T0322032

DARWIN
A COMPANION

Other titles by John van Wyhe

Charles Darwin in Cambridge: The Most Joyful Years
ISBN: 978-981-4583-96-1
ISBN: 978-981-4583-97-8 (pbk)

Dispelling the Darkness: Voyage in the Malay Archipelago and the Discovery of Evolution by Wallace and Darwin
ISBN: 978-981-4458-79-5
ISBN: 978-981-4458-80-1 (pbk)

DARWIN
A COMPANION

Paul van Helvert and John van Wyhe

Building on the work of R.B. Freeman

With iconographies by John van Wyhe

World Scientific

NEW JERSEY · LONDON · SINGAPORE · BEIJING · SHANGHAI · HONG KONG · TAIPEI · CHENNAI · TOKYO

Published by

World Scientific Publishing Co. Pte. Ltd.

5 Toh Tuck Link, Singapore 596224

USA office: 27 Warren Street, Suite 401-402, Hackensack, NJ 07601

UK office: 57 Shelton Street, Covent Garden, London WC2H 9HE

Library of Congress Cataloging-in-Publication Data
Names: Van Helvert, Paul, author. | Freeman, R. B. (Richard Broke) Charles Darwin, a companion. |
    Van Wyhe, John, 1971–   author.
Title: Darwin : a companion / Paul van Helvert and John van Wyhe ; building on the work of
    R.B. Freeman ; with iconographies by John van Wyhe.
Description: New Jersey : World Scientific, [2021] | Includes bibliographical references and index.
Identifiers: LCCN 2020038862 | ISBN 9789811208201 (hardcover) |
    ISBN 9789811229275 (paperback) | ISBN 9789811208218 (ebook for institutions) |
    ISBN 9789811208225 (ebook for individuals)
Subjects: LCSH: Darwin, Charles, 1809-1882. | Darwin, Charles, 1809-1882--Pictorial works. |
Naturalists--Great Britain--Biography.
Classification: LCC QH31.D2 V34 2021 | DDC 576.8/2092 [B]--dc23
LC record available at https://lccn.loc.gov/2020038862

British Library Cataloguing-in-Publication Data
A catalogue record for this book is available from the British Library.

© The portions of this work written by R.B. Freeman are copyright of The Charles Darwin Trust. All
rights reserved. For private academic use only. Not for republication or reproduction in whole or in
part without the prior written consent of The Charles Darwin Trust, 31 Baalbec Rd, London, N5 1QN
© The new introductions, bibliography, additions and corrections are copyright of John van Wyhe and
Paul van Helvert, 2020.
© Iconographies copyright John van Wyhe.

Copyright © 2021 by John van Wyhe

All rights reserved.

Cover illustrations:

1871 Albumen silver print of Charles Darwin by Oscar Gustav Rejlander. This has seldom been
reproduced in modern publications.

1886 Watercolour (13.5×9cm) by Julia 'Snow' Wedgwood (Darwin's niece) of Darwin's new study
as he left it after his death in 1882. Reproduced with permission of the Robert M. Stecher Collection,
Dittrick Medical History Centre, Case Western Reserve University. Published here for the first time.

Frontispiece:
Carte de visite of Charles Darwin in 1871 by George Charles Wallich. Very seldom reproduced.

Title illustration:
Carte de visite of Charles Darwin in 1869 by Elliott and Fry. Also very seldom reproduced.

For any available supplementary material, please visit
https://www.worldscientific.com/worldscibooks/10.1142/11497#t=suppl

Printed in Singapore

CONTENTS

LIST OF ILLUSTRATIONS

INTRODUCTION

Before Darwin's work appeared, the great majority of naturalists, and, almost without exception the whole literary and scientific world, held firmly to the belief that *species* were realities, and had not been derived from other species…But now this is all changed. The whole scientific and literary world, even the whole educated public, accepts, as a matter of common knowledge, the origin of species from other allied species by the ordinary process of natural birth. The idea of special creation or any altogether exceptional mode of production is absolutely extinct! … And this vast, this totally unprecedented change in public opinion has been the result of the work of one man, and was brought about in the short space of twenty years!

Alfred Russel Wallace, *Darwinism*, 1889.

In 1978 the zoologist, historian and bibliographer Richard Broke Freeman (1915-86) published his *Charles Darwin: A Companion*. Although it has been out of print for over forty years, it has remained one of the most unique, authoritative and wide-ranging reference works on all things Darwin. It is still cited by scholars, such as the editors of *The Correspondence of Charles Darwin*. It was a work of immense scholarship. Yet Freeman described his book as "a compilation, with almost nothing in it that has not appeared in print before." The present edition has not only completely revised and corrected Freeman's original work but supplemented it with new research and hundreds of discoveries never before published.

This is the richest single-volume reference work on Charles Darwin ever created. It is essentially an encyclopaedia of Darwin, his life, publications, family, contemporaries and many other features of his private and public life. No other single book provides so much information about Darwiniana. For example, nowhere else can one discover that 29 books were dedicated to Darwin (not 7 as previously thought), a list of all the scientific societies that elected him a member (92), charities that he supported, stamps, banknotes and coins depicting him (247), his finances, the names of his domestic staff, a photograph of his missing first coffin or details of hundreds of members of his extended family both ancestors and descendants. Only here can one find a list of more than 350 visitors, couples, families and groups that the Darwins entertained at their home over the years, rendering Darwin less of a recluse than often claimed. More than 150 recollections of Darwin and 38 theories to explain his illness are recorded. There is also the most detailed itinerary of Darwin's whereabouts throughout his life (656 entries, Freeman 1978 gave 68). Other new findings first published here include the fact that more than 70 institutions, 92 ships, streets etc. and 130 monuments around the world have been named after or dedicated to him. There is also a list of 50 family anniversaries such as births and deaths. The translation of his works into 64 languages is also recorded; Freeman was aware of 33 languages. No other man of science has had his publications translated into so many languages as Darwin. Even more surprising is the revelation that his name has been given to species of animals and plants at least 700 times, which exceeds even Alexander von Humboldt (c.400).

Also first published here is the most complete iconography (or list of portraits) of Darwin ever created. Freeman provided a list of about 50 depictions of Darwin in the first edition of *Companion*. This edition reveals for the first time that over 1,000 unique works have been created to portray Darwin including at least 48 19th-century caricatures, more than 213 paintings and drawings, at least 50 photographs, more than 590 printed portraits (up to 1982) and over 240 three-dimensional works such as busts, medallions and statues. Not just a list of portraits, further information is provided such as dates, artists, prices paid and quotations from Darwin or others about how a work was originally received. In addition to recording all known photographs of him, 227 printed variants of these have been identified and described. This iconography of Darwin is the most comprehensive ever created for any scientist, and quite possibly of any historical individual. While there is no possibility to provide reproductions of all of these in this volume, 300 representative, newly discovered and particularly interesting or unusual illustrations accompany the Darwin iconography and list of caricatures. There are also the first ever iconographies of HMS *Beagle*, his wife Emma Darwin and Down House and grounds. These too contain many new discoveries.

In 2007 a second online edition of Freeman's *Companion* was published in *The Complete Work of Charles Darwin Online* [http://darwin-online.org.uk/] (henceforth *Darwin Online*). It was prepared by Sue Asscher and John van Wyhe, mostly from Freeman's own large collection of unpublished corrections and additions, helpfully provided by The Charles Darwin Trust who purchased them and the copyright of his works from Freeman's widow. Together, these editions of *Companion* (1st and 2d) in *Darwin Online* have been accessed over 500,000 times since 2007. Clearly *Companion* has continued to be very useful.

A colossal amount of additional information and research about Darwin has become available, especially since 1985 when the publication of *The Correspondence of Charles Darwin* (henceforth CCD) began, the publication of *Charles Darwin's Notebooks, 1836-1844* in 1987 and other important works. A legion of scholars now collectively nicknamed the 'Darwin industry' has produced a vast array of modern scholarship and discoveries spread across hundreds of publications. In addition valuable writings with unique information on Darwin had already been appearing since the end of the 19th century. All of this means that the literature on Darwin is so immense and so varied and some of it is available only in now quite obscure publications that this presents a very daunting and often unknown body of material for any reader or researcher to attempt to grapple with. While no single volume could summarize or even list all of these sources, publications and findings, this *Companion* is the most complete guide to navigating the immense ocean of information about Darwin. But this book focuses more on primary sources than secondary works.

Ten years in the making, this new *Companion* has been checked, corrected and greatly enlarged from A to Z. We have added many entries on persons, places, events, manuscripts, servants at Down House, changes to the house and grounds, visitors to the Darwins, family pets, theories of his illnesses, the most extensive details of his library and manuscripts and so forth. There is a new map of the house and grounds. Over one million sources were consulted during the research for this volume, for example 16,880 references to Darwin in the British Newspaper Archive. So much new research and so many new findings and corrections to the existing literature make up this *Companion* that it can be considered an entirely new work. The first edition was c.100,000 words long. This edition has grown to 235,000 words including hundreds of corrections to Freeman's original entries.

The historian P. Thomas Carroll wrote as long ago as 1976 that "it seems certain that research and writing upon Darwin will continue, with scholars demanding access to ever more records of Darwin's life and work". (Carroll 1976, p. xv). This has certainly been borne out. However, the majority of the records of Darwin's life were not to appear until the publication of the CCD and the advent of the internet. By far the largest publication on Darwin is *Darwin Online* which began in 2002. This scholarly online edition contains, for the first time, *all* of Darwin's publications, including his shorter publications, the different editions of his books as well as foreign translations in thirty languages (both as images and high-quality transcriptions). Previously unknown publications continue to be discovered and added. It includes almost all of Darwin's c.20,000 pages of manuscripts and private papers as well as more than 2,000 publications relevant to the study of Darwin, his life, work and reactions to his theories. Many of these are cited in *Companion*, but not every item in *Darwin Online* is cited as such here, but many publications in *Companion*, whether by Darwin or not, are likely included there which will make it easy for a reader to consult them. *Darwin Online* is not an archive or library but a scholarly edition, with over 10,000 pages of newly transcribed manuscripts, the world's first and largest union catalogue of Darwin manuscripts, 200 new editorial introductions and over 4,900 new editorial footnotes to publications and manuscripts.

At the heart of *Darwin Online* is a heavily revised and enlarged edition of Freeman's other great work, *The works of Charles Darwin: An annotated bibliographical handlist* (2d edn. revised and enlarged, 1977) called the Freeman Bibliographical Database (http://darwin-online.org.uk/Freeman_intro.html). Freeman described Darwin's publications, their variants and translations in detail and assigned each a number. For example, he numbered the first edition of *Origin of species* as 373. This, with the prefix F (for Freeman), provides a standard reference to link the electronic text and images of that work with the database record for it. References throughout this book to these 'F numbers' can be found both in Freeman 1977 (up to his final number of F1805) and *Darwin Online* and are used in the scholarly literature on Darwin. The *Darwin Online* project has added over 760 newly discovered Darwin publications and translations to Freeman's original bibliography as well as making hundreds of corrections. Therefore, items that are cited here with little more than an 'F number' such as "Autobiography: 1936 Ukrainian (F2423)" can be easily found by consulting the Freeman Bibliographical Database and searching for the 'Identifier', F2423. This will bring up a complete bibliographical record for that publication. Freeman was a truly great bibliographer and this is reflected in the focus on publications and bibliographical detail still found in this *Companion*.

Editorial considerations

Several editorial priorities have informed the present edition. We have endeavoured to make the work accessible to a wide readership, from beginners to expert Darwin scholars. We have updated references to outdated selections from correspondence such as *Life and letters*, Stecher 1961 and Carroll 1976 and refer readers to the definitive edition of the letters, the CCD. Where letters have not yet been published in CCD we have maintained the original references. There is a wealth of unique information in many of the older secondary sources that is not familiar to most readers today so older references, not just to letters, have often been maintained. Space restrictions have obliged us to use many abbreviations and sometimes smaller fonts or to truncate an entry to save a line. In order to fit all of the information *Companion* contains into a single volume, difficult decisions about shortening had to be made.

No single book could contain all the information on Charles Darwin and his work. We have collected together more than any previous book which will enable readers to discover new facts and facets about Darwin and his work and pointers where to find more detailed information about him, his social and scientific world and context. As such we have followed Freeman's original intentions. Equally in agreement with Freeman we can quote an 1859 letter from Darwin to T.H. Huxley: "*The difficulty is to know what to trust.*" (CCD7:404)

John van Wyhe and Paul van Helvert, 2020

ACKNOWLEDGEMENTS

We are grateful to the late Mary Whitear, Randal Keynes and The Charles Darwin Trust for permission to reproduce the work of R.B. Freeman and to the Syndics of Cambridge University Library for permission to reproduce materials from the Darwin Archive, Cambridge University Library. Naturally we owe a very large and almost indefinable debt to the generations of scholars whose work has steadily made more and more of Darwin's life, work and context known and accessible and uncovered countless lost or forgotten documents and connections that throw light on this period in the history of science. John van Wyhe received important assistance from Kees Rookmaaker, J. David Archibald, Michael Matthews, Peter Worsley, Sue Asscher, Mark Pallen, Frederick Burkhardt, Duncan Porter, John Hayman and Henry von Wartenberg.

For assistance with the Darwin iconography research, van Wyhe is grateful to Gene Kritsky, Anthony Smith, Shannon Bohle, Juan Manuel Rodríguez Caso, Pedro Navarro, Vanessa Ong, Haiyan Yang, Elisabeth Leedham-Green, archivist of Darwin College, Cambridge, Paul Cox of the National Portrait Gallery, Clare Howe of the Torquay Museum, Peter C. Kjaergaard, Director of the Natural History Museum of Denmark, Wentao Song of the Department of Physiology, University of Cambridge, Danielle Czerkaszyn of the Oxford University Museum of Natural History, Daniel Lewis, Huntington Library, California, Jo Clarke, Plymouth Museums Galleries Archives, Cristina Cilli, Curator, Museo di Anatomia umana "Luigi Rolando" University of Turin, Jordi Fàbregas of the Jardí Botànic Marimurtra, Sara Belingheri, Wellcome Library, London, Bruce Marshall Rare Books, Emma Darwin, Cordula van Wyhe, Michael Barton, Ros Cameron, Galapagos Conservancy, Randy Moore, Olivia Fryman, curator of Down House Museum, Trevor Reynolds, Wendy Monkhouse, Francesa Barison, Biblioteca dell'orto botanico, Padua, Alexander O. Averianov, Victoria Clinton, Alice Pattullo, Hayley Kruger, Saffron Mackay, Royal College of Surgeons, Neil Gostling, Geoff Belknap, Alexander van Wyhe, Garrett Herman, Johannes Riütta and Andrew Eaves. Igor Fadeev, Dmitriy Olshanskiy and Antonina Nefedova of the State Darwin Museum, Moscow, provided important and extremely generous help and assistance including sending photographs and documents from their collections and archives. Samantha Evans of the Darwin Correspondence Project very generously provided details from the Down House account books regarding photographs and much other help. Christine Chua provided untiring assistance including invaluable help in identifying statues and busts of Darwin in China and beyond. Van Wyhe is also grateful for reproduction permission from Stefan Ståhle of the Moderna Museet, Stockholm. James Taylor, Gordon Chancellor and Robert Kuipers provided expert advice and information for the *Beagle* iconography. Very many others have contributed, indirectly, over the years to van Wyhe's research including William Huxley Darwin, Milo Keynes, Richard Keynes, Simon Keynes, Janet Browne, Jim Secord, Gordon Chancellor, Frank Sulloway, Nick Gill, Angus Carroll, Greg Radick, Jeff Ollerton, Tony Larkum, Antranig Basman, Cemil Ozan Ceyhan, Gene Kritsky, John Woram, Jon Hodge, Adrian Desmond, Gabriel David, Katrina Dean, and the staff of the Correspondence of Charles Darwin, especially Rosemary Clarkson and Samantha Evans; Adam Perkins, Godfrey Waller, Frank Bowles, Geoffrey Wallace, John Wallace, Emily Dourish and other staff of the Rare Books and Manuscripts Reading Rooms at Cambridge University Library, Judith Magee and Lorraine Portch of the library of the Natural History Museum, London, Gina Douglas and the staff of the library of the Linnean Society of London, the Master and Fellows of Christ's College, Cambridge, Geoffrey Thorndike Martin, Sir Peter Lachman; Candace Guite, John Wagstaff, Samantha Hughes and other staff of Christ's College Library, the staff of the library of Shrewsbury School, James Edmonson and the custodians of the Robert M. Stecher Collection at the Dittrick Medical History Centre, Case Western Reserve University, members of the Department of Biological Sciences, National University of Singapore and Gregory Clancey, the Master and Fellows of Tembusu College, National University of Singapore.

We are most particularly indebted to Christine Chua, Associate editor of *Darwin Online*, who generously and courageously helped check and improve *Companion* through very, very many revisions, especially with unique information from her transcription of *Emma Darwin's diary* and very many other sources. This book has been greatly enriched and improved by her tireless and cheerful assistance.

We are also grateful to Joy Quek and the staff at World Scientific Publishing for getting this book through the press. We apologise to any whose names we have inadvertently omitted. Naturally the errors remain our own.

ABBREVIATIONS

BAAS	British Association for the Advancement of Science.
BMNH	British Museum (Natural History) [Now Natural History Museum, London.]
CD	Charles Robert Darwin.
CCD	*The Correspondence of Charles Darwin.*
CDV	Carte de visite.
CUL	Cambridge University Library.
DAR	Darwin Archive, Cambridge University Library, e.g. DAR66-68.
d.s.p.	*Decessit sine prole*, died without issue.
ED	Emma Darwin.
EH	English Heritage, e.g. catalogue number EH88202653.
F	Freeman bibliographic number for a Darwin publication.
FGS	Fellow of the Geographical Society.
FLS	Fellow of the Linnean Society of London.
FRS	Fellow of the Royal Society.
MLNSW	Mitchell Library, State Library of New South Wales.
nd	no date.
NHM	Natural History Museum, London.
NMM	National Maritime Museum, Greenwich.
NPG	National Portrait Gallery, London.
PRS	President of the Royal Society
q.v.	*Quod vide*, which see.
qq.v.	*Quae vide*, which see plural.
RS	Royal Society.
s.p.	*Sine prole*, without issue.
SDMM	State Darwin Museum, Moscow.
UCL	University College London, London.

Many standard sources are abbreviated in the text using abbreviations already standard in the field, e.g., LL, ML, *Origin* etc. Books are listed alphabetically in the bibliography, once under the abbreviated title and again under the author or editor's name. Photographs, paintings and sculptures are cited by a 'creator date' abbreviation such as Richmond 1840, Edwards 1866a, Lock & Whitefield 1878 or Elliott & Fry 1881a. These can all be identified in the Iconography under the entry for Darwin, Charles Robert.

THE ENTRIES

"Abbety". 1879 Nickname used, with "Boo", "Mim", "Lenny". L. Darwin 1929. and "Dadda" (CD), by Bernard Darwin for members of the family. None of them is ED. ED used the nicknames in her Jan. 1879 letter to Henrietta. ED2:294-95.

Abberley, Anne. William Darwin's nursemaid 1839-52?. 1839 Aug. 5 ED recorded "Anne Abberley" came. 1840 gave Anne pounds of tea. *ED's diary*. CD made observations of William's interactions with her. William close to her and "shewed a decided wish to go to Anne". *Notebook of observations on the Darwin children*. (1839-56) DAR210.11.37, CCD4 Appendix III and transcribed in *Darwin Online*.

Abberley, John, 1840-50. Gardener at The Mount. Sent CD with observations on ants and bees. CD asked A to plant "single Peas, Kidney Bean & Bean, intertwined, without sticks" and observe. *Questions & experiments notebook*. DAR 206.11. CCD2.

Abbot, Dr Francis Ellingwood, 1836-1903. American Unitarian minister, liberal religious philosopher, writer, co-founder of the Free Religious Association. Editor (1870-80) of *The Index*, of the Free Religious Association (1867) of Cambridge, Massachusetts, USA. 1871 CD letters to on religion F1753 and CCD19. *Truths for the times* was a pamphlet which, as Abbot described: "is an effort to bring the truest science and the truest religion of the age into absolute harmony and mutual understanding. The supposed conflict between science and religion is superficial and unreal, when both are properly conceived." CD wrote A of his agreement with *Truths for the times* "I agree to almost every word". This quotation was published with the permission of CD in *The Index* from 1871-80 when CD asked his son William to write to A, who was no longer to edit the journal, to discontinue printing the endorsement in future. See LL1:304-6 and Browne 2002, pp. 391-2 and CCD19. 1876 thanked CD for a donation of £20 towards *The Index*. Abbot, 1882 Recollections of CD in DAR112.A1-A2, transcribed in *Darwin Online*.

Abbott, John. 1848 new tenant at Sutterton Fen, an estate willed by Dr Robert Darwin to CD. A held tenancy for thirty-six years. 1876 A renewed tenancy with CD at a rent of £62.10s. per annum. Worsley, *The Darwin farms*, 2017, p. 90.

Abinger Hall, West of Dorking, Surrey. House of Sir Thomas H. Farrer. 1873 Aug. 5 CD first visited, and often later, which he much enjoyed. Many records of visits are recorded in *ED's diary*.

Abrolhos, Arquipélago dos, Brazil. Coastal islands south of Salvador. Also spelt "Abrohlos". 1832 Mar. 27 *Beagle* visited and CD landed. 1835 Misspelt "Abrothos" in CD *Letters on geology*, 1835, pp. 4-5. (*Shorter publications*, F1) *Zoological diary*, pp. 31-35. *Geological diary*: (3.1832) DAR32.49-50, transcribed in *Darwin Online*.

Abstracts, of books, pamphlets, and articles from scientific journals. CD made notes and summaries of books he did not own and later of some that he did own. These range in length from a few lines to many pages. These are principally found in

DAR71-77 and also scattered throughout other sections of the Darwin Archive-CUL. Almost all of these are reproduced in *Darwin Online*.

Academia Caesarea Leopoldino-Carolina Naturae Curiosorum. 1857 CD Member under cognomen Forster. "Accipe...ex antiqua nostra consuetudine cognomen Forster". Named after Johann Reinhold Forster (1729-98); both Johann Reinhold F and his son Johann Georg Adam F (1754-94) on Cook's 2^d voyage.

Académie des Sciences de l'Institut de France. 1872 CD proposed for Zoologie section, but not elected. 1878 Elected in Botanique section. CD to Gray "It is rather a good joke that I shd be elected in the Botanical section, as the extent of my knowledge is little more than that a daisy is a compositous plant & a pea a leguminous one." CCD26:353. "He was in fact guilty of evolution but with extenuating botanical circumstances". Francis Darwin, *Annals of Botany*, 13, 1899, p. xi.

Acland, Sir Henry Wentworth, 1st Bart, 1815-1900. Physician. 1846 Married Sarah Cotton (d. 1878). 7 sons, 1 daughter, Sarah Angelina (1849-1930). 1980 Jan. 29 A copy of the 1865 offprint of *Climbing plants* (F835) inscribed to H. Acland in CD's own hand, Sotheby's sale, lot 345 sold for £37,500 in 2014.

Acton, Samuel Poole (Mr Acton), 1813-85. Postmaster, wine and spirit merchant at Bromley 1839-57. He would forward items to CD by messenger if necessary. CCD5:15.

Adler, Friedrich, 1857-1938. Austrian jurist, translator and writer of Bohemian origin. Composed a series of poems in honour of CD that were presented with the German and Dutch albums of photographs in 1877. CCD25.

Adventure **[I], HMS**, 1827-30. Command vessel, under Captain Philip Parker King, of first voyage of HMS *Beagle*.

Adventure **[II]**. Schooner, 170 tons, a sealer, originally built at Rochester as a yacht, had been used by Lord Cochran. 1833 Mar. Bought by FitzRoy on 2^d voyage from William Low or Lowe, at Port Louis, Falkland Islands, for $6,000 (nearly £1,300) with £403 for second-hand equipment from two ships wrecked on Falklands. Then named *Unicorn*. John Clements Wickham in command. 1834 Oct. Admiralty refused to reimburse FitzRoy, so sold at Valparaiso for $7,300 (nearly £1,400).

Agassiz, Alexander Emanuel, 1835-1910. Marine biologist. Swiss born but moved to the USA in 1849 with father and family. Son of Jean Louis Rodolphe A. Converted to belief in evolution by reading and corresponding with Fritz Müller. Fairly frequent correspondent with CD. 1869 Dec. 1 Visited Down House with wife.

Agassiz, Jean Louis Rodolphe (Louis), 1807-73. Born in Switzerland but emigrated to the USA in 1849. Ichthyologist and geologist. Biographies include: Elizabeth Cabot Cary Agassiz (2^d wife), *Louis Agassiz. His life and correspondence*, 1885; C.F. Holder, *Louis Agassiz. His life and work*, 1893; A.B. Gould, *Louis Agassiz*, 1901; J.D. Teller, *Louis Agassiz, scientist and teacher*, 1947. 1832-47 Prof. Natural History Neuchâtel. 1838 Foreign Member RS. 1847-73 Prof. Zoology and Geology Harvard. 1841 CD sent *Journal*. 1849 CD met at BAAS, Southampton. 1854 CD sent *Living Cirripedia*. 1859 CD sent 1st edn of *Origin*. 1860 Jan. Gray to CD "He says it is poor—

very poor!! (entre nous). The fact is he is very much annoyed by it". CCD8:16. 1860 Jul. "I shall therefore consider the transmutation theory as a scientific mistake, untrue in its facts, unscientific in its method, and mischievious in its tendency". *Silliman's Jrnl.*, 143. 1863 CD to Gray "I enjoy anything that riles Agassiz. He seems to grow bigoted with increasing years. I once saw him years ago and was charmed with him". CCD11. 1866 CD to Gray about an Amazonian glacier "We [CD and Lyell] were both astonished at the nonsense which Agassiz writes...his predetermined wish partly explains what he fancied he observed". CCD14. A continued against CD for the rest of his life and ML contains a number of examples of his attitude. Nevertheless, A and CD were on cordial terms in correspondence.

Aiken, 1829-31. Supplier of bird skins. CD recorded several transactions. CCD1.

Ainslie, Robert, 1810-95. Methodist minister and writer. 1845-58 Lived in Tromer Lodge, Down. 1845 CD wrote to Susan, his sister telling her about A altering the road illegally. CCD3:247. 1858 CD was angered by A not paying church rate. CCD7:92.

Ainsworth, William Francis, 1807-96. Physician, Wernerian geologist and traveller. Medical student at Edinburgh with CD and collected with him including Isle of May and Inchkeith. *Athenaeum*, May 13, p. 604, 1882. CD: "Knew a little about many subjects, but was superficial and very glib with his tongue". *Autobiography*, p. 48.

Airy, Dr Hubert, 1838-1903. Physician. Son of Sir George Biddell A. 1871 One of the people who pointed out the error in *Descent*, 1:19 that the platysma myoides (a muscle) cannot be brought into action voluntarily. Corresponded with CD 1871-76. Visited CD in 1872 Oct. 1. *ED's diary*.

Albanian. First editions in: 1976 *Journal* (F2067). 1982 *Origin* (F2389).

Albury, near Guildford, Surrey. 1871 Jul. 28 - Aug. 24 CD had a family holiday in a rented house, belonged to Henry Drummond, the Irvingite. Jul. 28, they came to Haredene. They were joined by Leonard on 29th and then to Leith Hill etc. on 31st. Then to Ewhurst and down to Albury on Aug. 8. William came on the 12th. Home on Aug. 25. 1882 Jun. 20, ED drove to Albury heath. *ED's diary*.

Alison, Robert Edward. English author and resident of Valparaiso and later managing director of a Chilean mining company who wrote on South American affairs. In the *St. Fe notebook*, Chancellor & van Wyhe, *Beagle notebooks*, 2009.

Alderson, Georgina [II], 1827-99. Daughter of Sir Edward Hall A and Georgina Drewe. Called "Georgy" by the Allen aunts. Married Marquis of Salisbury, Robert Gascoyne-Cecil (1830-1903). In 1840 Feb. 15, ED recorded "Went to Lady Alderson". 1840 Apr. 5 in a letter CD to ED, CD said he shared a carriage with an elegant female, "like a thin Lady Alderson" but did not venture to speak with her. 1856 Fanny A said she was an excellent ballet-mistress. 1882 A was on "Personal Friends invited" list for CD's funeral.

Alderson, Sir Edward Hall, 1787-1857. Judge, Baron of the Exchequer. 1823 Married Georgina Drewe. Daughter, Georgina [II] 1827-99. Lived Great Russell Street, London. "A most temperate man".

Allen, Frederick and Frances Allen. Residents at Down. 1868 Sept. Mr Robinson, Curate at Down, had been having a relationship with one of Mrs A's maids, Esther West. Brent 1981, p. 460 and CCD16.

Allen, Antoinette Caroline, 1768-1835. 3^d child of John Bartlett and Elizabeth Hensleigh. ED's aunt. 1793 Married Edward Drewe (1756-1810).

Allen, Catherine (Kitty) [I], 1765-1830. 2^d child of John Bartlett and Elizabeth Hensleigh. ED's aunt. "She could neither make herself or others happy". 1798 Married Sir James Mackintosh (1765-1832) as 2^d wife. 2 sons and 2 daughters.

Allen, Charles, 1842-55. 3^d child of Lancelot Baugh and Georgina Sarah Bayly.

Allen, Charles Grant (Grant) Blairfindie, 1848-99. Canadian naturalist. Chronically sick and often in financial difficulty. Not related to the other Allens. Biography: E. Clodd, 1900. 1877 CD to Romanes, asks to thank A for his book *Physiological aesthetics*. CCD25. 1879. [Extract from a CD letter]. In A., Colour in nature. *Nature* 19 (24 Apr.): 581, F2004. In the original letter CD referred to "the sombre aspect of nature" in the Galapagos and Patagonia. 1879 CD to Romanes, A was in some financial difficulty, CD subscribed £25, will send more if needed. CCD27. 1881 CD to Romanes relates to A's trouble, acknowledging cheque for £12.10s in 50% repayment of loan, and about giving a present of a microscope to. *Calendar* 13536. 1882 CD to Romanes, CD prefers to give the microscope now, rather than wait for the repayment of the other half of the loan. *Calendar* 13544, 13600. 1885 ED to Henrietta Litchfield about A's book *Evolutionist at large* (1881), "I do not like Grant Allen's book about your father. It is prancing and wants simplicity". ED2:272. 1885 A published one of the earliest biographies of CD.

Allen, Clement Francis Romilly, 1844-1920. 2^d child of Lancelot Baugh and Georgina Sarah Bayly. ED's first cousin. 1877 Married Edith Louisa Wedgwood (d. 1935). 2 sons and 3 daughters.

Allen, Edmund Edward, 1824-98. 2^d child of Lancelot Baugh and Caroline Romilly. ED's first cousin. 1848 Married Bertha Eaton, Dorothea Hannah's sister. Had 5 sons and 6 daughters.

Allen, Elizabeth Jessie Jane, 1845-1918. 3^d child of Lancelot Baugh and Georgina Sarah Bayly. ED's first cousin. Unmarried.

Allen, Emma Augusta, 1780-1866. 11th child of John Bartlett and Elizabeth Hensleigh. Unmarried. ED's aunt. ED named after her. 1842 her two widowed sisters Harriet and Fanny came to live with her and Fanny at Tenby.

Allen, Frances (Fanny), 1781-1875. 12th child of John Bartlett and Elizabeth Hensleigh. Unmarried. ED's aunt. 1842 her two widowed sisters Harriet and Jessie came to live with her and Emma at Tenby.

Allen, George Baugh, 1821-98. Barrister. 1st child of Lancelot Baugh and Caroline Romilly. ED's first cousin. 1846 Married Dorothea Hannah Eaton (1818-68), Bertha's sister. Had 8 children.

Allen, Gertrude Elizabeth, 1816-23. 3ᵈ child of John Hensleigh [I] and Gertrude Seymour. ED's first cousin. Died at seven.

Allen, Harriet (Sad), 1776-1845. 8ᵗʰ child of John Bartlett and Elizabeth Hensleigh. ED's aunt. 1799 Married Matthew Surtees (1756-1827). Had unhappy marriage. ED1:7. After death of husband lived with her sister Jessie at Chêne. ED1. 1842 returned to live at Tenby with Emma and Fanny. Jessie would move in later. 1845 Nov. 9 ED recorded "Aunt Harriet died".

Allen, Henry George, 1815-1908. 2ᵈ child of John Hensleigh [I] and Gertrude Seymour. Unmarried. ED's first cousin.

Allen, Isabella Georgina, 1820-1914. 5ᵗʰ child of John Hensleigh [I] and Gertrude Seymour. ED's first cousin. 1840 Married George Lort Phillips (1811-66). s.p.

Allen, Jessie, 1777-1853. 9ᵗʰ child of John Bartlett and Elizabeth Hensleigh. ED's favourite aunt. She described CD "Fresh and sparkling as the purest water". ED1:159. 1819 Married J.C.L. Simonde de Sismondi (1773-1842). After his death, went to live with her sisters, Emma, Fanny and Harriet, at Tenby. 1842 She burnt Sismondi's journals and her own. ED2:57.

Allen, Joan Bartlett, 1773-1801. 6ᵗʰ child of John Bartlett and Elizabeth Hensleigh. ED's aunt. Unmarried.

Allen, John, 1810-86. Friend of Edward FitzGerald and of Alfred Tennyson. 1836-46 School Commissioner. 1847-83 Archdeacon of Shropshire. Married 31 Jul. 1834 Harriet Higgin (d. 1904). 1 son, 9 daughters. 1847 Visited, with Jessie Sismondi and her sister Emma, the school at Caldy Island, which was paid for by Sarah Elizabeth Wedgwood [II]. ED2:107.

Allen, John Bartlett, 1733-1803. Of Cresselly, Pembrokeshire. CD's maternal great-grandfather. 1763 Married 1 Elizabeth Hensleigh (1739-1790). 2 sons, 10 daughters: 1 Elizabeth; 2 Catherine; 3 Antoinette Caroline; 4 John Hensleigh; 5 Louisa Jane; 6 Joan Bartlett 7 Lancelot Baugh; 8 Harriet; 9 Jessie; 10 Octavia; 11 Emma; 12 Frances. 1792 Married 2 Mary Rees (d.?1798) a coalminer's daughter. 3 daughters.

Allen, John Hensleigh [I], 1769-1843. 4ᵗʰ child of John Bartlett and Elizabeth Hensleigh. ED's uncle. 1812 Married Gertrude Seymour (1784-1825). 3 sons, 2 daughters: 1 Seymour Phillips; 2 John Hensleigh [II]; 3 Gertrude Elizabeth 4 Henry George; 5 Isabella Georgina.

Allen, John (Johnny) Hensleigh [II], 1818-68. 4ᵗʰ child of John Hensleigh [I] and Gertrude Seymour. Colonial Office. Worked much amongst the London poor. ED's first cousin. 1863 Married Margaretta Sarah Snelgar (c.1841-1919). s.p.

Allen, Lancelot Baugh (Baugh), 1774-1845. 7ᵗʰ child of John Bartlett and Elizabeth Hensleigh. ED's uncle. Married 1 Caroline Jane Romilly, 2 sons: 1 George Baugh; 2 Edmund Edward. 1842 Married 2 Georgina Sarah Bayly (?-1859), 2 sons, 1 daughter: 1 Charles Hensleigh; 2 Clement Francis; 3 Elizabeth Jessie Jane. 1811 Assistant Warden of Dulwich College. 1811-20 Master of Dulwich College. 1819-25 Solicitor, Police Magistrate. 1839 May 27 visited Darwins with Henslow et al.

Allen, Louisa Jane (Jane/Jenny), 1771-1836. 5th child of John Bartlett and Elizabeth Hensleigh. ED's aunt. 1794 Married John Wedgwood [IV] (1766-184). 7 children. Died at Shrewsbury when consulting Dr Robert Waring Darwin, CD's father.

Allen, Octavia, 1779-1800. 10th child of John Bartlett and Elizabeth Hensleigh. Unmarried. ED's aunt.

Allen, Sarah Elizabeth (Bessy) [I], 1764-1846. 1st child of John Bartlett and Elizabeth Hensleigh. CD's mother-in-law. 1792 Married Josiah Wedgwood [II], 4 sons and 3 daughters. 1833 suffered a stroke and damaged a foot, never to walk again. 1836-46 Bedridden and became mentally ill.

Allen, Seymour Phillips, 1814-61. 1st child of John Hensleigh [I] and Gertrude Seymour. ED's first cousin. 1843 Married Catherine Henrietta Fellowes (1821-1900). 6 sons, 2 daughters.

Allfrey, Charles Henry, 1839-1912. Physician of St. Mary Cray and Chislehurst. 1882 Attended to CD in his terminal illness. Signed CD's death certificate which was at the Register, Bromley; copy in DAR140.5. A was on "Personal Friends invited" list for CD's funeral. A came to Down House on Apr. 12 1882. *ED's diary*. Dr Moore was also in attendance. A was to see ED again later that year on Oct. 30, possibly coming to treat her for "windpipe" and a cold which began on Oct. 28.

Allingham, William, 1824-89. Irish poet. Visited Freshwater, Isle of Wight in Aug. 1868 recording CD in his diary: "Darwin expected, but comes not. Has been himself called 'The Missing Link.'" And in 1875: "I had not seen him for twenty years. He is a pleasant jolly-minded man (I thought this a very curious phrase), with much observation and a clear way of expressing it. Has long been an invalid. I asked him if he thought there was a possibility of men turning into apes again. He laughed much at this, and came back to it over and over again." Allingham & Radford eds. *A diary*, 1907, pp. 184-5, 239, 274, his recollection of CD is transcribed in *Darwin Online*.

Alvey, Elizabeth, 1666-1724. Daughter of Matthew A. Origin of forename Alvey in family. Married John Hill (1678-1717). CD's great-great-grandmother.

Alvey, Matthew. Son of William A and Frances Wymonsold. CD's ancestor in fifth generation. Married Susanna Rawlinson (1644-1724). Daughter Elizabeth A.

Alvey, William, ?-1649. Married Frances Wymonsold. Father of Matthew A. CD's ancestor in 6th generation.

Alwyne, Mrs. 1871 Played the organ in Downe church.

Amazon valley fauna. 1863 Contributions to an insect fauna of the Amazon valley, *Trans. Lin. Soc. of London*, 23: pp. 495-566, by Henry Walter Bates. 1863 Review of [unsigned] by CD, *Nat. Hist. Rev.*, 3: pp. 219-24 (*Shorter publications*, F1725). Anon review of Bates, *Naturalist on the river amazons*, mistakenly attributed to CD in the printed catalogue in the Department of Printed Books in the British Museum. *Calendar*, p. 584, CCD11.

American Philosophical Society, Philadelphia, USA. 1870 CD Honorary Member. For its holdings in CD letters etc., see P. Thomas Carroll, *An annotated calendar of the* *letters of Charles Darwin in the Library of the American Philosophical Society*, 1976, in *Darwin Online*, and the Darwin Online Manuscript Catalogue. Letter details of the *Calendar* available online: https://search.amphilsoc.org/collections/view?docId=ead/Mss.B.D25-ead.xml

Amharic. First edition in: 2003 *Origin* (F2452).

Ammonium carbonate. 1882 The action of carbonate of ammonia on the roots of certain plants, *Jrnl. of the Lin. Soc. of London (Bot.)*, 19: pp. 239-61 (*Shorter publications*, F1800). 1882 The action of carbonate of ammonia on chlorophyll bodies, *ibid.*, 19: pp. 262-84 (*Shorter publications*, F1801). Both read by Francis Darwin. Abstracts of these by Francis Darwin, *Nature*, 25: pp. 489-90. CD's notes are in DAR62.

Ampthill Park, Bedfordshire, 1826 Home of Sir James Mackintosh, lent to him by Henry Richard Vassal-Fox, 3ᵈ Baron Holland.

Andersson, Nils Johan, 1821-80. Swedish botanist who visited Galapagos in the frigate *Eugenie*, during its voyage from 1851-3. 1857 worked with Hooker at Kew. Assisted CD who perhaps sent him 1ˢᵗ edn of *Origin*. CCD7:350.

Anerley Gardens, near Sydenham, Kent. CD in 1855 Jul. 31 letter to Fox said he would go to a "great show of Poultry & Pigeons". Likely where CD met Tegetmeier. Entry in CD's Account book on Aug. 30 recorded the purchase of four breeds of pigeons. CCD5:422. 1855 Aug. 28-30 CD went to poultry show. Archibald, *Charles Darwin: A reference guide to his life and works*, 2019, p. xxiii.

Angulus Woolnerii. The infolded point of the human ear, also called Darwin's peak, Darwin's point or Darwin's tubercle. The name A. Woolnerianus was suggested by CD to Woolner jokingly, and never became the official name. LL3:140; *Nature*, Apr. 6, 1871. See also Woolner, Thomas.

Animal intelligence. 1882 George John Romanes, *Animal intelligence*, London, International Scientific Series XLI. Extracts from CD's notes throughout (F1416). First foreign editions: 1883 USA (F1419). 1887 French (F1429).

'The projecting point', from *Descent*.

Anne, 1865?-79 Domestic servant at Down House.

Ansell, Mark, 1851-? Groom at Down House in 1871 census.

Ansted, David Thomas, 1814-80. Prof. Geology at King's College London. 1844 FRS. 1860 wrote CD about *Origin* and *Geological gossip*, 1860, by A. CD to J. Torbitt 12 Apr. 1876 (CCD24:105) referred to an enclosed letter from A on using selection to produce a blight resistant potato which was not located by the CCD. The letter is in an item with CD letters in *Darwin Online* (F2542). To be included in CCD30.

Ants. 1873 Instinct: Perception in ants. *Nature*. 7: pp. 443-4 (*Shorter publications*, F1810) and Habits of ants, *Nature*. 8: p. 244 (*Shorter publications*, F1761); introducing letters from geologist and mining engineer James D. Hague.

Appleton & Co. New York publishing house, 1785-1948. American publishers of works by CD and Herbert Spencer.

Appleton, Mary (Molly), 1813-89? American spiritualist. Sister of Thomas Gold and Frances Elizabeth (Mrs Henry W. Longfellow). Married Robert Mackintosh.

Appleton, Thomas Gold, 1812-84. Writer, poet, artist and patron of the fine arts. Better described as wit, littérateur, interested in spiritualism. Brother of Mary and Frances Elizabeth (Mrs Henry W. Longfellow). A may have made an earlier call in 1849 Oct. 12 "Mr Appleton came". 1868 Jul. 17 A called on CD at Freshwater, Isle of Wight. Another record in 1874 Jun. 19. *ED's diary*.

Arabic. First edition in: 1918 *Origin* (F1928a).

Arding, Willoughby, 1805-79. Physician. Ashworth 1935 (in *Darwin Online*) identifies CD's Edinburgh naturalist friend "Hardie" as A, but CD says that Hardie died early in India. A was at Bombay and then Wallingford, Berkshire.

Argentina, CD on geology of. See Revista: Darwin in Argentina. *Revista de la Asociación Geológica Argentina* 64, no. 1 (Feb. 2009): 1-180, in *Darwin Online*. See also Sandra Herbert, *Charles Darwin, geologist*, 2005.

Argyll, 8th Duke of, see **Campbell, George John Douglas.**

Armenian. First editions: 1914 Biographical sketch of infant (F1310). 1896 *Earthworms* (F1402). 1936 *Origin* (F630). 1949 *Journal* (F169). 1959 *Autobiography* (F1510).

Armstrong, Robert. Physician at Royal Naval Hospital Plymouth and Inspector of Fleets (1847). 1833-4 CD sent boxes of fossils and specimens to A for forwarding to Henslow. CCD1. "Armstrongs" visited the Darwins in 1851 Jul. 22 and a Mrs Armstrong visited in 1871. *ED's diary*. Possibly an unrelated family.

Arsenic. CD began taking on 13 Mar. 1873 as a medication. *ED's diary*. At the time, low doses were believed to strengthen the heart.

Artizans' Dwelling Company, 1871 CD took 10 shares at £100 each from John Royle Martin. CCD19. 1881 CD did not then own them. Atkins 1976, p. 96.

Asborne, Emily, 1839-? Housemaid at Down House in 1861 census.

Asborne, Jane, 1836-? Nurse and housemaid at Down House in 1861-? Census 1861. Bernard Darwin was very fond of her and called her "Na". 1880 came for a visit to Down House. F. Darwin, *Story of a childhood*, 1920, p. 10 and 35.

Ascension Island, Atlantic Ocean. 1836 Jul. 19 *Beagle* arrived. 1836 Jul. 20 CD ashore. CD's notes on are in DAR40.93-96 and DAR38.936-953, transcribed in *Darwin Online*. Further notes (1856?) in DAR205.3.63. *Geological diary*. DAR38.936-953, transcribed in *Darwin Online*.

Ash, Edward John, 1799-1851. Bursar of Christ's College, Cambridge. Rector of Brisely and Vicar of Gateley, Yorkshire. 1831 Nov. 15 A failed to subtract furniture value from CD's final account with College. CCD1:177ff. 1836 or 1837 CD had dinner in A's rooms in Christ's College. See van Wyhe, *Darwin in Cambridge*, 2014.

Ashburner, Misses. Aunts of Sara Sedgwick. Their father was "the youth beloved" of Mrs John Opie's (née Amelia Alderson) poem "Forget me not". 1871 George and Francis stayed with them in USA. In 1880 Jul. 3 ED in her diary wrote "Miss Ashburner went" and an entry on Jul. 1, "Aunt Grace came". Called Aug. 20.

Ashworth, Emily, 1848 Married Edward Forbes. 1854 Widowed.

Athenaeum Club, Pall Mall, London. 1838 Before Aug. CD elected member, one of 40 new members called "The 40 Thieves", proposed by 4th Marquis of Lansdowne. CD used the Club often before marriage. See *Autobiography*, p. 35. In 1842 Feb. 9, ED wrote in her diary "Charles unwell & went to the Athenaeum".

Atkin, James Richard, 1867-1944. Born in Australia, eldest son of Robert Travers and Mary Elizabeth (née Ruck). Bernard Darwin's cousin. Known as Dick, came to Down House 1875 Jun. 1. *ED's diary* recorded "little Atkin". 1879 wrote to thank CD for a present. DAR159.122. 1891 Called to the bar by Gray's Inn. 1891-96 visited ED, last entry Jul. 4 1896, "Dick Atkin & wife". Bernard Darwin's *The world that Fred made*, is mentioned in *Lord Atkin*, p. 4. See Geoffrey Lewis, *Lord Atkin*, 1999. Bernard recollected "once I had tea in Magdalen with my cousin Dick Atkin, then an undergraduate and now a law lord." B. Darwin, *Pack clouds away*, 1941, p. 113.

Atkin, Robert Travers, 1841-72, Journalist and parliamentarian. Held a commission in the Shropshire militia. 1864 married Mary Elizabeth Ruck, 3 sons, eldest James Richard. Fell from his horse and never fully recovered, died at 30. 1868 Mrs A spent time with ED. ED2:220. 1879 Jun 17-Jul 21, Mrs A's sons stayed at Down House. 1880 Mrs A appeared to have had a long stay from Jun. 19-Jul. 2.

Auditory-Sac. 1863 On the so-called "auditory-sac" of Cirripedes, *Nat. Hist. Rev.*, 3: pp. 115-16 (*Shorter publications*, F1722). See CCD10:451-3.

Atkinson, Tindall, 1880 Apr. 30 drew Bernard Darwin's portrait. F. Darwin, *Story of a childhood*, 1920, pp. 28-9 and 70.

Aubertin, John James, 1818-1900. Spanish and Portuguese scholar. Was in Ilkley same time as CD in 1859. Corresponded with CD from 1863-72. Made a failed attempt to call on CD in 1871 Mar. 1 which both men regretted. A had sent CD's sons some postage stamps. Expressed sadness at the news from CD of Mary Butler's tragic death. Finally visited CD in Oct. 1 1871. *ED's diary*.

Audubon, John James, 1785-1851. Born in France as Jean-Jacques. American ornithologist, famed for his four volume, double elephant folio *Birds of America*, 1827-38. 1826 CD met and heard him lecture at Edinburgh. "Sneering somewhat unjustly at Waterton". *Autobiography*, p. 51. 1830 FRS.

Australia. 1836 Jan. 12-Mar. 16 *Beagle* was at. "Farewell Australia! you are a rising infant and doubtless some day will reign a great princess in the south, but you are too great and ambitious for affection, yet not great enough for respect. I leave your shores without sorrow or regret". *Beagle diary*, p. 413, *Journal*, p. 538. F. Nicholas, *Charles Darwin in Australia*. 2d edn 2009. P. Armstrong, *Charles Darwin in Western Australia*, 1985, in *Darwin Online*. Geological diary: DAR38.812-836, trans. in *Darwin Online*.

Autobiographical Fragment. This autobiography of CD's early years was written in 1838. DAR91.56-63. 1903 Printed first in ML1:1-5, in *Darwin Online*. Foreign editions: 1903 USA in stereo edition of ML. 1959 Russian, fragment alone. 2019 Portuguese, fragment alone, translated by Pedro Navarro, F2230, in *Darwin Online*.

Autobiography. 1876 Written between late May and Aug. 3 with later additions. MS title "Recollections of the development of my mind and character". DAR26. It has sometimes been claimed that CD never called this his "autobiography". However he did call it this in his Last will and testament (reproduced in *Darwin Online*). 1887 First printed in LL1:26-160, with omissions because of strong objections of ED and her daughters. 1892 Abbreviated version printed in *Charles Darwin: his life told in an autobiographical chapter, and in a selected series of his published letters*. 1958 Nora Barlow, ed. *The autobiography of Charles Darwin 1809-1882. With the original omissions restored*, (F1497): a re-transcription of the original manuscript, which lists, pp. 244-5, the more important omissions. See also Russian edn 1957 below. 1958 English braille edn based on Barlow (F1509). 2008-9 New transcriptions appeared, one by Anne Secord in J. Secord, *Charles Darwin: Evolutionary writings*, 2008 and one by Kees Rookmaaker in *Darwin Online*. First foreign editions: 1959 Armenian (F1510). 1959 Bulgarian (F1511). 1977 Catalan (F2495). 1917 Chinese (F1882). 1937 Croatian (F2420). 1910 Czech (F2304). 1909 Danish (F1512). 2000 Dutch (F1906). 1987 Finnish (F2028). 1888 French (F1514). 1887 German (F1515). 2007 Greek (F2049). 1948 Hebrew (F1520). 1955 Hungarian (F1521). 1919 Italian (F1522). 1891 Japanese (F2339). 1965 Korean (F1525). 1935 Latvian (F1526). 1959 Lithuanian (F1527), 2009 Malayalam (F2486). 1889 Norwegian (F1528). 1891 Polish (F1529). 2004 Portuguese (F2039). 1957 Romanian (F2296). 1896 Russian (F1533). 1937 Serbian (F1542). 1959 Slovak (F2426). 1959 Slovenian (F1543). 1902 Spanish (F1544). 1959 Swedish (F1546). 2000 Turkish (F2537). 1936 Ukrainian (F2423). 1908 USA (F1478). 1957 Russian (F1540) a transcription which precedes Barlow. See also: 1908 The education of Darwin, *Old South Leaflets*, vol. 8: 194, pp. 345-64 (F1478).

Aveling, Dr Edward Bibbins, 1849-98. Medical practitioner, freethinking atheist and crook. Took as common law wife Eleanor Marx, daughter of Karl Marx. See also Karl Heinrich Marx. 1880 Oct. 12 A to CD. A wanted to dedicate a book on free thought to CD. 1880 Oct. 13 CD declined. P.T. Carroll & R. Colp, *Annals Sci.* 33, pp. 387-94, 1976. 1881 A visited Down House with Ludwig Büchner, religion was discussed. LL1:317. 1881 *The student's Darwin*. 1882 *Darwinism and small families*. Recollection of CD in 1883 *The religious views of Charles Darwin*: "For, on further inquiry, [CD] told us that he had, when of mature years, investigated the claims of Christianity. Asked why he had abandoned it, the reply, simple and all-sufficient, was: 'It is not supported by evidence.'" p. 6. In *Darwin Online*. 1887 *Darwin made easy*.

Ayrton, Acton Smee, 1816-86. Barrister and M.P., served in Gladstone's cabinet, 1868-74, First Commissioner of Works and had a long-running feud with J.D. Hooker. (*Shorter publications*, F1937). See CCD.

d'Azara, Félix, 1746-1821. Spanish explorer and army officer who surveyed Spanish and Portuguese territories in South America. In the *Rio notebook*, Chancellor & van Wyhe, *Beagle notebooks*, 2009 and other *Beagle* manuscripts.

Azores, Atlantic Ocean. 1836 Sept. 20-25 *Beagle* anchored off Angra do Heroismo, capital of Terceira; CD visited Praya. 1836 Sept. 25 *Beagle* called at St. Michael for letters and left for England. *Beagle diary* pp. 437ff. *Geological diary*: DAR38.957-960. P. Armstrong, *Charles Darwin's last island*, 1992, both in *Darwin Online*.

"Babba". Bernard Richard Meirion Darwin's infant name for CD. B. Darwin, *Green memories*, 1928, p. 27 spells "Baba". As seen in a letter ED wrote to Henrietta in Jan. 1879, CD was also called "Dadda". ED2:294-95.

Babbage, Charles, 1791-1871. Mathematician, philosopher and inventor. CD regularly attended his "famous evening parties" in London. *Autobiography*, p. 108. 1816 FRS. 1828-39 Lucasian Prof. Mathematics Cambridge. 1839 Mar. 16, Apr. 20 and Jun. 22 visited CD. Again 1840 Apr. 25 and May 9. *ED's diary*.

Babington, Charles Cardale, 1808-95. Botanist, entomologist and archaeologist. At Cambridge B and CD competed in beetle collecting. 1837 Founded Cambridge Ray Club as a successor to Henslow's evening soirees. In May 1838 Babington exhibited *Dyticidae*, which CD had collected in South America. From 1842 editor of *Annals and Mag. of Nat. Hist.* 1842 Descriptions of the species of Dytiscidae collected by CD. (*'Only connect', learned societies in 19th century Britain*, William C. Lubenow, 2015). 1851 FRS. 1861 Prof. Botany Cambridge, succeeding Henslow.

Bacon, William. Tobacconist of Cambridge. The shop was located along Sidney Street, where Boot's the chemists now stands. 1828 Jan. 26- CD lodged over this shop in Sidney Street, "for a term or two". LL1:163. See van Wyhe, *Darwin in Cambridge*, 2014 for illustrations of building and further information.

Badel, Pauline, 1852-? Nurse from Switzerland at Down House. In 1881, ED wrote Mar. 29 "Pauline came". Took over from Mary Anne who was to marry Arthur Parslow. "A fine hitter" at cricket. F. Darwin, *Story of a childhood*, 1920, pp. 43 and 56.

Baer, Karl Ernst, Ritter von, Edler von Huthorn, 1792-1876. Embryologist. Born in part of Russia that is now Estonia. Of German parents who were Russian subjects. See J.A. Rogers, The reception of Darwin's *Origin of species* by Russian scientists, *Isis*, 64, 1973, pp. 488-93. 1834 Librarian Academy of Sciences, Prof. of comparative anatomy and physiology at the Medico-surgical Academy, St. Petersburg. 1860 Aug. B wrote to Huxley generally pro-*Origin*, although he never fully accepted CD's views. 1861 CD refers to B in Historical sketch of 3d edn of *Origin*.

Bagshaw's Directory for Kent. 1847 described CD as "farmer".

Bahia (Salvador, Brazil). 1832 Feb. 22-Mar. 18 *Beagle* at and CD ashore. 1836 Aug. 1-17 *Beagle* returned and CD ashore. *Zoological diary*, pp. 26-30, 91-106, 314-15. *Geological diary*: DAR32.3-14, both transcribed in *Darwin Online*.

Bahia Blanca, Argentina. A military outpost, known as Fort Antonio, separating the Pampas from Patagonia. 1832 Sept. 6-28 *Beagle* at. 1833 Aug. 25-Sept. 6 CD passed through on his journey from Rio Negro to Buenos Aires. His passport from the local government survives, signed 14 Aug. 1833. *Geological diary*: DAR32:41-48, 63-74 and 38:954-956, in *Darwin Online*.

Bain, Alexander, 1818-1903. Scottish philosopher and educationalist. Met CD at Moor Park Hydropathic Establishment in 1857. 1873 CD to about B's theory of spontaneity. CCD21:444. Recollection in *Darwin Online*. (F2024).

Bains, Edward, 1801-82. Fellow of Christ's College, Cambridge, who bet CD after dinner on 23 Feb. 1836 that CD could not guess the height of the Combination room ceiling. CD lost the bet. See van Wyhe, *Darwin in Cambridge*, 2014, p. 109.

Baily, John. "Baily the poulterer". CCD5-8. Dealer in fancy pigeons, poultry and rabbits in London. Mentioned many times in *Variation* and LL2. CD arranged tickets for him to attend a lecture by Huxley in 1860. He was trying to get a half-lop rabbit for CD as was Baker (see below). B was on CD's presentation list for *Origin*.

Baird, Spencer Fullerton, 1823-87. American naturalist and ornithologist. 1850-78 Assistant Secretary Smithsonian Institution. 1867 B showed CD's *Queries about expression* to George Gibbs. 1878- Secretary Smithsonian Institution.

Baker, either Samuel C. or Charles N. B, poultry dealers in Chelsea, London. They provided CD with different varieties of fowls. 1861 B was trying to get a half-lop rabbit for CD as was John Baily. CCD9:64.

Baker, Charles B., 1803-75. 1836 Dec. A missionary at Bay of Islands, New Zealand. CD was shown round by him. (*Shorter publications*, F1640).

Baker, Thomas, 1771-1845. Admiral, commanding the South American station, 1829-33. The South American station was operational base for the Royal Navy. In *Rio notebook*, Chancellor & van Wyhe, *Beagle notebooks*, 2009.

Balch, Charles Leland, 1840-72. 1871 Offered CD membership of the New York Liberal Club. CCD19. CD's reply published: *Shorter publications*, F1981.

Balfour, Sir Arthur James, 1st Earl of Balfour, 1848-1930. Cambridge friend of CD's sons. Statesman. 1882 Was on "Personal Friends invited" list for CD's funeral, with his sister. 1888 FRS. 1902-05 Prime Minister. 1916 OM. 1922 1st Earl, KG. ED recorded in her diary 1873 Jun. 21 "Arthur Balfour". c.1870 recollection of CD in Balfour, *Chapters of autobiography*, 1930, pp. 37-8, transcribed in *Darwin Online*.

Balfour, Francis Maitland, 1851-82. Embryologist. Younger brother of Sir Arthur James B. Strong personal friend of CD's sons at Cambridge. 1878 FRS. 1880 Jul. CD lunched with at Cambridge. 1881 Oct. B took tea with CD and ED at Cambridge. "He has a fair fortune of his own. He is very modest, and very pleasant, and often visits here [Down House] and we like him very much". LL3:251. B told George Darwin that he had never seen an experiment carried out except under anaesthesia. LL3:203. 1881 *A treatise on comparative embryology*, 2 vols. 1882 Prof. Animal Morphology Cambridge. 1882 B was on "Personal Friends invited" list for CD's funeral. Frequent visitor between 1872-81. *ED's diary*. 1882 Jul. Killed climbing on the Aiguille Blanche, part of the Mont Blanc massif in Switzerland.

Balfour, John Hutton, 1808-84. Botanist. 1841 Prof of Botany, Glasgow. 1845 Prof of Botany, Edinburgh. 1856 Regius Keeper of Royal Botanic Gardens. 1861 CD sent B Gray's *Natural selection not inconsistent with natural theology*. CCD9.

Bangor, Caernarvonshire, presently County Gwynedd, Wales. 1831 Aug. CD visited on geological trip with Sedgwick. 1843 Jun. CD visited.

Bar of sandstone off Pernambuco. 1841 On a remarkable bar of sandstone off Pernambuco, on the coast of Brazil, *The London, Edinburgh and Dublin Phil. Mag.*, 19: pp. 257-60 (*Shorter publications*, F266). Communicated by CD. Foreign editions: 1841 French (F267). 1904 Portuguese (F268). 1936 Russian (F270). 1959 Portuguese, English, French, as a pamphlet, (F269).

Barbier, Edmond, c.1834-80. Translator of CD's works into French (*Journal, Origin* 6th ed., *Variation* 2d edn, *Descent* 2d edn). 1880 ED in summer wrote to Sara Darwin telling her of having B and Francisque Sarcey to luncheon. ED2:308.

Barclay, Lucy, 1757-1817. Married Samuel John Galton. Mother of Samuel Tertius Galton.

Barellien, Mlle. 1865 B taught Elizabeth (Bessy) Darwin French at Down House.

Barlaston Lea, Staffordshire. Home of Francis Wedgwood, near Upper House. 1852 CD and ED visited on journey to Rugby, Betley and Shrewsbury. 1866 Home of Clement Wedgwood on marriage. 1878 Jun. CD and ED visited.

Barlow, Mrs. "My father used to quote an unanswerable argument by which an old lady, a Mrs Barlow, who suspected him of unorthodoxy, hoped to convert him:— 'Doctor, I know that sugar is sweet in my mouth, and I know that my Redeemer liveth'". *Autobiography*, p. 96.

Barlow, Nora, see Emma Nora Darwin.

Barmouth, Caernarvonshire, now Merioneth/Meirionnyd district, county Gwynedd. 1828 Summer, CD went on a study holiday under G.A. Butterton. 1829 Jun. CD visited with Frederick William Hope to collect beetles, but CD had to return home after two days owing to illness. 1829 Aug, CD visited with sisters. 1830 Aug, CD visited with Frederick William Hope and Thomas Campbell Eyton. 1831 CD visited alone after geological tour with Sedgwick. 1869 Jun. 10-Jul. 30 Darwin family holiday at Caerdeon, two miles east of, on north side of estuary. ED recorded in 1869 Jun. 14 "drove to Barm". Jul. 30 "slept at Stafford", returned home Jul. 31.

Barnacles. "Then where does he do his barnacles?" This story of a one of the Darwin children's misunderstanding that every father works on barnacles is Lubbock's. ML1:38. For CD's work on barnacles see *Cirripedia*.

Barnacles, for CD's publications on, see Cirripedia.

Barrande, Joachim, 1799-1883. French invertebrate palaeontologist. 1855 CD to Huxley, CD to Lyell, CD had proposed him for Foreign Member of Royal Society. He was not elected. ML1:81, ML2:231.

Barraud & Jerrard, (active 1873-80) later just "Barraud". Commercial London photography company. They photographed CD in c.1873-5 and 1881. See Darwin, Iconography. Barraud's most famous work is 'His Master's Voice'.

Barrow, Sir John, 1st Bart, 1764-1848. Civil Servant. 1805 FRS. 1830 Founding member of the Geographical Society, which received its Royal Charter in 1858. 1835 1st Bart. 1836 B communicated FitzRoy's paper on *Beagle* voyage to *Jrnl. Roy. Geogr.*

Soc., 6: pp. 311-43. Before May 1848 CD to Edward Cresy, CD considered that naval expeditions, especially those in search of missing vessels, were a waste of money. Barrow was much in favour of them. "That old sinner". CCD4:134.

Bartlett, Abraham Dee, 1812-97. 1859-97 Superintendent, Zoological Society's Gardens, Regent's Park, London. Renowned taxidermist. His shop was near the British Museum. In this capacity CD had contact with B. Among the animals B stuffed were the first gorilla ever shown in England (1858), and the famous elephant Jumbo owned by P.T. Barnum, of circus renown. Frequently helped CD by answering queries and sending materials. Letters with CD published in 1900 (F2183).

Basket, Fuegia, 1821?-1883? Woman, native name Yok'cushly, of the Alakaluf tribe from the western islands of Tierra del Fuego. See Keynes, *Fossil, finches and Fuegians*, 2002. 1830 Mar. After one of the *Beagle*'s boats was stolen B was captured as a hostage. She was named "Basket" to commemorate the return of the crew to the *Beagle* in a woven basket. Taken to England by FitzRoy, then aged about 9. 1833 Jan. 23 B returned in *Beagle* and aged only 12 married York Minster. She "daily increases in every direction except height". *Beagle diary*, p. 65. 1839 FitzRoy gives her name in Alikhoolip language as Yokcushlu. ?1843 "Captain Sulivan...heard from a sealer, that...he was astonished by a native woman coming on board who could talk some English. Without doubt this was Fuegia Basket. She lived (I fear the term bears a double interpretation) some days on board". *Journal*, 2ᵈ edn, p. 229. c.1872 T. Bridges saw her, and again in 1883 when she was old and "nearing her end". Two engraved portraits of B in *Narrative* 2, facing p. 324.

Basque. First edition in: 1994 *Origin* (F2147).

Bassett, North Stoneham, Southampton. 1862-1902 Ridgmount, home of William Erasmus D, sold on death of his wife Sarah. First record in *ED's diary* in 1872 Jun. 8.

Bassoon. Francis Darwin of CD: "Finding the cotyledons of Biophytum to be highly sensitive to vibrations of the table, he fancied that they might perceive the vibrations of sound, and therefore made me play my bassoon to it". LL1:149.

Bate, Charles Spence, 1819-89. FRS. Practiced dentistry in Swansea, then in Plymouth. One of CD's dentists. J. Hayman & J. van Wyhe, Charles Darwin and the dentists. *Jrnl. of the Hist. of Dentistry*, 2018, vol. 66, no. 1, pp. 25-35.

Bateman, James, 1811-97. Botanist and plant breeder especially of orchids. Sent CD plants of *Angraecum sesquipedale*, a native of Madagascar, which is now known to be fertilized by a sphingid moth, *Xanthopan morgani*, also known as Morgan's sphinx moth or Darwin's Hawk Moth, with proboscis about 20-35cm long. On the basis of the flower's anatomy CD predicted (in *Orchids*) that an insect with a proboscis of this length would exist; it was not discovered until 1903.

Bates, Henry Walter, 1825-92. Traveller and naturalist. Biography: G. Woodcock 1969; H.P. Moon 1977. 1861 Married Sarah Ann Mason. 3 sons, 2 daughters. 1861 CD sent B 3ᵈ edn of *Origin*. CCD9. 1863 CD was most impressed by *Naturalist on the river Amazons*, "the best work on natural travels ever published in England". B published an altered quotation from an 1840s letter by Wallace to incorrectly suggest the two had gone to the Amazon, not just to be collectors, but to "solve the problem of the origin of species", see John van Wyhe, A delicate adjustment: Wallace and Bates on the Amazon and "the problem of the origin of species". *Jrnl. of the Hist. of Bio.* (2014). 1863 Review of Amazons book, in *Nat. Hist. Rev.*, 3: pp. 385-9, is not by CD, although it is attributed to CD in early printings of Everyman edn of the book and from there by British Museum printed catalogue. 1863 Review of B's paper on insect fauna of the Amazon valley, which discusses Batesian mimicry, *Trans. Lin. Soc. of London*, 23: pp. 495-566, in *Nat. Hist. Rev.*, 3: pp. 219-24 is by CD. (*Shorter publications*, F1725) 1864-92 Assistant Secretary to Royal Geographical Society. 1871 FLS 1881 FRS. 1892. [Advice to B.] in: Anon. Obituary: *Proc. of the Roy. Geogr. Soc.* 14 (4): 245-57, pl. I, p. 251. F2162. 1862 Apr. 18 and 21 visited CD.

Bathurst, New South Wales, Australia. 1836 Jan. 20 CD visited from Sydney. 1949 A monument was erected to commemorate CD's 1836 visit. *Beagle diary*, p. 396; *Geological diary* DAR38.812-836 and *Sydney Mauritius notebook*. Chancellor & van Wyhe, *Beagle notebooks*, 2009.

Baxter, William W., 1829-1900. Dispensing chemist in Bromley. CD ordered spermaceti, and many other mixtures. Corresponded much with B between 1842-82.

Bayley, Major, [1832] May-Jun. CD to Emily Catherine Darwin mentions B as a Shrewsbury friend. CCD1:231ff. Possibly William Bayley, of the banking firm Rocke, Eyton, Campbell, Leighton and Bayley in Market Square, Shrewsbury. Misspelled Bagley in Barlow 1945, p. 69.

Bayly (or **Bayley**), **Georgina Sarah**, 1807-59. 1841 Married as 2ᵈ wife Lancelot Baugh Allen.

Beagle **[I]** His/Her Majesty's Ship, sometimes called His Majesty's Surveying Vessel. Third of the name (after the dog breed). Cherokee class 10 gun brig-sloop rigged as a brig. Built at Woolwich on the Thames. 1820 May 11 Launched. 1825 Rerigged as a barque. Displacement 235 tons; length of gundeck 90'; extreme breadth 24' 6"; keel for tonnage 73' 7 7/8"; light draught 7' 7" forward, 9' 5" aft.

Beagle laid ashore, River Santa Cruz. By Martens, from FitzRoy's *Narrative.*

Measurements differ slightly. no. 41 of a class of 107 ten-gun brigs which were nicknamed "coffins", or "half-tide rocks", from their ability to go down as sea swept over waist in bad weather. Guns varied, normally 7; 1x6lb carronade, 2x6lb fore guns, 2x6lb aft guns, 2x9lb, all brass. Heavily modified by FitzRoy before the second voyage. Much error has appeared in descriptions of the *Beagle*. See L. Darling, HMS *Beagle*, 1820-1870: voyages summarized, research and reconstruction. *Sea Hist.*, Spring, no. 31, 1984, 27-38; K. Thomson, *HMS Beagle*, 1995; K.H. Marquardt, *HMS Beagle survey ship extraordinary*, 1997 and J. Taylor, *The voyage of the Beagle*, 2008.

Sketch of the *Beagle* in Sydney harbour in 1839 by Conrad Martens. MLNSW, PXC 295.

Iconography: This is the first list of visual portrayals of HMS *Beagle*. Only contemporary or eye-witness depictions are recorded, 56 in total. The list contains some previously unknown illustrations of the ship.

1832 Oct. Sketch by CD, only 4 pencil lines, hull and three masts, "ship was thus" part of map of Punta Alta. *Rio notebook*. Chancellor & van Wyhe, *Beagle notebooks*, 2009, p. 93.

183? Sketch of *Beagle* at sea in full sail by John Clements Wickham, frontispiece in Murray's editions of *Journal* from 1901 (F97). The original drawing has not been found.

183? Sketch of floor plan of the poop cabin (by P.G. King?), annotated by CD. DAR44.16.

1832 Purported watercolour caricature of CD and *Beagle* crew entitled "Quarter Deck of a Man of War on Diskivery or interesting Scenes on an Interesting Voyage" attributed to Augustus Earle (20.5x34cm). Sold at Sotheby's in 2015 for £52,500. Published in black and white on cover of conference programme: Darwin's menagerie. Humanities West. San Francisco, 1998.

1831 View of the deck and skylight, "Crossing the line" by Augustus Earle, engraved on steel by Thomas Landseer for *Narrative* vol. 2.

1833 Aug. 3 Pencil sketch of *Beagle* at anchor by Conrad Martens signed "Monte Video August 3. 1833." Sketchbook I, CUL.

1833 Dec. 25 Watercolour "Slinging the monkey" by Conrad Martens. Sketchbook III f.27, CUL.

1834 Feb. 8 "Sedger River, Port Famine. Feby 8th 1834. Port Famine" with *Beagle* at anchor by Conrad Martens. Sketchbook IV, f.1. See Organ 1996.

1834? Pencil sketch "H.M.S. Beagle working up Santa Cruz river – on the coast of Patagonia" by Syms Covington, MLNSW, PXD 41 f5 d. A sketch based on this in an apparently modern style is in the NMM, reproduced in Darling 1984 and Keith Thomson, *HMS Beagle*, 1995.

1834? Scrimshaw depicting the *Beagle* rounding Cape Horn signed "J.A. Bute" with depiction of hauling boats up Santa Cruz on reverse. Western Australia Museum, Perth, CH1972.480.

1834 Scrimshaw "H M Sloop Beagle laid on shore to repair her Forefoot" signed "J A Bute". Sold at Bonhams in 2009 for £40,800. Very reminiscent of 'Beagle laid ashore'. See Bute, James Adolphus.

c.1834 Scrimshaw purportedly by J.A. Bute: "HBM Sloop Beagle laid on shore at Santa Cruz Eastn Patagonia, to repair her fore foot, in April 1834. J. A. Bute". Eldred's Lot #109 sold for $18,000.

c.1834 Scrimshaw purportedly by J.A. Bute: "HBM Sloop Beagle laid on shore at Santa Cruz Eastn Patagonia, to repair her fore foot, in April 1834. J. A. Bute" with depiction of hauling the boats up the Santa Cruz on reverse. Eldred's Lot #76 sold for $47,500.

1834 Feb. 24 Drawing of *Beagle* off "Cape Spencer and Cape Horn" by Conrad Martens, Sketchbook IV f.12, engraved by S. Bull for *Narrative* vol. 1.

1834 Mar. 3 Watercolour of *Beagle* in Beagle Channel by Martens (19.5x29.6cm). NMM PAF6232.

1834 Mar. 4 Watercolour of *Beagle* in Beagle Channel by Martens (26.5x41.9cm). NMM PAF6228.

1834 Apr. 16? Watercolour(?) *"Beagle* laid ashore, river Santa Cruz" by Conrad Martens, engraved on steel by Landseer for *Narrative* vol. 2. Not traced.

1834 Jun. 9 Watercolour of *Beagle* in front of Mount Sarmiento by Conrad Martens, engraved by Landseer for *Narrative*. Sketchbook IV, f.44. Basis for 1838 oil painting. See M. Organ, Conrad Martens' *Beagle* pictures, 1996.
https://ro.uow.edu.au/cgi/viewcontent.cgi?filename=0&article=1126&context=asdpapers&type=additional

1834 Jun. 10 Watercolour of *Beagle* at "Mount Sarmiento (from Warp bay)" by Conrad Martens (20.2x29.9cm) engraved by Landseer for *Narrative*. NMM PAF6229.

1834? Watercolour, the *Beagle* in the Magdalen Channel, by Conrad Martens. Sketchbook IV f.43. See M. Organ, Conrad Martens' *Beagle* pictures, 1996.

1834 Jul. 2 Pencil sketch "At Chiloe" by Conrad Martens. Sketchbook I f.20, CUL.

1834 Watercolour "Cove in Beagle channel (Portrait cove)" by Conrad Martens (19.4x29.5cm) engraved by Landseer for *Narrative*. NMM PAF6242.

1834 Watercolour "Murray Narrow — Beagle channel" by Conrad Martens (19.2x29.1cm) engraved by Landseer for *Narrative*. NMM PAF6238.

1834 Watercolour "The *Beagle* running into Berkeley Sound, East Falkland, March 1834" by Martens (19.5x29.5cm) NMM PAF6227, engraved by J.W. Cook for *Narrative*.

1834 Watercolour of *Beagle* at Mt Sarmiento, Lomas range by Conrad Martens. Sketchbook III f.32v.

1834 Aug. 17 Pencil sketch of *Beagle* "Almendrals, Valparaiso" by Conrad Martens. Sketchbook I.

1834 Sept. 1 Sketch of *Beagle* [at anchor in Valparaiso] by Conrad Martens. Dated "Sept 1./34-". MLNSW, DL PX 13.

1834 Nov. 10 Pencil sketches of *Beagle* at Valparaiso by Martens. MLNSW, DL PX 13.

1835 Watercolour of the *Beagle* in the Murray Narrow, Tierra del Fuego by Conrad Martens. CD purchased in Australia for £3 2s. Titled "Jemmy Button Sound, Tierra del Fuego." in 1909 Christ's College exhibition catalogue. Darwin Heirlooms Trust. On display in old study at Down House.

1835 Pencil sketch of *Beagle* anchored with Aconcagua in background by Martens. Sketchbook III f. 34.

1835 *Beagle* at "Albemarle island", Galapagos, by P.G. King, engraved by S. Bull, *Narrative*.

1835 *Beagle* at "Chatham Island (Watering place)", Galapagos, by P.G. King, engraved by S. Bull for *Narrative*.

1836 Feb. 7. Pencil sketch "View in Papieti Harbour, Tahiti." by Conrad Martens. Sketchbook I f.63.

1836 Pencil sketch *Beagle* at anchor "Near Bahia (St Salvador) Brazil" by Syms Covington, MLNSW, PXD 41.

1836 Watercolour "Distant view of Moorea from Tahiti" by Conrad Martens. Sketchbook I f.59.

nd Oil painting "Port Famine, Straits of Magellan, the anchorage of H.M. ships 'Adventure' and 'Beagle', off and on,1827-30." by Conrad Martens (19.7x25.5cm). Lawsons, 22 Feb. 1966, cat no. 304, $240. Not traced. See M. Organ, Conrad Martens' *Beagle* pictures, 1996.

1838 Oil on board "H.M.S. Beagle off Mount Sarmiento, Tierra Del Fuego 18 April 1838" by Martens (18x26cm). On verso: "Mount Sarmiento. Barbara Channel by Conrad Martens 1838. Drawn to order of Capt. Phillip P King R.N...". Stolen 1988. M. Organ, Conrad Martens' *Beagle* pictures, 1996.

1838? Pen and ink sketch of the *Beagle* in the Bass Straits, moored in East Cove, Kent Group, by Graham Gore (23.9x37.9cm). NMM PAH0067.

1838? Pen and ink sketch "'Beagle' at anchor under Circular Hd. Bass Straits", by Graham Gore (24.4x37.7cm). NMM PAH0068.

1838? Watercolour of *Beagle* in full sail "Black Pyramid, [Bass Strait]" by J.C. Wickham. MLNSW SSV/141. Bears a striking similarity to an unsigned watercolour in the NMM "H.M.S. Beagle rounding the Cape of Good Hope" dated 10 Jul. 1843. In Darling 1984.

1839? "North-west part of Magnetic Island" by Graham Gore, engraved in J.S. Stokes, *Discoveries in Australia...in the years 1837-38-39-40-41-42-43*, 1846, vol. 1, facing p. 338.

1839 Mar. 9 Pencil sketch of the *Beagle* in Sydney harbour by Conrad Martens, "H.M.S. Beagle Sydney" (29x39cm). MLNSW, PXC 295. Hammocks hang from the rigging. Reproduced above.

1841 Watercolour with ink "HMS *Beagle* off Fort Macquarie, Sydney Harbour" by Owen Stanley, signed (14x22.4cm). Inscribed on back "HMS 'Beagle' Sydney April 10 1841". NMM PAD8969. Similar to the 1841 Brierly painting below.

1841? "Coepang from the anchorage." Engraved in J.S. Stokes, *Discoveries in Australia*, 1846, vol. 2, facing p. 187.

1841? Sketch and watercolour of the *Beagle* at anchor in a calm, labelled "H M S Beagle" by Oswald Walters Brierly (1817-94), (11.4x19cm). NMM PAD9324.

1842? "Passing between Bald Head & Vancouver Reef." by Graham Gore. Engraved in J.S. Stokes, *Discoveries in Australia*, 1846, vol. 2, facing p. 222.

1842? "Dangerous situation of the Beagle", by A.J. Mason. Engraved in J.S. Stokes, *Discoveries in Australia*, 1846, vol. 2, p. 421.

1842 Pencil sketch by Henry I. Campbell "Farm Cove - H.M.S. Favourite & Beagle and the Revenue Cutters N.S.W. 1842" (26.5x18.4cm). MLNSW, PXC 291, 20. C.

1843 Jul. 10 Watercolour "H.M.S. *Beagle* rounding the Cape of Good Hope", unsigned. NMM. See 1838? watercolour by Wickham above. Reproduced in Thomson, *Beagle*, 1995, p. 131 who suggests this may be by J.L. Stokes. Reproduced also in Darling 1984.

1884? Sketch by B.J. Sulivan showing position of CD's hammock in *Beagle's* poop cabin. In letter to Francis Darwin. DAR107.42-47. Reproduced in Taylor, *Beagle*, 2008, et al.

1890 Sketch, "HMS Beagle 1832" starboard cutaway by P.G. King. Engraved in *Journal of researches*, F59. "prepared at the request of Mr Hallam Murray, who used one of the drawings in…[F59] 'from old drawings and recollections'". Barlow, *Beagle diary*, 1933, p. v. et al. MLNSW, A1977, p. 811.

1890 Sketch, "HMS Beagle Upper Deck" by P.G. King. Engraved in F59.

1890 Sketch of "HMS Beagle's Quarter-deck" by P.G. King. Engraved for but not printed in F59. MLNSW, A1977, p. 813. Copy in DAR221.4.255.

1890 Sketch, floor plan of the poop cabin by P.G. King. Engraved for but not printed in F59. Copy in DAR221.4.255.

1890 Sketch, "HMS Beagle Lower deck" by P.G. King. Engraved for but not printed in F59. Copy in DAR221.4.255.

The two 1980s oil paintings of the *Beagle* by John Chancellor are the result of meticulous research. The one of the *Beagle* at the Galapagos was made a UK stamp in 2019.

1826-1830 first surveying voyage: To South America, in company with HMS *Adventure* [I], Captain Philip Parker King who commanded the expedition. *Beagle* commanded by Lieut. Pringle Stokes. 1826 Aug.-Nov. Acting command of Lieut. Skyring. 1828 Aug. 12 Stokes committed suicide, thereafter commanded by FitzRoy. *Beagle* [I], First voyage. Patagonia and Tierra del Fuego. Extracted from a journal of the surveying expedition composed of His Majesty's Ships *Adventure* and *Beagle*. 1830 *United Services Jrnl.*, part 2: pp. 461-7 (Oct.), 671-79 (Nov.), 793-800 (Dec.).

1831-1836 second surveying voyage: To South America and round the world 1831 Dec. 27 to 1836 Oct. 2. Total time away from England 1,740 days (1835 Nov. 15 crossed date line, one day lost). Commanded by Commander FitzRoy, Captain 1835 Dec. On 2^d voyage carried two 9lb guns and four carronades; special fittings included upper deck raised 8-12", Lihou's rudder, Harris's conductors on all masts, 22 chronometers: 11 government, 6 FitzRoy, 4 on loan from makers, 1 Lord Ashburnham. Complement varied. CD wrote in the *Beagle diary*, p. 84, on 24 Jul. 1832:

I procured this evening a Watch-bill & as most likely our crew will for rest of the voyage remain the same.— I will copy it.—Boatswains mates, J. Smith & W. Williams:—Quarter-Masters, J. Peterson, White, Bennett, Henderson:—Forecastle Men, J. Davis; Heard: Bosworthick (Ropemaker); Tanner; Harper (sailmaker); Wills (armourer);—Foretopmen, Evans; Rensfrey; Door; Wright; Robson;

MacCurdy; Hare; Clarke;—Main top-men Phipps; J. Blight; Moore; Hughes; Johns B.; Sloane; Chadwick; Johns; Williams; Blight, B.; Childs;—Carpenters crew, Rogers; Rowe; J. May; James; Idlers, Stebbing (instrument mender); Ash gunroom steward; Fuller, Captains do; R Davis, boy do; Matthews, missionary; E Davis, Officers cook; G Phillips, ships cook; Lester, cooper; Covington, fiddler & boy to Poop-cabin; Billet, gunroom boy; Royal Marines.—Beazeley, sergeant; Williams, Jones, Burgess, Bute, Doyle, Martin, Middleton, Prior (midshipmens steward);— Boatswain, Mr Sorrell; Carpenter, Mr May; Midshipmen, Mrss Stewart, Usborne, Johnson, Stokes, Mellersh, King, Forsyth.— Hellyar, Captains clerk.—Mr Bino, acting surgeon—Mr Rowlett, purser.—Mr Chaffers, Master.— Mr Sulivan, 2d Lieutenant; Mr Wickham, 1st Lieutenant; R. FitzRoy, Commander.— There are (including Earl, the Fuegians & myself) 76 souls on board the Beagle.

Not mentioned are Robert McCormick, Peter Benson Stewart, 1808-? Mate (CD spelled "Stuart") and Harry Fuller, Captain's steward. FitzRoy gives lists of those on board at the start and end of the voyage in *Narrative* 2:19-21 and further details. See CCD1: Appendix III.

CD on board as supernumerary, a guest of FitzRoy, throughout voyage, but mostly on shore when *Beagle* was surveying. CD was not on board as a social companion to the captain but as official naturalist as approved by the Admiralty. See John van Wyhe, "my appointment received the sanction of the Admiralty": Why Charles Darwin really was the naturalist on HMS *Beagle*'. *Studies in Hist. and Phil. of Biol. and Biomedical Sciences* (Sept.), 2013. In *Darwin Online*.

Summary: Details of day-to-day positions and ports of call are given in *Narrative*, vol. II appendix and in the *Beagle diary*. The only complete day-by-day itinerary of CD's whereabouts during the entire voyage is by Kees Rookmaaker and is published in *Darwin Online*: http://darwin-online.org.uk/content/frameset?viewtype=text&itemID=A575&pageseq=1. The following is a greatly abbreviated extract from Rookmaaker's 90-page itinerary:

28 Dec. 1831 Devonport to Canary Islands.
16 Jan. 1832 Porto Praya on St Jago.
08 Feb. Cape Verde to Fernando Noronha.
20 Feb. 1832 Fernando de Noronha (Brazil).
21 Feb. 1832 At sea: Fernando to Bahia.
28 Feb. 1832 Bahia (Salvador da Bahia).
18 Mar. 1832 At sea: Bahia-Rio de Janeiro.
27 Mar. 1832 Abrolhos (Brazil).
31 Mar. 1832 At sea: Bahia-Rio de Janeiro.
04 Apr. 1832 Rio de Janeiro (Brazil).
05 Jul. 1832 At sea: Left Rio harbour.
26 Jul. 1832 Montevideo (Uruguay).
20 Aug. At sea: Montevideo-Bahia, Point Piedras.
06 Sept. 1832 Bahía Blanca (Argentina).
19 Oct. 1832 At sea: Baia Blanca.
26 Oct. 1832 Montevideo (Uruguay).
30 Oct. 1832 Sea: Montevideo to Buenos Aires.
02 Nov. 1832 Buenos Aires (Argentina).
10 Nov. 1832 Sea: Buenos Aires to Montevideo.
14 Nov. 1832 Montevideo (Uruguay).
27 Nov. 1832 At sea: Montevideo to San Blas, Santa Maria Cliffs.

03 Dec. 1832 San Blas (Bahia San Blas).
05 Dec. 1832 At sea: Patagonia Coast.
16 Dec. 1832 Tierra del Fuego, South of Cape St. Sebastian (Cape San Sebastian).
21 Dec. 1832 At sea: Tierra del Fuego, Near Bay of Good Success.
24 Dec. 1832 Wigwam Cove (Hermite).
31 Dec. 1832 At sea: Tierra del Fuego.
13 Jan. 1833 Behind False Cape Horn.
14 Jan. 1833 At sea: Tierra del Fuego.
15 Jan. 1833 Goree Sound (Navarino).
19 Jan. 1833 Towards Ponsonby Sound.
07 Feb. 1833 Goree Sound.
10 Feb. 1833 North of Orange Bay.
21 Feb. 1833 Good Success Bay.
26 Feb. 1833 At sea: To Falkland Islands.
01 Mar. 1833 Port Louis (Falkland Islands).
06 Apr. 1833 Sea: Patagonia Coast to Rio Negro.
26 Apr. 1833 Montevideo (Uruguay).
28 Apr. 1833 Maldonado (Uruguay).
08 Jul. 1833 At sea: Maldonado to Montevideo.
09 Jul. 1833 Montevideo (Uruguay).

13 Jul. 1833 Maldonado (Uruguay).
18 Jul. 1833 At sea: Near Maldonado.
21 Jul. 1833 Maldonado (Uruguay).
24 Jul. 1833 At sea: Maldonado-Rio Negro.
03 Aug. 1833 Rio Negro (Balneario Massini).
17 Aug. 1833 Bahia Blanca (Argentina).
20 Sept. 1833 Buenos Aires (Argentina).
02 Oct. 1833 St Fe (Santa Fe, Argentina).
05 Oct. 1833 Paraná, over river from Sante Fe.
21 Oct. 1833 Buenos Aires (Argentina).
02 Nov. 1833 Boat (Packet)-Monte Video.
04 Nov. 1833 Monte Video (Uruguay).
06 Dec. 1833 At sea: Montevideo to Port Desire.
23 Dec. 1833 Port Desire (Puerto Deseado).
04 Jan. 1834 At sea: Port Desire to Port St Julian.
09 Jan. 1834 Port St Julian (Argentina).
21 Jan. 1834 At sea: From Port Desire.
02 Feb. 1834 Port Famine (Puerto del Hambre).
10 Feb. 1834 At sea: Leaves Port Famine.
24 Feb. 1834 Wollaston Island (Chile).
26 Feb. Straits of Magellan, Wollaston Island.
28 Feb. 1834 East end of Beagle Channel.
02 Mar. Straits of Magellan to Ponsonby Sound.
04 Mar. 1834 Ponsonby Sound (Hoste).
06 Mar. Straits of Magellan to East Falkland.
10 Mar. 1834 Berkeley Sound.
07 Apr. 1834 At sea: Falkland to Santa Cruz.
13 Apr. 1834 Mouth of River Santa Cruz.
12 May 1834 At sea: Santa Cruz to Port Famine.
01 Jun. 1834 Port Famine (Puerto del Hambre).
08 Jun. 1834 At sea: Magdalen Channel.
28 Jun. 1834 S. Carlos in the island of Chiloe.
13 Jul. 1834 At sea: Chiloe to Valparaiso.
23 Jul. 1834 Valparaiso (Chile).
10 Nov. 1834 At sea: Valparaiso to Chiloe.
21 Nov. 1834 San Carlos, excursion in a boat.
11 Dec. 1834 At sea: Chiloe to Chonos.
13 Dec. 1834 Chonos Archipelago (Chile).
18 Dec. 1834 At sea: Chonos Archipelago.
21 Dec. 1834 Tres Montes (Península de Taitao).
28 Dec. 1834 At sea: Tres Montes.
29 Dec. 1834 Yuche Island (Chile).
05 Jan. 1835 At sea: Tres Montes to Chonos.
07 Jan. 1835 Chonos Archipelago (Chile).
15 Jan. 1835 At sea: Chonos to Chiloe.
18 Jan. 1835 S. Carlos (Ancud, Chile).
04 Feb. 1835 At sea: Chiloe to Valdivia.

08 Feb. 1835 Valdivia harbour (Chile).
22 Feb. 1835 At sea: Valdivia-Concepcion.
04 Mar. 1835 Concepcion, Quiriquina Island.
07 Mar. 1835 At sea: Concepcion to Valparaiso.
11 Mar. 1835 Valparaiso (Chile).
14 May 1835 Coquimbo (Chile).
22 Jun. 1835 Copiapò (Chile).
06 Jul. 1835 At sea: Copiapò to Iquique.
12 Jul. 1835 Iquique (Chile).
14 Jul. 1835 At sea: Iquique to Lima.
15 Sept. 1835 At sea: Chatham Island.
16 Sept. 1835 Chatham Island.
02 Oct. 1835 Sea: Albemarle (Isabella) to James.
08 Oct. 1835 James Island, Buccaneer Cove.
17 Oct. 1835 At sea: James Island (Santiago).
18 Oct. 1835 Along Albemarle Island.
20 Oct. 1835 At sea: Galapagos to Tahiti.
15 Nov. 1835 Tahiti (French Polynesia).
26 Nov. 1835 At sea: Tahiti-New Zealand.
21 Dec. 1835 Bay of Islands, Pahia (Paihia).
30 Dec. 1835 At sea: New Zealand-Sydney.
12 Jan. 1836 Sydney (Australia).
30 Jan. 1836 At sea: New South Wales to Hobart.
05 Feb. 1836 Hobart Town, Tasmania.
17 Feb. 1836 Tasmania to King George Sound.
06 Mar. 1836 King George Sound (Albany).
14 Mar. King George Sound to Keeling Islands.
01 Apr. 1836 Southern Keeling Islands.
12 Apr. 1836 At sea: Keeling to Mauritius.
29 Apr. 1836 Port Louis (Mauritius).
09 May 1836 Mauritius to Cape of Good Hope.
31 May 1836 Cape Town (South Africa).
18 Jun. 1836 At sea: Cape to St Helena.
08 Jul. 1836 St Helena (British Overseas).
14 Jul. 1836 At sea: St Helena to Ascension.
19 Jul. 1836 Ascension (Saint Helena).
23 Jul. 1836 At sea: Ascension to Bahia.
01 Aug. 1836 Bahia (Salvador da Bahia).
06 Aug. 1836 At sea: Bahia to Pernambuco.
12 Aug. 1836 Pernambuco (Recife, Brazil).
17 Aug. 1836 At sea: Brazil to St Jago.
31 Aug. 1836 Porto Praya, St Jago, Cape Verde.
04 Sept. 1836 At sea: St Jago to Terceira.
20 Sept. 1836 Azores (Terceira, Portugal).
25 Sept. 1836 At sea off St Michaels.
02 Oct. 1836 Falmouth, UK.

During the South American part of the voyage, FitzRoy used up to 7 inshore vessels: 4 schooners for inshore surveying work, *Adventure* [2], *La Liebre*, *La Paz* qq.v., and one, of 35 tons, whose name is not given, which was at first, 1835

Jun., loaned by Antonio José Vascunan of Coquimbo, when B.J. Sulivan surveyed parts of the Chile coast. It was later bought, and A.B. Usborne surveyed the whole coast of Peru after *Beagle* left for Galapagos Islands; finally sold at Paita, Peru. Barrow, [FitzRoy], *Jrnl. Roy. Geogr. Soc.*, 6: pp. 311-43, 1836.

1837-1843 third surveying voyage: To New Zealand and Australia. 1837-41 Under command of Captain John Clements Wickham until he retired due to ill-health. 1841-43 Captain John Lort Stokes. 1839 a harbour in north-west Australia named by Wickham "Port Darwin", now Darwin, capital city of the Northern Territory. 1843 Nov. 17 Finally paid off. Later history: 1845-70 Coastguard Watch Vessel on river Roach, near Pagglesham, Essex, with masts and all gear removed. 1863 Name removed and numbered W.V.7. 1870 May 13 Sold to Murray & Trainer (or T. Rainer) for scrap and towed to Thames estuary.

Beagle Channel, Tierra del Fuego. Divides Isla Grande to the north from Isla Hoste and Isla Navarino to the south. Surveyed and named on 1st voyage of *Beagle*.

Beagle diary. CD's "journal" account of the *Beagle* voyage. He sent it home in instalments during the voyage. The bound manuscript is now kept at Down House. Its 751 pages total c.190,000 words. It was later used to write *Journal* (1839) which totals c.223,000 words. First transcribed and published by Nora Barlow, *Charles Darwin's diary of the voyage on H.M.S. Beagle*, 1933 (F1566). A facsimile volume was published by Genesis Publications, 1979. The next edition was Richard Darwin Keynes ed., *Charles Darwin's Beagle diary*, 1988 (F1925), transcribed in *Darwin Online*. *Diary* transcribed again, complete, by Kees Rookmaaker ed. 2006, *Darwin's Beagle diary (1831-1836)* in *Darwin Online*.

Beagle field notebooks. Fifteen pocket notebooks used by CD mostly during his shore excursions during the voyage. They total 116,000 words and contain over 300 sketches and doodles. Much of the information from them fed into the *Beagle diary*, *Zoological diary* and *Geological diary*. The first substantial extracts published were in Nora Barlow, *Charles Darwin and the voyage of the Beagle*, 1945. A complete transcription was published as Gordon Chancellor and John van Wyhe eds. with the assistance of Kees Rookmaaker. *Charles Darwin's notebooks from the voyage of the Beagle*. Foreword by Richard Darwin Keynes, 2009. Also in *Darwin Online*. Rookmaaker also created an important Chronological register to the notebooks and a Concordance to Darwin's *Beagle* diaries and notebooks (and *Geological diary*), in *Darwin Online*.

Beagle Islands, Small islands in Galapagos group between James and Indefatigable Islands. 1892 Official Ecuadorian name. Another Beagle Island exists north of Tasmania, and one off the west coast of Australia.

Beagle **library**. Consisted of c.400 volumes, housed in the poop cabin where CD worked and slept. See CCD1: Appendix IV for a reconstructed catalogue. An expanded catalogue of the library and virtually all of its volumes have been recreated in *Darwin Online* (See also Library, CD's):
http://darwin-online.org.uk/BeagleLibrary/Beagle_Library_Introduction.htm

Beagle **specimens** (records of). CD and his servant Syms Covington prepared numbered lists of most of the specimens to act as catalogues of the different collections. The main catalogue for animals and plants is now at Down House in six notebooks. Three are labelled "Catalogue for Specimens in Spirits of Wine" with numbers 1-660, 661-1346, 1347-1529 respectively. The other three parts are for specimens not in spirits of wine labelled "Printed Numbers", 1-1425, 1426-3344 and 3345-3907. (Transcribed in Keynes, *Zoology notes*, 2000.) These numbers correspond to the labels attached to specimens and used in the *Zoological diary* and other notes. Beginning in the summer of 1836 CD and Covington prepared at least 13 collection specific catalogues to give, along with the respective collection, to an expert for identification and study. "Animals" (i.e. mammals) is in DAR29.1.A1-A49. It was transcribed and edited with an introduction by Richard Keynes in *Darwin Online*. There is also "Mammalia in Spirits of Wine" in DAR29.3.76-77. "Fish in Spirits of Wine" is in DAR29.1.B1b-B20. Transcribed in Pauly, *Darwin's fishes*, 2004. A corrected and enlarged version is in *Darwin Online*. The list of birds is in DAR29.2.1-85, transcribed and edited by Nora Barlow as 'Ornithological notes', 1963, in *Darwin Online*. Insect lists include "Insects in Spirits of Wine" DAR29.3.44 and Insect notes at the NHM. All are transcribed and edited in Smith, Darwin's insects, 1987 (in *Darwin Online*). "Shells in Spirits of Wine" are in DAR29.1.D1-D8, transcribed in *Darwin Online*. There is also a list of shells in DAR29.3.4-8 (transcribed in *Darwin Online*) which includes some annelids, barnacles, bivalves, corals and gastropods and "List of Mr Darwin's Shells" in DAR29.3.3 and DAR29.3.31-33 is a list of molluscs in spirits, including land slugs, snails, nudibranchs, and insects not in spirits. "Reptiles in Spirits of Wine" [and amphibians] is in the NHM (NHM-405052-1001) and is transcribed in *Darwin Online*. The surviving lists of plants have been transcribed and edited by Duncan Porter in 'Darwin's notes on *Beagle* plants', 1987, also in *Darwin Online*. DAR29.3.41-43 has lists of Algae and invertebrate specimens originally in spirits. There is also a brief 'List of fossil woods' in the NHM (NHM-408865-1001), transcribed in *Darwin Online*. On these see B.A. Thomas, Darwin and plant fossils. *The Linnean* 25, no. 2 (Apr. 2009). (The references to and quotations from CD's *Beagle field notebooks* in this article are from the transcriptions published in *Darwin Online*, although not credited.) There are four notebook catalogues of CD's mineralogical specimens at the Sedgwick Museum of Earth Sciences, Cambridge. Alfred Harker made a fair copy in 1907. Photographs and a transcription are in *Darwin Online*: 'Catalogue of the "Beagle" Collection of Rocks". Most of these diverse documents and their contents are explained in detail by Richard Keynes in *Zoology notes*, 2000, pp. 317ff (in *Darwin Online*) and Duncan Porter, The *Beagle* collector and his collections. In D. Kohn ed., *The Darwinian heritage*, 1985, pp. 973-1019. In P. Barrett, *Collected papers*, 1977, c.70 publications describing CD's specimens from the voyage of the *Beagle* were listed. *Darwin Online* lists and also provides over 180 such publications. http://darwin-online.org.uk/specimens.html. See also CD's notes on the preservation of specimens, DAR29.3.78, also transcribed in *Darwin Online*.

Beans. 1857 Bees and the fertilisation of kidney beans, *Gardeners' Chron.*, no. 43: p. 725 (*Shorter publications*, F1697). See CCD6:465-7. 1858 On the agency of bees in the fertilisation of papilionaceous flowers and on the crossing of kidney beans, *Gardeners' Chron.*, no. 46: pp. 828-9 (*Shorter publications*, F1701), reprinted in *Annals and Mag. of Nat. Hist.*, 2: pp. 459-65. Notes in DAR27.1. See CCD7.

Beaton, Donald, 1802-63. Plant breeder, working gardener and hybridizer. See Britten and Boulger, *A biographical index of British and Irish botanists*, 1893, 2^d edn, 1931 and CCD11. 1861 CD "I can plainly see that he is not to be trusted". CCD9. 1863 B's assertion against Carl Friedrich von Gaertner's work is controverted by CD in *Cottage Gardener* 29: p. 93 (*Shorter publications*, F1727a). B's reply to CD in *ibid.* 29: pp. 70-1, Influence of pollen on the appearance of seed. More information on B from *ibid.* 30: pp. 266, 385, 415. See CCD.

Beaufort, Rear Admiral Sir Francis, 1774-1857. Originator of the Beaufort Scale of wind speeds. Was a personal friend of FitzRoy. A. Friendly, *Beaufort of the Admiralty*, 1977. 1803 B visited CD's father at Shrewsbury re skin disease. 1814 FRS. 1829-55 Hydrographer to the Navy. 1831 B offered CD post of naturalist on *Beagle* through George Peacock. 1832-36 FitzRoy's letters to B, during 2^d voyage of *Beagle*, contain many comments on CD; extracts in F. Darwin, *Nature*, 88: pp. 547-48, 1912; Barlow, *Cornhill*, 72: pp. 493-510, 1932 (both in *Darwin Online*). 1848 KCB.

Bees. See also Humble bees. 1857 Bees and the fertilisation of kidney beans, *Gardeners' Chron.*, no. 43: p. 725 (*Shorter publications*, F1697). 1858 On the agency of bees in the fertilisation of papilionaceous flowers and on the crossing of kidney beans, *Gardeners' Chron.*, no. 46: pp. 828-9. (*Shorter publications*, F1701). CD was working on a document on bee's cells shortly before he received Wallace's letter from Ternate on 18 Jun. 1858. CD mentioned working on "Bees cells" in his 'Journal', for Apr. 1858. A MS headed "Last Sketch" is in DAR48.B67-B74, transcribed in *Darwin Online*. CD published his views on bee's cells in *Origin*, pp. 224-235. W. Swale sent CD a letter with four honeybees stuck to it. CD sent it to *Annals and Mag. of Nat. Hist.*, (Hive-Bees in New Zealand) 2: pp. 459-65 (in *Darwin Online*). 1860 Intercourse between common and Ligurian bees, *Cottage Gardener*, no. 24: p. 143 (*Shorter publications*, F1814). 1862 Bees in Jamaica increase the size and substance of their cells, *Jrnl. of Horticulture*, (15 Jul.): p. 70 (*Shorter publications*, F1826). 1862 Bee-cells in Jamaica not larger than in England, *ibid.*, (22 Jul.): p. 323 (*Shorter publications*, F1824). 1874 Recent researches on termites and honey bees, *Nature*, 9: pp. 308-9 (*Shorter publications*, F1768), introducing letter from Fritz Müller. Foreign translations: 1913 On the flight paths of male humble bees (Dutch) (F1583h). 1971 Russian (F1591). CD notes on bees in DAR46, DAR48 and DAR194.

Beesby, a village in Lincolnshire. 1845 CD bought a farm including 325 acres of land for £13,592 borrowed from his father; rent 1845 £377, 1877 £555 16*s*. 1845 Sept. CD visited the only time "to see a farm I have purchased". CCD3:258, Keith, *Darwin revalued*, 1955, p. 222. Atkins 1976, p. 100. Most of that time tenanted by

Francis Hardy. 1903 Sold by William Erasmus Darwin (for the Executors of CD, deceased) to the tenant, Edward Ash Young. See Worsley, *The Darwin farms*, 2017.

Beetles. 1828-32 CD collected avidly when at Cambridge, encouraged by William Darwin Fox. Some of his early collecting records are published in J.F. Stephens, *Illustrations of British entomology*, [1827-]28-36, suppl., 1846, about thirty records in first 5 vols. of *Mandibulata*. All available in *Darwin Online*, see van Wyhe, *Darwin in Cambridge*, 2014. A list of his beetles collected in 1828-9 is given in the *Edinburgh notebook*, DAR118, transcribed in *Darwin Online*. 1829 Feb. 20 Frederick William Hope gave CD specimens of about 160 species of British beetles in London. 1829 CD went on beetle collecting tour with Hope to Barmouth, but CD was ill and had to return to Shrewsbury after two days. A cabinet CD had made to store his collection while in Cambridge was passed down to Milo Keynes who lent it to Christ's College in 2009 for display in CD's rooms. See van Wyhe, *Darwin in Cambridge*, 2014 for photographs. Some CD Cambridge beetles survive in the Zoology Museum, Cambridge. 1859 Coleoptera at Down, *Entomologist's Weekly Intelligencer*, 6: p. 99 (*Shorter publications*, F1703), a note signed by Francis, Leonard and Horace, who were 10, 8 and 7 years old, clearly written by CD. LL2:140.

Behrens, **Georg Wilhelm Julius**, 1854-1903. German botanist. Assistant of Julius von Sachs in Würzburg. 1878 CD wrote to on fertilisation of plants by insects, praising Sprengel, and thanking B for sending his Geschichte der Bestaubungs-Theorie, *Progr. K. Gewerbschule zu Elberfeld*, 1877-78.

Belfast. 1827 CD visited on a spring tour.

Bell, **Lady Caroline**, 1836 "Lady Caroline Bell, at whose house I dined at the C. of Good Hope, admired Herschel much, but said that he always came into a room as if he knew that his hands were dirty, and that he knew that his wife knew that they were dirty". *Autobiography*, p. 107. Emma Wedgwood in 1837 wrote to Elizabeth Wedgwood that "Lady Bell was very sweet, engaging, and pretty". ED(1904)1:389.

Bell, **Sir Charles**, 1774-1842. Physician, anatomist and surgeon. 1812-36 Surgeon to Middlesex Hospital. 1826 FRS. 1831 Kt. 1833 Author 4th Bridgewater Treatise: *The hand, its mechanism and vital endowments as evincing design*. 1836-42 Prof. Surgery Edinburgh. CD had high admiration of his *Anatomy and philosophy of expression*, 1806, quoting in *Expression* from 3d edn 1844 which has B's latest corrections. "Admirable work on expression". *Autobiography*, p. 138.

Bell, **John**, 1782-1876. Army officer in South Africa. CD saw on 13 Jun. 1836. In the *Despoblado notebook*, Chancellor & van Wyhe, *Beagle notebooks*, 2009.

Bell, **Thomas**, 1792-1880. Physician, dental surgeon and zoologist. 1828 FRS. 1836- Prof. Zoology King's College London. 1842-43 B wrote *Reptiles* for *Zoology of the Beagle*, and delayed completion for nearly two years through procrastination and ill-health. An extensive list of CD's reptile specimens is now in the Natural History Museum and reproduced in *Darwin Online*. B was also entrusted with describing the crustaceans, but he never completed it. See Chancellor G., di Mauro A, Ingle R.,

King G., 1988. Charles Darwin's *Beagle* Collections in the Oxford University Museum. *Archives of Nat. Hist.* 15 (2): 197-231. Retired to The Wakes, Selbourne, Hampshire, Gilbert White's house. Often at Down House in the early years. In 1855 May 4 there was a diary entry by ED "went to Mr Bell with Etty, [Henrietta]". They came home the next day. In 1858 Apr. 28, Etty was taken to London to see Mr Bell and again on May 14 the same year. 1861 CD dined with B at Linnean Club, "Bell has a real good heart". CCD9:100.

Belloc, Louise Swanton, 1796-1881. French author. 1859 Dec. CD to Quatrefages, B considered translating *Origin*, but found it technically too difficult. CCD7. 1860 Translated part of *Journal* into French (F180).

Belt, Thomas, 1832-78. Engineer, geologist and naturalist. 1874 CD to Hooker, refers to *Naturalist in Nicaragua* 1874, about glacial period. 1874 CD to Hooker, again referring to *Naturalist in Nicaragua* "It appears to me the best of all the Nat. Hist. Journals which have ever been published." CCD22:169.

Bemmelen, Adriaan Anthoni van, 1830-97. Dutch ornithologist; Chairman of Netherlands Zoological Society. 1867- Director Blijdorp Zoological Gardens Rotterdam. 1877 sent a photograph album of 218 distinguished Dutch men for CD's 68th birthday. The album is now at Down House. (EH88202653) "To Charles Darwin, on his sixty-eighth birthday, as a proof of respect, offered by his Dutch admirers." See CCD25 Appendix V for a detailed discussion.

Benchuca Bug. A large house bug of South America (*Triatoma infestans*, Reduviidae). Vector of Chagas disease. 1835 Mar. 25 "It is most disgusting to feel soft wingless insects, about an inch long, crawling over one's body". CD mentions one of the officers of the *Beagle* feeding one at Iquique. *Beagle diary*, pp. 296-8, 315. In the *St. Fe notebook*, Chancellor & van Wyhe, *Beagle notebooks*, 2009, p. 152a.

Bengali. First edition in: nd *Descent* (F2394).

Bennett, Alfred William, 1833-1902. Botanist. 1874 CD to B, when B had ceased to be assistant editor of *Nature*, asking for return of wood blocks for *Climbing plants*, 1865. CCD22. B? 1872 Apr. 13 visited Darwins with Charles Crawley. *ED's diary*. 1869-76 B exchanged about twenty letters with CD.

Bennett, James, 1804-? Born in Devonport. Served on *Arrogant* with FitzRoy. 1830-31 Gunner's Mate of *Beagle* on first voyage. Remained with FitzRoy and looked after the four, later after the death of Boat Memory, three, Fuegians when they were in England. Acted as "Captain's Coxswain" on 2d voyage from time to time, although no such rank existed and quartermaster. On part of 3d voyage. "A most deserving and long tried companion in many difficulties". FitzRoy.

Bennett, Mrs. 1841 Mary Eleanor's wet nurse. 1842 Sept. 25 £2 7s paid "Fly for Mrs Bennett & travelling expenses". CCD2:335.

Bent, Eliza and Emma, 1824-53 Visited the Wedgwoods, then the Darwins. 1839 Jan. 11 ED wrote "Miss Bent 171 Regent".

Bentham, Mr. Of Holwood, Down. 1865 Sept. Called at Down House. Apparently a new neighbour. ED2(1904):208.

Bentham, George, 1800-84. Son of Sir Samuel B. Nephew of Jeremy B. Botanist. Biography: Jackson 1906, CD discussed evolution with before *Origin*. 1844 CD discussed flora of Sandwich Islands with. 1854 B presented his books and herbarium to Kew and worked there daily. 1858 Jul. 28 CD "I have ordered Bentham, for as Babington says it will be very curious to see a Flora written by a man who knows nothing of British plants!!!" CCD7:139. 1858 Jul. 30 "I have got Bentham and am charmed with it". These two quotations refer to *Handbook of the British flora*, 1858. 1859 Royal Medal Royal Society. 1859 B accepted evolution. 1862 FRS. 1862 B approved of *Orchids* in his Presidential address to Linnean Society. 1882 B was on "Personal Friends invited" list for CD's funeral. 1882 recollections of CD in DAR112.A5-A7, in *Darwin Online*.

Berkeley Sound, East Falkland Island. 1833 Mar. 1-Apr. 6, 1834 Mar. 10-Apr. 7 *Beagle* anchored at. CD there only in 1834.

Berkeley, Rev. Miles Joseph, 1803-89. Mycologist. Vicar of Sibbertoft, Northamptonshire. William Turner Thiselton-Dyer described B as "the virtual founder of British mycology". See Edible fungus from Tierra del Fuego. 1842 B published two papers describing fungi collected by CD during the voyage of the *Beagle*. 1862 Jun. 14 B reviewed *Orchids* in *London Rev.* (14 Jun.), pp. 553-54. 1868 CD thanks B for sending a copy of his Presidential address to Section D of BAAS at Norwich. CCD16. 1879 FRS. (*Shorter publications*, F1671, F2014).

Betley, Staffordshire, near Maer. Betley Hall. Home of George Tollet. CD and ED often visited in childhood. 1852 Apr. CD and ED visited.

Betsey, 1865?-79. Domestic servant at Down House. ED recorded in 1866 Apr. 21 "Betsey 3-15". No other mention in *ED's diary* until 1894 when ED saw her again and noted on 16 Sept. that B was ill. 1896 Jun. 19 saw B again and wrote to Henrietta two days later that B was looking much better.

Biddulph, Frances, née Owen, 1807-87 see Owen, Frances Myddelton Mostyn.

Biddulph, Robert, 1761-1814 Aug. 30. 1801 Dec. 24 Married Charlotte Myddelton.

Biddulph, Col. Robert Myddelton, 1805-72. Of Chirk Castle, Denbigh. Eldest son of Robert B. 1832 Married Frances ("Fanny") Owen. Had 3 sons, 3 daughters.

Big book. This term is used very often in the literature on CD for his unpublished work-in-progress on his theory of evolution interrupted by Wallace in 1858. Although constantly quoted, a source is never given. There are even misquotations such as "Big Species Book" in Glick & Kohn, *On evolution*, 1996, p. xvi, a phrase CD never used, despite the quotation marks. There seem to be only two sources. CD to W.D. Fox 3 Oct. [1856] "I find to my sorrow it will run to quite a big Book." CCD6:238 and CD to Lyell 10 Nov. [1856] "my big Book". CCD6:256. CD later used the phrase to refer to *Variation*, which was by far his longest book.

Billiard room, Billiards. 1859 CD to W.D. Fox "I find it does me a deal of good, & drives the horrid species out of my head." CCD7:269. See Alterations to house. CD played nightly games at Moor Park. Browne, *Power of place*, 2002, p. 64.

"Biographical Sketch of an Infant". 1877 *Mind*, 2: pp. 285-94 (*Shorter publications*, F1305). Observations made by CD 1839-41 on his first-born child William, written as a result of a paper on the same subject by Hippolyte Taine, a translation of which appeared in the previous number of *Mind*, pp. 252-9. First foreign editions: 1914 Armenian (F1310). 1958 Chinese (F1310a). 1877 French (F1311), German (F1312), Greek (F2059). Russian (F1314). 1956 USA (F1309). The German translation contains additional matter at the end by CD on the perception of colour not in the English original, a translation of a letter to Ernst Krause, 14 Jul. 1877, CCD25:260. 1881 ["On the bodily and mental development of infants"], *Nature*, 24: p. 565 (*Shorter publications*, F1797), report of a letter from CD to a social science meeting at Saratoga, New York, USA.

Birch, Samuel, 1813-85. Egyptologist and antiquary of the British Museum. B's information is cited in *Origin*, p. 27. B translated passages on fowls from a 1609 Chinese encyclopaedia for CD. *Variation* 1:238. CD visited B in 1856.

Bird, Edward Joseph, 1799-1881. Naval captain and arctic explorer. 1839 Jun. 6 Dined with the Darwins with Roberts and Herbert. *ED's diary*.

Bird, Mr. 1831 sent a fly to CD through Henslow. Barlow 1967, p. 27. CCD1:126.

Bird, Isabella L., 1831-1904. Traveller and japanophile. 1881 Married John Bishop. B told CD that *Origin* was translated into and much studied in Japan (*Autobiography*). 1896 First Japanese edition (F718).

Birds. CD ed. 1841. *Birds Part 3 of The zoology of the voyage of HMS Beagle* by John Gould [& G.R. Gray] Edited and superintended by Charles Darwin. London: Smith Elder and Co. On CD's work on birds see Clifford B. Frith, *Charles Darwin's life with birds. His complete ornithology*, 2016 and F.D. Steinheimer, Charles Darwin's bird collection and ornithological knowledge during the voyage of H.M.S. *Beagle*, 1831-1836. *Jrnl. of Ornithology* 145(4) (2004), pp. 300-20, (appendix [pp. 1-40]), in *Darwin Online*.

Birmingham, Warwickshire. 1829 CD visited with Wedgwoods for music meeting. CD visited for BAAS meetings 1839 Aug. 26-Sept. 11 and 1849 Sept. 11-21.

Bismuth (bizmuth). Bismuth subsalicylate, an antacid medication used to treat conditions such as diarrhoea, indigestion, heartburn and nausea. 1840 earliest record of CD taking B on Sept. 25. See CD to Susan Darwin 3[-4] Sept. 1845, CCD3. 1849 ED recorded on Jan. 26 that CD "began Bizmuth" and was to take it with other prescriptions. 1896 Jan. 12 was a final entry, on the cost of B being 5/-.

Bishop's Castle, Shropshire. 1822 Jul. CD had a holiday at with sister Elizabeth.

Blair, Rev. Robert Hugh, 1834/5–85. Principal, Worcester College for the Blind Sons of Gentlemen. 1871-2 B helped CD with expression in the blind. CCD.

Blair, Rueben Almond, 1841-1902. Amateur naturalist of Sedalia, Missouri, USA. 1877 wrote CD to about damaged goose wing and inheritance of similar damage by offspring. CCD25. 1881 CD to B about Mastodon remains and B's daughter's love of natural history, "I hope that the study of natural history may give your daughter a large share of the satisfaction which the study has given me". Carroll 1976, 593.

Blane, Robert, 1809-71. Officer in 2^d Life Guards. Cambridge friend of CD. One of the eight members including CD of the "Gourmet Club"/"Glutton Club". 1854-55 Assistant Adjutant General and Military Secretary. 1860 Colonel.

Blaxland, Gregory (or John), 1778-53. 1769-1845, landowners and merchants in Australia. In the *Sydney notebook*, Chancellor & van Wyhe, *Beagle notebooks*, 2009.

Blomefield, Leonard, see Jenyns.

Bloom. 1886 Francis Darwin. On the relation between the 'bloom' on leaves and distribution of the stomata, *Jrnl. Lin. Soc. Bot.*, 22: pp. 99-116 (F1805). Contains results obtained by Francis Darwin working as research assistant to CD in 1878. CD's notes on the subject are in DAR66-68.

Blue Pill, a mercury-based medicine that was taken by the family from 1852. CD usually took when he had been sick. ED took quite a bit of it in her old age, the frequency peaked in 1894. See CD to ED 25 Jun. 1846, CCD3.

Blunt, Thomas, 1803-74. Chemist at Wyle Cop in Shrewsbury with Joseph Birch Salter as Blunt and Salter. CD bought distilled water from him for his chemistry. Browne, *Voyaging*, 1995, p. 31. Pattison, *The Darwins of Shrewsbury*, 2009.

"Blunts". In 1861 May 22 "Blunts" visited Down House and again in 1881 Feb. 26, "Mr Blunt to lunch". *ED's diary*. Probably the Shrewsbury family of Thomas Porter Blunt (1842-1929). See CCD1 & CD to Thomas Blunt 5 Apr. 1867, CCD15.

Blyth, Edward, 1810-73. Zoologist. Neglected his druggist business at Tooting in favour of natural history and got into financial difficulties. LL2:315. CD discussed evolution with before *Origin*. Biographical note on: ML1:62. Wrote under pseudonyms "Zoophilus" and "Z". Helped greatly with *Variation*. 1835, 1837 His early views on causes reminiscent of natural selection maintaining fixity of species *Mag. Nat. Hist.* 8, 1835, pp. 40-53 and 1837 n.s. 1: pp. 1-9. L. Eiseley claimed that CD deliberately plagiarized the idea of natural selection from these articles. *Proc. of the American Phil. Soc.* 103: pp. 94-114, reprinted in Eiseley, *Darwin and the mysterious Mr. X*, 1979, pp. 42-80. Now an entirely discredited view. See R. Colp, Loren Eiseley and 'The Case of Charles Darwin and the Mysterious Mr. X.' *Jrnl. of the Hist. of Medicine*. (Jan.), 1981. 1841-62 Zoological Curator of Museum of Asiatic Society of Bengal, Calcutta. 1848 CD to Hooker, praising B's knowledge of Indian zoology, "he is a very clever, odd, wild fellow, who will never do, what he could do, from not sticking to any one subject." CCD4:139. 1855 Dec. 8 B drew CD's attention to Wallace's 'Sarawak Law paper' in *Annals and Mag. of Nat. Hist.*, 1855. CCD5:519-22. 1860 May B wrote to CD in favour of *Origin*. 1868 Mar. Visited CD at Down House. *ED's diary* 1868 Sept. 12 "Mr Blyth & J. Weir" (Joseph Jenner Weir). There are 60 known letters between B & CD spanning 1855 to 1882.

Blytt, Axel Gudbrand, 1843-98. Norwegian Botanist. Prof. Botany Christiania (Oslo). 1876 B sent CD his work on Norwegian flora, *Essay on the immigration of the Norwegian flora*. CD much approved of it. CCD24.

Boat Memory, ?-1830. Alakaluf man from Tierra del Fuego. "A great favourite with all who knew him...a pleasing intelligent appearance...quite an exception to the general character of the Fuegians, having good features and a well-proportioned frame". FitzRoy, *Narrative*, p. 10. Was, unusually, a good swimmer. 1830 Apr. Captured as hostage for stolen boat. 1830 Aged about 20 taken to England by FitzRoy. 1830 Nov. Died of smallpox in Plymouth Naval Hospital.

Bob, Bobby. 1870 A large half-bred black and white dog at Down House. See *Expression*, p. 64. 1870 Mar. 19 ED wrote to Henrietta "He looked so human, lying under a coat with his head on a pillow". ED2:230. CD was not very fond of.

Boehm, Sir Joseph Edgar, 1st Bart, 1834-90. Austrian of Hungarian descent. Sculptor and medallist. 1882 RA. 1883 B made statue of CD at BMNH. 1885 Jun. 9. 1887 B made the deep bronze medallion of CD in Westminster Abbey.

Bolton, Thomas. Dealer in chemicals at 146 High Holborn, London, and of Birmingham. Supplied CD with artificial sea salt for experiments on the longevity of seeds. Does sea-water kill seeds? *Gardeners' Chron.* (*Shorter publications*, F1682).

Boner, Charles, 1815-70, writer who lived most of his adult life in Germany. 1871. [Two letters by CD to B]. In Kettle, ed., *Memoirs and letters of Charles Boner*. 1:76-8. (*Shorter publications*, F1950). CD cited Boner's works in *Descent* 2: 245, 253, 259, 269.

Bonham-Carter, Elinor Mary, 1837-1923. Sister of Henry Bonham-Carter. 1866 William and Henrietta stayed with C in France. 1871 Provided CD information on expression in dogs through Henrietta. 1872 Married Albert Venn Dicey. Dined with the Darwins in 1855 Jul. 17 and paid many more visits after.

Bookselling Question. 1852 [Letter of 5 May 1852 by CD on the bookselling question.] In reply to Parker, *The opinions of certain authors on the bookselling question*, p. 27 (*Shorter publications*, F1912): "As an author of some scientific works, I beg to express strongly my opinion, that, both for the advantage of authors and the public, booksellers, like other dealers, ought to settle, each for himself, the retail price." John William Parker (1792-1870), London bookseller and publisher. See CCD5.

Boole, Mrs Mary Everest, 1832-1916. Mathematician. Widow of George B. 1866 B writes to CD to ask whether natural selection was consistent with the belief that each individual had the power to choose how far to yield to "hereditary animal impulses" and how far to follow "moral motives". CD replied: "it has always appeared to me more satisfactory to look at the immense amount of pain & suffering in this world, as the inevitable result of the natural sequence of events...rather than from the direct intervention of God". CCD14:426.

Boott, Dr Francis, 1792-1863. American-born physician and botanist in England from 1816. 1820 Married Mary Hardcastle of Derby. Her mother may have been an extramarital daughter of Erasmus Darwin. 1832-39 Secretary, Linnean Society of

London. 1856-61 treasurer. 1861-63 vice-president. 1838 Aug. CD dined with at Athenaeum. 1860 Mar. 8 CD to Gray, CD has had a long letter from B "full of the most noble love of truth and candour. He goes far with me but cannot swallow all. No one could until he had enlarged his gullet by years of practice, as in my own case". CCD8.

Bosnian. First editions in: 1878 *Origin* (F2498). 1922 *Descent* (F2499).

Bosquet, Joseph Augustin Hubert de, 1814-80. Dutch carcinologist of Maastricht. 1854 CD sent B copy of *Living Cirripedia*. 1856 B named *Chthamalus darwini*, a fossil barnacle from the Chalk, for CD and sent him specimen. 1856 CD to B who was apparently also interested in carrier pigeons. CCD6.

Bosworthick, John. Old shipmate of FitzRoy. Ropemaker on *Beagle*'s 2^d voyage.

Botanic Garden, Cambridge. New Botanic Garden, Trumpington Road. Owns CD's set of *Gardeners' Chron.* 1846 Opened, designed by J.S. Henslow.

Boucher de Crèvecoeur de Perthes, Jacques, 1788-1868. French geologist and archaeologist. Director of Customs, Abbeville. 1847-64 B, in *Antiquités Celtiques et Antediluviennes*, described flint artefacts with bones of rhinoceros and hyena at Abbeville. 1863 CD complains to Lyell that he had not done B justice in *Antiquity of man*, "Must be a very amiable man". CCD11:243.

Botafogo Bay, Rio de Janeiro, Argentina. Used as a base and address by CD.

Bournemouth, Hampshire. 1862 Sept. 1-27 CD on family holiday after visit to William at Southampton. "Came to Bournemouth" *ED's diary*, 1 Sept. 1862.

Bowcher, Frank, 1864-1938. Sculptor and engraver. Studied under Alphonse Legros in Paris. 1898 B designed medal for Linnean Society to commemorate Hooker's 80th birthday. 1908 B designed Darwin-Wallace medal for Linnean Society.

Bowen, Charles Synge Christopher, Baron Bowen, 1835-94. Judge. 1862 Married Emily Frances Rendel. 2 sons, 1 daughter, Ethel Kate B.

Bowen, Ethel Kate, 1869-1952. Poet. Daughter of Charles S. C. Bowen. 1894 Married Josiah Clement Wedgwood [IV] as 1st wife. 1913 Separated; 1919 Divorced.

Bowen, Francis, 1811-90. American theologian. 1853-89 Prof. Natural Religion, Moral Philosophy and Civil Polity, Harvard. 1860 Anti-*Origin* reviews in *Mem. American Acad. Arts Sci.* and *N. American Rev.*, as editor 1843-54.

Bowerbank, James Scott, 1797-1877. Distiller. A founder of London Clay Club. 1842 FRS. 1851, 1854 Secretary of the Palaeontographical Society when CD published *Fossil Cirripedia*. 1864-82 Best known work *British Spongiadae*.

Bowman, Sir William, 1816-92. Ophthalmic surgeon. 1841 FRS. 1868 CD called on him in London, but he was away. He had done some kindness to one of CD's sons. CCD16. 1868- Provided much information for *Expression*. *Expression*, pp. 160, 192. 1882 B was on "Personal Friends invited" list for CD's funeral. 1884 1st Bart.

Brace, Rev. Charles Loring, 1826-90. American philanthropist, social reformer and practical christian. 1872 Summer, visited Down House with his wife, Letitia Brace née Neill. ED recorded in 1872 Jul. 11 "Mr & Mrs Brace". B wrote to a friend from Down House 1872 Jul. 12, telling of the most cordial welcome he received from the Darwins. He recalled at dinner CD sitting "very erect, to guard his weakness" and was "simple and jovial as a boy". Recollections and letter of CD in B., *The life of Charles Loring Brace*, 1894. (F2086), transcribed in *Darwin Online*.

Bradford, Elizabeth. Housemaid at Down House in 1881.

Bradley, George Granville, 1821-1903. 1881-1902 Dean of Westminster Abbey. 1882 B's name is on admission cards for CD's funeral. He was abroad at the time and sent his consent for CD's burial in Westminster Abbey by telegram in French "Oui sans aucune hésitation regrette mon absence".

Bradshaw, Henry, 1831-86. Elected University Librarian, Cambridge in Mar. 1867. Fellow of King's College. CD in 1876 wrote to thank him for obtaining a translation by a learned rabbi of the Naphtali letter by Naphtali Lewy. 1872 Sept. 5 visited CD. 1884 May 18-25 dined and stayed. The Hookers were also staying 17-19 May.

Braille. English Braille editions of CD's works: 1916 *Journal* (F168). 1934 *Origin* (F629). 1962 *Autobiography* (F1509).

Brass Close. Darwin family estate at Kirton, Lincolnshire. 1722 Ann Darwin née Waring, bequeathed in her will, dated 1722 May 18, "the rents from Brass Close for four poor widows" who were to be provided with "4 grey coats" with a badge of red cloth "cut in the shape of Two Great Roman Letters A.D." 1879 Leonard Darwin visited Kirton when the piece of land was known as "Darwin's Charity".

Braun, Alexander Carl Heinrich, 1805-77. German botanist and plant morphologist. 1864 B was an early convert to CD's views on species. 1864 CD to Benjamin Dann Walsh mentioning B as convert. CCD12.

Brayley, Edward William, 1801-70. Geologist. A free-lance lecturer. 1854 FRS. 1865 Prof. Physical Geography London Institution. See next entry.

Brayley Testimonials. 1845 *Additional testimonials submitted to the Council of University College, London, By Edward William Brayley...a candidate for the Professorship of Geology*, London, Richard & John E. Taylor printed (*Shorter publications*, F324). CD's testimonial p. [7]. CD did not contribute to the testimonials, for the same chair, of 1841. The chair was not filled because the College could not find the salary. CCD3:133.

Brazil, Emperor of, Pedro II, 1825-91. 1878 Jun. Expressed a wish, whilst in England, to meet CD, but CD was away from home.

Bree, Charles Robert, 1811-86. Naturalist and anti-Darwinian. 1860 *Species not transmutable, nor the result of secondary causes*. CD's comments on. CCD8. 1860 CD to Hooker, "You need not attempt Bree", "He in fact doubts my deliberate word, and that is the act of a man who has not the soul of a gentleman in him". CCD8:444. 1872 *An exposition of the fallacies in the hypothesis of Mr. Darwin*. CD replied in: Bree on Darwinism, *Nature*, 6: p. 279 (*Shorter publications*, F1756).

Brehm, Alfred Edmund, 1829-84. German zoologist and writer on popular natural history. 1864-69 *Illustriertes Thierleben*, 6 vols., Hildburghausen. 1868 CD to the publishers about an English translation, not recommending it; one never appeared. CD used fourteen illustrations from it in *Descent*. See CCD16.

Brent, Bernard P. 1855 or 1856 A member of the Columbarian Society. CD in letter to William E. Darwin 1855 "Mr Brent was a very queer little fish". CCD5:508.

Bressa Prize. 1879 awarded to CD by Reale Accademia della Scienze. Turin. 12,000 lire. CD recorded £418 18*s.* 10*d.* for 'Bressa prize' on 17 Jan. 1880. Account books—banking account (Down House MS). CD gave £100 from it to Anton Dohrn's Zoological Station at Naples. See CCD27.

Bridge, Sir [John] Frederick, 1844-1924. Organist and composer. 1875-1918 Organist at Westminster Abbey. 1882 B composed and played anthem for CD's funeral, "Happy is the man that findeth wisdom, and the man that getteth understanding". *Proverbs* iii 13-17. In *Darwin Online*, A204. 1897 Kt. DAR215.4e.

Bridgen, George, 1840-? Footman at Down House in 1861 census as 21 years old.

Bridges, Esteban Lucas, christened **Stephen,** 1874-1949. Anglo-Argentine explorer and farmer in Tierra del Fuego. Second son of Thomas B. Born at Ushuaia and spent most of his life at Harberton. 1947 *Uttermost part of the earth*, New York, contains later information on the three Fuegians who returned home on 2^d voyage of *Beagle*. Chapter 1 is about *Beagle* voyages; also detailed information on Indian tribes, especially Yahgan.

Bridges, Thomas, 1842-98. Missionary and later farmer in Tierra del Fuego. See Esteban Lucas Bridges above, and Freeman and Gautrey, *Jrnl. for the Soc. of the Bibliography of Nat. Hist.*, 7: pp. 259-63, 1975. 1856 B arrived at Keppel Island Mission Station, West Falkland Islands. 1860 CD sent some preliminary queries about expression to—information from Admiral Sulivan about. CCD8. 1871 Oct. Set up home at Stirling House, Ushuaia, and mission to the Fuegians. See Keynes, *Fossils, finches and Fuegians*. 2002. 1887 Built farm at Harberton.

Briggs, Mark. Coachman to Robert Waring Darwin [II] and later to Susan Elizabeth Darwin until her death in 1866. 1832 Married Anne Latham, a laundrymaid at The Mount. B had sometimes to go before Dr Darwin to test the strength of the floors and stairs of his patients' homes. Pattison, *The Darwins of Shrewsbury*, 2009, p. 54. 1867 Mar. 3 Erasmus Alvey Darwin wrote to CD that he had promised B £20 per annum and that CD would continue the payment after his death. B would also stay in his house near The Mount rent free. ED(1904)2:13. CCD15:121. 1875 Alive.

Brighton, Sussex. 1853 Jul. CD visited on day trip from Eastbourne. In 1849 Aug. 29, Brodie and Anne went there. The Darwins went in 1855 Oct. 17- Dec. 17.

Brinton, William, 1823-67. Physician. Specialist on the stomach at St. Thomas's Hospital, London. 1863 CD saw, on the recommendation of George Busk, during his six months illness. 1864 FRS.

Brisbane, Matthew, 1787?-1833. Scottish Antarctic explorer. First British Resident at Falkland Islands. Was in employ of Louis Vernet who held Falkland Islands from Spanish Government in Argentina. 1833 Aug. 26 Murdered in an uprising of imported South American labour at Port Louis. 1834 CD, from Port Louis, to Edward Lumb, "Such scenes of fierce revenge, cold-blooded treachery, and villany in every form, have been here transacted as few can equal it". CCD1. In the *Falkland notebook*, Chancellor & van Wyhe, *Beagle notebooks*, 2009.

Bristowe, Mary Ann, 1800-29. Sister of W.D. Fox. 1827 CD to Fox mentions Fox's two charming sisters. CCD1. 1829 CD to F condoling on her early death.

British Association for the Advancement of Science (BAAS). 1831 Founded and first met at York. CD went to meetings at: 1839 Birmingham. 1846 Southampton. 1847 Oxford. 1849 Birmingham (at which he was a Vice-President). 1855 Glasgow (his last). 1860 Oxford; details of the Huxley/Wilberforce encounter at this meeting in LL2:320-323, ML1:156. There are many other versions of what was said, none of them verbatim. An excellent one in A.F.R. Wollaston, *Life of Alfred Newton*, 1921, pp. 118-21. See James 2005 for the most complete compilation of accounts. 1860 "When Professors lose their tempers and solemnly avow they would rather be descended from apes than Bishops; and when pretentious sciolists seriously enunciate follies and platitudes of the most wonderful absurdity and draw upon their heads crushing refutations from the truly learned". *Guardian*, (4 Jul.), p. 593. Francis Darwin 1892, pp. 236-42 gives an extended version. 1900 Tuckwell, *Reminiscences of Oxford*, p. 50. 1923 Huxley "There was inextinguishable laughter among the people, and they listened to the rest of my argument with great attention". *Nature*, p. 920. 1958 "The Bishop...had turned to Huxley and mockingly asked him whether he reckoned his descent from an ape on his grandfather's or on his grandmother's side? to which Huxley retorted 'If the question is put to me, would I rather have a miserable ape for a grandfather or a man highly endowed by nature and possessing great means and influence, and yet who employs those faculties and that influence for the mere purpose of introducing ridicule into a grave scientific discussion—I unhesitatingly affirm my preference for the ape'". Ellegård, *Darwin and the general reader*, 1958, p. 68. 1891 Huxley to Francis Darwin "When he turned to me with his insolent question, I said to Sir Benjamin [Brodie] in an undertone, 'The Lord hath delivered him into my hands'". Francis Darwin, 1892, p. 240. after 1860 Many Presidential Addresses and addresses by Presidents of Section D, after 1860, give an excellent summary of the progress of evolutionary thought.

British Museum, Trustees. 1858 *Enquiry by the Trustees of the British Museum*, (F345), contains letter from CD to Roderick Impey Murchison. CCD7.

British Museum (Natural History). Now the Natural History Museum, London. Holds many of CD's specimens from the *Beagle* voyage. 1838 Copy of a Memorial presented to the Chancellor of the Exchequer [Rice], recommending the Purchase of Fossil Remains for the British Museum. *Parliamentary Papers*, (27 Jul.): 1. Memorial signed by CD (*Shorter publications*, F1944). 1847 Copy of Memorial to the First Lord of the Treasury [Russell], respecting the Management of the British Museum. *Parliamentary Papers* (268), (13 April). Memorial signed by CD (*Shorter publications*, F1831).

1858 Memorial of the promoters and cultivators of science on the subject of the proposed severance from the British Museum of its natural history collections, addressed to Her Majesty's Government. *House of Commons Papers* (23 Jul.). Memorial

Base of the Skull of *Toxodon platensis* discovered by CD along a small stream entering the Rio Negro about 120 miles northwest of Montevideo. Plate I of *Fossil Mammalia.* The specimen illustrated at natural size required a fold-out sheet.

signed by CD (*Shorter publications*, F1942). 1858 [Letter to the Chancellor of the Exchequer on the removal of British Museum.] *Gardeners' Chron.*, 27 Nov. 1858, p. 861. CCD7, Appendix VI, pp. 525-30. 1859 [Letter on the herbarium of the British Museum.] *The Times* (18 Mar.): p. 12 (F1933). 1859 [Letter on the collections of the British Museum.] In Northcote, Further communications by architect and officers of British Museum on enlargement of building: plan. *Parliamentary Papers* (11 Mar.) (*Shorter publications*, F1934). 1866 *Memorial to the Chancellor of the Exchequer*, signed by CD (F869), 1873 [Letter from P.L. Sclater containing text of 1866 Memorial], *Nature*, 9: p. 41 (F870). 1875 BMNH established in South Kensington. 1885 Jun. 9 Statue of CD by Boehm unveiled.

Brodie, Jannet (Jessie), c.1790-1873. 1842-51 Scottish nurse at Down House aged 49 in 1842. Came from previous service with the Thackerays and Anne Thackeray (Mrs Richmond Ritchie). 1851 Left after death of Anne Elizabeth Darwin and returned to family home at Portsoy, Scotland. Continued to visit the Darwins. *ED's diary* mentions of Brodie in 1855 Jul. 7 "went to Hartfield with W(illiam), G(eorge), F(rank) & Brodie. 1862 Sept. 18 "drove out with Brodie", 1862 Sept. 25 "walked with Brodie" and 1865 Sept. 26, "drove out with Brodie". ED wrote to her often, but she had a monomania that she was forgotten. ED2:214. See R. Keynes, *Annie's box*, 2001. Her death on Jul. 3 1873 was recorded in *ED's diary*.

Brodie, Sir Benjamin Collins, 1st Bart, 1783-1862. Physician. 1810 FRS. 1853 ED consulted. 1855 PRS. 1860 Apr. CD went to reception at his house. 1860 Jun. B sat next to Huxley during Wilberforce's speech at Oxford BAAS meeting.

Bronn, Heinrich Georg, 1800-62. German palaeontologist and zoologist. 1822- Prof. Natural History Heidelberg. 1859 CD sent 1st edn *Origin*. 1860 B translated *Origin* into German, adding his own notes at CD's suggestion and slightly altering the text. CD was not pleased with the result. (F672), CCD9.

Brooke, Rajah Sir Charles Anthony Johnson, née Johnson, 1829-1917. Second British Rajah of Sarawak. 1868 B succeeded his uncle, Sir James B (1803-68). 1870 Nov. 30 B answered CD's *Queries about expression* from Sarawak.

Brooks, William, 1802-? Gardener at Down House. Census of 1861. Foulmouthed and morose. Lived in a cottage close to cow yard. Wife Keziah, son private in Guards. F. Darwin, *Springtime*, 1920, p. 57. Daughter Emily, same age as Anne Darwin; they were playmates. R. Keynes, *Annie's box*, 2001, p. 108.

Brooks, William Keith, 1848-1908. American zoologist. Studied anatomy and embryology of marine animals. 1883 published *The law of heredity*, with dedication "To the memory of Charles Darwin – from whose store of published facts I have drawn most of the material for this volume."

Brougham, Lord Henry, 1778-1868. Barrister and Whig politician, founder of the *Edinburgh Rev.* CCD8:664. CD read his *Dissertations*. CD noted in his reading notebooks "very good" and "profound". CD cited B in *Foundations*, pp. 3, 17, 117.

Brown. There is a Mr Brown in the *Red notebook*, p. 155 who Sandra Herbert 1980 suggests might have been Admiral William Brown (1777-1857) of Buenos Aires; an Irishman that CD met at Parish's house in 1837 (p. 71 of the *Red notebook* itself).

Brown(e), Andrew. Farmer and superintendent of Wallerawang, Australia. See Nicholas and Nicholas, *Charles Darwin in Australia*, 1989, pp. 44-5, *Beagle diary*, p. 401 and the *Sydney notebook*, Chancellor & van Wyhe, *Beagle notebooks*, 2009.

Brown, Jane, 1746-1835. Daughter of Joseph Brown of Swineshead, Lincolnshire. CD's great aunt-in-law. 1772 Married William Alvey Darwin [I].

Brown, Robert, 1773-1858. Botanist. First Keeper of Botany at British Museum. Dilatory in describing plants of first voyage of *Beagle*. CD gave him a list of specimens of fossilized wood from voyage of *Beagle*. List is in NHM, and transcribed with introduction in *Darwin Online*. B examined the specimens sometime between the end of Mar. to mid-May 1837. (See CCD2, letter to J.S. Henslow, 18 [May 1837]). Biography D. Mabberley, 1984. 1800 FLS. 1811 FRS. 1849-53 President Linnean Society. 1858 CD to Hooker, "I am glad to hear that old Brown is dying so easily". CD "I saw a good deal of". *Autobiography*, p. 103. 1858 Jul. 1 The Darwin/Wallace papers were read at Linnean Society special meeting because of B's death.

Brown-Séquard, Charles Édourd, 1817-94. French/Mauritian physiologist and neurologist, a pioneer endocrinologist and neurophysiologist. Met CD in 1858. 1862 Offered to review *Origin* in a French periodical. CD cited in *Variation* 1:105, 259.

Browne, Sir James Crichton-, 1840-1938. Physician. Director of West Riding Pauper Lunatic Asylum, Wakefield. Gave CD information and photographs for *Expression*. Sent CD Annual Reports of the Asylum, the run now being at CUL. See CCD21. 1870 FRSE. 1875-1922 Visitor in Lunacy. 1883 FRS. 1886 Kt.

Browne, William Alexander Francis, 1805-85. Physician of Stirling. Naturalist friend of CD at Edinburgh, and member of Plinian Natural History Society. Father of Sir James Crichton B. 1857-70 First Commissioner in Lunacy for Scotland.

Brullé, Gaspard Auguste, 1809-73. French zoologist. 1840- Prof. Zoology and Comparative Anatomy Dijon. 1864 Falconer to CD "He told me in despair that he could not get his pupils to listen to anything from him except à la Darwin!" CCD12:390.

Brummidge, Mrs. c.1890 Cook at Down House. CD once played B's hand at whist for her so that she could get on with meal preparation. "She was a superb cook in an uncompromisingly English manner and her roasts and her dumplings were remembered by the Darwin grandchildren for many years." Atkins 1976, p. 75.

Brunton, Sir Thomas Lauder, 1st Bart, 1844-1916. Physician. Consultant at St. Bartholomew's Hospital, London. B helped CD with experiments for *Insectivorous plants*. 1874 FRS. 1881 Nov. 19 CD to B about prosecution of Ferrier under the Vivisection act. CD wanted to be a subscriber if a subscription was got up to pay Ferrier's costs. CD met Ferrier at B's house, 50 Welbeck St. ML2:437. 1908 1st Bart.

Bryanston Street, London. No. 2. Richard Buckley and Henrietta Lichfield's house. First record of visit to B was in 1872 Jun. 5. Subsequently there would be many visits. ED recorded a "Bry. Sq. ball" in 1876 Feb. 18.

Bryce, James, 1838-1922. 1882 Jan. 22 came to luncheon with A.V. Dicey. 1909 B made a speech at a CD centenary celebration. *Proc. of the American Phil. Soc.* 48, no. 193 (Sept.): iii-xiv, transcribed in *Darwin Online*.

Buch, Christian Leopold von, 1774-1853. German geologist and palaeontologist. In *Geological diary* and the *Buenos Ayres notebook, Beagle notebooks*, 2009.

Büchner, Ludwig, 1824-99, German philosopher, physiologist, physician and atheist. 1881 visited Down House with Aveling. Recollection of CD in *Last words on materialism and kindred subjects*, 1901, pp. 147-8, transcribed in *Darwin Online*.

Buckland, Rev. William, 1784-1856. Geologist. Father of Francis Trevelyan B. "Though very good-humoured and good-natured seemed to me a vulgar and almost a coarse man". *Autobiography*, p. 102. 1812 Prof. Mineralogy Oxford. 1818 FRS. 1837 Discussed iguanas from the Galapagos with CD. 1845-56 Dean of Westminster.

Buckle, Henry Thomas, 1821-62. Self-educated historian. c.1842 CD met at Hensleigh Wedgwood's and discussed organization of facts. 1858 CD to Hooker "I was not much struck with the great Buckle". CD was reading B's *History of civilization* at the time. CCD7:31. "I doubt whether his generalisations are worth anything". *Autobiography*, pp. 109-110.

Buckley, Arabella Burton, 1840-1929. Natural historian and author. Secretary to Lyell. 1871 Mar. Visited Down House with the Lyells. 1871 *A short history of natural science*. 1876 Feb. 11 CD to B saying that he had enjoyed B's *A short history of natural science*. 1879 Alerted CD to Wallace's financial need, leading CD et al to successfully petition for a civil list pension for Wallace. CCD27:529. 1882 B was on "Personal Friends invited" list for CD's funeral. 1884 Married to become Mrs Fisher. Photograph of her in Samantha Evans ed., *Darwin and women*, 2017, p. 163.

Buckman, James, 1814-84. Agriculturist and geologist. 1848-63 Prof. Geology, Botany and Zoology Royal Agricultural College Cirencester. 1857 CD to B on varieties of domestic pigeon. CCD6:463-4.

Buenos Aires, (**Buenos Ayres**). 1832 Jul. 26-1833 Dec. 6 *Beagle* used mouth of La Plata river as a base for surveying trips. CD used B, Montevideo and Maldonado as bases for inland expeditions. [1833]. [Recollection of CD in Buenos Aires]. In *Reminiscences of diplomatic life*. F2097. CD's list of fossil shells in DAR43.2.A. *Zoological diary*, pp. 180-185. *Geological diary*: (11.1832) DAR32.75-76, all transcribed in *Darwin Online*.

Buist, George, 1805-60. Scottish journalist, naturalist and geologist. 1825-26 Studied at Edinburgh University; contemporary with CD. 1832-34 Editor *Dundee Courier*. 1839-59 Editor *Bombay Times*. 1845-46 Returned to England upon death of his wife. 1846 FRS. 1859 CD sent 1st edn *Origin*. 1860 Died at sea en-route to Calcutta.

Bulgarian. First editions in: 1906 *Descent* (F2504) 1928 *Origin* (F2503). 1959 *Autobiography* (F1511). 1967 *Journal* (F170).

Bult, Benjamin Edmund, (not Bull as in LL2:281). A London pigeon fancier and member of the Philoperisteron Society with CD. 1859 CD to Huxley "it was hinted that Mr Bult had crossed his Powters with Runts to gain size; & if you had seen the solemn, the mysterious & awful shakes of the head which all the fanciers gave at this scandalous proceeding, you would have recognised how little crossing has had to do with improving breeds, & how dangerous for endless generations the process was.— All this was brought home far more vividly than by pages of mere statements". CCD7:404-5. Described in *Variation* 1:208, as "the most successful breeder of Pouters in the world".

Bulwer, Sir Edward George Earle Lytton, Baron Lytton, 1803-73. Novelist and parliamentarian. A remote cousin of CD through Erasmus Earle (1590-1667). 1838 1st Bart. 1843 Added "Lytton" to his surname. 1858 In "one of his novels a Professor Long, who had written two huge volumes on limpets", was CD. *Autobiography*, p. 81. The novel was *What will he do with it?*, 4 vols., 1858, under pseudonym "Pisistratus Caxton". 1866 1st Baron Lytton.

Bunbury, Sir Charles James Fox, 8th Bart (1860), 1809-86. Palaeobotanist. Of Mildenhall, Suffolk. Lyell's brother-in-law. CD discussed evolution with before publication of *Origin*. Encouraged CD in pursuing species theory. 1844 Married Frances J. Horner, daughter of Leonard Horner. 1851 FRS. 1859 CD sent 1st edn *Origin*. Biography: [1894] by his wife. Recollections of CD in [K.M. Lyell], *The life of Sir Charles J. F. Bunbury*, 1906, recollections transcribed in *Darwin Online*.

Bunnett, Augustus Templeton, 1820-73. Australian. 1867 answered *Queries about expression*. He is acknowledged in the introduction of *Expression*, 1872, p. 20 (F1142).

Buob, Louise, 1822/3-? German-born governess and schoolmistress. CD's Classed account book (Down House MS) records two payments to B for 1863 Apr. and Aug. Lizzy went to B in 1863 Jan. 27, ED wrote "took Lizzy to Miss B". Miss B visited them twelve years later in 1875 Jan. 23. 1882 Recollections in: Darwin's Heim. *Ueber Land und Meer*. Transcribed in *Darwin Online* with an English translation.

Burchell, William John, 1781-1863. Explorer and naturalist. Travelled in South America and later in South Africa. CD knew in London after return of *Beagle*. *Red Notebook*, pp. 67 and 117. 1803 FLS.

Burdon-Sanderson, Sir John Scott, Bart, 1828-1905. Physician and physiologist. B helped CD with experiments for *Insectivorous plants*; CCD21, 22. 1867 FRS. 1874-82 Jodrell Prof. Physiology University College London. 1873 Jul. 4, Dr and Mrs Sanderson called on CD. 1875 S saw and agreed to Litchfield's sketch for vivisection bill. 1881 Feb. CD attended lecture by S at Royal Institution on plant movement. CD wrote to his son George Feb. 27 "We came up at this particular time that I might attend Burdon Sanderson's Lecture at R. Inst, on the movements of plants and animals compared. He gave a very good lecture."; audience applauded on CD's entrance. ED1:245. 1882-95 Waynflete Chair of Physiology Oxford. 1882 S was on "Personal Friends invited" list for CD's funeral.

Burgess, Thomas. Royal Marine on 2ᵈ voyage of the *Beagle*, became a Cheshire policeman. A scrimshaw he carved during the voyage sold for £6,000 at auction in 2010. See also Bute, James Adolphus.

Burnham Beeches. Fine woodland on Dunstable Downs. 1847 Jun. CD visited on a day trip from BAAS meeting at Oxford.

Busby, James, 1802-71. Viticulturist; introduced vine grape from Spain and France to New Zealand. 1832-40 First British Resident in New Zealand. 1835 Dec. CD met. *Shorter publications*, F1640, *Journal* 2ᵈ edn, p. 421 (spelt "Bushby").

Busk, George, 1807-1886. Surgeon and man of science. 1846 FLS. 1850 FRS. 1852, 1854 and 1875 published descriptions of marine Polyzoa collected by CD in British Museum catalogues. 1856-59 Hunterian Prof. Comparative Anatomy and Physiology Royal College of Surgeons. 1859 CD to Huxley, "I have heard that Busk is on our side in regard to species". 1863 B recommended Dr William Brinton to CD. 1871 CD to B, thanking him for pointing out an error about the supra-condyloid foramen in 1st issue of *Descent*. CCD19. 1871 Royal Medal Royal Society. 1878 Lyell Medal Geological Society. 1885 Wollaston Medal Geological Society. 1862 Apr. 6 ED recorded "J. Lubbock, Mr Busk & Mr Huxley." Visited Down with Sir John Lubbock on 1865 Dec. 17. And again on 1871 Jun. 18, 1872 Mar. 6 and 1873 Apr. 4. They called on him and Mrs Busk in 1875 Apr. 4.

Bute, James Adolphus. Royal Marine on 2ᵈ voyage of the *Beagle*. A pair of scrimshaw purportedly carved by B were sold at Bonham's for £40,800 in 2009. A rosewood and whalebone snuff box dated 1831 made by B was sold at Bonham's for £2,500 in 2018.

Butler, Miss Mary. Met CD at Moor Park. 1859 Sept. CD invited to stay at Ilkley in Oct. since he might not be able to take his family; "but if you were there I should feel safe and home-like". In the end he took his family. B did undergo treatment the same time as CD. She also met J.J. Aubertin. She then stayed at Down in 1860 Apr. 25-May 1. She died a "dreadful & prolonged death" according to a letter CD sent to Aubertin in 1871 Mar. 3. CCD19. See M. Dixon & G. Radick, *Darwin in Ilkley*, 2009.

Butler, Rev. Samuel [I], 1774-1839. Schoolmaster and priest. Father of Thomas B, grandfather of Samuel B [II]. 1798-1835 Headmaster of Shrewsbury School, including the time when CD was there. 1822 FRS. 1836-9 Bishop of Lichfield and Coventry. Pattison, *The Darwins of Shrewsbury*, 2009.

Butler, Samuel [II], 1835-1902. Author and controversialist. Son of Thomas B, grandson of Samuel B [I]. Biography: J.G. Paradis, *Samuel Butler, Victorian against the grain*, 2007. 1859 CD sent 1st edn *Origin*. 1872 May 19 CD invited B and the Woolners to Down House. 1872 May 13, CD wrote to Henrietta "I am dead tired. Woolners come on Sunday. I believe we shall ask S. Butler, author of *Erewhon*, and grandson of Dr Butler, my old master." 1880 B had a one-sided quarrel with CD over Krause's biography of Erasmus Darwin in its English version. For B's printed contributions see *Athenaeum*, Jan. 31, *St. James's Gazette*, Dec. 8. Also H.F. Jones, *Charles Darwin and Samuel Butler*, 1911. See F1992 for a letter by CD. 1880 Dec. 14 Romanes to CD, "[Butler] is a lunatic beneath all contempt. an object of pity were it not for his vein of malice". Romanes 1896, p. 104. 1881 Jan. CD to Romanes on Romanes's review of *Unconscious memory*, *Nature*, 23: pp. 285-7. B "will smart under your stricture", Romanes is right to attribute B's conduct to "the disappointment of his inordinate vanity"; CD thanks Romanes for saving him from B's "malignant revenge". 1881 Feb. CD to T.R.R. Stebbing thanking Stebbing for his letter to *Nature*, 23: p. 336 on the controversy. 1881 Apr. CD to Romanes, "I am extremely glad that you seem to have silenced Butler and his reviewers. But Mr. Butler will turn up again, if I know the man". 1881 Krause wrote a strictly accurate letter on the subject, *Nature*, 23: p. 288. *Autobiography*, gives references and reprints H.F. Jones's pamphlet in full. Butler's copy of *Erasmus Darwin*, with his manuscript notes, is in the British Library, B's books on evolution, a subject on which his knowledge was entirely theoretical, were 1879 *Evolution old and new*, 1880 *Unconscious memory*, 1887 *Luck or cunning*. *Erewhon* 1872 developed from "Darwin among the machines", *The Press*, Christchurch, New Zealand, 1863 Jun. 13; this was signed "Cellarius", a pseudonym. 1862 "Darwin on the origin of species", *The Press*, Dec. 20. H.F. Jones, *Charles Darwin and Samuel Butler*, 1911. 1872-82 recollections of CD in H. Breuer ed., *The note-books of Samuel Butler*, 1984, vol. 1, pp. 122-3, 129-31, 168, 204, 237, (F2103), transcribed in *Darwin Online*. Papers dealing with B's attack on CD, including fair copies of CD's first and second draft letters to *Athenaeum*, cuttings, reviews of B's *Evolution old and new*, and Butler's letter to *Athenaeum* are in DAR92. See also F2536.

Butler, Rev. Thomas, 1806-86. Son of Samuel B [I], father of Samuel B [II]. At St John's College, Cambridge, when CD was up. 1828 B was at Barmouth with a reading party in autumn with CD, under George Ash Butterton with John Lloyd. B and CD collected beetles together. 1834-76 Rector of Langar with Bamston, Nottinghamshire. 1839 B and CD travelled together in a stage coach from Birmingham to Shrewsbury, at end of BAAS meeting. H.F. Jones, *Life of Samuel Butler*, vol. 1, p. 13; Jones says that this is the last time that they met. 1868 Canon of Lincoln. 1872 CD to John Maurice Herbert, B has become "a very unpleasant old man". CCD16. 1882 recollections of CD as "we used to make long mountain rambles in which he inoculated me with a taste for botany which has stuck by me all my life." in DAR112.A10-A12 and in A. Silver ed., *The family letters of Samuel Butler, 1841-1886*, 1962, p. 209, both transcribed in *Darwin Online*.

Butterflies. 1880 The sexual colours of certain butterflies, *Nature*, 21 (8 Jan.): 237 (*Shorter publications*, F1787) and inherited instincts.

Butterton, George Ash, 1805-91. CD's tutor for classics and mathematics. CD "A very dull man". *Autobiography*. 1828 B took a reading party to Barmouth in autumn; CD and T. Butler were present. 1828-37 Fellow of St. John's College, Cambridge. 1839-45 Headmaster Uppingham. 1843 DD. 1847-59 Headmaster Giggleswick.

Button, c.1877 A stray minute female black and tan collie at Bassett, later thought to be a "special breed of dog from Thibet". ED2:287, Atkins 1976, p. 80.

Button, James, "**Jemmy**", 1816?-63. Boy from Yahgan Tribe, canoe people from southwest islands, Tierra del Fuego, different tribe from two of the other Fuegians on the *Beagle*. FitzRoy, *Narrative*, gives his name in Tekeenica (i.e. Yahgan) as Orundellico. FitzRoy says that he was bought for one mother-of-pearl button. E.L. 1830 Apr. Captured, "tied in a bag". FitzRoy Diary. 1830 Aged about 14, taken to England by FitzRoy. 1833 Jan. 23 Returned. 1858 Taken from home a second time to Falkland Islands mission station. 1863 He was alive and remained a bad lot; not mentioned later. His son, Threeboys B, visited England. Thomas Bridges calls him "Jimmy" and that the story about the button could not be true. *Jemmy Button: novela*, a novel by Benjamin Subercaseaux, 1950; USA translation by M. and F. de Villar, 1953; further abridged by Oliver Coburn, W.H. Allen 1955. *Jim og hans folk*, Danish children's book by S. Koustrup, 1978; Finnish translation *Tuliman Jim*, 1979. N. Hazlewood, *Savage. The life and times of Jemmy Button*, 2000. Two engraved portraits of B in *Narrative* 2, facing p. 324.

Button, Threeboys. Son of Jemmy B. 1865 Visited England with three other Fuegian youths. Died six months after return. Buried Port Stanley, Falkland Islands.

Byerbache, Mr. 1834 "a resident merchant" in Valparaiso, whom CD recorded in his 1840 paper 'On the connexion of certain volcanic phenomena in South America' (*Shorter publications*, F1656) informed him "that sailing out of the harbour one night very late, he was awakened by the captain to see the volcano of Aconcagua in activity". Also mentioned in Chancellor & van Wyhe, *Beagle notebooks* and in the *Geological Diary* DAR36.447-451. Possibly the Sardinian Consul Herr Bayerbach or Edward Beyerbache, United States Consul in Talcahuano.

Byerley, Thomas, 1747-1810. Josiah Wedgwood [I]'s partner at Etruria Works and his cousin. Son of Josiah's father's sister Margaret. Josiah II after the death of his father almost entirely entrusted the management of the potteries to B.

Bynoe, Benjamin, 1803-65. Born in Barbados. Assistant Surgeon on 1st, 2d and 3d voyage of *Beagle*; 1832 Apr.- acting surgeon on 2d voyage after departure of McCormick. 18 years on *Beagle*; gave first account of marsupial birth. CD probably met in London after return of *Beagle*. *Red Notebook*, p. 68. 1836 Surgeon. Later Medical Officer in charge of convicts. 1839 CD "Thanks...for his very kind attention to me when I was ill at Valparaiso". *Journal*, p. vii. 1844 FRCS.

Caddis-Flies. 1879 Fritz Müller on a frog having eggs on its back—on the abortion of hairs on the legs of certain caddis-flies, &c., *Nature*, 19: pp. 462-3 (*Shorter publications*, F1784); introducing a letter from Müller, *ibid*, pp. 463-4. Two woodcuts (a frog and caddis-fly hairs) accompanied this brief article in *Nature*.

Caerdeon, North Wales. Two miles east of Barmouth, on northern side of Barmouth estuary. 1869 Jun. 12-Jul. 30 CD had family holiday there. 1869 Jun. 10 the Darwins set out to London, arriving Caerdeon on 12.

Caernarvon, North Wales. 1842 Jun. CD visited and returned home 29.

Caird, Sir James, 1816-92. Agriculturalist. 1857-59 MP for Dartmouth. 1859-65 MP for Stirling Burghs. 1875 FRS. 1878 C subscribed, with CD and Sir Thomas Henry Farrer to keep Torbitt's experiments on potato disease going. CCD26.

Caldcleugh, Alexander, 1795-1858. Private Secretary to British Ambassador to Chile, later trader and plant collector in Santiago. Owner of copper mines at Panuncillo. 1825 *Travels in South America, During the years 1819-20-21*. 1831 FRS. In the *Buenos Ayres notebook*, Chancellor & van Wyhe, *Beagle notebooks*, 2009. 1834 CD stayed with at Santiago. 1835 CD to sister Susan Darwin "the author of some bad travels in South America...took an infinite degree of trouble for me". CCD1:446.

Caldwell, Mrs Anne Marsh, 1791-1874. Novelist. A friend of the Wedgwoods from childhood. Sister of Emma Holland. Family came from Linley Wood near Maer. 1817 Married Arthur Cuthbert Marsh, 1786-1849; 1858 Added Caldwell to surname. 1866 CD to C about her blind friend Mr Corbet. CCD7.

Callao, Peru. Seaport of Lima. 1835 Jul. 20-Sept. 7 *Beagle* at. Jul. 20 CD landed. *Geological diary*: DAR37.709-710, transcribed in *Darwin Online*.

Cambridge. Apart from his residence as an undergraduate, for which see Cambridge University, CD was in Cambridge on the following occasions: 1831 Sept. 2-4, 19-21 Staying with Henslow when preparing for *Beagle* voyage. 1836 Dec. 13-1837 Mar. 6 Staying with Henslow and in Fitzwilliam Street, sorting *Beagle* material. 1838 May 10-12 Visiting Henslow. ED went to the May eights in 1867 Jun. 22-25 with Amy Crofton. 1870 May 20-24 Visiting his sons, Francis, George and Horace, stayed at Bull Hotel. 1877 Nov. 16-19 CD visited with ED for award of Honorary LL.D. 1880 Aug. 14-18 CD and ED stayed with Horace in Botolph Lane. 1881 Oct. 20-27 CD and ED stayed with Horace. 1883 After CD's death, ED moved to The Grove, Huntingdon Road, for the winters. The latter Cambridge life for the Darwins is brilliantly depicted in Gwen Raverat's *Period piece*, 1952.

Cambridge Scientific Instrument Company, 1881-1968. Chairman Sir Horace Darwin, partner A.G. Dew Smith, Botolph Lane. First known as "The Shop". Made wormstone for Down House. Taken over by William G. Pye until 1898.

Cambridge Philosophical Society. Henslow and Sedgwick were the leading instigators. CD was never a member. 1819 Founded. 1835 Issued for private circulation CD's *Letters on geology* [Extracts from letters addressed to Prof. Henslow] (*Shorter publications*, F1). 1879 The members commissioned portrait of CD by William Blake Richmond, known sometimes as the red portrait, after ED's description.

Cambridge University. 1827 Oct. 15 CD entered at Christ's College, but did not come into residence until Lent term 26 Jan. 1828. CD was enrolled for an ordinary BA degree, which was a prerequisite for taking later divinity training and becoming ordained as a clergyman, something which was never realised. Contrary to common misconception, CD did not study divinity or theology at Cambridge. His studies were identical to all other undergraduates registered for an ordinary BA degree. See van Wyhe, *Darwin in Cambridge*, 2014. 1831 Jan. CD took degree examinations and kept two terms, leaving end of Jun. 10th in list of candidates who did not seek honours. 1831 Apr. 26 CD admitted BA. *Cambridge Chron.*, Apr. 29. He was "Baccalaureus ad Baptistam" and therefore included in 1832 list. LL1:163. 1831 Jun. CD left. 1836 MA. 1877 Nov. 17 Hon. LL.D. Public Orator, J.E. Sandys, ended "Tu vero, qui leges naturae tam docte illustraveris, legum Doctor nobis esto". LL3:222. 1877 Nov. 17 ED to William gives description of the scene with a monkey and a missing link lowered from the gallery by undergraduates. ED2:230, CCD25. At least two CDV photographs purportedly showing this monkey in cap and gown on copies of CD's books were produced by Hills & Saunders, Cambridge (one in profile). A copy seen has written on the verso "T Collisson placed this in the Senate House when Charles Darwin was receiving his honorary degree." CD et al. 1878. [Memorial to the Vice-Chancellor respecting the Examination in Greek]. *Cambridge University Reporter* (7 Dec.): 206-207. (*Shorter publications*, F1939)

Cambridge, Rev. Octavius Pickard-, 1828-1917. Arachnologist. 1868-1917 Rector of Bloxworth, Dorset. 1874 CD to C on natural selection and on spiders. CCD22:88. 1887 FRS.

Cameron, Rev. Jonathan Henry Lovett, 1807-88. Shrewsbury School and Trinity College. Cambridge friend of CD. Member of Gourmet Club. 1830 C was gulfed [to be in the gulf is said of an honours candidate who fails, but is allowed an ordinary degree]. 1831 B.A. 1860-88 Rector of Shoreham, Kent. 1882? recollections of CD in DAR112.A14, transcribed in *Darwin Online*. "At Cambridge I used to read Shakspere with him in his own room & he took great pleasure in these readings. He was also very fond of music, though not a performer".

Cameron, Julia Margaret, née Pattle, 1815-79. Photographer. Sister of Mrs Prinsep and Lady Somers. 1838 Married Charles Hay Cameron. 1868 CD with ED, Erasmus and Horace, visited C at Freshwater, Isle of Wight. Her house there, now a museum, is called Dimbola Lodge, and can be visited. C photographed CD, Erasmus and Horace, but not ED. A purported quotation from a letter from CD to C: "there are sixteen people in my house, and every one your friend." (Not in CCD.)

Lady Henry Somerset, [Review of H.H. Cameron, *Tennyson and his friends*.] *Woman's Signal*, (24 May 1894), p. 361. CD to Hooker: "She came to see us off & loaded us with presents of Photographs, & Erasmus called after her "Mrs Cameron there are six people in this house all in love with you". CCD16:692. CD "I like this photo graph very much better than any other which has been taken of me" a quotation from a lost letter, this was lithographed onto one of C's photographs of CD.

Campana, Cerro la, Chile. A peak 6,400 ft. high. Nowadays referred to as Darwin's Peak. Campana is Spanish for "bell". 1834 Aug. 16-17 CD climbed to summit, which now bears a plaque. *Journal* 2d edn, pp. 255-7.

Campbell, George John Douglas, 8th Duke of Argyll, 1823-1900. Statesman and geologist. 1847 8th Duke. 1851 FRS. 1862 C reviewed *Orchids* in *Edinburgh Rev*. 1864 C addressed Royal Society of Edinburgh anti-*Origin*. 1867 CD to Huxley about C's *Reign of law*, "Dukelets?, how *can* you speak so of a living real Duke?" CCD15:15. 1867 CD to Kingsley about *Reign of law*, "very well written, very interesting, honest & clever & very arrogant." CCD15:297. 1881 C "I wish Mr. Darwin's disciples would imitate a little of the dignified reticence of their master. He walks with a patient and a stately step along the paths of conscientious observation". ML1:396. 1882 C was Pallbearer at CD's funeral. Main works relating to evolution: 1867 *The reign of law*. 1884 *The unity of nature*. 1885 recollection of CD in 1882 in What is science? The substance of a lecture delivered in Glasgow. *Good Words*. (Apr.), pp. 243-4, (F2176), transcribed in *Darwin Online*. In it C said he had seen letters where writers "rejoiced in Darwin simply because they thought that Darwin had dispensed with God, and that he had discovered some process entirely independent of Design which eliminated altogether the idea of a personal Creator of the universe." But C felt sure that that was not the attitude of CD's own mind.

Camphill. House on Maer Heath, Staffordshire. 1827-47 Home of Sarah Elizabeth Wedgwood [I]. House was built for her. She moved in 1827. 1847 Sold with rest of Maer estate after Sarah Elizabeth Wedgwood [II] "Bessy" move to The Ridge, Staffordshire. Sarah Elizabeth Wedgwood [I] moved to Petley's, Down. The Darwins called in 1841 Jun. 23, 1845 Feb. 7-8, Oct. 18, 1853 Aug. 15-16.

Canary Islands. 1831 CD planned a trip there with Kirby and Ramsay, perhaps also Dawes and Henslow, before *Beagle* invitation came. See also Tenerife/Teneriffe.

Canby, Dr William Marriott, 1831-1904. Botanist of Wilmington, Delaware, USA. C provided information on *Dionaea* for *Insectivorous plants*. 1873 Feb. 19 CD to C describing *Dionaea* as "the most wonderful plant in the world". CCD21. CD 1874. [Memoranda on *Drosera filiformis*] in W.M. Canby, Observations on Drosera filiformis. *The American Naturalist* 8 (7) (Jul.): 396-397. (*Shorter publications*, F1932)

Candolle, Alphonse Louis Pierre Pyramus de, 1806-93. French botanist of Swiss origin. ED in 1839 May 27, recorded Mr Henslow, Uncle Baugh and "M. de Candolle". 1840 C dined at 12 Upper Gower Street to meet the Sismondis. 1841-50 Prof. Natural History Geneva, succeeding his father, Augustin Pyramus de Candolle

(1778-1841). 1855 C's *Géographie botanique raissonée*, 2 vols., Paris, was very important to CD in his study of cultivated plants. 1859 CD sent 1st edn of *Origin*. 1869 Foreign Member Royal Society. 1873 *Histoire des sciences et des savants depuis deux siècles*, Geneva. C used the same portfolio method of note reference as CD, independently evolved. LL3:333. 1880 Autumn C visited Down House. Recollection of the visit in: 1882 *Darwin considéré au point de vue des causes de son succès*, Geneva. In English: Bettany, 1887, pp. 148-150 (F2095), in *Darwin Online*. 1889 Linnean Medal Linnean Society.

Canestrini, Giovanni, 1835-1900. Italian naturalist and acarologist. C translated nine of CD's works into Italian. 1862-69 Prof. Zool. Modena. 1869-1900 Prof. Zoology and Comparative Anatomy Padua. 1877 *La teoria dell'evoluzione*.

Canning. Fishmonger at Down. C went to Billingsgate three times a week. His mother was unqualified midwife at Down. Atkins 1976, p. 104.

Cape Verde Islands, 1832 Jan. 17-Feb. 8 *Beagle* at Porto Praya, Santo Jago. CD landed. 1836 Aug. 31-Sept. 4 *Beagle* again at. CD landed. *Zoological diary*, pp. 4-21. *Geological diary*: (1-2.1832), transcribed in *Darwin Online*. It is one of the few differences between 6th edn *Origin* 1872, 11th thousand, and the altered 6th edn 1876, 18th thousand, that the name is changed from Cape de Verde to Cape Verde.

Capel Curig, 1831 Aug. CD on geological tour with Sedgwick. "At Capel Curig I left Sedgwick went in a straight line by compass map across the mountains to Barmouth". *Autobiography*, p. 70. 1842 Jun. CD visited.

Cape Town, Cape Colony, South Africa. 1836 May 31-Jun. 18 *Beagle* at. CD landed and made short excursion inland. CD met Sir John Herschel there. 1836 CD's first published work, with FitzRoy, A letter...moral state of Tahiti, New Zealand &c., *S. Afr. Christian Recorder*, 2: pp. 221-38, 1836 Sept. was published there (*Shorter publications*, F1640). *Geological diary*: DAR38.902-919, transcribed in *Darwin Online*.

Cardwell, Edward, 1st Viscount, 1813-86. Statesman. 1873 FRS. 1874 1st Viscount 1875 C was Chairman of Vivisection Commission, to which CD gave evidence. See *Shorter publications*, F1275.

Caricatures, see Darwin, Charles Robert, Iconography.

Carlisle, Cumberland. 1855 Sept. 11 ED records "Slept at Carlisle. 1855 Sept. 19 CD visited on return from BAAS meeting at Glasgow.

Carlisle, **Bishop of**, see Goodwin, Rev. Harvey.

Carlisle, Sir Anthony, 1768-1842. Surgeon. 1847 May CD to Hooker "Old Sir Anthony Carlisle once said to me gravely that he supposed Megatherium & such cattle were just sent down from Heaven to see whether the earth would support them & I suppose the coal was rained down to puzzle mortals." CCD4:86.

Carlyle, Thomas, 1795-1881. Essayist and historian. CD met several times at Erasmus Alvey Darwin's and at C's in London. 1836 Married Jane Baillie Welsh d.s.p. The Darwins drank tea in 1852 Mar. 16 with the Carlyles and visited them on three occasions in 1875. "I have been to see him, and a more charming man I have never met in my life." (1875) J. Tyndall, *New fragments*, 1898, p. 388. CD referred to C as "the crotchety old grouch", E.L. Griggs, A scholar goes visiting. *Quarterly*, 1934, vols. 40-1, p. 411, (F2167). 1853 recollection of CD in A. Carlyle, ed. 1904. *New letters of Thomas Carlyle*, vol. 2: 145-46, transcribed in *Darwin Online*. Notes on Carlyle's supposed attack on CD and refutation are in DAR132.

Carmichael, Dugald, 1772-1827. Scottish botanist and army surgeon. "Father of marine botany". Retired to Ardtur near Oban.

Carnarvon, Earl of, CD et al. 1877. [Memorial to Earl Carnarvon on the proposed South African Confederation by the Committee of the Aborigines Protection Society]. *The Times* (23 Jul.): 10.

Carpenter, William Benjamin, 1813-85. Physician and naturalist 1844 FRS. 1845 Fullerian Prof. Royal Institution London. 1856-79 Registrar London University. 1859 CD sent 1st edn of *Origin*. 1860 Jan. C reviewed *Origin* in *National Rev.*, Apr. in *Med. Chirurg. Rev.* 1861 Mar. 17 visited Down House with Huxley. 1861 Royal Medal Royal Society. 1874 *Principles of Mental Physiology*.

Carr, Colonel Ralph Edward, 1833-92. Of Hedley, Northumberland. 1870 Married Anne Jane Wedgwood. 1872 Lost first child. 2 sons, 1 Henry Arbuthnot C (1872 Sept. 2-1951 Jan. 22), 2 Martin Raymond C (1877-1914).

Carruthers, William, 1830-1922. Botanist. 1871 FRS. 1871-95 Keeper of Botanical Department, BMNH. 1871-1910 Consulting botanist to Agricultural Society. 1878 CD to James Torbitt in search of funds for potato blight work. C was against providing further money. CCD26.

Carter, Alice & Eliza, (sisters?) Alice, a partially blind Down cottager whom ED helped. A looked after old Mrs Osborn. 1872 May 29 A was invited to dine, 1873 May 5 came with Eliza in *ED's diary* entry as "Alice & E. Carter came P.M." and a few more times with them visiting together. 1876 Mar. 9, ED saw E. 1896 ED made two payments to A of £18.

Carter, James, Plumber and glazier at Down in 1855 directory.

Cartmell, James, 1810-81. 1849-81 Master of Christ's College, Cambridge. 1855-81 Chaplain to Queen Victoria. 1909 William's speech at Cambridge celebrations "He [CD] spoke to me with pride and pleasure of walking, dressed in his scarlet gown, arm in arm with Dr. Cartmell". ED2:171.

Carus, Julius Victor, 1823-1903. German zoologist. 1853- Prof. in Leipzig. 1860 Jun. was at BAAS meeting at Oxford. 1866 C revised 3d German *Origin*, which was published in 1867, from 4th English (*Origin* was translated by H.G. Bronn). "The connection was cemented by warm feelings of regard on both sides". C later translated twelve other of CD's works. 1868 C translated 1st German edn of *Variation*. 1876 Mar. 21 CD to C "I can assure you that the idea of anyone translating my books better than you never even momentarily crossed my mind". CCD24:81.

Carver, Miss Alice, Schoolmistress. 1907 acquired Down House and turned it into Downe House School with Miss Olive M. Willis (d.1964).

Cary, William, 1759-1825. Instrument maker of London. The firm maintained the name William Cary after his death in 1825 until 1890. His two sons George (c. 1788-

1858) and John Jr. (1791-1852) took over from their father. 1831 CD to Henslow about C making instruments for *Beagle* voyage. Barlow 1967, pp. 25, 41, CCD1.

Case, Rev. George Augustus, 1773-1831. Unitarian minister at Shrewsbury with a chapel in High St. 1797-1831 C was pastor at Shrewsbury. 1817 CD went for a year, with sister Emily Catherine, to an infant school run by C. *Autobiography*, p. 22. C's school was at The Old Parsonage, Claremont Hill. CD was there "up to the age of nine". 1959 Nov. 22 A special service was held when Alister Hardy, himself a Unitarian, gave an address. Arnold Broadbent, 1962, *The story of unitarianism in Shrewsbury*, 11 pp., Shrewsbury. Pattison, *The Darwins of Shrewsbury*, 2009.

Casks. 1879 [letter] Rats and water-casks, *Nature*, 19: p. 481, supporting one from Arthur Nicols (author of numerous popular zoological and travel books), *ibid.*, p. 433 (*Shorter publications*, F1785). CD replied to Nicols with a recollection of the experience of J.C. Wickham.

Caspary, Johann Xaver Robert, 1818-87. German botanist and director of Bonn herbarium. CCD15:545. Corresponded regularly with CD. 1859 Received a presentation copy of *Origin* bearing the inscription "Professor Caspary / Koenigsberg / from the author." Sold at auction in 2019 to an anonymous buyer for the record breaking price of US$500,075. 1868 Discussed *Variation* and 1877 *Different forms*.

Castor oil. From the first record in *ED's diary* in 1855, the family began to take. It was last mentioned in her diary in 1893.

Catasetum tridentatum. An orchid (*Catasetum macrocarpum*). 1861 CD to John Lindley. *C. tridentatum*, *Monachanthus viridis* and *Myanthus barbatus* are male, female and hermaphrodite flowers of the same species. 1862 "On the three remarkable sexual forms of Catasetum tridentatum, an orchid in the possession of the Linnean Society", *Jrnl. Proc. of the Lin. Soc. London (Botany)*, 6: pp. 151-7 (*Shorter publications*, F1718). 1863 French translation in *Annales des sciences naturelles* (F1967), with (F1723).

Catalan. First editions in: 1879 *Journal* (F1907), 1977 *Autobiography* (F2495). 1980 *Origin* (F2493). 2008 On the tendency (F2018).

Caton, John Dean, 1812-95. Chief Justice of Illinois, USA, and naturalist. 1855 Nov. 1 Mrs Caton paid a visit. 1868 CD thanks C for a paper on American deer. 1871 CD to C, George Howard and Francis Darwin are touring USA, please aid them and show "famous Deer-Park". CCD19.

Cats. Few of the names of Darwin family cats are recorded. A "rather fierce tabby cat, named Bullzig" belonged to Henrietta. The cat was later killed because "he took to killing the pigeons". Autobiographical fragment. (1926) DAR246. Transcribed by Richard Carter in *Darwin Online*. Mephistopheles, "Phisty" Henrietta's kitten in 1863 "a beautiful Persian kitten", ED(1904)2:202. Cinder, a kitten in 1871. CD patronized the national cat show at the Crystal Palace in 1873 and 1875. *London Evening Standard*, (22 Sept.). Jumbo, a black cat, 1883. F. Darwin, *Story of a childhood*, 1920.

Cattell, John, 1786/7-1860. Nurseryman of Westerham, Kent. CD often ordered seeds and plants from C. Many records in CD's Account book (Down House

MS). 1860 CD to Maxwell Masters, the nurseryman CD generally dealt with. CCD8:147.

Cavendish, Sir William, 7th Duke of Devonshire, 1808-91. Landowner and politician. 1845 Sept. or Oct. CD visited Chatsworth, the ducal seat, then of William C, 6th Duke. 1858 7th Duke. 1882 Pallbearer at CD's funeral, as Chancellor, University of Cambridge.

Caverswell Castle, 1878 Leased home of Godfrey and Hope Elizabeth Wedgwood. 1887 or 1888 they moved to Idlerocks to be nearer the factory.

Cecil, Robert Arthur Talbot Gascoyne, 3d Marquess of Salisbury (1868), 1830-1903. Son of Lady Mary. Statesman. 1857 Married Georgina Alderson [II].

Cecil, Lord Sackville Arthur, 1848-98. 5th son of 2d Marquess of Salisbury. Cambridge friend of CD's sons and neighbour in Kent. 1871 May. 20 C visited, *ED's diary*. 1882 C was on "Family Friends invited" list for CD's funeral.

Census Returns of England and Wales. Entries relating to Darwins for 1841, 1851, 1861, 1871, 1881 and 1891 are transcribed and reproduced in *Darwin Online*. http://darwin-online.org.uk/content/frameset?itemID=NationalArchivesCensus&viewtype=text&pageseq=1

Cerro Perico Flaco, Argentina. A hill near river Beguelo, a tributary of Rio Negro. 1833 Nov. 22-26 CD visited from estancia of Mr Keen and found skull of "Megatherium" [actually Toxodon]. The hill now bears an obelisk commemorating CD's visit and a nearby village is called Darwin. J.H. Winslow, *Jrnl. of Historical Geography*, 1, 1975, pp. 347-60, 1975.

Chaffers, Edward Main, 1806-45. Master and acted as Purser of *Beagle* during illness and after death of Rowlett. Master of *Beagle* on 2d voyage. In the *Sydney notebook*, Chancellor & van Wyhe, *Beagle notebooks*, 2009. Later Captain of New Zealand Association Ship *Tori*. Harbour Master Port Nicholson.

Chagas Disease. A trypanosomiasis of South America, spread to man by the house bugs *Triatoma infestans* and *Conorhinus magistus*. Long suggested, from 1959, that CD had the disease from being bitten by *T. infestans*, the benchuca bug, at Luxan, Argentina, 1835 Mar. 26. His symptoms were not those typical of the disease. This theory is no longer credited. See Darwin, Charles Robert, Health.

Chalk. 1849 Aug. 21 ED recorded giving C to Bessy. In 1854 Sept. 22 George was given C but it "did not do good". CD in 1864 Mar. 22 began lime water with chalk and magnesium. It was left off in Aug. and taken again in 1865 Jul. 14 after an "uncomf. mg." (uncomfortable morning). See W. Jenner to CD 14 Aug. 1864. CCD12.

Chambers, Robert, 1802-71. Edinburgh publisher, geologist and author. 1832 C founded *Chambers's Edinburgh Jrnl.* with his brother William C. In 1854 the title was changed to *Chambers's Jrnl. of Popular Literature, Sci. and Arts*. 1844, Anonymous author of *Vestiges of the natural history of creation*, and of *Explanations; a sequel*, 1845. See Secord *Victorian sensation*, 2000. 1844 CD to Hooker, "have been somewhat less amused at it, than you appear to have been". CCD3:108. 1845 CD to Hooker, CD calls him "Mr. Vestiges". CCD3:261. 1846 CD to Hooker, on *Explanations* and

Kerguelen cabbage. CCD3:199. 1847 CD to Hooker, "Somehow I feel perfectly convinced he is the author". CCD4:36. 1848 *Ancient sea-margins as memorials of changes in the relative levels of sea and land.* 1848 CD to Lyell, "if he be, as I believe, the Author of *Vestiges* this book [*Ancient sea margins*] for poverty of intellect is a literary curiosity". CCD4:152. 1860 C was at Oxford BAAS meeting and encouraged Huxley to attend the famous session where he tangled with Wilberforce. 1861 CD called "at his very nice house in St John's wood & had very pleasant half-hour's talk: he is really a capital fellow." CCD9:100. After the death of C, CD wrote to his daughter "Several years ago I perceived that I had not done full justice to a scientific work which I believed and still believe he was intimately connected with, and few things have struck me with more admiration than the perfect temper and liberality with which he treated my conduct." CCD19:208. 1884 Public acknowledgement of his authorship of *Vestiges* was not made until 12th edn 1884.

Chapman. Cambridge friend of CD. Possibly Benjamin Chapman, a fellow undergraduate at Christ's College. CCD1:73.

Chapman, Elizabeth, 1826-? Lady's Maid at Down House in 1861 census; Housemaid in 1871 census. 1880 Francis Darwin called her "the excessively staid housemaid". Took care of Bernard Darwin. F. Darwin, *Story of a childhood*, 1920, p. 39.

Chapman, Dr John, 1821-94. Physician and publisher, specialist in dyspepsia. 1851 Acquired the journal *Westminster Rev.*, which became important in the debate about evolution. 1865 Spring and summer, CD tried his ice-cure. ED marked in her diary "Dr Chap" in 1865 May 20. Another appointment was noted on May 28 1865.

Charities. CD frequently contributed to charities and liberal causes as well as donations raised to help individuals. In 1858 he donated £3 to the Field Lane refuge, a charitable housing and London Christian mission centre for the poor and unemployed (F1935 q.v.). In 1860 he gave £50 for the formation of a Bromley volunteer rifles corps. In 1861 he gave annual subscriptions of £5 to the Coventry Relief Fund and the Indian Famine Relief Fund. In 1862 £10 to the Lancashire and Cheshire Operatives Relief Fund. In 1864 £5 to the Ladies' Polish Relief Fund for sick and wounded Poles. In 1867 £5 to the East-End Central Relief Committee. In 1871 he supported the Voysey Establishment Fund on behalf of liberal theist preacher Charles Voysey and gave £50 to the Cresy Memorial Fund; he gave £3 to the Medical Education of Women Fund. In 1875 £100 to the Stazione Zoologica, Naples. In 1876 £15 to Viscountess Strangeford's Bulgarian Peasant Relief Fund. In 1877 £10 for the building of a new infant Sunday school in Frankenwell and £2 2s to the Sunday Lecture Society. In 1881 he subscribed to the Carlyle Memorial Fund. For some years near the end of his life CD paid an annual subscription to the South American Missionary Society for the orphanage at the Mission Station in Tierra del Fuego. In 1881 he gave 5 guineas to the women's Cambridge colleges, Girton and Newnham, for a physical and biological laboratory. Many more are recorded under

"Gifts and annual subscriptions" in CD's Classed account books (Down House MS).

Charlesworth, Edward, 1813-93. Geologist. 1838 CD to Lyell, "Charlesworth is to be pitied for many reasons". CCD2. 1842 CD to Lyell, discussing a controversy between C and Buckland, Lyell and Owen on the Crag, "it is not the wise who rule the universe, but the active rule the inactive and verily Charlesworth is...active". CCD2.

Chater, William, 1802-85. Nurseryman and landscape gardener of Saffron Walden, Essex. 1855 CD to Henslow [CD spelled Chator], on breeding of hollyhocks in which C specialized. CCD5:501, Barlow 1967, p. 189.

Chatsworth, Derbyshire. Seat of Dukes of Devonshire. 1845 Oct. CD visited. "was in transports with the great Hothouse". CCD3:258.

Cheek, Henry Hulme, 1807-33. A contemporary of CD at the University of Edinburgh. Elected member of the Plinian Society same day as CD. 1827 Was present when CD presented his two discoveries in Mar. 27. Minutes of the meeting transcribed in *Darwin Online*. Died young and buried at Saint Stephen's Church, Salford. B. Jenkins, Cheek and transformism: new light on Charles Darwin's Edinburgh background. *Notes and Records* (2015) 69, 155-171.

Cheeseman, Thomas Frederic, 1845-1923. New Zealand botanist. Curator Auckland Institute and Museum for 49 years. Described fertilization of orchids, especially *Pterostylis* to CD. See 2ᵈ edn *Orchids*, 1877, pp. 88, 90, 280. 1877 CD sent inscribed copy to C "with the author's compliments and respect".

Chêne, Vevey, Lac Léman, Switzerland. Home of J.C. L. Simonde de Sismondi.

Cherry blossoms. 1876 Cherry blossoms, *Nature*, 14: p. 28 (*Shorter publications*, F1772). CD had observed a squirrel biting off cherry blossoms.

Chester, Harry, 1806-68. Clerk in Privy Council Office. Novelist. Son of Sir Robert C, 1768-1848, DL, Hertfordshire. A personal friend of FitzRoy who was apparently invited to go on a planned private voyage to Tierra del Fuego to return the Fuegians. CCD1:143: "An inscription in volume one of a copy of Kirby and Spence 1828 [An introduction to entomology]...reads: 'Harry Chester | From his valued friend Robert FitzRoy'". See van Wyhe, "my appointment", 2013.

Chester, Colonel Joseph Lemuel, 1821-82. American genealogist. Worked on early history of the Darwin family. George's manuscript notes for C are in the Galton papers at University College London. 1855-58 Aide-de-Camp of James Pollock, governor of Pennsylvania, with rank of Colonel. 1858 C settled in London. 1879 Henrietta "My brothers had been having the pedigree of the Darwins made out by a certain Colonel Chester". ED2:237.

Chester Place, Regent's Park, London. 1868 no. 4, Sarah Elizabeth Wedgwood's [II] "little house". ED2(1904):217.

Chevening, Kent. Seat of 4ᵗʰ and 5ᵗʰ Earl Stanhope. 1849 Mar. 5 CD visited and 1855 Aug. 25 as recorded in *ED's diary*.

Chiloe Island, Chile. 1834 Nov. 10-1835 Feb. 4 and Jun. 28-Jul. 13 *Beagle* surveying around. CD much ashore, including visits to Chonos Archipelago to south of C. "Everyone was glad to say farewell to Chiloe". *Journal* 2ᵈ edn, p. 297. *Zoological diary*, pp. 234-8. *Geological diary*: DAR34.199-217 and DAR35.288-342, all transcribed in *Darwin Online*. There are hand-watercoloured maps of C in DAR35.306-309.

Chinese. First editions in: 1902 *Origin* (Historical sketch only) (F1846). 1903 Chapters 3 & 4 only. 1918 Whole work (F637). 1917 *Autobiography* (F1882), 1930 *Descent* (F1047a), 1935 *Expression* (F2254), 1941 *Journal* (F1841), 1958 *Beagle diary* (F1566a), 1965 *Orchids* (F817c), 1966 *Variation*. (F909c), 1972 *Power of movement* (F1341a), 1983 *Insectivorous plants*. (F1236a), 1983 On the tendency of species to form varieties. (F1844), 1995 *Earthworms* (F2276), 1995 *Climbing plants*. (F2272), 1996 *Different forms of flowers* (F2277), 1998 *Cross and Self Fertilisation*. (F2271).

Chloroform, used as anaesthetic, the Darwins adopted its use in 1848. ED was administered during the birth of Francis and later on the other births. In 1852, CD wrote to W.D. Fox "...I agree most entirely, what a blessed discovery is Chloroform". George Darwin had his teeth extracted under chloroform. See CCD4.

Chlorophyll. 1882 The action of carbonate of ammonia on chlorophyll bodies, *Jrnl. of the Lin. Soc. of London Bot.*, 19: pp. 262-84 (*Shorter publications*, F1801); abstract by Francis Darwin, who helped in the work, *Nature*, 25: pp. 489-90.

Chobham camp, Surrey. 1853 Aug. CD visited military camp for Crimean war. Henrietta wrote: "our visit was planned in order to see what we could of the camp with its mimic warfare. I well remember my father's intense enjoyment of the whole experience. Admiral Sulivan, his old shipmate on board the *Beagle*, showed us about and greatly added to our pleasure." ED2:154.

Choice, Felix, 1829-1915. Surgeon-dentist, London. 1873? Corresponded with CD in relation to tears being a sign of happiness, as discussed in *Expression*. CCD21:1.

Chonos, Archipiélago de los, Chile. 1834 Dec. 18-1835 Jan. 15 *Beagle* surveying off; CD ashore. *Zoological diary*, pp. 258-9, 274, 277-82. *Geological diary*: DAR35.233-258, both transcribed in *Darwin Online*.

Christ's College, Cambridge. 1827 Oct. 15 CD admitted, "Admissus est pensionarius minor sub Magistro Shaw", but did not go up until Lent term. Set in front court, G staircase, room 3, traditionally said to be the same as those of William Paley though this is almost certainly an apocryphal legend of recent date. The set now has a commemorative Wedgwood plaque and was restored to appear as in CD's student days in 2008-09. Darwin's student bills there are also reproduced in *Darwin Online*: (http://darwin-online.org.uk/EditorialIntroductions/vanWyhe_Christ%27s_College_student_bills.html) and in John van Wyhe, *Darwin in Cambridge*, 2014 with much additional information on student life at Christ's in CD's time.

Cinder. ED in her letter to Henrietta in Sept. 1871 talked about Cinder the kitten who was lost for two days. When it returned, "It caused a burst of indignation thro' the house; Jane [the housemaid] was sure she was starved..." ED2:205.

Cirripedia. 1856 [Typical list of Cirripedia.] In [On typical objects in natural history.] *Report of the twenty-fifth meeting of the British Association for the Advancement of Science*, p. 121 (*Shorter publications*, F1977). CD's Catalogue of parts of Cirripedes, mounted as microscopical slides. (1854-5). The University Museum of Zoology Cambridge, UMZC-Histories3.454. Published in *Darwin Online* with an introduction. 'Catalogue of the appendages and other parts of Cirripedes, mounted as microscopical slides.' (1854-5). As part of his barnacle researches, CD examined different parts of barnacles under his microscope. He prepared permanent slides which were labelled with a number which corresponded to a handwritten catalogue. On 20 Feb. 1897 Francis Darwin presented the slides and the catalogue to the University Museum of Zoology in Cambridge. Photographs of the entire catalogue are reproduced in *Darwin Online*. http://darwin-online.org.uk/content/frameset?itemID=UMZC-Histories3.454&viewtype=image&pageseq=1

CD's rooms at Christ's College, Cambridge, after restoration in 2009.
Previously unpublished photograph courtesy of Allison Maletz.

1863 On the so-called auditory-sac in cirripedes, *Nat. Hist. Rev.*, 3: pp. 115-6 (*Shorter publications*, F1722). 1873 On the males and complemental males of certain cirripedes, and on rudimentary structures, *Nature*, 8: pp. 431-2 (*Shorter publications*, F1762). See Marsha L. Richmond, Darwin and the barnacles. In M. Ruse ed., *Cambridge encyclopedia of Darwin and evolutionary thought*, 2013, pp. 807 and Love, Darwin and Cirripedia prior to 1846. *Jrnl. of the Hist. of Bio.* 35 (2002): 251-289.

Cirripedia, British Fossil. 1850 On British fossil Lepadidae, *Proc. of the Geol. Soc. of London*, 6: pp. 439-40, abstract. CD withdrew paper (F1679). 1851, 1854, 1858 *A monograph of the fossil Lepadidae, or pedunculated cirripedes of Great Britain...A monograph of* *the fossil Balanidae and Verrucidae*...[Index to vol. II 1858], Palæontographical Society vols. 5, 8 & 12 [index to vol. II], (F342), Facsimile 1966 (F343).

Cirripedia, Living. 1851, 1854 *A monograph of the sub-class Cirripedia...The Lepadidae; or, pedunculated cirripedes...The Balanidae (or sessile cirripedes), the Verrucidae*, 2 vols., Ray Society's Publ. nos. 21 and 25, London (F339). 1854 CD asks Huxley's advice on presentation copies; these were sent to Bosquet, Milne-Edwards, Dana, Louis Agassiz, Fritz Müller, Dunker; possibly also to Von Siebold, Lovén, d'Orbigny, Kölliker, Sars, Kröyer. 1936 Foreign edition: extracts only Russian (F341). 1964 Facsimile (F340, reprinted 1968). 1988 Facsimile (F1921).

Claparède, Jean Louis René Antoine Édouard, 1832-71. Swiss invertebrate zoologist. More generally referred to as René-Édouard, Édouard-René or Édouard C. Early convert to evolution. 1861 Articles on evolution in *Revue Germanique*. 1861 Offered to assist Clemence Royer with French translation of *Origin*. 1862- Prof. Comparative Anatomy Geneva.

Clapham, Abraham, 1810-88. Keen naturalist and horticulturalist specialising in the rearing of ferns. Corresponded with CD 1847-50. CCD4.

Clapham Grammar School. All CD's sons went there except William. Lawrence Ruck's sons made friends with CD's sons there. 1834 Headmaster and founder Charles Pritchard; George Howard and Francis educated by him. 1862 Headmaster Alfred Wrigley; Leonard and Horace educated by him. 1885 Closed.

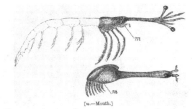

[*m.*—Mouth.]

Comparison of a crustacean and a lepas barnacle minus its shell, showing their similarities. From *Living Cirripedia*, vol. 1.

Clark, Sir James, 1st Bart, 1788-1870. Physician. 1837-60 Physician-in-Ordinary to Queen Victoria. 1837 1st Bart. 1837 Probably CD's physician in London. CCD2:51.

Clark, Sir Andrew, 1st Bart, 1826-93. Fashionable London physician. 1873 C first attended CD. In 1873 Oct. 8 ED diary entry for Dr Clark. Next mentions on 1876 Apr. 19 & 1880 Oct. 17. 1876 attended William at Down House for concussion after a riding accident in May. 1881 C saw CD in London, "some derangement of the heart". 1882 Mar. 10 C saw CD at Down House. 1882 Apr. C on "Personal Friends invited" list for CD's funeral. 1883 1st Bart. 1885 FRS.

Clark, John Willis, 1833-1910. Zoologist, archaeologist and Cambridge historian. 1866-91 Superintendent Zoology Museum Cambridge when CD got honorary LL.D. 1880 Aug. 18, ED recorded C in her diary. 1882 C was on "Personal Friends invited" list for CD's funeral. 1891-1910 Registrar Cambridge. 1909 C organised CD centenary celebrations at Cambridge.

Clark, Mary, 1750-1820. Daughter of Philip and Ann (née Wedgwood). 1783 Married Joseph Wedgwood.

Clark, William, 1788-1869. 1817-66 Prof. Anatomy Cambridge. 1826-59 Rector of Guisely, Yorkshire. 1836 FRS. 1860 May 18 CD to Lyell, says anti-*Origin*, but son J.W. Clark says not so.

Clarke, William Branwhite, 1798-1878. Priest and geologist. 1839 Emigrated to Australia. 1861-62 Corresponded with CD on geology, orchids and cirripedes.

Claus, Carl Friedrich, 1835-99. German zoologist. 1876 May 2 CD letter (in German) granting permission to C to dedicate his book to CD. *Neue freie Presse* (22 Apr.): 2 (F2286). See CCD24:146. 1871 recollection of CD in C, *Autobiographie*, 1899, p. 17. "Lubbock's amiable wife arranged my visit in Darwin's house." In *Darwin Online*.

Cleavage. 1846-7 CD's views on geological cleavage, with illustrations by CD. ML2:199-210 and in the Darwin Archive DAR41.59-77, and DAR42, transcribed in *Darwin Online*. These were never published as a paper.

Clemens, Samuel [Mark Twain], 1835-1910. American author. [1876]. [Recollection of CD] in 'Books, authors and hats. Address at the Pilgrims' Club Luncheon, Savoy Hotel, London, Jun. 25, 1907'. In Howell ed. *Mark Twain speeches*, 1907, pp. 33-5: C upon hearing from Norton that CD read his books before bed replied "I do regard it as a very great compliment and a very high honor that that great mind, laboring for the whole human race, should rest itself on my books. I am proud that he should read himself to sleep with them." (F2102).

Clement, William, 1763-53. Apothecary of Shrewsbury; "unflinching advocate of parliamentary reform and civil and religious liberty". Meteyard in Woodall 1884, p. 10. CD must have known as a child. Transcribed in *Darwin Online*.

Clemson, Richard, b.1806. Gunsmith of Shrewsbury. 1831 C made CD's gun and spare parts for *Beagle* voyage. CCD1:148. First name not previously identified.

Clifford, William, 1821-78. Book seller, mainly of religious material. Married Fanny Kingdon (1821-54). Friend of the Wedgwoods. House in Perristone, Herefordshire. 1839 Aug. 30-Sept. 2 Stayed with the Darwins at Gower Street. The Wedgwoods & Allens talked about encounters with C in their letters. ED2:10, 102-3, 122-4.

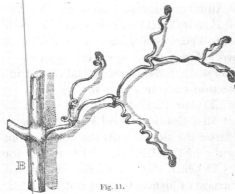

Fig. 11.

Tendrils of Virginia creeper (*Ampelopsis hederacea*, now *Parthenocissus quinquefolia*) from *Climbing plants*.

Clift, William, 1775-1849. Naturalist, keeper of the Hunterian Museum. Had examined some of CD's South American fossils before CD returned to England. 1792 Apprenticed to John Hunter. 1793-1844 Conservator Royal College of Surgeons Museum. 1823 FRS. 1835 His daughter married Richard Owen.

Climbing plants, 1865 "On the movements and habits of climbing plants", *Jrnl. of the Proc. of the Lin. Soc. of London*, 9, nos. 33 and 34, pp. 1-118 (F833. Commercially available offprint of the *Jrnl. of* *the Lin. Soc.* (F834), and available as author's offprint (F835); both in paper wrappers. 1866 Reprinted in *Flora*, 49. 1875 2^d edn *Climbing plants*, London (F836). Full title: *The movements and habits of climbing plants.* 1875 Nov. 3,012 copies sold. Drafts in DAR18 and proofs in DAR213. Notes for 2^d edn (1875) are in DAR69 and DAR157. 1882 2^d edn with appendix to preface by Francis Darwin, London (F839). Henrietta's copy, which was passed on to Margaret Keynes was sold in 2018 at Sotheby's for £37,500. First foreign editions: 1876 USA (F838), 1957 Chinese (F857a). 1876 German (F860). 1877 French (F858). 1938 Japanese (F863a). 1900 Russian (F865). 1970 Romanian (F864). List of presentation copies of 2^d edn is in CCD23 Appendix IV.

Clive, William, 1795-1883. Friend of the Darwins and Wedgwoods. An old friend of the Darwin family. A contemporary of Henslow at St John's College, Cambridge, 1813-18. 1829 Married Marianne, daughter of George Tollet. 1 daughter Marianne Caroline 1839 Mar. 23-25. ED wrote "M. Clive brought to bed". Two days later, the baby died. ED called in on Mar. 30. 1844-61 Archdeacon of Montgomery. 1841 Jan. 2 the Darwins "drank tea at Clives" and again two weeks later on Jan. 16. 1855 CD to Henslow, CD had seen C in London and he had enquired after Henslow. Barlow 1967, p. 174, CCD5:293.

Clough, Miss Anne Jemima, 1820-92. Promoter of higher education for women. Sister of Arthur Hugh Clough, poet. 1871-92 First Principal of Newnham College Cambridge. 1855 Feb. 3 the Darwins dined at Cumberland Terrace, Clough. 1883 C stayed at Down House. 1884 Feb. 16 C was ED's dinner guest and again in 1886, 1889 and 1891. ED noted her funeral as on Mar. 5, 1892.

Clowes, William, 1779-1847. Printer for John Murray. William Clowes and Sons.

Coal, origin of. CD and Hooker disagreed on the origin of coal. CD thought it came from plants in warm shallow seas. Hooker insisted that the plants had been terrestrial. 1846 CD to Hooker, four letters on the subject. CCD3. 1866 The matter resurfaced in correspondence without being resolved. CCD14.

Cobbe, Miss Frances Power, 1822-1904. Antivivisectionist. Editor of *The Echo* and *Zoophilist*. Reviewed *Descent* in *Theological Rev.* 1868 ED to her sister Elizabeth Wedgwood "I dined over the way [at Hensleigh Wedgwood's] Mar. 31, (and Charles also) to meet Miss Cobbe and Miss Lloyd. Miss Cobbe was very agreeable". ED2:189. 1872 *Darwinism in morals and other essays*. 1875-84 Secretary National Anti-Vivisection Society. 1881 C issued antivivisection circular which she sent to CD; letters by C to *The Times* Apr. 19 and 23, by CD Apr. 22 and by Romanes Apr. 25 relate. 1881 CD to Romanes "with the sweet Miss Cobbe—Good Heavens what a liar she is: did you notice how in her second letter she altered what she quoted from her first letter, trusting to no one comparing the two". LL2:203. 1894 C to ED for permission to publish correspondence from CD which she had altered and printed in *The Echo*, about what C considered a miscarriage of justice, but was not. ED2:302. Henrietta recalled how C's request harassed ED. 1894. Recollection and letters of CD in *Life of Frances Power Cobbe*. (F2099). 1894 May 13 ED wrote Snow "about Miss

Cobbe" and a few days later ED wrote in her diary "Miss Cobbe affair upset me." Photograph in Evans ed., *Darwin and women*, 2017, p. 155.

Cobbold, Thomas Spencer, 1828-86. Parasitologist. 1864 FRS. 1873, 1875 and 1886 C described CD's *Beagle* parasites in *Proc.* and *Jrnl. of the Lin. Soc. of London Zool.* See *Shorter publications*, F1974 also F1806.

Cockell, Edgar, b.1786. Surgeon and apothecary of Down. 1842 admitted to Royal College of Surgeons in London. Attended to ED's confinement for the birth of Mary Eleanor Darwin. CD's account book (Down House MS) recorded a payment of £5 5s for his services.

Cocos Keeling Islands, Indian Ocean. Coral atolls with lagoons. They had an important influence on CD's views on the origin of coral atolls. 1836 Apr. 1-12 *Beagle* at. 1836 Apr. 2-3 CD ashore on Direction Island. CD's notes are in DAR41.1-57, (including a fair copy) transcribed in *Darwin Online*. Captain John Clunies Ross, the owner, was away, and CD only met his assistant Mr Liesk. 1837 [Notes on Cocos-Keeling Islands plants.] In Henslow, Florula Keelingensis. An account of the native plants of the Keeling Islands. *Annals of Nat. Hist.*, 1 (Jul.), (*Shorter publications*, F1959). P. Armstrong, *Under the blue vault of heaven: A study of Charles Darwin's sojourn in the Cocos (Keeling) Islands*, 1991. *Zoological diary*, pp. 305-11. All transcribed in *Darwin Online*.

Cod liver oil. 1861 Feb. 15 CD began taking, Jul. 27 ED began taking. See CD to W.W. Baxter 26 Jan. [1862]. CCD10.

Coe, Henry, head gardener at Hampshire County Lunatic Asylum, Knowle. CD corresponded with 1857-58. CD published C's data in On the agency of bees in the fertilisation of papilionaceous flowers, 1858 (F1701); see *Origin*, p. 613.

Cohn, Ferdinand Julius, 1828-98. German bacteriologist and plant pathologist. 1859 Prof. Botany Breslau. 1876 Aug. C visited Down House. 1882 recollection of CD in Ein Besuch bei Charles Darwin. *Breslauer Zeitung* Apr. 23 (copy in DAR216.68c-69a) and 1901 *Blätter der Erinnerung*, both in *Darwin Online*.

Colburn, Henry, 1784-1855. Publisher of Great Marlborough Street, London. 1839 Published 1st edn of *Journal* (F11).

Colchicum, or dried sodium carbonate, an alkaline salt plant extract often prescribed as a treatment for acid dyspepsia and as a purgative in the treatment of gout. ED entry in 1863 Dec. 12 "Colchicum". CD took in 1863 and 1864 24 Mar.

Coldstream, John, 1806-63. Physician at Leith. Friend of CD at Edinburgh. 1823 Member of the Plinian Society, nominated CD for membership. 1826/27 examined marine animals on shores of Firth of Forth with CD and R.E. Grant. 1833-35 Wrote "Cirrhopoda" in Todd, *Cyclopaedia of anatomy and physiology*, 1, pp. 683-94.

Colenso, William, 1811-99. Botanist, ethnologist, missionary and printer. 1834-42 Missionary printer at Paihia, New-Zealand. 1835 CD spent Christmas Day with him. 1882 Eulogy of CD by C summarised in *Trans. & Proc. of the N.Z. Inst.* 15: p. 541 "a great and useful man". 1883 Hooker proposed C for Royal Society, asked J.F.J. von

Haast to sponsor C, saying that CD would gladly have signed. Tee, Darwin's contacts with New Zealand, *N.Z. Genetical Soc. Newsletter*, no. 4, 1978, p. 46. 1886 FRS.

Collier, Elizabeth, 1747-1832. Natural daughter of Charles Colyear, 2ᵈ Earl of Portmore. Mother was Elizabeth Collier, governess to the legitimate children. Married 1 Edward Sacheverell Chandos-Pole. CD's step-grandmother. Francis Galton's grandmother. 1781 Married 2, as second wife, Erasmus Darwin [I]. 6 children.

Collier, Hon. John (Jack) Maler, 1850-1934. Painter and rationalist. RA. Son of Sir Robert Porrett C, Baron Monkswell. 1879 Married 1 Marian Huxley, daughter of T.H. Huxley, who painted C's portrait in 1881. 1881 C painted CD in oils. CD sat for him in Aug. LL3:223. C recalled "I had to go to Mr. Darwin's house in Downe to paint him, for he was a chronic invalid. He told me that he had never recovered his health after the terrific seasickness he had suffered on board the Rattlesnake [sic] which provided so much of the material for the Origin of Species. I painted him in his study." *Singapore Free Press* (15 Aug. 1930), p. 18. C also recalled that CD "more than once" said to him "Huxley has always fought all my battles for me." *Straits Times* (8 Oct.), p. 17. (F1864) 1881 CD thanks for sending copy of "your Art Primer". "Everybody whom I have seen, and who has seen your picture of me is delighted with it. I shall be proud some day to see myself suspended at the Linnean Society [who commissioned it]". ML1:398. C's visits were recorded in *ED's diary* in 1881 Jul. 12, Jul. 19-22. 1882 C on "Personal Friends invited" list for CD's funeral. 1887 "Many of those who knew his face most intimately think that Mr. Collier's picture is the best of the portraits". LL3:223. Now at Linnean Society, Burlington House, London. Replica by the artist with the family, at Down House. 1889 Married 2 Ethel Gladys Huxley, daughter of T.H. Huxley. 1891 C painted T.H. Huxley.

Collingwood, Dr Cuthbert, 1826-1908. Naturalist and writer on religious subjects, anti-*Origin*. 1855 *On the scope and tendency of botanical study*. 1858-66 Lecturer Botany Royal Infirmary Medical School Liverpool. 1861 CD corresponded with C on evolution. CCD9:53. C sent CD two of his anti-evolution pamphlets which defended the obscure and abstract views of Agassiz. In the Darwin Pamphlet Collection-CUL. 1866-67 Surgeon and naturalist HMS *Rifleman* and HMS *Serpent* on voyage of exploration China seas. 1868 *Rambles of a naturalist on the shores and waters of the Chinese seas*.

Columbarian Society = Southwark Columbarian Society, for breeders of domestic pigeons, in which CD was much interested for *Variation*. See also Philoperisteron. CCD5:509. CD was a member since 1855 and attended meetings near London Bridge. LL2:51. 1859 CD to Huxley. "I have found it very important associating with fanciers & breeders...I sat one evening in a gin-palace in the Borough amongst a set of Pigeon-fanciers". CCD7:403. [1860] CD to Huxley sending him a card to admit him to a pigeon show at the Freemason's Tavern, London. CCD8.

Colyear, Charles (Beau), 2ᵈ Earl of Portmore, 1700-85. Natural father of Elizabeth Collier. CD's Step-great-grandfather in bastardy. Francis Galton's great-grandfather in bastardy. 1732 Married Juliana Hale, Dowager Duchess of Leeds.

Comfort, Joseph, c.1806-. c.1842-54. Gardener-coachman at Down House. See for example CD to W.D. Fox 6 Feb. [1849], CCD4:209.

Compton, Spencer Joshua Alwyne, 2nd Marquess of Northampton, 1790-1851. 1838-48 President of the Royal Society. ED mentioned social events sponsored by C in her diary in 1839 Mar. 23 "Lord Northampton Roy. Society" & Apr. 13 simply "Lord Northampton".

Concepcion, Chile. 1835 Mar. 4-7 *Beagle* at. Earthquake and tsunami of Feb. 20 had caused almost total destruction of the town and of its port Talcahuano. *Geological diary*: DAR35.354-370, transcribed in *Darwin Online*.

Condy's ozonised water. 1862 CD took for dyspepsia. CD to Hooker "with, I think, *extraordinary* advantage—to comfort, at least". CCD10:515. See G.C. Oxenden to CD 17 Sept. [1862]. CCD10, recommended C for scarlet fever.

Congreve, Mary, 1745-1823. From Shrewsbury, friend of the Darwin family. 1821 wrote to CD about London plays. CCD1.

Coniston, Lancashire. 1879 Aug. 2-27 CD had family holiday there. On Aug. 17 to Coniston Hall. "beaut row on lake to old Coniston Hall" *ED's diary*.

Constitucion. Small schooner, cost £400. Bought by FitzRoy. 1835 May Used to survey coasts of Chile and Peru by Sulivan and Usborne.

Conway, Caernarvonshire. 1831 Aug. CD visited with Sedgwick for geology.

Conway, Moncure Daniel, 1832-1907. American Unitarian clergyman. Abolitionist. 1863-84 Minister South Place Chapel, Finsbury, London. 1864 C abandoned theism upon the death of one of his sons. C and the congregation left fellowship with the Unitarian Church. South Place Chapel eventually became South Place Ethical Society. 1873 Jan. 24 Visited Down House. *ED's dairy*. 1873 Sent Col. T.W. Higginson's *Collected essays* to CD. Recollection of CD in 1867 in *Autobiography* (F2098), transcribed in *Darwin Online*.

Cooke, Robert Francis, 1816-91. Partner and cousin of John Murray. 1837 Joined firm. After 1845 much involved in publishing CD's books. CCD13.

Cookson, Montague Hughes, 1832-1913. Barrister and eugenicist. Cambridge friend of CD's sons. Later Montague Hughes Crackanthorpe. 1875 QC. 1882 C was on "Personal Friends invited" list for CD's funeral.

Cooper, James Davis, 1823-1904. Wood engraver and book illustrator, London. Cut woodblocks for *Expression*, *Insectivorous plants* and Wallace's *Malay archipelago*.

Cope, Edward Drinker, 1840-97. American palaeontologist and comparative anatomist. 1872 CD to Alpheus Hyatt about Hyatt and C's theories on evolution. 1876 CD to William Darwin, "He writes very obscurely, but is an *excellent* Naturalist. He looks, following [Jean Louis Rodolphe] Agassiz at a genus as something essentially distinct from a species, which I believe to be quite an error." CCD24. 1887 *The origin of the fittest*, New York. 1889- Prof. Zoology, Pennsylvania.

Cooper, Mary, 1834-? Nurserymaid from Staffordshire at Down House in 1851.

Coote, George, 1842-1908. Born in Bromley. 1871 census resident in Beckenham, gardener/domestic servant. Supposedly worked as a gardener at Down House. 1877 emigrated to USA, specifically Corvallis, Oregon, where he became a farmer. 1888 C accepted a position at Oregon Agricultural College, laying out the college grounds and greenhouses and subsequently filled the chair of Horticulture.

Copley medal. CD was awarded in 1864 for researches in geology, zoology, and botanical physiology. Falconer had spoken in the meeting for *Origin* to be considered on behalf of CD but to no avail. CD did not attend the dinner to receive the medal due to ill health and expressed only mild disappointment that the *Origin* was passed over. 1911, CD's son George was awarded for his researches on tidal theory, the figures of the planets, and allied subjects.

Copiapó, Chile. 1835 Jun. 22 CD reached C on expedition from Valparaiso, via Coquimbo. 1835 Jun. 26-Jul. 1 CD took a short expedition into cordilleras from C. 1835 Jul. 6 *Beagle* left C for Iquique. *Geological diary*: DAR36.597-610 and DAR37.611-676, transcribed in *Darwin Online*.

Coquimbo. 1835 May 14-Jun. 2 CD visited C on expedition from Valparaiso. Met FitzRoy there and stayed with Don Joaquin Edwards, whose silver mine at Arqueros they visited May 21. Small earthquake whilst they were there. See *Beagle notebooks*. CD's other notes are in DAR39.152. *Geological diary*: DAR36.550-591. 1840 [Letter to Basil Hall on the valley of C]. In: Hall, *Extracts from a journal, written on the coasts of Chili, Peru, and Mexico*. (F2163). *Coquimbo notebook*, Chancellor & van Wyhe, *Beagle notebooks*, 2009. All transcribed in *Darwin Online*. Further notes on plains and shells in DAR39.110, 115, 159-160.

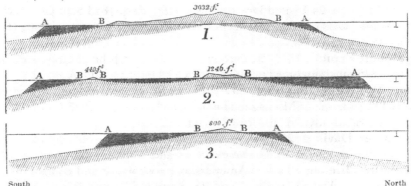

Three Pacific islands representing different stages of the formation of barrier reefs, from *Coral reefs*.

Coral islands. 1843 Remarks on the preceding paper in a letter from Charles Darwin, Esq. to Mr. Maclaren, *Edinburgh New Phil. Jrnl.*, 34: pp. 47-50 (*Shorter publications*, F1662); preceding paper was by Charles Maclaren, On coral islands and reefs as described by Mr. Darwin. 1962 Coral Islands, *Atoll Research Bull.*, no. 88, 20 pp., 1 map (*Darwin Online*, F1576); a transcript of CD's manuscript notes, with introduction by D.R. Stoddart, also transcribed in *Darwin Online*.

Coral Reefs. 1846 [Note on Sandstone and query on coral reefs.] In J.L. Stokes, *Discoveries in Australia*. 1:108, 331 (*Shorter publications*, F1915).

Coral reefs, presents CD's theory for the gradual formation of coral reefs and atolls. Part 1 of geology of the voyage of the *Beagle*. All English editions until 1889 published by Smith Elder. Copies of the 3d edn of 1889 published by Murray do exist however (F277a). 1842 *The structure and distribution of coral reefs*, London (F271). 1851 Same text in a combination volume with the other 2 parts (F274). 1874 2d edn (F275). 1889 3d edn (F277 and F277a). 1969 Facsimile (F306). First foreign editions, whole or part: 1875 Dutch (F2359). 1876 German (F311). 1878 French (F309). 1874 Italian (F2463)). 1940 Japanese (F2310). 1846 Russian (F320). 2006 Spanish (F2074). 1889 USA (F278). List of presentation copies of *Coral Reefs* 2d edn is in CCD22 Appendix IV. Notes for the 2d and 3d editions are in DAR69.

Corbet, Richard, A blind friend of Mrs Marsh Caldwell. 1866 CD to Mrs C enclosing note for C about diet. CCD14.

Corfield, Richard Henry, 1804-97. Son of Rev. Richard C (Barlow 1967, p. 96, n.2.) 1816-19 Shrewsbury School. 1829-68 In South America. School friend of CD living in Almendral, a suburb of Valparaiso. 1834 and 1835 CD stayed with. C cared for CD when he was very ill, perhaps with typhoid, in Sept. and Oct. 1834. In the *Galapagos notebook*, Chancellor & van Wyhe, *Beagle notebooks*, 2009.

Cornford, Frances, see Frances Crofts Darwin.

Cornford, Francis Macdonald, 1874 Feb. 27-1943 Jan. 3. Classical scholar and poet. Married Frances Crofts Darwin. Known in the family as "FMC" and his wife as "FCC". 1931 Laurence Prof. of Ancient Philosophy Cambridge. 1937 Fellow British Academy. 1908 Author of *Microcosmographia academica*.

Cote House, Westbury, Bristol. c.1795 A large country estate bought by John Wedgwood. A great social centre for young Wedgwoods and Allens. Gardens and greenhouses were famous. Painted in 1792 by J.M.W. Turner. See also *Ann Green of Clifton*, 1936. 1805 Sold because of John Wedgwood's financial troubles.

Cotton, Richard, A Shrewsbury naturalist. Judd, *The coming of evolution*, 1909, p. 340. 1822 "An old Mr. Cotton in Shropshire" had pointed out to CD the inexplicable bell stone, an erratic boulder in Shrewsbury. *Autobiography*, p. 52. Fragment remains.

Cotton, Mrs, 1855-56 CD acquired 91 pigeons from her. 1868 Aug. 1, a Mrs Cotton was visited but unclear if CD came along. *ED's diary*.

Covington, Syms, 1813-61. "Fiddler and boy to the poop cabin" on 2d voyage of *Beagle*. Boy 2d class, shoemaker. Drawing of Lima beauty, p. 289, and Napoleon's tomb, p. 362, in Keynes, *Beagle record*, in MLNSW. 1833 May 22 Became personal servant to CD at "under £60 per annum". Cost CD £30 because FitzRoy kept him on the books for food. 1834 Jul. 20 CD to his sister Catherine "my servant is an odd sort of person; I do not very much like him; but he is, from his very oddity, very well adapted to all my purposes." CD discussed evolution with before *Origin*. Barlow 1945, pp. 100-5, CCD1:392. C rearranged CD's notes on volcanic islands and many others. Remained in CD's employ as secretary-servant until 1839 Feb. 25, when

CD's accounts show "Present to Covington on leaving me £2". 1839 May 29 CD wrote testimonial for. 1839 C went to Australia working his passage as a cook. First employed at Australian Agricultural Co.'s coal depot in Sydney. c.1840 Married Eliza Twyford of Stroud. 6 sons, 2 daughters: eldest son Syms died 1923. From 1854 Employed at Pambula, running a store and postmaster, until death. Home The Retreat, Princes Highway, Pambula, Twofold Bay, New South Wales. 1859 CD continued to correspond with C. C sent CD large numbers of barnacles. Very deaf in later years, 1843 and 1860 CD sent new ear-trumpets. 1861 Death certificate says "21 years in this colony". 1884 Aug. 9 CD's letters to C published in *Sydney Mail*, 38: pp. 254-5, F2219. 1903 CD's pocket compass and sun-dial in C's possession at his death. Described in *Daily Telegraph & Courier*, (20 Oct.), p. 9. 1959 Reprinted in *Notes and Records Royal Soc.*, 14: pp. 14-27. Biography: J.B. Ferguson. *Syms Covington of Pambula*, 2^d edn 1988. Journal from the *Beagle* voyage in the MLNSW, transcribed and edited by Vern Weitzel, http://www.asap.unimelb.edu.au/bsparcs/covingto/contents.htm.

Cradock, Edward Hartopp, 1810-86. Childhood friend of CD in Shrewsbury. 1882: "I do not think that it was in his nature to be awkward to any one — never was a boy less set out for a bully or a school tyrant. Both in manner and in mind he was old for his age...I do not remember that he passed any distinction in ordinary school work or that he seemed to covet any — nor again that he was a hero of the cricket field or the river." DAR112.A16-A17, transcribed in *Darwin Online*.

Craik, Georgiana Marion (Mrs May), 1831-95. Novelist. 1858 C was a visitor to Moor Park Hydropathic Establishment. CD to ED "I like Miss Craik very much, though we have some battles & differ on every subject." CCD7:84. See Mike Dixon & Gregory Radick, *Darwin in Ilkley*, 2009. 1886 Married Allan Walter May, artist.

Crawfurd, John, 1783-1868. Scottish physician, colonial administrator and diplomat. 1818 FRS. 1856 CD to Hooker mentions C as being on selection committee of Athenaeum when Huxley was up for membership. 1859 C reviewed *Origin* in *Examiner*, hostile but free from bigotry.

Crawley, Charles, 1846-99. Cambridge friend of Francis D. 1871 Nov. 18 and 1872 Apr. 13 C visited Down House, together with Bennett (Alfred William B). Visited again on 1873 Jul. 5 and 1879 Jan. 18 visited with Dyers. (William Turner Thiselton-Dyer). 1882 C was on "Personal Friends invited" list for CD's funeral. C and his wife, Augusta, drowned while boating on river Wye.

Cresselly, Pembrokeshire. Home of John Bartlett Allen. From 1803 Home of John Hensleigh Allen. From 1843 Home of Seymour Phillips Allen.

Cresy, Edward [I], 1792-1858. Architect and civil engineer. Neighbour at Down. Father of Edward and Theodore. 1820 Fellow Society of Antiquaries. 1842 Advised CD on purchase of Down House. CD met C 1842 Aug. 12, C visited or met with CD on 1845 Aug. 19. More visits 1860-66. *ED's diary*. CD discussed evolution with before *Origin*. 1859 CD sent 1st edn *Origin*.

Cresy, Edward [II], 1825-70. Son of Edward C [I]. Architectural draftsman, 1849 Assistant surveyor under the commissioners of sewers; later engineer. 1858 Founder member of the Geologists' Association. 1859 Principal assistant clerk at the Metropolitan Board of Works. 1866. Architect to the fire brigade. Neighbour at Down. 1853 Oct. 1 "young Mr Cresy", 1860 Sept. 18? Visited with parents? "Cresy's & Frank W. came" and stayed 1867 Dec. 14-16s *ED's diary*. 1860 C helped CD with measurements for *Insectivorous plants*. CCD8.

Crichton-Browne, Sir James, see Browne, Sir James Crichton-

Crewe, Frances, 1786-1845. 1834 Married Robert Wedgwood as 1st wife.

Crick, Walter Drawbridge, 1857-1903. Businessman and palaeontologist. 1882 Feb. C to CD about dispersal of fresh-water bivalve molluscs by water beetles. LL3:252. See *Nature*, (Apr. 6 1882), pp. 529-30; (*Shorter publications*, F1802).

Croatian. First editions in: 1922 *Journal* (F2412.1). 1937 *Autobiography* (F2420). 1948 *Origin* (F2417). 2007 *Descent* (F2419).

Cripps Corner, Ashdown Forest, Sussex. 1900 Country home of Leonard Darwin when he married Mildred Massingberd. Raverat: "we always think of Uncle Lenny and Mildred...living in happy and uneventful solitude." *Period piece*, 1952, p. 197.

Crocker, Charles William, 1832-68. 1862 C had lately retired from being foreman at Kew. He was going to work on varieties of hollyhock. 1862 CD to Hooker on Crocker of Chichester, "he has real spirit of experimentalist, but has not done much this summer". CCD10:499.

Crofton, Amy, 1866 Jan. 4 visited Down. 1867 May 25. C was a family friend who went to May eights at Cambridge with ED and family. Stayed 4 days, left on 1867 May 28. Visited yearly 1868-71. *ED's diary*.

Crofts, Ellen Wordsworth, 1856-1903. Daughter of John C of Leeds. Fellow in English Literature and History Newnham College, Cambridge. Lecturer in English Literature. 1883 Sept. 13 Married as 2d wife Francis Darwin. 1 daughter, Frances.

Croll, James, 1821-90. Scottish man of science and geologist of Edinburgh. Correspondent of Lyell. 1869 CD to Lyell about C's estimates of geological time. CCD17. 1869 CD sent him 5th edn of *Origin*. 1875 *Climate and time, in their geological relation*.

Croquet. Many games were played and enjoyed by the Darwin family. ED recorded games from 1861 in her diary. 1861 "Etty played crocquet". There was a croquet party in 1893 and the grandchildren "at croquet" in the summer of 1894.

Cross and Self Fertilisation. 1876 *The effects of cross and self fertilisation in the vegetable kingdom* (F1249). Advertised in 1873 as "The evil effects of interbreeding in the vegetable kingdom." 1878 2d edn (F1251). 1891 3d edn, but really as 2d (F1256). CD's notes on the subject are in DAR76-79. Drafts are in DAR17.1. Notes for 2d edn are in DAR69. Drafts in DAR2-4 and DAR53. Proofs of the book are in DAR213. List of presentation copies is in CCD24 Appendix III. First foreign editions: 1959 Chinese (F1264a). 1877 French (F1265), German (F1266), 1878 Italian (F1269). 1940

Japanese (F2343). 1959 Polish (F1270), 1964 Romanian (F1271). 1938 Russian (F1272). 1877 USA (F1250).

Cross Breeding. 1856 Cross breeding, *Gardeners' Chron.*, no. 49: p. 806; p. 812 (*Shorter publications*, F1691, 1692). 1860 Cross bred plants, *Gardeners' Chron.*, no. 3: p. 49 (*Shorter publications*, F1704). 1861 [letter to D. Beaton] Phenomena in the cross-breeding of plants, *Jrnl. of Horticulture*, 1:112-13 (*Shorter publications*, F1713). 1861 Cross-breeding in plants, *Jrnl. of Horticulture*, 1:151 (*Shorter publications*, F1714).

Crüger, Dr Hermann, 1818-64. German botanist. 1857-64 Director of Botanic Garden, Trinidad. 1862 Mar. C helped CD with Melostomaceae. 1863 C observed fertilisation in *Catasetum* and *Coryanthes*. 1866 CD to Fritz Müller, "I am sorry to say Dr. Crüger is dead from a fever". CCD14:122.

Cudham Wood, Hangrove, not far from Down, one of the Darwin's favourite places for walks and rides. In Nov. 1842 Hensleigh's three children and Darwin's two got lost there. Snow brought William home after losing the other group. Elizabeth Harding, the nursery maid was with Erny, Brodie and Annie. They were found by CD and Parslow. In 1862 Apr. 9 ED wrote "went to Cudham on donkey with Lizzy". ED wrote to Henrietta in 1879 telling her of a delightful expedition Bernard Darwin took where "he talked all the way when he was not singing". ED may have taken her last walk there in 1896 Sept. 3 with an entry "went to Cudham Lane to Hangrove". *ED's diary.*

Cumberland Place, Regent's Park, London. 1868 no. 1, house of CD's cousin Hensleigh Wedgwood.

Cuming, Hugh, 1791-1865. Collector, especially of molluscan shells. 1829 Collected in Galapagos before CD. ML1:52. 1819 Sailmaker at Valparaiso. 1831 C returned to England. 1854 CD arranged and identified C's barnacles for him. 1866 His collection of nearly 83,000 specimens was sold to the BMNH for £6,000.

Cupples, Rev. George, 1822-91. Popular writer. 1869 CD letter to in F1874. 1873 CD to C, long letter of general nature about people. CD had recommended Mrs (Anne Jane) C's book *Tappy's chicks and other links between nature and human nature*, 1872, to Josiah Wedgwood [III]'s family, with whom CD was staying. CCD20.

Cut or uncut, 1867 Cut or uncut, *Athenaeum.* no. 2045: pp. 18-19. Reprinted in *The Times* (7 Jan. 1867): p. 8 (*Shorter publications*, F1815). CD supported a protest against books being sold with uncut pages.

Cypripedium. 1867 Fertilisation of cypripediums, *Gardeners' Chron.*, no. 14: p. 350 (*Shorter publications*, F1738). A genus of hardy orchids.

Cytisus scoparius. 1866 The common broom (*Cytisus scoparius*), *Jrnl. of the Lin. Soc. of London Bot.*, 9: p. 358; a note added to George Henslow's paper, Note on the structure of Indigofera etc., *ibid.*, 9: pp. 355-8 (*Shorter publications*, F1737). CD had corresponded with Henslow about the article and corrected it.

Czech. First editions in: 1906 *Descent* (F1048). 1910 *Autobiography* (F2304). 1912 *Journal* (F2300). 1914 *Origin* (F641). 1964 *Expression* (F1181).

Daisy, a donkey. ED wrote about several rides on a donkey by the family between 1857-62. Francis Darwin wrote in 1882 Apr. 15 Bernard told him Daisy the donkey was frightened of the thunder shower that day. F. Darwin, *Story of a childhood*, 1920, p. 55.

Dallas, William Sweetland, 1824-90. Zoologist. 1849 FLS. 1858-68 Curator Yorkshire Philosophical Society Museum. 1868 CD to Fritz Müller, "Prof. Huxley agrees with me that Mr. Dallas is by far the best translator" of *Für Darwin*. 1868 D compiled index to *Variation*, holding the publication up. 1868-90 Editor *Annals and Mag. of Nat. Hist.* 1872 D compiled glossary to 6th edn of *Origin*. 1879 Translated (for £21) *Erasmus Darwin* by Ernst Krause, which translation contains a long preliminary notice by CD (F1319). 1871 CD letter to in F2279 of 1902.

d'Alton, Josef Wilhelm Eduard, 1772-1840. German. Father of Johann Samuel Eduard. d'A. Vertebrate zoologist. Scientific illustrator. d'A is referred to in historical sketch to *Origin* as Johann Samuel Edward d'A, their names being persistently misprinted "Dalton". See *Book Collector*, 25: pp. 257-8, 1976. 1827- University Bonn.

Dana, James Dwight, 1813-95. American geologist and zoologist. CD discussed evolution with before *Origin*. Biography: Gilman, 1899. 1838-42 United States exploring expedition under Charles Wilkes to the Pacific as mineralogist and geologist. 1849 D sent CD his work on geology of US Expedition. 1849 CD to Lyell, "Dana is dreadfully hypothetical in many parts, and often as 'd d cocked sure' as Macaulay [was said to be by William Lamb]". CCD4:289. 1849-54 Provided CD with specimens and information for *Cirripedia*. 1850-92 Silliman Prof. Natural History and Geology Yale. 1854 CD sent copy of *Living Cirripedia*. 1859 CD sent copy of 1st edn of *Origin*. 1859 Dec. CD to Lyell, CD had had a letter from D saying that he is "quite disabled in his head" from overwork. CCD7:461. 1860 D to CD, from Florence, saying that his health was poor. 1863 CD to Lyell on D's classification of mammals in *Silliman's Jrnl.*, 25: pp. 65-71 and *Annals and Mag. of Nat. Hist.*, 12, "The whole seems to me to be utterly wild". CCD11:145. 1872 *Corals and coral islands*. 1877 Copley Medal Royal Society. 1881 Aug. CD to Hooker, says D was first to argue for permanence of continents. LL3:247. 1884 Foreign Member Royal Society.

Dangerous or **Low Archipelago** (Tuamotu Archipelago), Pacific island group, part of French Polynesia. Largest chain of atolls in the world. FitzRoy spells "Tuaamotu", with chart in appendix to vol. 2 of *Narrative*. 1835 Nov. 9-13 *Beagle* sailed through on way to Tahiti, charting two new islands, but did not stop.

Daniell, William Freeman, 1818-65. M.D., FLS. Army surgeon and botanist. 1856 sent CD skin of a dog from Sierra Leone. CD acknowledged and cited D in *Variation*, 1:132, 186, 184 and 237.

Danish. First editions in: 1870 *Journal* (F1834). 1872 *Origin* (F643). 1874-75 *Descent* (F1050). 1909 *Autobiography* (F1512). Introductions to several of the Danish translations were written as part of the Darwin in Denmark Project, in association with *Darwin Online* by Peter C. Kjærgaard, Stine Grumsen, Jakob Bek-Thomsen, Gry

Vissing Jensen, Marie Larsen, Lars Brøndum, Laura S. Thomasen and Hans Henrik Hjermitslev: http://darwin-online.org.uk/EditorialIntroductions/Kjaergaard_DarwinDenmark.html

Dapsy, László, 1843-90. 1873-74 translated *Origin* into Hungarian (F703). CD in 1873 Jun. 9 replied to express pleasure at the progress of the translation, asked for a copy. CCD21. Asked for permission to translate *Descent*, but did not complete it.

Darbishire, Alexander, ?-1841 Mate on *Beagle*, *Narrative* 2:19. CD spelled "Derbyshire". 1832 Apr. 25 CD to Caroline Darwin "is also discharged the service, from his own desire, not choosing his conduct, which has been bad about money matters to be investigated". CCD1:225.

Darby, Yvonne, 1910-84. 1931 1st wife of Sir Robert Vere Darwin.

D'arcy, M. Miss, visited the Darwins on three occasions, 1877 Sept. 15, 16 with Arthur and Eliz, 1879 Jan. 3-4. *ED's diary*.

Dareste de la Chavanne, Gabriel Madeleine Camille, 1822-99. French biologist. Held biological chairs in Paris. Specialist on monstrosities. 1863 CD to D, D was pro-*Origin*. 1869 CD to D on his application for chair of physiology in Paris.

Darwin Archive—CUL, see Darwin, Manuscripts.

Darwin census, 2009- a project between Angus Carroll, the Huntington Library and *Darwin Online* to record surviving copies of the first edn of *Origin* and, when possible, their whereabouts. As of Jul. 2020, 350 copies had been located around the world, or 28% of the 1,250 copies printed in 1859. There are surely many more.

Darwin, family of. See the family papers on ancestry, pedigrees etc. in DAR210.29. Burke 1888 gives by far the most detailed pedigree. Three pedigrees are given below: one, abridged from Burke 1888, shows the male Darwin line back to the 16th century, as far as he was able to trace it: a second shows CD's children and grandchildren, although the latters' marriages and the CD great-grandchildren are intentionally omitted: and thirdly one to show CD's relationship to ED. These pedigrees can be expanded, especially to the other 13 children of Erasmus Darwin [I], and to the Wedgwoods and Allens, by reference to the text. Before 1542 he traces the family in the male line back to William Darwin [I] of Marton, Lincolnshire, who died before 1542. The male descendants continue largely in that county. 1680 William Darwin [VI] married Ann Waring who inherited Elston Hall in the same county. The estate was inherited by their son Robert Darwin and is still held by the senior branch of CD's line of the family. 1849 Elston Hall passed to a distaff on the marriage of Charlotte Maria Cooper Darwin to Francis Rhodes in 1849. 1850 The latter changed his name to Darwin on inheriting Elston Hall under the will of his brother-in-law Robert Alvey Darwin, who had died in 1847. The headship of the family, in the male line, then passed back to the descendants of Erasmus Darwin [I] who was the younger brother of Charlotte Darwin's father William Alvey Darwin. 1847-48 Erasmus Darwin's [I] only surviving son Robert Waring Darwin, CD's father, held it briefly in 1847-48 and, on his death in the latter year, it went to his elder son Erasmus, CD's brother. 1881-82 He died in Aug. 1881, unmarried, and CD himself held it for a little over 6 months. 1882 From CD it went to his eldest son William who had no children. 1912 CD's second son, George died. 1914 His eldest son, Sir Charles Galton Darwin, became head on William Erasmus Darwin's death. 1962 On Sir Charles Galton Darwin's death in 1962, it passed to his eldest son George Pember Darwin. 1914, 1915 Less detailed pedigrees are printed in *Emma Darwin* and in *Life, letters and labours of Francis Galton*, vol. 1, 1914. 1952 There is also a brief one in Gwen Raverat, *Period piece*, 1952, which carries the pedigree one generation further into the 20th century.

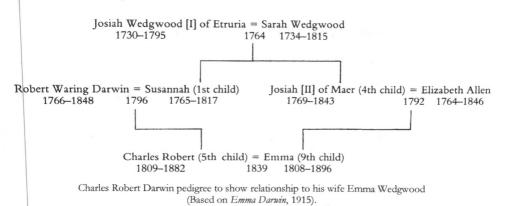

Charles Robert Darwin pedigree to show relationship to his wife Emma Wedgwood
(Based on *Emma Darwin*, 1915).

1.	2.	4.
1. Gwendolen Mary, 1885-1957.	Bernard Richard Meirion, 1876-1961.	1. Erasmus, 1881-1915.
2. Charles Galton, 1887-1962.		2. Ruth Frances, 1883-1973.
3. Margaret Elizabeth, 1890-1974.	3.	3. Emma Nora, 1885-1989.
4. William Robert, 1894-1970.	Frances Crofts, 1886-1960.	
5. A male died in infancy 1896.		

Table of Charles Robert Darwin's grandchildren

Darwin, family, head of: George Pember Darwin, 1928-2001 Jun. 18, was head of the family. He is succeeded by William Huxley Darwin.

Eponyms, list of forenames (all other eponyms are under **Darwin, Charles**):

Barlow, Erasmus Darwin, 1915 Apr. 15-2005 Aug. 2, named after his mother Emma Nora, Lady Barlow, née Darwin. Married Brigit Ursula Hope Black (known as Biddy). 1 son, 2 daughters. 1 Jeremy Barlow 1939. 2. Camilla Barlow 1942 and Phyllida Barlow 1944.

Fox, Edith Darwin, 1857 died an infant, named after father William Darwin F. Mother Ellen Sophia.

Fox, Rev. Samuel William Darwin, 1841-1913, named after his father Rev. William Darwin F. Mother Harriet Fletcher.

Fox, Victor William Darwin 1883-1915, named after his grandfather Rev. William Darwin F. Mother Harriet Fletcher.

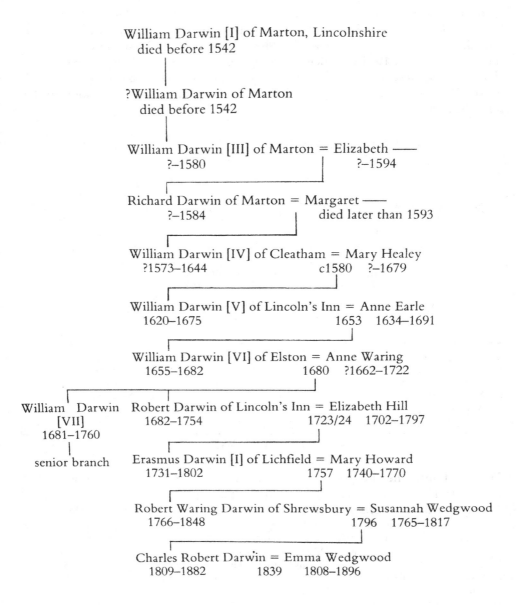

William Darwin [I] of Marton, Lincolnshire
died before 1542

?William Darwin of Marton
died before 1542

William Darwin [III] of Marton = Elizabeth ——
?–1580   ?–1594

Richard Darwin of Marton = Margaret ——
?–1584   died later than 1593

William Darwin [IV] of Cleatham = Mary Healey
?1573–1644   c1580   ?–1679

William Darwin [V] of Lincoln's Inn = Anne Earle
1620–1675   1653   1634–1691

William Darwin [VI] of Elston = Anne Waring
1655–1682   1680   ?1662–1722

William Darwin
[VII]
1681–1760

senior branch

Robert Darwin of Lincoln's Inn = Elizabeth Hill
1682–1754   1723/24   1702–1797

Erasmus Darwin [I] of Lichfield = Mary Howard
1731–1802   1757   1740–1770

Robert Waring Darwin of Shrewsbury = Susannah Wedgwood
1766–1848   1796   1765–1817

Charles Robert Darwin = Emma Wedgwood
1809–1882   1839   1808–1896

Skeleton pedigree of Charles Robert Darwin in the male line (from Burke 1888).

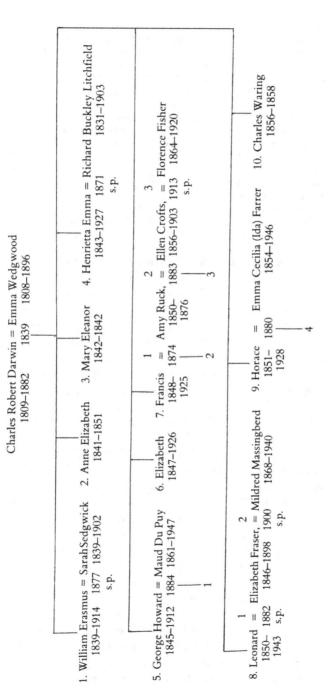

Charles Robert Darwin = Emma Wedgwood
1809–1882   1839   1808–1896

1. William Erasmus = Sarah Sedgwick 1877   1839–1914   1839–1902   s.p.

2. Anne Elizabeth 1841–1851

3. Mary Eleanor 1842–1842

4. Henrietta Emma = Richard Buckley Litchfield 1871   1843–1927   1831–1903   s.p.

5. George Howard = Maud Du Puy 1884   1845–1912   1861–1947
1

6. Elizabeth 1847–1926

7. Francis = Amy Ruck, = Ellen Crofts, = Florence Fisher
1848–   1   1850–   2   1883   3   1913
1925   1874   1876   1856–1903   1864–1920   s.p.
2   3

8. Leonard = Elizabeth Fraser, = Mildred Massingberd
1850–   1   1882   2   1900   1846–1898   1868–1940
1943   s.p.   s.p.

9. Horace = Emma Cecilia (Ida) Farrer
1851–   1880   1854–1946
1928   4

10. Charles Waring 1856–1858

Pedigree of Charles Robert Darwin's children (see table on p. 65).

Fox, Rev. William Darwin 1805 Apr. 23-80 Apr. 8, named after his mother Anne née Darwin. Married twice and had 18 children. Married 1 Harriet Fletcher (1799-1842), 1 d 1834 stillborn, 2 d Eliza Ann (1836-74) m Rev. Henry Martyn Sanders, 3 d Harriet Emma 1837 m Samuel Charles Overton, 4 d Agnes Jane (1839-1906), 5 d Julia Mary Anne 1840 m Samuel Everard Woods and 1 son Samuel William Darwin 1841-1913. Married 2 Ellen Sophia (1820-87) had 6 sons and 6 daughters.

French, Erasmus Darwin, fl. 1875, source of forenames unknown.

Galton, Darwin, 1814-1903, named after his mother Frances Anne Violetta née Darwin.

Galton, Violet Darwin 1862-?, named after her grandmother Frances Anne Violetta née Darwin.

Huish, Frances Violetta Darwin 1858-?, named after her grandfather Sir Francis Sacheverel Darwin.

Huish, Francis Darwin, 1850-1917, named after his grandfather Sir Francis Sacheverel Darwin.

Keynes, Richard Darwin, 1919 Aug. 14-2010 Jun. 12, named after his mother Margaret Elizabeth, Lady Keynes, née Darwin.

Overton, William Darwin, ?-1883, named after his great-great-grandfather William Alvey Darwin, through his grandfather Rev. William Darwin Fox.

Stowe, Darwin, fl. 1638, named after his great-grandfather Henry Darwin.

Swift, Francis Darwin, 1864-?, named after his grandfather Sir Francis Sacheverel Darwin.

Wilmot, Rev. Darwin, 1855-1935, named after his grandfather Sir Francis Sacheverel Darwin.

Wilmot, Sacheverel Darwin, 1885-1918, son of Rev. Darwin Wedgwood, q. v.

Arms and Crest of
Robert Waring
Darwin (bookplate).

Darwin, family of, Arms. Burke 1888. c.1573-1644. Records the arms of William Darwin [IV], c.1573-1644, as: Argent, on a bend gules between two cotises vert, three escallops vert. 1717 He illustrates the same coat for Robert Darwin of Lincoln's Inn in 1717, with a cadency crescent for second son. Erasmus Darwin [I] used them without cadency, although he was also a second son. His son, Robert Waring Darwin, shows a martlet for 4th son, although the pedigree gives him as 3d son. Crest in all these examples, a demi-griffin segreant vert, holding between the claws an escallop vert. Motto "E conchis omnia". Burke 1888 illustrates the arms of two of CD's sons, William and George in both of which the coat is quartered 2 and 3, vert a chevron argent, between 3 herons heads erased (for Waring of Elston Hall, Lincolnshire); crest the same; motto "Cave et aude". Fairbairn, *Book of crests of the families of Great Britain & Ireland*, c.1860, for four of CD's sons, records the crests as having in front of the griffin three escallops fesseway argent. The senior branch of the family had slightly variant arms: ermine a leopard's face jessant-de-lys between two escallops, all within two bendlets gules. In 1849 Francis Rhodes married Charlotte Maria Cooper Darwin, heiress of Elston Hall, the family seat. In 1850 he changed his surname to Darwin and was granted in the same year, by Queen Victoria, the Darwin arms quartering 2 and 3 those of Rhodes, per pale argent and azure, on a bend nebuly, a lion passant guardant, between two acorns slipped, all countercharged; twin crests, a demi-griffin segreant sable, semée of mascules or, resting the sinister claw upon an escutcheon argent, charged with a leopard's face jessant-de-lys gules (for Darwin). A cubit arm erect, vested of six argent and azure, cuffed gules, the hand holding in saltire an oak branch and a vine branch, both fructed proper (for Rhodes): Motto

"Cave et aude". See Susan Darwin to CD [c.24 Oct. 1839]. CCD1. There seems to be no record of CD using arms, although he did use a signet with the crest. CD to W.D. Fox 24 Oct. 1839 asking "what the motto to our crest is for I mean to have a seal solemnly engraved." CCD2 CD had one made with "Cave et aude" and used it on letters in 1840 but apparently not after. See: http://darwin-online.org.uk/EditorialIntroductions/Carroll_DarwinsSeal.html

Darwin, Lady, The following have borne the title as wives and some as relicts: 1. Maud du Puy, 1905-47, wife of Sir George Howard Darwin. 2. Florence Henrietta Fisher Darwin, 1913-20, wife of Sir Francis Darwin. 3. Emma ("Ida") Cecilia Farrer, 1918-46, wife of Sir Horace Darwin, was also The Hon. from 1893 when father became Baron. 4. Katharine Pember, 1925 Married Sir Charles Galton Darwin.

Darwin, Amy Richenda, see Ruck.

Darwin, Ann, 1727-1813. 4th child of Robert Darwin. CD's great-aunt. Unmarried.

Darwin, Anne [III], 1777-1859. Child of William Alvey Darwin [I]. CD's first cousin once removed. 1799 Married Samuel Fox. Children including Rev. William Darwin Fox.

Darwin, Anne Elizabeth (Annie), 1841 Mar. 2-1851 Apr. 23 midday. Second child of CD, born at 12 Upper Gower Street. Known as "Annie", "Anny" and "Kitty Kumplings". CD's favourite child. 1851 Died at Malvern of a fever (possibly tuberculosis). Death notice, by CD, in *The Times* (28 Apr.): 9 (F1999): "On the 23rd inst., at Malvern, of fever, Anne Elizabeth Darwin, aged 10 years, eldest daughter of Charles Darwin, Esq., of Down, Kent." A writing box with some of her things was preserved as well as a brooch containing a lock of her hair. All are transcribed in *Darwin Online*. See Desmond and Moore, *Darwin*, 1991 and R. Keynes, *Annie's box*, 2001. Contrary to common belief, her death did not lead to CD's loss of faith in Christianity which had already occurred in the years after the *Beagle* voyage. According to CD's *Autobiography*, p. 87, and elsewhere (see Aveling 1883), he gradually came to give up belief in Christianity because he found it was not supported by evidence and that "The rate was so slow that I felt no distress". The death of Anne was the most distressing event in CD's life. Also CD did not cease attending church as a result of A's death. CD's touching reminiscence of her as well as letters on her death are in DAR210.13.40 and published in LL1:132-4, Colp 1987, CCD5, Appendix II and transcribed in *Darwin Online*. Other manuscripts relating to Anne's death are kept in DAR210.13 including a reminiscence of her by ED, transcribed in *Darwin Online*. See also *Notebook of observations on the Darwin children.* (1839-56) DAR210.11.37, CCD4: Appendix III and transcribed in *Darwin Online*.

Darwin, "Annie", see Anne Elizabeth Darwin.

Darwin, Bernard Richard Meirion, 1876 Sept. 7-1961 Oct. 18. Went to Eton. Writer, mostly on golf. Only child of Sir Francis Darwin and Amy Richenda Ruck. CD's senior grandchild, the first of two born in CD's lifetime. Nicknames: "Babba", "Babsey", "Dubba", or "Dubsy" in infancy. Called himself "Abbedabba" and "Ubbadubba". Home Gorringes, Down. 1876-83 His mother died in childbed and he was brought up at Down House until his father remarried in 1883. First started playing golf in 1884. F. Darwin, *Story of a childhood*, 1920, p. 62. 1906 Married Elinor Mary Monsell. 1 son, 2 daughters: 1. Sir Robert Vere, 2. Ursula Frances Elinor Mommens, 3. Nicola Mary Elizabeth. 1908-53 Golf correspondent of *The Times*, Francis Darwin, *The story of a childhood*, 1920, privately printed. Contains extracts from letters from Francis Darwin to Mrs Ruck, née Mary Anne Matthews, his mother-in-law, about D, from birth to age 15. They were given back to Francis Darwin on Mrs Ruck's death, she died in her late 80s. 1925-27 wrote a two vol. illustrated children's book with Elinor *The tale of Mr. Tootleoo*. 1941 wrote introduction to the *Oxford dictionary of quotations*, 1941. 1947 *Fifty years of country life*. 1955 Autobiography: *The world that Fred made*, 1955, recollections of CD transcribed in *Darwin Online*. Photographs in DAR219 and DAR225. Fred Hill was a gardener at Down House.

Darwin, "Bessy", see Elizabeth Darwin [VI].

Darwin, Caroline Sarah, 1800 Sept. 14-1888 Jan. 5. Second child of Robert Waring Darwin. CD's sister. The only one of CD's siblings to outlive him. 1837 Married Josiah Wedgwood [III]. 1837 CD to William Darwin Fox "I never saw a human being so fond of little crying wretches (children) as she is". CCD2:29; Wedgwood 1980, p. 228. 1885 recollection of CD "I see from the Journal that Charles's recollection of his childhood was wrong. Instead of being "a naughty boy" he was particularly affectionate, tractable & sweet tempered, & my father had the highest opinion of his understanding & intelligence. My father was very fond of him & even when he was a little boy of 6 or 7, however bustled & overtired, often had C. with him when dressing, to teach him some little thing such as the almanack - & C. used to be so eager to be down in time. C. does not seem to have known half how much my father loved him." DAR112.A117, transcribed in *Darwin Online*. 1816 Pastel chalk drawing by Ellen Sharples. Darwin Heirlooms Trust. Down House.

Darwin, Catherine, see Emily Catherine Darwin.

Darwin, Charles, [I] 1758 Sept. 3-1778 May 15. 1st child of Erasmus Darwin and Mary Howard. Unmarried. CD's uncle and CD was named after him. Medical student, died from a dissecting room wound at Edinburgh. Andrew Duncan arranged his funeral. 1780 Publication of his thesis "Experiments establishing a criterion between mucaginous and purulent matter" in Lichfield, edited by his father.

Darwin, Sir Charles Galton, 1887 Dec. 19-1962 Dec. 31. 2d child of Sir George Howard Darwin. CD's grandson. Physicist. 1925 Married Katharine Pember. 4 sons, 1 daughter. 1922 FRS. 1923-36 Prof. Natural Philosophy Edinburgh. 1936-9 Master of Christ's College, Cambridge. 1938-49 Director National Physical Laboratory. 1927 D owned Down House when George Buckston Browne bought it. 1942 KBE. Had 4 sons and 1 daughter.

Darwin, Charles John Wharton, 1894 Dec. 12-1941 Dec. 26. Son of Charles Waring Darwin. Squadron Leader and Businessman. Head of senior branch of the family. Of Elston Hall, Nottinghamshire. CD's remote cousin. 1917 Married Sibyl Rose.

Darwin, Charles Robert, 1809 Feb. 12-1882 Apr. 19. Dates of birth, death, marriage and names of children are given first, followed by a few quotations to give some indication of CD's character. Other information is then given under the following heads:

Anniversaries.	Books, fiction.	Finances.	Manuscripts.
Appearance.	Books, statistics.	Funeral.	Medals.
Banknotes and Coins.	Books, translated.	Grave.	Order.
Books by.	Death and funeral.	Habits.	Prize.
Books, autobiographies.	Degrees.	Handwriting.	Religion.
Books, bibliographies.	Descendants.	Health.	Society.
Books, biographies.	Eponyms, e.g.	Homes.	membership.
Books, dedicated by.	animals, institu-	Iconography.	Stamps.
Books, dedicated to.	tions, places.	Itinerary.	

1809 Feb. 12 Sun. - 1882 Apr. 19 Wed. about 4 pm. Naturalist. 5th child of Robert Waring Darwin and Susannah née Wedgwood. Born The Mount, Shrewsbury. Died Down House, Downe, Kent. Nicknames: "Gas" (at Shrewsbury School). His siblings called CD "Bobby" and "Charley". "Postillion", by Frances Mostyn Owen. "Dear old Philosopher" (by officers on *Beagle*). "Flycatcher" (by all ranks on *Beagle*).

When CD was born he had only one grandparent living, Sarah Wedgwood, his maternal grandmother, who was ED's paternal grandmother. She died when CD was 5/6. His mother died when he was 7 and his father when he was 39. He had one brother and four sisters. Of his own ten children, three died in infancy or childhood, the rest outliving him. He had four grandsons and five granddaughters: two, Bernard Richard Meirion Darwin and Erasmus Darwin [III], were born in his lifetime. 1839 Jan. 29 Married Emma Wedgwood, by Rev. John Allen Wedgwood at St. Peter's Church, Maer, Staffordshire. 6 sons, 4 daughters: 1. William Erasmus, 2. Anne Elizabeth, 3. Mary Eleanor, 4. Henrietta Emma, 5. George Howard, 6. Elizabeth, 7. Francis, 8. Leonard, 9. Horace, 10. Charles Waring.

Quotations: [As a child] "I just remember him—a dullish apathetic lad, giving no token of his after-eminence". F.E. Gretton, *Memory's harkback*, 1889, p. 33.

1834 To Emily Catherine Darwin, from East Falkland Island, "there is nothing like Geology; the pleasure of the first day's partridge shooting or first day's hunting cannot be compared to finding a fine group of fossil bones, which tell their story of former times with almost a living tongue". CCD1:379.

1839 Jan. 29 "Uncle John [Wedgwood] believes one single turnip in a garden is enough to spoil a bed of cauliflowers". Entry made by CD on wedding day. *Notebook E: Darwin's notebooks*, 1987, p. 423.

1844 Aug. 29 CD to Horner, "I always feel as if my books came half out of Lyell's brain".

1856 CD to Thwaites, asking for information, "when a beggar once begins to beg he never knows when to stop". CCD6:54.

1859 CD to Lyell, "It is a pity he [FitzRoy] did not add his theory of the extinction of Mastodon etc., from the door of the Ark being made too small", about two letters to *The Times* signed "Senex". CCD7:413.

1860 Mar. CD to Joseph Leidy, "I have never for a moment doubted, that though I cannot see my errors, that much in my book [*Origin*] will be proved erroneous". CCD9:39.

1862 Dec. Hooker to B.H. Hodgson "First naturalist in Europe. Indeed I question if he will not be regarded as great as any that ever lived; his powers of observation, memory and judgement seem prodigious, his industry indefatigable and his sagacity in planning experiments, fertility of resources and care in conducting them are unrivalled, and all this with health so detestable that his life is a curse to him". Allan, *Darwin and his flowers*, 1977.

1863 CD to Hooker, "We are degenerate descendants of old Josiah W. for we have not a bit of pretty ware in the house." LL3:5, CCD11:8.

1863 CD to Hooker, "It is mere rubbish thinking at present of the origin of life; one might as well think of the origin of matter". CCD11:278.

1864 CD to Wallace, "It is an awful stretcher to believe that a peacock's tail was thus formed; but, believing it, I believe in the same principle somewhat modified applied to man". CCD12:248.

1875 CD to R.F. Cooke, "I sometimes think a man is a fool who writes books". CCD23:243.

1881 Jun. 15 CD to Hooker, "So I must look forward to Down graveyard as the sweetest place on earth". ML2:433.

Anniversaries: The following list gives the dates of 50 significant anniversaries in CD's immediate family circle during his lifetime.

Jan. 2 Charlotte, ED's sister, died 1862.
3 Horace Darwin, CD's son, married 1880.
23 Susannah Darwin born 1765.
29 CD and ED's wedding day 1839.
31 Sophia Mary Ann Wedgwood 1838, Jos Wedgwood & Caroline's first born died.
Feb. 2 Catherine Langton, sister, died 1866.
12 Charles Darwin born, 1809.
Mar. 2 Anne Elizabeth Darwin born 1841.
2 Charlotte, ED's sister, married, 1832.
11 Josiah, ED's brother, died 1880.
31 Elizabeth Allen, ED's mother, died 1846.
Apr. 4 Henry Allen Wedgwood born 1799.
7 Marianne Darwin, CD's sister, born 1798.
19 Charles Darwin died 1882.
23 Anne Elizabeth Darwin died 1851.
26 Francis Wedgwood married, 1832.
May 2 Emma Darwin born 1808.
10 Catherine, CD's sister, born 1810.
13 Horace Darwin born 1851.
30 Robert Waring Darwin born 1766.
Jun. 28 Charles Waring Darwin died 1858.
Jul. 8 Elizabeth Darwin born 1847.
9 George Howard Darwin born 1845.
11 Leonard Darwin married 1882.
12 Josiah [II], ED's father, died 1843.

15 Susannah Darwin died 1817.
18 Marianne Parker, CD's sister, died 1858.
Aug. 3 Susan, CD's sister, born 1803.
16 Francis Darwin, CD's son, born 1848.
20 Frances, ED's sister, died 1832.
26 Erasmus Alvey Darwin died 1881.
31 Henrietta Emma Darwin married 1871.
Sept. 7 Bernard, CD's grandson, born 1876.
11 Amy Richenda Ruck died 1850.
14 Caroline, CD's sister, born 1800.
14 ED moved into Down House 1842.
17 CD moved into Down House 1842.
23 Mary Eleanor Darwin born 1842.
25 Henrietta Emma Darwin born 1843.
Oct. 2 *Beagle* reached Falmouth 1836.
3 Susan, CD's sister, died 1866.
16 Mary Eleanor Darwin died 1842.
Nov. 7 Sarah, ED's sister, died 1880.
11 CD proposed marriage to ED 1838.
13 Robert Waring Darwin died 1848.
25 Francis, ED's brother, born 1800.
29 William, married Sarah Sedgwick, 1877.
Dec. 6 Charles Waring Darwin born 1856.
7 Erasmus, CD's grandson, born 1881.
27 *Beagle* sailed from Devonport 1831.
29 Erasmus Alvey Darwin born 1804.

Appearance: The only full description of CD's physical appearance and of his dress is Francis Darwin's reminiscences of his father in Chapter 3 of LL1, but he omits much and only treats of CD in his later years. The picture can be amplified from portraits. The only portrait in his childhood is the pastel by Ellen Sharples, when he was about 7 years of age. See **Iconography** below.

He was about six feet tall, sparely built with medium shoulders. According to R.W. Darwin's 'Weighing Account' book (Down House MS) CD weighed: Oct. 1836: 10 stone 8½ lb. Oct. 1837: 12 stone 5 oz. Sept. 1839: 11 Stone 4 lb. Apr. 1840: 10 stone 8 lb. Oct. 1843: 11 stone 1½ lb. In Francis Darwin's recollection CD had a tendency to stoop which increased with age; high forehead, much wrinkled in age, but his face otherwise unlined; wide-set eyes, iris bluish-grey according to Francis Darwin and the Sharples portrait but pale brown in the W.D. Richmond and Ouless portraits; eyebrows very bushy in age; nose straight; mouth small; chin neither prominent nor receding. All the portraits show a youthful face for his age, until he grew his beard, from which time he looked unchangingly old. His hair and side whiskers were light brown and the hairline started to recede before he was 30; by 60 he had only a fringe of hair at the back. His complexion was ruddy. His gait was springing and he walked with a stick which he banged on the ground. He used his hands a good deal in conversation, although the crossed arms and legs shown in the *Vanity Fair* cartoon were characteristic. He was right-handed (See 'A biographical sketch of an infant', p. 287). His laugh was a "free and sounding peal". LL1:111.

1832 Nov.-1834 Jul. He first grew a beard, as did everyone else, when the *Beagle* left Montevideo for the cold south, but they shaved when they returned to temperate waters. CD to his sister Emily Catherine "With my great beard". CCD1:382.

1835 "Whilst we all wore our untrimmed beards". *Journal* 2d edn, p. 209. No images of CD taken during the *Beagle* voyage exist although a purported caricature from the voyage was sold at Sotheby's in 2015.

The later portraits show that CD's dress was conventional and that of a man of his position, but in later years it became less so. He gave up wearing a tall hat even in London, wearing a soft black one with a rounded crown in winter and a big straw hat in summer. His clothes were dark and of a loose and easy fit.

1849 CD to Hooker, "Everyone tells me that I look quite blooming and beautiful; and most think that I am shamming". CCD4:344, LL1:111.

1862 CD finally grew beard and moustache after a case of eczema.

1864 May 28 CD to Gray, on sending a bearded photograph "Do I not look venerable". CCD12:212.

1866 Apr. 28 ED to Henrietta "He was obliged to name himself to almost all of them [people at a Royal Society soirée], as his beard alters him so". ED2:185.

c.1880 Outdoors he wore a cloak: the cloak and winter hat are well shown in the Elliott & Fry photographs of 1881. Indoors, he normally wore a shawl and "great loose cloth boots" over his indoor shoes. LL1:112.

1880 Jan. His sons bought him a fur coat. ED to Leonard Darwin "He has begun wearing it so constantly, that he is afraid it will soon be worn out". ED2:239.

He wore, for reading or close experiments, spectacles or more often pince-nez, visible on a string around his neck in several photographs, starting from the first in 1842. His hearing was unimpaired.

Banknotes and coins depicting CD: (27 recorded here)

1983 Darwin "tourist dollar", with bearded portrait, Australia.

1993 Cook Islands 20 dollar silver coin. CD after Richmond 1840 with Galapagos tortoise.

1996 Cook Islands 5 dollar silver coin. "Protect our world". Same as above.

1998 Gibraltar 1 Crown coin, "The Victorian age / Charles Darwin. Evolution of mankind."

1999 1 dollar coin depicting CD, Galapagos finch heads and the *Beagle*. Sierra Leone.

1999 CD after Maull & Polyblank 1857 with apes. 1000 Francs CFA silver coin, Republic of Congo.

2000-18 Bank of England £10 banknote with a portrait of CD on back. Often stated to be after a photograph by Julia Margaret Cameron but in fact after Elliott & Fry 1874a (reversed).

2001 1 Crown coin Isle of Man "The Victorian Age" after Collier 1881.

2003 Cocos (Keeling) Islands silver $10 coin commemorating the 1836 visit of the *Beagle*.

2003 Cocos (Keeling) Islands silver $100 coin as above.

2005 $1 silver coin loosely after Lock & Whitefield 1878. Cook Island.

2006 Jersey Great Britons Charles Darwin £5 Silver Proof Coin, the Royal Mint.

2007 Falkland Islands commemorative coins of CD. The Pobjoy Mint, in Cupro-nickel, 10,000 minted in sterling silver and 5,000 were struck in .999 gold.

2008 1 Crown 1/25oz. gold coin. Tristan da Cunha.

2009 1/64 Crown. Falkland Islands cupro-nickel & gold commemorative coins of bearded CD.

2009 Tuvalu three different one dollar coins of CD.

2009 The Perth Mint 1-ounce silver coin depicting CD young and old, the *Beagle* and his signature. Legal tender on the island of Tuvalu.

2009 1 Peso copper-nickel coin with three-quarter right profile and *Beagle*. Cuba.

2009 The Royal Mint £2 coin with head of CD and a chimpanzee facing each other by Suzie Zamit. The edge inscription is "ON THE ORIGIN OF SPECIES 1859". Also £2 gold proof coin.

2009 $10 silver coin after Richmond 1840 (reversed) but a poor likeness. Samoa.

2009 500 Sucres, three-quarter right profile bearded portrait. Galapagos Islands, Ecuador.

2009 1000 Sucres, portrait after Richmond 1840 (reversed). Galapagos Islands, Ecuador.

2009 $25 coin with right profile after Rejlander 1871c. Galapagos Islands, Ecuador.

2010 2500 Sucres, left profile after L. Darwin 1878a (reversed). Galapagos Islands, Ecuador.

2013 20 Dragones, with Collier 1881. Columbia.

2015 100 Kina. Bougainville Island.

2015 5 Numismas. Banco De Kamberra.

Books by CD: These and his publications in serials, are entered in the main sequence under brief titles. The following list gives full titles of his main books in strict alphabetical order, except for first articles, followed by the date of first appearance under that title and any needed cross reference. Several of his books appeared under more than one title. Works to which he contributed only an article, preface, or letter, have also not been included. Works printed from CD's manuscripts since his death have not been included, but will be found under the separate heading "Manuscripts" and under abbreviated titles in the main sequence.

CD wrote seventeen works in twenty-one volumes, or fifteen if the three volumes of geology of the *Beagle* are treated as one. They consist of more than 9,000 pages of text with a further 170 pages of preliminary matter. If the papers in serials are added, the total comes to well over 10,000 pages. This rough total does not consider the increase, or rarely decrease, in the length of the text in later editions, and represents about 230 pages on average a year for forty-three years. An important summary of CD's writings is M.T. Ghiselin, Darwin: A reader's guide. *Occasional Papers of the California Academy of Sciences* (2009): 185 transcribed in *Darwin Online*.

1 *The descent of man, and selection in relation to sex*, 2 vols., 1871 (F936).

2 *The different forms of flowers on plants of the same species*, 1877 (F1277).

3 *The effects of cross and self fertilisation in the vegetable kingdom*, 1876 (F1249).

4 *Erasmus Darwin. Translated from the German by W.S. Dallas, with a preliminary notice by Charles Darwin*, 1879 (F1319). Text by E. Krause, but CD's preliminary notice is longer.

5 *The expression of the emotions in man and animals*, 1872 (F1141).

6 *The formation of vegetable mould through the action of worms, with observations on their habits*, 1881 (F1357).

7 *Geological observations on coral reefs, volcanic islands, and on South America*, 1851 (F274). Combination volume of Nos. 8, 9 and 27, from the same sheets.

8 *Geological observations on South America. Being the third part of the geology of the voyage of the Beagle, under the command of Capt. Fitzroy, R.N. during the years 1832 to 1836*, 1846 (F273).

9 *Geological observations on the volcanic islands visited during the voyage of H.M.S. Beagle, together with some brief notices of the geology of Australia and the Cape of Good Hope. Being the third part of the geology of the voyage of the Beagle, under the command of Capt. Fitzroy, R.N. during the years 1832 to 1836*, 1844 (F272).

10 *Insectivorous plants*, 1875 (F1217).

11 *Journal and remarks 1832-1836*, 1839 (F10 part). vol. 3 of no. 18, first issue of no. 12.

12 *Journal of researches into the geology and natural history of the various countries visited by H.M.S. Beagle etc.*, 1839 (F11).

13 *Journal of researches into the natural history and geology of the countries visited during the voyage of H.M.S. Beagle round the world, under the Command of Capt. Fitz Roy, R.N.*, 1845 (F13). 2^d edn of no. 12.

14 *The life of Erasmus Darwin...Being an introduction to an essay on his scientific work*, 1887 (F1321). 2^d edn of no. 4, same text but new preliminaries.

15 *A monograph of the fossil Lepadidae, or pedunculated cirripedes, of Great Britain. A monograph of the fossil Balanidae and Verrucidae of Great Britain*, 2 vols. and index, 1851, 1854[1855], 1858 (F342).

16 *A monograph of the sub-class Cirripedia, with figures of all the species*, 2 vols. 1851, 1854 (F339).

17 *The movements and habits of climbing plants*, 1875 (F836). 2^d edn of no. 20.

18 *Narrative of the surveying voyages of his Majesty's ships Adventure, and Beagle*, 3 vols. and appendix to vol. 2, 1839 (F10). Edited by Robert FitzRoy. vol. 3 is CD's volume, titled *Journal and remarks*, =No. 11, 1st edn of no. 12.

19 *A naturalist's voyage. Journal of researches etc.*, 1879 (F34). An unchanged reprint of no. 13.

20 *On the movements and habits of climbing plants*, 1865 (F833 and F834).

21 *On the origin of species by means of natural selection, or the preservation of favoured races in the struggle for life*, 1859 (F373).

22 *On the various contrivances by which British and foreign orchids are fertilised by insects, and on the good effects of intercrossing*, 1862 (F800).

23 *The origin of species by means of natural selection, or the preservation of favoured races in the struggle for life*, 1872 (F391). 6th edn of no. 21. Starting with the 6th edn the title ran "The Origin...", omitting "On". The 1876 issue, 18th thousand, was the final text as CD left it.

24 *Movement in plants*, 1880 (F1325).

25 *Queries about expression*, [1867] (F871, F872 and F873).

26 *Questions about the breeding of animals*, [1839] (F262).

27 *The structure and distribution of coral reefs. Being the first part of the geology of the voyage of the Beagle, under the command of Capt. Fitzroy, R.N. during the years 1832 to 1836*, 1842 (F271).

28 *The variation of animals and plants under domestication*, 2 vols., 1868 (F877).

29 *The various contrivances by which orchids are fertilised by insects*, 1877 (F801). 2ᵈ edn of no. 22.

30 *The voyage of the Beagle*, 1905 (F109). Unchanged reprint of no. 13.

31 *The zoology of the voyage of H.M.S. Beagle, under the command of Captain Fitzroy, during the years 1832 to 1836. Published with the approval of the Lords Commissioners of Her Majesty's Treasury*, 19 numbers making 5 parts, 1838-43 (F8). Edited by CD.

Books, autobiographies: 1876 The original publication of CD's autobiography is in Chapter 2, LL1:26-107 (1887), but CD's description of his father, which is in the manuscript, is printed in Chapter 1, LL1:11-20, instead of in its correct place. It was written in 1876, between May 28 and Aug. 3, with some additions and alterations in 1878 and 1881. The manuscript is headed "Recollections of the development of my mind and character". This version was expurgated by Francis Darwin after a dispute with ED and his sisters, "passages should occur which must have to be omitted". One omitted passage, about CD's mother, was printed in ML1:30.

1838 A further autobiographical fragment of his first eleven years, written in 1838, was printed in ML1:1-5, CCD2 and Secord, *Evolutionary writings*, 2008 and Colp, 'I was born a naturalist': Charles Darwin's 1838 notes about himself, *Jrnl. of the Hist. of medicine and allied sciences*, 35, 1980.

1957 The first full transcription of the original manuscript of the autobiography appeared in Russian translation by S.L. Sobol' (F1540).

1958 Nora Barlow's version of it, which was independently transcribed "with original omissions restored", with an important appendix (F1497).

1974 de Beer edited an edition of the Barlow transcription, with slight modifications after the manuscript had been re-examined by James Kinsley, in *Charles Darwin, Thomas Henry Huxley, autobiographies*. This edition also contains the fragment of 1838 (F1508).

2008 A new transcription by Anne Secord, was published in James Secord's *Charles Darwin: Evolutionary writings*.

2009 A new transcription by Kees Rookmaaker, was published in *Darwin Online* alongside images of the manuscript.

Books, bibliographies: There is no full bibliographical work even of the first editions of CD's books; Freeman's 1977 bibliographical handlist remains the best source available in print, now heavily corrected and greatly supplemented in the Freeman Bibliographical Database in *Darwin Online*:

http://darwin-online.org.uk/Freeman_intro.html.

1959 *Origin* has been surveyed in great detail by Morse Peckham in his variorum edition of 1959. (F588) He covers all English editions and issues up to 1890, and his descriptions include paper, type and binding cases, as well as summaries of John Murray's accounts. Another variorum is available in *Darwin Online* by Barbara Bordalejo:

http://darwin-online.org.uk/Variorum/index.html

1964 Harrison D. Horblit, in the Grolier Club volume *One hundred books famous in science*, 1964, gives another description of the 1st edition.

1954 A full description of *Living Cirripedia* is given in R. Curle, The Ray Society a bibliographical history, 1954, pp. 48-9.

There are several handlists:

1883 F.W. True, A Darwinian bibliography, *Smithson. Misc. Coll.*, 25:92-101.

1887 John P. Anderson, pp. i-xxxi, in G.T. Bettany, *Life of Charles Darwin*, a good list which also contains list of early darwiniana and of reviews.

1887 Francis Darwin, LL3, pp. 362-72, not so useful as Anderson.

1977 R.B. Freeman, *The works of Charles Darwin. An annotated bibliographical handlist*, 2d edn. *Addition and correction to the 2nd edition* were privately published in 1986 in an edition of 25 copies. [Incorporated into and superseded by the Freeman Bibliographical Database.]

1977 P.H. Barrett, *The collected papers of Charles Darwin*, 2 vols., contains a large collection of CD's works in serials, with their references, but with few notes and many omissions.

2006 Freeman Bibliographical Database, the most comprehensive bibliography of CD's writings ever published. http://darwin-online.org.uk/Freeman_intro.html

2009 John van Wyhe ed., *Charles Darwin's shorter publications, 1829-1883*, contains almost all CD's shorter publications with 70 not present in Barrett 1977, fully edited and annotated.

Books, biographies, including letters. Biographies of CD are very numerous and include ODNB and those listed here all contain general biographical matter as well as considerations of his work and theories. Many more, which are concerned with darwinism from the biological, ethical or sociological viewpoints, contain some facts about his life, but usually nothing new, these are not listed here.

1882 *Charles Darwin, memorial notices*, Nature Series. Six obituaries from *Nature*.

1883 L.C. Miall, *The life and work of Charles Darwin*.

1883 J.M. Winn, *Darwin*.

1884 E. Woodall, *Charles Darwin*.

1886 J.T. Cunningham, *Charles Darwin; naturalist*.

1887 G.T. Bettany, *Life of Charles Darwin*.

1887 Francis Darwin ed., *Life and letters of Charles Darwin*, 3 vols. In 1887 and 1888 four further issues and editions appeared, all with slight changes and corrections. Many later issues were published, not printed by Murray.

1889 W. Mawer, *Truth for its own sake: the story of Charles Darwin, written for young people*.

1891 C.F. Holder, *Charles Darwin*.

1892 Francis Darwin, *Charles Darwin. His life told in an autobiographical chapter, and in a selected series of his published letters*. An abridged version of *Life and* Letters, 1887, with some additions.

1894 E.A. Parkyn, *Darwin his work and influence*.

1903 Francis Darwin & A.C. Seward eds., *More letters of Charles Darwin*, 2 vols.

1904 H.E. Litchfield ed., *Emma Darwin, wife of Charles Darwin*, 2 vols., privately printed edn. 1915 *Emma Darwin*, 2 vols., published edn.

1909 E.B. Poulton, *Charles Darwin and the Origin of species*.

1921 Leonard Huxley, *Charles Darwin*.

1923 Karl Pearson, *Charles Darwin, 1809-1882*.

1927 Henshaw Ward, *Charles Darwin*.

1937 Geoffrey West [H.G. Wells], *Charles Darwin, the fragmentary man*.

1950 P.B. Sears, *Charles Darwin, the naturalist as a cultural force*.

1955 William Irvine, *Apes, angels and Victorians*.

1955 Dorothy Laird, *Charles Darwin. Naturalist*.

1959 Arthur Keith, *Darwin revalued*.

1963 Gavin de Beer, *Charles Darwin, evolution by natural selection*.

1966 Julian Huxley & H.B.D. Kettlewell, *Darwin and his world*.

1970 P.J. Vorzimmer, *Charles Darwin: the years of controversy*.

1970 A.J. Marshall, *Darwin and Huxley in Australia*.

1973 D.L. Hull, *Darwin and his critics*.

1977 Mea Allan, *Darwin and his flowers*.

1981 Peter Brent, *Charles Darwin*.

1981 J.J. Parodiz, *Darwin in the New World*.

1982 Wilma George, *Darwin*.

1982 Jonathan Howard, *Darwin*.

1985 R.W. Clark, *The survival of Darwin*.
1985 D. Kohn ed., *The Darwinian heritage*.
1985 F. Burkhardt and S. Smith eds., *A calendar of the Correspondence of Charles Darwin, 1809-1882*. Reprinted with supplement in 1994.
1985- F. Burkhardt and S. Smith et al eds., *The correspondence of Charles Darwin*. [By far the most important, indeed definitive, edition of CD's letters.]
1991 Adrian Desmond & James Moore, *Darwin*.
1995-2002 Janet Browne, *Charles Darwin. voyaging. Volume I of a biography*. And, *Charles Darwin. The power of place. Volume II of a biography*.
1996 Tort, P. ed., *Dictionnaire du darwinisme et de l'évolution*. Paris: PUF, 3 vols., 5000 pp.

2000 Richard Keynes ed., *Charles Darwin's zoology notes & specimen lists from H.M.S. Beagle*.
2001 Randal Keynes, *Annie's box*.
2005 Sandra Herbert, *Charles Darwin, geologist*.
2006 Janet Browne, *Darwin's Origin of species*.
2008 John van Wyhe, *Darwin*. [2019, 2020].
2009 G. Chancellor & J. van Wyhe eds., *Charles Darwin's notebooks from the voyage of the 'Beagle'*. Foreword by Richard Darwin Keynes.
2009 M.T. Ghiselin, Darwin: A reader's guide.
2009 M. Dixon and G. Radick. *Darwin in Ilkley*.
2013 John van Wyhe, *Dispelling the darkness*.
2014 John van Wyhe, *Charles Darwin in Cambridge*.

The torrent of books on CD to appear around the 2009 bicentenary and since are too numerous to list here. Most are derivative and provide no new information. The largest list of the 2009 outpourings is the one compiled by John van Wyhe in http://darwin-online.org.uk/2009.html. Very far from complete, the list records c.630 events, books, periodicals and articles, radio/podcasts, films, installations, documentaries and performances.

Books, dedicated by CD:

1845 *Journal of researches*, 2d edn 1845 and later to Charles Lyell.
1854 *Living Cirripedia*, to Henri Milne Edwards.
1877 *Forms of flowers*, to Asa Gray.

Books, dedicated to CD: (29, Freeman 1978 listed 7)

1854 Hooker, J.D., *Himalayan journals*, 2 vols.
1856 Wollaston, T.V. *On the variation of species*.
1858-82 Mueller, F., *Fragmenta phytographia australie*. (vols. 1 and 5).
1861 Grant, R.E., *Tabular view of the primary divisions of the animal kingdom*.
1866 Haeckel, E., *Generelle Morphologie*.
1869 Wallace, A.R., *The Malay archipelago*, 2 vols.
1870 Orton, J., *The Andes and the Amazon; or across the continent of South America*.
1870 Wallace, A.R., *Contributions to the theory of natural selection*.
1871 [White, R.W.], *The fall of man: or, the loves of the gorillas. A popular scientific lecture upon the Darwinian theory of development by sexual selection. By a learned gorilla*.
1872 Bennett, C.H., *Character sketches, development drawings, and original pictures of wit and humour*.
1873 Sievert, K.S., *Der Communisten-Staat: Cultur-Historische Studie*.
1873-4 Kovalevsky, V.O., *Monographie der Gattung Anthracotherium Cuv. und Versuch einer natürlichen Classification der fossilen Hufthiere*.
1875 Goodacre, F.B., *A few remarks on hemerozoology; or the study of domestic animals*.
1876 Claus, C.F., *Untersuchungen zur Erforschung der genealogischen Grundlage des Crustaceen-Systems*.
1877 Ludwig, R.A.B.S., *Fossile Crocodiliden*.
1877 Semper, C.G., *Über Sehorgane von Typus der Wirbelthieraugen auf dem Rücken von Schnecken*.
1877 Adler, F., (dedicated a number of poems to CD, see CCD25).
1878 Hoare, C., *Dogma, doubt, and duty: a poem in five cantos*.
1878 Nash, W., *Oregon: there and back in 1877*.
1879 Moseley, H.N., *Notes of a naturalist on the "Challenger"*.
1879 *What Mr. Darwin saw in his voyage round the world in the ship "Beagle"*.
1881 Wise, J.R. de C., *The first of May, a fairy masque*.
1881 Mengozzi, G.E., *Nuova classificazione degli esseri naturali e saggio sulla generazione degli animali*.

1882 Dixie, F., *In the land of misfortune.*
1883 Brooks, W.K., *The law of heredity.*
1883 Berger, M., *Der Materialismus im Kampfe mit dem Spiritualismus und Idealismus.*
1897 Newman, E., *Pseudo-philosophy at the end of the nineteenth century.*
1909 Thomson, J.A. *Darwinism and human life.*
2000 Keynes, Richard, *Charles Darwin's zoology notes & specimen lists from H.M.S. Beagle.*

Books, fiction:
1867 [Waugh, E.; Benjamin Brierley et al.], *The Lancashire wedding or Darwin moralized.* (a play).
1936 Baker, Ethel Winifred, *Miss Ann Green of Clifton.* (a novel).
1980 Stone, Irving, *The Origin: a biographical novel of Charles Darwin.*
1982 Ward, Peter, *The adventures of Charles Darwin: a story of the Beagle voyage.*
2002 Drayson, Nicholas, *Confessing a murder.*
2003 Hosler, Jay, The Sandwalk Adventures
2005 Thompson, Harry, *This thing of darkness.*
2005 Darnton, John, *The Darwin conspiracy.*
There are very many others as well as children's books, especially since 2009.

Books, statistics: CD reckoned that he had made £10,248 from his books by the end of 1881. His Murrays totalled 94,000 copies sold at the time of his death, of which 15,000 were *Journal* in which he had no copyright. He made about 2s 6d per copy sold, excluding *Journal.*

Books, translated: No single scientist has had his publications translated into so many languages as CD. As of 2020, there are 64 known languages: Albanian, Amharic, Arabic, Armenian, Basque, Bengali, Bosnian, Bulgarian, Catalan, Chinese, Croatian, Czech, Danish, Dutch, English, English Braille, Esperanto, Estonian, Finnish, Flemish, French, Galician, Georgian, German, Greek, Hebrew, Hindi, Hungarian, Icelandic, Indonesian, Italian, Japanese, Kannada, Kazakh, Korean, Latin, Latvian, Lithuanian, Macedonian, Malay, Malayalam, Maltese, Mongolian, Norwegian, Persian, Polish, Portuguese, Punjabi, Romanian, Russian, Sanskrit, Serbian, Serbo-Croat, Slovak, Slovene, Spanish, Swedish, Tamil, Tibetan, Turkish, Ukrainian, Urdu, Vietnamese and Yiddish.

CD's first coffin. *The Sketch*, (15 Apr.), 1896, p. 528.

Death and funeral: On 23 Mar. 1882 ED recorded "Ch. out a good deal." Drs Moore & Allfrey came on Apr. 12. On Apr. 17, two walks in the orchard and a little work were recorded. On Apr. 18, ED wrote "Ditto" perhaps referring to CD going out to the orchard again and doing a little work.

The following day, Apr. 19, Henrietta recounted (ED2:251), that she arrived in the morning and found her father being supported by her mother and Frank. ED left for a little rest. During that time CD said to Frank and Henrietta, "You are the best of dear nurses" and shortly after ED and Bessy had to be sent for. In their presence, he was said to have "peacefully passed away" at half-past three. ED later recorded in her diary for that day "fatal attack at 12".

Although he had left no instructions, CD's family intended a burial at Downe cemetery. The first coffin was made by John Lewis [II], the village carpenter, for two years a page at Down House. Lewis put CD into it. After a burial in Westminster Abbey was arranged, CD was transferred to a white oak one in which he was buried. The original coffin's brass plate read "CHARLES ROBERT DARWIN, / Died 19 April, / 1882, / Aged 73 years." *Maidstone Jrnl. and Kentish Advertiser*, (22 Jun.), p. 5. The coffin was sold to "a young chap that kept a beerhouse out at Farnborough". "Darwin laid in that coffin thirty-one and a half hours exactly. I put him in myself". *Zoologist*, 1909, p. 120, from *Evening News*, (12 Feb.), 1909. See also S. Maxwell, *Just beyond London*, 1927, pp. 105-6. Maxwell relates a tale of an old man of 87 who had helped to put CD into the first coffin and transferring him to the second by "fitful moonlight". The beerhouse was The New Inn, Rocks Bottom, Farnborough. Colp, Darwin's coffin, and its maker, *Jrnl. of the Hist. of Medicine* 35: pp. 59-63, 1980. 1925 The coffin, with six brass handles and "much elaborate brass work" was re-discovered, forgotten, in the former coach house of the New Inn. "Inquiries have elicited…the coffin was made from old timber on the estate belonging to" CD. *Belfast News-Letter*, (1 Apr.), 1925, p. 9. Its re-discovery was widely reported in the newspapers. 1925 The coffin was purchased by Leonard Darwin and burned at the wishes of the family. *Daily Herald* (29 Jun.), p. 9. The second coffin 'shell' was enclosed in a lead coffin, this again being enclosed in the coffin of white oak. On this, the outermost coffin, was a plate bearing the simple inscription: 'Charles Robert Darwin. Born February 12, 1809, Died April 19, 1882." *Nottinghamshire Guardian*, (28 Apr.), p. 4. The coffin was covered "with numerous wreaths, which, it was observed included not merely the customary combination of milk-white flowers and delicate ferns, but also wreaths of blossoms of many hues, suggestive of the re-awakening of nature in the spring." *Derby Mercury*, (3 May 1882), p. 6.

1882 Apr. 19 CD died. CD was the first and only naturalist to be buried in Westminster Abbey. A copy of his Last will and testament, dated 21 Sept. 1881, from the York Probate Sub Registry is transcribed in *Darwin Online*.

Apr. 21 Letter to the Dean, G.G. Bradley, on House of Commons paper: "Very Rev. Sir, We hope you will not think we are taking a liberty if we venture to suggest that it would be acceptable to a very large number of our countrymen of all classes and opinions that our illustrious countryman Mr. Darwin should be buried in Westminster Abbey, We remain your obedient servants", signed by Lubbock and nineteen other MPs. The Dean was abroad and replied by telegram "Oui sans aucune hésitation regrette mon absence".

Apr. 25 Tuesday, afternoon. CD's body was carried from Down House, in a hearse drawn by four black horses, accompanied by Francis, Leonard and Horace. Vigil in St. Faith's Chapel, where they were joined by William and George. The undertakers were T. & W. Banting. *The Times*, (26 Apr.).

Apr. 26 Wednesday at noon, the mourners invited for 11 am. Service conducted by Canon George Prothero, Senior Canon. Pallbearers, to left of body: Lubbock, Huxley, J.R. Lowell (as American Ambassador), Duke of Devonshire (as Chancellor University of Cambridge), Wallace, to right of body: Canon Farrar (Rector of St. Margaret's Westminster), Hooker, W. Spottiswoode (as President of Royal Society), Earl of Derby, Duke of Argyll. Chief Mourner: William Erasmus Darwin, followed by thirty-one relatives, including all surviving children, servants Parslow and Jackson at rear followed by representatives of scientific bodies. ED not present. Queen Victoria in Council was represented by Earl Spencer, the President. Ambassadors of France, Germany, Italy, Russia and Spain were present. There are manuscript lists by George at CUL including one of "Personal Friends invited" with 74 named individuals "and other old servants and inhabitants of Down". (DAR215.3c, transcribed in *Darwin Online*) There is a printed list, one copy of which is marked by George Darwin (DAR215.3d) "very erroneous". Many names there are not on his manuscript and vice versa. Anthem specially composed by Sir Frederick Bridge "Happy is the man that findeth wisdom, and the man that getteth understanding". Proverbs iii 13, 15-16 omitting 14. May 1 Memorial Service: Westminster Abbey, sermon by Harvey Goodwin, Bishop of Carlisle. The Archbishop of Canterbury, A.C. Tait, had withdrawn at short notice. H.D. Rawnsley, *Harvey Goodwin*, 1896, pp. 223-5.

1915 Nov. 1 Memorial plaque to Wallace was placed next to that for CD, Westminster Abbey. A marble memorial to Hooker by Frank Bowcher is nearby.

Degrees: 1831 Apr. 26 Cambridge, B.A., 10th in list of 178 candidates who did not seek honours. 1837 Cambridge, M.A. 1862 Breslau, Germany, Hon.D.Med. and Chirurg. 1868 Bonn, Germany, Hon.D.Med. and Chirurg. [1870 Jun. 17 Oxford, CD declined Hon. DCL, on grounds of ill health. (F1940)] 1875 Leiden, the Netherlands, Hon. Doctorate (Doctor *honoris causae*). 1877 Nov. 17 Cambridge, Hon. LL.D.

Descendants: CD had 29 great-grandchildren. Erasmus Darwin Barlow, *Zoo Newsletter*, Autumn 1980, on his appointment as Secretary of the Zoo, p. 1, lists 27 great-grandchildren. Those that are known are listed in order of their parents seniority:
1. Gwendolen Mary, (1885-1957) daughter of Sir George Howard Darwin, married Jacques Pierre Paul Raverat (1885-1925), had two daughters.
    1. Elisabeth (1916-2014). 1940 Married Edvard Hambro (1911-77). 4 children: 1 Anne (1941-?), Carl J (1944-?), Christian (1946-?), Linda (1948-?)
    2 Sophie Jane (1919-2011). 1940 Married 1 Mark Pryor (1919-1970). 4 children: Emily (1942-?), William (1945-?), Lucy (1948-?), Nelly (1952-?). 1973 2 Married Henry Gurney (1913-1997), 1 son, 2 daughters.
2. Sir Charles Galton Darwin (1887-1962), son of Sir George Howard Darwin, had 4 sons 1 daughter.
    1 Cecily Darwin, 1926-, X-ray crystallographer, married John Littleton of Philadelphia in 1951. She inherited the chair CD used in his study to write. In 1989 she presented the chair to the Academy of Natural Sciences of Drexel University. She was alive in 2009.
    2 George Pember Darwin, 1928-2001 Jun. 18, was eldest son and head of family.
    3 Henry Galton Darwin, 1929-92, married Jane Sophie Christie, 3 daughters.
    4 Francis William Darwin, 1932-2001, of Kings College London, zoologist.

5 Edward Leonard Darwin, 1934-.

3. Margaret Elizabeth, daughter of Sir George Howard Darwin, 1890-1974, married 1917 Sir Geoffrey Langdon Keynes 1887-1982, FBA, 1981, 1 daughter, 4 sons:

1 Harriet Frances K, 1918-18.

2 Richard Darwin K, 1919-2010, FRS 1959, married 1945 Hon. Anne Pinsent Adrian. 4 sons (1 deceased by 1979).

3 Quentin George K, 1921-2003.

4 William Milo K, 1924-2009.

5 Stephen John K, 1927-2017, married 1955 Mary Knatchbull-Hugesson. 3 sons, 2 daughters.

4. William Robert, 1894-1970, son of Sir George Howard Darwin, married Monica Slingsby. 2 sons, 1 daughter.

5. Bernard Richard Meirion Darwin., 1876-1961, son of Sir Francis Darwin by 1st marriage, Elinor Monsell (1879-1954). 1 son, 2 daughters:

1 Sir Robert (Robin) Vere Darwin, 1910-74, twice married, s.p.

2 Ursula Frances Elinor Darwin 1908-2010, married 1 Julian Trevelyan, 1910-1988 (one son, Philip Trevelyan, 1943) and 2 Norman Mommens, 1922-2000.

3 Nicola Mary Elizabeth Darwin, 1916-76.

6. Frances Crofts Darwin., 1886-1960, poet, daughter of Sir Francis Darwin by second marriage, married Francis Macdonald Cornford, 1874-1943. 3 sons, 2 daughters. One of whom was Francis Cornford, poet.

7. Ruth Frances Darwin, 1883-1972, daughter of Sir Horace Darwin, married W. Rees-Thomas (1887-1978), s.p.

8. Emma Nora Darwin., 1885-1989, daughter of Sir Horace Darwin, married 1911 Sir James Alan Noel Barlow, Bart (1881-1968). 4 sons, 2 daughters:

1 Joan Helen Barlow, 1912-54.

2 Sir Thomas Erasmus Barlow, 1914-2003, RN retired, DSC, DL, 3d Bart. 1968, married 1955 Isabel Body (1915-2005). 2 sons, 2 daughters:

    1. Sir James Alan Barlow, 4th Bart. 2003, 1956-

    2. Dr Monica Ann Barlow, 1958-

    3. Philip Thomas Barlow, 1960-

    4. Teresa Mary Barlow, 1963-

3 Erasmus Darwin Barlow, 1915-2005, physician, psychiatrist, married 1938 Brigit Ursula Hope Black, 1916-2004. 1 son, 2 daughters:

    1. Thomas Jeremy Erasmus Barlow, 1939-, married 1962 Jane Marian Hollowood. 1 son: Josiah Bernard Barlow, 1973-.

    2. Camilla Ruth Barlow, 1942-, married 1 Martin Christopher Mitcheson 1965 diss. 1973, 1 son: Luke Thomas Mitcheson, 1966-. Married 2 1974 Stuart Anthony Whitworth-Jones. 1 daughter: Eleanor Gwen Whitworth-Jones, 1975-.

    3. Gillian Phyllida Barlow, 1944-, married Fabian Peake, has children.

4 Andrew Dalmahoy Barlow, 1916-2006, married Yvonne Tanner. 1 son, 1 daughter:

    1. Martin Thomas Barlow, 1953-

    2. Claire Barlow, 1954-.

5 Hilda Horatia Barlow, 1919-2017, married 1944 John Hunter Padel (1913-99).

    1. Ruth Sophia Padel, 1946-, wrote *Darwin: A life in poems*, 2009.

    2. Oliver James Padel, 1948-

    3. Nicola Mary Padel, 1951-

    4. Felix John Padel, 1955-

    5. Adam Frederick Padel, 1958-.

6 Horace Basil Barlow, 1921-, FRS 1969, married 1 1954 Ruthala Chattie Salaman, diss. 1970, 4 daughters:

    1. Rebecca Nora Barlow, 1956-

2. Natasha Helen Barlow, 1958-
3. Naomi Jane Barlow, 1963-
4. Emily Anne Barlow, 1967-
Married 2 1980 Miranda Weston-Smith, 1 son, 2 daughters; Oscar, Ida and Pepita Barlow.

Eponyms: Gathered under this heading are an anatomical feature, animals, institutions, monuments, places and plants in which "Darwin" referring to CD occurs. The plant genus *Darwinia* relates to Erasmus Darwin [I].

Anatomical feature: Tubercle, = Tuberculum Darwini = Darwin's peak; a cartilaginous prominence on fold of pinna of human ear in some—Jessie Dobson, *Anatomical eponyms*, 2d edn, 1962, p. 52. CD called it "the projecting point". *Descent*, 1:22.

Animals named after CD: (523) This list records animal names, scientific and common. The current validity of the names is not considered, only when CD's name has been invoked to name animals. Therefore the number here does not equate to the number of animals named after CD, but the number of times his name has been used, since sometimes his name has been given more than once to the same animal or group and some animals bear his name both in the formal scientific name and in a common name. The same is true of the list of plants named after CD below. Sources include K.G.V. Smith, Darwin's insects. *Bulletin of the BMNH Historical Series*, 1987, F1830, in *Darwin Online*, Daniel Pauly, *Darwin's fishes*, 2004, FishBase https://www.fishbase.org, the Integrated Taxonomic Information System https://www.itis.gov/, MolluscaBase, Grant & Estes, *Darwin in Galapagos*, 2009, Frith, *Charles Darwin's life with birds*, 2016 and especially Miličić et al. How many darwins? - list of animal taxa named after Charles Darwin. *Natura Montenegrina*, Podgorica (2011) 10(4):515-532, and many other sources:

Ablechrus darwinii Waterhouse, 1877 Coleoptera.

Acarterus darwini Sinclair, 1996 Diptera.

Achrysocharis darwini Girault, 1917 Hymenoptera, Maryland.

Achryson galapagoense darwini Linsley & Chemsak, 1966 Coleoptera.

Actia darwini Malloch, 1929 Diptera.

Adelopsis darwini Jeannel, 1936 Coleoptera, Maldonado, Uruguay.

Aedimorphus darwini Taylor, 1914 Diptera.

Agarista darwiniensis Butler, 1884 Lepidoptera.

Agonostoma darwiniense Macleay, 1878 Chordat.

Agonum darwini Van Dyke, 1953, a ground beetle; synonym *Platinus darwini*, Galapagos.

Agrilus darwinii Wollaston, 1857 Coleoptera, Madeira.

Alectorobius darwini Kohls, Clifford & Hoogstraal, 1969.

Alleloplasis darwini Waterhouse, 1839, a bug of the family Derbidae, the planthoppers, Australia.

Amasa darwini Bright & Skidmore, 2002 Coleoptera.

Amblyomma darwini Arachnida, Galapagos.

Amblyomma darwini Hurst & Hurst 1910, an ixodid tick from St. Paul's Rocks.

Ameghinomya darwini Philippi, 1887. Fossil mollusc.

Amphisbaena darwini darwini, Duméril & Bibron, 1839, a reptile.

Amphisbaena darwinii Duméril & Bibron, 1839, a legless lizard.

Anaploderma darwini Lameere, 1902 Coleoptera, Cerambycidae, Brazil.

Anastatus darwini Girault, 1915 Hymenoptera, Queensland.

Anaulocomera darwinii Scudder, 1893 Orthoptera, Galapagos.

Anipo darwini Evans, 1942 Hemiptera, Australia.

Antarctodarwinella Zinsmeister, 1976. A mollusc.

Anthophora darwini Cockerell, 1910 Hymenoptera.

Aplodes rubrofrontaria var. *darwiniata* Dyar, 1904 Lepidoptera, British Columbia.

Aprostocetus darwini Girault, 1913 Hymenoptera.
Aprostocetus darwinianus Boucek, 1988 Hymenoptera.
Apteronemobius darwini Otte, D. & R.D. Alexander, 1983 Orthoptera.
Archophileurus darwini Arrow, 1937 Coleoptera, Maldonado.
Argynnis darwini Staudinger, 1899 Lepidoptera, Tierra del Fuego.
Astarte darwinii Forbes, 1846, a bivalve mollusc.
Atheta Acronota darwini Cameron, 1943 Coleoptera.
Atropos darwini Duméril, 1854 = *Trimeresurus strigatus*, a venomous pitviper in southern India.
Attus darwini White, a jumping spider.
Aulonodera darwini Champion, 1918 Coleoptera, Chiloe.
Australoeucyclops darwini Karanovic & Tang, 2009, a copepod.
Austroliotia darwinensis Laseron, 1958. A mollusc.
Belomicrus darwini R. Bohart, 1994. Apoid wasp.
Benthocardiella darwinensis Middelfart, 2002. A mollusc.
Berthelinia darwini Jensen, 1997. A mollusc.
Bombinator darwinii Duméril & Bibron, 1841.
Boursinidia darwini Staudinger, 1899 Lepidoptera.
Brachidontes darwinianus d'Orbigny, 1842. A mollusc.
Brachyleon darwini Banks, 1915, Australian insect.
Brachyleon Macronemurus darwini Banks, 1915. Neuroptera.
Bracynemorus darwini Stange, 1969 Neuroptera, Galapagos.
Bulimus darwini Pfeiffer, 1846, a land snail. Synonym *Bulimulus darwini* Pfeiffer.
Caberea darwinii Busk, 1884, Bryozoa.
Caerostris darwini, an orb-weaver spider, Madagascar. Produces the toughest known biomaterial.
Calantica darwini Jones & Hosie, 2009, a barnacle.
Callimicra darwini Hespenheide, 1980, a buprestid beetle.
Calosoma darwinia van Dyke, 1953, a ground beetle, Galapagos.
Camponotus darwinii darwinii Forel, 1886 Hymenoptera.
Camponotus darwinii Forel, 1886 Hymenoptera.
Camponotus darwinii themistocles Forel, 1910 Hymenoptera.
Camponotus radovae radovaedarwinii Forel, 1891.
Cancellaria darwini Petit, 1970. A mollusc.
Cantharis darwiniana Sharp, 1867, Coleoptera.

Canthidium Eucanthidium darwini, Kohlmann & Solís, 2009, dung beetle from Costa Rica.
Capsus darwini Butler, 1877 Hemiptera, Galapagos.
Carabus darwini Hope, 1837, a ground beetle, Chiloe.
Cardium darwini Mayer, 1866. A mollusc.
Carenum darwiniense Macleay, 1878 Coleoptera.
Carios darwini Kohls, Clifford & Hoogstraal, 1969.
Cavernulina darwini Hickson, 1921, a sea pen, Galapagos.
Centrodora Aphelinus darwini Girault, 1913 Hymenoptera.
Cephaloplatus darwini Distant, 1910 Hemiptera.
Ceratina darwini Friese, 1910 Hymenoptera, South America.
Chelonoidis nigra ssp. darwini Van Denburgh, 1907, a tortoise. Originally named *Testudo darwini*, renamed in 1907. James Island, Galapagos.
Chile Darwin's Frog, *Rhinoderma rufum* Philippi, 1902.
Chionelasmus darwini Pilsbry, 1907, marine invertebrate.
Chlamydopsis darwiniensis Lea, 1918.
Chrysopa darwini Banks, 1940 Neuroptera.
Chthamalus darwini Bosquet, 1857. An allegedly Late Cretaceous barnacle = *C. stellatus*.
Citharichthys darwini Victor & Wellington, 2013, a flatfish.
Clathrina darwinii Haeckel, 1870. A sponge.
Clathropsis darwinensis Laseron, 1956. A mollusc.
Clivina darwini Sloane, 1916 Coleoptera.
Cnetispa darwini Maulik, 1930, a leaf beetle.
Coelioxys albolineata var. *darwiniensis* Cockerell, 1929 Hymenoptera.
Coelioxys darwiniensis Cockerell, 1929 Hymenoptera.
Coelostoma darwini Blair, 1933 Coleoptera, Galapagos.
Coenonica darwini Cameron, 1943 Coleoptera.
Coenonympha cephalidarwiniana Verity 1953 Lepidoptera.
Coenonympha darwiniana Staudinger, 1871, a pearly heath, Satyrinae, European Alps.
Coenonympha insubridarwiniana Verity, 1927 Lepidoptera.
Coenonympha philedarwiniana Verity, 1927 Lepidoptera.
Columbella darwini Angas, 1877 = *Zafra darwini*. A mollusc.
Colymbetes darwini Babington 1842, a water beetle, Tierra del Fuego.

Copemania darwini, a nematode.
Cormodes darwini Pascoe, 1860 Coleoptera.
Corynura darwini Cockerell, 1901 Hymenoptera, Brazil.
Corythaica darwiniana Drake & Froeschner, 1967 Hemiptera, Culpepper, Galapagos.
Cosmocalanus darwinii Lubbock, 1860 syn. *Calanus darwinii*. Zooplankton, South Atlantic Ocean.
Crocisa caeruleifrons Kirby var. *darwini* Cockerell, 1905 Hymenoptera.
Crocodilus darwini originally *Alligator darwini* Ludwig, 1877 a tertiary fossil crocodile.
Cruria darwiniensis Butler, 1884 Lepidoptera.
Cryptocercus darwini Burnside, Smith & Kambhampati, 1999. Blattodea.
Cryptocheilus darwinii Turner, 1910 *Hymenoptera*.
Ctenispa darwini Maulik, 1887 Coleoptera, Bahia.
Cubinia darwinii Gray, 1845, a gecko.
Cylindrolepas darwiniana Pilsbry, 1916.
Cyrtobill darwini Framenau & Scharff, 2009, an orb-weaving spider.
Cyrtophium darwini Bate, 1860, an amphipod crustacean = *Platophium darwini* Bate = *Podocerus variegatus*. Leach.
Dagbertus darwini Butler, 1877, plant bugs in the family Miridae.
Darwin Blind Snake=*Anilios tovelli* Loveridge, 1945.
Darwin Carpet Python=*Morelia spilota variegata* Gray, 1842.
Darwin Grouper=*Epinephelus darwinensis*.
Darwin Herring=*Herklotsichthys gotoi* Wongratan 1983.
Darwin Silky-Oak=*Grevillea pteridifolia* Knight.
Darwin Sole=*Leptachirus darwinensis*.
Darwin Stringybark=*Eucalyptus tetrodonta* F. Muell.
Darwin Termite=*Mastotermes darwiniensis*.
Darwin Tyrann=*Colorhamphus parvirostris* Gould & Gray, 1839.
Darwin Wallace Poison-Frog=*Epipedobates darwinwallacei*.
Darwin Woollybutt, *Eucalyptus miniata* A. Cunn. ex Schauer.
Darwin's Band-rumped Storm-Petrel, *Hydrobates castro* (bangsi) Nichols, 1914.
Darwin's Bark Spider=*Caerostris darwini*.
Darwin's Cactus Ground Finch=*Geospiza scandens* Gould, 1837.
Darwin's Caracara=*Phalcoboenus albogularis* Gould 1837.

Darwin's Cocos Island Finch=*Pinaroloxias inornata*. Gould, 1843
Darwin's Crake=*Coturnicops notatus*, Gould, 1841.
Darwin's Cubinia=*Cubinia darwinii*.
Darwin's Finches; sub-family Geospizinae, family Fringillidae, Galapagos.
Darwin's Flycatcher=*Pyrocephalus nanus* Gould, 1839.
Darwin's Fox=*Pseudalopex fulvipes* Martin, 1837. Syn. *Lycalopex fulvipes*.
Darwin's Frog=*Ingerana charledarwini*, Das, 1998.
Darwin's Frog=*Rhinoderma darwinii*.
Darwin's Heath=*Coenonympha darwiniana*.
Darwin's Iguana=*Diplolaemus darwinii*.
Darwin's Koklass Pheasant=*Pucrasia macrolopha darwini*.
Darwin's Large Cactus Finch=*Geospiza conirostris* Sharpe 1888.
Darwin's Large Fan-throated Lizard=*Sarada darwini*.
Darwin's Large Ground Finch=*Geospiza magnirostris* Gould, 1837.
Darwin's Large Tree Finch=*Camarhynchus psittacula* Gould, 1837.
Darwin's Leaf-Eared Mouse=*Phyllotis darwini*.
Darwin's Leiolaemus=*Leiolæmus darwinii*.
Darwin's Mangrove Finch=*Camarhynchus heliobates* Snodgrass & Heller, 1901.
Darwin's Medium Ground Finch=*Geospiza fortis* Gould, 1837.
Darwin's Medium Tree Finch=*Camarhynchus pauper* Ridgeway, 1890.
Darwin's Mouse=*Mus Phyllotis darwinii*.
Darwin's *Papilio feronia*. A cracker butterfly.
Darwin's Racer=*Pseudalsophis darwini*.
Darwin's Rail=*Coturnicops notata*, Rallidae, Guyana to southern Argentina.
Darwin's Redfish=*Gephyroberyx darwini*.
Darwin's Rhea=*Pterocnemia pennata*.
Darwin's Rice Rat=*Nesoryzomys darwini*.
Darwin's Ringed Worm Lizard=*Amphisbaena darwini*.
Darwin's Roughy=*Gephyroberyx darwini*
Darwin's Sanddab=*Citharichthys darwini*.
Darwin's Sharp-beaked Ground Finch=*Geospiza difficilis* Sharpe, 1888.
Darwin's Sheep=*Ovis ammon darwini*.
Darwin's Slimehead=*Gephyroberyx darwinii*.
Darwin's Small Ground Finch=*Geospiza fulignosa* Gould, 1837.

Darwin's Small Tree Finch=*Camarhynchus parvulus* Gould, 1837.

Darwin's Storm-Petrel=*Oceanodroma bangsi* Nichols, 1914.

Darwin's Tanager=*Thraupis bonariensis darwini.*

Darwin's Taraguira=*Tropidurus torquatus.*

Darwin's Tree Iguana=*Liolaemus darwinii.*

Darwin's Vegetarian Finch=*Platyspiza crassirostris* Gould, 1837.

Darwin's Wall Gecko=*Tarentola darwini.*

Darwin's Warbler Finch=*Certhidea olivacea* Gould, 1837.

Darwin's Woodpecker Finch=*Camarhynchus pallidus* Sclater & Salvin, 1870.

Darwin's Worm Lizard=*Amphisbaena darwinii.*

Darwinea Bate, 1856, amphipod crustacean, nomen nudem = *Darwinia* Bate, 1857.

Darwinella amaroides Enderlein, 1912 Coleoptera.

Darwinella australiensis Carter, 1885 Sponge.

Darwinella corneostellata Carter, 1872 Sponge.

Darwinella dalmatica Topsent, 1905 Sponge.

Darwinella duplex Topsent, 1905 Sponge.

Darwinella gardineri Topsent 1905 Sponge.

Darwinella intermedia Topsent, 1893 Sponge.

Darwinella muelleri Schultze, 1865 Sponge.

Darwinella oxeata Bergquist 1961 Sponge.

Darwinella rosacea Hechtel, 1965 Sponge.

Darwinella simplex Topsent, 1892 Sponge.

Darwinella tango Poiner & Taylor, 1990 Sponge.

Darwinella viscosa Boury-Esnault, 1971 Sponge.

Darwinella warreni Topsent, 1905 Sponge.

Darwinellidae Merejkowsky, 1879 Sponge.

Darwinhydrus Sharp, 1882, dytiscid water beetles. Coleoptera.

Darwinhydrus solidus Sharp, 1882 Coleoptera.

Darwinia Dybowski, 1873. Coelenterate. 1874, fossil anthozoan coelenterates. =*Prodarwinia.*

Darwinia Pereyaslawzew, 1880, tubellarian flatworms.

Darwinia Schultze, 1865, fossil sponges.

Darwinia subtilis Graff, 1882.

Darwinia variabilis Pereyaslawzewa, 1892.

Darwinida, a barnacle.

Darwiniella angularis, a species of Cirripedia in the family Pyrgomatidae.

Darwinilus, a genus of staphylinid or rove beetles.

Darwininitium Budha & Mordan, 2012. A mollusc.

Darwiniothrips, a genus of thrips.

Darwiniphora Schmitz, 1953, phorid fly.

Darwinites, a genus of Palaeozoic mollusc.

Darwinius masillae, Franzen et al, 2009, a primate-like fossil species of the genus *Adapiformes.*

Darwinius, a genus of Eocene primates.

Darwinivelia Andersen & J. Polhemus, 1980, water strider.

Darwinivelia angulata Polhemus & Manzano, 1992. Hemiptera.

Darwinneon crypticus Cutler, 1971, Arachnida, a jumping spider, Galapagos.

Darwinocoris, a genus of lygaeid bug.

Darwinomyia J.R. Malloch, 1922, muscid dipterans.

Darwinoplectanum is a genus of Platyhelminthes in the family Diplectanidae, flatworms.

Darwinopterus, a genus of extinct long-tailed middle Jurassic pterosaurs from China.

Darwinornis Moreno & Mercerat, 1891, fossil bird.

Darwinornithidae Moreno & Mercerat, family of fossil birds for *Darwinornis.*

Darwinula, T.R. Jones, 1885, genus of ostracod crustaceans, mostly Pleistocene fossils, one living *D. stevensoni*, for *Darwinella* Brady & Robertson, 1872.

Darwinulidae, Brady & Norman, 1889, ostracod crustaceans, mostly Pleistocene. Family.

Darwinulocopina Sohn, 1988 Infraorder.

Darwinuloidea Brady & Norman, 1889 Suborder.

Darwinylus marcosi Peris, 2016, an Early Cretaceous oedemerid beetle found in Spanish amber.

Darwinysius marginalis Ashlock, 1967 Hemiptera, Galapagos.

Darwinysius marginalis Dallas, 1852 Heteroptera, Galapagos.

Delphacodes darwini Muir, 1929 Hemiptera, Chiloe.

Demandasaurus darwini, a rebbachisaurid sauropod dinosaur from Spain.

Diastichus darwini Cartwright, 1970 Coleoptera, Galapagos.

Dichelaspis darwinii Filippi, 1897. A barnacle.

Dichotomius (Selenocopris) darwini, Nunes, Rafael & Vaz-De-Mello, Fernando, 2016. Coleoptera.

Diplatys darwini Bey-Bienko, 1959 Dermaptera.

Diplax frequens var *darwiniana* Selys, 1883 Insecta.

Diplolaemus darwinii Bell, 1843 Darwin's Iguana.

Dischistodus darwiniensis Whitley, 1928 Chordat.

Distigmoptera darwini Scherer, 1964 Coleoptera, Uruguay.

Docema darwini Mutchler, 1924, a beetle of the family Hydrophilidae. Coleoptera, Galapagos.

Doleromyrma darwiniana darwiniana Forel, 1907 Hymenoptera.

Dorcus darwinii Hope, 1841, a stag beetle, Chile. = *Sclerognathus femoralis* Guérin, 1887

Duplominona darwinensis Martens & Curini-Galletti, 1989.

Dusicyon australis darwinii Thomas, 1914 canid.

Dynamene darwinii. Crustacea.

Ectemnostega darwini Hungerford, 1948 Hemiptera, St Cruz, Patagonia.

Elimia darwini Mihalcik & F. G. Thompson, 2002. A mollusc.

Enochrus darwini Knisch, 1922 Coleoptera.

Epinephelus darwinensis Randall & Heemstra, 1991 Chordat. Darwin grouper.

Epipedobates darwinwallacei Cisneros-Heredia & Yánez-Muñoz, 2011.

Epitrix darwini Bryant, 1942 Coleoptera, Maldonado.

Epuraea darwinensis Blackburn, 1927 Coleoptera, Australia.

Erechthias darwini Robinson, 1983 Lepidoptera, St Paul's Rocks.

Eremaeozetes darwini Schatz, 2000.

Eudicella darwini Kraatz, 1890 Coleoptera.

Eudicella darwiniana flavoclypealis Devecis, 1997 Coleoptera.

Eudicella darwiniana Kraatz, 1840 Coleoptera.

Euryglossina Euryglossella darwiniensis Exley, 1974 Hymenoptera.

Euthria darwini Monteiro & Rolán, 2005. A mollusc.

Euvallentinia darwinii. Crustacea.

Felis darwini Martin=*F. yaguarundi* Desmarest. Jaguarondi, a race of *Felis Herpailurus yagouaroundi*, South America to Texas.

Fissurella darwini Reeve, 1849, a keyhole limpet.

Fissurella darwinii. Mollusca.

Foenus darwini Westwood, an ichneumonid wasp, Australia.

Foenus darwinii. An insect.

Forcipomyia darwini Marino & Spinelli, 2003, Diptera.

Gabaza Wallacea darwini Hill, 1919 Diptera.

Galagete darwini Landry, 2002 Lepidoptera.

Galapagodacnum darwini Blair, 1933, a plant beetle of the family Chrysomelidae.

Galapagoleon Brachynemurus darwini Stange, 1969 Neuroptera.

Gebiopsis darwinii. Crustacea.

Geochelone darwini van Denburgh, a giant tortoise, James Island, Galapagos = *Testudo darwini.* Synonym *Chelonoidis.*

Geochelone nigra ssp. darwini van Denburgh, 1907 Testudines, James Island subspecies of Galápagos tortoise.

Geophagus darwini Keller, 1887.

Geoplana darwini Hallez, 1893.

Geospiza conirostris darwini Rothschild & Hartert, 1899.

Gephyroberyx darwini Johnson, 1866. A fish.

Gephyroberyx darwinii Johnson, 1866. A fish.

Gerygone chloronota darwini Mathews, 1912.

Gibbosaverruca darwini Pilsbry, 1916.

Glaphyromorphus darwiniensis darwiniensis Storr, 1967 reptile.

Globocornus darwini Espinosa & Ortea, 2010. A mollusc.

Gnathaphanus darwini Blackburn, 1888 Coleoptera, Australia.

Gonatocerus darwini Girault, 1912 Hymenoptera, Queensland.

Gossypium darwinii. A virus.

Grania darwinensis Coates & Stacey, 1997.

Graphelmis darwini Ciampor & Kodada, 2004.

Graviceps darwini Whitley, 1958. A fish.

Gryllus darwini Otte, D. & Peck, 1997 Orthoptera.

Gryphaea darwini, Forbes, 1846 in d'Orbigny, a fossil oyster = *Ostraea darwini.*

Gymnodactylus darwinii Gray, 1845, a reptile.

Haeckeliana darwiniana Haeckel. Zooplankton of the South Atlantic Ocean.

Halictus Evylaeus darwiniellus Cockerell, 1932 Hymenoptera, Australia.

Halictus eyrei darwiniensis Cockerell, 1929 Hymenoptera.

Haliophasma darwinia Poore & Lew Ton, 1988.

Halobates darwini Herring, 1961 Hemiptera.

Haplodelphax darwini Fennah, 1965 Hemiptera.

Haptoncura darwinensis Blackburn, 1903 Coleoptera.

Helix (Hadra) darwini Brazier, 1872. A mollusc.

Herpailurus darwini Martin =*Felis darwini,* a race of *F. yagouaroundi.*

Hesperomys darwini Wagner in Schreber, a cricetine rodent. Obsolete genus, now *Calomys.*

Heterodiomus darwini Brèthes, 1924 Coleoptera, Brazil.

Heteronyx darwini Blackburn, 1889 Coleoptera, Australia.

Holacanthus darwiniensis Saville-Kent, 1889 Chordat.

Homonota darwinii Boulenger, 1885 reptile.

Horouta darwini McCaffrey & Harding, 2009, a leafhopper, family Cicadellidae.

Hugoscottia darwini Knisch, 1922 Coleoptera, South America.

Hydraena darwini Perkins, 2007.

Hydrelaps darwiniensis Boulenger, 1896, Seasnake.

Hydrometra darwiniana J. Polhemus & Lansbury, 1997.

Hydroporus darwini Babington, 1842 a water beetle. Superseded by *Necterosoma darwini*, King George's Sound, Australia.

Hyperaspis arrowi Brèthes var. *darwini* Brèthes, 1925. Coleoptera, Maldonado, Uruguay.

Hypolimnas alimena darwinensis Waterhouse & Lyell, 1914 Lepidoptera.

Hypolimnas darwinensis Waterhouse & Lyell, 1914 Lepidoptera.

Idiocephala darwini Saunders, 1843 a chrysomelid beetle, Australia.

Ingerana charlesdarwini, Das, 1998, a frog of the Andaman islands, =*Rana charlesdarwini*.

Ischnodemus darwini Slater, 1964 Hemiptera, South Africa.

Ischnothyreus darwini Edward & Harvey, 2014, an Australian goblin spider.

Isomyia Strongyloneura darwini Curran, 1938 Diptera.

Issoria darwini Staudinger, 1899 Lepidoptera.

Kaapia darwini Theron, 1983 Hemiptera, South Africa.

Kalotermes darwini Light, 1935 Isoptera, Galapagos.

Kynotus Geophagus darwini Keller, 1887.

Labidocera darwini Lubbock 1853, a calanid copepod crustacean.

Labidocera darwinii, Lubbock, 1853. A calanid copepod.

Lasiacantha darwini Cassis & Symonds, 2011.

Lasioglossum darwiniellum Cockerell, 1932 Hymenoptera.

Lates darwiniensis Macleay, 1878 Chordat.

Leda darwini E. A. Smith, 1884. A mollusc.

Leiolæmus darwinii Bell, 1843.

Lepidiota darwini Blackburn, 1888 Coleoptera, Australia.

Leptachirus darwinensis Randall, 2007.

Leptocera Limosina darwini Richards, 1931 Diptera, Concepcion.

Leskia darwini Emden, 1960 Diptera, South Africa.

Leucosolenia darwinii Haeckel, 1870 a sponge.

Limnonectes charlesdarwini Das, 1998.

Limosina Leptocera darwini Richards, 1931 Diptera.

Liolaemus darwinii Bell, 1843, Iguana, Argentina.

Longitarsus darwini Bryant, 1942 Coleoptera, Maldonado.

Loxostege darwinialis Sauber 1904 Lepidoptera.

Mactra darwinii Sowerby in CD, a bivalve mollusc.

Mallada Chrysopa darwini Banks, 1940 Neuroptera.

Manzonia darwini Moolenbeek & Faber, 1987. A mollusc.

Marilyna darwinii Castelnau, 1873. A fish. "Dedicated to the greatest naturalist of the age".

Masasteron darwin Baehr, 2004.

Mastotermes darwiniensis, termite of Australia. The most primitive living termite.

Mastotermes darwiniensis. A bacterium.

Medon Hypomedon darwini Cameron, 1943 Coleoptera.

Megachile darwiniana Cockerell, 1906 Hymenoptera.

Megalomus darwini Banks, 1924 Neuroptera, Galapagos.

Melanococcus Melancoccus darwiniensis Williams, 1985 Hemiptera.

Melizoderes darwini Funkhouser, 1934 Hemiptera, Chiloe.

Mesocyclops darwini Dussart & Fernando, 1988.

Microdarwinula inexpectata Pinto, Rocha & Martens, 2005.

Microdarwinula zimmeri Menzel, 1916.

Microdontomerus darwini Grissell, 2005 Hymenoptera.

Migadops darwinii Waterhouse, 1842 carabid beetle, Tierra del Fuego; now *Pseudomigadops darwini*.

Mimacraea darwinia apicalis Smith & Kirby, 1890 Lepidoptera.

Mimacraea darwinia Butler, 1872 Lepidoptera, West Africa.

Minihippus darwini Jago, 1996 Orthoptera.

Monophora darwini Agassiz, 1874 a fossil sea urchin; now *Monophoraster.*

Moruloidea darwinii Cunningham, 1871.

Mus darwini=Phyllotis darwini.

Mus Phyllotis darwinii Waterhouse, 1839 a cricetine rodent, South America.

Mycale darwini Hajdu & Desqueyroux-Faúndez, 1994, a sponge.

Mycale darwini Hajdu & Desqueyroux-Faúndez, 1994.

Mylodon darwini Owen, a fossil giant sloth. South America. Owen 1842 =*Grypotherium darwini.*

Mysteria darwini Lameere, 1902 Coleoptera.

Mytilus darwinianus d'Orbigny, 1842 a fossil mussel.

Naesiotus darwini synonym of *Bulimulus darwini*. A Galapagos land snail.

Natica darwinii Hutton, 1885. A mollusc.

Navicula darwiniana Seddon & Witkowski, Kingdom Chromista.

Necterosoma darwinii Babington, 1841.

Nemoria darwiniata Dyar, 1904 Lepidoptera.

Neobidessodes darwiniensis Hendrich & Balke, 2011.

Neobrachypterus darwini Jelínek, 1979 Coleoptera, Patagonia.

Nephopullus darwini Brèthes, 1885 Coleoptera, Brazil.

Nesoryzomys darwini Osgood, 1929, a cricetine rodent, Academy Bay, Indefatigable Island, Galapagos. Darwin's rice rat; now extinct.

Netastoma darwinii Sowerby, 1849. A mollusc.

Nettastomella darwini Sowerby, 1849.

Nitela darwini Turner, 1916 Hymenoptera, Galapagos.

Nocticanace darwini Wirth, 1969 Diptera, Galapagos.

Nomia darwinorum Cockerell, 1910 Hymenoptera.

Nothura darwinii darwinii Gray, 1867. Darwin's Nothura.

Nothura darwinii salvadorii Hartert, 1909.

Nuculana (Saccella) darwini E. A. Smith, 1884. A mollusc.

Nuculana darwini E. A. Smith, 1884. A mollusc.

Nyctelia darwinii Waterhouse, 1841 a heteromeran beetle from Port Desire.

Odontoscelis curtisii darwinii G.R. Waterhouse, 1840 a pentatomid bug Carabidae, *Cnemacanthus*, Patagonia.

Ogcocephalus darwini Hubbs, 1958 Chordat. Red-lipped batfish, Galapagos.

Ogcodes darwini Westwood, 1876 Diptera. Australia.

Oncopterus darwinii Steindachner, 1874 Chordat. Described as *Rhombus* in *Fish*. Remo flounder.

Onthophagus auritus darwini Paulian, 1937 Coleoptera.

Ontiscus darwini Hamid, 1975 Hemiptera, King George's Sound and Sydney, Australia.

Ophiocara darwiniensis Macleay, 1878 Chordat.

Ophiothrix darwini Bell, 1884 Echinoderm.

Opistognathus darwiniensis Macleay, 1878 Chordat, Darwin jawfish.

Orchestia Darwinii Chelorchestia darwinii. Müller, 1864.

Ornithodoros darwini Kohls, Clifford & Hoogstraal, 1969.

Orthosia ? darwini Staudinger, 1899, Uschuaia.

Orynipus darwini Brèthes, 1924 Coleoptera, Chiloe.

Ovis ammon darwini, Przewalski, 1883. Northern China & central Mongolia.

Oxycantha darwini Surekha & Ubaidillah, 1996 Hymenoptera.

Oxytelus Anotylus darwini Cameron, 1843 Coleoptera.

Pachycondyla darwinii Forel, 1893 Hymenoptera.

Pacifigorgia Gorgonia darwinii Hickson, 1928, a sea fan, Galapagos.

Paenibacillus darwinianus. A bacterium.

Palpibracus darwini Soares & Carvalho, 2005 Diptera.

Paludicola darwinii Bell, 1843.

Pandalosia darwinensis Laseron, 1956. A mollusc.

Pantinia darwini China, 1962 Hemiptera, Chiloe.

Parachernes darwiniensis Beier, 1978.

Parachlus darwini Brundin, 1966, Chironomidae, Chile.

Paracleis darwini Parent, 1933 Diptera, Australia.

Parahelops darwini Waterhouse, 1875 Coleoptera, Tierra del Fuego & Valparaiso.

Paralastor darwinianus Perkins, 1914 Hymenoptera.

Paraliochthonius darwini Harvey, 2009 a pseudoscorpion.

Paraliparis darwini Stein & Chernova, 2002 Chordat. A fish.

Paraplatoides darwini Waldock, 2009 a spider of Australia.

Parochlus darwini Brundin, 1966 Diptera.

Parvaponera darwinii Hymenoptera.

Pecten darwinianus d'Orbigny, 1842 a scallop, *Pecten darwinii* Sowerby 1846.

Pediobomyia darwini Girault, 1913 Hymenoptera, Australia.

Pelecorhynchus darwini Ricardo, 1900 Diptera, Chiloe.

Pelycops darwini Aldrich, 1934 Diptera, Tierra del Fuego.

Periclimenes darwiniensis Bruce, 1987.

Periophthalmus darwini Larson & Takita, 2004 Chordat. Darwin's mudskipper.

Perissopmeros darwini Rix, Roberts & Harvey, 2009 an Australian spider.

Phalacrus darwini Waterhouse, 1877 Coleoptera, Galapagos.

Phlyctaenodes darwinalis Sauber, 1904 Lepidoptera, Central Asia.

Pholas darwini Sowerby, 1849 a piddock bivalve mollusc.

Phthiracarus darwini Mahunka, 1980.

Phyllodactylus darwini, Taylor, 1942 Galapagos. Darwin's leaf-toed gecko.

Phyllotis darwini Waterhouse, 1837, a leaf-eared mouse from South America.

Phymaturus darwini Núñez et al, 2010. An iguana.

Phytosus darwinii Waterhouse, 1879 Coleoptera, Falklands.

Pieris napi bryonia subsp. *darwiniana* Stichel, 1910 Lepidoptera, Europe.

Pimelometopon darwini, Jenyns, 1842 a wrasse, *Cossyphus darwini* synonym, Galapagos.

Plagithmysus darwinianus Sharp, 1896 Coleoptera, Hawaii.

Platinus darwini=Agonostoma darviniense.

Platophium darwini Bate 1860, an amphipod crustacean = *Cyrtophium darwini* Bate 1860 = *Podocerus variegatus* Leach.

Platydecticus darwini Rentz & Gurney, 1985 Orthoptera.

Platytomus darwini Cartwright, 1970 Coleoptera.

Pleurodema darwini Bell, 1843 a tree-frog of South America. Synonym for *Pleurodema bibroni.*

Plotopuserica darwiniana Brenske, 1900 Coleoptera, Madagascar.

Polybothris darwini Théry, 1912. Coleoptera, Buprestidae). Madagascar.

Polycladus darwini Diesing, 1862 a flatworm.

Polylobus darwini Bernhauer, 1935 Coleoptera, Chiloe.

Polynema darwini Girault, 1913 Hymenoptera, Australia.

Polystyliphora darwini Ax & Ax, 1974, a flatworm.

Pomacentrus darwiniensis Whitley, 1928. Banded Damselfish, Australia.

Pristhesancus darwinensis Miller, 1958 Hemiptera.

Proctotretus darwinii Bell, 1843. =*Liolaemus darwinii.*

Psephenus darwinii Waterhouse, 1880 Coleoptera, Brazil.

Pseudalsophis darwini. Zaher, 2018. A species of Galapagos snake.

Pseudechis darwiniensis Macleay, 1878 reptile.

Pseudobankesia darwini Stengel, 1990 Lepidoptera.

Pseudococcus darwiniensis Williams, 1985 Hemiptera.

Pterocnemia darwini Gould, 1837, junior synonym for *P. pennata* d'Orbigny 1834.

Pteronema darwini Haeckel, 1879 Hydrozoa.

Ptilocnemus darwinensis Malipatil, 1985.

Pucrasia macrolopha darwini Swinhoe, 1872.

Puijila darwini, Rybczynski, Dawson & Tedford, 2009, a Miocene mammal.

Pulvinaria darwiniensis Froggatt, 1915 Hemiptera.

Radinacantha darwiniis Moir, 2009 a lacebug.

Rana charlesdarwini Das, 1998 amphibian.

Regioscalpellum darwini Hock, 1883. A barnacle.

Rhea darwinii Gould, 1837=*Pterocnemia darwini.*

Rhinoderma darwinii Duméril & Bibron, 1841 a

dwarf frog. Southern Darwin's frog.

Risius darwini Fennah, 1962 Hemiptera, South Africa.

Ropalidia darwini Richards, 1978 Hymenoptera.

Ruggieria darwinii. Crustacea.

Saccella darwini E. A. Smith, 1884. A mollusc.

Salius darwinii Turner, R.E., 1910 Hymenoptera.

Sapphirina darwini Haeckel, 1864. A species of parasitic copepod.

Sarada darwini Deepak et al, 2016. Lizard.

Sarcophaga darwiniana Kano & Lopes, 1979 Diptera.

Scenopinus darwini Kelsey, 1969 Diptera.

Sclerostomus darwini Burmeister, 1847. Synonym for *Apterodorcus bacchus* Hope & Westwood; a stag beetle.

Scolytogenes darwini Eichhoff, 1878 Coleoptera, Burma.

Scolytomimus darwini Schedl, 1951 Coleoptera.

Sebastes darwini Cramer, 1896 Chordat. A fish.

Selitrichodes darwini Girault, 1915 Hymenoptera, Queensland.

Semicossyphus darwini Jenyns, 1842 Chordat, fish, Pacific red sheephead, Galapagos. Original name *Cossyphus darwini.*

Senoculus darwini Holmberg, 1883. A spider species found in Argentina.

Solen darwinensis Cosel, 2002. A mollusc.

Sphex darwinensis Turner, 1912 Hymenoptera.

Spirifer darwini Morris, 1845 in Strzelecki, a fossil brachiopod.

Spondylus darwini Jousseaume, 1882. A mollusc.

Spurlingia darwini Brazier, 1872. A mollusc.

Stagira darwini Distant, 1905 Hemiptera, Mauritius.

Stenaelurillus darwini Wesolowska & Russell-Smith, 2000. A jumping spider species, Tanzania.

Stenoconchyoptera darwini Muir, 1931 Hemiptera, South Africa.

Stictospilus darwini Brèthes, 1924 Coleoptera, Chile.

Strongyloneura darwini Curran, 1938 Diptera.

Stylatula darwini Kölliker, 1870. A sea pen.

Swolnpes darwini Framenau, 2009 a trapdoor spider.

Sympetrum darwinianum Selys, 1883 Insecta.

Tabanus darwinensis Taylor, 1917 Diptera.

Tanagra darwini Gould, 1841 Darwin's tanager.

Tanaoneura darwini LaSalle, 1987 Hymenoptera.

Tapetosa darwini Framenau et al, 2009 a wolf spider.

Taraguira darwinii Gray, 1845 an iguana. Synonym for *Tropidurus torquatus* Wied-Neuwied, 1820.

Tarentola darwini Joger, 1984, Darwin's Wall Gecko.

Telephorus darwinianus Sharp, 1866 Coleoptera, Scotland.

Testudo darwini van Denburgh, 1907 = *Geochelone darwini*, a giant tortoise, Galapagos Islands, James Island. Now a subspecies *Geochelone elephantopus darwini*.

Testudo darwini Van, 1907 Testudines.

Tetraodon darwinii Castelnau, 1873. A fish.

Tetrastichus darwini Girault, 1913 Hymenoptera, Australia.

Tetrodon darwinii Castelnau, 1873, now *Marilyna darwinii*.

Thecacera darwini Pruvot-Fol, 1950, a sea slug.

Thraupis bonariensis darwinii Gould, 1841 Darwin's tanager; *Tanagra darwinii* is a junior synonym.

Thynnus darwinensis Turner, 1908 Hymenoptera.

Torresitrachia darwini Willan, Köhler, Kessner & Braby, 2009. A mollusc.

Toxodon darwinii Burmeister, 1866.

Tramea cophysa darwini Kirby, 1889. Galapagos insect.

Tramea darwini Kirby, 1889 Insecta, Galapagos.

Trapania darwini Gosliner & Fahey, 2008. A mollusc.

Trechisibus darwini Jeannel, 1927 Coleoptera, Argentina.

Trichoniscus darwini Vandel, 1938.

Trichopteryx darwinii Matthews, 1889 Coleoptera, Brazil.

Trochocopus darwinii Johnson, 1866 A fish.

Turbonilla darwinensis Laseron, 1959. A mollusc.

Undinula darwinii Lubbock, 1860.

Upucerthia darwini Scott 1900, Scaly-throated Earthcreeper.

Upucerthia dumetaria darwini, Scott, 1900. A bird.

Utetheisa pulchelloides darwini Staudinger, 1899 Lepidoptera, Keeling.

Valdivia darwini Shannon, 1927. Syrphidae, *Valdiviomyia* Chile.

Valdiviomyia Valdivia darwini Shannon, 1927 Diptera.

Venus darwinii Römer, 1858. A mollusc.

Vexillum darwini R. Salisbury & Guillot de Suduiraut, 2006. A mollusc.

Wallacea darwini Hill, 1919 Diptera.

Xanthomelon darwinense Köhler & Burghardt, 2016. A mollusc.

Xylcopa darwinii Cockerell, 1926. Hymenoptera, carpenter bee. Only native bee in Galapagos.

Xyleborus darwini Schedl, 1972 Coleoptera.

Xylocopa darwini Cockerell, 1926. Insect.

Zafra darwini Angas, 1877. A mollusc.

Zalmoxis darwinensis Goodnight & Goodnight, 1948.

Zelleriella darwinii. Protozoa.

Institutions: (73)

Carlos Darwin College, Mexico.

Centro Educacional Charles Darwin, High School, Vitória, Brazil.

Charles Darwin advanced laboratory at the Botany School, University of Sydney. 1927.

Charles Darwin Elementary School, Chicago.

Charles Darwin Elementary School, Milwaukee.

Charles Darwin Foundation Library, Galapagos.

Charles Darwin Foundation. A USA organization, founded 1959, which runs the Darwin Research Station, see Galapagos.

Charles Darwin Primary School, Northwich.

Charles Darwin Prize, Mater Christi School, Burlington, Vermont. 2015?

Charles Darwin prize, Tribeca Film Festival, USA. 2013.

Charles Darwin Professor of Anthropology at Rutgers University. -2015.

Charles Darwin Professorship of Animal Embryology, endowed by the Bles Fund, School of Biology, University of Cambridge. 2008?

Charles Darwin School, Biggin Hill, Kent, UK.

Charles Darwin School, Wellington, Telford, UK.

Charles Darwin School, Westerham, UK.

Charles Darwin Trust. British educational charity, part of a campaign with other organizations to back the unsuccessful bid to make Down House a World Heritage Site. 1999.

Charles Darwin University, Australia, 2003.

Charles-Darwin-Schule (Gymnasium), Berlin.

Charles-Darwin-Schule, Chemnitz, Germany.

Colégio Darwin, Private school in Fortaleza, Brazil.

Darwin Award, presented to the member who has made the most significant contribution to coral-reef studies. International Society for Reef Studies. 1986-?

Darwin Central School, São Bernardo do Campo, Brazil.

Darwin Centre for Biology and Medicine, Cardiff.

Darwin College, Cambridge: 1964 Jul. 28 founded for postgraduate and post doctorate students. First buildings were conversions of Newnham Grange and the Old Granary, home of Sir George Howard Darwin.

Darwin College, University of Kent, at Canterbury; a student residence opened 1970.

Darwin College, University of Kent.

Darwin Community College, Australia, 1974.

Darwin Day, since the late 1990s 12 Feb. has been commemorating by various organisations around the world.

Darwin Fund of the Royal Society. 1963.

Darwin Hall of Invertebrate Zoology, American Museum of Natural History. 1909.

Darwin Institut (institutea), of "Heieiei" (German), "Hy-yi-yi" (English), an imaginary country, destroyed Oct. 1957, in "Harald Stumpke" *Bau und Leben der Rhinogradentia,* Stuttgart 1964; "Stumpke" is pseudonym for Gerolf Steiner, Heidelberg University.

Darwin Institute for Scientific Research and Development, Phnom Penh, Cambodia.

Darwin Lecture, in human biology, under the auspices of Eugenics Society and Institute of Biology, London; annually 1960-.

Darwin Leisure Centre, Biggin Hill.

Darwin Medal, Royal Society; first struck 1890. Effigy reduced from a medallion by Allan Wyon. First awarded 1890, "in the field in which Charles Darwin himself laboured". Biennial with British or foreign recipients. Awarded to Wallace 1890, Hooker 1892, Huxley 1894. In 1885 the Committee of the Darwin Memorial Fund transferred to the Society the balance of the fund in trust. *Yearbook* 1968. Now awarded annually.

Darwin Memorial Fund: Committee set up 1882 May 16, with William Spottiswoode, PRS, in Chair. 1883 Huxley took over the Chair as PRS on S's death. 1888 Printed Report, Spottiswoode, London, lists about 700 subscribers; £5,128 raised; £2,100 paid to Boehm for a statue at BMNH, and a further £150 for the medallion in Westminster Abbey; £9.0.6 paid to Edward Whymper for a woodcut of a bust which illustrates Report. £2,608.8.8 remained, after expenses, some of which, although the Report does not refer to it, went to funding the Darwin medal.

Darwin mission, European Space Agency project to find Earth-like planets in our galaxy.

Darwin Montessori School, Chennai, India.

Darwin Press, Princeton, New Jersey.

Darwin Prize (1880), to include a Darwin Medal. Midland union of natural history societies.

(See *Shorter publications,* F1993) Still being awarded in 1895.

Darwin Prize (for Historical Geology), Department of Geology, State University of New York and College at Cortland. 1998-

Darwin Prize for best use of an Evolutionary framework. London evolutionary research network. 2009.

Darwin Prize for Biological Anthropology, University of Kent. 2014?

Darwin Prize for Science, awarded for outstanding achievements and contributions in Science, Keele University. -2012?

Darwin Prize Fund, Hampshire Community Fund. 2017?

Darwin Prize in evolutionary biology, Centre for Ecological and Evolutionary Synthesis, Department of Biosciences, University of Oslo, Norway. 1988-

Darwin Prize Visiting Professorship, Edinburgh University. 1990s?-

Darwin Prize, Christ's College, Cambridge. 1885.

Darwin Prize, Department of Biology at Queens College, the City University of New York. 2009.

Darwin Publications, Sherman Oaks, California.

Darwin Publishing Company, Detroit, Michigan.

Darwin Regatta, held each year at Darwin, Northern Territory, Australia; the craft being built of empty beer cans. 1974.

Darwin School of Business, Guwahati, India.

Darwin School, Education Management, Hayward, California.

Darwin School, Winnipeg, Canada.

Darwin Shipping Company Ltd. Owners of R.M.V. *Darwin.*

Darwin Society, Christ's College, Cambridge. 1937.

The Darwin Tree of Life project, part of a global initiative to sequence all complex life on Earth.

Darwin Trust (Colchester). Established to study causes of amentia. 1931.

Darwin's chrysanthemum and horticultural society, UK. c.1920?

Darwineum. Zoologische Garten Rostock, Germany. 2012.

Darwin-Wallace Medal, Linnean Society of London, first struck 1906, designed by Frank Bowcher. 1908, to Wallace, Hooker, Haeckel, Weismann, Strasburger, Francis Galton, Ray

Lankester, in that order. Only awarded on the 50th (1908), 100th (1958), and 150th (2008) anniversary of the reading of the Darwin-Wallace paper for "major advances in evolutionary biology." Two replicas were exhibited at the BMNH in 1909.

Doctor Darwin Prize, Italian Society for Evolutionary Biology. 2017.

Ecole Darwin, Algiers, Algeria.

Escola Darwin, a business school, Brazil.

Escuela Carlos Darwin, Progresso, San Cristobal, Galapagos.

Escuela Primaría "Carlos Darwin", Mexico City.

IISS Charles Darwin, Rome, Italy.

Liceo Scientifico Statale di Rivoli Charles Darwin, Rivoli, Italy.

Primaria Charles Darwin, Pachuca de Soto, Mexico.

Royal Society Darwin Trust Research Professorship. 2016?

Scholarships (5) in natural science in the name of Darwin. St. Petersburg, Russia. 1884.

State Darwin Museum, Moscow. 1907.

Szkoła Darwin, Poland.

Young Darwin Prize, a competition to find the best videos that describe a conservation project. Open Air Laboratories (OPAL) network, UK, organised by the NHM. 2010.

Monuments: (132) Excluding those which are solely a representation of CD such as a bust or statue, for which see Darwin, Iconography.

1882 Marble gravestone "CHARLES ROBERT DARWIN / BORN 12 February 1809 / DIED 19 April 1882", Westminster Abbey.

1882? Tree planted in CD's memory in Cambridge. Reported in newspapers as "stolen" in 1887.

1883 Marble memorial in Shrewsbury Unitarian chapel. "To the memory of CHARLES ROBERT DARWIN, author of 'Origin of species', born in Shrewsbury Feb. 12th, 1809. In early life a member and constant worshipper in this church. Died April 19th, 1882". Woodall 1884, p. 12. Pattison, *The Darwins of Shrewsbury*, 2009.

1884 "DARWIN" in stone above a window, Brantford Public Library, Canada.

1885 Wedgwood green jasperware plaque signed "T. Woolner. SC" by Thomas Woolner in CD's set, Christ's College, Cambridge. "SC"= "sculpsit", Latin for he sculpted it. A separate rectangular jasper panel below reads: "CHARLES ROBERT DARWIN 1829-31". The date 1829 is a mistake for 1828. A label on the wooden frame reads: "Erected by G. H. Darwin. Plumian Professor 1885".

1887 Circular bronze portrait wall medallion by Sir Joseph Boehm, Westminster Abbey, London. Boehm was paid £150 for it. Memorial to Alfred Russel Wallace placed next to it 1915 Nov. 1.

1888 Round wall tablet, now vanished, 11 Lothian Street, Edinburgh, at the expense of Ralph Richardson. The building was demolished by 1913.

1889 The Darwin Frieze (in stone) by Johannes Benk (1844-1914) "an ape pointing to its own chest and holding a mirror out to a boy, who is covering his eyes because he does not want to see his own reflection and the ape. A monkey behind the boy is holding an open book entitled 'Darwin. The Descent of Man'". Upper Dome Hall, Natural History Museum, Vienna, Austria.

1895? "DARWIN" inscribed on base of bronze statue, A muse for the sciences, by Bela Lyon Pratt, Boston Public Library.

1902 "DARWIN" in façade of Warrington Technical School, Palmyra Square, Warrington, Cheshire.

1904 Blue plaque erected Dec. 13, "CHARLES DARWIN / NATURALIST / LIVED HERE / 1838-1842", first date was wrong, should be 1839. 110 Gower Street (formerly 12 Upper Gower Street), Greater London Council. House bombed during the war (Sept. 1940). Photograph after bombing in Barlow 1945, facing p. 258.

1906 "blue encaustic ware" plaque on 110 Gower Street. "LCC / CHARLES DARWIN / 1809-1882 / Lived Here / 1839-1842." *Morning Post*, (27 Feb.) et al. Photograph in *The Sphere*, (13 Feb.), p. 148.

1909? Stone plaque "Charles Darwin lived here 1836-7". 22 Fitzwilliam Street, Cambridge.

1913 "DARWIN" on façade of Multnomah County Central Library, Portland, Oregon.

1916 "DARWIN" in frieze, Killian Court building, Massachusetts Institute of Technology (MIT).

1925 CD's name on tablet listing justices of the peace for the past 100 years. Bromley Court House.

1925 "DARWIN" carved on the shield of a gargoyle. Others for Mendel and Hofmeister. Botany School, University of Sydney.

1927 "DARWIN" on façade of Pasadena Central Library, California.

1929 Stone plaque set in front garden wall of Down House. "HERE DARWIN THOUGHT AND WORKED FOR FORTY YEARS AND DIED 1882". Words by Sir Buckston Browne.

1935 Bronze tablet inscription by Leonard Darwin on obelisk supporting bust: "Charles Darwin landed on the Galapagos Islands in 1835 and his studies on the distribution of animals and plants thereon led him for the first time to consider the problem of organic evolution. Thus was started the revolution in thought on this subject which has since taken place". Ecuadorian Naval Base, Puerto Baquerizo Moreno, San Cristobal Island, Galapagos. Copies of the plaque appear on the plinths of the busts at the University of Guayaquil, Ecuador and at Quito. The latter is translated into Spanish.

1935 Bronze plaque on a boulder commemorating CD's ascent of Bell mountain, Cerro La Campana, by the Scientific Society of Valparaiso, Chile. "Homenaje a Carlos Darwin. 'Pasamos el día en la cima del monte, y nunca me ha parecido el tiempo más corto, Chile se extiende a nuestros pies como un panorama inmenso limitado por los Andes y el océano Pacífico.' Carlos Darwin ('Mi viaje alrededor del mundo') 17 de Agosto 1834".

1936 Darwin Tree or Oak. English oak planted by President of the Field Naturalists' Club of NSW at Wentworth Falls, New South Wales, Australia, to commemorate CD's visit there 1836 Jan. 17.

1949 Tablet erected to commemorate CD's visit. Bathurst, New South Wales, Australia.

1953 Bronze plaque by South African National Monuments Council, Sea Point, Cape Town, South Africa. Reported as stolen in 2005. "The rocks between this plaque and the sea reveal an impressive contact zone of dark slate with pale intrusive granite. ...Notable amongst those who have described it was Charles Darwin who visited it in 1836."

1958 Brass plaque commemorating Darwin-Wallace reading to the Linnean Society, Meeting Room of the Linnean Society, Burlington House, London. "Charles Darwin and Alfred Russel Wallace made the first communication of their views on the origin of species by natural selection at a meeting of the Linnean Society on 1st July 1858. 1st July 1958."

1959, 2009 The Darwin plaque (Darwin-Plakette) "to provide recognition for an outstanding contribution that develops the Darwinian ideas and throws light on the main problems of evolution" by The Academy of Sciences Leopoldina, Germany.

1959 Stone and cement? tablet. "Homenaje en el centenario de la publicacion de el Origen de las especies". Agua de la Zorra, near Uspallata, Mendoza, Argentina. Now Parque Paleontológico Araucarias de Darwin. By The National University of Cuyo. The 1959 tablet was replaced in 2009.

1961 Ceramic blue plaque at Biological Sciences Building, UCL, (1982 renamed Darwin Building), site of 110 Gower Street, Bloomsbury, London. "CHARLES DARWIN 1809-1882 Naturalist lived in a house on this site 1838-1842". 1838 is a mistake for 1839.

-1966 Plaque at the Mount, Shrewsbury. No longer extant?

1960s? Plaque with left profile bust in modernist style. "Charles Darwin 1809-1882". Zoologischer Garten, Halle, Germany.

-1976 CD mentioned three times on bronze plaque in honour of H.W. Bates and A.R. Wallace. Leicestershire Museum & Art Gallery, Leicester. Presented by Frank Preston.

1976 "Darwin was right" temporary but now famous plaque by Brian Rosen, on Eniwetok atoll. See Darwin was right. q.v.

1981 Bronze plaque: "CHARLES DARWIN At daybreak on 27-12-1831 HM Survey ship BEAGLE, commanded by Captain H. FitzRoy, departed its anchorage in Barn Pool, opposite this point. Darwin's 'Origin of the Species' resulted directly from experiences gained during this voyage to South America" Devil's Point, Stonehouse, Plymouth. [2017 moved to a lone stone from original wall.]

1981 Vertical sundial in south wall of tower with inscription below "This sundial is in memory of CHARLES DARWIN 1809-1882 who lived and worked in Downe for 40 years He is buried in Westminster Abbey". St Mary's Church, Downe, Kent.

c.1983 Plaque. Castle Gates Library, Shrewsbury. "Charles Darwin [was] educated here."

1986 Enamelled plaque, Bathurst, Australia. "Charles Darwin passed this way in 1836, remembered by his friends in 1986."

1986 Metal sign mounted on rock: "CHARLES DARWIN PASSED THIS WAY 1836". Charles Darwin Walk, Jamison Creek, Australia.

1986 Enamelled sign with unique portrait and explanatory text. Charles Darwin Walk, Jamison Creek, Australia.

-1990 CD mentioned on plaque in honour of Wallace. Neath 'Mechanics' institute', Wales. "…In his life he collaborated with Charles Darwin in the study of the law of Natural Selection. And with him presented the first paper on the subject in 1858."

c.1991 Bronze street plaque, Writers Walk, Sydney, Australia. "This is really a wonderful Colony; ancient Rome, in her imperial grandeur, would not have been ashamed of such an offspring. Letter from Charles Darwin (1836) Charles Darwin, the British naturalist, spent two months in Australia in 1836. His voyages helped shape his theories of evolution and natural selection. Propounded in *Origin of species* (1858)[sic]. NSW Ministry for the Arts…"

1997 Bronze plaque set in a boulder outcrop. Cerro Santa Lucía, Castillo Hidalgo, Santiago, Chile: "Santiago 27 de Augusto 1834 Una inagotable fuente de placer es escalar el Cerro Santa Lucia, una pequeña colina rocosa que se levanta en el centro de la ciudad. Desde allí las vista es verdaderamente impresionante y única". From *Journal*, 2d edn, p. 262: "A never-failing source of pleasure was to ascend the little hillock of rock (St. Lucia) which projects in the middle of the city. The scenery certainly is most striking".

19?? Two plaques. Cambridge lodgings on Sidney Street, rebuilt as branch of Boots the Chemist, a brass plaque: "CHARLES DARWIN 1809-1882 KEPT IN LODGINGS ON THIS SITE IN 1828 WHILE AND UNDERGRADUATE OF CHRISTS COLLEGE" and a newer blue plaque: "CHARLES DARWIN LIVED IN A HOUSE ON THIS SITE 1828".

19?? CD's name in stained glass in the northeast window of St John's Church, Peterborough.

nd Wooden sign in the Darwin Building, University College London, erected by Richard Keynes.

2000 Medallion plaque. Darwin Gardens Millennium Green, Ilkley. "1809-1882 Charles Darwin".

2000? Quotation from p. 61 of *Origin* in window of The Henry Wellcome Building, Cambridge.

2001 Darwin Square [China] (达尔文广场).

2001 Painted panel commemorating the Darwin-Wallace reading to the Linnean Society, Reynolds room, Burlington House, London. "Charles Darwin and Alfred Russel Wallace made the first communication of their views on the origin of species by natural selection at a meeting of the Linnean Society on 1st July 1858. 1st July 1958."

2001 Blue plaque. Wells House, Ilkley. "…In 1859 Charles Darwin was a patient [here]."

2002 Bronze plaque "On this site Charles Darwin (1809-1882) author of *The Origin of Species* lodged at 11 Lothian Street whilst studying medicine at the University of Edinburgh 1825-1827." Rear wall of National Museum of Scotland, South College Street, Edinburgh.

2002 Engraved slate plaque by Lynne Davies with quotations by CD and Evan Roberts in English and Welsh. CD: "Neither of us saw a trace of the wonderful glacial phenomena around us, …a house burnt down by fire did not tell its story more plainly than did this valley. — Charles Darwin 1831". Ogwen Cottage, Llyn Ogwen, Wales, UK.

-2004 Portrait after Elliott & Fry 1881a and name on village sign, Downe, Kent.

2004 Bronze plaque, The Darwin Gate, Shrewsbury.

2004 Plaque: "This residence was home to Britain's most famous naturalist and geologist during six weeks of the summer of 1861…[CD] traveller, adventurer and writer b.1809-d.1882 author of Origin of Species published 1859". 2 Meadfoot House, Hesketh Crescent, Torquay.

-2006 Village sign "Welcome to Shrewsbury Home of Charles Darwin". Shrewsbury.

2006 Bronze plaque with platypus by Tim Johnman & Philip Sparks. Lake Wallace, Australia.

2006 Bronze plaque commemorating CD's two day visit to the Wallerawang area in 1836. Australia.

2006 Blue plaque commemorating J.S. Henslow at Cambridge Botanic Garden. "1796-1861 Professor, Churchman, Botanist and Geologist Innovator in universal education A guiding light for his student Charles Darwin. He created this Botanic Garden".

-2007 Quote from *Journal* 2^d edn, p. 437, carved on a large standing stone. Echo Point Blue, Mountains, New South Wales, Australia.

-2007 Stone marker "Carl von Linné Charles Darwin Alfred Wallace Henry Bates". Colorado University.

-2008 CD mentioned on stone plaque for Charles Lyell by Bruce Walker, Kirriemuir, Scotland. "He was mentor and friend of Charles Darwin."

2008 Stone and cement obelisk with marble tablets to P.P. King and CD. Santa Ana, Chile. "Homenaje al eminente naturalista Charles Darwin 1809-1882…"

2008 Steel-framed signs "Caminhos de Darwin", with reproduction of Richmond 1840 and map placed in 11 locations along CD's route in Brazil.

2008 Fake CD quotation: "It is not the strongest of the species that survives, nor the most intelligent, but the most responsive to change." in the floor of the California Academy of Sciences. This widely repeated quotation was first revealed as not genuine in John van Wyhe, Misconceptions. *The Guardian* [Saturday pamphlet] (9 Feb. 2008).

2008 Stone monument commemorating 175th anniversary of CD's stay with Sebastián Pimienta 18-20 May 1833. Pan de Azúcar, Uruguay. "…el sabio y naturalista ingles…"

-2009 Plaque commemorating CD's visit. Machattie Park, Bathurst, Australia.

2009 "Caminos de Darwin", [The Darwin Trails] map and portrait (after Richmond 1840) in glazed tiles placed in 12 locations along CD's route including Maldonado and Montevideo, Uruguay and 12 in Brazil: Maricá, Saquarema, Araruama, São Pedro d'Aldeia, Cabo Frio, Barra of São João, Macaé, Conceição de Macabu, Rio Bonito, Itaboraí and Niterói.

2009 Blue plaque commemorating the wedding of CD and ED in Maer, Newcastle Civic Society. "Charles Darwin & Emma Wedgwood were married in St. Peter's, Maer on 29th January 1839…"

2009 Plaque by the Western Province Branch of the Geological Society of South Africa. Sea Point, Cape Town, South Africa.

2009 Granite plaque. Agua de la Zorra, near Uspallata, Mendoza, Argentina by the Centro Regional de Investigaciones Científicas y Tecnológicas of Mendoza. Designed by Eugenia Zavattieri.

2009 Stone plaque with portrait after Elliott & Fry 1881a by in the gardens of CCT-CONICET-Mendoza, Argentina. Designed by Eugenia Zavattieri.

2009 In metal lettering on a wall "DARWIN 1809-1882", Boston Museum of Science.

2009 Marble monument, Founder's alley of the Botanic Gardens, Balchik, Bulgaria. By the Sofia University St. Kliment of Ohrid.

2009 Marble plaque, Wulaia Bay, Chile. "Charles Darwin 1809-1882. Disembarco en esta caleta de Wulaia, centro del territorio Yamana el 23 de enero de 1833. Su estadia en Chile entre 1832 y 1835, durante su viaje a bordo de la "Beagle", contribuyo a la elaboracion de sus idea cientificas. Homenaje en el bicentenario de su natalicio." Instituto de Commemoracion Historica de Chile.

2009 Metal plaque with 225-word inscription commemorating CD's visit to Gardners Inn, Blackheath, Australia.

2009 'The face of Darwin.' Ceramic mosaic after Collier 1881 with tiles representing themes from CD's life by students of FRS 004. Storer Hall, University of California, Davis.

2009 Mural by Bruce Williams after Richmond 1840 with evolution motifs, Bromley, Kent.

2009 Darwin's Tree of Life Mural by students from Bickley Park School, Bromley, Kent.

2009 Sculpture and plaque. Darwin Memorial Geo-Garden Quantum Leap, Mardol Quay Park, Shrewsbury.

2009 Way-marker or stone street plaque with iguana. Shrewsbury Library. "12th February 2009 Shrewsbury library Darwin200". Shrewsbury. The series sponsored by the Royal Mail.

2009 Stone street plaque with beetle. "The Bellstone Darwin200 12th February 2009". Shrewsbury.

2009 Stone street plaque with newts. "St. Chads church Darwin200 12th February 2009". Shrewsbury.

2009 Stone street plaque with the *Beagle*. The Lion Hotel, Shrewsbury.

2009 Stone street plaque with a trilobite. The Unitarian Church, Shrewsbury.

2009 Stone street plaque with a fossil fish. The Mount, Shrewsbury.

2009 Stone street plaque with fossil ammonite. 13 Claremont Hill, Shrewsbury.

2009 Illustrated sign "Darwin's Shrewsbury River Walk." Shrewsbury.

2009 Enamel signs with QR codes and portrait after Rejlander 1871c "Darwin's Shrewsbury". Different colour examples located at The Bellstone, The Library, St. Chad's and The Lion Hotel.

2009 "Tree", by Tania Kovats, a 17 metre-long installation across the ceiling of the mezzanine gallery, Natural History Museum, London.

2009 Malvern Civic Society blue plaque outside house: "In 1851 Charles Darwin stayed here with his daughter Anne Elizabeth who was being treated by pioneer of the Malvern Water Cure Dr James Manby Gully." Montreal House, Worcester Road, Malvern.

2009 Stone plaque. Fish Strand Hill, Falmouth. "CHARLES DARWIN 1809-1882 arrived at Falmouth aboard 'The Beagle' on 2nd October 1836 and departed on the Royal Mail coach from this site".

2009 Blue plaque. "Hillside. Charles Darwin & family stayed here (Wells Terrace) while awaiting the publication of 'On the Origin of Species' in autumn 1859." Ilkley, Yorkshire.

2009 Enamel plaque after 1840 Richmond sketch, Princes Park, Hobart Town, Tasmania.

2009 Enamel plaque, Royal Tasmanian Botanical Gardens, Hobart Town, Tasmania.

2009 Enamel plaque, Mount Wellington Reserve, Tasmania.

2009 Illustrated information pane. Marimurtra Botanical Garden, Spain.

2009 "HMS Beagle Ship Bell Chime" by Anton Hasell "Featuring a series of cast bronze bells and a replica HMS Beagle ship's bell, cast in brass" with a different species of Australian bird atop each bell in bronze, painted. Civic Park, near Civic Centre and public library, Darwin, Australia.

c.2009 Reproduction of writing in *Notebook B*, p. 228, with added full signature, in stone on the wall of the entrance to the Institute of Zoology, Chinese Academy of Sciences, Beijing.

nd Woodburned outdoor wall portrait after Edwards 1866a. Galapagos National Park Headquarters.

-2010 Explanatory sign with portrait after Elliott & Fry c.1880b, Peninsula Valdes, Chubut, Argentina.

2010 Monument stone. "On 30 June 1860 Thomas Henry Huxley Samuel Wilberforce and others debated Charles Darwin's Origin of Species in the museum 1860-2010". Oxford University Museum of Natural History.

2010 Information board of laminated synthetic material erected by City of Cape Town, Sea Point, Cape Town, South Africa. In three languages: English, Afrikaans and isiXhosa. Reported vandalized.

2010 Bronze plaque with bas-relief bust (after Richmond 1840) & *Beagle* commemorating CD's visit on 12 July 1835, with second plaque with further text in front of large anchor on a plinth. In front of the former Customs building or Rímac Palace, Iquique, Chile.

2010 Eleven stone plaques with CD quotations and a sign. The Darwin Trail. Caledonian Park Islington, London.

-2011 Metal silhouette after Richmond 1840. Partition wall along the Darwin Waterfront, Australia.

-2011 Metal sign on rock "Mudstone Rock Measured by Charles Darwin 1831". Offa's Dyke Path, Llanymynech. English/Welsh border.

2011 Mural with image of CD with fake hand added, Sean Williams's wall in Avondale Point's Alycia Alley, behind Avondale Wine & Cheese.

2011 'Darwin', Varnished stainless steel modernist abstract sculpture by Phillip King, UK.

2012 Blue plaque: "THE RAMSGATE SOCIETY CHARLES DARWIN 1809 - 1882 Naturalist stayed here in 1850", 8 Paragon Street, Ramsgate.

-2012 Plaque at The Mount, Shrewsbury. "Charles Darwin was born here 12th February 1809".

2014 "Interpretation" panel on the connection between CD et al and Keston. Keston Common, Bromley, Kent.

-2015 Outline portraits of CD and Robert Falcon Scott cut from sheet metal and erected near a bench at North Cross, Plymouth.

2015 Human Nature - Darwin's Wonderland, street mural. Bristol, UK.

-2016 Bronze wall plaque "Berberis darwinii Charles Darwin". Near Avrille Airport, Pays de la Loire, France.

-2016 Bronze ape holding a human skull on stack of books, enlarged version of Rheinhold statuette. Spelling on spine of a book "DARWEN". Sanya Park, Hainan Island, China.

2016 CD's work in stained glass window (not a portrait). Fulton Library, Utah Valley University.

2016 Seven interpretive signs commemorate CD's visit to Albany in 1836. Western Australia.

nd Sign at Darwin Tree, Wentworth Falls, New South Wales, Australia.

nd Brass plaque: "The Old Parsonage or Manse...Charles Darwin along with his sister were pupils here in 1817 under the Reverend Augustus Chase before he went on to The Schools, Shrewsbury (now the Library)." 13 Claremont Hill, Shrewsbury.

nd Stone plaque "Darwin was here August – 1834", base of Bell mountain. Hacienda San Isidro, Chile.

nd CD mentioned in blue plaque in honour of Conrad Martens. Library, Exeter Road, Exmouth. "From 1833 to 1834, he was the ship's artist on the Beagle and corresponded with Charles Darwin for many years."

20?? Stone plaque, CD's head facing a chimpanzee. Devonport Heritage Trail marker.

There are very many more.

Places named after CD: (284)

1991 Darwin, a stony Florian asteroid.

Avenida Charles Darwin (road). Puerto Ayora, Santa Cruz Island, Galapagos.

Cerro Darwin, see Mountain, Albemarle Island, Galapagos.

Charles Darwin Building, Bristol Polytechnic.

Charles Darwin Building, University of Worcester.

Charles Darwin Centre, Downe, Kent, 2001.

Charles Darwin Centre, office tower, Darwin, Australia.

Charles Darwin Drive, Monto, Australia.

Charles Darwin Exhibition Hall, Puerto Ayora, Santa Cruz Island, Galapagos Islands.

Charles Darwin House, Roger Street, London.

Charles Darwin National Park, Northern Territory, Australia.

Charles Darwin Park, Wallerawang, Australia.

Charles Darwin Research Station (Estacion Cientifica Charles Darwin), Puerto Ayora, Santa Cruz Island, Galapagos Islands, 1964-

Charles Darwin Research Station, Galapagos Islands.

Charles Darwin Reserve, Western Australia.

Charles Darwin Road, Plymouth.

Charles Darwin Trail, Bellerive, Tasmania.

Darwin Sports Centre, Biggin Hill.

Charles Darwin Trail, Leura, Australia.

Charles Darwin Trail, Wentworth Falls, Australia.

Charles Darwin Walk, Wentworth Falls, New South Wales, Australia, 1986.

Charles Darwin-Strasse, Hamburg, Germany.

Charles-Darwin-Platz. Zentrum-Bozner Boden-Rentsch, Italy.

Charles-Darwin-Straße, Erfurt, Germany.

Charles-Darwin-Straße, Hanau, Germany.

Charles-Darwin-Strasse, Windischholzhausen, Germany.

Cordillera Darwin Metamorphic Complex, a geological unit in Tierra del Fuego.

Darvinsky Zapovednik (Дарвинский заповедник, Darwin Nature Reserve), North Russia, 1945-.

Darwin (settlement), Falkland Islands.

Darwin, village, Cerro Perico Flaco, Argentina.

Darwin (polling district, Bromley).

Darwin Arch, Culpepper Island/Isla Darwin, Galapagos.

Darwin Avenue, Buxton, UK.

Darwin Avenue, Chesterfield, UK.

Darwin Avenue, Christchurch, UK.

Darwin Avenue, Derby, UK.

Darwin Avenue, Green Bay, USA.

Darwin Avenue, Maidstone, UK.

Darwin Avenue, Worcester, UK.

Darwin Bank, an oil field in Azerbaijan.

Darwin Bar, Darwin College, Cambridge.

Darwin Bar, Queen's Head public house, Downe.

Darwin Bay, coast of Chonos Archipelago, Chile.

Darwin Bay, southwest side of Tower Island/Isla Genovesa, Galapagos Islands.

Darwin Bicentennial Oak, planted by Richard Carter, 2009, UK.

Darwin Bend, a bend in the Tasman glacier, New Zealand, where it goes round Mount Darwin.

Darwin Boulevard, Edison, NJ, USA.

Darwin Boulevard, Louisville, KY, USA.

Darwin Boulevard, Port Saint Lucie, FL, USA.

Darwin Building, Faculty of Education, Newman University.

Darwin Building, Keele University, UK.

Darwin Building, La Sabina, Spain.

Darwin Building, Royal College of Art, Kensington.

Darwin Building, Science Park, Cambridge.

Darwin Building, UCL, Biological Sciences block, renamed 1982, see Darwin Lecture Theatre.

Darwin Building, University of Central Lancashire.

Darwin Building, University of Edinburgh.

Darwin Canyon, Kings Canyon National Park.

Darwin Centre, a regional 15-bedded inpatient unit, Penkull, UK.

Darwin Centre, a Tier 4 adolescent inpatient unit for young people, Cambridge.

Darwin Centre, Netherlands Institute of Ecology.

Darwin Centre, NHM, UK.

Darwin Channel, leading to Port Aysen, Chile.

Darwin Close, Burntwood, UK.

Darwin Close, Cannock, UK.

Darwin Close, Colchester, UK.

Darwin Close, Ely, UK.

Darwin Close, Hemel Hempstead, UK.

Darwin Close, Heron Ridge, Nottingham, UK.

Darwin Close, Hinckley, UK.

Darwin Close, Huntington, York, UK.

Darwin Close, Lichfield, UK.

Darwin Close, London N11, UK.

Darwin Close, Medbourne, UK.

Darwin Close, Milton Keynes, UK.

Darwin Close, New Southgate, UK.

Darwin Close, Orpington, UK.

Darwin Close, Reading, UK.

Darwin Close, Spalding, UK.

Darwin Close, Stapenhill, UK.

Darwin Close, Taunton, UK.

Darwin Close, Thorpe Astley, UK.

Darwin Close, Wakeley, NSW, Australia.

Darwin Close, Walsgrave-on-Sowe, UK.

Darwin Close, Worcester, UK.

Darwin Close, York, UK.

Darwin Cordillera (Cordillera Darwin), Chile.

Darwin Court, Crail Row, London, UK.

Darwin Court, Gloucester Avenue, London, UK.

Darwin Court, Greenville, NC, USA.

Darwin Court, Grimsby, UK.

Darwin Court, Laurel, MD, USA.

Darwin Court, Lichfield, UK.

Darwin Court, Margate, UK.

Darwin Court, Orange Park, FL, USA.

Darwin Court, Plaistow, London, UK.

Darwin Court, Rochester, UK.

Darwin Court, Salisbury Heights, Australia.

Darwin Court, San Jose, CA, USA.

Darwin Court, Southwark, Greater London, UK.

Darwin Court, Whitchurch, UK.

Darwin crater, western Tasmania, named in 1972.

Darwin Crescent, Beechboro WA, USA.

Darwin Crescent, Deep River, Canada.

Darwin Crescent, Dunclair Park. Efford, UK.

Darwin Crescent, Loughborough, UK.

Darwin Crescent, Maraenui, New Zealand.

Darwin Crescent, Morley WA, USA.

Darwin Crescent, New Plymouth, New Zealand.

Darwin Crescent, Newcastle upon Tyne, UK.

Darwin Crescent, Perth, Australia.

Darwin Crescent, Plymouth, UK.

Darwin Crescent, Queenstown, Tasmania.

Darwin Crescent, Spotswood, New Zealand.

Darwin Crescent, Torquay, UK.

Darwin Discovery Park in Yichuan, New Village Kindergarten, Shanghai, China.

Darwin District, Zimbabwe.

Darwin Drive, Amherst, NY, USA.

Darwin Drive, Brookside, Ohio.

Darwin Drive, Cambridge, UK.

Darwin Drive, Cheektowaga, NY, USA.

Darwin Drive, E. West Lafayette, IN, USA.

Darwin Drive, Fremont, CA, USA.

Darwin Drive, Leyland, UK.

Darwin Drive, Llanarth, NSW, Australia.

Darwin Drive, Machesney Park, IL, USA.

Darwin Drive, New Ollerton, Newark-on-Trent.

Darwin Drive, Oceanside, CA, USA.

Darwin Drive, Osage Beach, MO, USA.

Darwin Drive, Tonbridge, Kent, UK.

Darwin Drive, Wilmington, NC, USA.

Darwin Drive, Yeovil, UK.

Darwin Forest, an in situ Middle Triassic forest in the Paramillo Formation of Argentina.

Darwin Garden, Christ's College, Cambridge, renamed Charles Darwin Sculpture Garden.

Darwin Gardens Millennium Green, Ilkley, UK.

Darwin Gate, Mardol Head, Shrewsbury, UK.

Darwin Glacier, Antarctica.

Darwin Glacier, New Zealand, flows from Mount Darwin into Tasman glacier.

Darwin Glacier, Kings Canyon National Park, California, USA.

Darwin Guyot, a seamount in the Pacific Ocean.

Darwin Hall, Flanders meeting and convention centre, Antwerp, Belgium.

Darwin Hall, later Darwin Hall of Evolution, former name used for a hall in the American Museum of Natural History.

Darwin Hall, Sonoma State University, California.

Darwin Harbour, Choiseul Sound, East Falkland.

Darwin Island, Danger islands, Antarctica.

Darwin Island/Isla Darwin, official Ecuadorian name of Culpepper Island, Galapagos group.

Darwin Junction, a town in California, USA.

Darwin Laboratories, three at Shrewsbury School. Opened by Sir Francis Darwin, 1911 Oct. 20.

Darwin Lake, Isla Isabel, Galapagos.

Darwin Lane, Austin, TX, USA.

Darwin Lane, Crosspool, Sheffield, UK.

Darwin Lane, Highlands Ranch, CO, USA.

Darwin Lane, Kokomo, IN, USA.

Darwin Lane, NE Palm Bay, FL.

Darwin Lane, North Brunswick, NJ, USA.

Darwin Lane, Oak Ridge, TN, USA.

Darwin Lane, Remuera, Auckland, Australia.

Darwin Lane, Walpole, MA, USA.

Darwin Lane, Wheaton, Illinois, USA.

Darwin Lecture theatre, University College London. Botany theatre renamed by Richard Darwin Keynes 1982, Apr. 19; on site of no. 12 Upper Gower Street.

Darwin Library, King's Buildings, Edinburgh. Demolished 2018.

Darwin Mounds, NW Cape Wrath, Scotland, 2002.

Darwin Mountain, Antarctica, above Beardmore Glacier, Ross Dependency.

Darwin Mountain, California, King's Canyon National Park; named 1895.

Darwin Mountain, New Zealand, South Island. 18km northeast of Mount Cook. 2561 m. Named by J.F.J. von Haast, who corresponded with CD. See also Glacier.

Darwin Mountain, on the Moon. Midway between Mare Orientale and Mare Humorum.

Darwin Mountain, Peru.

Darwin Parkway, Asheville North Carolina, USA.

Darwin Parkway, Atlanta, Georgia, USA.

Darwin Parkway, Birmingham AL, USA.

Darwin Parkway, Gómez, Panama.

Darwin Parkway, New Mexico, USA.

Darwin Parkway, Wisconsin, USA.

Darwin Place, Bracknell, UK.

Darwin Place, Derby, UK.

Darwin Place, Newmarket Road, Cambridge, UK.

Darwin Rise, Dartford, UK.

Darwin Road, Gloucester, UK.

Darwin Road, in the Essex town of Tilbury, UK.

Darwin Road, London W5, UK.

Darwin Road, Long Eaton, Nottingham, UK.

Darwin Road, Madison, Wisconsin, 1890s-.

Darwin Road, Pudong District, Shanghai, China.

Darwin Road, Welling, UK.

Darwin's Charity. A piece of land Darwin owned at Kirton, Lincolnshire.

Darwin Shopping Centre, Pride Hill, Shrewsbury.

DarwIN Shrewsbury Trail, Shropshire, UK.

Darwin Sound, Canada.

Darwin Sound, Tierra del Fuego, Chile, continuing northwest arm of Beagle Channel.

Darwin Spring, see town, California, USA.

Darwin's Peak, Campana, Cerro la, Chile.

Darwin station, a railway station in Madison, Wisconsin from the 1890s-1950s.

Darwin station, an Early/Lower Miocene fossil shell locality in Patagonia, Argentina.

Darwin Straße, Baden-Wurttemberg, Germany.

Darwin Straße, Berlin, Germany.

Darwin Straße, Charlottenburg, Germany.

Darwin Straße, Debschwitz, Germany.

Darwin Strasse, Dresden, Germany.

Darwin Strasse, Heidelberg, Germany.

Darwin Straße, Herbrechtingen, Germany.

Darwin Straße, Leipzig, Germany.

Darwin Straße, Mörfelden-Walldorf, Germany.

Darwin Straße, Paniówki, Poland.

Darwin Strasse, Sibiu, Romania.

Darwin Strasse, Vienna, Austria.

Darwin Street, Birmingham, UK.

Darwin Street, London SE, UK.

Darwin Street, Northwich, UK.

Darwin Street, Shrewsbury, "a short street of new houses near St. George's church has been called 'Darwin Street'. Woodall 1884, p. 12.

Darwin Street, Ushuaia, Argentina.

Darwin Terrace, Dudley Park, Australia.

Darwin Terrace, Shrewsbury, UK.

Darwin Terrace, The Villages, FL, USA.

Darwin Tower, University of Edinburgh.

Darwin Town, Choiseul Sound, East Falkland.

Darwin Township, Clark County, Illinois, USA.

Darwin Trail, Banos, Ecuador.

Darwin Trail, cycling route, Azores.
Darwin Trail, Genovesa Island, Galapagos.
Darwin Trail, La Campana National Park, Chile.
Darwin Trail, Sea point, Cape Town.
Darwin Trail. Caledonian Park Islington, London.
Darwin Village, Uruguay, on river Beguelo, near Cerro Perico Flaco.
Darwin volcano, Isla Isabela, Galapagos.
Darwin Walk, Northampton.
Darwin Way, Bunbury, WA, USA.
Darwin Way, Citrus Springs, FL, USA.
Darwin Way, College Grove, WA, USA.
Darwin Way, El Dorado Hills, CA, USA.
Darwin Way, Ellesmere Port, Cheshire, UK.
Darwin Way, Knoxville, TN, USA.
Darwin Way, Matawan, NJ, USA.
Darwin Way, Meadowridge, South Africa.
Darwin Way, Mount Juliet, TN, USA.
Darwin Way, North Brunswick, NJ, USA.
Darwin Way, Pakenham VIC, Australia.
Darwin Way, Rosemount, MN, USA.
Darwin Way, San Diego, USA.
Darwin Way, San Jose, CA, USA.
Darwin Way, Wollert, Australia.
Darwin, a city in Minnesota, USA.
Darwin, a former coal mining town near Laredo, Texas, founded in 1882.
Darwin, a former settlement in Fresno County, CA, USA.
Darwin, a ghost town near Death Valley, USA.
Darwin, a lunar crater.
Darwin, a Martian crater, 1973.
Darwin, an Australian Electoral Division in Tasmania, 1903-55.
Darwin, an electoral ward in Bromley, UK
Darwin, an unincorporated community in central Bedford Township, Meigs County, Ohio, USA.
Darwin, an unincorporated community in Dickenson County, Virginia, USA.
Darwin, an unincorporated community in western Pushmataha County, Oklahoma, USA.
Darwin, Río Negro, a municipality in Río Negro Province, Argentina.

Darwin's Gate, modern art piece, Shrewsbury.
Darwin's Oak planted in 1936, Wentworth Falls, Australia.
Darwin's Rocks, Sea Point, Cape Town.
Darwin's Trail, Brazil.
Darwin's Trail, mountain bike trail, Alta Lake, British Columbia.
Darwin's Walk, Wentworth Falls, Australia.
Darwin Ward, Bromley Hospital.
Darwinstraße, Emsbüren, Germany.
Darwin-Strasse, Frankfurt, Germany.
Darwin-Straße, Gera, Germany.
Darwin-Straße, Halberstadt, Germany.
Darwin-Straße, Kharkiv (Charkiw), Ukraine.
Darwin-Straße, Königsbrücker, Germany.
Darwin-Straße, Luxemburg.
Darwin-Straße, Penza, Russia.
Darwin-Straße, Schwaigern, Germany.
Darwin-Straße, Tscheljabinsk, Russia.
Darwin Villas, Downe, Kent.
Jardin Darwin, Municipality of Santiago, Chile.
Karola Darwin, (street), Katowice, Poland, 1960.
Mount Darwin [a town], Zimbabwe.
Mount Darwin, Isla Grande, Tierra del Fuego, Chile, west of Ushuaia on Beagle Channel. 1834 CD to Catherine, "Mount Sarmiento, the highest mountain in the south, excepting!! Darwin!!". CCD1:381. Sarmiento is the higher.
Mount Darwin, Tasmania.
Mount Darwin, Zimbabwe.
Mount Darwin. Sierra Nevada, Kings Canyon National Park, California.
Parque Paleontológico Araucarias de Darwin, Argentina.
Plaça Charles Darwin, (building), Barcelona.
Port Darwin, East Falkland Island.
Port Darwin, Northern Territory, Australia.
Rua Charles Darwin, São Paulo, Brazil.
Rue Darwin. Paris, named in 1884, near Rue Lamarck.
Rue Darwin/Darwinstraat, Brussels, Belgium.
Senda Darwin Biological Station, Chiloe island, Chile, established 1993.

Plants named after CD: The following list of 221 plants and fungi is based on B. Daydon Jackson, *Darwiniana*, 1910, (with altered orthography) the International Plant Names Index https://www.ipni.org/, The Plant List http://www.theplantlist.org/, AlgaeBase https://www.algaebase.org/, Mycobank http://www.mycobank.org/ and other sources.

Abutilon darwinii Hooker 1871. Malvaceae, Brazil. Shrub sent by Fritz Müller to CD from Brazil.

Acmella darwinii D.M. Porter R.K. Jansen 1985.
Allochlamys darwinii Moq., [A.P. de Candolle] 1849.

Amphiroa darwinii. Harvey 1849. Algae.
Arthrocardia darwinii (Harvey) Weber Bosse, algae.
Asterina darwinii Berkeley, Fungi, Chiloe, Chile.
Asteromphalus darwinii Ehrenberg, 1844, Algae,
  Antarctica.
Aulacodiscus darwini Pantocsek, Algae (Diatom),
  fossil.
Baccharis darwinii Hook. & Arn., Compositae,
  Patagonia, Argentina. Known as "chilca".
Berberis darwinii W.J. Hooker, Berberidaceae,
  Chiloe, a garden plant. Discovered by CD.
Bonatea darwinii Weale, Orchidaceae. Darwin's
  barberry.
Botryocladia darwinii Schneider & Lane, algae.
Buellia darwini J. Steiner & Zahlbr.
Calceolaria darwinii Benth., Scrophulariaceae, Pata-
  gonia, Argentina. Synonym for *Calceolaria uni-*
  flora. Darwin's Slipper Flower.
Callianthe darwinii Hook.f. 1871.
Carex darwinii Boott 1845. Cyperaceae.
Catasetum darwinianum Rolfe, Orchidaceae.
Chaetomorpha darwinii. Hook.f. marine algae.
Charles Darwin, "a rich brownish crimson [rose],
  quite a new tint in colour." 1880.
Charles Darwin, a late-flowering tulip, named by
  Jacob Heinrich Kreglage, Haarlem, 1889.
Charles Darwin, a purple chrysanthemum. York
  Florist's Chrysanthemum Show, 1884.
Charles Darwin, a scarlet Azalea hybrid of *A.*
  mollis and *A. sinensis,* 1901.
Charles Darwin, English Rose by D. Austin 2003.
Charles Darwin, rich purple pelargonium, 1894.
Cheilosporum darwinii De Toni, Algae, Chile.
Chiliotrichum darwinii J.D. Hooker, Compositae,
  synonym for *Nardophyllum darwinii.*
Cladonia darwinii S. Hammer 2003. Fungi.
Clinopodium darwinii Kuntze 1891.
Coldenia darwini Gürke, *Coldenia dichotoma* Leh-
  mann, Boraginaceae, Floreana, Galapagos.
Colobanthus subulatus var. *darwinii* Hook. 1845.
Conferva clavata var. *darwinii* Hooker & Harvey,
  algae.
Cortinarius darwini Spegazzini, 1887 Fungi, Pata-
  gonia.
Croton scouleri var. *darwinii* G.L. Webster 1970,
  Galapagos.
Cruoriopsis darwinii Rosenvinge. Algae.
Cryptothecia darwiniana Bungartz & Elix 2013. Fun-
  gi.
Cytarria darwinii Berkeley 1842. Fungi, Tierra del
  Fuego. Eaten by natives.

Darwin auricula 1882 C. Turner named an alpine
  auricula strain "Charles Darwin" at Royal Ag-
  ricultural Society's show. *The Times,* Apr. 26.
Darwin Black Wattle=*Acacia auriculiformis* A.
  Cunn. ex Benth.
Darwin clematis 1887 Apr. 25 A clematis strain
  "Darwin in memoriam" at Royal Agricultural
  Society Show. *The Times,* Apr. 26.
Darwin potato, *Solanum maglia,* Solonaceae, in
  Chonos archipelago, Chile. Named "Charles
  Darwin's potato" by G. Nicholson, *Illustrated*
  dictionary of gardening, 1885-89.
Darwin tulip 1889 E.H. Krelage asked Francis
  Darwin if he might name the strain in honour
  of CD. A drawing by Herbert Bone is in the
  Royal Collection Trust.
Darwin's darklings (Coleoptera), Galapagos.
Darwin's dream. *Astilbe arendsii.*
Darwin's fern bamboo, native to India.
Darwin's fungus=*Cytarria darwinii.*
Darwin's orchid, *Angraecum sesquipedale.*
Darwin's snow sprite. *Astilbe simplicifolia.*
Darwin's surprise. *Astilbe chinensis.*
Darwinara hort. 1980. Orchidaceae.
Darwini othamnus, "Darwin's daisies", Galapagos.
Darwinia acerosa Fitzg. Shrub of Western Australia.
Darwinia apiculata N.G. Marchant 1984.
Darwinia axillaris F. Muell. 1882.
Darwinia biflora (Cheel) B.G. Briggs 1962. Shrub
  of Eastern Australia.
Darwinia brevifolia F. Muell. 1882.
Darwinia brevistyla Turcz. 1847.
Darwinia briggsiae Craven & S.R. Jones 1991.
Darwinia camptostylis B.G. Briggs 1962.
Darwinia capitellata Rye 1983.
Darwinia carnea C.A. Gardner 1928.
Darwinia chapmaniana Keighery 2009.
Darwinia ciliata F. Muell. 1882.
Darwinia citriodora Benth. 1865.
Darwinia collina C.A. Gardner 1923.
Darwinia decumbens Byrnes 1984.
Darwinia diminuta B.G. Briggs 1962.
Darwinia diosmoides Benth. 1865.
Darwinia divisa Keighery & N.G. Marchant 2002.
Darwinia drummondii Turcz. F.Muell. 1847.
Darwinia endlicheri F. Muell. 1882.
Darwinia exaltata Raf., now invalid.
Darwinia fascicularis subsp. fascicularis Rudge 1961.
Darwinia ferricola Keighery 2009.
Darwinia fimbriata Benth. 1865.
Darwinia flavescens A. Cunn. ex Schauer 1841.

Darwinia foetida Keighery 2009.
Darwinia forrestii F. Muell. 1878.
Darwinia glaucophylla B.G. Briggs 1962.
Darwinia gracilis F. Muell. 1882.
Darwinia grandiflora Benth. 1917.
Darwinia helichrysoides Benth. 1865.
Darwinia heterandra F. Muell. 1882.
Darwinia homoranthoides J.M. Black 1926.
Darwinia hookeriana Benth. 1865.
Darwinia hortiorum K.R. Thiele 2010.
Darwinia hypericifolia Domin 1923.
Darwinia intermedia A. Cunn. ex Schauer 1841.
Darwinia laxifolia Schauer 1841.
Darwinia leiostyla Domin 1923.
Darwinia leptantha B.G. Briggs 1962.
Darwinia luehmannii F. Muell. & Tate 1896.
Darwinia macrostegia Benth. 1865.
Darwinia masonii C.A. Gardner 1964.
Darwinia meeboldii C.A. Gardner 1938, anglice.
Darwinia megalopetala F. Muell. 1882.
Darwinia meisneri (Kippist) Benth. 1865.
Darwinia micropetala Benth. 1865.
Darwinia mollissima Marchant & Keighery.
Darwinia neildiana F. Muell. 1875.
Darwinia nubigena Keighery 2009.
Darwinia oederoides Benth. 1865.
Darwinia oldfieldii Benth.
Darwinia oxylepis Marchant & Keighery 1980.
Darwinia pauciflora Benth. 1865.
Darwinia peduncularis B.G. Briggs 1962.
Darwinia pimelioides Cayzer & F.W. Wakef. 1922.
Darwinia pinifolia Benth. 1865.
Darwinia polycephala C.A. Gardner 1924.
Darwinia polychroma Keighery 2009.
Darwinia porteri C.T. White 1932.
Darwinia procera B.G. Briggs 1962.
Darwinia purpurea Benth. 1865.
Darwinia quinqueflora Dennst.
Darwinia radinophylla F. Muell. 1875.
Darwinia repens A.S. George 1967.
Darwinia rhadinophylla F. Muell. 1875.
Darwinia salina Craven & S.R. Jones 1991.
Darwinia sanguinea Benth. 1865.
Darwinia saturejifolia Turcz. 1852.
Darwinia schuermannii (F. Muell.) Benth. 1865.
Darwinia sect. Genetyllis (DC.) Benth. 1865.
Darwinia speciosa Benth.
Darwinia squarrosa (Turcz.) Domin 1922.
Darwinia taxifolia subsp. *macrolaena* Briggs 1961.
Darwinia terricola Keighery 2012.
Darwinia thomasii Benth. 1865.

Darwinia thryptomenioides D.A. Herb. 1922.
Darwinia thymoides Benth. 1865.
Darwinia turczaninowii F. Muell. 1882.
Darwinia uncinata F. Muell. 1882.
Darwinia verticordina Benth. 1865.
Darwinia vestita Benth. 1865.
Darwinia virescens Benth. 1865.
Darwinia virgata F. Muell. 1875.
Darwinia whicherensis Keighery 2009.
Darwinia wittwerorum Marchant & Keighery 1980.
Darwiniella antarctica Speg. 1888. Fungi.
Darwiniella globulosa Sacc. 1891. Fungi.
Darwiniella gracilis Speg. 1902. Fungi.
Darwiniella orbicula Syd. 1911. Fungi.
Darwiniella, a genus of Fungi.
Darwiniera, a genus of orchids.
Darwinilithus, a genus of Mesozoic alga.
Darwiniothamnus alternifolius Lawesson & Adsersen.
Darwiniothamnus Gunnar Harling, for *Erigeron lancifolium* J.D. Hooker, Isabela, Galapagos.
Darwiniothamnus lancifolius (Hook.f.) Harling.
Darwiniothamnus tenuifolius (Hook.f.) Harling.
Dendryphion darwinii Siboe, Kirk & Cannon 1999. Fungi.
Dictyonema darwinianum Dal-Forno, Bungartz & Lücking 2017. Fungi.
Distephanopsis crux subsp. *darwinii* (Bukry) Desikachary & Prema, algae.
Eschscholzia minutiflora var. *darwinensis*, now invalid.
Eugenia darwinii Hook.f. 1846, Myrtaceae, Chile.
Eustephia darwinii Vargas 1960.
Exobasidium darwinii Piatek & M. Lutz 2012. Fungi.
Fagelia darwinii (Benth.) Kuntze 1891. Current name *Calceolaria darwini*.
Frullania darwinii Gradst. & Uribe, 2004, liverwort.
Galapagoa darwini J.D. Hooker, 1847.
Gentiana magellanica var. *darwinii* Speg. 1902.
Gentiana patagonica var. *darwinii* Griseb. = *Gentianella magellanica* (Gaudich.)
Gomphonema darwinii Bahls, algae.
Goniolithon darwinii (Harvey) Foslie, algae.
Gossypium darwinii Watt, 1907 Malvaceae, Galapagos; a cotton species.
Hebe darwiniana Colenso, Scrophulariaceae, New Zealand.
Herbertia darwinii Roitman & Castillo 2008. A bulb plant of Argentina, Uruguay and Brazil.
Hirsutella darwinii Evans & Samson 1982. An Entomogenous fungi from the Galapagos.
Hoya darwinii Loher 1910.

Hymenophyllum darwinii Hook.f. 1861, Fern, Chile.
Hypocopra darwinii Spegazzini, Fungi, Patagonia.
Isora uniflora var. *darwinii* Benth. Witasek ex Reiche
Laboulbenia darwinii Thaxter, 1902. Fungi.
Laeliocattleya darwiniana × hort. Orchidaceae.
Lecocarpus darwinii Adsersen 1980, Galapagos.
Lippia darwini Spegazzini, Verbenaceae = *Neosparton darwinii* Benth. & Hook.
Lithophyllum darwinii Harvey Foslie, Algae, South Australia. = *Melobesia darwinii* Harvey.
Lithothamnion darwinii (Harvey) Areschoug, algae.
Luticola darwinii Witkowski, Bak, Kociolek, Lange-Bertalot & Seddon, algae.
Maihueniopsis darwinii (Hensl.) F. Ritter 1980.
Meliola darwiniana Mibey 1997. Fungi.
Melobesia darwinii Harvey, algae.
Micromeria darwinii Benth. 1848, Patagonia, Argentina = *Clinopodium darwinii*.
Myrtus darwini (Hook. f.) Barnéoud, Myrtaceae, Chile. Synonym of *Amomyrtus luma*, a tree occurring in Chile and Argentina.
Nardophyllum darwinii A. Gray, Compositae, Patagonia =*Chiliotrichum darwinii*.
Nassauvia darwinii O. Hoffm. & Dusén 1900.
Neobarya darwiniana Etayo 2008. Fungi.
Neomolina darwinii (Hook. & Arn.) Hellw. 1993.
Neosparton darwinii Benth. & Hook. 1876.
Opuntia darwinii Henslow, Cactaceae, Patagonia, Argentina. Synonym of *Maihueniopsis darwinii*.
Oxalis darwinii Ball 1884.
Panargyrum darwinii W.J. Hooker & Arnott, Compositae; synonym of *Nassauvia darwinii*.
Panargyrus darwinii Hook. 1836.
Pediastrum darwinii Haeckel, algae.
Pertusaria darwiniana A. Yánez-Ayabaca, F. Bungartz 2015. Fungi.
Phaseolus coccineus subsp. *darwinianus* Hern.-Xol. & Miranda-Colin.
Pisonia darwinii Hemsley, Nyctaginaceae, Fernando Noronha.

Platichthys darwiniana Seddon & Witkowski, Kingdom Chromista, a diatom.
Pleuropetalum darwinii Hook.f. 1846. Amaranthaceae, Galapagos.
Poa darwiniana Parodi 1937, a grass.
Polygala darwiniana A.W. Benn, Polygalaceae, Patagonia, Argentina. A subshrublet.
Pseudocaryophyllus darwinii Hook.f. 1941.
Puccinia berberidis-darwinii H.S. Jacks. & Holw. 1931. Fungi.
Puccinia caricis-darwinii Speg. 1909. Fungi.
Satureja darwini (Benth.) Briquet, Labiatae.
Scalesia darwini J.D. Hooker, Compositae, Santiago Island, Galapagos. "Darwin finches of the plant world".
Senecio darwini W.J. Hooker & Arnott, Compositae, Tierra del Fuego, Chile. A ragworth.
Sideroxylon portus-darwini O. Schwarz 1927.
Siegfriedia darwinioides C.A. Gardner, 1934, Australian shrub.
Solanum galapagense S.C. Darwin & Peralta, a wild Galapagos tomato.
Spilanthes darwinii D.M. Porter 1978.
Stelis darwinii Luer & R. Vásquez 2018.
Stylidium darwinii Punekar & Lakshmin 2010.
Tephrocactus darwinii (Hensl.) Frič 1934.
Thelypteris darwinii L.D. Gómez 1986.
Tiquilia darwinii (Hook.f.) A.T. Richardson 1976, Galapagos.
Torula darwini Spegazzini, Fungi, Tierra del Fuego, Chile. A yeast. Currently *Candida utilis*?
Torula darwinii Speg. 1888. Fungi.
Tripodiscus darwinii, (Pantocsek) Kuntze, 1898, a fossil diatom. = *Aulacodiscus darwinii*.
Tulip hybrid. Crosses between Darwin tulips and *Tulipa fosteriana*.
Ulota darwinii Mitten, moss, Patagonia, Argentina.
Urtica darwinii Hook.f. 1846. Chonos Archipelago.
Veronica darwiniana Colenso, Scrophulariaceae. Possible synonym *Hebe darwiniana*.
Zinnia darwinii Haage & Schmidt 1875.

Other: (92) Businesses, restaurants, products and slang and popular culture references etc. named after CD are not recorded.

ANL Darwin Trader, cargo ship built in 2008.
Betheal.Darwin, Indian fishing vessel in 2011.
Carlos Darwin, trawler built in 1955.
Charles Darwin, a yacht named in 1878, USA.
Charles Darwin, a ship in 1984.
Charles Darwin, a ship in 2011.
Charles Darwin, catamaran; built 1990.

Charles Darwin, dredger, Paita, PE.
Charles Darwin, a Super voyager diesel train (221121), UK, 2001/2.
Charles Darwin, hopper dredger built in 2011.
Charles Darwin, Luxembourg, dredging or underwater operations vessel.

Charles Darwin, recreational vessel; Milwaukee, WI; built 2014.

Charles Darwin Scholar Program, Charles Darwin University, 2013.

Conti Darwin, container ship built in 1999.

Charles Darwin Research and Innovation Medal, established to recognise individuals who have made an outstanding contribution to research and innovation in a variety of fields across the Northern Territory, Australia. From 2009-.

Darwin (ADL), an architecture description language.

Darwin (programming game).

Darwin (programming language).

Darwin 1, Chilean owned vessel in 2019.

Darwin 3, French sailing vessel in 2019.

Darwin Awards, satirical website, 1994-, originating in Usenet groups c. 1985.

Darwin buddy, passenger (cruise) ship in 2011.

Darwin Core, an extension of Dublin Core for biodiversity informatics.

Darwin Fellowship Award, UK government.

Darwin Four, Canadian 5 tonne vessel in 2019.

Darwin Glass, of volcanic origin, occurs abundantly around Mount Darwin, Tasmania.

Darwin Information Typing Architecture, an XML data model for publishing.

Darwin Initiative, a UK Government funding program that assists countries with rich biodiversity but poor financial resources.

Darwin Karindo Indah, cargo ship, Indonesia.

Darwin Lagrangian, a concept in physics.

Darwin Medal of the Midland Union of Natural History Societies, from 1880.

Darwin Medal of the Royal Society, from 1890.

Darwin Now Award. British Council.

Darwin pilot, Australian pilot vessel in 2019.

Darwin princess, vessel sunk near Australia in 1974.

Darwin sound, Canadian sailing vessel in 2019.

Darwin star, 11 tonne vessel, UK in 2019.

Darwin star, cargo vessel, Singapore 2019.

Darwin Streaming Server (DSS).

Darwin, 15/5m sailing vessel, Chile in 2019.

Darwin, a computer operating system, 2000.

Darwin, a houseboat.

Darwin, a sailing bark, 1885.

Darwin, a ship in 1901.

Darwin, a ship in 1957.

Darwin, a ship in 1977.

Darwin, a ship in 1996.

Darwin, a ship in 2001.

Darwin, a unit of evolutionary rate of change proposed by J.B.S. Haldane in 1948.

Darwin, barquentine; built 1884, Summerside Prince Edward Island.

Darwin, cargo ship built in 1984.

Darwin, cargo ship, built 1884, at Newcastle.

Darwin, cargo ship, built 1901, at Sunderland.

Darwin, cargo? in Ecuadorian waters in 2019.

Darwin, container ship built in 1996.

Darwin, Danish pleasure craft in 2019.

Darwin, Dutch cargo ship in 2019.

Darwin, Dutch sailing vessel in 2019.

Darwin, passenger vessel, Saugatuck, MI; 2011.

Darwin, Polish cargo ship in 2019.

Darwin, recreational vessel; San Diego, CA; 1998.

Darwin, sailing ship, 1887.

Darwin, sailing ship, Charlottetown, Prince Edward Island, in 1884.

Darwin, Singapore registered vessel in 2019.

Darwin, steamer in 1886.

Darwin, steamer, 1892.

Darwin, steamer, London, 1885.

Darwin, steamer, Stanley, Falklands, in 1957.

Darwin, UK pleasure craft in 2019.

Darwin, UK, container ship.

Darwin, warship in 2019.

DarwinD, USA sailing vessel in 2019.

Darwinian Demon, a hypothetical organism that would exist if there were no biological constraints on the evolutionary process.

Darwinian puzzle, a trait that appears to reduce the fitness of individuals that possess it.

Darwinism, a term used to refer to CD's evolutionary theories or simply evolution per se.

Empire Darwin, on Lloyd's Register 1940-45.

Knight-Darwin Law, "nature abhors perpetual self fertilisation". ML2:250.

KM.Darwin Samudra, Indonesian cargo ship 2019.

M/V Darwin, dry cargo vessel. Antigua & Bermuda, 1984.

Matilha Darwin, Brazilian sailing vessel in 2019.

MV Darwin, built in 1930, sunk in 1978.

Mv Flex Darwin, cargo ship in 2013.

My Darwin, USA pleasure craft in 2019.

Nautilus Darwin, Australian sailing vessel in 2019.

NDS Darwin, British sailing vessel in 2019.

Odyssea Darwin, USA ship in 2019.

RRS Charles Darwin, was a Royal Research Ship belonging to the British Natural Environment Research Council. 1988-89 Antarctic expedition. Since 2006, she has been the geophysical survey vessel, RV Ocean Researcher.

RT *Darwin*, tugboat built in 2011.

The Darwin, 1958-73, Royal Mail Vessel plied between Port Stanley, Falkland Islands and Buenos Aires, Montevideo. 1958-63 Falkland Islands Trading Co. Ltd, 1963-73 Darwin Shipping Co. Ltd.

The Darwin, 1984, Royal Research Ship. Belonged N.E.R.C. for geological research. First cruise 1985 Aug. *New Scientist* Feb. 23, 1984.

The Darwin, a barque, copper ore carrier, late 1960s. Probably had wooden figurehead by a Mr Thomas.

XDarwin, a display server supporting the X Window System (X11) ported to run on the Mac OS X and Darwin operating systems.

Finances: CD's finances at Cambridge are detailed in John van Wyhe, *Darwin in Cambridge*, 2014, with the complete College bills and the following summaries:

Tutors' Accounts & Students Bills		The itemized quarterly bills added together			
55.16.06	22 Apr. 1828	15.3.0	Barber	20.11.0	Library
106.05.0	1 Jul. 1828	12.1.6	Bedmaker	8.5.0	Painter
15.18.06	21 Nov. 1828	1.7.0	Bookseller	16.19.9½	Porter
39.15.03	5 Mar. 1829	0.8.6	Carpenter	1.13.6	Scullion
26.15.03	29 May 1829	57.11.0	Chamber	0.1.6	Seamstress
58.05.07	Jul. 1829	0.4.0	Chimney sweep	5.3.6	Shoeblacker
70.18.07	8 Mar. 1830	28.0.1	Coals	15.12.6	Shoemaker
49.10.00	4 Jun. 1830	78.0.0	Cook	0.8.6	Smith
[53.09.11]	1830 Midsummer	5.11.0	Eve Joiner	125.16.7	Steward
122.05.09	2 May 1831	4.12.6	Glazier	35.11.6	Taylor
33.12.6½	9 [Jun. 1831]	39.17.0	Grocer	37.10.0	Tuition
636.0.9½		12.19.6	Laundress	40.5.6	Woollen draper

On the *Beagle* voyage, apart from kitting-out expenses, CD drew bills on his father's account through Robarts, Curtis & Co. CD reported a total of £735 to his father in letters to his sisters. He was on the ship's books for victuals, but paid £50 per annum to FitzRoy towards the expenses of his table, £250 in all, leaving £485 for his personal expenses whilst travelling on land. The wages of his servant from July 1833, Syms Covington, were about £30 per annum, Covington being on the books for messing.

CD kept detailed accounts from the time of his marriage, as did ED, for household expenditure. Francis Darwin recalled: "In money and business matters he was remarkably careful and exact. He kept accounts with great care, classifying them, and balancing at the end of the year like a merchant. I remember the quick way in which he would reach out for his account-book to enter each cheque paid, as though he were in a hurry to get it entered before he had forgotten it." LL1:120.

The accounts are in two series, arranged by year, the Account Books and Classed Account Books. The latter were details copied from the Account Books into categories such as science, gardens, personal, household etc. See Down House, Household expenditure, Annual Household Expenditure 1867-81. Janet Browne describes: "[The Annual Ledger which] was dedicated to recording income against the cheques and cash sums he paid out to his wife, shopkeepers, salaried staff, investment companies, and suchlike, the historical equivalent of modern bank statements." Browne, The natural economy of households: Charles Darwin's account books. In M. Beretta et al eds., *Aurora Torealis*, 2008, pp. 87-110. Extracts are given in Keith, *Darwin revalued*, 1955, pp. 221-23, and Atkins 1976, pp. 95-100. The editors of *The Correspondence of Charles Darwin* provide the following excellent summary of CD's account books (CCD25:830-1):

CD's Account books (Down House MS). This series of seventeen account books begins on 12 February 1839. A fortnight after CD and Emma's marriage, and ends with CD's death. The books contain two sets of accounts. From the start, CD recorded his *cash account* according to a system of double-entry book-keeping. On each left-hand page he recorded credits (i.e., withdrawals from the bank, either in the form of cash paid to himself or cheques drawn for others), and on each right-hand page he recorded debits (i.e., cash or cheques paid to others). CD also recorded details of his *banking account* from the start, but only noted them down in a single column at the bottom of the left-hand page of his cash account. In August 1848, however, he began a system of detailing his banking account according to double-entry book-keeping, in a separate chronological section at the back of each account book. On the left, he recorded credits to the account in the form of income (i.e., investments, rent, book sales, etc.). On the right, he recorded debits to the account (i.e., cash or cheque withdrawals).

CD's Classed account books (Down House MS). This series of four account books, covering the years 1839—81, runs parallel to CD's Account books. For each year, September-August (after 1867, January-December). CD divided his expenditure into different classes; in addition, he made a tally for the year of his income, expenditure, cash in hand, and money in the bank. From 1843, CD also compiled at the back of each book a separate account of the total expenditure under the various headings in each year, and from 1844 he added a full account of his income in each year, and of capital invested and 'paid' up.

CD's Investment book (Down House MS). Records for each of CD's investments the income received during the period 1846-81.

Until his father's death in 1848 CD was wholly dependent on him, except for ED's marriage settlement and £150 which he received for the sale of his copyright in *Journal* in 1845. In his early manhood years he received £400 per annum which was increased to £500 on marriage.

1839- ED's dowry of £25,000 brought £400 per annum. From his father CD received a marriage bond (a trust fund) of £10,000.

A summary analysis of the Classed Accounts, 1842-3 to 1881 from Chris Miele, *Darwin's garden: The estate and gardens at Down House, Bromley*, 1996, pp. 23-25:

Year/s	Household	Repairs	Furniture	Menservants	Gardens	Fields etc.	Stables
1842/43	380	742.16	256.6.3	77.10	88.01.6	69.07.12	158.12
1843/44	435	486.11	112.0.6	80.10	193.02.12	37.14.6	46.18.0
1844/45	365	32.9.8	60.15.6	77.0.0	111.02.0	34.04.6	29.8.0
1845/46	447	390.13.3	11.4.0	80.15	149.04.6	35.19	59.14.9
1846/47	525	196.7.8	110.5.8	82.8.6	58.17.6	60.08	47.0.6
1847/48	580	47.11.7	62.1.3	87.12.6	45.16	42.12.1	42.15.0
1848/49	655	109.8.10	39.14.8	90.6	47.10.1	47.7.7	75.18.0
		Furniture and Repair					
1849/50	575	150.8.7		101.17	73.8.6	38.13.1	57.17.8
1850/51	615	112.16.0		108.16.6	51.1.0	34.4	109.10.6
1851/52	600	258.2.10		108.16.6	56.3.6	43.3.10	34
1852/53	625	206.10		124.0.6	80.13.6	42.15.8	61.18.9
1853/54	735	208.18.10		127.10.6	69.18	53.11.7	79.2.3
1854/55	737.5.7	187.2.5		110.18.6	67.5.6	65.13.6	37.8

Year/s	Household	Furniture and Repair	Menservants	Gardens	Fields etc.	Stables
1855/56	750.8	99.13.2	125.19.6	65.6.6	67.6.9	29.4
1856/57	749.13	75.10.6.	123.10.6	68.7.10	65.18.3	100.13.8
1857/58	833.2.6	280.15.2	121.3.6	71.15	54.6.6	70.9.10
1858/59	834.6.4		154.11	56.10.2	42.17	229.3.4
1859/60	855.3.8	185.0.6	145.8	61.6.4	99.2	276.0.3
1860/61	961.8.7	156.3.10	153.9.6	68.12	145.2.3	211.0.10
1861/62	899.14.9	117* *118.16 cowshed & stable entered here, not stable	158.14.3	76.11.3	84.14.4	360.3.9
1862/63	1009.8.2	164.1.6	165.6	98.4.2	60.6.9	129.13
1863/64	823.18.9	182.3.1	160.10.6	124.2.6		12.9.1
1864/65	952.16.1	113.11.4	164.18.0	120.2.5	72.8.6	142.5.6
1865/66	887.2.0	257.18.9	162.0.7	142.13.6	94.12.8	170.19.5
1866/67	906.7.7	214.4.7	196.15	159.16	94.4.9	329.1.9
1868	1065.13.2	280.6.3	243.3.0	144.1.11	85.7.6	198.13.6
1869	1176.19.4	250.14.3	224.6.0	154.19.1	120.2.10	152.0.1
1870	1073.6.5	444.18.1	236.17.8	125.1.1	80.4	209.9
1871	1108.16.11	257.0.5	230.5.6	133.16.7	11.6.9	166.6.4
1872	1050.3.5	150.17.6	260.15.7	128.5.7	86.2.7	154.1.3
1873	1168.19.5	399.19.12	252.13.6	146.0.6	83.17.7	141.16.3
1874	1083.6.9	275.15.7	265.18	153.0.6	80	150.11.6
1875	1080.2.6	185.15.4	242.19.7	142.0.6	70.4.6	205.7.4
1876	1171.6.3	429.11.10	241.9	177.18.9	111.18.11	246.17.10
1877	1103.8.4	1008.18.2	215.8.11	193.8.3	92.19.10	351.4.0
1878	1173.13.4	475.10.5	290.17.3	181.1.7	85.11.6	242.7.7
1879	1057.2.9	681.6.3	251.14.3	201.4.10	140.14.4	116.18.7
1880	1137.8.11	319.19.4	254.11.6	292.0.2	159.6.1	141.17.10
1881	1159.6.5	388.8.5	260.10.11	256.10.11	70.3.6	95.9.3

1839 CD had saved and invested a little, so that his total income was £1,244.

1848 His father left him probably more than £40,000 on his death.

1859-81 his books brought in £10,248, an average of about £465 per annum.

1845- His farm, at Beesby, Lincolnshire, was bought in 1845 for £13,592 borrowed from his father at interest of £461 16s 10d, about 3 per cent. At that time the rent was £377, but by 1877 it had increased to £555 16s.

In 1854 CD's total income was £4,603. By 1871 it had risen to around £8,000, and it continued at this level until his death. See Worsley, The Darwin farms, 2017.

His investments, which were partly looked after by his banker son William, were largely in railways and government bonds. CD's investment portfolio in 1865 is given below (from Atkins 1976, p. 96): The Great Northern Railway, The Lancashire and York Railway, Maryport and Carlisle Railway, The Monmouth Railway, The Great North of Scotland Railway, Penarth Railway, The Mid Kent Railway, The Lancashire and Carlisle Railway, The North Eastern Railway, The Massachusetts Railroad, The Pennsylvania Railroad, The Shropshire Union Canal, The East & West India Docks, The London Docks, Etruria (the Wedgwood factory), The Beesby Estate in Lincolnshire, Interest on mortgage to Major Owen, and a loan to Mrs Ransome. Totalling £5483 per annum.

1873 He was able to save a considerable sum each year, the highest being £4,819 in 1873. One of CD's banks was Union Bank of London. Sotheby's, 1979 Jun. 18, lot 467, a £50 cheque to Sydney Sales. Another £100 "to self" (21 Mar. 1871) was discovered inside a picture frame at Christ's College in 2006. Reproduced in *Darwin Online* and van Wyhe, A Darwin manuscript at Christ's College, *Christ's College Mag.*, (2007) no. 232, pp. 66-8.

1875-81 CD's accounts with (from) his publisher John Murray recording sales of books for 1875-80, are in DAR210.14. Accounts for 1878-79 in DAR139.1.1 and DAR210.11.15. Accounts for 1878-80 are in DAR210.11.18. Accounts for 1881 are in DAR139.1.1.

Annual Income and Surplus 1854-80 (Atkins 1976, p. 97):

Year	Surplus for Investment	Income	Year	Surplus for Investment	Income
1854	2127	4603	1868	2157	6779
1855	2270	4267	1869	2274	6839
1856	2250	4048	1870	2031	7186
1857	2173	4187	1871	2558	7897
1858	1174	4963	1872	2402	8107
1859	1378	5187	1873	4819	8475
1860	2734	5079	1874	790	8708
1861	2073	5797	1875	4658	8530
1862	1981	4743	1876	3053	8259
1863	1249	5847	1877	1543	7916
1864	2196	5553	1878	3157	7997
1865	1952	5828	1879	802	7650
1866	1418	5483	1880	499	7937
1867	1897	5737			

1881 investment portfolio: North Eastern Railway, London and North Western Railway, London and South Western Railway, South Eastern Railway, Great Northern Railway, North Eastern Railway, Leeds and York Railway, Great North of Scotland Railway, Pennsylvania Railroad, Albany and Boston Railroad, New South Wales Bonds, Metropolitan Consols, Consols, Holborn Viaduct Co., East and West India Docks, St. Katharine's Docks, United States Funded Loan, Penarth Harbour Co., Blackfriars Bridge Stock, Leeds Corporation Stock, The Beesby Estate and the interest on the mortgage to Major Owen. Totalling about £8000 per annum.

Rates and taxes were always small: in the 1860s a little over £60 per annum, in the 1870s over £70. His highest income tax was £52 in 1872.

1881 On the death of his brother Erasmus, CD inherited half of his fortune, perhaps the £9,354 19s 6d. shown as extraordinary receipts in his summary of income for 1881. In that year CD had an income of £17,299 1s 4d., a bank balance of £2,968 and £165 19s 4d. in hand. His expenses were £4,880 16s 6d.; he invested £10,218 6s 6d. and gave £3,000 to each of his children. William wrote to his father 1881 Sept. 8 that the total estate was about £282,000 and that, calculated at 7 to 12, his daughters would inherit about £34,000 and sons £53,000. See also Down House, household expenditure.

1882 Apr. At time of his death his worth was calculated on his last will and testament to be £146,911 7s 10d.: CGPLA Eng. & Wales, 1882.

Grave: 1882 Westminster Abbey, "north-east corner of the nave next to that of Sir John Herschel", 7ft deep in a coffin of white oak. *The Times*, Apr. 27 1882. "A few feet from the grave monument of Sir Isaac Newton". LL3:361. The pale Carrara marble slab reads "CHARLES ROBERT DARWIN / BORN 12 February 1809 / DIED 19 April 1882". 1887 Memorial wall plaque by Sir Joseph Boehm.

Habits: The most detailed account of CD's day-to-day pattern of life is in Francis Darwin's reminiscences of his father. LL1:108-160. This stems from his middle and later years when he had developed a rigid pattern, seldom changed even when there were visitors in the house. His own autobiography tells little about his habits, except something of his hobbies and enthusiasms. A typical day for CD at Down House may be summarized as follows:

7am Rose and took a short walk. 7.45am Breakfast alone.

8-9.30am Worked in his study; he considered this his best working time.

9.30-10.30am To drawing-room and read letters, followed by reading aloud of family letters.

10.30-12 or 12.15pm Returned to study, which he considered the end of his working day.

12 noon Walk, starting with visit to greenhouse, then round the sandwalk, the number of times depending on his health, usually alone or with a dog.

12.45pm Lunch with whole family, which was his main meal of the day. After lunch read *The Times* and answered his letters.

3pm Rested in his bedroom on the sofa and smoked a cigarette, listened to a novel or other light literature read by ED.

4pm Walked, usually round sandwalk, sometimes farther afield and sometimes in company.

4.30-5.30pm Worked in study, clearing up matters of the day.

6pm Rested again in bedroom with ED reading aloud.

7.30pm Light high tea while the family dined. In late years never stayed in the dining room with the men, but retired to the drawing-room with the ladies. If no guests were present, he played two games of backgammon with ED, usually followed by reading to himself, then ED played the piano, followed by reading aloud.

10pm Left the drawing-room and usually was in bed by 10.30, but slept badly.

Even when guests were present, half an hour of conversation at a time was all that he could stand, because it exhausted him and he would feel sick later.

Alcohol: Francis Darwin records that CD "drank very little wine, but enjoyed and was revived by the little he did drink". LL1:118. However, CD told him that "he had once drunk too much at Cambridge" as his enthusiastic membership of the Gourmet Club perhaps indicates. "Darwin had once told him [Hooker] that he had got drunk three times in early life, and thought intoxication the greatest of all pleasures". M.E. Grant Duff, *Victorian vintage*, 1930, p. 144. CD's accounts show a considerable consumption of brandy and of beer at Down House, but the former was probably for guests and the latter for growing sons and the staff. The alcoholic beverages in Down House at the time of CD's death are detailed in Lea, *Down House...Notes on the inventory made following the death of Charles Darwin...*1996, p. 22: "Champagne quarts 14, [Champagne] pints 15, Best Sherry 33, Port 33, [Port] pints 42, [Port] from London 25, Cheap [Port] 11, Burgundy quarts 22, [Burgundy] pints 19, Best Claret 22, Claret from London 20, Common Claret 6, [Common Claret] pints 11, Hock [=German white wine] quarts 35, [Hock] pints 12, Marsala [wine] 61, Quarter cask Sherry Say 10 gallons".

Hobbies and pastimes: As a boy he enjoyed fishing. He was not good at ball games such as cricket, although he records that he enjoyed bat-fives whilst at Shrewsbury School. His beetle collecting whilst at Cambridge seems to have been little more than collecting. Not uncommonly, CD not only tried to collect as many species as possible but also multiple specimens of a single species and arranged them to show intraspecific variation. He played Van John (Vingt-et-un) at Cambridge a lot, but does not seem to have played cards later. In his youth, he was an enthusiastic shot, especially when visiting Maer and the Owens at nearby Woodhouse. He shot for the pot and for scientific need during the *Beagle* voyage, but during the voyage this became largely the responsibility of Covington and CD gave it up entirely on his return. He rode for pleasure in his youth and as the only way of covering ground on inland trips from the *Beagle*. He took up riding again in 1866 for health reasons on his quiet cob Tommy, on the recommendation of Dr Bence Jones, but rode less frequently after he had been rolled on in 1869 Apr. 9. 1858 he purchased a billiard table like the one he had enjoyed at Moor Park. He enjoyed watching his family play lawn tennis. His evening recreation, other than reading, being read to and listening to ED play the piano, was backgammon. He and ED played two games every evening when they were alone. He won most games, she most gammons. 1876 Jan. 28 CD to Gray records 5285 games played; ED won 2490 games and CD "hurrah, hurrah" 2795. ED2:221, CCD24:32.

Diet: A recipe book compiled by ED is in DAR214, transcribed in *Darwin Online*. One recipe "To Boil Rice" is in the handwriting of CD. More detail in J. Browne, *Power of place*, 2002, and Mike Dixon & Gregory Radick, *Darwin in Ilkley*, 2009.

Tobacco: CD started taking snuff when he was a student at Edinburgh and continued to do so, finding it a stimulant. See LL1:122. He smoked a few cigarettes when travelling with gauchos in South America, and restarted late in life when he was relaxing. The mock coat of arms by fellow undergraduate Albert Way from 1828 depicts crossed tobacco pipes, meerschaum pipes and cigars. (DAR204.30). See van Wyhe, *Darwin in Cambridge*, 2014. In 1849 at Malvern CD wrote to Henslow complaining that Dr Gully was a "cruel wretch" who "has made me leave off snuff" and calling it a "chief solace of life" CCD4:235. CD reminisced after the death of Anne how she "used sometimes to come running downstairs with a stolen pinch of snuff for me". Two of his snuff jars were returned to Down House in 1929 "one the size of a pound jar, the other equal to a 4lb size; both contain a brown snuff left by Darwin. The larger jar bears the name 'Heynon and Stocken at the Highlander, London.'" *Todmorden & District News*, (2 Aug. 1929). According to Leonard Darwin, CD never smoked a pipe. *The Times* (15 Aug.), p. 11. 1883. CD letter on wine and tobacco. In Reade, *Study and stimulants*. (F1988).

Handwriting: CD's handwriting, even at its best, is notoriously difficult to read especially in later years. Francis Darwin comments of rough notes that they "were almost illegible, sometimes even to himself". LL1:119. "I defy anyone, not familiar with my handwriting & odd arrangements to make out my M.S. till fairly copied" CD to John Murray, 31 Mar. [1859]. CCD7:273. Final manuscript for the press was, for many years, transcribed by the Down schoolmaster, Ebenezer Norman, and long letters were dictated, often to ED and later to Francis Darwin. He was considerate to foreign correspondents, remarking to Francis Darwin "You'd better try to write well, as it's to a foreigner". LL1:119. His formal signature was "Charles Darwin", as in the example given above, but on letters or autographed photographs he usually signed "Ch. Darwin". He seldom used his middle initial. One such exception being: CD 1877. [Letter of thanks] in Harting, Testimonial to Mr. Darwin—evolution in the Netherlands. *Nature* 15 (8 Mar.): pp. 410-12. (*Shorter publications*, F1776).

Charles Darwin's signature from 1876.

Health: A great deal has been written on CD's ill-health, much of it guesswork based on ambiguous descriptions of his symptoms in letters, *ED's diary*, his autobiography and on a few remarks by Francis Darwin in LL1:Ch. 3. See Health diary. The Darwin family Bible (Down House MS) lists many of the children's illnesses. As for CD, no case notes his physicians have ever been found, nor was an autopsy carried out at his death. The number of proposed diagnoses is very great. The following list has been kindly provided by pathologist Dr John Hayman who has himself argued in recent years for a mitochondrial disorder to explain CD's symptoms. See also John Bowlby, *Charles Darwin*, 1990 and R. Colp, *Darwin's illness*, 2008.

1901 neurasthenia.
1903 refractive error.
1920 psychogenic- repressed hostility father.
1920 psychogenic- latent homosexuality.
1929 pyorrhoea.
1943 psychoneurosis.
1958 brucellosis.
1959 depressive psychosis.
1959 Chagas' disease.
1963 malaria.
1963 diaphragmatic hernia.
1965 paroxysmal tachycardia (+).
1965 psychogenic- unresolved grief mother.

1971 arsenic poisoning.
1971 mercury poisoning.
1997 psychogenic- repressed hostility wife.
1989 syphilis.
1990 allergy.
1991 psychogenic- fear of species theory effects.
1987 pyroluria.
1994 adrenal insufficiency.
1997 lupus erythematosus.
1997 panic disorder.
1997 Ménière's disease.
1998 psychological- father-son bonding.
2000 atopic dermatitis.

2002 obsessive compulsive disorder.
2002 excessive flatulence.
2005 lactose intolerance ('systemic').
2005 Asperger's syndrome.
2007 Crohn's disease.
2009 cyclic vomiting.
2009 helicobacter infection.
2012 irritable bowel syndrome.
2013 Candida overload.
2014 mitochondrial disorder.
2016 PTSD (stress disorder).
2018 chronic borreliosis (Lyme disease).

Scholars from the humanities tend to favour psychosocial causes.

It is routinely claimed that CD worried a great deal that having married his first cousin made his children susceptible to ill-health. Despite constant repetition, there is no evidence for this view. In fact, *all* of CD's statements about concern for his children's congenital health explicitly refer to his fear that they might inherit *his own* weak constitution. See his statement in *Cross and self fertilisation*, p. 461:

> With respect to mankind, my son George has endeavoured to discover by a statistical investigation whether the marriages of first cousins are at all injurious, although this is a degree of relationship which would not be objected to in our domestic animals; and he has come to the conclusion from his own researches and those of Dr. Mitchell that the evidence as to any evil thus caused is conflicting, but on the whole points to its being very small. From the facts given in this volume we may infer that with mankind the marriages of nearly related persons, some of whose parents and ancestors had lived under very different conditions, would be much less injurious than that of persons who had always lived in the same place and followed the same habits of life.

Another common claim, also without evidence, is that CD lied about his health in order to avoid supposedly unwelcome invitations and responsibilities. It has to be remembered that CD's ill-health caused him to miss many things he dearly wished to do or attend and it cannot be doubted that his health was often precarious. Therefore one should not conclude that because CD pleaded ill health on an occassion, that it was untrue.

1831 In 1831 Oct.-Dec., just before the *Beagle* sailed "I was also troubled with palpitation and pain about the heart, and like many a young ignorant man, especially one with a smattering of medical knowledge, was convinced that I had heart disease." *Autobiography*, p. 80.

1834 During the voyage, apart from a few minor accidents, some mild fever and continuing and severed sea-sickness, he had only one serious illness. This was at Valparaiso, 1834 Sept. 19 until the end of Oct. He "was there confined to my bed till the end of October". *Journal*, 2^d edn, pp. 268-9. Possibly typhoid. For most of the voyage he was fit and lived an extremely energetic life.

1839-42 During his residence in London, "I did less scientific work", "This was due to frequent recurring unwellness, and to one long serious illness". *Autobiography*, p. 98. Again he gives no symptoms. At Down House, he explained that after entertaining company "my health almost always suffered from the excitement, violent shivering attacks and vomiting being thus brought on". *Autobiography*, p. 115. This condition continued for the rest of his life, although the attacks seem to have been less frequent or less violent in his later years.

1839-82 Records of medicines CD took are in *ED's diary* and the 'Receipts and memoranda book' book of prescriptions at Down House (transcribed in Colp, *To be an invalid*, 1977): aloe, arsenic, bismuth, bitters, blue pill, calomel (a mercury based purgative), carbonate of ammonia, castor oil, chalk, cod liver oil, colchicum, condy's ozonised water, croton, grey powder, iron, lime water, logwood, magnesia, nitre, nitroglycerin, pepsin, podophyllin peltatum, phosphate of iron, potassium bicarbonate and sulphuric acid. ED also recorded his fluctuating weight, pulse, what he ate and what foods disagreed with him.

In 1845-46 he tried electrotherapy, in a letter to J.D. Hooker [5 or 12 Nov. 1845], CD wrote that he was trying a popular therapeutic treatment or galvanisation by passing "a galvanic stream through my insides from a small-plate battery for half an hour". CCD3.

In 1865 CD tried ice therapy on the recommendation of John Chapman. It had no effect.

1873, Aug. 26. CD had a partial loss of memory for twelve hours.

1881, 1882 During Dec. 1881 CD began to suffer anginal pains which became more frequent in Feb.-Mar. 1882. He had a severe attack with fainting on Apr. 18. Francis Darwin records his father's last words, on the 18th, as "I am not the least afraid to die". LL3:358. See Health diary.

Homes: Until 1836 CD's home was his father's house, The Mount, Shrewsbury, until after his return from the *Beagle* voyage in Oct. 1836. He was however away for much of the year whilst an undergraduate student at Edinburgh and Cambridge, and for almost five years when on the *Beagle*. On his return, he stayed in Cambridge with Henslow and then took lodgings at 22 Fitzwilliam Street, and in London with his brother Erasmus at 43 Great Marlborough Street.

1837 Mar. 13, he took furnished rooms at 36 Great Marlborough Street with Syms Covington: this house can perhaps be regarded as his first personal home.

1838-42 After his engagement to ED, he rented a furnished house, 12 Upper Gower Street, into which he moved in 1838, Dec. 31, and where he and his bride took up residence the day after their wedding, 1839, Jan. 30. They lived there until 1842, Sept. 16. The so-called Macaw Cottage.

1842-82 On Sept. 14th ED moved to Down House and CD followed on 17th. There they lived for the rest of their lives, from 1882 ED spent winters in Cambridge.

The following list summarizes CD's homes and dates:

1809 Feb. 12-1837 Mar. 13 The Mount, Shrewsbury. One might also count boarding at Shrewsbury School.

1825 Oct. 22-1827 Apr. 23 11 Lothian Street, Edinburgh, in term time.

1827 The Mount, Shrewsbury.

1828 Jan. 26, Sydney Street, Cambridge.

1828 Oct. 31- Jun. 16 1831 Christ's College, Cambridge, in term time.

1831 Dec. 10-1836 Oct. 2 HMS *Beagle*.

1836-7 22 Fitzwilliam Street, Cambridge.

1837 Mar. 13-1838 Dec. 30 36 Great Marlborough Street, London.

1838 Dec. 31-1842 Sept. 16 12 Upper Gower Street, London.

1842 Sept. 17-1882 Apr. 19 Down House, Down, Kent.

Russian sculptor Vasily Vatagin (1884-1969) preparing his large statue of CD in 1927. Note the sitter on the right and the small maquette on the left. The sitter is zoopsychologist and comparative psychologist Alexander Feodorovich Kohts (1889-1963), the founder of the SDMM in 1907. Photograph courtesy of the State Darwin Museum Moscow (SDMM). The statue is still on display at the SDMM.

Iconography: This iconography is by far the most complete ever made of CD and is probably the largest for any scientific figure in history. The previous lists of likenesses of CD include Freeman 1978 with c.50 items and the ODNB (2004) with 55 entries, although some refer to multiple items in a collection. The present list reveals that there have been at least 210 unique oils, watercolours and drawings, more than 590 printed, at least 50 photographs and over 240 three-dimensional works such as statues and medallions. CD must surely be the most prolifically illustrated scientist in history. Except for the photographs, illustrations are provided only for selected items. These have been chosen for being either particularly important, not previously known in the literature or as representative of a particular type. Portraits taken from life are marked *.

The most comprehensive exhibition of portraits and related material was that at Christ's College, Cambridge in 1909. A similar exhibition, with some of the same material, was held at the BMNH in that autumn. There are printed catalogues of both (reproduced in *Darwin Online*).

Three dimensions: There are many works in three dimensions (240 are recorded here) ranging from full-scale statues to bas-relief busts for small medallions. It is difficult to identify which photograph a three-dimensional work might be based on. However, the two most influential two-dimensional portraits for sculptors have certainly been Richmond 1840 and Collier 1881. One surprising revelation is that there are so many busts and statues of CD in China (45 have been found). Portraits that are illustrated here have their date underlined to help link image with reference.

Woolner 1870. Woolner 1870? von Hildebrand 1873. Moore 1880.

-1866 Medallion after W.E. Darwin 1864a. Henrietta to George Darwin: "[Ernst Haeckel] told us that there are over 200 medallions of Papa made by a man from W^ms photo in circulation amongst the students in Jena' (DAR 245: 269)." CCD14:xviii. Not seen.

1868 Nov. Clay model for the bust below by Thomas Woolner.* Visible in photograph of the artist in his studio in *Thomas Woolner, R.A., sculptor and poet*, 1917, facing p. 325, reproduced in Browne, *Power of place*, 2002, facing p. 313. Possibly that given by Woolner to the Museum and Picture Gallery in Sunderland in 1879. *Sunderland Daily Echo* (18 Dec.), p. 3.

1870 Marble bust by Thomas Woolner.* "CHARLES DARWIN" on base. 1868 Nov. CD sat for at home for several days. CD wrote to Hooker on 24 Nov. "I ... am undergoing the purgatory of sitting for hours to Woolner, who, however, is wonderfully pleasant & lightens, as much as man can, the penance.— As far as I can judge he will make a fine Bust, & I tell my wife she will be proud of her old husband". CCD16:862. Finished in 1870. One critic described how the "eyes lie hidden beneath their overhanging brows...[the bust displays] calm repose and contemplation undisturbed by outward things." *London Evening Standard*, (3 May 1870), p. 6. Francis Darwin commented: "It has a certain air, almost of pomposity, which seems to me foreign to my father's expression". LL3:106. Listed as from 1870 in Woolner's biography. Cast in Herbarium, Sainsbury Laboratory, Department of Plant Sciences, Cambridge. Cast "bought in Germany by the late Mr. Terrero, a grandson of General Rosas. Restored by Charles L. Hartwell". BAAS, *Historical and descriptive catalogue of the Darwin Memorial at Down House*, 1969. A cast also at Shrewsbury School.

<u>1870?</u> White on sage green jasper oval portrait medallion, signed "T. WOOLNER. SC", by Thomas Woolner, produced by Wedgwood. Copies in CD's set at Christ's College, Cambridge (See van Wyhe, *Darwin in Cambridge*, 2014); Wedgwood Museum, Stoke-on-Trent. Copy at American Philosophical Society has "CHARLES DARWIN" just below the bust as part of the ceramic, attributed to A.H Bentley. Another version has a beige rather than green background, as the copy in the Birmingham Museum of Art, UK; a copy in the Victoria and Albert Museum has a blue background.

<u>1873</u> Bust in plaster with "CH. DARWIN" on base by Adolf von Hildebrand. Stazione Zoologica Anton Dohrn, Naples, Italy. Photograph in F. Fehrenbach in Groeben, *Art as autobiography*, 2008, pp. 93-104.

Legros 1881.   Lehr 1883.   Boehm 1883.   Echteler 1883-4.

<u>1880</u> "The Darwin medal" of gold or bronze after L. Darwin 1878a by Joseph Moore (1817-92) for the Midland Union of Natural History Societies. Signed "JOS.F MOORE F:" (See van Wyhe, editorial note in F1993). On verso "the Darwin medal / Awarded to [...] / Founded by the Midland Union of Natural History Societies 1880" with a branch of coral. A bronze copy (EH88202657) and its wax impression are at Down House.

<u>1881</u> Bronze left profile portrait medallion by Alphonse Legros, signed "A.L." with "CHARLES DARWIN 1881" (11.5cm) "wrought from a rough sketch on an envelope at a meeting of the Royal Society. That powerful noble head made a deep impression on [Legros]." T. Okey in Attwood, The medals of Alphonse Legros. *The medal*, no. 5, 1984. "Somewhat after the manner of the Italian medallists of the fifteenth century." *Shields Daily Gazette*, (10 Nov. 1881) Only a few copies of this medallion are known to have been struck. Exemplars in Christ's College Library, Fitzwilliam Museum, Cambridge, Manchester City Gallery and the British Museum.*

1881 Plaster(?) portrait medallion by Alphonse Legros. Copy of the above.

-1882 "a bust by Mrs. Mica Heideman". Mentioned in *Proc. of the Biol. Soc. of Washington*, vol. 1 (1882), p. 43. Whereabouts not traced.

c.1880s Whitby jet brooch with left profile of CD.

1882 A wax figure by Gustave Castan, for Castan's Panopticum (a wax figure museum 1869-1922). "A wax figure of the late Mr Darwin has been added to the Panopticon, a kind of Berlin Madame Tussaud's." *Glasgow Evening Citizen*, (29 Apr. 1882), p. 2 and *Führer durch Castan's Panopticum*. Berlin, c. 1900.

1882 Plaque in bronze by Allan Wyon after Rejlander 1871c? (reversed), (22cm). Signed "Allan Wyon SC". An electrotype of the wax model from which the Royal Society's Darwin Medal was reduced was exhibited at the BMNH in 1909; the die made in 1890. Wax sculpture at Down House, EH88202270.

1883 Bust by Christian Wilhelm Jacob Lehr. "Charles Darwin." on base. Museum für Naturkunde in Berlin. Plaster copies in Keynes Room, CUL. (This copy was displayed in a 2018 exhibition at CUL and described as the 1902 Montford bust, transferred from Down House in 1982); other copies at Botanical Institute of Bucharest and the RKD (Rijksdienst voor Kunsthistorische Documentatie. https://rkd.nl/nl/collecties/explore). A copy was on display at the American Museum of Natural History in 1932. A CDV of the bust was produced by A. Naumann, Leipzig.

1883 Plaster statue, painted to look like bronze, seated, by Sir Joseph Boehm, signed on back "Charles Darwin / E. Boehm fecit 1883". Exhibited at Christ's College in 1909: "Lent by Sir George H. Darwin, K.C.B., F.R.S. Study, about half-size, for the statue in the British Museum...Purchased at Sir Edgar Boehm's death by the Countess of Derby, and given by her daughter, the Lady Margaret Cecil, to Sir George H. Darwin, K.C.B." Now on display on the first floor landing at Down House, EH88202495.

1883 White marble statue, seated, by Sir Joseph Boehm, with "CHARLES DARWIN. J. E. BOEHM, FECIT" on base, (statue 191x86, plinth 22x91cm). 1885 Jun. 9 unveiled by T.H. Huxley in presence of Prince of Wales; Admiral Sulivan and Parslow were also present. *ED's diary* "Unveiling at S K" [South Kensington]. The event illustrated in *The Graphic* (20 Jun. 1885), p. 621. Boehm was paid £2,100 for it. In 1927 it was replaced in that location with an elephant and then with the 1896 bronze statue of museum founder Sir Richard Owen by Sir Thomas Brock. In 1971 the Boehm statue was transferred to a low plinth and moved to the North Hall. *Royalist and realist: The life and work of Sir Joseph Edgar Boehm*, 1988, p. 414. In May 2008 the statue was moved back to the landing of the Central Hall in preparation for the CD bicentenary of 2009. 1883? A plaster copy was in the library of the Cambridge Philosophical Society, later in the Department of Physiology, Cambridge, destroyed by a flood in the mid-20th century. "The Darwin family have presented the University the cast of the model executed by the late Sir J. E. Boehm, R.A. for his statue of Mr. Charles Darwin in the British Museum, and it is placed in the Lecture-room of Comparative Anatomy." *East Anglian Daily Times*, (4 Nov. 1891).

1883 Bust in terracotta by Sir Joseph Boehm (61cm). National Portrait Gallery, UK.

1883-4 Plaster bust painted to imitate terracotta, "CHARLES DARWIN" on base by Joseph Anton Echteler (1853-1908). Given by artist to Metropolitan Museum of Art, USA 1887.

1884 Painted plaster bust by Robert Stark (1853-1931), Torquay sculptor and painter. Marked underneath "R. Stark 84" and "R. Stark Sculptor". Torquay Museum.

1887? Life-size maquette for the below by Joseph Boehm. Of the family home, Nash Mills House, Dame Joan Evans wrote: "The long passage that led from the side-door to the garden entrance was light and airy, but rather encumbered with a...life-size maquette of Sir Edgar Boehm's great roundel of Darwin made for his tomb in Westminster Abbey." *Prelude and fugue: an autobiography*, 1964, p. 16.

<u>1887</u> Deep bronze wall medallion bust with "DARWIN" underneath by Joseph Boehm, near CD's grave in Westminster Abbey. Boehm was paid £150 for it.

1888 Bust in centre façade by Léopold Noppius. Aquarium et musée de zoologie, Liège, Belgium.

c.1888 Bust in stone by Victor de Pol(?). One of several busts depicting the founders of modern science on the façade of the Natural Sciences Museum in La Plata, Argentina.

1889 Large bust in stone by Léopold Nopius in the centre of the central pediment of the façade of the Édouard Van Beneden Institute of Zoology, Liège, Belgium.

1889 Stone bas-relief bust by Josef Beyer, Burgring, right façade, Natural History Museum, Vienna, Austria.

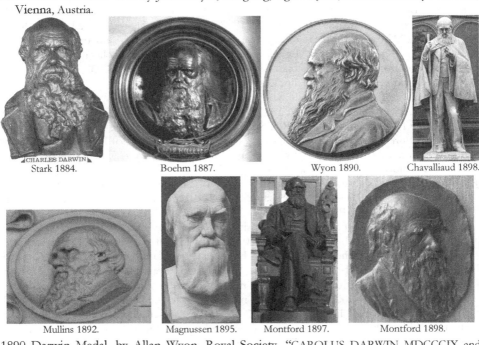

Stark 1884.   Boehm 1887.   Wyon 1890.   Chavalliaud 1898.

Mullins 1892.   Magnussen 1895.   Montford 1897.   Montford 1898.

<u>1890</u> Darwin Medal, by Allan Wyon, Royal Society. "CAROLUS DARWIN MDCCCIX and MDCCCLXXXII" on verso. Wallace was first recipient. See 1882 plaque by Wyon above.

-1892 Bas-relief bust by John Rhind (1828-92), Royal Scottish Museum.

<u>1892</u> Stone roundel bas-relief portrait after L. Darwin 1878a (reversed) by Edwin Roscoe Mullins(?) Façade of Croydon Public Library, Katharine Street.

1894 Marble bust by Jane Nye Hammond (66cm tall). Providence Athenaeum, Rhode Island.

<u>1895</u> Bust by German sculptor Harro Magnussen (1861-1908).

1897 "Bronze miniature of statue by Horace Montford, outside the Museum, Shrewsbury." Back of the base is signed "H Montford". BAAS, *Historical and descriptive catalogue of the Darwin Memorial at Down House*, 1969. EH88202496. Copies were for sale in 1909. *Report of British Association*, 1938. A copy is on display at the Rare Book Room, Mertz Library, New York Botanical Garden.

1897 "Sculptor's model of above." by Horace Montford. Down House EH88204496.

1897 Bronze statue, seated with manuscripts on lap, by Horace Montford, outside Old School, Shrewsbury (then the Free Library). "1809 1882" on base of statue. "DARWIN" on pedestal. Broad & Son. Founders London. "Erected by the Shropshire Horticultural Society 1897" which paid £1,086. The 137x142cm moulded and polished emerald granite pedestal was made by Messers Henderson & Webster, Aberdeen, from a single 6 tonne block. There is a life-size plaster cast of this statue. "Those who knew the late Charles Darwin…speak in warm terms of praise of the likeness". *Liverpool Mercury*, (6 Apr.), p. 5.

1897 Small white round medallion in plaster(?) by Horace Montford, framed at Shrewsbury school library. Another copy at Down House, BAAS, *Historical and descriptive catalogue of the Darwin Memorial at Down House*, 1969. EH88204610.

1897 Silver-gilt medallion with CD in right profile (vaguely after Rejlander 1871c) by Elkington, 39mm. Manchester & North of England Orchid Society.

1898 Marble statue by Léon-Joseph Chavalliaud, after Collier 1881, contemplating a flower bud. Palm House, Sefton Park, Liverpool. On base: "CHARLES DARWIN Born of Shrewsbury 1809 buried in Westminster Abbey 1882 naturae minister et interpres". Latin from Bacon's *Novum Organum* (1620), meaning servant and interpreter of nature.

| Montford 1898? | Montford 1898. | Lambeaux 1898. | Hope-Pinker 1899. |

1898? Bust in terracotta by Horace Montford (71x57x35cm). *London for the literary pilgrim*, 1949, p. 76, NPG, copy of the above.

1898 Marble bust of CD in old age by Horace Montford (69.1cm). On a socle. Presented to Royal Society in 1959. Royal Society, 6-9 Carlton House Terrace, London.

1898? Plaster copy of statuette in bronze by Montford. Copy was in UCL, Zoology 1982-?

c.1898 Medallion in bronze by Horace Montford (46.9x38.4cm). Copy on display in the Great Gate of Christ's College, Cambridge. Another copy was at Down House in the 1990s.

1898 Bronze bust together with full-figure female nude by Jef Lambeaux (1852-1908), Antwerp Zoo, Belgium, a gift of English painter and sculptor William Fiott Barber.

1899 Maquette for the below by Henry R. Hope-Pinker. At Down House. Presented by E.B. Poulton. Down House EH88204496.

1899 Statue in stone after Collier 1881, "CHARLES DARWIN" on base, by Henry R. Hope-Pinker, at University Museum, Oxford. Presented by E.B. Poulton. Unveiled 14 Jun. 1899. See: Anon. Unveiling the Darwin statue at the museum. *Jackson's Oxford Journal* (17 Jun. 1899), p. 8 with unique recollections of CD. (In *Darwin Online*, F2169).

late 19th century. Bust of pine on a socle inscribed "Charles Darwin" on the base (67.3cm high). Sold at Christie's in 2000 for £3,680.

c.1900 Porcelain bust by Robinson & Leadbeater (20cm). Museum Applied Arts & Sciences.

18?? Terracotta bust of CD (13.2cm), Rex Nan Kivell Collection, Australia.

late 19th century? Small bronze desktop shell (for paperclips?) with CD from shoulders up, left arm extended contemplating an ivory(?) human skull with quill pen in right hand.

1902 Two busts in bronze by Horace Montford commissioned for Andrew Carnegie "based on the marble bust made by the same sculptor for the late Lord Farrer. One of the busts will be placed in the [Pittsburgh] Museum, and the other will go to Skibo Castle." *Knowledge*, vol. 25, p. 15, with photograph. 1909: a bronze bust of CD was presented to Kew Gardens by Montford. *Kew Bulletin*, vol. 23, p. 315. A bust is still at Kew.

1902? Plaster cast of above by Horace Montford. On display at Christ's College, 1909. Copies at the Grant Museum of Zoology and another at Down House EH88202848. 2014 Resin copy on display at the Darwin Building, University College London.

Hampton 1906. Pegram 1907. Battle Library 1908. Bowcher 1908.

Bowcher 1908. Couper 1909. Goetz 1909.

1905 Medallion full face portrait "C. Roberto Darwin", one of a series "edition artistique" including Victor Hugo. Photograph on a French postcard by "E.A.T.M. 526".

c.1906 Marble bust by William J.S. Webber (c.1843-1919), Mercer Art Gallery, Harrogate, UK.

1906 CD as standing figure in bronze relief panel with thirteen other men of science by Herbert Hampton (1888-1927). West Frieze Queen Victoria Memorial, Lancaster, UK.

1907 Statue in stone after Collier 1881 by Henry Pegram. Frieze in the main façade of the Great Hall, University of Birmingham, UK.

1908 Stone bas-relief bust, façade of The Battle Library, originally West End Public Library. Reading, Berkshire, UK.

1908 Darwin-Wallace Medal of the Linnean Society of London, once in gold (for Wallace), otherwise in silver after Rejlander 1871c? (reversed) by Frank Bowcher. CD on recto with "LINN: SOC: LOND: 1858-1958 DARWIN" and "LINN: SOC: LOND: 1858-1958 WALLACE" on verso.

1908 Bronze medallion, "DARWIN" signed "F. Bowcher F 1908" (55cm), at Linnean Society and V&A. Plaster casts at NHM and NPG. Portrait only almost identical to Darwin-Wallace Medal of the Linnean Society of London.

1908 Plaster bust by William L. Couper (1853-1942). American Museum of Natural History.

1908 Bust in bronze, twice life size, by William Couper, inscribed on pedestal, "DARWIN" signed on the cut-off left "Wm Couper / N.Y." and "GORHAM Co. FOUNDERS". Couper was paid $1000 for it. Unveiled at the American Museum of Natural History (New York) for the opening of its Darwin Hall of Invertebrate Zoology on 12 Feb. 1909. Now at the New York Academy of Sciences, USA. Copy by Couper at Christ's College, Cambridge, presented by USA delegates to June 1909 centenary celebrations. See van Wyhe, *Darwin in Cambridge*, 2014. Two plaster casts were originally made, J.W. Goodison, *Catalogue of the portraits in Christ's, Clare and Sidney Sussex Colleges*, 1985, p. 11. Probably the most widely imitated or reproduced bust of CD.

-1909 Medallion in bronze by William Rothenstein, shown at Christ's College, Cambridge anniversary exhibition, 1909.

1909 Bronze medallion portrait with a monkey contemplating a human skull on verso by Karl Goetz, 36mm. c.2018 Re-struck in 2oz silver by Piedfort.

1909 Medals. Darwin, Charles Robert * 1809 Shrewsbury, +1882 London, English naturalist.

1909 One-sided bronze badge roughly after Collier 1881 by Mayer and Wilhelm, Stuttgart (5.7x3.84cm).

1909 Soviet Union struck medal to commemorate *Origin* centenary(?).

1909 Circular bronze plaque by Franz Kovnitzky. Science Museum, London.

1909 Marble tablet with bronze bas-relief profile portrait and inscription: "A Darwin Los escolares médicos en el primer centenario de su nacimiento 1909". Historical-Medical Museum of the Faculty of Medicine of the University of Valencia, Spain.

1918 Bust of CD in old age by Vasily Vatagin. (Василий Алексеевич Ватагин, also spelled in the Roman alphabet Watagin) The first sculpture of CD in Russia. SDMM.

-1921 Porcelain bust "gift of Thomas and Martha Lennard, 1921". Museum of Applied Arts & Sciences, Australia.

1923 Bronze bas-relief figure, part of "The Progress of Science" by Lee Oskar Lawrie (1877-1963). National Academy of Sciences, Washington, D.C., USA.

1924 Bas-relief figure, seated left profile, contemplating an ape skull on his lap by Bertram Goodhue. Bronze door panel of The National Academy of Sciences, Washington, D.C.

c.1924 Bust in stone on black granite pedestal, inscribed "Charles Darwin 1809-1882". Sun Yat-sen School of Life Sciences, Guangzhou, China. (中山大学生命科学院, 广州市).

c.1924 Terra cotta niche figure by Allan Clark (c.152.4x45.7cm). Suzzallo Library, University of Washington, Seattle, Washington.

1927-8 Larger than life seated gypsum statue by Vatagin. SDMM. (See photograph above.)

1928 CD one of four plaster medallions in curved niche behind colonnade (Richard Wagner, Karl Marx, CD, and William Morris) Bethnal Green Library, London, UK.

1929 Bronze bust by Charles Leonard Hartwell (1873-1951). Commissioned by Joseph Leidy. Inscription: "Presented by Dr. Joseph Leidy II of Philadelphia, to the British Nation in memory of those American naturalists who came to the support of Charles Darwin upon the publication of 'The origin of species' in 1859". 1929 transferred to Down House from the Royal Academy where it had been exhibited. The Japanese government ordered a copy in 1929. Photograph of face of bust in Michael Neve, Charles Darwin: Down House, Downe, Kent. In K. Marsh ed. *Writers and their houses*, 1993, p. 153.

c.1930s Granite statue by Gustav Vigeland (1869-1943), Vigeland Park, Oslo, Norway.

1930 Stone bust corbel, Convocation Hall, McMaster University, Canada.

<u>1933</u> Stone bas-relief bust "CH·R·DARWIN/1809—1882", by Szilard Konstantin Szody (1878-1939), signed "SZODY", (50x80cm). Façade of the former Hungarian Society of Natural Sciences, Budapest.

<u>1935</u> Galapagos centennial medallion (Victor von Hagen). On verso the *Beagle* and "The Galapagos Centennial / "The voyage of the Beagle has been the most important event in my life and determined my whole career." Copy at Down House, EH88202658.

National academy of sciences 1924.

Suzzallo Library 1924.

McMaster 1930.

Hartwell 1929.

Szody 1933.

Galapagos medal 1935.

Bezpalov 1935.

<u>1935</u> Bust in bronze on red granite pedestal with signatures on the bust section from the back: "I. Bezpalov / V - 35 g .; on the shoulder cut to the right: Ex. Miglinnik." By Innokenty Fedorovich Bezpalov, Institute of Experimental Medicine, St. Petersburg, Russia.

c.1935 Bust in bronze, mounted on a column, by Innokenty Fedorovich Bezpalov. In front of the Old Laboratory, Pavlov village, Vsevolozhsk District, Leningrad Region, Russia.

1935 Concrete bust, based on a mould from a plaster cast of the Couper bust in New York by E. Gargani & Sons in New York, under direction of Victor Wolfgang von Hagen by Luis Mideros. Ecuadorian Naval Base, Puerto Baquerizo Moreno, San Cristobal Island, Galapagos. A copy, also arranged by von Hagen, at General Library of the Central University, Quito, Ecuador. See John Woram, Portraits in the Round: Busts of Charles Darwin. http://galapagos.to/TEXTS/COUPER.HTM and John Woram, *Charles Darwin slept here.*

1935 Bust by Alfredo Palacio, related to the von Hagen campaign, but not a copy of the Couper bust. University of Guayaquil campus at Calle Chile & Chiriboga, Ecuador.

1935 Bas-relief portrait of beardless CD on bronze plaque for bust at Puerto Baquerizo Moreno. Signed by Luis Mideros, Director of la Escuela de Bella Artes in Quito who oversaw the creation of the concrete Couper bust copy in 1935.

1939 Plaster bust of Young CD by Vatagin after Richmond 1840. Perhaps the finest bust of a young CD. SDMM OF7667/1776. Copy in NHM. A handwritten date on a photograph from SDMM of this in CUL dates it to 1917.

1939 Plaster bust of middle aged beardless CD by Vatagin. SDMM OF7667/1775. 1959 a copy of this and a matching bust of Wallace were made and sent to the BMNH.

1939 Plaster bust of older CD by Vatagin. SDMM OF7667/1777. Copy in NHM.

1941 Bronze façade monument with bust, Russian State Library, Moscow, Russia.

1947 Bust of young CD by D. Gofman after Maull & Polyblank 1857. SDMM OF7667/1790.

1951 Sculpture by Keith Godwin (1916-91) exhibited at the Festival of Britain, London, South Bank Exhibition, 1951 in the section "The Living World". *Guide to the exhibition of architecture, town-planning and building research*, p. 111. Not seen.

| Vatagin 1939. | Vatagin 1939. | Vatagin 1939. | Gofman 1947. |

1951 Stone bas-relief sculpture of CD with skull by Duarte Angélico at the Library of the University of Coimbra, Portugal.

1954 Terracotta bust of CD with skull by Sergey Timofeyevich Konenkov (1874-1971) (Сергей Тимофеевич Коненков). SDMM NVF9140.

1955 Bust by Tatyana Kirpichnikova, signed and dated. Zoological Museum of the Zoological Institute of the Russian Academy of Sciences, St. Petersburg, Russia.

nd Plaster sculpture. National Museum of American History, Smithsonian Institution.

nd Plaster bust on a socle in the Copenhagen Zoological Museum, Denmark. "C. DARWIN" on base. Purchased commercially, there is no information on its date or maker.

1958 Bust in plaster by Vatagin. SDMM. Copy presented to the BMNH. Described as "Darwin in middle age, head and chest, facing forward, wearing coat, stock and necktie." John C. Thackray, *A catalogue of portraits, paintings and sculpture at the Natural History Museum, London*, 1995. The copy is still at the NHM but not on display.

1958 Plaster bust of young CD by Victor Evstafiev (1916-89) [sometimes spelled Eustaphieff]. After Richmond 1840. A pair with a bust of ED. SDMM NVF1264/1192. 1960 copy made and sent to Down House, EH88204646.

1959 Rectangular left profile portrait medallion after Maull & Polyblank 1857, signed "M" with "CHARLES DARWIN / ON THE ORIGIN OF SPECIES – 1859" on recto and "DEUTSCHE AKADAMIE DER NATURFORSCHER LEOPOLDINA FÜR BESONDERE VERDIENSTE UM EVOLUTIONSFORSCHUNG UND GENETIC 1959" on verso.

<u>1959</u> Plaster bust of older CD by Evstafiev. SDMM OF7667/1878. 1960 copy sent to Down House, not now known in that collection. Cast on display at Darwin College, Cambridge.

1959? Bronze(?) three-quarter length statue after Collier 1881. Museum of Natural Sciences Budapest. Depicted on a contemporary postcard of the "Darwin memorial".

c.1959 Statuette of seated CD, gift of the SDMM to Down House. Visible in a photograph of the 'Russian room' at Down House, early 1960s in P. Simpson, A Cold War curiosity?: The Soviet collection at the Darwin memorial museum... *Jrnl. of the Hist. of Collections*, vol. 30, issue 3, (Nov. 2018): 487-509. Not now found in Down House holdings.

<u>1959</u> Portrait medallion by G.S. Shklovsky after Rejlander 1871c. Leningrad Mint for the Soviet Academy of Sciences, (6.5cm) 1,261 copies. The *Beagle* on verso.

1961 White on sage green jasper portrait medallion, without border, Wedgwood, (15.8x12.4cm). Not by Woolner. A copy, pale blue jasper with laurel frame (11.2x8.5cm) was produced in 1987. Many variations have been produced.

1964? Bronze bust by Senta Baldamus (1920-2001) in Tierpark, Berlin, Germany.

1965? Bust (concrete?) at Escuela Carlos Darwin, Progreso, San Cristobal, Galapagos.

1972 Bronze medallion with bearded CD bas-relief bust, 5 insect outlines on verso by Christiane Cochet. 2000 donated anonymously to American Psychological Association.

Evstafiev 1959.   Leopoldina 1959.   Shklovsky 1959.   Copenhagen nd.

1973 Plaster bust by Helmut Karl Wimmer based on but not a cast of the Couper bust. On display at the American Museum of Natural History. The name on the pedestal was changed to "Ch. Darwin" in a facsimile of CD's handwriting.

c.1974 Silver medallion by John Pinches "Charles Darwin voyage of discovery...amassed scientific data which led to publication of 'Origin of Species'". Mountbatten medallic history of Great Britain & the sea.

nd Bust formerly on display at the Evolution Gallery at the NHM. A plaster or plastic copy of a bust not otherwise seen. The NHM was unable to trace it in 2019.

nd Miniature bronze bust. Copy in the library of the Linnean Society of London "presented by S. Baldwin F.L.S."

<u>1980?</u> Resin statue of seated and reading CD attributed to G.J. High. On display (2020) in the library of the NHM (formerly in store).

1982 Bronze medallion, 55mm by 73mm by Nicola d'Alton Moss. Produced by the Medallic Art Company of Danbury, Connecticut, 2,500 pieces.

1982 Three-quarter right profile bust relief loosely after Maull & Polyblank 1857. "The Coalbrookdale Company Remembers in 1982 / Charles Darwin A Shropshire Lad 1809-1882" cast iron commemorative medallion (18.5cm).

1982 Bronze medallion loosely after L. Darwin 1878a by Jose de Moura.

1985 Medallion by I.A. Daragan. Issued by the Leningrad Mint for the 175th anniversary of the birth of CD (6cm) 850 copies.

1986 Bronze bust by Frank Diettrich. Botanischer Garten, Gera, Germany.

1988? Painted statue of seated and bearded CD in cloak. Intendencia Parque Nacional Los Glaciares, El Calafate, Argentina.

1988? Painted statue of beardless CD with (mock) *Beagle* anchor. Intendencia Parque Nacional Los Glaciares, El Calafate, Argentina.

1989 Statue in stone with curved walking stick by Rhyl Hinwood. Left of entrance to Goddard Building, Great Court, University of Queensland, St. Lucia, Brisbane, Australia.

1989 Bronze Medal. Barely recognizable abstract CD portrait, verso a giant tortoise. The Society of Medallists 119th Issue.

1989 Seashell-shaped medal that features CD and a Galapagos tortoise from the Society of Medallists. Designed by Nicola Moss. Mintage was limited to 2,500.

1992? Porcelain medallion (3.9cm) "150 Jahre Orden Pour le Mérite", by Meissen, Germany.

c.1993 Bust at the Daejeon Expo, Daejeon, South Korea.

after 1993 Bust in bronze on red plinth, by Xikun Yuan (袁熙坤) inscribed in Chinese, "The discoverer of the Theory of Evolution, Charles Robert Darwin, English naturalist". Beijing International Sculpture Park, China. (北京国际雕塑公园西园).

High 1980?   Daragan 1985.   Diettrich 1986.   Wouda 1998.

1997 Bust in stone(?) "One side of the head is Charles Darwin, the other side is God as depicted in the Sistine Chapel" by Mel Chin. The bust has been nicknamed "Godwin". University of Georgia, Athens, USA.

1998 Bronze statue of young CD in top hat riding a Galapagos tortoise by Marjan Wouda. Garden of Heroes and Villains, Dorsington, UK.

-2000 Bronze bust. Charles Darwin Research Station, Santa Cruz Island, Galapagos.

2000 Silver medallion roughly after Collier 1881, head only. The Millennia Collection Inc.

2000 Statue in stone, half body, holding an open book by Jukui Lai (赖聚奎). Shenzhen West Waterland Park, Guangdong, China. (深圳西部海上田园).

2000 Bronze maquette for below by Jemma Pearson, sold at Sotheby's in 2017 for £6,875.

2000 Bronze statue of young CD on the Galapagos by Jemma Pearson, unveiled by Sir David Attenborough, in front of Shrewsbury School's main building. Shrewsbury.

2000 Bronze miniature sculpture of above by Jemma Pearson (38x20cm). Edition of 40. A copy at the Royal Society of London, UK.

2000 Bust in bronze, after Couper 1909? Anhui Chinese Sculpture, China. (安徽/华派雕塑).

c.2000 Statue in bronze, life-size, seated and resting right elbow on rock with open book, with bird perched on left hand. Shanghai Oriental Land, China.

2001 Carved wooden bas-relief portrait, Bristol Zoo, UK.

2001 Medallion plaque with seated bas-relief figure. "Darwin Gardens Trust..." Darwin Gardens Millennium Green, Ilkley, Yorkshire, UK.

2002 Engraved slate plaque after Maull & Polyblank 1857 by Lynne Davies and unveiled by the Countryside Council for Wales. Photograph here is an extract.

2002 Portrait plaque after Richmond 1840 by Vincent Butler on rear outer wall of the National Museum of Scotland, South College Street, Edinburgh: "On this site Charles Darwin (1809-1882) author of *The Origin of Species* lodged at 11 Lothian Street whilst studying medicine at the University of Edinburgh 1825-1827."

2004 Statue in bronze, seated, with left hand on a globe, head of an ape on the right by Qiguangqi Advertising Communication (GZ) Co., Ltd. South China Botanical Garden, Guangzhou, China. (中国科学院华南植物园, 广洲市).

Pearson 2000.   Davies 2002.   South China 2004.   Baquerizo M 2009.

Ramos -2008.   Hasell 2009.   Smith 2009.   Smith 2009.   Eduardo 2010.

-2005 Metal sign with outline portrait. Zhangjiang Hi Tech Park, China.

-2006 Wax figure. Madame Tussaud's San Francisco, California, USA.

2006 Bronze statue, seated at a table with outstretched left hand by Miguel Barranco. Science Park, Granada, Andalucia, Spain.

2006 Pavement portrait of CD in salt and iron filings on gauze by Esther Solondz. Boston.

-2007 Wax figure. Madame Tussaud's Wax Museum, London.

2007 Bronze miniature. The Devil regarding the bust of Darwin, by Paul Dibble.

2007 Head of older CD with hummingbird and image of the CD £10 note. 50mm cupro-nickel medal issued as part of a British Banknotes medal series.

2007 Bust in RP resin, cast copper and stone, Hebei, China.

-2008 Fibreglass standing statue at purported site of CD's first landing, signed by M. Ramos. Las Tijeretas, San Cristobal, Galapagos.

2008 Royal Doulton Figurine of standing full-figure bearded CD by Robert Tabbenor.

2009 Bronze bust of a Cambridge student age CD by Anthony Smith, Sedgwick Museum of Earth Sciences, University of Cambridge.

2009 Maquette in bronze for below in different pose, looking left, by Anthony Smith.

2009 Bronze statue, seated on a wooden garden bench, of a 22-year-old CD, 10% larger than natural size, with books by Humboldt, Herschel and Paley and a beetle on one of the books, by Anthony Smith, Christ's College, Cambridge. Unveiled by the Duke of Edinburgh on CD's 200th birthday in the First Court of Christ's College. Now on display in The Charles Darwin Sculpture Garden, Christ's College.

2009 Bronze bust of older CD by Anthony Smith. First casting at the Science department of Winchester College, UK.

2009 Bronze bust after Richmond 1840, Puerto Baquerizo Moreno, San Cristobal, Galapagos.

2009 Wax figure of older CD at the exhibition Darwin observador, Darwin naturalista in CosmoCaixa, Barcelona, Spain.

2009 "Caminos de Darwin", portrait of CD after Richmond 1840 with map in glazed tiles placed in 12 locations along CD's route including Maldonado and Montevideo, Uruguay.

2009 120-tonne sand sculpture by Jamie Wardley and Nicola Wood, Bradford, UK.

2009 Bronze bust by Francisco Peralta Fabi, in front of the Science Faculty of the National Autonomous University of Mexico, Mexico City, Mexico.

2009 Bronze bust of young CD after Richmond 1840 by Anton Hasell. Civic Park, near Civic Centre and public library, Darwin, Australia.

2009 Bronze bust, copy (not a cast) of Couper 1909 bust by John Milner & Assoc. commissioned by National Academy of Sciences, for display in the Great Hall, Washington, D.C.

2009 Bronze medal, CD with beard "theory of the natural selection". UK.

2009 Bronze medal after Lock & Whitefield 1878a. Charles Darwin Research and Innovation Medal, established to recognise individuals who have made an outstanding contribution to research and innovation in fields across the Northern Territory, Australia.

2009 Clay bust by Daniel Fairbanks. Sculpted in one hour at Brigham Young University.

2009 Plaster/pigment bust by Suzie Zamit.

2009 Privately commissioned plaster bust, private collection, Italy.

2009 Silver medal with gold marching figures from monkeys to men. UK.

2009 Silver-plated copper "Greatest Britons" commemorative medallion, after Lock & Whitefield 1878a? A poor likeness with pointy nose. "CHARLES DARWIN 1809-1882 NATURALIST".

2009 Statue in bronze, life-size, nude, seated with iguana across lap feeding from open right palm and a monkey sitting below on the left side, and other animals, by Anthony Stones. China. Copy -2012, Shenyang Normal University, China (沈阳师范大学).

2009 Plaster original £2 coin by Suzie Zamit (30cm).

2009 Wax figure of an eighteen-year-old CD by Élisabeth Daynès, Calouste Gulbenkian Foundation, Galeria da Biodiversidade, Lisbon, Portugal.

2009 Wax figure of CD riding a giant tortoise in the exhibition "Darwins rEvolution". Natural History Museum, Vienna, Austria.

2009? 1.3oz silver medallion with full-figure CD on Galapagos touching nose of an iguana. *Beagle* in background. Medallic Art Co.(?)

-2010 Arch of Darwin, with bas-relief bust reminiscent of Collier 1881. Adjacent to the cemetery near the Galapagos National Park and the Charles Darwin Foundation, Puerto Ayora, Santa Cruz Island, Galapagos.

-2010 Statue in stone seated on rock with walking stick hooked to right arm. China.

-2010 Painted bust (concrete?) Barrio La Cascada. Erected on a private house. Puerto Ayora, Santa Cruz Island, Galapagos.

2010 Plaque monument commemorating CD's landing at Iquique on 12 July 1835. Iquique, Region I de Tarapacá, Chile.

2010 Bronze bust by Robert R. Toth (30cm). Commissioned by Columbia Pictures for 'Salt'.

2010 Bronze statue of young CD with walking stick by Pablo Eduardo. Cold Spring Harbor Laboratory, New York.

2010 Bust, wearing a soft hat, Academy of Sciences, Santo Domingo, Dominican Republic.

2010 Sandstone bust. Base inscribed "Darwin 达尔文" with small figures of an ape and upright man beneath by Wang Yun-ming. (210cm high). Nanjing, China. (南京十九山雕塑院).

2010 Bronze bust on black plinth (80x100cm). Shenzhen, China.

c.2011? Standing statue after Collier 1881 with CD holding an open book. In the central niche of the recently redeveloped Buckston Browne Farm building, Downe, Orpington. Originally the niche held the statue of a monk or friar.

2011 Bronze bust, looking downwards, by Gabriel Navas Vinelli. Charles Darwin Research Station, Puerto Ayora, Santa Cruz Island, Galapagos.

2011 Commemorative plaque (cement?) with bas-relief bust by Szilárd Sződy affixed on the front of the former Headquarters of the Hungarian Society for Natural Sciences.

-2011 Metal bas-relief bust, The Border View, Llanymynech Rocks, Offa's Dyke Path, Wales.

2011 Terracotta bust of older CD by C. Corbet. SDMM NVF14695.

2011 Sculpture in clay by Yana Amaru, recreating L. Darwin 1878b.

-2012 Bronze bust with modern neck-tie on pink granite plinth with inscription in blue "达尔文 1809-1882", Yuying Lake, Shandong Province, China. (育英湖 山东).

-2012 Cement bust with inscription on plinth: "Charles Darwin, The father of evolution". Northern Thai High School of Noen Maprang, Phitsanulok, Thailand.

-2012 Abstract outline bust "1831 Charles Darwin visits", bust inscribed panorama featured at The Border View at Llanymynech Rocks on Offa's Dyke Path, Shropshire, UK.

2012 Bronze statue of CD as a monkey "Anis del Mono" by Susana Ruiz. Badalona, Spain.

2012 Sand sculpture in right profile. The Sand Museum, Tottori Dun, Japan.

2012 Sand sculpture. Festival white lights. Perm, Russia.

-2013 Wax statue of *Beagle*-age CD. China Science and Technology Museum, Beijing, China.

-2013 Slender statue in bronze, life-size, standing in long coat, arms crossed looking at a bronze monkey, seated below. Monkey looking up at CD. After *Truthseeker* 1904? China.

2013 Sand sculpture bust. Burgas Sand Fest. Burgas, Bulgaria.

-2014 Bust in roughly worked bronze on black stone plinth, brass plaque inscribed "查尔斯罗伯特达尔文 1809-1882." China.

-2014 Statue in bronze, holding a book with an iguana on plinth. Zhejiang Gongshang University, China. (浙江工商大学).

-2014 Statue in bronze, life-size, cloaked, folded right arm clutching a book. China.

-2014 Wax sculpture seated at desk holding a quill pen. China.

2014 Bronze bas-relief of CD in hat and cloak after Elliott & Fry 1881a. "Knowledge is power" with thirty other figures by Zenos Frudakis. Rowan University, New Jersey, USA.

2014 Bronze statue of young CD with lens around neck, holding notebook, by Patricio Ruales. Charles Darwin Research Station, Puerto Ayora, Santa Cruz Island, Galapagos.

-2015 Monumental bust in white jade. Wuhan, China. (武汉).

2015 Life-size silicone statue by Csam Wheatley in the Freedom from Religion Foundation Library, Madison, Wisconsin.

-2016 Outline sculpture of CD with notebook, finch on shoulder. Location untraced.

-2016 Bust at Puerto Villamil, Isabela Island, Galapagos.

-2016 Bronze bust with base inscribed "Darwin" in his handwriting. National Zoological Museum of China. (国家动物博物馆).

2016 Sand sculpture bust by Ferenc Monostori. Binz, Ruegen island, Germany.

2016 CD as a scarecrow. Wray's Annual Scarecrow Festival, Wray, Lancashire, UK.

2016 Bronze bust of young CD by Elizabeth O'Kane. Commissioned to celebrate the life and work of Frank Noble, former Vice-Principal and Head of Biology Department at Ballymena Academy Grammar School, Northern Ireland, UK.

| Zamit 2009. | Dominican 2010. | Ruales 2014. | O'Kane 2016. | Clendining 2019. |

-2017 Bust in stone or cement(?) National Botanical Garden of Iran.

-2017 Statue in bronze, seated in long coat. Hong Kong Wetland Park, China. (香港湿地公园).

-2017 Bust in bronze on red stone plinth, inscribed in gold "达尔文 (1809-1882)". China.

-2017 Bust in white stone or cement(?), on black granite plinth, gold border with inscription "达尔文" (Darwin), Gu Guo-he Middle School, Zhejiang, China. (宁波市北仑区顾国和中学).

-2017 Bas-relief stone sculpture of CD and procession of apes walking to become man and woman with inscription in English "The Origin of Species Charles Darwin". China.

-2017 White bust on white stone plinth, carved "达尔文" (Darwin). China.

2017 Painted bronze bust by Guy Maestri, Lake Macquarie, Australia.

-2018 Bust in bronze, head only no plinth. China.

-2018 Bust in stone on stack of 12 books. Guangxi Medical University, China. (广西医科大学).

-2018 Statue in bronze(?) in cloak and hat as in Elliott & Fry 1881 photographs, iguana eating from his hand. Sculpture Park, Hebei, China. (河北雕塑公园).

-2018 Statue in stone, half body on grey granite cube with metal plaque with long inscription. Sun Yat-sen University, Guangzhou, China. (中山大学, 广洲市). It was originally planned to name the university "Darwin university" but this was rejected when it became known that CD had married his first cousin.

-2019 Bust in stone on a stack of three books. Darwin Discovery Park in Yichuan, New Village Kindergarten, Shanghai, China. (宜川新村幼儿园的达尔文探索园).

2019 Clay model for the statue below by David Clendining. Now at Ottawa, Ontario.

<u>2019</u> Bronze statue of young CD, seated with Galapagos finch on shoulder and *Notebook B* in hand (which he only started after returning home) by David Clendining. Summit Studios, in Smithsonian's National Museum of Natural History, Washington D.C., USA.

2019 Metal statue, head of CD on gorilla-like metallic body, "Darwinism" by Luke Kite.

2019 Silver finish and gold-plated seven-sided 'coin' with bust after L. Darwin 1878a with march of evolution on reverse. Commemorative Coin Company, UK.

nd Sterling silver commemorative medallion, 19.8 Grams. CD relief bust after Maull & Polyblank 1855 with *Beagle* and crow(?) On verso: "Charles Darwin 1809-1882 His theory of evolution by natural selection was one of the most important milestones in science. Long the subject of controversy, his theory today is widely accepted."

nd Bas-relief on information centre/observation deck, Malecón, Shipwreck Bay in Puerto Baquerizo Moreno, Galapagos.

<u>nd</u> Bronze bust on red plinth with gold inscription. Shenyang Normal University, China. (沈阳师范大学).

Sun Yat-sen c.1924.   Shenyang Normal University nd.   Stones 2009.   Zhejiang 2014.

nd Bronzed flat back wall medallion with impressed lozenger date registration mark, (10x7.6cm) Listed on an online antique sale.

nd Bust in bronze on grey granite plinth, inscribed in gold "达尔文, 1809-1882." China.

nd Bust in bronze on stone plinth, carved "达尔文...1809-1882". China.

nd Bust in bronze on tall tapering pedestal, inscribed "达尔文 (1809-1882)" with quotation in Chinese "The way to complete a task is to cherish every minute". China.

nd Bust in stone on black plinth, Chongqing, China. (重庆).

nd Bust in white stone on grey plinth with biography and quotation in English and Chinese. Changsha University of Science and Technology (est. 1957), China. (长沙理工大学).

nd Painted statue in cement(?) of young CD. Puerto Baquerizo Moreno, San Cristóbal Island, Galapagos.

nd Statue in white, seated with arms folded, on dark grey stone plinth. China.

nd Statue in stone, kneeling on one leg, holding handful of earth. China.

2019- Life-size bronze bust by Zenos Frudakis. USA.

2019- Monumental bas-relief head in bronze (c.182cm high) by Zenos Frudakis. USA.

The many recent commercially available busts are not included here.

Oils to 2009: (71 recorded here)

1862 Half-length oil on canvas by Eliphalet Frazer Andrews (1835-1915) "signed and dated", (75x62cm). Cited in *Index to the miscellaneous documents of the House of Representatives...*1884 and *Smithsonian Institution Bulletin: National Gallery of art,* 1909, p. 80. National Museum of American History, Smithsonian Institution, Object number 12568. Not seen.

1869 Aug. Oil on canvas, left profile, by Laura Russell (1816-85), (17.7x25.4cm). Painted inscription on back: "23d. Augt. 1869 / Down / from / nature by Laura Russell". In private collection, USA, exhibited Cambridge 2009, reproduced as frontispiece to CCD17.*

1871 Oil painting "The youthful Darwin expounding his theories" by William Holbrook Beard (1824-1900). American Museum of Natural History. See plate in CCD20:290.

1875 Oil on canvas, left three-quarter profile, by Walter William Ouless (1848-1933), signed and dated 1875 (59.4x54.6cm). CD sat for in Feb.-Mar. Darwin Heirlooms Trust (now at Darwin College, Cambridge). Registered for copyright by Ouless 22 Mar. 1875. 1883 copy by the artist in the Hall of Christ's College, Cambridge signed "W. W. Ouless 1883 replica". Etching by Paul Rajon 1877. CD "I look a very venerable, acute melancholy old dog,— whether I really look so I do not know." CCD23:124. Francis Darwin's opinion: "Mr. Ouless's portrait is, in my opinion, the finest representation of my father that has been produced". LL3:195.*

1879 Half-length portrait in oils. CD as honorary LL.D., wearing the red robes, by William Blake Richmond (George Richmond's son), (125.1x99.7cm). Subscribed for by members of the University of Cambridge, £400 being raised to preserve a permanent memorial of CD at the University. CD sat for in Jun., 1879. 1908 A report that during the painting CD was asked "'What were the most important years in the education of a child's life?' He said unquestionably the first three." *Morning Post,* (24 Jul.), p. 9. Initially hung in the Library of the Cambridge Philosophical Society, now in the Department of Zoology. Copy by the artist, Darwin Heirlooms Trust. Cambridge Philosophical Society. A contemporary review described a "small but powerful unquiet eye of intense observation." (*The Standard,* 1880) ED's opinion in 1881 Oct. "The red picture, and I thought it quite horrid, so fierce and so dirty". Francis Darwin's opinion "according to my own view, neither the attitude nor the expression are characteristic of my father". LL3:222.* One contemporary said it "looked like a scarlet ape." *Leicester Chron.,* (8 May 1880), p. 3.

1881 Three-quarter length portrait, standing, full-face, hat in hand, in cloak by Hon. John Collier. CD sat for in Jul., finished by Aug.* At Linnean Society and commissioned by them. Collier's recollections of the sitting are cited under Collier q.v. Francis Darwin's opinion "many of those who knew his face most intimately think that Mr. Collier's picture is the best of the portraits, and in this judgment the sitter himself was inclined to agree." At least one watercolour portrait study for the 1881 Linnean Society portrait exists. LL3:223. Copies include:
1883 copy by the artist, signed "John Collier replica 1883", presented in 1896 to the NPG by William Erasmus Darwin, (125.7x96.5cm).
1905 copy oil on canvas of the head and shoulders, (74x61cm), by John Lewis Reilly (c.1835-1922), signed on the reverse "John L. Reilly, 1905", numbered "65" on the canvas. On display at Down House, dining room, EH88202623.
1906 a photograph of it was registered for copyright by William Axon Mansell and George Fluck.
1912 copy oil on canvas by Mabel Beatrice Messer (127x99.5cm), for the Royal Society.

1922 aquatint engraving after the 1883 portrait was made by G. Sidney Hunt for the BMNH, reproduced as a colour mezzotint by the NPG in the same year. See *The literary and cultural reception of Charles Darwin in Europe*, pp. 140-1.

1929 copy oil on canvas by Collier (c.127x96.5cm), commissioned by George Buckston Browne, for Down House. EH88202263.

2000 copy oil on canvas by Philip Craig (b.1951), (73x55cm), sold at Sotheby's in 2017 for £3,250 from the collection of Garrett Herman.

1881 Grisaille painting, three-quarter profile, by Ernst Hader (1866-1910). As CDV, "E. Hader pinxit. /Gesetzlich geschützt. / Photographie und Verlag Sophus Williams, Berlin W." Later as a postcard. A very poor likeness. Engraved by Mayerhofer(?). Re-drawn as frontispiece to *Origin* (historical sketch only, in Chinese), 1902, F1846. Often re-printed, e.g. as a German postcard c.1900 "Collection das grosse Jahrhundert" and on stamps.

-1882 Oil on canvas(?) "a large portrait of Darwin, painted by Henry Ulke [1821-1910]" who named his son Darwin Ulke, mentioned in *Proc. of the Biol. Soc. of Washington*, vol. 1 (1882), p. 43. Whereabouts not traced.

1880s? Oil on canvas(?), three-quarter left profile. South African National Gallery, Cape Town, South Africa.

-1884 Three-quarter right profile head and shoulders, part of "the great painting of Mr M. C. Tiers", a New York portrait painter. Reproduced as a cabinet card for editor A. Wilford Hall of The scientific arena featuring a large number of figures with CD in the top centre.

Richmond 1879.   Ouless 1883.   Baroni 1890 (extract).   Workman 1930-5.   Diego 1934.

1890 Oil on canvas, lunettes panel by Pasquale Baroni (b.1831), after L. Darwin 1878a. Human Anatomy Museum of the University of Turin, Italy.

1898 "A small portrait" of CD by Janet A. Boyd was exhibited at the Royal Academy.

18?? Bearded miniature portrait in black round frame. Dated by Christie's to 19th century.

1909 Oil on canvas, three-quarter left profile. Not from life nor a copy of a particular photograph, by Adolf Dagobert Libeski (50.8x69.85cm). Sold on eBay 200?.

1920 Oil on canvas, CD with pigeons behind Down House by Vatagin. SDMM. OF7667/460.

1922 Oil on canvas, CD in Tierra del Fuego by M. Ezuchevski. SDMM. NVF1321.

1922 Oil on canvas, CD on Galapagos Islands by M. Ezuchevski. SDMM. NVF1263/736.

1923 Oil on canvas, CD excavating fossils in South America (with Glyptodon skull) by M. Ezuchevski. SDMM. NVF1263/735. Copy sent by the SDMM to Down House in 1959.

1925 Oil on canvas, CD ponders the relationships in nature, with a bee, mouse, flower and a cat by M. Ezuchevski (150x200cm). SDMM. NVF1263/1025.

1925 Oil on canvas, CD counts the fertility of mice by M. Ezuchevski. SDMM. NVF1263/1026.

c.1925 Oil on canvas (42x32cm); collection E.J.O. Kompanje. (Otto van Duijn).

1926 Oil on canvas, CD with Anne Elizabeth (Annie) Darwin on 24 Apr. 1851 by M. Ezuchevski. SDMM. NVF1263/731. Daguerreotype of her obviously not seen by artist.

1926 Oil on canvas, Procession of scientists (CD, Wallace, Huxley, Haeckel and ten others) by Victor Evstafiev. SDMM. NVF1263/734.

1930 CD at his desk in the old study by S. Uranova. Tretyakov Gallery, Moscow, Russia.

1930-5 Oil with gold leaf, on canvas mounted on plywood. Ely Mural Panel by David Tice Workman (1884-1971). On loan to Washington Elementary School, Ely, Minnesota. Lent by Ely-Winton Historical Society, Ely, Minnesota, USA.

1934 CD pointing at a monkey, detail in the mural 'Man at the crossroads' by Diego Rivera (1886-1957), Palace of Fine Arts, Mexico City.

1935 Oil on canvas. CD in cloak in garden behind Down House with dog by Victor Vatagin. SDMM. OF7667/110. Black and white photograph in DAR238.14:94.

-1938 Oil portrait by E. Pailthorp acquired by Buckston Browne for Down House. *Report of British Association*, 1938. Not now in records at Down House.

-1940 "an original painting of" CD by Michalowshi [sic] "the painting is hanging in the biological library at present" donated to the University of California in 1940 by Lee A. Stone. *Madeira Tribune* (22 Apr.)

-1945 Oil on canvas after Lock & Whitfield 1878a. By anon. Russian artist, online sale 2019.

1920 CD & pigeons. 1922 on Galapagos.   1926 CD with Annie.   c.1958 "early Spring". 1960 CD in his library.

1948 Oil painting on board of beardless CD in the old study, by Victor Evstafiev, (40x52cm). Sent by the SDMM, to the BMNH, later lent on indefinite loan to Down House, 1959. P. Simpson, A cold war curiosity?: The Soviet collection at the Darwin memorial museum, Down House, Kent. *Jrnl. of the Hist. of Collections*, vol. 30, issue 3, (Nov. 2018): 487-509. Down House, EH88204429.

1950 "Painting of Charles Darwin." Festival of Britain. National Archives. WORK 25/202/D1/FOB2597.

1954 Painting after L. Darwin 1878a by Ria Novosti (Russia).

1950s Oil on cardboard. "Darwin's departure from England" on the *Beagle* by S. Uranova after Ezuchevski 1926, (39x28cm). Spelled "Beagle" on the life preserver. Sent to Down House by the SDMM in 1960. Down House EH88204427.

1950s Oil on laminated paper board. CD excavating fossils in South America by Victor Evstafiev after Ezuchevski 1923 (43x32cm). Sent to Down House by the SDMM. Down House EH88204458.

1950s Oil on laminated paper board. CD's first encounter with a Fuegian by Victor Evstafiev after Ezuchevski 1920 (30x40cm). Sent to Down House by the SDMM. Down House EH88204459.

1950s Oil on cardboard. CD contemplating a giant tortoise on the Galapagos Islands by S. Uranova after Ezuchevski 1920 (40x30cm). Sent to Down House by the SDMM in 1960. EH88204456.

1950s Oil on cardboard. "Ch. Darwin with Orchids" in his hothouse by S. Uranova after Ezuchevski 1923 (42x30.5cm). Sent to Down House by the SDMM in 1960. EH88204426.

1950s Oil on cardboard. CD with pigeons behind Down House by S. Uranova after Vatagin 1920 (44x33cm). Sent to Down House by the SDMM in 1960. EH88202276.

1950s Oil on paper. CD in his greenhouse by Victor Evstafiev after Ezuchevski 1923 (42x30.5cm). Sent to Down House by the SDMM. Down House, EH88204426.

1950s Oil on cardboard. CD greets Ernst Haeckel at the door of Down House by S. Uranova after Ezuchevski 1926. (Frame size 47x36.5cm). Sent to Down House in 1960. EH88202275 (Haeckel often misidentified as Timiraziev).

1959 Oil on cardboard. CD with Anne Elizabeth (Annie) Darwin on 24 Apr. 1851 by Victor Evstafiev after Ezuchevski 1926 (50x70cm). Sent to Down House by the SDMM.

1959 Oil on card. CD reading a sample chapter of *Origin* to Lyell and Hooker in the old Study by Evstafiev, after Evstafiev 1948 (44.5x64cm). Sent to Down House by the SDMM. EH88202271. This painting has been often reproduced but modern writers, e.g. Tim Berra, *Charles Darwin*, incorrectly describing it as CD holding Wallace's Ternate essay.

c.1958 "Early Spring", oil painting showing young CD on verandah and ED at her piano by Evstafiev(?). Sent by the SDMM to Down House. Photograph sent by the SDMM to Sir Charles Darwin in 1963, now in DAR238.14.64.

c.1958 Oil painting, "Late Autumn" by Evstafiev. ED at her piano playing for elderly CD (frame size 36x47cm). Sent to Down House by the SDMM. Down House EH88202062.

1958 Oil on canvas of beardless CD in his old study writing with pen by Victor Evstafiev, signed and dated (82.5x109cm). Sent to Down House by the SDMM. EH88202272.

nd Three-quarter profile in oils, at San Diego State University, 2009. Private collection.

1960 Oil on canvas, Prof. Henslow advising young CD on the voyage of the *Beagle* by Victor Evstafiev, initialled and dated. SDMM. NVF1263/1190.

1960 Oil on laminated paper of bearded CD at his library in the old study by Victor Evstafiev (64.5x47cm). Initialled and dated. Sent to Down House by the SDMM. EH88204432. No original or copy is in the SDMM.

1972 Oil on canvas. Seated CD outdoors with pigeons and cat by Maus Slangen.

1978 Oil on canvas after L. Darwin 1878a by Gergel(?) (67x87cm). USSR.

1980 Oil on wood by Stephen Nash, (152.5x213.3cm). UCL Art Museum, London, UK.

1980s Oil on canvas, CD meeting K.A. Timiraziev in the hothouse by N. Navashina. SDMM. OF12434.

1986 Oil painting by Fruls Tilp (156.5x122cm). Moderna Museet, Stockholm, Sweden.

c.2003 Right profile oil painting of beardless CD looking at painting of a bird with a bird stuffer in background by Amod Damle. Ohio State University, University Libraries.

2009 Full-face half-length original portrait in oils by Victoria Clinton (61x50.8cm).

2009 Full-length oil portrait of young CD in the Galapagos by Gordon Chancellor.

2009 Three-quarter left profile in oils after Cameron 1868b? by Victoria Clinton (35.5x25.5cm).

2009 Collage of older CD and young CD, after Richmond 1840 with Christ's College, Down House, *Beagle* and other CD scenes by Fiona Macpherson. Christ's College, Cambridge.

2009 Full-length dancing saint icon (with halo) by Mark Dukes. St. Gregory of Nyssa Episcopal Church, San Francisco.

2009 Oil on cradled board by Fred Bell.

2009 Oil on board by Francesco Masci.

nd Oil painting by unknown artist. Smithsonian Institution, National Museum of American History, Washington, D.C.

nd Bearded portrait at Special Collections and University Archives Reading Room Library and Information Access, San Diego State University.

There are innumerable further works of more recent date.

Watercolours, drawings and stained glass: (142 recorded here)

Many more watercolours and other artworks have been created than are recorded here as many were made to be transferred to print, such as the numerous illustrated books and magazines of the early and mid-20th century e.g. the Ladybird books and *Look and Learn* magazine. Some of these original artworks apparently survive but have not been traced. Those that are reproduced here have their dates underlined.

Christ's 1882.   P.R. Montford 1897.   Dukes 2009.   Bratislava 2009.   Providence 2013.

1816 Pastel chalk drawing of CD with his sister Emily Catherine by Ellen Sharples (1769-1849), (33.7x37.5.1cm). Darwin Heirlooms Trust. On display at Down House with three others of the Darwin family.* Sepia-toned photograph, (28.5x33.5 frame size), at Down House, EH88202289.

1828? Two humorous ink sketches of CD riding giant beetles by Albert Way, with captions "Darwin & his hobby." and "Go it Charlie!". DAR204.19.

1830 Sketch(?) by Fanny Owen Not found. See F. Owen to CD 4 Oct. [1830], CCD1:108.

1832 Purported full-length watercolour caricature of CD and *Beagle* crew entitled "Quarter Deck of a Man of War on Diskivery or interesting Scenes on an Interesting Voyage" attributed to Augustus Earle. The speech bubble for the character said to be CD reads: "Observe its legs are long, and the palpi are strongly toothed on the inner side. I think the whole insect appears of a dark chestnut brown colour with a yellowish cast on the abdomen its history is but little known but there can be no doubt of its being of a predacious nature." This is in fact a modified quotation from G. Gregory ed., *A new and complete dictionary of arts and sciences*, 1819, vol. 3 on *Phalangium*, a genus of harvestmen. This work is not known to have been in the *Beagle* library. 2015 sold at Sotheby's for £52,500.

1836 July Pencil sketch full-length self-portrait by CD as a 'stick man' on the cliffs of St. Helena, "gale of wind to hand not to man", *Despoblado notebook*, p. 78b. In Chancellor & van Wyhe, *Beagle notebooks*, 2009, p. 583.

1840? Pencil sketch, half-length, by George Richmond (38x27cm). On display in Manuscripts Reading Room, CUL. Label in tooled leather: "This drawing of Charles Darwin,
presumed to be a preliminary sketch by George Richmond for his watercolour of c.1840,
was discovered in the basement of the Botany School, Cambridge 1929. Deposited in the
University Library by the family of Nora, Lady Barlow, September, 1987."*

1840 Mar. Watercolour by George Richmond (20.3x25.4cm).* Note on back of frame reads
"Charles Robert Darwin age 31 March 1840." Marriage portrait, a twin with that of ED
by Richmond. Both hung at The Mount until the death of Susan Darwin in 1866 when
they went to Erasmus Darwin in London and after his death to Down. Darwin Heirlooms Trust. On display at Down House. Down House LI.2008.0046.2. Copy: Watercolour on paper (33x23.5cm). The Charles Darwin Trust. Down House EH88202068.

nd Watercolour "Within sight of the Andes" the boats with FitzRoy and CD ascending the
Santa Cruz river, 1833. Attributed to O.W. Brierly when exhibited at Christ's College,
1909. "Lent by H. N. Sulivan, Esq. (son of Admiral Sir B. Sulivan". Not traced.

1853 Full-face chalk drawing by Samuel Laurence (1812-84).* Botany School, Cambridge.
Study for the below? Copy in DAR225.127.

1853 Left profile coloured chalk drawing, signed and dated by Samuel Laurence. Darwin
Heirlooms Trust. On display at Down House, Billiard Room.

1855 Half-length pencil (and charcoal?) sketch, full face, by Harriet Caroline Lubbock né
Hotham (1810-73), wife of John William Lubbock.* In the family.

1868 Ink caricature sketch by T.H. Huxley of CD as a pope with "pangenesis" on his mitre
and "selection" and "variation" on his staff. CD gestures a blessing to a kneeling figure
swinging a censer of incense before him, viz. Wilhelm Friedrich Kühne who, as Huxley
humorously put it: "wants to know whether there is any possibility of his paying his devotions at the Shrine of Dr Darwin". Signed "Huxley". The letter and sketch are reproduced
in CCD16:634-5. Photocopy in DAR221.4.254. Original said to be in a private collection.

1871 Original watercolour and pencil lithographic outline of the *Vanity Fair* caricature by
James (Jacques) Tissot (31.7x19.7cm). Sold at Christie's in 1912, Lot 197, the drawing was
partly drawn in outline on lithographic stone in black, one copy printed, then completed
and coloured by Tissot. Now at Old Library, Christ's College, Cambridge.

after 1873 Pen and ink drawing of CD in chair, legs crossed, stroking beard thoughtfully, by
Harry Furniss (1854-1925), (45.4x30.2cm). Printed in Furniss, *Some Victorian men*, 1924,
facing p. 25. (From life?)

c.1875 Washed India ink, right profile, by Louisa A'hmuty Nash (1837-1926). In possession
of a Nash descendant, USA.* Photograph at Down House EH88204919.

1878 Full-face, half-length pencil sketch by Marian Huxley (1859-87), (18.1x12.4cm). Signed
in ink with monogram "MH". NPG.* See under Iconography, Prints, Himmelfarb 1959.

1880 Watercolour of Down House from the south-west by Albert Goodwin. According to
Shipley & Simpson, *Darwin centenary: the portraits...exhibited at Christ's College*, 1909: "The
small figures in the verandah are very faithful portraits of Charles Darwin, Mrs Charles
Darwin and their grandson Bernard; with the dog Polly." EH88202055.

1881 Watercolour on paper on canvas, inscribed and signed below: "Charles Robert Darwin
1809-1882/ Study by John Collier" (30x23cm). Sold at Sotheby's in 2009 for £13,125.*

1881 "a rough sketch on an envelope at a meeting of the Royal Society." by Alphonse
Legros. T. Okey in Attwood, The medals of Alphonse Legros. *The medal*, no. 5, 1984.

c.1881 Drawing, three-quarter portrait, goldpoint, heightened with white on pink prepared paper, by Alphonse Legros, Fitzwilliam Museum, Cambridge.

1882 Stained glass window, full-length in LL.D. gown, holding small ape skull, "Carolus Darwin" by Burlison and Grylls in west oriel of the Hall of Christ's College, Cambridge.

c.1882? Drawing in ink after Elliott & Fry c.1880b by Thomas G. Johnson (1844-1904). Reproduced in J. David Archibald, *Origins of Darwin's evolution*, 2017, p. [xvi].

Furniss 1873-. © NPG.   RHS Lindley Library.   Pengov 1950s.

-1894 Portrait sketch by Edwin Tryon Billings (1824-1893). *Billings Memorial Exhibit*, 1894.

1895 Wall mosaic portrait in left profile. La Biblioteca Pública Arús, Barcelona, Spain.

-1897 A "pastel portrait" of CD was sold in 1897 for 45s. *Shrewsbury Chron.*, (4 Jun.), p. 6.

1897 Drawing by Paul Raphael Montford (1868-1938) in R. Lydekker, Biological progress in the Victorian era. *Knowledge*, (1 May), facing p. 110.

c.1901 Pencil drawing, three-quarter right profile, by Henri Lévy, signed (21.5x16.5cm). Artist's cache de vent shows work was sold in 1901. Sold at Sotheby's in 2017 for £1,750.

-1902 Miniature painting on mantelpiece of the room of the assistant secretary of the Royal Society, Burlington House, UK. *St James's Gazette*, (18 Sept.), p. 17.

nd Graphite and watercolour sketch of CD with seated in a chair pointing at a manuscript. Unknown artist. Royal Horticultural Society (RHS), UK, Lindley Library LIB0006240.

nd Pen and ink drawing (16.7x16.2cm) after Ouless 1875. Wellcome Library no. 2368i UK.

nd Portrait pencil drawing. Wellcome Library, GALTON/1/1/3/7 UK.

early 20th century? Colour painting or lithograph after Collier 1881, signed by an Italian artist. Look and Learn History Picture Archive (https://www.lookandlearn.com/about/cookies.php).

1910 Pen and ink drawing after Elliott & Fry c.1880a by Geska(?) For sale in 2019 with Kotte Autographs GmbH, Roßhaupten, Germany.

1920 Pastels on paper, CD contemplating a giant tortoise on the Galapagos Islands by M. Ezuchevski. SDMM. OF7667/529.

1920 Pastels on paper, CD in pigeon house by M. Ezuchevski. SDMM. OF7667/447.

1920 Pastels on paper, CD with fossils by M. Ezuchevski. SDMM. OF7667/450.

1920 Pastels on paper, CD's first encounter with a Fuegian by M. Ezuchevski. SDMM. NVF12305. A copy was sent by the SDMM to Down House in 1948-58.

1923 Pastels on paper, "Ch. Darwin with Orchids" in his hothouse by M. Ezuchevski. SDMM. OF7667/448.

1926 Pastels on paper, "Darwin's departure from England" on the *Beagle* by M. Ezuchevski. SDMM. OF7667/508 'Beagle' on a life preserver in the painting spelled "Beegle".

1926 Pastels on paper, CD greets Ernst Haeckel at the door of Down House by M. Ezuchevski. SDMM. OF7667/503.

1930-5 Portrait miniature by Percy Buckman. Royal College of Surgeons. RCSSC/P 383.

1930-37 Pencil drawing of aged CD on a slip of paper sent to Buckston Browne, (17.5x11.4cm), with a note "Dear Browne, From the adjoined photo out of the 'Morning Post' it appears one of Darwin's descendants has been caught and put in the zoo. Poor thing! yours always, W Logsdair". The news clipping (EH88202939) referred to an albino monkey. Down House, EH88202938.

1931 Stained glass window in the Shove Memorial Chapel, Colorado Springs, Colorado.

1934 Photograph of drawing of CD by Lawrence Baker (with frame 43.5x37.5cm). Down House EH8820443.

1945 Pastels on paper, CD in cloak near Down House, K. Flerov. SDMM. NVF1265/427.

1948 Pastels on paper, adolescent CD watching birds by Victor Evstafiev (1916-89). SDMM. NVF1265/1110.

1948 CD and Catherine Darwin.　　1948 CD with John Edmonston.　　1948 Henslow and CD.

1948 CD and FitzRoy by Evstafiev.　1948 (1957) CD hunting rheas on the pampas.　1923 CD with orchids.

1948 Pastels on paper, boyhood CD goes for a walk by Evstafiev. SDMM. NVF1265/1079.

1948 Pastels on paper, CD and Catherine as children fishing in the river Severn by Evstafiev. SDMM. NVF1265/1078 A copy was sent by the SDMM to Down House in 1959.

1948 Pastels on paper, CD and Catherine as children in front of The Mount, CD carrying a fishing rod, by Evstafiev. SDMM. NVF1265/1076. A copy was sent to Down House by the SDMM in 1959.

1948 Pastels on paper, CD and Catherine near bird nest by Evstafiev. SDMM. NVF1265/1077.

1948 Pastels on paper, CD and Catherine on horseback in North Wales by Evstafiev. SDMM. NVF1265/1107.

1948 Pastels on paper, CD and Henslow on tour by Evstafiev. SDMM. NVF1265/1108.

1948 Pastels on paper, CD as a child collecting beetles by Evstafiev. SDMM. NVF1265/1083. A copy by Evstafiev was sent by the SDMM to Down House in 1959.

1948 Pastels on paper, CD as a child reading a book in a bay window of The Mount by Evstafiev. SDMM. NVF1265/1075. A copy sent by the SDMM to Down House in 1959.

1948 Pastels on paper, CD in Edinburgh learning taxidermy from John Edmonston by Evstafiev. SDMM. NVF1265/1084. A copy sent by the SDMM to Down House in 1959.

1948 Pastels on paper, CD in a fisherman's boat at sea [near Edinburgh] by Evstafiev. SDMM. NVF1265/1080.

1948 Pastels on paper, CD collecting sea creatures by Evstafiev. SDMM. NVF1265/1082.

1948 Pastels on paper, CD geologizing by Evstafiev. SDMM. NVF1265/1109.

1948 Pastels on paper, CD on the hunt by Evstafiev. SDMM. NVF1265/1081.

1948 Pastels on paper, CD at night on the pampas by Evstafiev. SDMM. NVF1265/1184. A copy painted in 1957. Reproduced here.

1948 Pastels on paper, CD hunting rheas with gauchos on the pampas by Evstafiev. SDMM. NVF1265/1186. A 1957 copy by Evstafiev given by SDMM to Down House in 1959.

1948 Pastels on paper, CD on a boat in a rainforest river by Evstafiev. SDMM. NVF1265/1185.

1948 Pastels on paper, CD reading a sample chapter of *Origin* to Lyell and Hooker in the old Study by Evstafiev. SDMM. NVF1265/1189.

1948 Pastels on paper, Evening at Maer by Evstafiev. CD with ED and Robert Waring Darwin. SDMM. NVF1265/1085. Copy sent by the SDMM to Down House in 1959.

1948 Pastels on paper, Prof. Henslow advising young CD on the voyage of the *Beagle* by Evstafiev. SDMM. NVF1265/1086.

1948 Pastels on paper, CD and FitzRoy meeting, by Evstafiev. SDMM. NVF1265/1087.

c.1950s Pen and ink three-quarter right profile by Slavko Pengov (1908-66), (11.9x9.7cm).

nd Watercolour on paper. Formerly belonging to grandson William Robert Darwin (d.1970). In original gilt frame, and decorative window mount. Down House, EH88202621.

1960 Monochrome watercolour portrait by Czech artist Burian Zdeněk (1905-81).

1960s Watercolour portrait by Burian Zdeněk, (11x10.5cm).

1965 Ink on paper, after Maull & Polyblank 1857 by Peter Blake.

1971 Watercolour on paper. CD as a schoolboy being admonished by a teacher standing at a lectern by Robert Broomfield. 1 in a series of 21. For a work entitled The voyage of the *Beagle*? (25x30cm). Down House EH88302414.

1971 Watercolour on paper. CD standing in a drawing room being admonished by his father. 2 in a series of 21 by Robert Broomfield. (65x50cm). Down House EH88302416.

1971 Watercolour on paper. CD as a teenager doing a chemistry experiment in a laboratory. Three other schoolboys look on and one of them is holding his own nose. Reference to "gas". 3 in a series of 21 by Robert Broomfield. (55x49cm). Down House EH88302417.

1971 Watercolour on paper. CD as a young man reading a letter that has just arrived that invites him for the *Beagle* voyage. He sits at the breakfast table with his father. 4 in a series of 21 by Robert Broomfield. (67x50cm). Down House EH88302418.

1971 Watercolour on paper. CD saluting FitzRoy in introduction. They stand next to a carriage where sailors are unloading trunks. Ship can be seen on the water in the background. 5 in a series of 21 by Robert Broomfield. (64x50cm). Down House EH88302419.

1971 Watercolour on paper. CD entering the poop cabin carrying his books - he bangs his head on the lantern. FitzRoy is sitting at the chart table inspecting a chronometer. 6 in a series of 21 by Robert Broomfield. (62x50cm). Down House EH88302420.

1971 Watercolour on paper. FitzRoy on the deck of HMS *Beagle* using a surveying instrument. 7 in a series of 21 by Robert Broomfield. (56x50cm). Down House EH88302421.

1971 Watercolour on paper. CD in a small boat, landing on a beach and meeting some Fuegians. 8 in a series of 21 by Robert Broomfield. (62x50cm). Down House EH88302422.

1971 Watercolour on paper. CD sitting in the poop cabin writing in his journal. 9 in a series of 21 by Robert Broomfield. (23x33cm). Down House EH88302424.

1971 Watercolour on paper. CD catching insects with a net. 11 in a series of 21 by Robert Broomfield. (68x50cm). Down House EH88302425.

1971 Watercolour on paper. CD packing up his specimens. An assistant hammers a nail into packing crate. 12 in a series of 21 by Broomfield. (58x51cm). Down House EH88302426.

1971 Watercolour on paper. CD standing with hammer beside a Megatherium skeleton. 13 in a series of 21 by Robert Broomfield. (59x50cm). Down House EH88302429.

1971 Watercolour on paper. CD sitting on an upturned bucket reading a book, two Galapagos tortoises in the foreground, on the deck of the *Beagle*. 16 in a series of 21 by Robert Broomfield. (60x49cm). Down House EH88302430.

1971 Watercolour on paper. CD returning from voyage with men loading his trunk onto a carriage. 17 in a series of 21 by Broomfield. (59x49cm). Down House EH88302431.

1971 Watercolour on paper. CD embraced by his sisters with his father looking on. 18 in a series of 21 by Robert Broomfield. (65x49cm). Down House EH88302432.

1971 Watercolour on paper. CD sitting at a desk writing his book. 19 in a series of 21 by Robert Broomfield. (62x48cm). Down House EH88302433.

1971 Watercolour on paper. CD sitting in the new study at Down House. 20 in a series of 21 by Robert Broomfield. (64x50cm). Down House EH88302416.

1972 Watercolour over graphite with blue and red pencil on wove paper by Jack Cowin. "Darwin and the Beagle Series" (Above: Hunting Ostriches' Eggs with the Bolas). (56.5x76.3cm). National Gallery of Canada.

1974 Pencil on paper, "Darwining." Study for below(?) by Aldo Mondino.

1974 Pastel(?) drawing. "Darwining." Abstract portrait after Collier 1881 by Aldo Mondino.

1978 Three-quarter profile watercolour by Gösta Werner, (132x176cm). Moderna Museet, Stockholm, Sweden.

1978 Three-quarter left profile caricature in pen and ink by Lucinda Levine, (20x22cm).

1970s? Three-quarter right profile, gouache, pen and ink with chimpanzee by Robert Engle.

1980s? Three-quarter right profile watercolour by Jack Coughlin.

1981 Three-quarter right profile crayon drawing by Freda Reiter.

1981 Three-quarter right profile crayon drawing with apes in background by Freda Reiter.

1984 Half-length full-face gouache on paper portrait by Song Ren (China), (63x45.2cm).

1986 Full-face modernist portrait, colour etching, drypoint, by Mauricio Lasansky.

-1998 Watercolour(?) after Richmond 1840 depicting CD and the *Beagle*. Cover of *Charles Darwin's letters: a selection*, 1998 et al. Mary Evans Picture Library/Jeffrey Morgan.

1992 Three-quarter left profile pencil drawing by Edward Forster.

1999 Acrylic on board by Geoffrey Appleton (120x151cm). Commissioned by the Faculty of Science, Staffordshire University, UK.

2000 Acrylic abstract realism portrait by Mike Wehner.

2000 Pen and ink three-quarter left profile by Kevin Sprouls for the *Wall Street Journal*.

2005 Pen on paper. Darwin Statue (Survival of the phattest) by Tim Maxwell.

2007 Graphite pencil three-quarter right profile after a photograph (Elliott & Fry c.1880b?) by Victoria Clinton, (42x29.7cm).

2008 Pen and ink and watercolour three-quarter left profile by Andreas Noßmann.

2008 Darwin's garden. Etching of CD as outline in gaps of vegetation by István Orosz.

nd Bearded portrait of CD in stained glass at St Giles & All Saints Church, Orsett, UK.

2009 Cambridge 800: an informal panorama. With bearded CD reading a book on a Galapagos tortoise, watercolour frieze by Quentin Blake. Addenbrooke's Hospital, Cambridge.

2009 Stained-glass window by "F.R.", Faculty of Medicine, Charles University, Bratislava.

2009 Left profile after Rejlander 1871c (reversed) by William Edwards. Fine bone china commemorative plate. Charles Darwin Bicentenary Collection.

2009 Three-quarter right profile on lid of Bilston & Battersea Enamels oval trinket box.

2009 Full-figure portrait roughly after Richmond 1840. Lid of Halcyon Days trinket box.

2009 Watercolour. Darwin Blue by Daniel Bosler after Cameron 1868a.

2009 Watercolour by 'gemiee'.

2009 Watercolour by Adolfo Gutiérrez after Lock & Whitfield 1878.

nd Portrait after Richmond 1840 by George Frazer. Fine bone china thimble. Number 38 in a series of 50. History of Britain series.

nd Portrait after Richmond 1840, commemorative beaker. Staffordshire Enamels Old Hall.

-2010 Composite drawing of CD after L. Darwin 1878a by Sarah King.

-2010 Mural of bearded CD with a Galapagos finch perched on his finger, a ship in the background. Puerto Ayora, Santa Cruz Island, Galapagos.

2010 Graphite sketch of bearded CD by Jason Cottle.

2010 Bearded portrait by Ernest Pignon-Ernest.

-2011 Color etching, aquatint and woodcut with embossing by Mauricio Lasansky.

-2011 Bearded portrait in blue tiles, façade of Berlin-Kreuzberg, Körtestrasse 8.

2011 Acrylic painting by Eric Dee, after the 2009 NHM exhibition poster with a hand added to Elliott & Fry c.1880a, with a finger held to the lips.

2011 Study for "Self-portrait as Charles Darwin" after L. Darwin 1878a acrylic, ink and collage on paper by Adrian Ghenie.

c.2012 Full figure (head after L. Darwin 1878a) dressed as a conquistador in a forest by Cuban artists Reynaldo Pagán and Orestes Campos (d.2016).

2012 Wall mural of bearded CD by street artist Rocket01. Sidney Street, Sheffield.

2013 Stained glass window. Fiondella Great Room, Ruane Center for the Humanities, Providence College, Rhode Island, USA.

2015 Watercolour three-quarter left profile by Fabrizio Cassetta.

2015 Four-colour screen print of young CD with *Beagle* by Alice Pattullo.

2016 Roots of knowledge. Stained glass window, with CD amongst thousands of other figures, by Tom Holdman, Utah Valley University.

2016 Acrylic painting "C.D." by Jonathan McAfee, after 2009 NHM exhibition poster.

2016 Watercolour by Kan Srijira, after 2009 NHM exhibition poster with fake finger to lips.

-2017 "Charles Darwin with Atari joystick" by Avril E. Jean.

2017 Acrylic painting by Victor Ovsyannikovnd "Origin of Species".

2017 Colour pencil drawing (and a copy) after Lock & Whitfield 1878a by Dave Duffy, USA.

2018 Watercolour portrait by Jim Stovall.

nd Graphite drawing of bearded CD by David R. Steward.

nd Watercolour "tea with Darwin..." by Will Bullas.

nd Pencil drawing of bearded CD by Omar Bautista.

There are very, very many more of recent date.

Prints to 1982: Original depictions of CD in print are extremely numerous and there are no doubt hundreds more not in the list of 596 below. English-language sources are the most complete here. Publications from the following countries and languages are included: Argentina, Armenian, Australia, Belgium, Canada, China, Cocos (Keeling) Islands, Czechoslovakia, Denmark, East Germany, Finland, France, Germany, Hungary, Italy, Japan, Latvia, The Netherlands, Norway, Portugal, Romania, Slovenia, Spain, Switzerland, UK, USA, USSR and Yiddish. Many more in other languages exist. The scale and range revealed here is surprising and significant from many perspectives. The earliest prints were expensive engravings, special publications and periodicals. After CD's death they tended to be frontispieces, cover art or illustrations to his own books and then to books about him and newspaper illustrations. The early woodcuts and engravings were widely pirated throughout the 19th century, though often an artist would make a new work only based on or imitating an earlier model. Photographs were often the model for printed portrayals, sometimes very closely followed, other times only loosely so. The two most popular have been Lock & Whitfield 1878a and L. Darwin 1878a, followed closely by Maull & Polyblank 1857, Rejlander 1871c and Elliott & Fry c.1880b. Slightly less popular were Elliott & Fry 1874a, Edwards 1866a, Elliott & Fry c.1880a, Elliott & Fry 1881a and Elliott & Fry 1874b. With only a handful were Cameron 1868a, and Elliott & Fry 1881b, W.E. Darwin 1866, Elliott & Fry 1871a, L. Darwin 1878b, Elliott & Fry 1874d and Barraud 1881b. And apparently only being imitated once were Edwards 1866d (1868), Cameron 1868b (1875), Cameron 1868c (1982), Rejlander 1871d (1882) and Elliott & Fry 1874c (c.1900?). Of the paintings, the clear favourite was Collier 1881 with at least 27 (1883-1979) but its popularity seems to have waned. The next most imitated has been Richmond 1840, but only since the mid-20th century (1955-81) when it became known in the public domain. The only other painting being significantly imitated was Ouless 1875 with at least 10 (1877-1957), also having apparently lost popularity.

In the following list, exact reproductions of an earlier work, whether mechanical, electrotype or photographic are not recorded. If they were, the list would be as long as this entire volume. Items which are illustrated here are indicated in the list by the date being underlined. Original printed portraits not recorded here include postage stamps. Over 100 depicting CD are known. Many are original artworks, although often based on 19th-century photographs or paintings.

<u>1849</u> Lithograph by Thomas Herbert Maguire (1821-95), signed "T H Maguire 1849". Printed by M. & N. Hanhart ("M & N Hanhart Impt."), (29x24cm, on sheet 45.5x34cm). Lithographed signature "Charles Darwin" below and raised blind stamp of Ipswich Museum with arms bottom right. Ipswich Museum BAAS Portraits. One of more than 60 lithographs paid for by George Ransome, chemist and druggist in Ipswich. See CCD4:250-51. CD is seated in a Down House study chair. See CD to Henslow 20 Nov. [1849] "My portrait has been taken". CCD4:282. CD paid the publishers of the series, Lovell Reeve, £1 6d, on 17 Nov. 1849. 8 Jan. 1850 "Subscription, per. L. Reeve for Portrait of G. Ransome for a sum of 10s. 6d". 5 Aug. 1852 CD paid "Ipswich Museum" £1 1s Classed account book (Down House MS). CD to Henslow: "My wife says she never saw me with the smile, as engraved, but otherwise that it is very like." CCD4:303. See also George Ransome.*

<u>1865</u> Woodcut (with very un-Darwinian nose) after Maull & Polyblank 1857 by August Johann Daugell (1830-99). Frontispiece to *Journal*, [in Russian] F2383.1.

1866 Tinted lithograph by Vincent Brooks (1814-85) after W.E. Darwin 1864a, *Quarterly Jrnl. of Science.* 3, no. 10, facing p. 151. Mistakenly described in CCD14:xviii as after photograph by E. Edwards.

1868 Woodcut after Edwards 1866d, signed "Cesch"(?) *Illustrierte Zeitung* [Leipzig], (9 May), p. 324. A very similar one in *American Phrenological Jrnl.*, (Oct. 1868), vol. 48, no. 4. p. 121.

1870 Engraving after W.E. Darwin 1864a with standing imaginary body added, frontispiece to *Origin* [in German], F675 and [in Hungarian] 1873, F703 et al "Stich u. Druck v. Weger, Leipzig."

1871? Engraving after Elliott & Fry 1871a. Stich & Druck von [August] Weger [1823-92], engravers, Leipzig. A copy sent by CD to Daniel Oliver in Nov. 1871 sold at Christie's in 2010 for $4,375.

[1871] Lithograph after Elliott & Fry 1871a by N.J.W. de Roode, P. Blommers, The Hague.

1871 Line and stipple engraving after Elliott & Fry 1871a by George E. Pernie. Frontispiece to *Eclectic Magazine*, (New York) (Jun.).

1871 Engraving by "ST" after Pernie 1871. *Appleton's Jrnl.*, vol. 3, pp. 439-41. Reproduced in *Popular Sci. Monthly*, 1873, vol. 2, pp. 497-98.

Maguire 1849.   Daugell 1865.   Brooks 1866.   *Illustrierte Zeitung* 1868.

Weger 1870.   Pernie 1871.   *Illus. Lond. News* 1871.   'ST' *London Jrnl.* 1872.   Austin 1873.

1871 Etching, after Edwards 1866a (reversed), *Illustrated London News*, (11 Mar.), pp. 243-4. Reproduced in *Every Saturday*, (Apr.) 1871, *Harper's Weekly*, (8 Apr.), and widely elsewhere. Reproduced again, head and shoulders only, in *Illustrated London News*, (29 Apr. 1882), p. 416.

1871 Woodcut after above, head and shoulders only. *Australian Town and Country Jrnl.* (22 Jul.), p. 9.

1871 Woodcut after *Illustrated London News* 1871. *Illustreret Tidende* [Denmark], pp. 341-43.

1871 Woodcut after Rejlander 1871c. *The Illustrated Review*, (15 Nov.), vol. II, p. 289.

1872 Engraving after Rejlander 1871d, head only, by 'ST'(?). *The London Journal*, (8 Jun.), p. 357.

1873 Drawing loosely after Maull & Polyblank 1857. *Austin American-Statesman*, (22 Mar.), p. 19.

1873 Woodcut after Edwards 1866a (reversed), *Die Illustrierte Welt*, no. 51.

1873 Woodcut portrait (with Lamarck, St. Hilaire and Haeckel) by Adolf Neumann, captioned 'The four main representatives of Darwinism'. *Die Gartenlaube*, no. 43, p. 711.

1874 Engraved portrait. *Pictorial world*, (6 Jun.), p. 228. Reprinted in *Pokrok*, 1876, *Schoolhive*, 1877.

1874 Steel stipple engraving by Charles Henry Jeens (1827-79), after Rejlander 1871c. "…Presented to the subscribers to Nature No. 240 June 4th 1874…" Asa Gray, Charles Robert Darwin. *Nature*, vol. 10 (4 Jun.), pp. 79-81; Reproduced in *Popular Science Monthly*, vol. 5, (1874), pp. 475. *The Graphic* 1875, Frontispiece, *Charles Darwin memorial notices*, 1882; 1892 *Expression* [in Spanish] F1214, *Galerie hervorragender Aertze und Naturforscher*. (16x24.5cm), J.F. Lehmann, Munich, 1909 and CCD22 et al.

1875-78 Gold embossed cover portrait after W.E. Darwin 1866. Carus ed., *Ch. Darwin's Gesammelte Werke*. F1602.

1875 Woodcut after Wallich 1871a by Moritz Klinkicht, *Gardeners' Chron.*, (6 Mar.), p. 309.

1875 Woodcut loosely after Elliott & Fry 1874b. *Neue Alpenpost* [Zurich], (13 Feb.), vol. 1, no. 7.

1875 Woodcut after Cameron 1868b by E. Ade, Stuttgart. *Bulletin d'insectologie agricole: journal mensuel de la Société centrale d'apiculture & d'insectologie*, p. 784. France.

1876 Woodcut after Rejlander 1871c right profile by R. Taylor in [W. Robinson,] Charles Darwin. *Garden, an Illustrated Weekly Jrnl.*, 8 (Supplement, 1 Jan.), 1876, pp. xi-xii

1876 CDV of drawing of J.S. Mill, C. Lamb, C. Kingsley, H. Spencer, J. Ruskin, and CD. With a statue of a monkey above the standing figure of CD with arms behind his back. London: Hughes & Edmonds (22.5x16cm). Some sold with a green mount and "With best wishes for a happy Christmas and a bright new year" in gold lettering.

Pictorial World 1874.　　Jeens 1874.　　Klinkicht 1875.　　Ade 1875.　　Hughes 1876.

Rajon 1877.　　Simms 1879.　　Closs 1880.　　Legros c.1881.　　Meyer 1882.

1876 Woodcut after Rejlander 1871c (reversed), signed "P". *Neue Illustrierte Zeitung*, (1 Oct.), p. 1.

1876 Woodcut after Elliott & Fry 1874a. *Popular Science Monthly* vol. 2, facing p. 497, 1882 *ibid* p. 261.

1877 Copper etching by Paul-Adolphe Rajon (1843-88), (40x60cm) after Ouless 1875. On Whatman paper £5 5s, Japanese paper £6 6s, "a few early proofs, with remarques" £15 15s. There is a proof with marginal dry point sketches at the American Philosophical Society. One version has a lithographed signature "Ch. Darwin" another has eight surround vignettes, one being CD loosely after Wallich 1871a. Rajon considered it his greatest work. By 1893 this highly-praised engraving was selling for $300. Reproduced in Feb. 1909 in Bookman.

1877 Drypoint etching by Georges Pilotell (1845-1918), (16.3x11.8cm). British Museum.

-1878 Etching by George Paul Chalmers. A remarque proof on vellum sold for £19 19s in 1878.

1879 Engraving after Elliott & Fry 1874d by Vincent, Brooks and Co. with biographical notice. *The Examiner*, (11 Oct.) issue 3741, facing p. 1312. Registered for copyright 11 Oct. 1879.

1879 Woodcut loosely after a photograph to exemplify the phrenological organ of Observativeness (Large). Joseph Simms, *Nature's revelations of character; or physiognomy illustrated*. New York, pp. 192-3.

1880 Woodcut loosely after Elliott & Fry 1874a. Frontispiece to *Origin* [in Spanish] F771.

1880 Woodcut after W.E. Darwin 1864a by Adolf Closs (1846 - 1894) in *Nutiden* [Denmark], p. 59.

1880 Woodcut after Rejlander 1871c (reversed). L.N. Fowler, Charles Darwin: A phrenological delineation. *The Phrenological Magazine*, vol. 1 (Apr.), p. 89.

1880 Woodcut after Elliott & Fry 1874a. *The Leisure Hour*, (11 Sept.), 1498, p. 585.

c.1881 Drypoint etching by Alphonse Legros, three-quarter left profile, (21x15cm).

1881 Engraving loosely after Lock & Whitfield 1878. Frontispiece to *Descent* [in French] F1061.
1881 Engraving after L. Darwin 1878a, signed and dated by Worthington Smith for *Gardeners' Chron.*
1881 Engraving after Elliott & Fry 1874a. *L'Illustration Belge* no. 37, (12 Aug.), cover.
1881 Three-quarter right profile woodcut in Chr. G. Hottinger, *Die Welt in Bildern.*
1881 Outline drawing of CD's hand (no. 20). Claude Warren, *The life-size outlines of the hands of twenty-four celebrated persons.* London: Modern Press. "Darwin is rather a large man, and his hand is not a large one in proportion. It is hard and rough, (the right one most so), very spatulous, and rather hairy, with knotty fingers; the lines are numerous and confused. It is a very interesting hand to those studying 'fingerology', for instance to compare it with that of an artist, or, of a man devoid of reasoning power."
1882 Engraving by Moritz Klinkicht (1845-1912?). "(From the Medal by Alphonse Legros, Exhibited at the Royal Academy, 1882.)" *The Magazine of Art*, vol. V.
1882 Woodcut after L. Darwin 1878a in Otto Zacharias, *Vom Fels zum Meer* 2: 348-53.
1882 Woodcut after Elliott & Fry 1874a (reversed), longer beard. *Illustrated Police News*, (29 Apr.), p. 4.

Klinkicht 1882.   Loewenstein 1882.   Tilly 1882.   Klíč c.1882.   Johnson 1883.

Flameng 1883.   Kruell 1884.   Hader 1885.   Nicholson 1886.   Kruell 1887.

1882 Three-quarter right profile with huge beard. *Rivista Illustrata* [Milan], front page.
1882 Unique three-quarter right profile woodcut. *Il Secolo* [Milan], 22-23 April.
1882 Engraving after Lock & Whitfield 1878. Signed H. Ease(?). *Le Monde Illustré* [France], (6 May).
1882 Woodcut by Henri Meyer after Lock & Whitfield 1878. *Le Journal illustré* [France], (14 May).
1882 Woodcut after Rejlander 1871c by R. Loewenstein. *La Ilustracion* [Argentina], (7 May).
1882 Engraving after Elliott & Fry 1874b "A.G. / Smeeton Tilly". *La illustration* [France], (29 Apr.).
1882 Chromolithograph bearded portrait by Karl Václav Klíč (1841-1926) after Elliott & Fry 1874a(?) (reversed). Published in Vienna. Österreichische Nationalbibliothek, #3476835 - Pg 1100: II (1).
1882 Steel engraving "after a photograph" by John Gilbert. *Magasin Pittoresque* [France], p. 392. A very stern and poor likeness. Another version is signed "A. Gusman"(?)
1882 Woodcut after Wallich 1871a in *Popular Science Monthly*, vol. 21.
1882 Engraving after L. Darwin 1878a by M. Weber.
1882 Green and black drawing after Richmond 1840 with *Beagle*. UK anniversary stamp issue card.
1882 Engraving after Klíč c.1882 but with pointy nose. From an untraced Spanish publication. Printed as a postcard in Spain c.1900.
c.1882? Three-quarter right profile engraving after Lock & Whitfield 1878 by R. Bong. Untraced.
c.1882? Engraving after Wallich 1871a, *Il mese agricolo* [Italy]. Similar to *Gardeners' Chron.* 1875.
c.1882? Russian engraved portrait after Edwards 1866a (reversed), surrounded by laurel wreath.
1880s? Three-quarter left profile not after a photograph. Domela Nieuwenhuis, Leiden.
1880s Half-length three-quarter left profile woodcut. L. Zehl's Verlag, Leipzig.

nd Engraving after Elliott & Fry 1874b, signed "S.T. sc". Publication untraced, French?

1883 Engraving after L. Darwin 1878b by Thomas Johnson. *Century Mag.*, (Jan..), vol. XXV, no. 3, facing p. 323 and frontispiece to LL2. Proof exhibited New York Academy of Sciences, Feb. 1909.

1883 Copper engraving by Léopold Flameng, after Collier 1881. Copies dated 10 Mar. Fine Art Society (Limited) London, with engraved signatures of artist & engraver.

1883 Woodcut after Lock & Whitfield 1878. *Westermanns illustrierten deutschen Monatshefte.*

1883 CDV of "Krao" with an apparition of CD floating above by Farini (William Leonard Hunt). In Nadja Durbach, *Spectacle of deformity: freak shows and modern British culture*, 2010, p. 92.

1880s-90s Lithograph after Lock & Whitfield 1878a. Added body with velvet collar. As CDV in at least 4 variants. 1. with "DARWIN" and publisher's text on recto mount (illeg. in only copy seen). 2. "DARWIN" in white on lower left of image. 3. "DARWIN." in black on recto mount. 4. On mount: "Bruckmann's Collection" / "Geo. Kirchner & Co., New York" / "DARWIN". Frontispiece to *Descent* [in Polish], vol. 2, 1960, F1103.2. Often mistaken for a photograph. Reproduced in *Popular Science Monthly*, vol. 74, (1909), p. 334. The editor, thinking it a unique photograph, stated that the source was unknown to him.

1880s Etching signed "E.W. Andrews fecit" after Elliott & Fry 1874d.

1880s? Unsigned lithograph after Barraud 1881b.

Andrews 1880s.   *St. Louis* 1887.   Whymper 1888.   L&F 1878a 1880s-90s.   *Humoristické* 1889.

1884 Woodcut after Maull & Polyblank 1857 by Gustav Kruell, *Harper's Mag.*, (Oct.). LL2 frontispiece, F1452 and widely reproduced thereafter. In 1889 Kruell, then in the USA, sold a limited edition of 200 mounted, signed and numbered prints of this and the Kruell 1887 woodcut for $20.

1884 Engraved portrait after Elliott & Fry 1874a? (reversed). Walford, *Greater London*, vol. 2, p. 120.

1884 Cabinet card for American editor A. Wilford Hall of *The scientific arena* featuring a large number of figures with CD in the top centre. The card purports to be "from the great painting of Mr M. C. Tiers", a New York portrait painter.

1885 CDV of painting by Ernst Hader. Photographed by Sophus Williams (Berlin) with lithographed signature of CD. An extremely poor likeness with pointy nose, reproduced as frontispiece to 1889 *Origin* [in Dutch] F650 and many other publications.

1885 Photogravure after Maull & Polyblank 1857 by Meisenbach in Ernst Krause ed., *Charles Darwin und sein Verhältniss zu Deutschland.*

1885 Photogravure after Barraud 1881c by Meisenbach in *ibid.*

1886 Gold embossed cover art after Elliott & Fry 1874a. H.A. Nicholson, *Natural history: its rise and progress in Britain as developed in the life and labours of leading naturalists.*

1886 Engraving after Elliott & Fry 1874a, frontispiece to *ibid.*

1886 Three-quarter left profile engraving by R. Bong after Tilly 1882. Adams, W.H. Davenport, 'Charles Darwin' In: *Master minds in art, science, and letters*, London, pp. 251-76.

1887 Engraving by Gustav Kruell after Elliott & Fry 1881b. LL3 frontispiece, later many others.

1887 Woodcut after Elliott & Fry 1874a. *El Museo Popular*, Madrid.

1887 Engraving of CD "on his usual walk". *Illustrated London News*, (10. Dec.), p. 686.

1887 Engraving of CD (drawn in scale far too small) in the hothouse. *Ibid.*

1887 Engraving of CD sitting in the new study (drawn in scale far too small). *Ibid.*

1887 Woodcut after Lock & Whitfield 1878a. Signed JHast(?) *St. Louis Post-Dispatch*, (4 Dec.), p. 21.

1888 Wood engraving by Edward Whymper (1840-1911) of Boehm statue in and for the Report of the Darwin committee and some later publications.
1889 Woodcut after Elliott & Fry 1874a. *Appletons' Cyclopædia of American Biography*, vol. 6, p. 677.
1889 Woodcut after Maull & Polyblank 1857, frontispiece to *Life and letters* [in Danish] F1528.
1889 Woodcut after L. Darwin 1878a, frontispiece to *Life and letters* [in Danish] F1528.
1889 Woodcut after Elliott & Fry 1881b, frontispiece *Life and letters* [in Danish] F1528 (poor likeness).
1889 Lithograph after Lock & Whitfield 1878a(?) in *Humoristické listy* [in Czech].
1890 Etching by Gustave Mercier (1858-98) after Rajon 1877, signed "G. Mercier", (40x30.8cm). Published by Robert M. Lindsay (USA).
1890 Engraving, with etching of a study including a profile portrait of CD in a roundel on the wall after Rejlander 1871c by Charles William Sherborn. Wellcome Library.
1890 Woodcut after Lock & Whitfield 1878a, frontispiece to *Journal.* F60.
c.1890 Woodcut after Edwards 1866a? "Illustrationsprobe aus dem 'Baer=Hellwaldchen Werke.'"
c.1890 Lithograph(?) bearded portrait by V.A. Heck, Vienna.
c.1890? Engraving after Cameron 1868a by "RL", Boston.
c.1890? Woodcut after Lock & Whitfield 1878a? by Holter.

Nugent 1891. *Inter-Ocean* 1894. Cook 1897-8. *Herald* 1897.

c.1891 Lithographed left profile by Johannes Müller, on box of "Das Darwin-Spiel". Leipzig.
1891 Left profile woodcut. *The Independent* [Kansas], (9 Feb.).
1891 Woodcut after Elliott & Fry 1881d. Frontispiece to *Autobiography* [in Japanese], F2339.
1891 Woodcut "Darwin and the squirrels.", by Meredith Nugent. In C.F. Holder, *Charles Darwin: his life and work*. New York: Putnam. See story in LL1:115. Also the following woodcuts by Nugent: "Darwin finding a vampire bat biting a horse."; "Darwin testing the speed of an elephant tortoise (Galapagos Islands)." The clothing and beard style are very much of the 1890s. Often reproduced since.
1891 Woodcut depiction of the (imagined) eyes and nose of CD as representative of "the observing eyelid", "overhanging folds of flesh in the outer corner of the upper eyelid, such as are shown in the picture of [CD], are common in persons noted for accurate observers." Character in eyes. *The Evening World*, [New York], (2 Jan.), p. 2.

Evening World 1891.

1892 Engraving after Lock & Whitfield 1878. *Les Contemporains* [Paris], no. 11, (25 Dec.).
1893 Woodcut after Elliott & Fry 1874b. Rev. John Vaughan, Boyhood of Charles Darwin. *Boys Own Paper*, (8 Apr.), p. 445.
1893? Woodcut after Barraud 1881b, frontispiece to *Origin*. New York: National Book Co. F2057.
1894? Woodcut after Elliott & Fry c.1880b? frontispiece to *Origin*. New York, F433 & F443.
1894 Woodcut "Mr. Darwin in his garden." Louise Lyndon, *Inter-Ocean* (4 Nov.), pp. 5-6.
1894 Woodcut after Edwards 1866a (reversed) in A. Geikie, *Great men and famous women*, vol. 4.
1894 Gold embossed half-length, three-quarter right profile bearded portrait after a photograph. Cover art to *Lives and labours of leading naturalists*. Edinburgh: W&R Chambers.
1895 Steel engraving after L. Darwin 1878a. *Le Magasin Pittoresque*, [France] p. 317.
1896 Engraving(?) after Elliott & Fry c.1880a. Frontispiece to *Origin* [in Japanese] F718.
1896 Left profile woodcut loosely after Cameron 1868a? *The Atlanta Constitution*, (20 Aug.), p. 8.
1896 Woodcut loosely after Barraud 1881c (reversed). A.R. Spofford et al eds., *The library of historic characters & famous events*. Philadelphia, p. 46.

1897 Outline woodcut after P.R. Montford 1897. *Tamworth Herald*, (12 Jun.), p. 6, et al.

1897 Unique engraving after a photograph. *The Herald*, [Los Angeles, California], (13 Jun.), p. 23.

1897 Unique woodcut of the Montford statue in Shrewsbury. *The Cardiff Times*, (7 Aug.), p. 7.

1897 Small chromolithograph portrait roundel in right profile by Walter Besant, "Her Majesty's glorious jubilee", *Illustrated London News*.

1897 Photogravure after Elliott & Fry c.1880b. *Harper's Monthly*, (Nov.), p. 936.

1897 Engraving after Ouless 1875. Cambridge engraving Co. Frontispiece *Origin*, John Murray, F453.

1897 Three-quarter left profile, no resemblance to a photograph. *Austin Daily Statesman*, (31 May), p. 5.

1897-8 Half-length steel engraving after Elliott & Fry c.1880a by C. Cook (10.7x13.4cm). The body, seated in chair with walking stick, was added by Cook, only the head is after the photograph. In James Taylor, *The Victorian Empire*. Mackenzie: London, Edinburgh & Glasgow; later many others.

1899 Engraving after Rejlander 1871c. *Aftonbladet* [Sweden], (2 Oct.).

1899 Woodcut after L. Darwin 1878a in *Blackie's Modern Cyclopedia*. Blackie & Son, Ltd.

1899 Engraving after Elliott & Fry 1881b (head only). Cover art to *Origin* [in German] F684 and other works by CD with the same publisher, Schweizerbart of Stuttgart.

1899 Engraving loosely after Collier 1881. *The Salt Lake Herald*, (31 Dec.), section two, p. 11.

F60 1890.   Hollyer 1900.   Bauer c.1900-10.   Hech c.1900?   Ogden's 1901.

1890s? "Memorial of Christ's College, Cambridge" with montage of alumni bust portraits with CD after Ouless 1883. Lithograph by Benyon & Co., Cheltenham, (78x55cm).

c.1890s Cigar box colour lithograph after Barraud 1881a, New York

c.1900? Lithograph loosely after Elliott & Fry 1874c by V.A. Hech. Severin Worm-Petersen, Vienna.

c.1900? CDV engraving after Collier 1881. Wesenberg & Co. Russia.

c.1900 Engraving after Ouless 1875. Frontispiece to *Descent*. Chicago: Thompson & Thomas. F996.

1900 Woodcut loosely after Elliott & Fry c.1880b, *Appletons' Cyclopædia of American Biography*, v. 7, p. 82.

1900 Engraving after Elliott & Fry c.1880b by Samuel Hollyer (1826-1919), (14x21.6cm).

1900 Small roundel portrait, loosely after Elliott & Fry c.1880b? The fifty greatest men of the nineteenth century. *The San Francisco Call*, (30 Dec.), p. 6.

1900 Postcard. Lithograph after Elliott & Fry c.1880a. Stengel & Co. (1885-1945) Dresden.

1900 Inverted shadow drawing after L. Darwin 1878a. Ten master minds of the century. *The Salt Lake Herald*, (30 Dec.), p. 26. A widely syndicated set of small portraits.

c.1900-10 Lithographed portrait by Karl Bauer (1868-1942), signed "Karl Bauer gez." (54x44cm), Künstlerbund Karlsruhe.

c.1900-10 Lithographed postcard by Karl Bauer, three-quarter left profile, dark collar. Under image "Bauer. Charles Darwin [CD in Cyrillic] 96" in white.

c.1900? Original painting of CD in three-quarter left profile, eyes directed upwards. Pince-nez visible. A poor likeness. Seen on two postcards: on recto "Charles Darwin" in Cyclic and "GG/C°/2097" all in white. And "Darwin." "GG/C°/5130" also in white text.

c.1901 Woodcut after Elliott & Fry c.1880b. "Leaders in science". *Cassell's History of England*.

c.1900? Printed drawing loosely after Elliott & Fry 1874a by Heinrich Scheu. Zürich.

1901 Colour portrait after Collier 1881. Ogden's Guinea Gold Cigarette Tobacco Card No. 290, UK.

1901 Engraving after Elliott & Fry c.1880b. Robert Wilson, *The life and times of Queen Victoria*.

1902 Woodcut loosely after Elliott & Fry c.1880b? Frontispiece to *Origin* F485.

1902 Redrawing of Hader 1885. Frontispiece (along with J.S. Mill) to *Origin* [in Chinese]. F1846.

1902 Colour portrait after Collier 1881. Ogden's Guinea Gold Cigarette Tobacco Card No. 337, UK.

1902 Lithograph(?) after Rejlander 1871c (reversed). Frontispiece to *Expression* [in Spanish] F1214 & *Descent* F1124.

1903 Woodcut after Elliott & Fry c.1880b. Frontispiece to *Origin*. London: Watts. F491.

1905 Lithographed sketch after Lock & Whitfield 1878a. Frontispiece to E. Hubbard, *Little journeys to homes of great scientists*. New York: The Roycrofters.

1906 Outline woodcut bearded portrait. *Aberdeen Herald*, [Montana, USA], (17 Sept.), p. 7 and others.

1907 Image of a bust of CD after Elliott & Fry c.1880a, UK.

1907 Godfrey Phillips cigarettes "Busts of Famous People". Imaginary CD bust after Barraud 1881a.

1907 Woodcut after L. Darwin 1878a. Frontispiece to P.M. Gander, *Darwin und seine Schule*.

1907 Artwork after L. Darwin 1878a. Boyhood of Darwin. *New-York Daily Tribune*, (24 Nov.), p. 5.

1908-09 Steel engraving after L. Darwin 1878a. *Meyers grosses Konversations-Lexikon*, 6th edn.

1908-15 Postcard, black and white, after Collier 1881. USA.

1909 Woodcut loosely after Elliott & Fry c.1880b? Cover art to *Origin* [in Danish] F644.

Hubbard 1905.   Poulton 1909.   Huebsch 1909.   Planck 1909.   *Marion* 1909.

1909 Right profile bust in gold embossed on cover of Poulton, *Fifty years of Darwinism*,

1909 Full-face drawing by Gertrude Huebsch after Elliott & Fry c.1880a? *The marvellous year*, p. [58].

1909 Lithograph loosely after Rejlander 1871c? by Willy Planck. Cover art to *Charles Darwin Gedenkschrift zur Jahrhundertfeier seiner Geburt*. Stuttgart. Also issued as postcard. "Kosmoskarte Nr. 1".

1909 Unique engraving of CD, three-quarter left profile, by Bergy(?). *Marion Weekly Star* [Ohio], (13 Feb.), p. 11. The same in several other American newspapers in 1909.

1909 Chromolithograph after L. Darwin 1878a on illuminated Celebratory address from the University of Rome, on the occasion of the Darwin Centenary Conference.

909 Engraving after Klíč c.1882. Georg Thieme. Leipzig, Germany.

1909 Lithograph(?) very loosely after Maull & Polyblank 1857, head and chest only. *The Open Court* [Chicago magazine], vol. XXIII. Reproduced in *American Review of Reviews*.

1909 Woodcut after Elliott & Fry c.1880b. *The New Era* (Lancaster PA), special feature page.

1909 Woodcut after Collier 1881. *Social-Demokraten* [Norway], (12 Feb.), p. 2.

1909 Engraving after Lock & Whitfield 1878. *Landsbladet* [Norway], (12 Feb.), p. 1.

1909 Woodcut after Collier 1881 (head only) in *Folkelaesning* [Denmark], (21).

1909 Woodcut after Collier 1881 in *Hoejskolebladet* [Denmark], pp. 223-230.

1909 Woodcut after Maull & Polyblank 1857, frontispiece to *Foundations*, F1555.

1909 Engraving in ornate frame with two monkey heads after Lock & Whitefield 1878. *The Hickman Courier*. [Kentucky], (4 Feb.). Also in many other American newspapers of the year.

1909 Engraving after Elliott & Fry c.1880b in *Frems Aarbog over Ny Viden og Virken* [Denmark], p. 90.

1909 Woodcut loosely after Lock & Whitfield 1878 by "WK". *The Province* [Vancouver], (9 Jan.), p. 4.

1909 Darwin's centenary. *Mining Journal* [Maryland], (6 Feb.), p. 1. Woodcuts by 'Parker': CD shooting birds in Shrewsbury; CD hearing about the Bell stone; Elderly CD walking on sandwalk in top hat and overcoat, very much in style of 1909. Polly trots alongside; Elderly CD in shawl, standing, looking through magnifying glass and portrait after Lock & Whitfield 1878. Widely syndicated in American newspapers in 1909.

1910 Half-length, full-face lithograph, not after a photograph. *Illustrated London News* (16 Jul.), p. 94.

1910 Illustration of an unknown bust. Frontispiece to *Origin of species* [in Russian], F757.

1910 Modern-style engraving after a photograph, Lock & Whitfield 1878a? Frontispiece to *Origin*. Ward Lock & Co. Not in Freeman 1977. Now F1878.

1910 Engraving loosely after Rejlander 1871c (reversed). Cover art to *Origin* [in Spanish] F2436.

1912 Original drawing of beardless "Darwin watching an ant-lion capturing its prey" in 20th-century hat and clothing by C.M. Sheldon. *Children's Mag.*, vol. 4, p. 495.

1912 Lithograph(?) of CD in hat and cloak with walking stick behind Down House, by E.F. Skinner. *Children's Mag.*, vol. 4, p. 495. Still appearing in newspapers in the 1930s.

1912 Woodcut loosely after Elliott & Fry c.1880a. *Kongsberg Dagblad* [Norway], (9 Oct.), p. 2.

Aarbog 1909.   *Province* 1909.   *Mining Journal* 1909.   Sheldon 1912.

Skinner 1912.   Klaatsch 1919.   *Daily. H.* 1920.   Hoover 1922 (extract).   Gallieni 1925.

c.1913 Postcard with engraving after Elliott & Fry c.1880a. Germany.

1913 Embossed profile after Wyon 1882 medallion (reversed). Cover art to *Origin* [in Dutch] F652.

1914 Woodcut after Lock & Whitfield 1878a, frontispiece to 1914-15 *Origin* [in Latvian] F736.

1914 An original three-quarter left profile colour painting(?) of bearded CD. Cover art to *Journal*. London: Thomas Nelson (Nelson's Classics). F126.

1919 Black, white and yellow woodblock half-length portrait after a photograph. Cover art to Hermann Klaatsch, *Grundzüge der Lehre Darwins*. Mannheim: Bensheimer.

1920 Woodcut after L. Darwin 1878a. *Daily Herald*, (20 Oct.), p. 7.

nd Postcard with facsimile line "notwithstanding that Leibnitz formerly accused Newton of introducing 'occult qualities & miracles into philosophy / Charles Darwin". A line (somewhat altered) that first appeared in *Origin* 4th USA printing, 1860, F380. With engraving after Barraud 1881a.

1921 Lithograph after Maull & Polyblank 1857 but with fierce expression, signed "Chandler".

1922 CD as a cigar-smoking ghost passing through the body of reporter Thomas L. Masson by Ellison Hoover. *New-York Tribune*, (18 Jun.), p. 6.

1922 Chromolithograph after Collier 1881 issued by BMNH, 400 copies, signed by Collier and G. Sidney Hunt.

1923 Original three-quarter left profile by Frenckell. *Floréal* [French illustrated magazine].

1924 Ogden Cigarette card, after Collier 1881, (6.5x3.5cm). "Leaders of Men" series 15, UK.

1924 Portrait loosely after Collier 1881 "Darwin/Descent of man" with an ape and human skull. Millhoff cigarette card "Men of Genius" series, UK.

c.1924 Cigarette card loosely after Collier 1881. "Leaders of Men" series. Imperial Tobacco Company.

1924 Woodcut, three-quarter left profile portrait loosely after Ouless 1875, in *Journal* [Finnish] F179a.

1924 CD in hat and neckerchief with fossil in South America. *Chillicothe Gazette* [Ohio], (8 Jul.), p. 6.

1925 Left profile woodcut by G.E. Gallieni. "Darwin Carlo." A very poor likeness.

1925 Three-quarter left profile woodcut, looking downwards. *Het volk* [Netherlands], (14 Feb.), p. 1.

1925 Engraving after Elliott & Fry 1871a? in *Atlas van de geschiedenis der geneeskunde*. Amsterdam.

1925 Three-quarter right profile after Elliott & Fry c.1880b? G. Overton, *Portrait of a publisher*, p. 46.

1926 Engraving(?) after L. Darwin 1878a. Frontispiece to *Origin* [in Yiddish] F1139.

1926 Negative-like drawing after Elliott & Fry 1881a. Cover art to Gamaliel Bradford, *Darwin*.

1928 Full-face portrait after Collier 1881 by S. Carlisle Martin. *St. Louis Post-Dispatch*, (8 Feb.).

1929 Portrait after Ouless 1875 (reversed). *The Age* [Melbourne], (19 Jan.), p. 4.

1930 "Charles Darwin finds an interesting specimen", showing it to FitzRoy with *Beagle* in the background, by Archibald Webb. Harold Wheeler ed., *Cassell's romance of famous lives*. London: Cassell.

1931 Drawing after Collier 1881. CD "did not contend that man descended from monkeys" by John Hix. *The News* [Paterson, New Jersey], (21 May), p. 34. Widely syndicated in the USA.

1932 Full-face lithograph sketch of CD by Rosalind Thornycroft. E. & H. Farjeon, *Kings and Queens*.

1932 Engraving loosely after Lock & Whitfield 1878. *Het volk* [Netherlands], (19 Apr.), p. 6.

1934 Engraving after L. Darwin 1878a. *De grondwet* [Netherlands], (3 Feb.).

1934 Unique three-quarter right profile portrait. *Der Abend* [Vienna], (10 Feb.), p. 4.

c.1930s Cigarette card after Collier 1881. "Turf" cigarettes "50 celebrities of British history N° 11". The same image, though in colour not blue monochrome, sold by Carreras Ltd., London.

c.1930s Cigarette card after Collier 1881, signed by artist. "Carlo Darwin" in red below image.

1935 Unique woodcut frontispiece, signed "T", in *Origin*, London: Odhams Press, F553.

Der Abend 1934. *Origin* F554 1936. Travelers insurance 1940. *Danville* 1941.

1935 Three-quarter left profile by John Hix. *Greenville News* [South Carolina], (19 Oct.), p. 4.

1935 Three-quarter right profile by John Hix. *Bradford Evening Star* [Pennsylvania], (17 Jul.), p. 6.

1935 Three-quarter left profile by Bill Venn. Furbay, Debunker. *Atlanta Constitution*, (12 Feb.), p. 11.

1935 After Collier 1881, head only. "Celebrities of British History" Carreras Cigarette card, UK.

1935 Printed bookmark after L. Darwin 1878a by Olleschau, a paper company of Prague.

1935 Engraving after Lock & Whitefield 1878 (reversed). *From ape to man* [in Polish]. Universum, 18.

1935 Re-drawing of Richmond 1840 (reversed). Kaempffert, *New York Times*, (15 Sep.), p. 10.

1936 Cover art woodcut of bearded CD resting head in his hand, thinking. In *Origin*. New York: Modern Library (World's Best Books). F554.

1936 Three-quarter right profile sketch by William Ferguson. *Plattsburgh Daily Press*, (6. Aug.), p. 4.

1936 Tiny drawing of CD on his knees, gardening. Center of evolution wars. *The Bismarck Tribune*. [North Dakota], (10 Oct.), p. 3.

1938 Three-quarter left profile by John Hix. Darwin's nose. *Bradford Evening Star* [Penn.], (12 May), p. 6.

1938 Chromolithographed cigarette card portrait, (6x8.8cm), by Cigarette Oriental de Belgique, to be assembled in an album: *Honderd portretten van beroemde mannen*. Brussels.

1938 Re-drawing of L. Darwin 1878a with wrinkled forehead. *Atheist* [Soviet newspaper] (1 Jul.), p. 2. Later used as cover art to Gurev, *Charles Darwin and Atheism* [in Russian], 1975. USSR.

1939 Woodcut of CD's head, three-quarter right profile, after Ouless 1875 (reversed). Cover art to *The living thoughts of Darwin*. London: Cassell and Co.

1939 Three-quarter right profile pen and ink sketch of bearded CD. Cover art to *Calendar of the letters of Charles Robert Darwin to Asa Gray*. Boston, USA.

1939 Three-quarter right profile cover art vignette to *Cross and Self Fertilisation* [in Russian], F1273.

1939 Woodcut roughly after Elliott & Fry c.1880b. *New Zealand Herald*, (30 Dec.), p. 8.

1940 Sketch portrait after Ouless 1875 by Nat Falk. *Cape Vincent Eagle*, (11 Apr.), p. 5.
1940 Full face head after Collier 1881. Cover art to *Journal* [in Spanish] F2434.
1940 Chromolithograph of CD at desk with a skull. Advert, The travelers insurance company [Conn.].
1941 Outline drawing after Maull & Polyblank 1857. *Danville Morning News*, (26 Mar.), p. 5.
1942 Half-length three-quarter portrait by John Hix. *Greenville News* [South Carolina], (10 Jan.), p. 4, etc.
1943 Woodblock right profile. Cover art to *Descent* [in Spanish], F1129.
1943 Abstract portrait of seated CD with beard, right profile. Henshaw Ward, *Charles Darwin*.
1945 Silhouette right profile bust. Cover art to *Kurs darvinisma* [Course of Darwinism] Moscow.
1945 Woodcut roughly after Elliott & Fry 1874a. Great Masters. Willow Blend Tea [advert extolling "careful & observant study and perseverance"]. *Forfar Dispatch*, (26 Jul.), p. [3]. Scotland.
1946 Engraving after Barraud 1881c (reversed). Cover art to *Memorias y Epistolario Intimo*. Argentina.
1946 Lithograph(?) of bearded CD, by Gordon Ross. In *Living biographies of famous men*. New York.
1946 Three-quarter left profile cover art vignette *Darwin and his teachings* [in Bulgarian], N.V. Turbin.

Willow 1945. Rostand 1947. *New Yorker* 1948. Jahns 1949. *St. Louis D.* 1950.

Origin F633 1950. *Origin* F2538 1951. Newnes c.1950? Shukhvostova 1955. Vladimirovich 1957.

1947 Plato-like bust portrait, cover art for Jean Rostand, *"Les figures" Charles Darwin*. Gallimard.
1947 Small three-quarter right profile woodcut. World's great thinkers. *Chicago Tribune*, (9 Nov.), p. 228.
1948 Left profile drawing of CD in a four-engined propeller-driven airliner looking at engine through the window. "Evolution keeps our wings growing, Mr. Darwin." 72,000 "miles of routes…without a missing link, sir!" British overseas airways corporation. Advertisement in the *New Yorker*.
1949 Three-quarter left profile drawings of CD and A.R. Wallace by George Jahns. Dick Kirby, As it seems. *The York Dispatch* [Pennsylvania], (30 Mar.), p. 23.
1949 Modernist-style colour painting after Elliott & Fry 1881a. Cover art to Julian Huxley, *Darwin: Edizione integrale*. Mondadori, Italy.
1950 Full-face woodcut. Evolution re-examined. *St. Louis Post-Dispatch*, (6 Jul.), p. 18.
1950 CD and a servant in "Brazilian forest". CD firing a rifle, by Papé. For our young readers. Uncle Ray's corner. *Ottawa Citizen*, (30 Jan.), p. 8.
1950 Three-quarter left profile cover art vignette to *Origin* [in Bulgarian], F633.
1950 CD, Einstein and James Joyce portraits, from a painting? *Courier-Journal*, (19 Mar.), p. 59.
c.1950? Black and white drawing of CD with butterfly net. Charles Darwin: A naturalist in the forest in Brazil. *Newnes Pictorial Knowledge*.
1950 Very crude three-quarter right profile. Cover art K. Timiriazew, *Darwin's theory* [in Polish].
1950 Woodcut of seated CD loosely after Collier 1881 used in adverts for pharmacies and companies in New Jersey for "artificial breast" surgery. *The Herald-News* [New Jersey], (9 Oct.) 1950, p. 2.
1951 Engraved three-quarter left profile, not after photograph. Cover art to *Journal* [in Polish] F2538.

1952 Woodcut "My grandfather on his horse Tommy." by Gwen Raverat. *Period piece*, 1952.

1952 Three-quarter left profile by Elsie Hix. Strange as it seems. *The Messenger*, (28 Apr.), p. 8.

1952 Original three-quarter left profile bearded portrait, with gorilla in the background. "Scientist Card # 124". Look 'n See "No. 7 of 16 Famous Americans 2nd series."

Courier 1950.　　*Herald-News* 1950.　　Hix 1952.　*Journal* F1842 1955.　*Saturday Rev.* 1955.

Reinfeld 1956.　　Himmelfarb 1959.　　Horrabin 1959.　　Reurich 1959.　　Vestal 1960.

1953 CD on horseback in South America. Cover art to E. Cheesman, *Charles Darwin and his problems*.

1955 A fine engraving of CD looking out to sea from the bulwark of the *Beagle* in 1831 by O. Shukhvostova and T. Kochubey. D.V. Panfilov, *Путешествие в страну нектара*. Moscow.

1955 Drawing(?) after Richmond 1840 by Chinese artist. Cover art and frontispiece to *Journal* [in Chinese] F1842. Also in (reversed from F1842) F1843, F1566a and F1883a, heavily retouched in latter.

1955 After Barraud 1881a? Delbert Clark, What makes a genius? *Saturday Rev.*, (Nov.), pp. 9-10.

1956 Head only portrait of young CD with sideburns and *Beagle* in green, white and black. Cover art to Fred Reinfeld, *Young Charles Darwin*. New York: Sterling Publishing Co.

1956 Churchman cigarettes card with portrait after Collier 1881 in roundel and CD with pick unearthing fossil mammal with two helpers in South America. "Pioneers set". UK.

1957 CD leans against bulwark of *Beagle*, looking inboard and perhaps the most dashing depiction of CD, after Richmond 1840. Cover art to K.G. Vladimirovich, *Charles Darwin* [in Russian].

1957 Chromolithograph postcard after Lock & Whitfield 1878. Publisher: IZOGIZ, USSR.

1957 Full-face sketch of bearded CD. *Chicago Tribune*, (26 Jul.), p. 29.

1957 Colour lithograph postcard after Lock & Whitefield 1878. Verso with short biography. USSR.

1957 Engraving(?) after Maull & Polyblank 1857? in *Origin* [in Romanian] F747.

1957 Black outline drawing of CD after Ouless 1875. *The Age* [Australia], (5 Dec.), p. 13.

1957 Three-quarter left profile of CD in a line with Napoleon, Newton, Voltaire and Victor Hugo by Elsie Hix. "All were premature babies!" Strange as it seems. *Valley Times* [California], (4 Dec.), p. 46.

1958 Engraving of bearded CD in right profile, possibly after a 19th-century portrait. Cover art to Loren Eiseley, *Darwin's century*.

1958 Two-tone image after Elliott & Fry c.1880b? Cover art to *A book that shook the world*. Pittsburgh.

1958 Left bearded profile sketch of CD conjoined with Marx (centre) and Wagner by Leonard Baskin. Cover art to Jacques Barzun, *Darwin, Marx, Wagner: critique of heritage*.

1958 Small full-figure sketch of CD with walking stick behind an imaginary Down House with three dogs by Ed Stevenson(?) CD "included many *dogs* among the animals he *studied* on his estate, Down House, Kent, England." *Richland Beacon-News* [Louisiana], (29 Mar.), p. 3.

1958 Redrawing of Richmond 1840 as frontispiece to *Journal* [in Romanian] F225.

1958 Woodblock illustration after L. Darwin 1878a. Cover art to *Origin*. F583.

1958 Two-tone outline after Lock & Whitfield 1878a? Cover art to N. Barlow, *Autobiography*, F1497.

1959 Chromolithograph three-quarter right profile, not after a photograph. Pointy nose. Trading card of celebrities "14" by Guterman, Belgium. Verso has biography in French and Flemish.

1959 Etching after Maull & Polyblank 1857. *Zemědělská ekonomika*, vol. 5:11 (Nov.), p. 851.

1959? Postcard with drawing of CD after Richmond 1840 and *Beagle* in brown. Museum of Natural History, Bucharest. A copy seen has a postmarked 1959 CD stamp.

1959 Sketch by Marian Huxley (1878). Cover art to G. Himmelfarb, *Darwin and the Darwinian revolution*.

1959 Woodcut by Frank Horrabin after Maull & Polyblank 1857 (reversed). *Discovery* [Kew gardens] no. 20. Republished without provenance in John Chancellor, *Darwin*, 1973.

1959 Outline re-drawing of Maull & Polyblank 1857, cover art to Rolf Reurich trans., *Charles Darwin: Autobiographie* [in German] F1519.

  Vestal 1960.   Simpson 1961.   Low 1961.   *Der Spiegel* 1962.   Wilson 1962.

1959 Cover portrait drawing after Maull & Polyblank 1857. *Life and letters* [in German] F1519.

1959 Postcard with outline portrait after L. Darwin 1878a (reversed). Slovenia.

1959 Dark two-tone cover portrait after L. Darwin 1878a. Gerhard Heberer, *Charles Darwin*.

1959 Lithograph(?) by Royle(?) advert with quote from *Origin* for The Rand Corporation.

1959 Original half-length bearded left profile. Cover art to *Il Contemporaneo* [Italy], (Dec.) no. 20.

1959 Корсунская, *Великий натуралист чарлз дарвин* (In Russian: *Great naturalist Charles Darwin*). Woodcuts after the following earlier portraits: Richmond 1840; Sharples 1816; Claudet 1842; Maquire 1849; Maull & Polyblank 1857 and L. Darwin 1878b.

1959 Minimalist sketch of CD's head after L. Darwin 1878a. Cover art to G. Kazimierz & T. Tadeusz, *Charles Darwin: Three anniversaries 1859, 1839, 1809.* [in Russian].

1959 Engraving (head and shoulders) after Edwards 1866c by Sam Kweskin. *Saturday Rev.* (14 Nov.).

1959 Crude sketch after Elliott & Fry 1881. Flyer for 'Time will tell' (musical). University of Chicago.

1959 Lithograph after L. Darwin 1878a. Calendar of Politizdat (Political Publishing House), USSR.

1960 Full-figure three-tone drawing of CD with long sideburns, in hat with rifle, pencil, notebook and jodhpurs with various *Beagle* animals. Cover art to Millicent S. Selsam, *Around the world with Darwin*.

1960 Earth, stars and man (9) Voyage of the 'Beagle' by Don Oakley and John Lane. *Lincoln Journal Star* [Nebraska], (19 Jul.), p. 19. CD astride a giant tortoise in "the Galapagos—the birthplace of new species"; CD on a horse on the Pampas; CD after Richmond 1840 sitting before a bookcase studying a shell. "The key: natural selection…" Widely syndicated in USA.

1960 Earth, stars and man (10) Battle of the century by Don Oakley and John Lane. *News Journal* [Delaware], (28 Oct.), p. 21. Small comic strip-style CD head and hand with Wallace's Ternate essay; Linnean meeting "the double discovery is announced"; "the battle line is drawn"; concluding with three-quarter left profile of bearded CD "there is grandeur in this view of life." Widely syndicated in USA.

1960 Full-figure painting(?) of CD in straw hat holding Toxodon skull. Syms Covington stands nearby with rifle by H. Vestal. Dust jacket art to Philip Eisenberg, *We were there with Charles Darwin on H.M.S. Beagle.* Further illustrations: CD inspects the ruins of Concepcion after the 1835 earthquake; Pen and ink drawing of CD with net. "Enmeshed in the net was the cuttlefish" Cover art to *ibid.*; CD at the top of a mountain striking a heroic pose. "Darwin stared with awe and amazement".

1961 В. **Корсунская** [Vera Korsunskaya], *Рассказы о чарлзе дарвине* [In Russian: *Tales of Charles Darwin*], Illustrated by **М.Ц. Рабинович** [M. TS. Rabinovich]. Detsky Mir Publishing. Lithographs of: CD as a child in a tree; CD beats a puppy; Catherine "repeats after CD"; CD ponders colours of primroses; CD tricked into taking pastries; CD plays with Catherine near the Mount; "A few minutes later Charles left the house, head bowed low."; CD called a *Poco curante*; CD by seashore is told "It is not a profession to collect worms!"; CD learns taxidermy; CD uses a microscope; CD reads a paper at the Plinian Society; CD asked by his father about missing lectures and running from the operating theatre; CD fires an unloaded pistol at a candle in his rooms at Christ's; CD on geological tour in North Wales; Wickham tells CD to clear deck of his collection; CD unearths fossils; CD stands on a tortoise; Three of his children come to CD in study for trifles; Beardless CD walks with Polly & pigeons; CD in top hat queries a dog breeder; CD shows a colleague a collection of animal skeletons, his "friends"; CD consoles ED after death Charles Waring; Bearded CD in chair takes notes on a plant on windowsill; Outline three-quarter, head only with beard. Chromolithographs: CD and Erasmus in home laboratory; CD rides a horse through the [Brazilian] forest; CD with gauchos around campfire.

A selection from Rabinovich drawings in Korsunskaya, *Tales of Charles Darwin*, 1961.

1961 Original woodcut bearded bust, looking rather angry. Frontispiece to *Origin* [in Portuguese] F745.

1961 Original three-quarter left profile face of young CD from an ink drawing(?) Cover art to *Journal* (Everyman Library, Science no. 104). F153.

1961 Two-tone blue and black abstract left profile. Cover art to *Charles Darwin's autobiography…with introductory essay by George Gaylord Simpson*. Collier Books. F1504.

1961 Cover art, abstract portrait of bearded CD. J.C. Greene, *Charles Darwin and the modern world view*.

1961 Abstract half-face pen and ink by Joseph Low. Cover art to Alex Comfort, *Darwin and the naked lady: discursive essays on biology and art*.

1961 Drawing after Lock & Whitfield 1878a? (reversed). *Philadelphia Daily News*, (19 Oct.), p. 31.

1962 Woodcut after L. Darwin 1878a. Profiles in science. *The Troy Record* [New York], (7 Apr.), p. 28.

1962 Frontiers of science: Evolution—parts 1-5. *Times Mirror Syndicate*. Widely syndicated in USA. 1: CD as a boy reading Gilbert White with 1960s hairstyle; CD upset watching cadaver dissection in Edinburgh. CD in cap and gown at Cambridge. 2: CD netting a butterfly in South America, "The profusion of life Darwin found in South America…"; CD finds "skeletons of animals long extinct, but resembling living species…"; CD ponders "why had some species survived and others had not?" 3: On the Galapagos "Darwin found creatures that existed nowhere else on Earth!"; "Similar variations in giant tortoises convinced Darwin that all living species were undergoing continual, gradual change". Measuring carapace of a giant tortoise; Back in England, CD ponders a stuffed bird. "What causes these variations?" 4: Watching chickens. "Darwin knew that every offspring was fractionally different from its parents, and from each other…And that *man* could *select* the best for breeding—but what happened with *wild creatures?*"; CD reading papers at desk. "Strangely Darwin did not publish his theory for twenty years". 5: CD as bearded sage. "While today the process of evolution continues endlessly, by the laws of 'natural selection'…and through Darwin's insight, man now knows, alone of all living things, how he has evolved."

1962 Frontiers of science…The first struggles—part 1. *Times Mirror Syndicate*. Widely syndicated in USA. Bearded CD at desk writes with [anachronistic] quill pen.

1962 Full-length lithograph of CD in Galapagos with iguanas, remarkably bald for his age, by N.G. Wilson. Cover art to E. Royston Pike, *The true book about Charles Darwin*. Further woodblocks: Half-length portrait after Maull & Polyblank 1857; Student CD "One day he spotted two rare beetles"; CD in hat with butterfly net, again looking old; CD trying to make fire with a stick in Tahiti; CD watches dance of aborigines in Australia; CD behind Down House in soft hat and cloak; CD with beard writing "at his desk"; CD on sandwalk in soft hat and cloak.

1962 CD head and beard, tree of life and a monkey. Cover art of *Der Spiegel*, no. 52, (26 Dec.).

1962 Engraved bust portrait by Zdenek Burian. *Grosse Entdeckungen.*
1962 Unique three-quarter left profile in colour with dinosaurs. Teacard #40 Coopers & Co.
1963 Redrawing or painting of Elliott & Fry c.1880a. Frontispiece to *Origin* [in Armenian] F631.
1963 Full face sketch of straight lines by Fortun. *Origin* [in Spanish] F1130.
1963 Woodblock after L. Darwin 1878a. Cover art to *Origin* (World's Classics no. 11). F985.
1963 Woodcut of CD's bearded head, three-quarter left profile, above a tree. Cover art to S.E. Hyman, *Darwin for today: The essence of his works.* New York: Viking Press.
1963 Drawing of imaginary busts of CD and Huxley. Cover art to W. Irvine, *Apes, angels & Victorians.*
1964 Pen and ink drawing of CD loosely after Richmond 1840 (head only) looking very stern. Cover art to H.E.L. Mellersh, *Charles Darwin: Pioneer in the theory of evolution.* New York: Praeger.
1964 Pen and ink redrawing of Collier 1881. Cover art to Gavin de Beer, *Charles Darwin: evolution by natural selection*, 1st US printing. New York: Doubleday.
1965 Colour three-quarter left profile. "R.J." cigar band.

Frontiers of science: Evolution—parts 1-5 (selection of the CD illustrations). *Times Mirror Syndicate.* 1962.

   Hyman 1963.   Irvine 1963.   Mellersh 1964.   Cooper 1966.   Wilson 1967.

1960s Colour three-quarter left profile, quite different from above. Alvaro cigar band.
1965 Original bearded portrait of CD with ape, cave man and modern man. Bancroft Tiddlers cards.
1965 Illustration after Collier 1881. *Tampa Bay Times*, (7 Aug.), p. 40.
1965 Rough sketch, left profile, head only, by Janusz Klechniowski. Cover art to S. Skowron, *Narodziny wielkiej teorii.* [Polish: *The birth of a great theory*].
1966 Ink sketch of CD after Maull & Polyblank 1857, shaking hands with Wallace, Lyell looking on. Ink sketch of CD after Maull & Polyblank 1857, reading Wallace's Ternate essay, by John Kaufmann. J. Cottler, *Alfred Wallace: explorer-naturalist*, Boston: Little, Brown, pp. 135 & 197.
1966 Unrecognizably poor likeness of bearded CD in 'Frontiers of science: synthetic life—part 1'. *The Journal News* [New York], (11 Jul.), p. 27. Widely syndicated in USA.
1966 Chromolithograph(?) of beardless CD at microscope by George Allen Cooper. Cover art for George Allen Cooper, *Charles Darwin: voyager-naturalist.* New York: Macmillan. Also: CD on deck of

Beagle. CD as a schoolboy contemplating a dandelion. CD as a schoolboy with Erasmus in home laboratory. Topless CD collecting shells. Topless CD raising rock to kill a snake. CD contemplates a Toxodon skull. CD experiences an earthquake. CD on horseback in the Andes. CD with geological pick. CD at microscope.

1966 Simple three-quarter left profile lithograph. Cover art to B.Z. Jones, *The golden age of science*.

1966 Black and white painting (head only) after Richmond 1840. Cover art to Jean. F. Leroy, *Charles Darwin* (in Spanish).

1967 Great men of science: Parts 1-4. *Star-Phoenix* [Canada], (5-26 Aug.). 1. CD as a boy showing butterfly collection to his father; CD as a boy; sits reading by shed; CD collecting specimens on shore near Edinburgh. 2. CD with net in Brazilian forest; CD with a tortoise with the *Beagle* in the background; CD writing in a *Beagle* cab-in with Captain FitzRoy. 3. CD in 1836 London lodgings showing specimens to "scientific colleagues"; CD in top hot shows ED the back of Down House; Beardless and rather portly CD in the old study, in his horsehair chair, with writing board; Beardless CD in study smiles at his children at play outdoors. 4. Bearded CD walks his dog with a farmer ploughing in the background; Bearded CD sits while ED plays the piano; CD reads accounts to Down Friendly Society; CD visited by Gladstone, "He talked to me as if I were an ordinary man like him-self."

Great men of science. parts 1-4 (selection of the CD illustrations). *Star-Phoenix* [Canada], 1967.

1967 Unique half-length bearded full-face portrait, woodblock(?) with tree of life sketch from *Notebook B*. Cover art to R.J. Wilson, *Darwinism and the American intellectual: a book of readings*. Illinois: Dorsey.

1967 Sketch portrait loosely after Richmond 1840 (reversed) by Harry Toothill. Cover art to John Meehan, *With Darwin in Chile*. London: Muller.

1967 Engraving after Rejlander 1871c, head only. Cover art to Walter von Wyss ed., *Charles Darwin: Ausgewählte Schriften*. F1637.

1967 Abstract bust of CD as a chess piece. Cover art to Wagar, W. Warren ed., *European intellectual history since Darwin and Marx*. New York: Harper Torch Books.

1967 Drawing of Huxley meeting CD in his study. *Star-Phoenix* [Canada], (10 Nov.), p. 39.

1967? Lithograph after L. Darwin 1878a. *Great Soviet Encyclopedia*. USSR.

1968 Young CD in hat with a tapir by David Hodges. Cover art to Carla Greene, *Charles Darwin*.

1969 Drawing of beardless CD in dressing gown at microscope at home. *Mexia Daily News* [Texas], (26 Mar.), p. 6. A widely syndicated image part of the series "World almanac: facts".

1969 Drawing of CD walking ashore with *Beagle* in background. *The Times Recorder* [Ohio], (27 Feb.), p. 12. "World almanac: facts".

1969 Original bearded portrait with series of ape-men in background. Explorer Card "Famous Peo-ple". Brooke Bond Tea Trade Card.

1970 From an original watercolour: "Charles Darwin carried out scientific experiments on the pollina-tion of primulas." The flowers that bloom. *Look and Learn*, no. 463, (28 Nov.).

1970 CD, head after Elliott & Fry 1881a, and orchids. The flowers that bloom. *Look and Learn*, no. 463, (28 Nov.). An illustrated UK weekly children's educational magazine.

1970? Colour lithograph of CD in hat inspecting a fossil leg bone by Peter Jackson. Once upon a time... discovering the secrets of the Earth. *Treasure* [magazine] no. 319. UK.

1970s Unique three-quarter right profile with Galapagos finch and Megatherium skeleton. Panini Trading Card (UK and Europe), Prehistoric animals series number 54.

1970s "Charles Darwin on the Galapagos Islands" CD coming ashore amidst sea lions. *Look and Learn*.

1970s From a watercolour "Charles Darwin when travelling on The Beagle" [taking notes on a Galapagos iguana] by Pat Nicolle. *Look and Learn*.

1970s From a watercolour "Charles Darwin in his study", CD with microscope, by Pat Nicolle. [loosely after L. Darwin 1878a] *Look and Learn*.

1970s Chromolithographs by Andrew Howat. Charles Darwin—The man whose ideas angered the world. *Look and Learn*. "Darwin wondered why the giant tortoises of the Galapagos Islands differed on each island."; "Darwin witnessed the great earthquake that wrecked the city of Concepcion." Sepia only; CD climbs through South American forest with the *Beagle* in the background; "Darwin noticed that the natives of Tierra del Fuego had adapted to the dreary climate."; "Darwin was interested in collecting fossils." In a cave with a lantern; "Many years later Darwin wrote his book The Origin of Species." CD in his study with a quill pen.

1970s Bearded CD and The Mount, Shrewsbury, by Angus McBride. Famous Rivers - The Severn – Part 2: Highwayman's Haunt. *Look and Learn*, p. 36.

Mexia 1969.   *Times Recorder* 1969.   Great 1972.   Hall 1973.   Francavilla 1973.

1971 Three-quarter left profile drawing by Robert Holzapel(?). *Paladium-Item* [Indiana], (18 Jul.), p. 6.

1971 Drawing after Elliott & Fry c.1880a. *Frontiers of Science*. (508).

1972 Full figure four-tone drawing of CD standing with cloak. Cover art to *Great thinkers*, Warsaw.

1972 Pen and ink re-drawing of CD's face after Richmond 1840. Cover art to Roy. A. Gallant, *Charles Darwin: The making of a scientist*. New York: Doubleday.

1972 Drawing(?), half-length three-quarter left profile of seated and bearded CD. Cover art to T. Glick ed., *The comparative reception of Darwinism*. University of Texas Press.

1972 Original drawing (woodcut?) loosely after Elliott & Fry 1881, face and brim of hat only. Cover art to Sprague, de Camp & de Camp, *Darwin and his great discovery*. Macmillan.

1972 Re-drawing(?) of Ouless 1875. Cover art to J.G. Crowther, *Charles Darwin* (Brief lives).

1973 Chromolithograph of young CD in *Beagle* poop cabin with FitzRoy by Roger Hall. Cover art to L. Du Garde Peach, *Charles Darwin*. Lives of the Great Scientists Series 708. Ladybird Books. Also: CD as a child "found robbing a nest"; CD fishing as a child "sometimes found ponds more interesting than people"; CD with John Edmonston; CD on cliff with hammer, *Beagle* below "To study rock, Darwin took samples"; CD meets FitzRoy; CD "pleaded for permission to land on Teneriffe"; CD on St. Paul's rocks; CD "In the forest, overtaken by a tropical storm"; CD with insect net in swamp "despite the danger of fever"; CD riding with gauchos; CD & FitzRoy greeted on Falkland Islands; CD "saw a volcano in action"; CD with "mule train...high in the Andes"; CD with Galapagos iguana; CD with aborigines in Australia; CD with a fossil leg bone; CD confronted by "the Church and others"; CD in soft hat and cloak behind Down House with walking stick.

1973 Drawing after L. Darwin 1878a with a section of beard missing by James Francavilla. *The Los Angeles Times*, (15 Apr.), p. 113.

1973 From a watercolour "Charles Darwin sails towards the Galapagos Islands." CD (after Richmond 1840) with FitzRoy reading from the Bible inside the *Beagle*. They broke the rules: the man who shocked a nation. *Look and Learn*, no. 574, (13 Jan.).

1973 Original drawing of CD in red loosely after L. Darwin 1878a. Cover art to Paul Ekman ed., *Darwin and facial expression: A century of research in review*. New York: Academic Press.

1973 Drawing loosely after Maguire 1849. *Orlando Sentinel* [Florida], (16 May), p. 76.

1973 Painting after Collier 1881. Cover art to *Descent* [in Spanish] F1135a.

1974 Re-drawing of Richmond 1840, head only, in three descending sizes. Cover art to Gruber & Barrett, *Metaphysics, materialism & the evolution of mind*. Chicago: University of Chicago Press.

1975 Simple line sketch, three-quarter right profile, by "Cunningham". *Greeley Daily Tribune* [Colorado], (19 Mar.), p. 11.

1975 From a watercolour, CD after Maull & Polyblank 1857 and A.R. Wallace from a photograph, route of *Beagle* voyage across a globe, by Harry Green. *Look and Learn*, no. 688, (22 Mar.).

1976 Aquatint(?) frontispiece portrait by R. Sholl(?) after Collier 1881. Darwin, *Origin*. Easton Press. 100 Greatest Books Ever Written series. Norwalk: Connecticut. F1865.

1976 Cartoonish bearded CD with giant head. Cover art to Daniel-H. Bouanchaud, *Charles Darwin et le transformisme*. Paris: Payot. Appeared also on several later works to 1982.

1976 Unique outline silhouette of bearded CD. Cover art to Thomas P. Carroll ed., *Annotated Calendar of the Letter of Charles Darwin in the Library of the American Philosophical Society*.

1977 Nude CD as terminal figure in a 'march of progress' line of nine hominids. *Sydney Morning Herald* [Australia], (8. Mar.), p. 8.

1977 Drawing after Barraud 1881c? Cover art to *Charles Darwin: Autobiographia*. Spain. F2443.

Easton 1976.　　Bouanchaud 1976.　　*Sydney* 1977.　　Drysdale 1979.　　Zhou 1979.　　Genesis 1979.

1977 Young CD with sideburns and spectacles reading in the *Beagle* and further illustrations by Christopher Spollen in Irwin Shapiro, *Darwin and the enchanted isles*. New York: Coward, McCann.

1978 After Richmond 1840, head only. Ripley's believe it or not. *Times* (Louisiana), (31 May), p. 23.

1978 Abstract re-drawing of Rejlander 1877b. Cover art to J.M. Montero Pérez, *Charles Darwin*. Madrid.

1979 Half-length felt pen sketch of CD in top hat by Russell Drysdale in P. Haefliger, *Duet for dulcimer and dunce*. Reproduced in *The Age* [Melbourne, Australia], (10 May 1980), p. 27.

1979 Painted portrait after Elliott & Fry c.1880a by Zhou Ruizhuang 周瑞庄, poster (78.7x109.2cm) with a quotation from Lenin, CD "put an end to the belief that the animal and vegetable species bear no relation to one another, except by chance, and that they were created by God, and hence immutable." Shanghai People's Fine Arts Press, first print run 50,000. Still in print in 1981.

1979 Chromolithograph(?) (black and green) of CD in hat with beard amidst vegetation and wildlife with the *Beagle* in the background. Frontispiece to *The journal of a voyage in H.M.S. Beagle*. Guildford: Genesis Publications. F1583b.

1979 Black-and-white drawing of CD's bearded face, three-quarter left profile. Cover art to Jean. F. Leroy, *Darwin: la vita il pensiero le opere*. Italy.

1979 Watercolour of CD (head only) after Collier 1881. Cover art to Paul Clark, *Famous names in science*.

1979 Postcard. Full-length chromolithograph(?) of CD after Richmond 1840 standing with hand to his head with Toxodon skull. USSR.

1979 Aquatint(?) bearded head frontispiece portrait by Fritz Kredel(?) *Descent*. Easton Press. 100 Greatest Books Ever Written series. Norwalk: Connecticut. F1866.

1979 Sketch of CD in hat with stick on Galapagos islands. *Frontiers of Science*. (905).

1979 Pen and ink drawing after Elliott & Fry c.1880a by Lapsley. Cover art to Loren Eiseley, *Darwin and the Mysterious Mr. X*. New York: E.P. Dutton.

1980 Full-face sketch loosely after Elliott & Fry c.1880a. *Chicago Tribune*, (16 Mar.), p. 18.

1980 Illustrations of CD's life for children by Jutta Petzold in Eberhard Hilscher, *Mister Darwin macht eine Entdeckung*. East Germany.

1980 Nude full-figure in right profile, final figure in three in typical 'evolution of man' procession. Cover art to Edmund O'Connor, *Darwin*. Greenhaven Press, USA.

<u>1980</u> Three-quarter left profile drawing by Ray Mellen. *Atlanta Constitution*, (21 Sept.), p. 94.

<u>1980</u> Drawing in red of bearded CD, right profile, with unusually trimmed and smooth beard. Cover art to C. Chant & J. Fauvel eds., *Darwin to Einstein: Historical studies on science & belief*. Also in black in N.G. Coley & V. M.D. Hall, *Darwin to Einstein: Primary sources on science & belief*, 1982.

Mellen 1980.　　Chant 1980.　　*Vancouver* 1981.　　Kelly 1981.　　Gould 1982.　　Jahn 1982.

<u>1981</u> Three-quarter left profile drawing of bearded CD. *Vancouver Sun*, (30 Sept.), p. 111.

<u>1981</u> Original drawing of CD with very long straight beard, three-quarter right profile. Cover art to Alfred Kelly, *The descent of Darwinism: The popularization of Darwinism in Germany, 1860-1914*.

1981 Fine re-drawing of Richmond 1840. *The Baltimore Sun*, (27 Jan.).

1981 Embossed young CD with a native, *Beagle* & armadillo. Cover art to *Journal* [in Spanish] F2535.

1981 Comic book with numerous illustrations and cover depicting CD and the voyage of the *Beagle*. *Grandes heroes, el descubrimiento del mundo*. Madrid: Planeta.

1981 Coloured drawing based on L. Darwin 1878b. Cover to Cocos (Keeling) stamp issue.

1981 Coloured drawing based on Richmond 1840. Cover to Cocos (Keeling) stamp issue.

<u>1982</u> From an unknown colour painting, half-length, right profile, CD holding hat. Ilse Jahn, *Charles Darwin*. Urania Verlag [Germany].

1982 Three-quarter left profile sketch with face of a chimpanzee on the back of CD's head by Greg McHenry. *Courier Post* [New Jersey], (25 Apr.), p. 65.

1982 Small caricature-style CD holding a skull. *Star Tribune* [Minnesota], (26 Sept.), p. 112.

1982 Pen and ink portrait, full face with very round nose and re-drawing of Adam from Sistine chapel emerging from his left eye. Cover art to N.C. Gillispie, *Charles Darwin and the problem of creation*.

1982 Original full-length drawing of CD after Elliott & Fry 1881, walking on cracking pavement. Cover art to Brian Leith, *The descent of Darwinism*. London: Collins.

1982 Colour artwork of CD's bearded head, three-quarter right profile, with primate and finches in beard. Cover art to *Natur – Das Umweltmagazin*. No. 4. Munich, Germany.

1982 Lithograph half-length beardless CD by P. Slot issued with Centenary UK stamps.

1982 Right profile portrait. Stamp sheet for São Tome and Principe centenary stamps.

1982 Two-tone re-drawing of Maull & Polyblank 1857, cover art to B.G. Gale, *Evolution without evidence*.

1982 Watercolour illustrations by Annabel Large. Peter Ward, *The adventures of Charles Darwin: A story of the Beagle voyage*. Cover: CD in hat in Brazilian forest, stalked by a jaguar. Other illustrations: CD in forest with straw hat; Beardless CD seated in a chair; CD meets Captain FitzRoy; CD falls from his hammock on the *Beagle*; CD during crossing the line ceremony; CD standing in frock coat in the *Beagle*; CD in carnival at Bahia; CD in the forest with straw hat and butterfly net; Sketch of CD running to catch a butterfly with net; CD trekking with hat, poncho and walking stick; CD around campfire with gauchos; CD in straw hat and poncho watching armadillo hunt; CD watches from boat as Tahitians catch a sea turtle.

1982 Abstract minimalist line drawing of CD's bearded head, three-quarter right profile. Cover art to *Revista Anthropos*, no. 16-17. Spain.

1982 Outline drawing of CD's head, left profile. Cover art to *Darwin, 100 anys*. Barcelona, Spain.

1982 Rough charcoal(?) sketch of CD, right profile. Bookmark. Madrid: Primavera.

1982 Painted(?) portrait after Elliott & Fry 1881b. Conference poster. Universidad Complutense, Madrid, Spain.

1982 Colour cover art painting after L. Darwin 1878b. R. Flos & M. Serrahima, *Darwin*. Spanish.

1982 Abstract drawing of CD's head, full face, scraggly beard. Cover art to *Novelle Ecole*, no. 37.

1982 Modernist-style re-drawing in blank ink of Maull & Polyblank 1857. Cover art to Benjamin Farrington, *What Darwin really said*. New York: Schocken.

1982 Engraving after L. Darwin 1878a. Cover art for *Autobiography* [in Chinese] F1510l.

1982 Engraving(?) after Cameron 1868c. Cover art for *Autobiography* [in Chinese]. Zhou Bang-li, *Life of Darwin*.

1982 Engraving after Elliott & Fry c.1880b in *Autobiography* [in Chinese]. *Ibid.*

1982 Engraving(?) after Cameron 1868a in *Autobiography* [in Chinese]. *Ibid.*

1982 c.60 caricatures of CD from young to old, often based on historic paintings and photographs, by Borin van Loon in Jonathan Miller, *Darwin for beginners*. Current title: *Darwin: a graphic guide*.

Photographs: Like all the iconographies provided here, no claim is made to being definitive. A very great deal of further information and discussion would be included in a full iconography. Nevertheless, this is by far the most complete and accurate list of photographs of Darwin ever published. The list includes several that have never been seen by virtually any reader interested in Darwin as well as seven unknown photographs found during the research for the present study.

The first commercially available was Maull & Polyblank 1855 but this was only for members of a subscription club. Maull & Polyblank 1857 was produced in a large format and CDV but apparently in small numbers. Photographs from the Edwards 1865 sitting were mass produced. It is well known that CD declined a request in 1869 by A.B. Meyer that CD might be photographed together with Wallace to illustrate a German translation of the 1858 papers. Meyer to CD 24 Nov. 1869 CCD17:497. CD replied that Meyer was welcome to include a photograph "But I am not willing to sit on purpose; it is what I hate doing & wastes a whole day owing to my weak health; and to sit with another person would cause still more trouble & delay ... PS I am very sorry to be disobliging about the Photographs, but I cannot endure the thought of sitting again, and I have refused 3 or 4 Photographers lately." 27 Nov. [1869] CCD17:505. Wallace agreed, as he wrote to CD on 4 Dec. "It is of course out of the question our meeting to be photographed together, as Mr. Meyer coolly proposes." CCD17:514. Despite CD's oft-expressed aversion to sitting for photographs, from 1865 he would be professionally/commercially photographed every year or alternate year for the remainder of his life except for 1875-77. It was common practice at the time, to sit for a more up-to-date CDV to send to friends and correspondents. In comparison, ED was photographed much less.

Dating Victorian photographs is particularly difficult as they were almost never dated at the time. They were often reproduced for many years by the same photographic company (and also pirated by others) but often differently cropped and edited. It must be remembered that hundreds of thousands of photographs were produced in the 19th century and even more in the first two decades of the 20th century. Many traditionally accepted dates for CD photographs derive from much later annotations on them. Some of these are clearly very inaccurate. Also, CD's dress is almost invariably very similar with a double-breasted waistcoat (often with a similar spotted design), silk cravat and heavy, loose, dark woollen jacket of plain weave. Of the photographs that reveal this, only Claudet 1842, Wallich 1871 and Rejlander 1871a-b show him wearing a single-breasted waistcoat. The latter two are so similar that they might even be the same suit. His personal appearance was also very consistent after the 1860s with a mostly bald head and full, bushy white beard. Therefore some of the dates adopted here might be further revised in future. There are probably further exposures from sittings already known.

A careful reading of the list will make it clear that many errors about dates and photographers have gone unnoticed for many years in the literature and some may still remain in the present list. Details are given not only for each photograph, but of the various printings or variants that were produced, or rather those that have been seen. It is unknown how large these print runs were. This reveals much about the extent to which CD's image was circulated, and in what forms, and which photographs were most popular in different periods. Such details will also be of interest to collectors. None given here have been cropped in any way and preference has always been given for the most complete, i.e. least cropped variant seen.

Of all those whose records have been found, Julia Margaret Cameron was the most active in registering photographs of CD for copyright. There was a rush by some photographers to register their photographs of CD for copyright shortly after his death. The first time (so far as seen) each photograph was engraved is noted to illustrate when a particular likeness became widely reproduced and available in print. Gene Kritsky has for many years been collecting and studying photographs of CD. He has discovered many photographs and likely some not recorded here. His findings are to be much anticipated. See also Janet Browne, "I could have retched all night": Charles Darwin and his body. In Lawrence & Shapin eds., *Science incarnate* (1998); Browne, Looking at Darwin: portraits and the making of an icon. *Isis* (Sept. 2009), 542-570 and Phillip Prodger, *Darwin's camera: Art and photography in the theory of evolution*, 2009.

1842 Aug. 23 Seated half-length three-quarter right profile daguerreotype with first child William on his lap by Antoine-François-Jean Claudet (1797-1867), 18 King William Street, Strand and Coliseum, called The Royal Adelaide Gallery. Only known daguerreotype of CD and the only 'photographic' image of him with another person. See Keynes, *Annie's box*, 2001, pp. 59ff. A London studio portrait with painted drop background. These were advertised as "EXCLUSIVELY HIS OWN [Claudet's], representing Landscapes, Interior of apartments, &c. which produces the most PICTURESQUE EFFECT". CD chose a landscape with neoclassical motifs. Modern writers have claimed that the sitting took so long that CD appeared in it perhaps only to hold William steady. But Claudet's advertisements boasted that "the sitting generally occupies less than one second, by which faithful and pleasing likenesses are obtained." The *Bradford Observer* (21 Apr. 1842) noted that this rapidity of the exposure method of Claudet "obviated in a great degree" the "objections made to their [i.e. photograph's] stern and gloomy expression". Thus some Victorians themselves were displeased with what has become the modern stereotype of Victorian photographic portraits because other technologies required long exposure times. According to Evelleen Richards, *Darwin and the making of sexual selection*, 2017, "William is wearing a dress, as was conventional for Victorian boys who were not 'breeched' until around five years of age". The original (8.3x9.4cm) is at Down House (EH88202861) mounted in a leatherette velvet lined case. Label on glass inscribed "Aug 23 1842 Ch and William 3 ¾". Paper label on rear of box "WED", William Erasmus Darwin, who lent it for display at Christ's College in 1909. Claudet made a daguerreotype copy of the 1853 drawing of CD by Laurence, (9.2x11.5cm). Annotated on back of case "Nov 1853". Down House EH88202863.

Claudet 1842.   Maull & Polyblank 1855.   Maull & Polyblank 1857.   W.E. Darwin 1861.

1855 Seated half-length, full face in embroidered waistcoat, by Maull & Polyblank for the Literary and Scientific Portrait Club. The Club was "instituted for the purpose of attaining a uniform set of portraits of the literary and scientific men of the present age at a moderate cost." (Contemporary advert.) The Club was joined by paying a fee of 10s 6d to the secretary, J.S. Bowerbank, and being photographed in the studio of Henry Maull (1829-1914) and George Henry Polyblank (b.1831), 187a Piccadilly, London. By joining the club one would receive one print and could purchase a photograph of any other member for 3s. Copies were available to members only. Francis Darwin first saw the photograph in 1909 and suggested it might date to 1854. CD paid Maull & Polyblank £3 1s. 6d. on 31 Dec. 1855. Classed account book, Down House. This suggests that CD purchased c.14 additional prints, though not necessarily all of himself. The portraits in the series are rather uniform. Some depict other men in the exact same pose as CD, almost none of them use props such as a microscope or fossil. The series of portraits was issued without text from 1855-c.1858. The Linnean Society has a set of 95, but there were many more. CD to Hooker 27 May 1855: "if I really have as bad an expression, as my photograph gives me, how I can have one single friend is surprising." CCD5:339. And to Hooker in 1860: "It makes me look atrociously wicked." CCD8:532. Asa Gray to CD 1856 Aug. 20 "[Francis] Boott lately sent me your photograph, which (tho. not a very perfect one) I am well pleased to have." CCD6:195. Boott was also a member of the Club. See: http://darwin-online.org.uk/EditorialIntroductions/vanWyhe_MaullandPolyblankPhoto.html.

1. Arched albumen silver print (20x15cm) with 2 gold border lines. Printed on recto mount "MAULL & POLYBLANK, Photo. 55 Gracechurch Street". Note no "London" in address on this variant. Verso blank. Christ's College, Cambridge. No copy in CUL.

2. Arched albumen silver print (20x15cm) with 2 gold border lines. Printed on recto mount "MAULL & POLYBLANK, Photo. 55 Gracechurch Street, London". NPG P106(7).

3. Photogravure (slightly cropped on all sides). Printed on recto mount "Maull & Fox, photographers Emery Walker Ph sc" with a lithographed signature "Charles Darwin" followed by "1854 / Published by Maull & Fox, 187 Piccadilly London, & by Emery Walker Limited. 16 Clifford's Inn in the City of London. 19th day of April, 1912." DAR140.1.25.

1857 Almost full-length seated left profile, checked trousers, waistcoat and cravat, by Maull & Polyblank whose partnership was 1854-65. This photograph is always attributed to "Maull & Fox" (after name on later prints) but their partnership was 1879-85. The editors of the CCD note "An entry in CD's Account book (Down House MS) in February 1858 suggests that CD may have sat for the photograph in the summer of 1857." CCD8:531. See also CCD8:488, 10:141, 11:22 and 16:16. Frontispiece to *Origin* 2d [in German] (1863, F673). Engraved in *Journal*, vol. 1 [in Russian] (1865, F2383.1.) et al. Head and shoulders engraved by G. Kruell, 1884. CD sent a copy to J.V. Carus in 1866. Variants seen:

1. Arched albumen silver print (32x27cm). Down House EH88204428.

2. CDV head and shoulders only. Printed on recto mount "MAULL & POLYBLANK LONDON". On verso "MAULL & POLYBLANK / Photographers / 55, GRACECHURCH STREET, / 187a, PICCADILLY, / and / TAVISTOCK HOUSE FULHAM ROAD / LONDON / N°" with photographer's handwritten number on both copies in CUL (DAR225.110-111) "33326". A copy in CUL (DAR225.110) is signed on verso "W.E Darwin". Down House EH88207472.

3. Arched print (c.17x10cm). Printed on recto mount "Photographed by Messers. Maull & Fox. London." Printed on verso: three royal coats of arms and "MAULL & FOX / PHOTOGRAPHERS TO THE ROYAL FAMILY / Portrait and Miniature Painters / 187A. PICCADILLY / London" and handwritten "4516". DAR225.175, Down House EH88207471.

4. An enlargement (54.5x41cm) on display at Down House, has image heavily edited with a chair back added, a printed label on the back of the frame "Messrs. MAULL & FOX, / (FORMERLY MAULL & POLYBLANK), / Photographers and Portrait Painters to the Royal Family. / 187A, PICCADILLY, W., / Name [Charles Darwin Esq] / No [162898.] On ordering Copies please to quote the Name and Number as above." The bracketed text written by hand. With "Belonged to William & later to Horace Darwin." written on another label. Down House EH88204460.

5. CDV cropped to almost head only. No printing on mount. Houghton Library.

6. Photogravure. Cropped to waist and above wrist exactly like woodcut by Kruell 1884, (11x9.3cm). Printed below image "Maull & Fox, photo. Walker & Cockerell, ph.sc. / *Charles Darwin. / Cir-1854.*" ML2, facing p. 204.

7. Photogravure in K. Pearson, *Life, letters and labours of Francis Galton*, (1914), vol. 1, pl. XXXVII with the unusual caption "From a photograph by Maull and Fox [sic], touched up by Mrs Darwin and now in the possession of Mr William E. Darwin." CD's trousers have been heavily scored on the negative, obscuring the checked pattern. No chair back.

1861 Apr. 11 Half-length right profile photograph by William Erasmus Darwin. Gray Herbarium, Harvard University and DAR225.113. CD to Asa Gray: "P.S. I enclose a little

Photograph made this morning by my eldest Son". CCD9:88-89. The last known photograph without beard. CD later referred to "a small dark closet, in which my son formerly photographed" CD to G.C. Wallich 18 Apr. [1869] CCD17:185. There are several records for photographic expenses for William in CD's Account book (Down House MS) from Jul. 1857. Not engraved or published during the 19th century.

W.E. Darwin 1864a.   W.E. Darwin 1864b.   Leonard Darwin c.1866.

1864 Apr. Two photographs by William Erasmus Darwin. The first photographs with beard.
a. Three-quarter left profile. Produced as CDV by S.J. Wiseman, Southampton, where William lived. Wiseman (1831-1914) started photography in 1863. William may also have taken the photographs of Henrietta and George Darwin that were produced by Wiseman which was dated by the editors of the CCD as c.1862. Lithographed by Brooks, *Quart. Jrnl. of Sci.* (1866), facing p. 151. Frontispiece to CCD12. See CCD24:464. Variants seen:
a.1. CDV cropped to head only in an oval, no printing on recto or verso. A copy sold at Bonham's in 2009 is signed and dated on the verso by CD "Ch. Darwin 1864".
a.2. CDV with more of chest and shoulders visible. Printed on verso: maker's monogram and "S.J.WISEMAN / 9, Bernard Street / SOUTHAMPTON". The copy in CUL (DAR225.112) has written on the verso "Photo^d by W.E.D." and on recto mount "1866 / 1864?" The latter date has been added since microfilming in the 1960s.
b. Seated full-figure, three-quarter right profile, draped in cloak with blanket over knees and legs. DAR257.3. This is a collodion positive, on a small glass plate (c.5.8x7cm), with black varnish on the back. Reproduced in Browne, *Power of place*, 2002, facing p. 441, with the caption "one of the last portraits of" CD. The CUL catalogue suggests 1870s. However, the collodian process was superseded by tintypes during the 1860s. CD's hair and unusually short beard in this photograph appear identical with the preceding photograph by W.E. Darwin. By permission of the Syndics of Cambridge University Library. It probably remained unprinted/reproduced because it was of poor quality.
1865 Nov. Three photographs by Ernest Edwards. Taken in London. CD paid £1 for "E. Edwards Photo" on 2 Mar. 1866. Classed account book, Down House. See CCD13:37. Photographs from this sitting are sometimes incorrectly stated to have been taken by the London Stereoscopic Company. An 1867 photograph by Edwards of Alexander Tweedie, in the same studio chair, shows the legs and floor. Just visible behind the chair is the heavy metal base of a stand that held the head of the sitter steady during the long exposure.
a. Three-quarter left profile seated in a chair with curled arms. Frontispiece to CCD13. Variants seen:

a.1. CDV knees cropped away in an oval frame with "CHARLES DARWIN, M.A., F.R.S." / "STEREOSCOPIC C[OY]. COPYRIGHT" printed on recto mount in red with thin border line around image. Verso photographer's imprint.

a.2. CDV barely cropped with "CHARLES DARWIN, M.A., F.R.S." / "Stereoscopic Co. Copyright" printed on recto mount in green. Printed in red on verso: ornately framed publisher's logo with coats of arms, symbols for medals from Berlin, London, Dublin and New York and "PHOTOGRAPHERS TO H.R.H. THE PRINCE OF WALES, / TO H.R.H. THE DUKE OF EDINBURGH. / SOLE PHOTOGRAPHERS TO THE INTERNATIONAL EXHIBITION 1862 / THE LONDON/ STEREOSCOPIC & PHOTOGRAPHIC COMPANY / 110 & 108. REGENT STREET. / AND / 54, CHEAPSIDE./ THIS NEGATIVE IS PRESERVED FOR FURTHER COPIES. / ENLARGEMENTS CAN AT ANY TIME BE PRODUCED WITHOUT ANOTHER SITTING."

Edwards 1865a.   Edwards 1865b.   Edwards 1865c.   Allen c.1866-71.

a.3. CDV, image not cropped. "STEREOSCOPIC C[OY]. *Copyright*" printed on recto mount in red. Verso, identical to a.2. Least cropped variant. Reproduced above.

a.4. CDV with seat of chair and legs slightly blended away. No printing on recto mount but a large blind stamp on the lower right corner of the photograph with "LONDON STEREOSCOPIC C[OY]. COPYRIGHT" in a shield.

a.5. CDV with image in oval, hands and chair arm blended away.

b. Same as above but almost full face. CDV uncropped with "Edwards & Bult, 20 Baker S[t.] W.", printed in black on recto mount (a partnership active only in 1868-69).

c. Three-quarter left profile almost indistinguishable from a. but in this one part of CD's shirt and collar are visible as is his right shoulder. NHM.

c.1866 CD on his cob Tommy in front of Down House, by Leonard Darwin. Sometimes dated to 1866 (when Tommy was acquired) or 1867 and very often to 1868, based on the annotation on the verso of the copy in CUL. No contemporary evidence has been seen for any of these dates. However CD looks much more like his 1866 photographs here. CD was still riding Tommy in 1870. Exhibited at Christ's College in 1909, lent by WED.

1. Albumen photograph (19.5x19.5cm). A copy on display at Down House (EH88202292) is framed with a note by CD: "Hurrah — no letters to day C. D." A copy at CUL (DAR225.116) is annotated on recto mount "Charles Darwin" and on verso "H.E. Litchfield (c.1868)". An annotation in another hand is "N[o] 2". Another copy is inscribed on verso "Charles Darwin on Tommy". EH88207651/DAR225.138.

c.1866-71 Half-length left profile by unknown photographer. Unusual light-coloured waistcoat. Apparently unknown and never reproduced in any work on CD.

1. CDV recto blank, verso publisher's emblem "ALLEN. N[o]. 6 TEMPLE PLACE, BOSTON." Edward L. Allen (1830-1914) had an establishment at this address only from 1866-71 (*The*

Boston directory (1866) and *The Boston almanac and business directory* (1867-71)). This does not indicate when the photograph was taken, but does establish an end date. It is unknown how Allen acquired a negative or photograph of CD not seen elsewhere. It could have been taken by Rejlander who took photographs of CD in 1871. No other photographs sold by Allen of foreign notables have been found. He did reproduce drawings by the artist S.W. Rowse such as one of Ralph Waldo Emerson.

Edwards 1866a.   Edwards 1866b.   Edwards 1866c.   Edwards 1866d.

1866 Apr. 24 Four photographs by Ernest Edwards. Taken in London. CD paid Edwards £3 8s. 6d. on 5 Sept. 1866. Classed account book, Down House. Janet Browne, *Power of place*, 2002, p. 363, noted that during 1866 CD "paid out a total of £14 in small sums for photographs, nearly doubling his overall costs for "Science" that year". It has been claimed that stereoscopic photographs of CD were produced and that one of these is such a photograph. (*Scientific American*, vol. 253, no. 2, Aug. 1985, p. 66.) This would mean that two photographs were taken simultaneously from slightly different angles. When such dual photographs are viewed through a stereoscope, they produce a three-dimensional effect. However, no actual stereoscopic photographs of CD, which are two images on the same card, have ever been found. Instead, standard CDVs by Edwards were sold by the London Stereoscopic & Photographic Company (1854-1922), as printed on the recto of the CDVs, which have been mistaken for stereoscopic photographs. However, the Company produced and sold standard photographs routinely, only some of their sales being of actual stereoscopic views. They produced these cartes of CD until c.1881.

a. Left profile seated, right hand on left with fingers exposed, chair not visible, pince-nez visible. Verso blank. A large (28.3x24cm) print in CUL (DAR232.2) is inscribed in the negative (but reversed in the positive print) "Mr Darwin - 24-4-66". One of very few dated photographs. The copy in the NPG (x1500) has written on the recto mount "C. Darwin. Photo. E Edwards." Etching in *Illustrated London News* in 1871. Apparently not produced for sale.

b. Left profile seated, right hand on left with fingers in sleeve, chair back visible, pince-nez not visible. CD to Wallace 5 Dec. 1869 "I like best the profile of Ernest Edwards". CCD17:515.

b.1. Albumen silver print in variants of Reeve & Walford eds., *Portraits of men of eminence in literature, science, and art, with biographical memoirs. The photographs from life, by Ernest Edwards*. London: Lovell Reeve & Co., 1866, vol. 5, facing p. 49. Legs cropped out.

b.2. CDV on recto mount "Ernest Edwards 20, Baker St. W."

b.3. CDV with no printing on recto mount.

c. Three-quarter left profile, top of chair back not visible, fingers but not thumb in left sleeve. A light patch on the floor beneath the chair is probably where the stand to hold the sitter's head steady has been edited out.

c.1. Albumen silver print in variants of Reeve & Walford eds., *Portraits of men of eminence*, 1866, vol. 5, facing p. 49. Legs of chair visible in some copies but not others. One copy seen has the chair legs cropped but reveals more of an umbrella(?) handle at bottom left than any other seen.

c.2. Albumen silver print (9x6cm) in E. Walford ed., *Representative men in literature, science and art: The photographic portraits from life by Ernest Edwards*. London: Bennett, 1868, n.p. See CCD13:314-5. Different copies of the work have slightly differently cropped images. The one reproduced here is the least cropped variant seen. A cane or umbrella handle is just visible bottom left.

c.3. CDV with "E. Edwards 20 Baker St. W." printed on recto mount. Photograph is blind stamped in the bottom right-hand corner "E. Edwards 20 Baker St." A copy in the collection of van Wyhe is signed on verso in Edward's handwriting "Mr Ernest Edwards / 20 Baker Street. / London W."

c.4. CDV as above but not blind stamped.

c.5. CDV part of lower leg and chair leg cropped with "E. Edwards 20 Baker St. W." printed on recto mount.

c.6. Same as above but cropped just above seat of chair.

c.7. CDV with lower legs and chair legs cropped, in oval frame, with green printing on recto and verso identical with Edwards 1865a.2.

c.8. CDV, uncropped, with "STEREOSCOPIC Cᵒʸ. *Copyright*" printed on recto mount in black. Verso has green printing similar to Edwards 1865a.2. but with statement about negatives preserved etc. absent.

c.9. CDV, cropped just below right knee, in an oval. On recto mount "CHARLES DARWIN, M.A., F.R.S." / "STEREOSCOPIC Cᵒʸ. COPYRIGHT" in green. Printed text on verso.

c.10. CDV with no printing on recto mount but a large blind stamp on the lower right corner of the photograph with "LONDON STEREOSCOPIC Cᵒʸ. COPYRIGHT" in a shield.

c.11. CDV cropped to head and chest in an oval. Entomological Society of America.

d. Same as above but legs not visible, right-hand fingers not in left sleeve. Engraved in 1868.

d.1. Albumen silver print (9x7cm) in variants of Reeve & Walford eds., *Portraits of men of eminence*, 1866, vol. 5, facing p. 49.

d.2. CDV with hands slightly more visible. "E. Edwards 20 Baker St. W." printed on recto mount. On verso "MESSʳˢ EDWARDS & BULT / 20, Baker Sᵗ· / W." Blind stamp in lower right corner of photograph "E. Edwards 20 Baker St."

d.3. CDV with hands almost entirely visible, thus least cropped variant. Used here. On recto mount "EDWARDS & BULT 20, Baker Sᵗ· W." Blind stamp and verso as d.2. above.

d.4. CDV cropped to almost head and shoulders only with "STEREOSCOPIC CO. COPYRIGHT" on recto mount in green.

d.5. CDV with hands and lower body cropped, in an oval. "CHARLES DARWIN, M.A., F.R.S." / "STEREOSCOPIC Cᵒʸ. COPYRIGHT" on recto mount in red with border line around image. Verso as Edwards 1865a.2.

d.6. CDV with hands and lower body cropped and "CHARLES DARWIN, M.A., F.R.S." / "STEREOSCOPIC Cᵒʸ. COPYRIGHT" on recto mount in red and border line around image. Verso like Edwards 1865a.2. but with "PRIZE MEDAL FOR PORTRAITURE / VIENNA

EXHIBITION 1873." and two medallions at the top. A copy seen has written on verso "London. Oct. 19th 1876."

d.7. CDV as d.4. above but with CD in oval frame.

d.8. CDV as above but with printed text in green and verso identical to d.2.

d.9. CDV with hands and lower body cropped in oval frame with "CHARLES DARWIN. M.A., F.R.S. / LONDON STEREOSCOPIC COMPᵞ COPYRIGHT." on recto mount in red.

d.10. CDV identically cropped as above, not in oval frame, with faint arched frame outline and "STEREOSCOPIC COᵞ *Copyright.*" on recto frame in green.

d.11. CDV cropped as above, not in oval frame, with "STEREOSCOPIC CO. COPYRIGHT." on recto mount in green.

d.12. CDV cropped (faded or blended out) below elbows with no printing on recto mount but a large blind stamp on the lower right corner of the photograph with "LONDON STEREOSCOPIC Cᵒᵞ. COPYRIGHT" in a shield.

d.12. Magic lantern slide (reversed) by George Washington Wilson. See K.D. Wells, Lincoln-Darwin bicentennial 1809-2009. *Magic Lantern Gazette*, vol. 20, no. 4, 2008, pp. 3-12.

Cameron 1868a. Cameron 1868b. Cameron 1868c. Cameron 1868d.

1868 Jul.-Aug. Four photographs by Julia Margaret Cameron; taken at Freshwater, Isle of Wight in two sittings. Several writers have claimed that Cameron took three photographs of CD. He recorded £4 7s. for photographs on 19 Aug. 1868, and later made other payments for further copies. Classed Account Books, Down House. Cameron also photographed Erasmus and Horace Darwin during this visit. Cameron registered CD photographs for copyright but some of her descriptions, National Archives, Kew, are not clear enough to unambiguously assign. Those not assigned by J. Cox & Colin Ford, *Julia Margaret Cameron: The complete photographs*, 2003, are below.

"Photograph of Charles Darwin, ¾ face, bust. No 1". 3 Aug. 1868. COPY 1/14/571.

"Photograph of Charles Darwin, nearly profile, bust. No 2". 3 Aug. 1868. COPY 1/14/572.

"Photograph of Charles Darwin, bust, nearly profile, with long beard. No 1". 10 Aug. 1868 COPY 1/14/582.

"Photograph of Charles Darwin, ¾ face, bust, with long beard. No 2". 10 Aug. 1868. COPY 1/14/583

"Photograph of Charles Darwin profile with part of second eyebrow showing". 6 Feb. 1874. COPY 1/24/211.

"Photograph of profile portrait of Charles Darwin only second eyebrow shown". 6 Feb. COPY 1/24/212.

"Photograph of Charles Darwin nearly ¾ face with second eye and second eyebrow showing". 6 Feb. 1874. COPY 1/24/213.

Ten years later prints were still commercially available, see CCD26:280. Cameron's 1868 catalogue gave the price range of 7s. 6d. to c.14s. Some prints of 1868a carried the lithographed line by CD "I like this photograph much better than any other which has been taken of me. / Ch. Darwin" on the recto mount. This is often mistaken for a handwritten inscription. LL3:92. Purportedly a copy in a private collection in the UK is actually inscribed thus by CD. Multiple signatures of CD were lithographed onto different photographs and variants by Cameron. Prints of 1868a are very variable. Some have a lithographed line by Cameron, in others the text is hand written by her. Glass negatives (almost certainly not Cameron's own) are at Down House, EH88202894 and EH88202895; as well as two framed prints (not identified in the catalogue), (frame size 49x38.5cm) EH88204450 and (frame size 55x47.5cm) EH88204436. See also Cox & Ford, *Julia Margaret Cameron*, 2003, p. 317, CCD16 and J. Browne, "I could have retched all night", 1998 and *Power of place*, 2002, pp. 298ff.

a. Half-length, left profile. Copyright registration: "Photograph of Charles Darwin, profile". 24 Aug. 1868. COPY 1/14/611; carbon print copyright 18 Oct. 1875. Cox 645.

a.1. Albumen silver print. On recto mount "From Life Registered Photograph Copy right Julia Margaret Cameron". On verso "From glass negative at Down House P.J. Gautry Feb '78". DAR225.139.

a.2. Albumen silver print. Print in the Norman Album (compiled by Cameron 1864-9). Recto mount inscribed by Cameron in pencil: "Ch: Darwin. He likes this photo better than any other that has been taken of him". Album dedicated 7 Sept. 1869.

a.3. Albumen silver print (28.5x22.7cm). On recto mount "From Life Registered Photograph Copy right Julia Margaret Cameron Fresh Water / I like this Photograph very much better than any other which has been taken of me. / Ch. Darwin". The lithographed lines by CD are in black. Blind stamp of P.D. Colnaghi's London gallery where her photographs were for sale. The first of her photographs of CD was on display there by 27 Jul. 1868. Down House EH88204438.

a.4. Albumen silver print (28.5x22.7cm). On recto mount "From Life Registered Photograph copy right Julia Margaret Cameron, 1868 / I like this Photograph very much better than any other which has been taken of me. / Ch. Darwin".

a.5. Albumen silver print (28.5x22.7cm). On recto mount "From Life Registered Photograph copy right Julia Margaret Cameron. / Ch. Darwin". Blind stamp of Colnaghi.

a.6. Albumen silver print. On recto mount "I like this Photograph very much better than any other which has been taken of me. / Ch. Darwin" with no Cameron text. Kew.

a.7. Albumen silver print (28.5x22.7cm). Reversed. On recto mount "From life Copyright Julia Margaret Cameron". Gernsheim Collection, Harry Ransom Center.

a.8. CDV on recto mount lithographed signature in black "Ch. Darwin" and below in red: "From life. Copy right. Julia Margaret Cameron." with gold lithograph border.

a.9. CDV on recto mount "From life. Copy right. Julia Margaret Cameron." in red. Below "Ch. Darwin" in black with gold lithograph border around card. These two CDVs have different CD signatures.

a.10. CDV on recto mount a different lithographed signature "Ch. Darwin" only.

a.11. Cabinet card (11.2x7.7cm) with handwritten on recto mount (not lithographed) "From Life registered Photograph Julia M. Cameron" signed in black ink: "Ch. Darwin" and linear blind-stamp "REGISTERED PHOTOGRAPH" by print-seller William Spooner. Harvard Art Museums/Fogg Museum.

a.12. Carbon print from 1874, produced by the Autotype Company (26.7x21cm) with no printing. Sometimes bearing the Company's blind stamp on the bottom right corner. Prints sold for 7*s* 6*d*.

a.13. Carbon print (reversed) cited in Cox & Ford, *Julia Margaret Cameron*, 2003, p. 317.

a.14. Photogravure. Printed beneath image *"Charles Darwin. / (from life)"* (27x21.4cm). *Alfred, Lord Tennyson and his friends*, 1893.

a.14. Carbon print (27.5x20.6cm). Handwritten on recto mount "from Mrs. Julia Margaret. Cameron's negative / 543". "This is not an. original Cameron. Modern print from her negative". George Eastman House.

b. Albumen silver print, half-length, three-quarter left profile. Engraved by E. Ade in 1875. Not included in Cox & Ford, *Julia Margaret Cameron: The complete photographs*, 2003.

b.1. Albumen silver print. On recto mount "From Life . Copy right Julia Margaret Cameron." Chicago Art Institute et al.

c. Head and shoulders, three-quarter right profile. Beard looks pointy. A blanket is just visible draped over shoulders. Copyright registration: "Photograph of Charles Darwin, ¾ face, bust". 23 Jul. 1868. COPY 1/14/530. Cox 644.

c.1. Albumen silver print (34.5x25.3cm). On recto mount "From Life Registered Photograph (copyright) Julia Margaret Cameron / Ch. Darwin". Blind stamp of Colnaghi. A copy sold at S. Maas & Co. 1973 had on verso "E. Darwin / Traverston / West Road / Cambridge".

c.2 Albumen silver print. On recto mount "From Life Registered Photograph copyright Julia Margaret Cameron / Ch. Darwin". Note lack of parenthesis around "copyright". Linear blind-stamp "REGISTERED PHOTOGRAPH" by William Spooner.

c.3 Albumen silver print. On recto mount "From Life Registered Photograph (copy right) Julia Margaret Cameron". No lithographed signature of CD unless cropped off. Colnaghi stamp.

c.4. CDV according to Cox & Ford, *Julia Margaret Cameron*, 2003, p. 317.

d. Albumen silver print, half-length left profile with cloak or blanket over shoulders, eyes directed upwards. Cox 646.

d.1 Albumen silver print (33x25.6cm). On gilt-ruled card mount. On recto mount "From Life Registered Photograph Julia Margaret Cameron. / Ch. Darwin". Note absence of "copyright". Blind stamp of Colnaghi. A copy was given to Joseph Parslow on his retirement in 1875 and remained in the family for many years. Private collection.

d.2. CDV on recto mount in red "From life. Copy right. Julia Margaret Cameron." No lithographed signature of CD.

d.3. CDV, no markings. Special Collections, Fine Arts Library, Harvard College Library.

Elliott & Fry 1869a.   Elliott & Fry 1869b.   Barraud & Jerrard. c.1873 a & b.

1869 Two CDVs by Elliott & Fry. Joseph John Elliott (1835-1903) and Clarence Edmund Fry (1840-97) established a studio at 55 Baker Street, London, in 1863 and remained in business long after the death of CD. From Feb. 1881 they sometimes added 56 Baker

Street to their address in newspaper adverts. CD recorded payments to Elliott & Fry on 25 Jul. 1869 and 5 Apr. 1870. Account books-banking account, Down House. CD to A.B. Meyer 27 Nov. [1869]: "Messʳˢ Elliot & Fry are the last who have taken me, & they came down here on purpose". CCD17:506 and CCD19:334.

a. Half-length left profile. Bottom edge of pince-nez just visible under jacket in both a. and b. CD wrote to Wallace on 5 Dec. 1869 "I like best the profile of Ernest Edwards, or a 3/4 face vignette which Messrs Elliott & Fry (Baker St) & which is a strong likeness & pleasing". CCD17:515. Variants seen:

a.1. CDV with arched top. Shows CD's right hand.

a.2. CDV, with CD's right hand cropped. Printed on recto mount "CHARLES DARWIN. / BORN FEB.12.1809 DEID [sic] April 19.1882. / ELLIOTT & FRY, Copyright 55&56, BAKER Sᵗ LONDON. W", verso: the Sovereign's coat of arms and "ELLIOTT AND FRY / Photographers / 55 & 56, Baker Street, London / W." and in small print: "Marion, Imp, Paris", one of the largest firms to sell photographs, photographic paper and the whole range of photographic equipment. A copy in CUL (DAR257.10) has "HE Litchfield abt ~~1870~~ 1869" written on the verso in her handwriting.

b. Full face, three-quarter length looking at camera with legs crossed, wearing corduroy trousers. Cabinet card with "CHARLES DARWIN. / BORN FEB 12 1809 DIED APRIL 19 1882. / ELLIOTT & FRY Copyright 55&56, BAKER Sᵗ LONDON. W" on recto mount. Verso identical to a.2. above. Copy in CUL (DAR225.117) has "HE Litchfield ~~abt 1870~~ 1869" on verso.

Rejlander 1871a.   Rejlander 1871b.   Rejlander 1871c.   Rejlander 1871d.

1871a-b Two photographs by Oscar Gustav Rejlander. 1 Albert Mansions, Victoria Street, London. These two have almost never been reproduced. CD to Elliott & Fry 23 Apr. [1871] "I lately was several times with Mʳ Rejlander, who was assisting me on a scientific subject, &, who took so much trouble for my sake that I gladly complied with his request to take several photos: of me, and these I imagine he intends to sell to any purchasers". CCD19:334. This suggests that both sittings took place during CD's stay in London and visits to Rejlander from 1-5 Apr. 1871. CD recorded the following payments to Rejlander in his Classed account books (Down House MS): £5 5s on 2 May 1871; £7 7s on 30 Oct. 1871 under the heading "Gifts and annual subscriptions"; £7 7s on 14 Mar. 1872 for "Photos"; and £1 1s on 22 Aug. 1872. There must have been another payment in c. Aug. 1871 of £1 1s, and this was clearly for CDVs of CD. See Rejlander to CD 11 Nov. 1871 thanking CD for "the cheque" and "I have sent a bundle of cards for the £1.1 and thank you". CCD19:680. Some of these payments include photographs provided by Rejlander for CD's work on *Expression*, which included photographs from and even of Rejlander.

DAR53.1 & DAR225. In 1874 CD paid him another £2 for photographs. Rejlander also photographed W.E. Darwin and R.B. Litchfield. Darwin Archive CUL.

a. Seated, legs not crossed, resting right arm on a desk with a pencil in right hand, pince-nez visible on single-breasted corduroy waistcoat.

a.1. Albumen silver print. Least cropped variant. Private collection. Cover of this book.

a.2. Albumen silver print. As above but slightly cropped up to CD's knees (19.5x14cm). Moderna Museet, Stockholm, ex-collection of Helmer Bäckström.

b. As above but legs crossed and facing more to the left, pince-nez fully visible, but reversed from a. Reproduced in Prodger, *Darwin's camera*, 2009, p. 159, where CD's relationship with Rejlander is discussed.

b.1. Albumen silver print (19.4x13.5cm). Moderna Museet, Stockholm.

b.2. CDV on recto mount "O. G. REJLANDER. N°. 1 ALBERT MANSIONS. / VICTORIA STREET. S.W." Photographer's imprint on verso. Cropped to head and shoulders only.

1871c-e Three photographs by Oscar Gustav Rejlander. CD's hair, beard, clothing and prop furniture are different from the other sitting with Rejlander. CD to ? 10 May [1871]: described c. and d. as "two of the best photographs which have been made of me". CCD19:375. CD gave signed copies as gifts. See also the letter to an unknown photographer in CCD19:475.

c. Right profile, head and shoulders. Engraved in *Nature*, 1874; *The Graphic* 1875 and 1882.

c.1. CDV with no printing on recto or verso. Extends slightly further down the torso than those below but CD's back slightly cropped. A copy in CUL (DAR225.114) has Rejlander's name and address written on the verso in his own handwriting. A copy for sale by Brick Walk Bookshop (Hartford, CT) in 2013 is signed by CD, has written on the verso Rejlander's name and address in his own handwriting and "Received June 12, 1871 at Toledo, O. from Mr. Darwin in a private letter dated May 27." by recipient F.E. Abbott. See CD to Abbot 27 May [1871] CCD19:400.

c.2. CDV as above but cropped just below the shoulder. CD sent a copy signed "Charles Darwin" on recto mount to the Academia Nacional de Ciencias of Argentina. See CD to Hendrik Weyenbergh 18 Mar. 1879 CCD27:117.

c.3. A carbon print cabinet card in CUL (DAR257.4) is blind stamped on the bottom right of the photograph "AUTOTYPE COMPANY" which was at 35 Rathbone Place, London (est. 1868). Printed on verso: "Copyright / Marion, Imp Paris. Déposé". CD's back has not been cropped out. This print is very fine, apparently not cropped, and is the one given here. Reproduced by permission of the Syndics of Cambridge University Library.

d. Left profile to knees, seated, hands clasped on lap, pince-nez visible. Only this one was registered for copyright, on 12 Apr. 1871, described as: "Photograph of Charles Darwin left side profile, light on the face, seated, one hand enclosing the other". National Archives, COPY 1/17/244. Engraved in *The London Journal* in 1872. Frontispiece CCD19.

d.1. CDV with no printing on recto or verso. Right chair arm cropped out. Rejlander often wrote his name and address on the verso of his CDVs. Copies in the collection of van Wyhe and CUL (DAR257.14) have "O.G. Rejlander / 1 Albert Mansions / Victoria Street SW." in his handwriting.

d.2. CDV on recto mount "O. G. REJLANDER. N°. 1 ALBERT MANSIONS. / VICTORIA STREET. S.W. / COPYRIGHT". Right chair arm just visible.

d.3. A carbon print cabinet card blind embossed on the bottom right of the photograph "AUTOTYPE COMPANY". On verso "Autotype Company Copyright Marion, Imp Paris

Deposé". A copy in CUL (DAR225.118) is very fine and much less cropped than those published by Rejlander. This is therefore the version reproduced above by permission of the Syndics of Cambridge University Library. A copy sold at Bonhams is signed and dated by CD "Charles Darwin / May 1. 1881.—" in blue ink on the recto mount.

e. CDV, half-length, head and shoulders, full face, looking left.

Elliott & Fry 1871a.   Elliott & Fry 1871b.   Wallich 1871a.   Wallich 1871b.

1871 Two Elliott & Fry CDVs. CD to Elliott & Fry 23 Apr. [1871]: "If you really think it worth while to come down, I shall be happy to give you a sitting & aid in any way…If you do come, be so kind as to let me know a day or two before." CCD19:334 And a postcard in the collection of Gene Kritsky from CD to Elliott & Fry postmarked 2 Aug. 1871: "Many thanks for the Photographs which are very good." CCD19:523. See also CCD19:475. The signature on this postcard seems to be the very one used for the lithographed signature on the Elliott & Fry 1874 CDVs below.

a. Three-quarter right profile, sharply trimmed moustache, speckled waistcoat with thick buttons, pince-nez and cord not visible. Hands not visible. Engraved by Weger, Leipzig, in 1871. No clearly posthumously produced cards have been found.

    a.1. CDV on recto mount "ELLIOTT & FRY Copyright 55, BAKER Sᵗ PORTMAN SQᴱ", on verso "Elliott & Fry / TALBOTYPE GALLERY / 55, Baker Street / PORTMAN SQUARE / LONDON. / Registered No.........[blank] ". Chair back not visible. A copy seen is stamped on verso by photographer P.E. Chappuis, 69, Fleet Street, London.

    a.2. CDV on recto mount "ELLIOTT & FRY Copyright 55, BAKER ST. PORTMAN SQUARE", on verso "ELLIOTT & FRY / 55, Baker Street. PORTMAN SQUARE.W." Chair back just visible.

    a.3. CDV all as above but verso lacks "W." after "Square".

    a.4. CDV on recto mount "ELLIOTT & FRY Copyright 55, BAKER ST. PORTMAN SQUARE", on verso an E&F monograph "ELLIOTT & FRY / 55, Baker Street. PORTMAN SQUARE.W." Chair back not visible.

    a.5. CDV on recto mount "ELLIOTT & FRY Copyright 55, BAKER Sᵗ PORTMAN SQᴱ", on verso "ELLIOTT & FRY / 55, Baker Street. PORTMAN SQUARE. W." Chair back not visible.

    a.6. CDV with no printing on recto or verso. A copy in the collection of van Wyhe has poor resolution and was most likely pirated. Chair back not visible.

b. As above, looking slightly upward, pince-nez almost fully visible. On recto mount "CHARLES DARWIN. / BORN FEB.12.1809 DEID [sic] April 19.1882. / ELLIOTT & FRY, Copyright 55&56, BAKER Sᵗ LONDON. W", same layout (also verso) and typographical error as in Elliott & Fry 1869a.2. A copy in CUL (DAR257.11) has "HE Litchfield abt ~~1870~~ 1869" on verso. Such late cartes sometimes had fading negatives retouched, as here, with rather crude markings to darken or lend texture to CD's hair and beard.

1871 Three photographs by George Charles Wallich, Trevor House, 2 Warwick Gardens, Kensington. Registered for copyright 3 Jul. 1871. CD was in London during parts of Jun.-Jul. and Wallich visited Down in May. CD declined for reasons of ill health to go to London to be photographed by Wallich in 1869 for inclusion in his *Eminent men of the day*. Scientific series, 1870. CD to Wallich 18 Apr. [1869] CCD17:185. In the copyright registration descriptions below, "a quarter plate" refers to glass plate negatives divided into four quarters so that each plate could take four exposures. This is why most of the photographs of CD are in groups of two or four per sitting. None of the Wallich photographs are preserved in the National Archives with the copyright registration. Rarely reproduced.

a. "Photograph of Charles Darwin Esq. FRS on a quarter plate, head and shoulders, ¾ face looking towards his left". National Archives, COPY 1/17/499. Heavy shadows around eyes, pince-nez visible. Woodcut in *Gardeners' Chron.*, (6 Mar. 1875). Variants seen:

a.1. CDV on recto mount: "Dr. Wallich, Trevor House, 2, Warwick Gardens, Kensington." See CD to O. Kratz 8 Sept. [1871] CCD19:562, n2. A copy signed by CD sold (with a letter and other photographs) at Christie's in 2005 for £7,800.

a.2. CDV on recto mount: "W. & D. DOWNEY COPYRIGHT". On verso "W & D DOWNEY, / PHOTOGRAPHERS TO HER MAJESTY [in 1862] 9, Eldon Square, / NEWCASTLE ON TYNE. / LONDON STUDIO, 61 EBURY St EATON SQUARE / Portraits taken by appointment." with seven coats of arms. Less cropped than the Wallich version above, hence this is the variant reproduced here. The firm of William and Daniel Downey began around 1855 and opened a studio at Eldon Square, Newcastle in 1863 and in 1872 at 57 & 61 Ebury Street, London, the address printed on the verso of these cards.

b. "Photograph of Charles Darwin Esq. FRS on a quarter plate, head and shoulders, rather more than ¾ face looking towards his left".

b.1. CDV on recto mount "W. & D. Downey Copyright". A copy in the Bancroft Library, University of California, Berkeley, has written on verso "Bought N.Y. Jan. 1874".

c. "Photograph of Charles Darwin Esq. FRS on a quarter plate, head and shoulders, front face, eyes directed a little towards his right".

c.1873 Two(?) half-length three-quarter left profile photographs by Barraud & Jerrard, 96 Gloucester Place, Portman Square, London (active 1873-80). CD's beard looks thin in the centre. Essentially unknown. Possibly taken at the same time as the undated photograph of ED by Barraud & Jerrard (DAR225.77 and Huntington Library).

a.1. CDV with parts of CD's face, eyebrows and beard retouched. No pince-nez string visible. In an album so mount not visible. Bibliothèque nationale de France.

a.2. CDV as above but pince-nez string visible and right eye appears to be cross-eyed to the right- probably from editing of the negative to enhance a poor exposure. On verso "BARRAUD & JERRARD / 96 Gloucester Place / Portman Square. W."

a.3. CDV as above but cropped lower down so handkerchief just visible in jacket pocket.

b. CDV same as above, except a pince-nez string and handkerchief are visible (lacking in a.) and the image is more closely cropped to head and shoulders than a. Eyes appear to be directed downwards.

1874 Four Elliott & Fry CDVs. CD wearing spotted waistcoat. Probably taken 10-17 Jan. in London. Some cartes seen of a. and d. have the identical lithographed signature "C. Darwin" on the recto mount, often mistaken for (and advertised or sold as) handwritten.

a. Half-length left profile. Pince-nez cord and handkerchief visible. Engraved in *Popular Science Monthly* in 1876. An engraved woodblock of this from W. & R. Chambers is in the

National Museum of Scotland, T.2011.56.162. This photograph (reversed) was the model for the engraving on the Bank of England's £10 note from 2000-18 (not one by Julia Margaret Cameron as often claimed).

a.1. CDV cropped to head and chest only in oval frame, slight retouching to CD's hair. On recto mount "ELLIOTT & FRY, 55, BAKER ST W".

a.2. CDV on recto mount "ELLIOTT & FRY, Copyright 55, BAKER ST. / PORTMAN SQE.".

a.3. CDV, image with arched top. On recto mount lithographed signature "C. Darwin" and printed "ELLIOTT & FRY, Copyright. 55, BAKER ST.". On verso the Sovereign's coat of arms and "ELLIOTT & FRY, / 55, BAKER STREET / PORTMAN SQUARE / LONDON / W / No......" A copy seen has "Got for 1s High St., Kensington on May 27th /79".

a.4. CDV as above but with CD's hair lightly retouched.

a.5. CDV as above but without arched top or retouching and slightly less cropped.

a.6. CDV as above but CD's beard and hair heavily retouched with grey streaks.

a.7. CDV cropped to little more than head and chest. On recto mount lithographed signature "C. Darwin" and "ELLIOTT & FRY, Copyright 55&56, BAKER ST. LONDON W".

Elliott & Fry 1874a.8.   Elliott & Fry 1874b.   Elliott & Fry 1874c.   Elliott & Fry 1874d.

a.8. CDV high-quality Woodburytype from original negative, less cropped than other variants. This is the version reproduced above. No printing on recto. On verso, ornate printed design and text in red "PORTRAIT FROM LIFE / COPYRIGHT. / CHARLES D. [sic] DARWIN. / JOHN G. MURDOCH, / Publisher, / LONDON, EDINBURGH & GLASGOW." A copy for sale on eBay is with a letter by CD which reads "May 4th [1875- added in another hand] / Dear Sir / I have much pleasure in sending you one of the best photographs, in my opinion, which has been made of me." CCD:30. If this is the photograph originally enclosed with the letter, this would date the Murdoch printings to as early as 1875. The verso of this example is the identical printing but the colour is not red but a pale maroon. This may be the result of fading.

a.9. CDV cropped to upper-chest. On recto "THE LATE" then identical text as Elliott & Fry 1869a.2 and Elliott & Fry 1871b including "deid" and identical text on verso. CD's hair and beard crudely touched up with grey spots and streaks. A copy in CUL (DAR257.12) has "HE Litchfield abt ~~1870~~ 1869" written on verso.

a.10. CDV with no printing on recto mount. Back of CD's hair retouched.

a.11. CDV cropped to head only, even end of beard gone. No printing on recto or verso.

a.12. CDV cropped to head only in oval frame. Image has clipped corners. On recto mount "Newnham & Field. Bournemouth." On verso N&F monogram and "NEWNHAM & FIELD. / ARTISTE PHOTOGRAPHERS. / 2, WELLINGTON TERRACE, / BOURNEMOUTH. /

THIS OR ANY OTHER PORTRAIT ENLARGED / UP TO LIFE SIZE AND PAINTED IN OIL OR / WATER COLORS."

a.13. CDV cropped to head only. On bottom of image: "*Дарвинъ. Darvin.*" Russia.

b. As above but facing more to the front, both eyes clearly visible, pince-nez cord not visible. No lithographed signature on any variant seen. Engraved in *Neue Alpenpost* in 1875.

b.1. CDV with unique thin red outline around photograph. Only copy seen, in the William L. Clements Library, University of Michigan. It is signed on the recto mount "Ch. Darwin March 7th 1874." No printed text on recto mount. Reproduced in Morris & Wilson, *Down House*, 1998, p. 48 et al.

b.2. CDV on recto "ELLIOTT & FRY (COPYRIGHT) 55, BAKER STREET / PORTMAN SQUARE / W." On verso "Elliott & Fry / TALBOTYPE GALLERY / 55, Baker Street / PORTMAN SQUARE / LONDON. / Registered No.........." Identical with Elliott & Fry 1871a.1 text.

b.3. CDV on recto "ELLIOTT & FRY Copyright 55, BAKER St / PORTMAN SQE". On verso "ELLIOTT & FRY, / 55, Baker Street, / PORTMAN SQUARE, W."

b.4. CDV with thin red line around photograph. On recto mount in black "Darwin. / 1391." on verso "P" monogram in red.

b.5. CDV cropped almost to head only. Small label on photograph "Charles Darwin".

b.6. CDV with thin red border line around photograph with "Darwin. / Collection Dèsmarets. Déposé." on recto mount in red. Verso blank.

c. As above but facing even more forward. Shirt collar, pince-nez cord and handkerchief not visible. No lithographed signature seen on any variant. Engraved by V.A. Hech c.1900?

c.1. CDV on recto mount "ELLIOTT & FRY, Copyright 55, BAKER ST. / PORTMAN SQUARE". On verso "ELLIOTT & FRY, / 55, BAKER STREET, / PORTMAN SQUARE."

c.2. CDV on recto mount "ELLIOTT & FRY Copyright 55, BAKER St. / PORTMAN SQE".

c.3. CDV cropped near uppermost jacket button with no printed text on recto mount. Victoria & Albert Museum etc.

c.5. CDV with thin red border line around photograph. On recto mount "PHOTOGRAPHIE WESENBERG & Co." Saint Petersburg, Russia. On verso, elaborate logo and Cyrillic text.

c.6. CDV with thin red border line around photograph. On recto mount "PHOTOGRAPHIE [monogram] WESENBERG & Co." Verso as above.

c.7. CDV thin red outline around photograph. On recto mount in red "WESENBERG" and address in Cyrillic.

c.8. A print in the Arbejdermuseet, Copenhagen, Denmark, has had the eyes touched up so poorly that it resembles a figure from Easter Island.

d. As above but facing more to the left, arms folded. Engraved in *The Examiner* in 1879. The extent that CD's hands and shoulders are cropped varies considerably, even amongst the same printed variant. The image reproduced above is the least cropped seen.

d.1. CDV with "ELLIOTT & FRY, 55, BAKER St W." on recto mount. On verso "ELLIOTT & FRY / 55, BAKER STREET, PORTMAN SQUARE, / LONDON. W. / №"

d.2. CDV "ELLIOTT & FRY (COPYRIGHT) 55, BAKER STREET / PORTMAN SQUARE / W." on recto of mount uniform with Elliott & Fry 1874b.2.

d.3. CDV "ELLIOTT & FRY Copyright 55, BAKER St. / PORTMAN SQE" on recto mount, uniform with Elliott & Fry 1871a.1. On verso "ELLIOTT & FRY / 55, Baker Street / PORTMAN SQUARE. W."

d.4. CDV on recto mount lithographed signature "C. Darwin" "ELLIOTT & FRY, Copyright 55, BAKER Sᵀ·" and verso identical with Elliott & Fry 1874a.3. A copy seen has an orange paper label pasted on the verso from photographer M.M. Couvée, The Hague.

d.5. CDV as above but image has arched top.

d.6. CDV on recto "CHARLES DARWIN. D.C.L., F.R.S. / ELLIOTT & FRY, Copyright. 55, BAKER Sᵗ· LONDON. W." CD did not receive the D.C.L. from Oxford in 1870. CD was awarded an LL.D. from Cambridge in 1877. Hair on back of head lightly retouched.

d.7. CDV with no printing on recto mount.

d.8. CDV cropped so that arms not visible with "Darwin." at bottom of image. On recto mount "Emerson NEW YORK." Verso blank.

d.9. CDV cropped to head in oval frame. No printing on recto. Hair, beard and eyes so heavily retouched as to be almost unrecognizable. On verso "D. M. BENNETT…N.Y." and lengthy adverts. A secular publisher.

d.10. CDV as above but not retouched and printed on recto of mount "CHAS. DARWIN. / Published by GEO. W. THORNE, 60 NASSAU St, N.Y."

d.11. Postcard. Shoulders largely absent but chest visible. On recto *"Darwin."* c.1909.

Leonard Darwin 1878a.   Leonard Darwin 1878b.   Lock & Whitfield 1878a.6.   Lock & Whitfield 1878b.

1878a Three-quarter right profile, seated in a Down House chair, by Leonard Darwin. 1881 Engraved in *Gardeners' Chron.* 1881 Letter to Henri de Saussure, CD regarded this as the best photograph of himself. *Calendar* 13088f. See also a letter from ED to an unknown London photographer (*Calendar* 13792), dated by the editors of the CCD as 1874 or later.

a.1. Woodburytype in *The University Mag.*, vol. 2, (Aug. 1878), facing p. 154, with lithographed signature below "Charles Darwin".

a.2. Woodburytype (9.5x13cm) no printed text. A copy at Down House (EH88207475) is annotated "Charles Darwin taken in 1878 by me Leonard Darwin 1933". A copy in CUL (DAR319.12.27) annotated in blue ink written with a nib on recto mount: "by his son Leonard Darwin 1878?". Another copy (DAR225.119) is annotated on the verso in ballpoint pen: "Photograph taken in 1878 by Leonard Darwin" and another copy (DAR225.120) is annotated on the verso "C Darwin 1880?" again in ballpoint pen.

a.3. Woodburytype (10.4x7.8cm) with blind stamp on bottom right of image "Autotype Company". Down House EH88207476. A copy seen is signed on the recto mount "Charles Darwin". Patrick Pollak Antiquarian and Rare Books, 2009.

a.4. Woodburytype as frontispiece to Woodall 1884 with lithographed signature "Ch. Darwin" below. Reproduced above.

a.5. A print available in the Wellcome Collection has "Elliott & Fry." in white on the lower left corner of the image. Attributed to the Galton Institute.

a.6. Postcard with "Charles Darwin" in Cyrillic below image. c.1900?

a.7. Postcard with monograph and "GF1659 and "Charles Darwin" in Cyrillic on image bottom left, and bottom right "BOP". All in white. c.1900?

a.8. Photogravure (21.2x15.8cm). Die Photographische Gesellschaft, Berlin, 1910.

a.9. CDV-size print, heavily re-touched. On recto mount: "Carlo Darwin".

a.10. c.1930s-1940s a colour-tinted lantern slide was produced by Beavis Photographic Studio, Sydney. Museums Victoria.

1878b Full-length left profile, seated in a basket chair on the verandah at Down House by Leonard Darwin. Engraved in *Century Mag.* in 1883. In LL2 (1887), Francis Darwin dated this as "1874?". A copy in CUL (DAR140.1) has written on the verso "?1874" and a framed copy at Down House is also dated on a label as 1874. These are not three independent datings, but are all Francis Darwin's estimate. Leonard Darwin's 1909 estimate was "between 1872 and 1878". Harmer & Ridewood eds., *Memorials of Charles Darwin*, 1910, p. 20. (In *Darwin Online*) Although Leonard presumably must have visited at some point, there are no records in *ED's diary* or the correspondence of Leonard visiting Down House during 1874 and in June he departed for New Zealand for the remainder of the year. Whereas there are seven visits recorded in *ED's diary* in 1878 when he certainly did photograph CD. For this and other reasons, 1878 seems more likely than 1874. Variants:

b1. Albumen photograph in CUL (DAR140.1), (20x15cm), is annotated on verso "Frontispiece L&L, vol II ?1874". Another copy (DAR225.1) is annotated on verso "Photograph of C. Darwin taken in the Verandah of Down House c. 1880". A copy at Down House (EH88207473) is annotated "C D in the verandah at Down". Another copy in gold card mount, EH88207474.

b2. Albumen photograph (31x26cm) framed and on display at Down House, EH88202824.

b3. A print (frame size 47x38.5cm) cropped just above the knees, on display at Down House (EH88204448?), bears a label on the frame: "COPY OF PHOTOGRAPH TAKEN BY HIS SON LEONARD DARWIN IN 1874 AND PRESENTED TO THE ROYAL SOCIETY". A label "9" is on the bottom left of the frame.

b4. Three-quarter length photograph in frame (65x50cm). Down House EH88204731.

b5. Oval enamel miniature cropped just above knees (c.8.5x6cm). These were produced for the family and found in a box with the words "Father's portrait". Down House.

b6. Smaller oval (c.6x4cm) enamel miniature cropped to head and chest only. As above.

b7. Smallest oval (c.4x3cm) enamel miniature cropped a little more than the above.

b8. A differently cropped version is at the Royal Geographical Society.

b9. CDV cropped to wrists just and below seat of chair. No markings on CDV. Special Collections, Fine Arts Library, Harvard College Library.

1878 Two photographs by Lock & Whitfield. Samuel Robert Lock (1822-81) and George Corpe Whitfield (1833-1904) established a partnership and studio in 1856 at Regent Street, London, and another studio at Kings Road, Brighton in 1864. Usually dated 1877 but with no evidence provided. It was first published in Dec. 1878.

a. Half-length three-quarter left profile.

a.1. Woodburytype (11.4x9cm) in oval border in Lock & Whitfield, *Men of mark: a gallery of contemporary portraits of men distinguished in the senate, the church, in science, literature and art, the army, navy, law, medicine, etc. photographed from life by Lock and Whitfield with brief biographical notices by Thompson Cooper*. London: Sampson Low, Marston & Co., vol. 3, 1878, opposite p.

36. Below image: "CHARLES ROBERT DARWIN." / "*Lock and Whitfield. Woodbury Process.*" Preface: "The photographs, taken from life, expressly for this publication, are produced in an absolutely permanent form by means of the Woodbury process". The work was issued in monthly parts with three portraits for 1*s* 6*d*. CD appeared in part 36 in Dec. 1878. *Publishers' Circular*, (18 Dec.), p. 1202. The 3ᵈ vol. in the series was advertised in *Publishers' Circular* from 16 Nov. 1878, p. 917. It sold for £1 5*s*. Engraved as frontispiece to *Descent* [in French] F1061 in 1881. Whitfield registered for copyright on 24 Apr. 1882.

a.2. CDV cropped to little more than head and beard. In an oval.

a.3. CDV with "Darwin." printed at bottom of image. No printed text on recto mount.

a.4. CDV as above but with red border line around image and rounded corners to card.

a.5. CDV cropped to head only. On recto of mount "H. RIISE / Amagertorv 6 KØBENHAVN." Denmark.

a.6. After Apr. 1882 a Woodburytype cabinet card by photographic and fine art publisher and print dealer William Luks (1840-1911) with a thin red border. On recto mount [in black:] "The late / CHARLES DARWIN / [in red:] WILLIAM LUKS COPYRIGHT. LONDON. / [in black:] Gesetzlich Deponirt". Verso blank. As the least cropped variant, this is the photograph reproduced here. NPG (CC BY-NC-ND 3.0), et al.

b. Almost identical to above, but with CD facing slightly more to the left.

b.1. CDV cropped to head only with "DARWIN." on recto of mount.

b.2. CDV slightly more cropped than b.1. Thin red border line around image.

Elliott & Fry c.1880a.   Elliott & Fry c.1880b.

c.1880 Two Elliott & Fry cabinet cards. Some modern works claim 1879, 1880 or 1881 or that these are the last photographs of CD. No contemporary datings have been found. Both were registered for copyright on 30 Mar. 1882 and again on 1 Jun. 1896 described as "large head". Cards with the printed address 7 Gloucester Terrace date from 1887-93. c.1880a was reproduced in Seward ed., *Darwin and modern science*, 1909, a work that was assisted by Francis Darwin, and it was there dated "circ. 1880." This is the earliest known dating of the photograph and therefore used here. Francis clearly did not regard these as the last photographs of his father. No record of payments for these has been found.

a. Half-length, full-face, looking at camera. Very widely reproduced after CD's death. Engraved as frontispiece to *Origin* [in Japanese], 1896, F718. This is the photograph (reversed) that was altered for the 2009 NHM exhibition poster with a hand 'photoshopped' onto it making the gesture 'shh'. Tragically, this faked photograph has become one of the most popular and widely reproduced images of CD. It has appeared on the covers of books, the *Times Literary Supplement*, postage stamps, postcards, in paintings, drawings, throughout the internet and has even been tattooed onto many avid Darwin admirers. This altered image perpetuates the myth that CD kept his theory a secret, which he did not. This erroneous view arose only in the mid-20ᵗʰ century. The fact that a Victorian photograph modified to reflect a 20ᵗʰ/21ˢᵗ-century version of history should be preferred by modern audiences to a more austere Victorian original should not be surprising.

a.1. Cabinet card with "ELLIOTT & FRY 55, BAKER STREET. W." on recto mount.

a.2. Cabinet card with "CHARLES DARWIN." in red and "ELLIOTT & FRY 55, BAKER STREET. / LONDON. W." in black on recto of mount.

a.3. Cabinet card with "CHARLES DARWIN. / ELLIOTT & FRY, Copyright 55 & 56, BAKER S^t. LONDON W". on recto mount.

a.4. Cabinet card with "The Late / CHARLES DARWIN. / ELLIOTT & FRY, Copyright. 55 & 56, BAKER S^T. LONDON. W." recto mount. On verso a royal coat of arms and "ELLIOTT AND FRY / Photographers / 55, BAKER STREET. / LONDON. W." On verso: royal coat of arms and "ELLIOTT AND FRY / Photographers / 55 & 56, Baker Street. / London. / W."

a.5. Cabinet card. Mount recto: "THE LATE / CHARLES DARWIN. / BORN FEB 12. 1809 DIED April 19. 1882. / ELLIOTT & FRY, Copyright 55&56, BAKER S^T. LONDON. W".

a.6. Cabinet card with "ELLIOTT & FRY Copyright 55, BAKER STREET. W. / AND AT 7, GLOUCESTER TERRACE. S.W." on recto of mount.

a.7. Cabinet card. On recto in Cyrillic Saint Petersburg: Vezenberg i Ko, [1900s].

a.8. CDV with Wesenberg & Co., Saint Petersburg in Cyrillic on recto and verso in red.

a.9. Postcard. "Darwin" in bold type on recto. Verso blank. [c.1900].

a.10. Postcard. As above but verso "B.K.W.I. 2111." Copies postmarked 1903 and 1927.

a.11. Postcard, on recto in white "C. Roberto Darwin 2191". Italy, 1904.

a.12. Postcard with "B.102." and "Darwin." on recto of mount. Verso blank. Copy seen is postmarked 1909(?) with a Romanian stamp.

a.13. Postcard, slightly cropped, with "DARWIN." in black across his chest.

a.14. Postcard. Below image "'ROTOGRAPH' SERIES / B26 / DARWIN".

a.15. Postcard by Balierini & Co., Florence.

a.16. Cabinet card with "Darwin." and "A / 3.102." on recto of mount. [c.1909].

a.17. Postcard from Poland, 1909, "252 K-M.W." on recto.

a.18. Postcard. Polish/Russian by "A.r.D. #119".

a.19. Postcard. "Charles Darwin" in Cyrillic below image and "GG/C9/2065". c.1900?

a.20. Postcard. "Charles Darwin" in Cyrillic and "413." and monograph, verso blank.

a.21. Postcard. Recto "Darwin" and "397".

a.22. Postcard Recto "C. DARWIN." Verso "CARTE POSTALE" also French and Russian.

b. Three-quarter right profile. Often engraved in later publications, never dated. Engraved in *Harper's Monthly* in 1897.

b.1. Cabinet card. On recto of mount "ELLIOTT & FRY 55, BAKER STREET. W."

b.2. CDV cropped to head. Recto of mount with same lithographed signature "C. Darwin" and "ELLIOTT & FRY, Copyright 55&56, BAKER S^T. LONDON. W".

b.3. Cabinet card. Recto of mount "DARWIN." in red and in black: "ELLIOTT & FRY Copyright. 55, BAKER STREET. W. / AND AT 7, GLOUCESTER TERRACE. S.W." on recto of mount. On verso royal coat of arms and "ELLIOTT AND FRY / Photographers / 55, BAKER STREET, / LONDON. W. / AND AT 7, GLOUCESTER TERRACE. S. W." in pale green.

b.4. Cabinet card. On recto of mount "THE LATE / CHARLES DARWIN. / ELLIOTT & FRY Copyright. 55, BAKER STREET. W. / AND AT 7, GLOUCESTER TERRACE. S.W."

b.5. Cabinet card. On recto of mount "THE LATE / CHARLES DARWIN. / BORN FEB 12. 1809 DIED April 19. 1882. / ELLIOTT & FRY, Copyright 55&56, BAKER S^T. LONDON. W". uniform with Elliott & Fry c.1880a.5. A copy in CUL, DAR257.13.

b.6. Postcard c.1906 by Rotograph Series No. A26, New York, etc.

1881 Four Elliott & Fry photographs. This well-known sitting includes the only known photographs of CD standing. The BMNH exhibition of 1909 showed four photographs from this sitting, dating them 1882. Sometimes dated by modern writers to 1880. LL3 frontispiece (1887), ML2 (1903), ED(1904):2 and *Order of the proceedings at the Darwin celebration 1909* all date to 1881. No record of payments for these photographs has been found. Registrations for copyright for a. and "3/4 face" and d. were made on 30 Mar. 1882, the same date as the two c.1880 photographs.

Elliott & Fry 1881a.   Elliott & Fry 1881b.   Elliott & Fry 1881c.   Elliott & Fry 1881d.

a. Half-length full face standing on verandah at Down House, leaning against one of the pillars, in velvet collar cloak and soft hat with round crown, looking at camera. A cloth is apparently draped behind so that no background details, such as the house, are visible. DAR140.1, DAR225.123, F1473, etc.

a.1. Cabinet card. On recto mount "ELLIOTT & FRY 55. BAKER STREET. W."

a.2. Cabinet card. On recto mount "ELLIOTT & FRY Copyright. 55, BAKER STREET. / LONDON. W." On verso a royal coat of arms and "ELLIOTT & FRY / Photographers / 55, BAKER STREET, / LONDON. W". DAR.257.8.

a.3. Cabinet card. On recto mount "The late CHARLES DARWIN." in dark red. And in black: "ELLIOTT & FRY 55, BAKER STREET. W."

a.4. Cabinet card. On recto mount "CHARLES DARWIN. / BORN FEB 12 1809 DIED APRIL 19 1882. / ELLIOTT & FRY, Copyright 55&56, BAKER ST. LONDON. W". On verso a royal coat of arms and "ELLIOTT AND FRY / Photographers / 55 & 56, Baker Street, London / W." and "Marion, Imp Paris". A copy in CUL (DAR225.123) is annotated on verso in pencil in the handwriting of Henrietta "(c.1880)". Another (DAR225.124) is annotated by her on verso "Charles Darwin at Down".

a.5. Cabinet card. On recto mount "CHARLES DARWIN, / Born, Feb., 12th, 1809; Died, April 19th, 1882." in reddish brown. In green: "ELLIOTT & FRY, Copyright. 55, BAKER STREET. W. AND AT 7, GLOUCESTER TERRACE. S.W." The copy in Manchester University Library is inscribed on the verso, probably by Meta Gaskell, indicating that the photograph was received by her on 4 Jul. 1892 from William Erasmus Darwin.

a.6. Cabinet card size image on an A4 mount in CUL (DAR140.1.32). On recto "COPYRIGHT." And hand-written "Elliott & Fry". On verso "Elliott & Fry, / 55, Baker Street, W. / No.......... / COPYRIGHT." Verso annotated in pencil "1881 Frontispiece L&L vol. III".

a.7. Cabinet card. A copy in Glasgow University Library is unlike any other seen, with a bookplate on the verso showing a figure in 18th-century dress holding a parchment reading "Elliott & Fry Baker St W.1" and embossed on the photograph bottom right hand corner, in red, "Elliott & Fry". Presumably a very late printing.

a.8. Large framed print (frame size 70x61cm) cropped just below hands clutching cloak. On display at Down House, EH88204454.

a.9. Undated enlargement (40.5x48.5cm) annotated "Charles Darwin". Down House EH88205819.

b. Same as above but looking slightly to the left. Engraved by G. Kruell 1887 yet the least commonly seen photograph of the set.

b.1. Cabinet card. On recto mount "C. DARWIN. / ELLIOTT & FRY Copyright. 55&56, BAKER Sᵀ· LONDON. W".

b.2. Cabinet card. On recto mount lithographed signature "C. Darwin" and "ELLIOTT & FRY Copyright 55&56, BAKER Sᵀ· LONDON. W".

b.3. No text on recto mount. Slight touching up to eyebrow and moustache.

c. Same as above but turned even further to the left. Unlike the above, no cloth obscures the background and other verandah columns are visible.

c.1. Cabinet card. On recto mount "ELLIOTT & FRY 55, BAKER STREET. W."

c.2. Cabinet card. Photograph covers entire mount. Blind embossed at the bottom of the photograph "ELLIOTT & FRY. LTD. COPYRIGHT." No printing on verso.

c.3. Cabinet card. On recto mount "THE LATE / CHARLES DARWIN. / BORN FEB 12. 1809 DIED APRIL 19. 1882. / ELLIOTT & FRY Copyright 55&56, BAKER Sᵀ· LONDON. W". On verso a royal coat of arms and "ELLIOTT AND FRY / Photographers / 55 & 56, Baker Street, London / W." A copy in CUL (DAR225.125) is annotated in the handwriting of Henrietta "H.E. Litchfield (c. 1880)". A copy seen is embossed on the bottom right corner of recto mount "WILLIAM LUKS / LONDON".

d. Seated, half-length, full face eyes to left, in the same cloak and hat. Verandah columns also visible in background, though at a different angle.

d.1. Cabinet card. Cropped to head and chest. On recto mount lithographed signature "C. Darwin" and "ELLIOTT & FRY Copyright 55&56, BAKER Sᵀ· LONDON. W".

d.2. Cabinet card. On recto mount "THE LATE / CHARLES DARWIN. / BORN FEB 12. 1809 DIED APRIL 19. 1882. / ELLIOTT & FRY Copyright 55&56, BAKER Sᵀ· LONDON. W". On verso a royal coat of arms and "ELLIOTT AND FRY / Photographers / 55 & 56, Baker Street, London / W." and "Marion, Imp Paris". A copy in CUL (DAR219.12.29) has "Carbon Print H.E. Litchfield" written on verso in black ink.

Barraud 1881a.6.   Barraud 1881b.   Barraud 1881c.   Barraud 1881d.

1881 Four photographs by Herbert Rose Barraud (1845-96). CD made a payment for photographs of £8 16s. to Barraud on 6 Jul. 1881. Classed account books, Down House. Barraud announced issuing photographs from a sitting "about ten months ago" in Apr. 1882. *The Globe*, (29 Apr.), p. 6. A payment was recorded on 3 Sept. 1881 for "Woodbury Photographs £1 12s" but it is not specified which photographs these were. The cabinet card of ED by Barraud, possibly done on the same day, is dated 1881 in ML1:xv (1903) and ED2 1904:[x]. Barraud's cards may be dated by the address printed on them. Until 1883 his studios were located at 96 Gloucester Place, Portman Square. From 1883-91 they were located at 263 Oxford Street and from 1893-96, 73 Piccadilly. No cards have been found printed with an address after Oxford Street. All of the copies printed by Barraud seen are cropped to little more than the head only, as in b-d above.

a. Head and shoulders, full face, eyes looking into camera, torso directed somewhat left.

a.1. Cabinet card. On recto mount "BARRAUD LONDON" (14x9.6cm).

a.2. Cabinet card. Thin red line around photograph. less cropped than the above. On recto mount in dark red "BARRAUD LONDON". On verso "Mr BARRAUD / 96, Gloucester Place / PORTMAN SQUARE / LONDON. W. Marion, Imp, Paris" in reddish brown. DAR257.6.

a.3. Cabinet card. Recto as above but text in black and backing card has rounded corners.

a.4. Cabinet card. On recto mount "COPYRIGHT / Barraud 263 Oxford St. London / A few doors West of The Circus.'" On verso "No...... / Mr. BARRAUD / 263, OXFORD St. W / (REGENT CIRCUS) / LONDON" followed by paragraphs of advertisement text.

a.5. Postcard (6.5x13.5cm) by the "ROTARY PHOTOGRAPHIC SERIES" (printed on verso). A much clearer print from Barraud's original negative and less cropped so that it is almost half-length and clearly shows a handkerchief in the inside jacket pocket and the top ring of the pince-nez, with "2308 ROTARY PHOTO. E.G. CHARLES DARWIN" in white on the bottom left of the photograph. The Rotary Photographic Co. (1901-16) used rotary presses to mass-produce postcards of popular portraits. They published other photographs by Barraud such as Huxley who was number 2315. There is a copy postmarked and addressed 20 Feb. 1909 in the Huntington Library, California. Rotary published another postcard with a photograph of the rear of Down House (not otherwise seen) with this photograph of CD, cropped and in an oval frame, in the upper right corner.

a.6. Postcard. Even less cropped than the Rotary above, extending down to show the whole right lens of the pince-nez. Just below the lens is "BARRAUD" in black. On recto mount "90 CHARLES DARWIN. J B & CO". 90 was the number in the publisher's series. The initials refer to J. Beagles & Co., postcard publishers, London (c.1901-09, thereafter J. Beagles and Co. Ltd.), founded by John Beagles (1844-1909). Because it is the least cropped image known, this is the one reproduced here. A postcard portrait of someone else in the NPG by the company and 94 in their series is dated 1902.

a.7. CDV, cropped to head and chest only, with "PROFESSOR DARWIN" on image itself.

a.8. As above but with no full stop after "DARWIN".

b. Full face, eyes looking slightly to the right, lips slightly parted. Engraved as frontispiece to *Origin*, New York, F2057, c.1893?

b.1. Cabinet card. On recto mount "BARRAUD LONDON". Cropped to head only. Pince-nez not visible.

b.2. As above but the text in thinner font and full stops after Barraud and London.

b.3. As b.2. but CD's beard has been heavily retouched on the negative.

b.4. Cabinet card. On recto mount, lithographed signature "Charles Darwin" and printed in reddish brown "BARRAUD LONDON" with border line around photograph. Extending down almost to waist with half of pince-nez clearly visible. Verso same as a.2. Reproduced above.

c. Right profile. DAR140.1.31, NPG, etc. Engraved by Meisenbach in 1885.

c.1. Cabinet card. On recto mount in dark brown "BARRAUD. LONDON." Verso blank.

c.2. Cabinet card. Thin red border around photograph. On recto mount "BARRAUD LONDON" in red. On verso "MR. BARRAUD / 96 Gloucester Place / PORTMAN SQUARE / LONDON. W. Marion, *Imp. Paris.*" in red. A pre-1884 copy seen is inscribed on the verso "Pronounced by his wife to be a very good likeness." WP Watson Antiquarian Books.

c.3. Cabinet card. On recto mount "BARRAUD LONDON". On verso "Mᵣ BARRAUD / 96, GLOUCESTER PLACE / PORTMAN SQUARE / LONDON. W. Marion, Imp, Paris". Followed by blind stamp "MARION & CO. LONDON".

c.4. Cabinet card. Thin red line around photograph. On recto mount, lithographed signature "Charles Darwin" and "BARRAUD LONDON" in red. Verso uniform with a.3 "263 Oxford Sᵗ" etc. Suggesting lithographed signature copies are later than those without. One copy the least cropped seen with the top of chair back just visible. Used above.

c.5. Cabinet card with "COPYRIGHT / Barraud 263 Oxford Sᵗ London / A few doors West of The Circus". on recto. Verso as a.3.

c.6. Cabinet card with "PROFESSOR DARWIN." bottom of photograph. Verso blank.

d. As above but head turned slightly to the right to a three-quarter right profile.

d.1. Cabinet card. Thin red line around photograph. On recto mount, lithographed signature "Charles Darwin" and "BARRAUD LONDON" in red and blind stamp "WILLIAM LUKS / LONDON".

Caricatures: There are many caricatures of CD and no complete list has ever appeared. Probably the largest collection is that in the Darwin Archive-CUL (DAR140.4, DAR141, DAR225 and DAR251). The following list cannot be complete, but it is the largest ever compiled and includes some not previously known. The ODNB, for example, lists ten. The present list records 48 from the 19th century alone. The far larger number of caricatures, not of CD himself, but referring to him or his theories and publications, are not included here. The list makes it clear that published caricatures of CD erupted very suddenly with the publication of *Descent* in 1871, rather than in the twelve years that had elapsed since the appearance of *Origin*. James Duncan Hague recalled speaking to CD about caricatures in 1881:

> The humorists have done much to make Mr. Darwin's features familiar to the public, in pictures not so likely to inspire respect for the author of *The Descent of Man* as they are to imply his very close relation to some slightly esteemed branches of the ancestry he claims; but probably no one has enjoyed their fun more than he.... 'Ah, has *Punch* taken me up?' said Mr. Darwin, inquiring further as to the point of the joke, which, when I had told him, seemed to amuse him very much. 'I shall get it to-morrow,' he said: 'I keep all those things. Have you seen me in the *Hornet?* (*Harper's New Monthly Mag.*, 1884, in transcribed *Darwin Online.*)

Those illustrated here have the date underlined in the list. See Janet Browne, Darwin in caricature. *Proc. of the American Phil. Soc.*, (Dec. 2001); revised in B. Larson & F. Brauer eds., *The art of evolution; Darwin, Darwinisms, and visual culture*, 2009, pp. 18-36.

<u>1871</u> Chromolithograph "Natural Selection", CD seated on a hassock in a small armchair, laughing, by James Jacques Tissot (often misattributed to Carlo Pellegrini). Men of the Day no. 33, *Vanity Fair*, (30 Sept.). For sale in two sizes, 31cm and 18cm, the former much better coloured. John Murray persuaded CD to appear there.*? Original drawing at the library of Christ's College, Cambridge.

<u>1871</u> Woodcut of CD's head on gorilla body addressing "my hairy hearers", by Stephene(?). In [Richard Grant White], *The fall of man: or, the loves of the gorillas. A popular scientific lecture upon the Darwinian theory of development by sexual selection. By a learned gorilla.* New York: G.W. Carleton, p. 7.

The Hornet 1871.   Fun 1871.   Harper's Weekly 1871.   The Graphic 1871.

Vanity fair 1871.   Figaro 1871.   The fall of man 1871.

Cruikshank 1871? (extract from upper left corner)   "AB" nd.

<u>1871</u> CD in monkey-like posture with handkerchief looking like a tail before a blackboard. "A little lecture by Professor D—n on the development of the horse." *Fun*, (22 Jul.), p. 38. CD's copy in DAR140.4.5.

<u>1871</u> "Mr. Bergh to the rescue." At the door of the "Society for the prevention of cruelty to animals". "The Defrauded Gorilla. 'That *Man* wants to claim my Pedigree. He says he is one of my Descendants.' Mr. Bergh. 'Now, Mr. Darwin, how could you insult him so?'" by Thomas Nast. *Harper's Weekly*, (19 Aug.), p. 776. Henry Bergh (1813-88) was the founder of the American society for the prevention of cruelty to animals.

1871 "Darwin eclipsed". An apelike CD as the sun with a dark bust of William Thomson, later Lord Kelvin, rolling in to cause an eclipse. The British Association at Edinburgh.— Humours of science. *The Graphic*, vol. 4:91 (26 Aug.), p. 5.

1871 CD with ape body "A venerable Orang-Utan. A contribution to unnatural history." *The Hornet*, (22 Mar.), (39.9x27.9cm). CD purportedly said "The head is cleverly done, but the gorilla is bad: too much chest; it couldn't be like that." In Hague 1884 cited above.

1871 "A Darwinian hypothesis". *Figaro*, (28 Oct.). CD's copy in DAR140.4.6.

1871 "The youthful Darwin expounding his theories", oil painting by William Holbrook Beard. A young anthropomorphic ape shows two older apes a line of animals from fish to amphibian. Prints were published by the American Photoplate Printing Company. See CCD20 for further details.

c.1871 Watercolour caricature of CD inspecting a painting of an ape-like man with large brow ridges in "Gallery of Ancestors" by Georges Montbard (Charles Auguste Loye). It was sent to CD. Now in DAR225.178.

1871? "Darwin's origin of species" [anthropomorphised animals dressed in clothing greet CD in the garden of Eden under a serpent] by Alfred Henry Forrester (1804-72), signed "Crowquill del". Published by W.H. Mason. DAR221.4.267.

Once a week 1872.   *Fun* 1872.   *Fun* 1872.

c.1871? "Comparative anatomy a la Darwin" by George Cruikshank jnr. (1792-1878). An ape-like CD shakes hands with an orangutan, top left corner. CD's copy in DAR225.181.

1870s? Anonymous print, CD with an ape-man pointing to a shouting ape. DAR140.4.24.

1870s? Watercolour caricature of CD in tails walking on all fours like an ape, signed "With Compliments of the Season", in DAR225.179. The mount is printed recto and verso with "Harding, 157 Piccadilly W". Signed on verso "From E. Kensley Aug 1939".

nd Watercolour caricature of CD standing to left, lecturing and holding a scroll entitled "...SPECIES". Beyond is the silhouette of a monkey. Inscribed: "Science Proffr Darwin" (25.4x18cm). The Royal Collection Trust. RL 28408 (RCIN 928408).

1870s? Mirrored silhouettes of CD as a monkey with quill pen in ear and looking in a hand mirror. DAR225.180 with a note "Possibly by Albert Bryan. Obtained by E. Kensley among a collection of the period. GL Keynes Dec. 1937."

1872 CD with Huxley et al looking at a gaping hippopotamus mother and baby at the zoon by Ernest Griset. "Emotional!" *Fun*, (23 Nov.), p. 209. Caption: "'An open countenance denotes a gentle and good-tempered character.' Still we think if Messrs. D—n, B—d, B—t, and Professor H—Y were to act on that theory and enter the den, their emotion would not be one of unmixed pleasure." CD's copy is in DAR140.4.8. A baby hippo had been born in the Zoological Garden of London a short time before.

1872 "Natural selection", CD with walking stick, by Frederick Waddy, *Once a Week*, (8 Jun.).

1872 CD as a monkey taking the pulse of a lady by Edward Dalziel, "That troubles our monkey again" *Fun*, (16 Nov.), referring to *Expression*. Caption: "Female descendant of Marine Ascidian:—'Really, Mr. Darwin, Say what you like about man; but I wish you would leave my emotions alone!'" Ascidian is a reference to *Descent* vol. 1, p. 206.

1873 CD with ape body and walking stick holding copy of *Origin*. A Hoen & Co. Varieties Theatre, Richmond, Virginia.

[1873] CD as monkey (inset) beckoning to Huxley and Tyndall. "Our National Church. The aegis of liberty, equality, fraternity. 'A House divided against itself cannot stand.'" Attributed to Gordon E. Flaws, London: E. Appleyard, 1*d*. (51x37cm). The scene with CD is at the top left corner. The biblical quote linked to CD is Genesis XXVII, 11: "Behold, my brother is a hairy man, and I am a smooth man." The 1883 version of the print states that this one was published ten years before it. Darwin family copies are in DAR141.10-11.

Our national church [1873].　　　　　　Our national church 1883.

Figaro 1874.　　*Fun* (extract) 1874.　　　*Punch* 1875.　　　　*La Lune Rousse* 1878.

1874 "The wedding procession". CD one of c.42 figures. "6. Dr. Darwin and our distinguished ancestor" [an ape or ape-man marches with CD]. *Fun*, (25 Mar.), p. 126.

1874 Chromolithograph "Prof. Darwin. This is the ape of form," by Faustin Betbeder (1847-c.1914) (Not "Betlader" as in ODNB). London Sketch Book of Celebrities. *Figaro*, (18 Feb.), (21.7x13.7cm). Artist's marked up proof in DAR225.177 annotated in an unknown hand: "The world has changed its mind since this was a 'joke'."

1874 Cover image to satirical song and sheet music. "Too thin; or, Darwin's little joke." A clean-shaven man (CD?) is surrounded by a circle of dancing monkeys. "Music by O'Rangoutang." New York: W.A. Pond.

1875 "Darwins Theorie", copper engraving by Max Klinger (1857-1920). Staatliche Museen zu Berlin. An ape hands a bearded man (CD?) a baby, with other evolution motifs.

1875 CD as an ape smoking a pipe in a tree by Linley Sambourne. "Suggested illustration. 'Dr. Darwin's Movements and habits of climbing plants.'...*We had no notion the Doctor would have been so ready to avow his connection with his quadrumaneous ancestors—the tree-climbing Anthropoids—as the title of his work seems it imply." *Punch*, (11 Dec.), p. 242. CD's copy in DAR140.4.15.

1877 Feb. 19 "Drawing of Professor Darwin and the gorilla" by Clement William Smith. Registered for copyright on this date. Artwork not found.

1877 Punch to Dr. Darwin. *Punch*, (1 Dec.), p. 241. Full face head and arm of CD illustrative of an accompanying comic poem.

Harper's Weekly 1878.   *La petite Lune* 1878.   *Punch* 1881.

Punch 1881.   Goedecker 1881.   *The Wasp* 1882.

1878 A simian looking "gentleman" labelled "Darwin" observes a "chimpanzee" who says "How are you Darwin?" Sketches in the New York Aquarium, by Frank Bellew. *Harper's Weekly*, (11 May), p. 368.

1878 Chromolithograph of CD as a monkey hanging from the "arbre de la science" by André Gill. *La petite Lune*, no. 10 (Aug.), cover. CD's copy in DAR140.4.20.

1878 Chromolithograph "L'homme descend du singe". CD as monkey leaps through circus hoops labelled credulity, superstition, errors and ignorance held by medical writer and populariser of A. Comte, Émile Littré. By André Gill, *La Lune Rousse* [Paris], (18 Aug.).

1881 "Charles Robert Darwin." Punch's fancy portraits. no. 54. CD contemplates an earthworm in the shape of a question mark. By Linley Sambourne. *Punch*, (22 Oct.), p. 190.

1881 "Man is but a worm." By Linley Sambourne. Punch's Almanack for 1882. *Punch*, (6 Dec.), n.p. This and the previous item refer to the recently published *Earthworms*.

1881 CD grins at a very similar looking monkey by Franz Goedecker who drew for *Vanity Fair*. The monkey sits on a copy of *Origin* which rests on a copy of "Erdenwhermer". Appeared in the UK and an offprint was published by Carl Simon, Berlin. Entirely re-drawn copy in *Morgen-Post* (3 Mar. 1885), cover.

1882 Chromolithograph "Reason against unreason." By Joseph Ferdinand Keppler, the "Light of Reason" containing many thinkers, with CD foremost, on a beam of light from the sky against the cowering religious orthodox. *Puck*, vol. 11, no. 261, (8 Mar.).

1882 Chromolithograph portrait (very poor) of the recently deceased CD held by two orangutans by George Frederick Keller. *The Wasp* [USA], (28 Apr.), vol. 8., cover.

1882 Chromolithograph "A sun of the nineteenth century". CD as a shining sun, dispelling the clouds of religion and superstition. *Puck*, (3 May).

1882 Lithographed cartoon of imaginary bust of CD to poke fun at Domingo Faustino Sarmiento by Henry Stein(?). *El Mosquito* [Argentina], (21 May), front page.

Untraced 18??.   *Puck* 1882.

1883 Jun. Chromolithograph "Our National Church The aegis of liberty, equality, fraternity. A house divided against itself—?." Signed "ION", drawn by Francis Carruthers Gould. Manchester & London: John Heywood. Different examples cost 6*d* to 1*s*. (The entire large print (45x55cm) is given as in insert in the top right corner here.) A bust of CD stands on the summit of a hill. A monkey bearing a flag marked "Darwinism" leads Huxley, Tyndall and Spencer up the stairs (named after geological eras) towards "protoplasm" and the rising sun, seemingly referring to enlightenment and the origin of life, with the statement "This way to daylight my sons." Their group was defined on the print as "Agnosticism". On this print and the 1873 original, see Janet Browne, *Power of place*, 2002, pp. 380-1. Another version (priced 6*d*) was split into 15 parts and glued on backing that allowed the print to be folded into quarter size in black buckram book covers with the title "National Church." One in a private collection (USA) has contemporary bookplates "Rev. Richard Covey, M.A." and "Rev. W.W. Covey Crump".

18?? CD as the figure at the end of an evolutionary sequence starting from marine creatures in water. (Reminiscent of "Man is but a worm" *Punch*, 1881.) Source untraced. Dutch?

1883 Chromolithograph "The universal church of the future—from the present religious outlook." Central portrait of CD overlooks 'congregation' from the wall. *Puck*, (10 Jan.).

1885 Re-drawing of Goedecker 1881 caricature. *Morgen-Post*, (3 Mar.), p. 1, "C Ohnesorg".

1885 Chromolithograph "The old attempt", "Mr. Beecher is trying to bridge the chasm between old orthodoxy and science with his little series 'Evolution sermons.'" Other figures

beside CD (the predominant figure) as giant heads carved into a mountainside are Spino-
za, Copernicus, Huxley, Spencer, Tyndall and Haeckel. *Puck*, (3 Jun.).

1885 Chromolithograph "Sheol". "This business is removed to Sheol, opposite." CD as one
of many respectable non-Christian figures moved from the old hell to the pleasant water-
ing-place known as Sheol. By Joseph Ferdinand Keppler. *Puck*, vol. 17, no. 429, (27 May).

1890 CD hands a bespectacled man dressed as a pharaoh a copy of *Origin*. 'Pharaoh' holds a
document labelled "races of man" behind his back, with a chimpanzee and ibis in foreground. Publication untraced. Getty images.

1890s? Caricature of CD in woodcut-style with monkey in the background on a playing card
by Multum In Parvo, London.

1890s? Chromolithograph of monkeys at a table celebrating Christmas with a painting of CD
on the wall "The toast of the evening. / Wishing you and yours a merry Christmas."

c.1893 Bronze statuette of an ape holding a human skull and sitting on a pile of books, one
of which has "Darwin" on the spine, by German sculptor Wolfgang Hugo Rheinhold
(1853-1900). 'Affe einen Schädel betrachtend' cast by Gladenbeck & Sohn, Berlin. The
statuette was not intended as a caricature, but has been widely reproduced and imitated
many times as such.

nd CD with long beard and a tail protruding from under his coat captioned "Natural history
repeating itself". Signed "A.B." untraced.

| *Puck* 1885. | *Truthseeker* 1904. | *Simplicissimus* 1909. |

1903 Colour cartoon: "Professor Darwin Doenutt studies infantile expression." by Walter
McDougall. *Chicago Record-Herald*, (25 Jan.).

1904 [CD as referee] *The Truthseeker*, (Oct.). Caption: "Great prize fight for the champion-
ship of the world / Between G. Hovah, the Celestial Crusher, and B L. Zebub, the Pet of
Pandemonium." See S. Paylor, Edward B. Aveling The people's Darwin. *Endeavour*, 2005.
The newspaper *Blue-Grass Blade* [Lexington, Ky.], (3 Jan.), p. 1, claimed that the *Truthseeker*
editor G.W. Foote was charged with blasphemy over this cartoon.

1909 "Zu darwins hunderstem Geburtstag." [CD embraces a chimpanzee and orangutan in a
tree. Haeckel offers a halo] by Thomas Theodor Heine. *Simplicissimus*, (15 Feb.), p. 792.

1913 CD bearing the term "eugenics". Forcible feeding. [John Bull being force-fed via a
stomach pump, by a queue of people; representing Britain's numerous political problems]
by Stanger Pritchard. *Truth*, (25 Dec.).

1925 "Darwin's theory reversed" by John M. Baer. CD with pointer at monkey to man images. "It took millions of years, Darwin claimed, to make a man out of a monkey". In
other panel a "company 'union' boss" has a monkey sign a contract: "While a company
union can make a monkey out of a man in a few minutes". *Labor* [USA] (18 Jul.).

1927 CD as a small child with his mother on the street with a monkey and musician nearby. "The childhood of great men. Little Charles Darwin begins to consider." by George Morrow. *Punch* (27 Apr.).

1929 There was an explosion of caricatures surrounding the so-called Scopes Monkey trial and subsequent high-profile trials regarding the teaching of evolution in American public schools. Not recorded here.

1930 CD with monkey taking coffee by Lluís Bagaria. *The Graphic*, (1 Feb.), p. 22.

Very many more 20th-century and more recent caricatures exist but are not recorded here.

Itinerary: Where CD was at any one time in his life is usually well documented except for the earliest years. For these the autobiographical fragment (DAR91.56-63), printed in ML1:1-5, CCD2, Appendix III, Secord, *Evolutionary writings*, 2008 and in *Darwin Online*, is the most helpful; this was written in 1838 when he started his personal 'Journal' (DAR158). The 'Journal' is reproduced year-by-year in the appendices to the CCD interleaved with additional information by the editors. The complete 'Journal', transcribed by the editors of the CCD and edited by John van Wyhe, is in *Darwin Online*: (http://darwin-online.org.uk/content/frameset?viewtype=side&itemID=CUL-DAR158.1-76&pageseq=3).

The 'Journal' contains only a little on the *Beagle* voyage, but the *Beagle diary* and Fitz-Roy's vol. 2 of the *Narrative* give details. There is a complete day-by-day itinerary of CD during the *Beagle* voyage by Kees Rookmaaker, 'Darwin's itinerary on the voyage of the *Beagle*' in *Darwin Online*:

(http://darwin-online.org.uk/content/frameset?viewtype=text&itemID=A575&pageseq=1).

See a greatly condensed version of this under *Beagle*, 1831-1836 second surveying voyage, Summary, above.

After 1838, all important visits from home are noticed in his 'Journal', except that brief trips to London for a night or so were omitted, or else he does not say where he stayed. Very many dates and details have been found by the editors of the CCD and by Christine Chua in *ED's diary*.

1842-81 After his move to Down House in 1842 CD was away from home for a considerable part of each year. Much of the time was spent at hydropathic establishments, but there were also holidays and journeys for scientific business. From 1842 to 1881 he was away for a total of about 2,000 days, exceeding 50 days in 23 of these 40 years.

1809-12 The Mount, Shrewsbury.

1813 Family summer holiday at Gros, Abergele, North Wales.

1814-16 No information, Shrewsbury.

1817 Spring, CD with his sister Catherine to Mr G. Case's day school in Shrewsbury.

1818 Summer, CD to Shrewsbury School as a boarder, stayed seven years.

Jul. CD to Liverpool with Erasmus Darwin.

1819 Jul. Summer holiday at Plas Edwards, Towyn, North Wales.

1820 Jul. riding tour with Erasmus to Pistyll Rhaeadr, North Wales.

1822 Jun. Downton, Wiltshire, with Caroline.

Jul. To Montgomery and Bishop's Castle, Shropshire, with sister (Elizabeth) Susan.

1824 Jun. 19 Susan & CD visited Maer.

Jul. 30 Emma Wedgwood "met with the Darwins" at Matlock, not specifying CD. CD left 1 Aug., returned 3 Aug. "The Darwins came in the evenings". *ED diary*.

Dec. 18 CD Shrewsbury to Maer.

1825 Jun. 17 left Shrewsbury School.

Oct. 22 matriculated Edinburgh University. Lodged at 11 Lothian St. Oct. 26 First lecture.

1826 At Edinburgh all year in term time.

Jun. 15 North Wales, walking tour with N. Hubbersty, climbed Snowdon.

Oct. 30 Parle, North Wales, with Caroline.

1827 Apr. c.24 finally left Edinburgh, toured Dundee, St. Andrew's, Stirling, Glasgow, Belfast, Dublin with Josiah Wedgwood [II].

May, end of, visited Paris with Josiah Wedgwood [II] and Catherine Darwin.
Autumn, many visits to Woodhouse, Shropshire, especially for shooting.
Sept. at Maer and visited J. Mackintosh.
1828 Jan. 26 Cambridge, lodging in Sydney Street.
Jul. 1 to Barmouth, North Wales with J.M. Herbert & T. Butler for private coaching by G.A. Butterton.
Aug. 27 to Maer for shooting.
Sept. at Maer and then Osmaston Hall.
Oct. 31 Cambridge.
Dec. 20 returned to Shrewsbury.
1829 Feb. 19 two days in London to talk about beetles with F.W. Hope.
Feb. 24 in Christ's College, Cambridge.
Jun. 8 to London & Shrewsbury
Jun. mid. to Barmouth with F.W. Hope.
Jun. 29? back to Shrewsbury with bad lips.
Jul. Maer one week.
Aug. to Barmouth with sisters.
Oct. Birmingham with Wedgwoods for music meeting.
Oct. 16 to Cambridge. [Oct. 12 according to Christ's College records.]
1830 Jan. 1 at Cambridge.
Jun. 3 out of College.
Aug. to North Wales, beetles and fishing. In Barmouth with F.W. Hope and T.C. Eyton.
Oct. 7 in College.
1831 Jan. to Cambridge for three months to keep terms, stayed with J.S. Henslow.
Jun. 16 left Cambridge end of May term.
Aug. to Llangollen, Ruthin, Conway, Bangor, Capel Curig, with Sedgwick for geology, then alone to Barmouth.
Sept. 1 Maer for shooting.
Sept. 2-4 Cambridge.
Sept. 5 London, 17 Spring Gardens.
Sept. 9 left by Packet with FitzRoy.
Sept. 11 arrived at Plymouth to see *Beagle*.
Sept. 11-13 sailing.
Sept. 13-16 Devonport.
Sept. 17-19 London.
Sept. 19-21 Cambridge.
Sept. 22 Shrewsbury.
Oct. 2 London, 17 Spring Gardens.
Oct. 21 Shrewsbury.
Oct. 24 Plymouth.
Dec. 10 *Beagle* sailed but put back.
Dec. 21 sailed but put back.

Dec. 27 sailed.
(For more detail on CD's *Beagle* voyage itinerary, see *Beagle*, 1831-1836 second surveying voyage, Summary, above.)
1832 Jan. 16-Feb. 8 Cape Verde Islands.
Feb. 16-17 St. Paul's Rocks.
Feb. 16-17 *Beagle* crossed equator.
Feb. 20 Fernando de Noronha.
Feb. 29 Landed at Brazil.
Mar. 27 Abrolhos archipelago.
Apr. 5-Jul. 5 Rio de Janeiro.
Jul. 26-19 Aug. Montevideo.
Sept. 6-Oct. 17 Bahia Blanca.
Nov. 2-26 Montevideo.
Dec. 16 Tierra del Fuego.
1833 Feb. 26 Tierra del Fuego.
Mar. 1-Apr. 6 Falkland Islands.
Apr. 28-Jul. 23 Maldonado.
Aug. 3-Dec. 6 Rio Negro & Montevideo.
Dec. 23-Port Desire.
1834 Jan. 4 Port Desire.
Jan. 9-Jan. 19 Port St. Julian.
Jan. 29-Mar. 7 Straits of Magellan via Falklands.
Mar. 10-Apr. 7 Falkland Islands.
Apr. 13-May 12 Santa Cruz River.
Jun. 28-Jul. 13 Chiloe.
Jul. 23-Nov. 10 Valparaiso.
Nov. 10 Valparaiso to Chiloe.
Dec. 13 Chonos Archipelago.
1835 Feb. 4 Chiloe.
Feb. 8-22 Valdivia.
Mar. 4-7 Concepcion.
Mar. 11-Jul. 6 Valparaiso-Copiapó.
Jul. 12-14 Iquique.
Jul. 19-Sept. 7 Callao for Lima.
Sept. 16-Oct. 20 Galapagos Islands.
Nov. 15-26 Tahiti.
Dec. 21-30 Bay of Islands, New Zealand.
1836 Jan. 12-30 Sydney.
Feb. 5-17 Hobart Town, Tasmania.
Mar. 6-14 King George's Sound.
Apr. 2-12 Cocos Keeling Islands.
Apr. 29-May 9 Mauritius.
May 31-Jun. 18 Cape of Good Hope.
Jul. 8-14 St. Helena.
Jul. 19-23 Ascension.
Aug. 1-6 Bahia Blanca.
Aug. 12-17 Pernambuco.
Sept. 4-8 Porto Praya, Cape Verde.
Sept. 20-25 Terceira, Azores.
Oct. 2 Falmouth, Cornwall.

Oct. 4 Shrewsbury.
Oct. end of, Greenwich unloading *Beagle.*
Nov. 6 London, 43 Gt. Marlborough St.
Nov. 12 Maer, 16th Shrewsbury.
Dec. 2-13 London.
Dec. 13- Cambridge, J.S. Henslow's home and 22 Fitzwilliam Street.
1837 Jan. 4 trip to London.
    Feb. 27 Phil. Soc. Meeting.
    Mar. 6 Cambridge.
    Mar. 6-12 London, 43 Marlborough St.
    Mar. 13-Jun. 25, 36 Great Marlborough St.
    Nov. 21 Isle of Wight visit to W.D. Fox.
    Nov. 23 London.
1838 May 10-12 Cambridge with Henslow.
    Jun. 23 London to Leith, Edinburgh for one day, Salisbury Crags, Loch Leven, Glen Roy for eight days, Glasgow, Liverpool.
    Jul. 12 Overton-on-Dee, Flintshire one night.
    Jul. 13-31 Shrewsbury and Maer.
    Aug. 1 to London.
    Oct. 25 Windsor for two days of rest.
    Nov. 9 Maer; Nov. 11 proposed to Emma Wedgwood.
    Nov. 12 Shrewsbury.
    Nov. 17 Maer.
    Nov. 20 to London.
    Dec. 6 Emma Wedgwood came to London.
    Dec. 21 to Maer.
    Dec. 31 slept at 12 Upper Gower St.
1839 Jan. 11 to Shrewsbury.
    Jan. 12-13 Half Moon St., London.
    Jan. 15 to Maer.
    Jan. 18 to Maer to London.
    Jan. 25 to Shrewsbury.
    Jan. 28 to Maer; Jan. 29 CD married Emma.
    Jan. 30 to 12 Upper Gower St.
    Apr. 26-May 6 Maer.
    May 7-8 Seabridge, Etruria, Fenton.
    May 8-12 Gower St.
    May 13-19 Shrewsbury.
    May 20 Came home to Gower St.
    Jun. 18 Windsor
    Aug. 23 to Maer.
    Aug. 26-31 to Birmingham for BAAS.
    Sept. 12-28 to Shrewsbury.
    Oct. 2 to London.
    Nov. 30 "Royal Soc. 1 o'clock"
1840 Jan. 31. Lecture Royal Institution.
    Apr. 3-10 to Shrewsbury.
    Jun. 10 to Maer and Shrewsbury.

Nov. 10 to London.
1841 Mar. 7-17 Shrewsbury.
    May 28 to Maer.
    Jun. 23 Camphill, Staffordshire.
    Jun. 30 to Shrewsbury. 1 Jun. in Account Book (Down House MS).
    Jun. 23 Camphill, Staffordshire.
    Jul. 23 to London.
1842 Feb. 21-23 Eastbury Park, Dorset.
    Mar. 7-17 Shrewsbury.
    May 18-Jun. 14 Maer.
    Jun. 15 to Shrewsbury.
    Jun. 18 Capel Curig, Caernarvon, 10 days.
    Jun. 29 returned from Wales.
    Jul. 15 returned from London in Account Book (Down House MS).
    Jul. 18 to London. 'Journal'.
    Jul. 22-23 CD and ED first saw Down House, slept at inn.
    Jul. 28-29 went to Down.
    Sept. 14 ED slept at Down House.
    Sept. 17 CD slept at Down House.
    Nov. 16 London.
1843 Jan. 17-19 went to London.
    Apr. 4 went to London.
    May. 3 went to London.
    Jul. 8 Maer and Shrewsbury one week.
    Jul. 16-17 unwell in London.
    Oct. 12 London then Shrewsbury ten days.
    Oct. 20-21 to London with ED.
    Nov. 30 London.
    Dec. 14 London.
1844 Mar. 20-21 to London with ED. Geo. Soc.
    Apr. 17 meeting of the Geological Society.
    Apr. 23 to Maer and Shrewsbury.
    May. 16-25 Shrewsbury.
    May 30 return to Down House.
    Jun. 12 meeting of the Geological Society.
    Jul. 17 to London with ED, ancient lectures.
    Jul. 18 to Kew with ED & returned home.
    Oct. 18-29 Shrewsbury. ED wrote 16th.
    Nov. 20 meeting of the Geological Society.
    Dec. 18 meeting of the Geological Society.
1845 Jan. 8 meeting of the Geological Society.
    Jan. 31 to London.
    Feb. 1 to Maer.
    Feb. 5 meeting of the Geological Society.
    Feb. 7-8 Camphill, Staffordshire.
    Mar. 12 meeting of the Geological Society.
    Apr. 2 meeting of the Geological Society.
    Apr. 10 April to London.

Apr. 29-May 10 London & Shrewsbury.
May 11 Down House.
May 28 meeting of the Geological Society.
Jun. 11-12 meeting of the Geological Society.
Jul. 2 meeting of the Geological Society.
Sept. 15 Shrewsbury, Beesby, Manchester to visit W. Herbert, Walton Hall to visit C. Waterton, Chatsworth, Camphill to visit S.E. Wedgwood [I]. Slept at Rugby.
Sept. 16 came to Shrewsbury.
Oct. 2 CD "came from his tour."
Oct. 16 Camphill, Staffordshire.
Oct. 26 to Down House.
Nov. 11-15 Chester Terrace.
Nov. 19-24 London, Geological Society.
Dec. 10 meeting of the Geological Society.
1846 Jan. 7 meeting of the Geological Society.
Feb. 4 meeting of the Geological Society.
Feb. 20 meeting of the Geological Society.
Feb. 21-Mar. 2 Shrewsbury. 3rd home.
Mar. 11 in London.
Mar. 25 meeting of the Geological Society.
Apr. 21 meeting of the Geological Society.
Apr. 23 in London.
May 20 meeting of the Geological Society.
Jun. 3 meeting of the Geological Society.
Jun. 17 meeting of the Geological Society.
Jun. 29 London.
Jul. 31-Aug. 8 Shrewsbury. 9 Aug. home.
Sept. 9-17 Southampton for BAAS with ED.
Sept. 12 visited Portsmouth and Isle of Wight.
Sept. 13 Winchester and St. Cross.
Sept. 14 Netley Abbey and Southampton Common.
Sept. 22 day at Knole Park, Sevenoaks with ED and Susan Darwin.
Sept. 29 London with ED.
Oct. 19-23 London ten days in two visits.
Nov. 18 meeting of the Geological Society.
Dec. 2 meeting of the Geological Society.
1847 6 Jan. meeting of the Geological Society.
Feb. 19-Mar. 5th Shrewsbury.
Apr. 28 meeting of the Geological Society.
May 12 meeting of the Geological Society.
May 26 meeting of the Geological Society.
May 29 London.
Jun. 9 meeting of the Geological Society.
Jun. 22-Jun. 30 Oxford for BAAS, visited Newnham Courtney, Dropmore, Burnham Beeches.
Sept. 18 London.

Oct. 22-Nov. 5 Shrewsbury.
Nov. 17 meeting of the Geological Society.
Dec. 1 meeting of the Geological Society.
1848 Jan. 5 meeting of the Geological Society.
Feb. 2 meeting of the Geological Society.
Feb. 18 to London, back same day.
Mar. end of to London.
Apr. 7-12 to the Hermitage.
Apr. 19 meeting of the Geological Society.
May 17 meeting of the Geological Society. To Shrewsbury.
May 31 meeting of the Geological Society. To Shrewsbury.
Jun. 1 Return to Down.
Jun. 14 meeting of the Geological Society.
Jul. 22 week at Swanage by Wareham and Corfe Castle.
Jul. 29 to Poole in Sir William Symonds's yacht and New Forest.
Oct. 10-24 to Shrewsbury.
Oct. 25 to Down.
Nov. 17-26 at Shrewsbury with Erasmus.
Nov. 26 to Down.
Dec. 13 to London.
Dec. 16 London(?)
1849 Jan. 31- Feb. 1 to London with ED.
31st meeting of the Geological Society.
Mar. 5 to Chevening, Lord Stanhope.
Mar. 8 to London with ED.
Mar. 9 to Malvern.
Mar. 10-Jun. 29 Malvern, slept at Chester Terrace. Malvern Wells, family and servants.
Jun. 30 returned home.
Sept. 11-21 Birmingham for BAAS, day visit to Malvern on 16. (ED "came home" 20th).
Nov. 7 meeting of the Geological Society.
Dec. 19 meeting of the Geological Society.
1850 Feb. 6 meeting of the Geological Society.
Apr. 10 meeting of the Geological Society.
Jun. 1-5 Hermitage.
Jun. 5 London.
Jun. 11-18 Malvern Wells.
Aug. 10-16 Leith Hill Place, visit Josiah Wedgwood [III], with 3 eldest and baby Lenny.
Sept. 5-6 to London then to Bruce Castle.
Oct. 14-21 Hartfield, Sussex, The Ridge, visit S.E. Wedgwood [II]. Took William.
Oct. 18 Ramsgate for the day.
Oct. 22 went home with the rest but ED stayed with Annie.

Nov. 15 with ED to Mitcham & brought home William.

Dec. 18 meeting of the Geological Society.

1851 Feb. 13 London to see Dr Paget.

Mar. 24-31 Malvern with Annie.

Apr. 16-24 Malvern with Annie who died there on Apr. 23. 16th CD went to London.

Jul. 30-Aug. 9 London, 7 Park St to see Great Exhibition. (ED records Aug. 10 came home.

Nov. 20-21 to London. Saw Sir. B.B.

Dec. 17 attended Geological Society Club.

1852 Jan. 29 to London & brought back William.

Mar. 24 Rugby 1 day to see W.E.D., then Barlaston, Betley and Shrewsbury to sister Susan with Henrietta & George.

Apr. 1 return to Down.

Jun. 2-5 to London.

Jun. 24-25 London.

Jul. 14-17 to Hartfield with William, Henrietta & George. (CD not mentioned) *ED's diary.*

Sept. 11-16 Leith Hill Place, home via Godstone and Reigate. George, Henrietta, Lizzy & Horace.

Nov. 8-12 to London.

1853 Feb. 1-3 to London to visit Susan, Catherine and Erasmus.

Apr. 4-7 in London. (*Health diary*) 6th meeting of the Geological Society.

Apr. 28-May 3 to London.

May 7 Attended Lord Rosse's Roy. Soc. party.

Jun. 1 meeting of the Geological Society.

Jul. 14-Aug. 4 Eastbourne with family, to Brighton & Hastings on day. visits.

Aug. 13-16 The Hermitage near Woking to visit Henry Allen Wedgwood with ED, George and Henrietta, visited military camp for Crimean war at Chobham. (ED records home Aug. 17).

Oct. 6 to Queen Anne St. (CD not specified by ED) *ED's diary.*

Nov. 28 to London

Dec. 10 London.

1854 Jan. visited London.

Feb. 23-25 to London with ED, Henrietta & Leonard.

Mar. 13-17 The Ridge, Hartfield, Sussex with ED, came home leaving George & Franky who returned with ED on 20th.

Mar. 29-Apr. 1 to London with Henrietta & Lizzy (CD not named).

May 1-3 to London. Linnean Society meeting.

May 12 to Westerham, home by Knockholt (CD not named).

May 20-22 to Hartfield (CD not named).

May 24 to London. Philosophical Club of the Royal Society.

May 27 to London with Etty to Mr Robinson (CD not specified).

May 31 Leith Hill Place (CD not named).

Jun. 10 Crystal Palace opening at Sydenham.

Jun 21-23 London. Philosophical Club.

Jul. 13-15 The Ridge, Hartfield, Sussex.

Oct. 9-14 Leith Hill Place, all but Horace.

Oct. 23-Nov. 3 to London with ED.

Nov. 2-4 London with ED.

Nov. 30-Dec. 1 London for anniversary meeting of the Royal Society.

Dec. 1 in London for breakfast.

1855 Jan. 18-Feb. 15 London, 27 York Place, Baker St. Royal Society & Philosophical Club.

Feb. 3 dined at Cumberland Terrace.

Mar. 20-24 to London.

Apr. 21-23 meetings of the Royal Society.

Apr. 30-May 5 London with ED.

Jun. 14 Crystal Palace.

Jun. 28 council meeting of the Royal Society.

Jul. 7-10 to Hartfield with William, George, Franky & Brodie. (CD not named).

Jul. 12 to Hartfield with Henrietta & Lizzy. (CD not named)

Aug. 7-9 London.

Aug. 25 to Chevening with ED & George.

Aug. 28-30 Anerley for Poultry show.

Sept. 10-18 Glasgow for BAAS with ED. Slept. 11 at Carlisle, arrived Glasgow on 12th. Attended Col. Rawlinson's lecture on 17th.

Sept. 19 slept at Carlisle.

Sept. 20 to Shrewsbury by Rugby.

Sept. 22 to Barlaston then Down.

Sept. 25 home, seeing William at Rugby.

Sept. 28 dined at Holwood.

Oct. 11 dined at Holwood.

Oct. 17 went to Brighton.

Oct. 25 council meeting of the Royal Society.

Nov. 8 meeting of the Royal Society.

Nov. 29 meeting of Columbarian Society.

Nov. 30 meeting of the Royal Society.

Dec. 19 to London to Mr Robinson.

Dec. 20 council meeting of the Royal Society & Philosophical Club.

1856 Jan. 8 meeting of Philoperisteron Society.

Jan. 31 meeting of the Royal Society.

Feb. 9-12 to London with Etty and Willy (CD not named). *ED's diary.*
Feb. 23 to Hartfield, children since Feb. 21.
Feb. 28 drove to East Grinstead.
Feb. 29 walked to Hartfield.
Mar. 1 returned home with George.
Mar. 10-14 to London, spent 10th with Cumberland, 11th to Exeter. Royal Society on 13th.
Mar. 18-19 to London.
Apr. 10 to London.
Apr. 17 meeting of the Royal Society.
May 5-8 to London. Royal Society.
May 16 to Hartfield.
May 20 drove to Buckhurst.
May 29 to London with Etty and George.
Jun. 12 meeting of the Royal Society.
Jun. 18-21 meeting of the Royal Society.
Jul. 16 to Chevening.
Jul. 29 Anerley poultry show.
Aug. 14 to London.
Aug. 25 to London with Etty.
Sept. 8 dined at Holwood.
Sept. 13-18 Leith Hill Place.
Sept. 19 returned home.
Oct. 15-18 London. Philosophical Club.
Nov. 13 London. Philosophical Club.
1857 Jan. 13-16 took William to London.
Mar. 4-7 London, brought George home.
Apr. 9 to Hastings.
Apr. 22-May 6 Moor Park.
May 27 to Westerham.
Jun. 6 to Leith Hill Place.
Jun. 16-30 Moor Park.
Jun. 23 all to Barlaston.
Jun. 26 picnic in Trentham.
Jun. 27 visited Selborne to see Bessy.
Jun. 29-30 to Shrewsbury.
Jul. 18-21 to Moor Park.
Aug. 10 Crystal Palace, poultry show.
Sept. 21 CD as magistrate, Bromley.
Sept. 24-27 to Moor Park.
Oct. 16 to Withyham.
Oct. 29-30 to Queen Anne St.
Nov. 5-12 Moor Park.
Nov. 16 CD as magistrate, Bromley.
Nov. 16-20 London.
Dec. 12 to Crystal Palace with Miss Pugh. (CD not named). *ED's diary.*
1858 Jan. 18 CD as magistrate, Farnborough.
Feb. 15 CD as magistrate, Bromley.
Feb. 16-20 London, joined by children 17th.
Mar. 15 CD as magistrate, Bromley.

Apr. 19 CD as magistrate, Bromley.
Apr. 20-May 4 Moor Park.
May 15 London visit.
May 17 CD as magistrate, Bromley.
May 20 London visit. Philosophical Club.
Jun. 5-9 to Leith Hill Place.
Jun. 21 CD as magistrate, Bromley.
Jul. 9-13 The Ridge, Hartfield.
Jul. 16 to Portsmouth with Etty.
Jul. 17-26 via Portsmouth, Sandown, Isle of Wight, King's Head Hotel with family.
Jul. 26-Aug. 12 Norfolk House, Shanklin, Isle of Wight.
Aug. 16 CD as magistrate, Bromley.
Aug. 19 ED: "came home".
Sept. 15 Moor Park with Etty. (CD not named) *ED's diary.*
Sept. 20 CD as magistrate, Bromley.
Oct. 6-7 to London with Lizzy. (CD not named) *ED's diary.*
Oct. 18 CD as magistrate, Bromley.
Oct. 19 visited London.
Oct. 25-31 Moor Park.
Nov. 1 returned from Moor Park.
Dec. 14-16 London. Philosophical Club.
Dec. 20 CD as magistrate, Bromley.
1859 Jan. 5-7 to London with Lizzy.
Jan. 17 CD as magistrate, Chelsfield.
Jan. 20 to London with William.
Feb. 5-18 Moor Park.
Feb. 21 CD as magistrate, Bromley.
Mar. 21 CD as magistrate, Bromley.
Mar. 26 to Crystal Palace.
Apr. 1-4 to London with Etty.
Apr. 18 CD as magistrate, Bromley; to Moor Park.
May 16 CD as magistrate, Bromley.
May 21-28 Moor Park.
Jun. 7 to Hartfield with Etty.
Jun. 20 to Crystal Palace with Etty.
Jul. 19-26 Moor Park.
Aug. 11-15 to Bromley, to London.
Aug. 15 CD as magistrate, Bromley.
Aug. 20-23 Leith Hill Place.
Sept. 10 to Bromley.
Sept. 19 CD as magistrate, Farnborough.
Oct. 2-Dec. 7 Wells Terrace, Ilkley.
Oct. 17 ED: "came to Ilkley".
Dec. 8-9 London.
Dec. 19 CD as magistrate, Bromley.
Dec. 31 Lord and Lady Stanhope, Sevenoaks.

1860 Jan. 16 CD as magistrate, Farnborough.
Jan. 24-27 London. Philosophical Club.
Jan. 30-Feb. 11 to London with William to Cambridge.
Feb. 10 Huxley lecture at Royal Institution.
Feb. 20 CD as magistrate, Bromley.
Feb. 27-28 London.
Mar. 5-6 London.
Mar. 19 CD as magistrate, Farnborough.
Mar. 28 to London.
Apr. 14 to London.
Apr. 21 to London. Philosophical Club.
Jun. 28-Jul. 6 Sudbrook Park, Petersham.
Jul. 7 visited Hooker at Kew.
Jul. 10-Aug. 2 The Ridge, Hartfield.
Aug. 14 to London to see Robinson. (CD not named) *ED's diary*.
Aug. 18 to London. (CD not named).
Aug. 20 CD as magistrate, Farnborough.
Aug. 21 to London.
Sept. 17 CD as magistrate, Farnborough.
Sept. 19 to Hartfield.
Sept. 22-Nov. 10 15 Marine Parade, Eastbourne due to Etty illness.
Dec. 31 Stanhope visit, Sevenoaks.
1861 Jan. 3 to Baston. (CD not named)
Jan. 28 to London with George.
Feb. 21-22 to London.
Apr. 1-4 London, Queen Anne St.
Apr. 16-20 to London.
Jul. 1-Aug. 27 1 slept at Reading, 2 Hesketh Terrace, Torquay.
Aug. 26 left Torquay, 27th home.
Sept. 18 to Westerham with ED.
Nov. 5-9 St. Leonards. (CD not named).
Nov. 21-23 London.
Nov. 28 dined at Holwood House with ED.
Dec. 16 CD as magistrate, Farnborough.
Dec. 27-31 to St. Leonards, Edenbridge.
1862 Feb. 17 CD as magistrate, Farnborough.
Feb. 19 to London, saw dentist.
Apr. 2-4, London, Queen Anne St.
Apr. 23 to London with 3 boys and Lizzy to a play. (CD not named) *ED's diary*.
May 6-9 to London. International Exhibition.
May 13 to Clapham.
May 15-22 Leith Hill Place. 17 drove with ED to Ockley. Home on 22nd.
Jun. 3-12 to Southampton.
Aug. 7 day to London to see children.
Aug. 9-10 to London hearing Etty was ill.
Aug. 12-31 1 Carlton Terrace, Southampton.

Sept. 1-27 Cliff Cottage, Bournemouth.
Sept. 29 London, Queen Anne St. Slept at 62 Gower St., returned home 30.
Oct. 16 to London. Exhibition & Adelphi.
Nov. 8-12 to Cumberland Terrace.
Nov. 25-Dec. 9 to London.
Dec. 17-20 to Chester Place.
1863 Jan. 22-23 London for ball, Crystal Palace.
Feb. 4-14 London, Queen Anne St. with Henrietta & Horace. Lizzy joined on 7th. To Kew on 11th. Queen Anne St. on 13th.
Apr. 14 to Westerham. (CD not named).
Apr. 27-May ?6 Hartfield. 4th to Buckhurst.
May 6-13 Leith Hill Place.
Sept. 2-3? London.
Sept. 3-Oct. 12 Malvern Wells.
Oct. 14 from London to home.
1864 Aug. 25-1 Sept.? 4 Chester Place.
1865 Nov. 8-18? London, Queen Anne St.
1866 Apr. 21-May 2 London, Queen Anne St. 28 soirée at Roy. Soc. ED: "came home" May 1.
May 29-Jun. 2 Leith Hill Place.
Oct. 3-4 to Wardour Castle. (CD not named)
Nov. 22-29 London, Queen Anne St.
Dec. 18-22 London. (not specified).
1867 Feb. 13-21 London, Queen Anne St.
Jun. 17-24 London, Queen Anne St. 20 to Royal Academy.
Oct. 18-24 London, Queen Anne St.
Nov. 28-Dec. 7 London, Queen Anne St.
1868 Feb. 21-24 to Leith Hill Place.
Mar. 3-10 London, Queen Anne St.
Mar. 10-31 4 Chester Place, London. 30 or 31 Kew to see Hooker. Returned home Apr. 1.
Jul. 16 Bassett, Southampton to Isle of Wight.
Jul. 17-Aug. 20 Freshwater, Isle of Wight.
Jul. 27 to Alum Bay.
Aug. 20 slept at Southampton, home Aug. 21.
Nov. 7-16 London, Queen Anne Street.
1869 Jan. 21-23 to Leith Hill Place.
Feb. 16-24 London, Queen Anne St.
Jun. 10 Shrewsbury on way to Barmouth.
Jun. 11-Jul. 29 Caerdeon, Barmouth, North Wales, to recuperate from fall from Tommy.
Jul. 30-31 Stafford on way home.
Nov. 1-9 London, Queen Anne St. Nov. 6 to House of Commons. Met Haweis.
1870 Mar. 5-10 London, Queen Anne St. (ED recorded home on 12th.)
May 20-24 Bull Hotel, Cambridge.
Jun. 24-Jul. 1 Queen Anne St.

Aug. 13-26 Bassett, Southampton. Aug. 21 to
Bishopstone.
Oct. 13-20 Leith Hill Place. 17 Ightham Mote.
Dec. 8-14 London, Queen Anne St.
1871 Feb. 23-Mar. 2 London, Queen Anne St.
Lunch with Wallace. Went to pictures Mar. 1.
Apr. 1-5 London, Queen Anne St.
May 11-19 Bassett, Southampton.
Jun. 24-30 London, Queen Anne St. Royal
Academy on 29.
Jul. ?-13 Queen Anne St., Crystal Palace.
Jul. 28-Aug. 25 Haredene, Albury, Guildford,
family holiday.
Nov. 3-10 Leith Hill Place.
Nov. 14 to London. (CD not named).
Dec. 7 to London. (CD not named).
Dec. 12-22 London, Queen Anne St. ED: 14
"came to London".
1872 Feb. 13-Mar. 21 London, 9 Devonshire St.,
a rented house. ED: 16 "came up to London".
27th to see procession at St. Paul's cathedral.
Mar. 15 to Crystal Palace.
Jun. 8-20 Bassett, Southampton.
Aug. 13-21 Leith Hill Place.
Oct. 5-26 Sevenoaks Common (Horace lodg-
ings in Sevenoaks).
Dec. 17-23 London, Queen Anne St. (ED
recorded home on 22.)
1873 Mar. 15-Apr. 10 London, 16 Montague St.,
a rented house.
Jun. 4-12 Leith Hill Place.
Aug. 5-9 Abinger Hall visiting Farrer.
Aug. 9-21 Bassett, Southampton.
Nov. 8-18 London, 4 Bryanston St. visiting
Richard Buckley Litchfield. Nov. 9 at Kew.
1874 Jan. 10-17 London, Queen Anne St. Jan. 16
séance of 20 persons at Erasmus's house.
Apr. 21-29 London, 4 Bryanston St. Royal
Society soirée on 22nd.
Jul. 25-30 Abinger Hall. 28 to Leith Hill Place.
Jul. 30 to Southampton.
Jul. 31-Aug. 24 Bassett, Southampton. 6 drove
to Southampton.
Dec. 3-12 London, 4 Bryanston St.
1875 Mar. 31-Apr. 12 London, Queen Anne St.
Apr. 6 to Bryanston St.
Jun. 3-Jul. 6 Abinger Hall. Jun. 6 to Bishops
Cross. Jun. 7 to Leith Hill Place. Jun. 21 at
Combe Bottom. Jul. 6 returned.
Aug. 28-Sept. 11 Bassett, Southampton.

Nov. 3-5 London, Queen Anne St., for Vivi-
section Commission.
Dec. 10-20 London, Bryanston St.
1876 Jan. 25 to Bryanston St.
Feb. 3-5 London, Queen Anne St.
Apr. 27-May 3 London, Queen Anne St.
May 6-Jun. 6 Hopedene, Dorking (home of
Hensleigh Wedgwood).
Jun. 1 to Abinger Hall.
Jun. 7-9 Hollycombe, Midhurst (home of Sir J.
Hawkshaw). Returned home Jun. 10.
Oct. 4-6 Leith Hill Place.
Oct. 7-20 Bassett, Southampton.
Dec. ? London to Royal Society.
1877 Jan. 6-15 London, Bryanston St.
Apr. 12-28 London, Bryanston St. on 20th,
then Queen Anne St. on 24th.
Jun. 8-12 Leith Hill Place. ED recorded going
to Leith Hill place on 6th.
Jun. 13-Jul. 3 Bassett, Southampton, visited
Stonehenge for earthworm research.
Jul. 4 back home from Bassett.
Aug. 20-25 Abinger Hall. 21 Leith Hill Place.
Oct. 26-29 London, Queen Anne St.
Nov. 16-19 Cambridge for LL.D degree.
1878 Jan. 17-23 London, 6 Queen Anne St. Jan.
22 CD at Kew.
Feb. 27-Mar. 5 London, Bryanston St. on
Mar. 4
Apr. 27-May 13 Bassett, Southampton. ED
recorded home on May 13th.
Jun. 7-?14 Leith Hill Place and Abinger Hall.
Jun. ?15 Barlaston.
Aug. 7-22 to Leith Hill Place. To Abinger on
Aug. 12. Barlaston on Aug. 15.
Nov. 19-27 Bryanston St. on 19. 20-22 Queen
Anne St. ED wrote "came home" on Dec. 4.
1879 Feb. 27-Mar. 5, No. 6 Queen Anne St.
May 6-7 Worthing to see A. Rich on 7th.
May 8-20 Bassett, Southampton.
May 21-26 Leith Hill Place.
Jun. 26 London. Crystal Palace. 27 Queen
Anne St.
Jun. 28-30 West Hackhurst, Abinger Hammer,
home of L.M. Forster. Jul. 1 home.
Aug. 1 London, No. 6 Queen Anne St.
Aug. 2-27 Coniston, Lake District, Grasmere
on 14. Drove up to Tilberthwaite farm on 16.
to Furness on 21. Drove to How Tarn on 23.
Dec. 2-12 London, 5 days Bryanston St., 5
days Queen Anne St.

1880 Mar. 4-8 London, No. 6 Queen Anne St.
  Apr. 8-13 Abinger Hall with Horace & his
  wife Emma Cecilia (Ida) Farrer.
  May 25-Jun. 8 Bassett, Southampton. Jun. 5 to
  Salisbury.
  Aug. 14-18 Cambridge to Trinity Chapel,
  Botolph Lane to visit his sons. Drove to Jesus
  College on 16. Visited King's Chapel on 17 &
  Botanic Gardens on 18th.
  Aug. 19- 21 London, No. 6 Queen Anne St.
  Oct. 20-Nov. 2 London. Bryanston St.
  Dec. 7-10 London, No. 6 Queen Anne St.
  Dec. 11- 15 Leith Hill Place.

1881 Feb. 24-Mar. 3 London, Bryanston St.
  Jun. 2-Jul. 5 Glenrhydding House, Patterdale,
  Ullswater. Jun. 2 slept at Penwith. Jul. 4 to
  Penrith. Home on Jul. 5th.
  Aug. 3-5 London, No. 6 Queen Anne St.
  Aug. 20 drove to Orpington with ED.
  Aug. 27 to No. 6 Queen Anne St. on the
  death of Erasmus. Back the same day.
  Sept. 8-10 West Worthing Hotel, Worthing,
  Sussex, visiting Anthony Rich.
  Oct. 20-27 Cambridge, stayed with Horace.
  Dec. 13-20 London, Bryanston St.
1882 CD did not leave Down. He died Apr. 19.

Manuscripts: Much material which was left in manuscript at CD's death has been published since. The autobiographical manuscripts have been considered above and published letters will be found in the main sequence. Other manuscript material which has been published will also be found in the main sequence under brief title, but is summarized here in date order of first publication. The majority of his papers are now in the Darwin Archive at Cambridge University Library which Sydney Smith rightly described as "incomparable in richness". An important Handlist outlining these holdings was published in 1960 (transcribed in *Darwin Online*). The history of how CD's manuscripts and private papers came to be divided between Down House and CUL is described in Sydney Smith, Historical preface, In *Darwin's notebooks*, 1987. *Darwin Online* includes the 'Darwin Manuscript Catalogue', the first and largest union catalogue of CD's manuscripts and private papers ever published. Intended to record all CD manuscripts in the world, it contains c.76,000 records from over 25 institutions and private collections. Based primarily on the detailed catalogue of the Darwin Archive at Cambridge University Library by Nick Gill. http://darwin-online.org.uk/MScatintro.html The other large collection is at Down House, which includes the *Beagle diary*, field notebooks, and documents on domestic matters, financial accounts, garden, poultry, pigeon houses and so forth.

1882 In G.J. Romanes, *Animal intelligence*, contains extracts from CD's notes on behaviour, published with his permission and in press before his death (F1416).

1883 In G.J. Romanes, *Mental evolution in animals*, contains an appendix from chapter 10 of the 2d part of CD's intended 'big book', on evolution (F1434). See also *Natural selection* (F1583), for a transcript of CD's intended 'big book'.

1885 E. Krause, Über die Wege der Hummel-Männchen, in *Gesammelte kleinere Schriften*, 2: pp. 84-8 (F1584, F1602). See also Freeman 1968, below.

1909 Francis Darwin, *The foundations of The origin of species, a sketch written in 1842*. Printed for private distribution (F1555).

1909 Francis Darwin, *The foundations of The origin of species, Two essays written in 1842 and 1844*. Published edn. The sketch of 1842 is from the same setting of type as previous (F1556).

1933 Nora Barlow, *Charles Darwin's diary of the voyage on H.M.S. Beagle*. (F1566).

1935 S. L. Sobol', Journal. [in Russian] Translated by D.L. Weiss. Moscow: Academy of Sciences, U.S.S.R. pp. 423-564.

1945 Nora Barlow, *Charles Darwin and the voyage of the Beagle, unpublished letters and notebooks.* (F1571).

1959 G. R. de Beer, Darwin's journal, *Bulletin of the BMNH, Hist. Series.*, 2: pp. 1-21. (F1573).

1960 P.H. Barrett, ed. A transcription of Darwin's first notebook [B] on 'Transmutation of species'. *Bull. of the Museum of Comparative Zoology, Harvard.* (F1575)

1960-7 G. R. de Beer, M.J. Rowlands and B. Skramovsky, Darwin's notebooks on transmutation of species, *Bulletin of the BMNH, Historical Series.*, 2: pp. 23-200; 3: pp. 129-76. *Notebooks B-E* (F1574).

1962 D.R. Stoddart, Coral islands, *Atoll Research Bulletin.*, no. 88. (F1576).

1963 Nora Barlow, Darwin's ornithological notes, *Bulletin of the BMNH, Historical Series.*, 2: pp. 201-78. (F1577).

1963 R.C. Olby, Darwin's manuscript of pangenesis, *British Jrnl. for the Hist. of Sci.*, 1: pp. 251-63. (F1578).

1968 R.B. Freeman, Charles Darwin on the routes of male humble bees, *Bulletin of the BMNH, Historical Series.*, 3: pp. 177-89. Translation of 1885 German paper above, with transcription of field notes by (F1581).

1974 H.E. Gruber, *Darwin on man*, contains transcription of M & N notebooks on behaviour, with other manuscripts, by Paul H. Barrett (F1582).

1975 R.C. Stauffer, *Charles Darwin's Natural selection*, transcribed from what was intended by CD to be Part 2 of his 'big book', *Variation* being Part 1 (F1583).

1980 S. Herbert, Charles Darwin's red notebook, *Bulletin of the BMNH, Historical Series.*, 7. Contains CD's earliest notes on evolution, covering the period Jun. 1836-Jun. 1837.

1985- F. Burkhardt, S. Smith et al. eds., *The Correspondence of Charles Darwin.* 27 vols.-

1987 D.M. Porter, Darwin's notes on *Beagle* plants. *Bulletin BMNH Historical Series* (F1827).

1987 K.G.V. Smith, Darwin's insects: Charles Darwin's entomological notes, with an introduction and comments. *Bulletin BMNH Historical Series* (F1830).

1987 P.H. Barrett, P.J. Gautrey, S. Herbert, D. Kohn and S. Smith, *Charles Darwin's notebooks, 1836-1844; geology, transmutation of species, metaphysical enquiries.* (F1817).

1988 R.D. Keynes, *Charles Darwin's Beagle diary.* (F1925).

1990 M.A. di Gregorio with the assistance of N.W. Gill, *Charles Darwin's marginalia.*

1990 G. Chancellor, Charles Darwin's St. Helena model notebook. *Bulletin BMNH Historical Series* (F1839).

1995 S. Herbert, From Charles Darwin's portfolio: An early essay on South American geology and species. *Earth Sciences Hist.* (F1956).

1999 D.M. Porter, Charles Darwin's Chilean plant collections. *Revista Chilena de Historia Natural.* (F2114).

2000 R.D. Keynes, *Charles Darwin's zoology notes & specimen lists from H.M.S. Beagle.* (F1840).

2008- The vast majority of the Darwin Archive at CUL and manuscripts of many other institutions are published online as electronic images, as well as hundreds of newly prepared transcriptions in J. van Wyhe ed. 2002-. *The complete work of Charles Darwin online*: http://darwin-online.org.uk/manuscripts.html.

2009 G. Chancellor and J. van Wyhe eds. with the assistance of K. Rookmaaker, *Charles Darwin's notebooks from the voyage of the "Beagle".* Foreword by R.D. Keynes. (F2044).

2012 K. Rookmaaker, G. Chancellor and J. van Wyhe, *Geological diary* from the voyage of the *Beagle*. CD's longest previously unpublished manuscript. (http://darwin-online.org.uk/EditorialIntroductions/Chancellor_GeologicalDiary.html)

Medals: 1853 Royal Medal (Royal Society). 1859 Wollaston Medal (Geological Society), from 1846-60 it was made of palladium. 1864 Copley Medal (Royal Society), CD was proposed in 1862 but failed. 1879 Baly Medal (Royal College of Physicians).
Order: 1867 Pour le Mérite, Prussia.
Prize: 1879 Premio Bressa [Bressa Prize], Reale Accademia della Scienze, Turin. CD first recipient. 12,000 lire (£418 18*s* 10*d*) for his discoveries in plant physiology. CD gave £100 from it to the Zoological Station at Naples.
CD was elected in 1872 to the Lord Rectorship of Aberdeen University, but declined because of ill health. In 1874 he was proposed for the Lord Rectorship of Edinburgh University and Lord Rectorship of St. Andrews University, also declined.
Religion: CD was raised as a Unitarian and then an Anglican Christian, but eventually gave up his belief in Christianity and a personal god shortly after the *Beagle* voyage. CD was however never, so far as is known, an atheist, but on occasion regarded himself an agnostic although in fact believing in a non-intervening deistic creator for most of his life, thus he was a theist not an agnostic. It has to be remembered what enormously negative connotations were attached to the word atheist at the time. Only the most extreme thinkers would be so reckless as to think of or call themselves atheists. CD's religious views are summarized in LL1:304-317. The most important source is CD's discussion of "Religious Belief" in his *Autobiography*, pp. 85-96. Francis Darwin states "My father spoke little on these subjects, and I can contribute nothing from my own recollection". LL1:317. CD considered religious views to be a personal matter and took pains not to offend ED. Entries throughout this volume will show how CD actively supported the village church and its activities as well as religious charities- despite being a non-believer. See John Hedley Brooke, Darwin and Victorian Christianity. In Jon Hodge & Gregory Radick eds., *The Cambridge companion to Darwin.* 2ᵈ edn, 2009, pp. 197-218; van Wyhe & Pallen, The Annie Darwin hypothesis: Did the death of his daughter cause Darwin to "give up Christianity"? *Centaurus* 54 (2012), pp. 1-19 and van Wyhe, Was Charles Darwin an atheist? *The Public Domain Review* (28 Jul. 2011).

1809 Baptism, Nov. 17 (not 15ᵗʰ) at St. Chad's, Shrewsbury, by Rev. Thomas Stedman. "Darwin Chasˢ. Robᵗ. Son of Dr. Robᵗ. & Mʳˢ. Susannah his wife/born Febʳ. 12ᵗʰ". Confirmation: no evidence available from Shrewsbury School, the sacrament perhaps being neglected at the time, although Dr Samuel Butler was an appointed catechist. CD's christening mug was bequeathed to his son W.E. Darwin.

1832-36 "Whilst on board the Beagle I was quite orthodox, and I remember being heartily laughed at by several of the officers (though themselves orthodox) for quoting the Bible as an unanswerable authority on some point of morality..." *Autobiography*, p. 85.

1836-39 "But I had gradually come, by this time, [Oct. 1836 to Jan. 1839] to see that the Old Testament from its manifestly false history of the world, with the Tower of Babel, the rainbow as a sign, etc., etc., and from its attributing to God the feelings of a revengeful tyrant, was no more to be trusted than the sacred books of the Hindoos, or the beliefs of any barbarian." *Autobiography*, p. 85.

1876 CD discussed his gradual and distress-free loss of faith in Christianity in his *Autobiography*, pp. 85-96. He came to realise that there was no evidence to support revelation or the stories of miracles in Christianity. The death of his daughter Anne occurred long after his loss of faith and he did not stop attending church because of her death.

1879 CD to John Fordyce, "In my most extreme fluctuations I have never been an Atheist in the sense of denying the existence of a God. I think that generally (and more and more as I grow older), but not always, that an Agnostic would be the more correct description of my state of mind". LL1:304, *Aspects of scepticism*, 1883.

1879 CD to N.A. von Mengden (b.1862): "Science has nothing to do with Christ; except in so far as the habit of scientific research makes a man cautious in admitting evidence. For myself I do not believe that there ever has been any Revelation. As for a future life, every man must judge for himself between conflicting vague probabilities." F1998, also F1973. Mengden wrote to CD on 2 Apr. 1879 asking if a believer in his theory could also believe in God. A reply in the affirmative was written by ED. Mengden wrote again stating that Haeckel disbelieved in the supernatural, what did CD think? This letter was his response.

1879 Of his grandfather, CD wrote: "Dr. Darwin has been frequently called an atheist, whereas in every one of his works distinct expressions may be found showing that he fully believed in God as the Creator of the universe." (F1319) The same could be said about his own writings, perhaps he wrote this intentionally.

1880 CD to G.E. Mengozzi: "I do not believe that any organic form of life exhibits evidence of design." *Roma Etrusca* no. 2 (15 Jul.): 10, (*Shorter publications*, F1970)

1880 CD to E.B. Aveling "though I am a strong advocate for free thought on all subjects, yet it appears to me (whether rightly or wrongly) that direct arguments against christianity & theism produce hardly any effect on the public; & freedom of thought is best promoted by the gradual illumination of men's minds, which follows from the advance of science. It has, therefore, been always my object to avoid writing on religion, & I have confined myself to science. I may, however, have been unduly biassed by the pain which it would give some members of my family, if I aided in any way direct attacks on religion."

1880 CD to Frederick McDermott "I am sorry to have to inform you that I do not believe in the Bible as a divine revelation & therefore in Jesus Christ as the son of God." Letter was marked "private".

1881 CD discussed his views with E.B. Aveling who published an account in *The religious views of Charles Darwin*, 1883. Christianity, CD told Aveling, Christianity "is not supported by evidence." In *Darwin Online*.

For CD's imaginary deathbed conversion to a fundamentalist Christianity see Atkins 1976, pp. 51-2 and especially J. Moore, *The Darwin legend*, 1994.

Society membership: As was customary, CD joined those London societies whose meetings might be of interest to him, although, after he left London in 1842, his attendance at their meetings was infrequent. He used the periodical publications of all these societies, except those of the Shropshire, Entomological and Ethnological Societies, for his own papers. CD was an Honorary member of societies in the Americas and Europe. Most of these are listed by countries in LL3:373-6, but their titles are sometimes given in English. The following list is in the original languages.

1826-27, 1861 Royal Medical Society, Edinburgh. Member 1826-27, Honorary Member 1861.

1826 Plinian Natural History Society. Member.

1831, 1839 Zoological Society of London. Corresponding Member 1831, Fellow 1839. 1850, 1855 CD served on the Council.

1833 Made a founding member of the Entomological Society in 1833.

1836 Geological Society of London. Fellow. 1838 Feb. 16 to 1841 Feb. 19. Publications Secretary. 1843 Elected Vice-President.

1837 Société Géologiques, Paris, France. Life Member.

1838 Royal Geographical Society. Fellow.

1838 Athenaeum Club. Member.

1839 St Andrews Literary and Philosophical Society. Honorary member 1839 Jan. 7.

1839 Royal Society, London. Fellow 1839 Jan. 24.

1840 The Shropshire and North Wales Natural History and Antiquarian Society. Honorary Member.

1854 Linnean Society of London. Fellow.

1855 Southwark Columbarian Society. Member.

1855 Philoperisteron Society. Member.

1857 Academia Caesarea Leopoldino-Carolina Naturae Curiosorum, Germany. Honorary Member, cognomen Forster.

1860 Academy of Natural Sciences, Philadelphia, USA. Correspondent.

1860 Kongliga Vetenskaps-Societeten, Uppsala, Sweden. Fellow.

1860 Sociedad de Naturalistas Neogranadinos, Colombia. Honorary Member.

1861 Ethnological Society of London. Fellow.

1861 Royal Medical Society of Edinburgh. Honorary Member.

1861 Royal Irish Academy. Honorary Member.

1862 Anthropological Society, London. Honorary Fellow from its foundation.

1863 Canterbury Philosophical Institute, New Zealand. Honorary Member.

1863 Manchester Scientific Students' Association. (See F2174) Honorary Member.

1863 Société des Sciences Naturelles, Neuchâtel. Corresponding Member.

1863, 1878 Königlich-Preussische Akademie der Wissenschaften, Berlin, Germany. Corresponding Member 1863, Fellow 1878.

1865 Kongliga Svenska Vetenskaps-Akadamien, Sweden. Foreign Member.

1865 Royal Society of Edinburgh. Fellow.

1866 Royal Irish Academy, Dublin, Ireland. Honorary Member.

1867 Academia Scientiarum Imperialis Petropolitana (Imperatorskaya Akademiya Nauk), Russia. Corresponding Member.

1867 Kaiserliche-Königliche Zoologische-Botanische Gesellschaft, Austria. Honorary Member.

1868 Manchester Literary and Philosophical Society. Honorary Member.

1868 Medico-Chirurgical Society of London. Honorary Member.

1868 Royal Medical and Chirurgical Society of London. Honorary Fellow.

1869 American Philosophical Society, Philadelphia, USA. Member.

1870 Académie Royale des Sciences, des Lettres et des Beaux-Arts de Belgique. Associate.

1870 Società Geografica Italiana, Florence, Italy. Honorary Member.
1870 Societas Caesarea Naturae Curiosorum (Société Imperiale des Naturalistes), Moscow, Russia. Honorary Member.
1871 Asiatic Society of Bengal, Calcutta, India. Honorary Member.
1871 Kaiserliche Akademie der Wissenschaften, Vienna, Austria. Corresponding Member.
1871 New York Liberal Club.
1871 Société d'Anthropologie, Paris, France. Foreign Member.
1871 Wellington Philosophical Society, New Zealand. Honorary Member.
1871 West Kent Horticultural Society, as Vice-president.
1872 Anthropologische Gesellschaft, Vienna, Austria. Honorary Member.
1872 California Academy of Sciences, USA. Honorary Member.
1872 Koninklijke Akademie der Wetenschappen, Amsterdam. Honorary Fellow.
1872 Magyar Tudomanyos Akademia, Budapest, Hungary. Member.
1872 Società Italiana di Antropologia e di Etnologia, Italy. Honorary Member.
1873 American Academy of Arts and Sciences, Boston, USA. Foreign Honorary Member.
1873 Boston Society of Natural History, USA. Honorary Member.
1873 Reale Accademia della Scienze, Turin, Italy. Honorary Member.
1873 Senkenbergische Naturforschende Gesellschaft, Frankfurt-am-Main. Corresponding Member.
1874 Sociedad Zoológica Argentina, Cordova, Argentina. Honorary Member.
1874 Société Entomologique, Paris, France. Honorary Member.
1875 Kaiserliche Akademie der Wissenschaften, Vienna, Austria. Honorary Foreign Member
1875 Reale Accademia dei Lincei, Rome, Italy. Foreign Member.
1875 Società dei Naturalisti in Modena, Italy. Honorary Member.
1875 Society of Naturalists of the Imperial Kazan University, Russia. Honorary Member.
1876 Birmingham Natural History Society. Honorary Member.
1876 Physiological Society. Honorary Member.
1877 Berliner Gesellschaft für Anthropologie, Germany. Corresponding Member.
1877 California State Geological Society, USA. Corresponding Member.
1877 Koninklijke Hollandsche Maatschappij der Wetenschappen. Foreign Member.
1877 Institución Libre de Enseñanza (ILE), Madrid, Spain. Honorary Professor.
1877 Siebenburgische Verein für Naturwissenschaften, Hermannstadt. Honorary Member.
1877 Sociedad Cientifica Argentina, Buenos Aires. Honorary Member.
1877 Sociedade de Geographia de Lisboa, Portugal. Corresponding Member.
1877 The Watford Natural History Society and Hertfordshire Field Club, UK. Honorary Member.
1877 Zeeuwsch Genootschap der Wetenschappen te Middelburg, the Netherlands. Foreign Member.
1878 Academia Nacional de Ciencias de la Republica Argentina. Honorary Member.
1878 Franklin Literary Society, Indiana, USA. Honorary Member.
1878 Institut de France, Académie des Sciences, France. Corresponding Member, Section of Botany, after several failed elections.
1878 Königlich-Bayerische Akademie der Wissenschaften, Munich. Foreign Member.
1878 Medizinisch-Naturwissenschaftliche Gesellschaft zu Jena. Honorary Member.
1878 Schlesische Gesellschaft für Vaterländische Cultur, Breslau. Honorary Member.
1878 Société Royale des Sciences Médicales et Naturelles, Brussels. Honorary Member.
1879 The English Spelling Reform Association. Member.
1879 Gabinete Portuguiz de Leitura, Pernambuco, Brazil. Corresponding Member.
1879 Kongeligt Dansk Videnskabernes Selskab, Copenhagen, Denmark. Fellow.
1879 Naturforschende Gesellschaft zu Halle, Germany. Honorary Member.
1879 New York Academy of Sciences, USA. Honorary Member.
1879 New Zealand Institute, New Zealand. Honorary Member.
1879 Royal Society of New South Wales, Sydney, Australia. Honorary Member.
1879 Deutsche Akademie der Naturforscher Leopoldina. Foreign Member.

1880 Birmingham Philosophical Society. Honorary Member.
1880 Koninklijke Natuurkundige Vereeniging in Nederlandsch-Indië, Batavia. Corresponding Member.
1880 Società La Scuola Italica Pitagorica, Rome, Italy. Presidente Onorario.
1880 Epping Field Club, later the Essex. Honorary Member.
1881 Société Royale de Botanique de Belgique, Belgium. Associate Member.

Darwin anniversary stamps issued by Gibraltar in 2009. Illustrations supplied by *Darwin Online*.

Stamps: Many countries and territories have issued stamps (88 are listed below, totalling more than 220 stamps) commemorating CD or certain aspects of his life, work and the voyage of the *Beagle*. The list below is hardly complete. First day covers also exist in great numbers. Many of the stamps and their accompanying flyers contain original artwork depicting CD, usually based on 19th-century depictions. See a large collection of over 100 CD stamps in *Darwin Online* mostly from the collection of Cemil Ozan Ceyhan: http://darwin-online.org.uk/DarwinStamps/Darwin_stamps.html.

General: Antigua & Barbados, Ascension, Austria, Bashkiria, Bosnia and Hercegovina, British Antarctic, British Indian Ocean Territory, Buenos Aires, Bulgaria, Cambodia, Cape Verde, Cayman Islands, Chad, Chile, Cuba, Cocos (Keeling) Islands, Comores Islands, Congo, Cuba, Czech Republic, Czechoslovakia, Djibouti, German Democratic Republic, Ecuador, Falkland Islands, Gabon, Galapagos Islands, Gambia, German Democratic Republic, Great Britain, Grenada, Gibraltar, Guernsey, Guinea, Guinea-Bissau, Ecuador, India, Ireland, Israel, Italy, Ivory Coast, Jersey, Kiribati, Liberia, Macedonia, Madagascar, Malawi, Maldives, Mali, Marshall Island, Madagascar, Mauritius, Micronesia, Mozambique, Moldova, Mongolia, Montserrat, Mozambique, Nevis, Niger, North Korea, Palau, Paraguay, Pitcairn Islands, Poland, Portugal, Romania, Rwanda, Sakhalin, San Marino, São Tome and Principe, Serbia, Seychelles, Sierra Leone, Slovenia, Solomon Islands, South Georgia, St. Helena, Soviet Union, Spain, Togo, Turks and Caicos, Tuvalu, Uganda, Ukraine, United Kingdom, Uruguay, Vanuatu, Vietnam, Yakutiya, and Zaire.

Special anniversaries:

1935 (Commemorating the centenary of CD's visit): Ecuador, 2, 5, 10 and 20 centavos, with map, marine iguana, giant tortoise and head of CD respectively.

1958 German Democratic Republic. "Charles Darwin 100 Jahre Abstammungslehre".

1958 (Centenary of Linnean Society publication): Great Britain, no CD stamps before 1982, but cancel, called special slogan, London, South Kensington, SW7. used Jul. and Aug. 1958 only "1958 / CENTENARY OF / DARWIN & WALLACE / EVOLUTION THEORY / 1958. D.W. Tucker, *Gibbons Stamp Monthly* 1958 Jul.

1959 (Centenary of publication of *Origin*): Czechoslovakia (designed by Jindra Schmidt), German Democratic Republic, Poland, Soviet Union, Yugoslavia.

1981 (Sesquicentennial of start of *Beagle* voyage): Cocos (Keeling) Islands.

1982 (Centenary of death of CD): Antigua & Barbados, Ascension, Czech Republic, Ecuador, German Democratic Republic, Falkland Islands, Mauritius, Mongolia, Poland, Sierra Leone, St. Helena, São Tomé and Príncipe, Sierra Leone, St. Helena, UK, Vietnam.

1983 Ecuador (150 years after acquiring sovereignty).

1986 (Sesquicentennial of visit of CD): Australia, Cocos (Keeling) Islands.

2006 (175 years after start of *Beagle* Voyage): Ascension Island, Congo, Kiribati. United Kingdom Millennium Series; 2006, National Portrait Gallery series).

2007 (125 years after death of CD): British Indian Ocean Territory.

2008 (150 years after Linnean Society Darwin-Wallace papers publication): Bosnia and Hercegovina, Cayman Islands.

2009 (Bicentennial of birth of CD, Sesquicentennial of publication of *Origin*): Antigua & Barbados, Ascension Island, Austria, Bashkiria, Bosnia and Hercegovina, Buenos Aires, Bulgaria, Cape Verde, Chad, Cocos (Keeling) Islands, Comoros Islands, Congo, Cuba, Czech Republic, Djibouti, Ecuador, Falkland Islands, Gabon, Czechoslovakia, Galapagos Islands, Gambia, Germany, Gibraltar, Grenada, Guernsey, Guinea, Guinea-Bissau, Ireland, Israel, Italy, Ivory Coast, Liberia, Macedonia, Mali, Micronesia, Moldova, Montserrat, Mozambique, Nevis, Nigeria, North Korea, Palau, Pitcairn Islands, Paraguay, Portugal, Romania, Rwanda (2x), São Tome and Príncipe, Serbia, Sierra Leone, Spain, Turks and Caicos Islands, United Kingdom, Uruguay, Vanuatu, Vietnam, Yakutiya.

2012 Guinea-Bissau, Ivory Coast, Mozambique, Togo.

-2019 "Down House Stamp Collection", English Heritage international postcard stamps.

Very many Darwin stamps have been issued since.

"Darwin was right", the caption for a well-known photograph, by Brian Rosen, on Eniwetok atoll in 1976 on the rediscovered site of the bore hole where the US Atomic Energy Commission drilled in 1952 in preparation for a hydrogen bomb test to discover the distance to the volcanic bedrock. If CD's theory of the formation of coral atolls was wrong there would be a thin layer of coral and then volcanic bedrock. The drill passed through 1,267 meters (4,158 feet) before striking bedrock.

"Darwin's bodysnatchers", a widespread myth popular amongst modern creationists that CD and his followers sought the skulls of the aboriginal peoples of Tasmania for research on evolutionary theory. The story is traced to its source and exposed as a myth without foundation in John van Wyhe, Darwin's body-snatchers? *Endeavour* (Dec.), 2016. See Tasmanians.

"Darwin's bull-dog" or **"Darwin's bulldog."** Despite universal belief for the past century that T.H. Huxley was widely known by this during the 19th century, in fact the nickname was never used in print or in public during his lifetime. Coined in Henry Fairfield Osborn, A student's reminiscences of Huxley. *Biological lectures delivered at the Marine Biological Laboratory of Wood's Hole*, 1896, recalled Huxley saying "You know I have to take care of him—in fact, I have always been Darwin's bull-dog." John van Wyhe, Why there was no 'Darwin's bulldog': Thomas Henry Huxley's famous nickname. *The Linnean*, vol. 35 (1) (Apr.), 2019, pp. 26-30.

Darwin's Farm, at Beesby, Lincolnshire. Bought with a loan from Dr Darwin. CD was also promised more funds to erect new buildings. Pattison, *The Darwins of Shrewsbury*, 2009, p. 93. See Worsley, *The Darwin farms*, 2017.

Darwin's Finches. 1838 John Gould described the finches collected by CD and others on the Galapagos islands and identified them as 13 distinct species. Decades after CD's death a legend arose that the finches or their beaks had inspired CD's theory while in the Galapagos. This was overturned by Frank Sulloway, Darwin and his finches: The evolution of a legend. *Jrnl. of the Hist. of Bio.* 15 (1982), pp. 1-53. 1935 The phrase 'Darwin's Finches' was coined by Percy Lowe. See P.R. Lowe, The finches of the Galapagos in relation to Darwin's conception of species, *Ibis*, (1936), pp. 310-21 and John van Wyhe, Where do Darwin's finches come from? *The evolutionary review* 3, 1 (2012): 185-195. 1942 The sub-family Geospizinae of the Galapagos Islands used by Robert T. Orr, *Bulletin N.Y. Zool. Soc.*, 45: pp. 42-5, 1942. 1947 Used by David Lack, *Proc. of the Zool. Soc. of London*, pt. 5, no. 53, p. 49, 1944, and title of his 1947 book. It was the work of Lack, not CD, to identify that the beaks of the birds are specially adapted for specific food or niches. Whereas CD had observed (perhaps near a settlement with artificially abundant food) "all the other species of this group of finches, mingled together in flocks, feed on the dry and sterile ground of the lower districts." *Journal* 2nd edn, p. 379.

Darwin's Peak. Another name for Angulus woolnerii, see also *Nature*, Apr. 6 1871. Other name for Cerro la Campana, mountain in Chile, between Valparaiso and Santiago. Visited by CD.

"Darwin's True Knight", Hooker's description of A.R. Wallace in his presidential address to the BAAS meeting in Norwich, given on 19 Aug. 1868.

Darwin's window, Window in Hooker's retirement house at Sunningdale. CD suggested its insertion on seeing the plans, to improve the view of the garden.

Darwin, Colonel Charles Waring [I], 1855 Aug. 28-1928 Aug. 1. CD's remote cousin. Head of the senior branch of the Darwin family, of Elston Hall 1894 Married Mary Dorothea Wharton. 3 sons; 1 Charles John Wedgwood Darwin (1894-1941), 2 Francis Wedgwood Darwin (1896-1972), 3 Gilbert William Lloyd Darwin (1899-1979). C. Wedgwood died in Wiesbaden, Germany.

Darwin, Charles Waring [II], 1856 Dec. 6-1858 Jun. 28. 10th and last child of CD. Died of scarlet fever?, had Down's syndrome. "He had never learnt to walk or talk". ED2:162. Death notice in *The Times*, (28 Jun.): 1. The date of his death probably contributed to the fact that CD did not attend the meeting of the Linnean Society on 1 Jul. 1858 where the papers by Wallace and Darwin on the transmutation of species were read. *ED's diary*: 1858 Jun. 23 "Baby taken ill" On 27, "Baby worse" and on 28 "Death". CD's private memorial is in DAR210.13.42, transcribed in CCD7 Appendix V and *Darwin Online*.

Darwin, Charlotte Maria Cooper, 1827 May 4-1885 Jun. 22. Eldest surviving daughter of William Brown Darwin. 1849 Married Francis Rhodes, later Darwin.

CD's remote cousin. Last of the senior branch of family. Elston Hall, the family seat, was left to her husband.

Darwin, Edward, 1782-1829. 1st child of Erasmus Darwin [I] and Elizabeth Pole. CD's half uncle. Officer in 3d Dragoon Guards. Lived at Mackworth, Derbyshire.

Darwin, Edward Levett, 1821 Apr. 12-1901 Apr. 23. Second son of Sir Francis Sacheverel Darwin and Jane Harriet Ryle. CD's half first cousin. 1858 Author on sporting matters under pseudonym "High Elms"; *The game-preservers manual*, 1858. High Elms was the mansion of the Lubbocks. 1858 CD of *The game-preservers manual*, "shows keen observation and knowledge of various animals". Woodall 1884, p. 4.

Darwin, Elizabeth [I], 1725-1800. Second child of Robert Darwin. CD's great-aunt. 1751 Married Rev. Thomas Hall, Rector of Westborough, Lincolnshire.

Darwin, Elizabeth [II], see Collier.

Darwin, Elizabeth [III], 1763-64. 3d child of Erasmus Darwin [I] and Mary Howard. CD's aunt.

Darwin, Elizabeth, 1820-35. Eldest child of William Brown Darwin. Died aged 15 after a protracted illness.

Darwin, Elizabeth [VI], 1847 Jul. 8-1926 Jun. 8. 6th child of CD. Oct. 10 Baptized at St Mary's, Down, 1847. Unmarried. Known as "Bessy" or "Lizzy". Bernard Darwin called her "Ubbady", F. Darwin, *Story of a childhood*, 1920, p. 46. A book offered in a recent auction: H. Myrtle, *A day of pleasure...for young children*. 1853, is inscribed "Lizzy Darwin Dec. 1852" with the plates coloured by her. "Very stout and nervous...not good at practical things...and she could not have managed her own life without a little help and direction...but she was shrewd enough...and a very good judge of character". Raverat, *Period piece*, 1952, pp. 146-7. "If family legend be true, my aunt Bessy when young had looked into the drawing-room at Down and flounced out again with the words 'Nothing but nasty, beastly boys'". B. Darwin, *Green memories*, 1928, p. 40. Photographs in DAR225.

Darwin, Emily Catherine (Catherine), 1810 May 10-1866 Feb. 2. 6th and last child of Robert Waring Darwin. CD's sister. Known as "Catty" or "Katty". "Had neither good health nor good spirits". ED2:180. "Failed to work out her capabilities either for her own happiness or that of others (perhaps)". ED2:184. CD's sisters, after their mother's death, ran an Infant School in the grounds of Millington's Hospital, Frankwell. 1863 Married Charles Langton as 2d wife (1st wife Charlotte Wedgwood), d.s.p. In 1866 Feb. 2, ED wrote in her diary "death of C. L."

Darwin, Emma, see Wedgwood, Emma [I].

Darwin, Emma. 1904 H.E. Litchfield ed., *Emma Darwin, wife of Charles Darwin. A century of family letters*, 2 vols., (F1552), 250 copies printed for family and friends. In 2007 Milo Keynes lent his unbound copy to be scanned for *Darwin Online* and remarked "the sheets of which were got from Aunt Etty's house when she died by my mother for me". 1915 *Emma Darwin. A century of family letters*, 1792-1896, 2 vols., (F1553), text as 1904 with some alterations, The 1904 edn contains a portrait of

"Mrs John Wedgwood" from a watercolour by Linnell absent from F1553. 1915 USA from stereos of English edition (F1554).

Darwin, Emma Georgina Elizabeth, 1784-1818. 3d child of Erasmus Darwin [I] and Elizabeth Pole. Unmarried. CD's half aunt.

Darwin, Emma Nora, [Nora Barlow] 1885 Dec. 22-1989 May 20. 3d child of Sir Horace Darwin. CD's granddaughter. Known as Nora. 1911 Married Sir James Alan Noel Barlow, 2d Bart. 2 daughters, 4 sons: 1. Joan Helen, 2. Thomas Erasmus (later 3d Bart.), 3. Erasmus Darwin, 4. Andrew Dalmahoy, 5. Hilda Horatia, 6. Horace Basil. Grandchildren Phyllida and Ruth Padel amongst others. 1933 Editor *Diary of the voyage of the Beagle* (F1566). 1945 *Charles Darwin and the voyage of the Beagle* (F1571). 1958 *The autobiography of Charles Darwin 1809-1882. With the original omissions restored* (F1479). 1963 Darwin's ornithological notes, *Bulletin of the BMNH, Historical Series* 8 (F1577). 1967 *Darwin and Henslow* (F1598). She visited ED with her sister Ruth (Boofy) in 1891 Jul. 4-13, 1892 Apr. 19, Aug. 26, and last record 1895 Oct. 7. Portrait by Yvonne Rosalind Barlow hangs in CUL near MS reading room.

Darwin, Erasmus [I], 1731-1802. Physician and man of science. 4th child of Robert Darwin. Born at Elston Hall, Nottinghamshire. CD's grandfather. Biography: Seward, 1804; Dowson, 1861; Krause and CD, 1879; Pearson, 1930; King-Hele, 1963, 1977, 1999; Fara, 2012. -1756? Educated at Chesterfield Grammar School, St. Johns College Cambridge and University of Edinburgh Medical School. 1756 Physician at Nottingham. 1757-81 Practised at Lichfield. 1757 Married 1 Mary Howard (1740-70). 4 sons, 1 daughter: 1. Charles, 2. Erasmus, 3. Elizabeth, 4. Robert Waring, 5. William Alvey. 1761 FRS. 1765- Member Lunar Society with a.o. Josiah Wedgwood [I]. 1781 Married 2 Elizabeth Chandos Pole, née Collier (1747-1832). 4 sons, 3 daughters. 1. Edward, 2. Frances Anne Violetta, 3. Emma Georgina Elizabeth, 4. Francis Sacheverel, 5. John, 6. Henry, 7. Harriet. He also had two illegitimate daughters by the governess Mary Parker of the children from his marriage to Mary Howard, 1 Susanna Parker (1772-1856), 2 Mary Parker Jr. (1774-1859). 1781-83 Radburn Hall, Derby. 1783-1802 Full St, Derby. 1802 Breadsall Priory, Derby, where his relict continued to live until her death in 1832. Oil painting of the house is on display at Down House. Main works: 1791, 1789 *The botanic garden*. Published anonymously. Part I was actually published in 1792, but has 1791 on the title page. 1794, 1796 *Zoonomia; or, the laws of organic life*. 1800 *Phytologia, or the philosophy of agriculture and gardening*. 1803 *The temple of nature; or, the origin of society*. Portraits: three in oils, one by Joseph Wright of Derby from 1770 in NPG, another by Joseph Wright from 1792 and one by Rawlinson of Derby in Derby Museum. Of all three portraits several copies were made. Silhouette in Huxley & Kettlewell, *Charles Darwin and his world*, 1956, p. 8 and Chancellor, *Charles Darwin*, 1973, p. 12 etc. Medallion in Lichfield Cathedral after Wright portrait. His commonplace book is now at Down House. 1813 The genus *Darwinia* Rudge, 1813, was named for D, (Myrtaceae) about thirty-seven species of Australian heath-like shrubs. *Darwinia* Rafinesque 1817 and *Darwinia* Dennstedt 1818 are junior homonyms.

Darwin, Erasmus [II], 1759-99. 2d child of Erasmus [I] and Mary Howard. Unmarried. CD's uncle. Solicitor and genealogist. Committed suicide by drowning.

Darwin, Erasmus [III], 1881 Dec. 7-1915 Apr. 25. 1st child of Sir Horace Darwin. Unmarried. CD's grandson, the second of the two born in CD's lifetime. Director Cambridge Instrument Co. Obituary by B. Darwin, *Green memories* 1928, in ED1:xi-xvi, 1915. Born the year his great uncle Erasmus Alvey Darwin died. First recorded visit to ED in 1883 Jan. 5. Came again on Jan. 26. His 8th birthday was also noted. Received a letter from ED when he was 10. The last record in *ED's diary* was in 1896 Apr. 29 "Eras to school." 1915 killed in action at Ypres.

Darwin, Erasmus Alvey, 1804 Dec. 29-1881 Aug. 26. 4th child of Robert Waring Darwin. Unmarried. CD's only brother. Known as "Ras", "Eras" and "Uncle Ras". Took a medical course at the University of Cambridge from 1822, and trained as a physician at Edinburgh in 1825 but never practised. Nicknamed "Bones" at school because tall, thin and delicate. Brent 1981, p. 28. Also known as "John" and "Strol" at school for unknown reasons. CCD1:10. 1835 Autumn, took 43 Great Marlborough Street house in London. Also at 24 Regent Street, 7 Park Street, 6 Queen Anne Street. 1849 Trustee Bedford College, University of London from its foundation, see *Bedford College Mag.*, 1902 Jun. Census 1851 had address at St George, Hanover Square, Westminster. Had servants John Griffith, 30 and Emma Hund, 32. Both from Shrewsbury. CD was visiting when census was taken. *Darwin Online.* 1859 Nov. D to CD regarding *Origin* "In fact the a priori reasoning is so entirely satisfactory to me that if the facts wont fit in, why so much the worse for the facts is my feeling". CCD7:391. Census 1871 taken at 6 Queen Anne Street, had staff Elizabeth Pearce (Servant), Elizabeth Squire (Cook), William Petit (Butler), James Texter (Footman). Anne Parslow was there as the Darwins were visiting when census took place. 1881 census Elizabeth Pearce (Housekeeper), Elizabeth Squire (Cook), Frederick Surman (Butler), ED and Horace D were visiting. 1881 CD to Sir Thomas Farrer, "He was not I think a happy man". ML1:395. "He had something of original and sarcastically ingenious in him, one of the sincerest, naturally truest, and most modest of men". Thomas Carlyle, *Reminiscences*, 1882, vol. 2, p. 208. 1881 Buried 1 Sept. in Down Churchyard. ED wrote in her diary in 1881 Aug. 22 "Eras taken ill" and four days later "died 11pm". His estate was valued at £158,000; all of his properties and three-sixths of his money were left to CD. He was remembered very fondly by his nephew William, who wrote to his mother on Aug. 27, "To me there was a great charm in his manner that I never saw in anybody else." 1816 Pastel chalk drawing by Sharples. Darwin Heirlooms Trust. Down House. 1853 Chalk drawing by Samuel Laurence, signed and dated. Down House. Photographs in DAR225 and DAR257.

Darwin, Florence Henrietta Fisher, 1864?-Jan. 31-1920 Mar. 5. Author of *Six plays*, Cambridge, 1921. 1886 Married 1 Frederic William Maitland. 1913 Married 2 Sir Francis Darwin as 3ᵈ wife s.p.

Darwin, Frances Anne Violetta, 1783-1874. 2ᵈ child of Erasmus [I] and Elizabeth Pole. CD's half great-aunt. 1807 Married Samuel Tertius Galton. 7 children: 1 Darwin Galton, 2 Erasmus Galton, 3 Francis Galton, 4 Elizabeth Anne (m. E. Wheler), 5 Lucy Harriot (m. J. Moilliet), 6 Millicent Adele (m. Rev. R. S. Bunbury), 7 Emma Sophia (unmarried).

Darwin, Frances Crofts, 1886 Mar. 30-1960 Aug. 19. Poet. Only child of Sir Francis Darwin and Ellen Crofts. CD's granddaughter. 1909 Married Francis Macdonald Cornford. Mother of John Cornford, the poet. 5 children.

Darwin, Sir Francis [II], 1848 Aug. 16-1925 Sept. 19. Botanist. 7ᵗʰ child of CD. Known as "Backy", "Frank" and "Franky". Assisted CD with his botanical work, including drawing figures of *Aldrovanda* and *Utricularia* for *Insectivorous plants*. 1860 Educated Clapham Grammar School. 1867 Trinity College, Cambridge. First studying mathematics, then natural sciences. Graduated in 1870. 1875 Qualified as a physician at St. George's Medical School, London. Did not practice. 1882 FRS. 1874 Married 1 Amy Richenda Ruck. 1 son Bernard Richard Meirion, lived at vicarage Down. After Amy's death in 1876, moved into Down House with infant son. 1882 after CD's death, moved to a house on Huntingdon Road, Cambridge, that was to be called the "Darwinery". *Nature*, 1925, vol. 116. 1883 2 Married Ellen W. Crofts (d. 1903). 1884 Dec. moved to Wychfield, 80 Huntingdon Road. F. Darwin, *Story of a childhood*, 1920, p. 59. 1 daughter Frances Crofts born there. 1887 Edited *Life and letters* (F1452). 1888-1904 Reader in Botany, Cambridge. 1894 With E.H. Acton, *Physiology of plants*. 1895 Main work: *The elements of botany*. 1903 Edited, with A.C. Seward, *More letters* (F1548). 1904 after death of Ellen, sold Wychfield and moved to London. 1905 returned to Cambridge for his research. 1909 Edited *Sketches of 1842 and 1844* (F1555, F1556). 1913 Kt. 1910 had a house built in Madingley Road for his daughter Frances and his son-in-law Francis Cornford. 1913 Married 3 Florence Henrietta Fisher s.p. During his 3ᵈ marriage, spent spring and summer at a converted farmhouse at Brookthorpe, Gloucestershire. It was on waste land which had belonged to her 1ˢᵗ husband Frederic William Maitland. After death of 3ᵈ wife (d. 1920 Mar. 5), moved to 10 Madingley Road. (Photograph in DAR257) *Nature*, 1925, vol. 116. 1917 Published *Rustic sounds*. 1917-20 In *Springtime* contains lists of plants and birds observed at Brookthorpe (1917), *Recollections* with reminiscences of CD, pp. 51-69. (1920). 1920 printed privately, *The story of a childhood*, contains extracts of letters from him to Mrs Lawrence Ruck (née Matthews), his mother-in-law, about her grandson Bernard's childhood up to age 15. The letters were to him on Mrs Ruck's death. Transcribed in *Darwin Online*. Photographs in DAR225 and DAR257.

Darwin, Sir Francis Sacheverel, 1786 Jun. 17-1859 Nov. 9. Physician and traveller. 4ᵗʰ child of Erasmus Darwin [I] and Elizabeth Pole. CD's half uncle. 1815 Married

Jane Harriet (Harriot) Ryle (d. 1866). Eldest son Reginald Darwin. 1820 Knighted by King George IV. Buried in All Saints Churchyard, Derbyshire, England.

Darwin, Violetta Harriot, 1826-80. Daughter of Sir Francis Sacheverel. Her drawing of Elston Hall as it was before 1754 was reproduced in *Erasmus Darwin*, p. 3.

Darwin, Sir George Howard, 1845 Jul. 9-1912 Dec. 7. Mathematician. 5th child of CD. Sometimes called "Georgy" or "Gingo". Interested in heraldry in youth; "the young herald". CD to Huxley CCD16:49. Drew the figures of *Drosera* and *Dionaea* for *Insectivorous plants*. Trained as a barrister but never practised. The only remaining male line of CD's family comes through him. 1884 Newnham Grange, Cambridge. Biography: Francis Darwin edited the *Scientific papers by George Howard Darwin*, 5 vols. 1916. 1856 16 Aug. Educated Clapham Grammar School. 1865 St. John's College but moved to Trinity College, Cambridge. 1868 Placed 2d for Smith's Prize. 1868 2d Wrangler, Cambridge. 1868 Fellow, Trinity College, Cambridge. 1872 Barrister. 1879 FRS. 1882 Inherited Down House. 1883-1912 Plumian Prof. of Astronomy and Experimental Philosophy, Cambridge University. 1884 Jul. 22 Married at Erie, Pennsylvania, USA, Maud du Puy. 3 sons, 2 daughters. 1. Gwendolen Mary, 2. Charles Galton, 3. Margaret Elizabeth, 4. William Robert. Another son died in infancy. A biographical sketch was reprinted by F. Darwin in *Rustic sounds*, pp. 152-194. 1899 Main work: *The tides*. 1905 KCB. 1911 Copley Medal Royal Society. Photographs DAR219 and DAR225.

Darwin, Georgina Elizabeth, 1823-1902. 7th child of Sir Francis Sacheverel Darwin. 1862 Married Rev. Benjamin Swift (1820-82). 6 children. Mother of Francis Darwin Swift. CD's half cousin.

Darwin, Gwendolen Mary, (Gwen Raverat) 1885 Aug. 27-1957 Feb. 11. 1st child of Sir George Howard Darwin. Known as "Gwen", and as "The Genie" from boarding school days. Married Jacques Raverat in 1911. 2 daughters, Elisabeth and Sophie Jane. CD's granddaughter. Artist, trained at Slade School, University College London. Engraved on wood. 1939 Illustrated published edn of *The bird talisman*. 1952 Main work: *Period piece* with recollections of life at Down House and family.

Darwin, Harriet, 1790-1825. 7th child of Erasmus Darwin [I] and Elizabeth Pole. 1811 Married Admiral Thomas James Maling, d.s.p at Valparaiso.

Darwin, Henrietta Emma, 1843 Sept. 25-1927 Dec. 17. 4th child of CD. Was sickly as a child. Helped CD with editing *Descent*. ED2:196, CCD. 1879 Did some editing of CD's part of *Erasmus Darwin*, advising the removal of some sensitive passages. Ironically passages mentioned that Krause had revised his work, that led to offending Samuel Butler. King-Hele 1977. 1861 CD to Hooker, "Poor H...has now come up to her old point & can sometimes get up for an hour or two twice a day." CCD9:20. 1865 Known as "Body", "Budgy", "Harriot" (she tried to use this name in 1865, ED objected "the pertest of names"), "Rhadamanthus minor" or just "Rhadamanthus" by Huxley, "Mr. Huxley used to laugh at for the severity of her criticisms". ML1:238 note 1. (In Greek mythology Rhadamanthus is the son of Zeus and Europa; renowned for his justice on earth, the gods made him one of the judges of the underworld.) "Trotty Veck" or Trotty (a character in Dickens' *The chimes*), most often however known as "Etty". 1871 Aug. 31 Married Richard Buckley Litchfield d.s.p. CD's only married daughter. Lived in 31 Kensington Square. 1903 On death of husband moved to Burrow's Hill, Gomshall, Surrey. Raverat, *Period piece*, 1952, chapter 7 gives a description of her valetudinarian habits and remembered her fondly "how easy it is to draw your absurdities, how difficult to show your lovableness; yet we were all your children, and coming to Burrow's Hill was always coming home". Her journal for 1871 is in DAR247 and transcribed in CCD19 Appendix VI. A 1926 autobiographical fragment is in DAR246 and is transcribed in *Darwin Online*. 1904, 1915 Editor *Emma Darwin*, (F1552 and F1553). The last thing ED wrote in her diary in 1896, Sept. 28 was "wr H E L". She left behind numerous recollections of CD: DAR262.23.6-11, DAR112.A79-A82, DAR112.B99 and *Richard Buckley Litchfield*, 1910, all are transcribed in *Darwin Online*. Numerous photographs of her are in DAR219 and DAR225.

Darwin, **Henry**, 1789-90. 6th child of Erasmus Darwin [I] and Elizabeth Pole. CD's half uncle.

Darwin, **Henry Galton**, 1929-92. Son of Sir Charles Galton Darwin. Barrister Foreign Office. CMG. 1958 Married Jane Sophia Christie. 3 daughters.

Darwin, **Sir Horace**, 1851 May 13-1928 Sept. 22. 9th child of CD. ED wrote in her diary "Birth of Horace" 1851 May 13. Known as "Jemmy", "Jim", "Jimmy" or "Skimp". 1880 Jan. 3 Married Emma ("Ida") Cecilia Farrer. 1 son, 2 daughters: 1. Erasmus [III], 2. Ruth Frances, 3. Emma Nora. 1881 Built The Orchard on land to the east of The Grove, Cambridge. Spalding, *Gwen Raverat: Friends, family and affections*, 2001, p. 30. The site now occupied by New Hall. 1885 Founder and Director of Cambridge Scientific Instrument Co., Botolph Lane, Cambridge. 1896-97 Mayor of Cambridge in year of Queen Victoria's Diamond jubilee (1897). 1903 FRS. 1918 KBE. Photographs in DAR225.

Darwin industry, a term coined by philosopher of science Michael Ruse in the title of a review essay in 1974. Ruse did not define the term, noting simply "Books and papers on Charles Darwin, his theory, his predecessors, his successors, his supporters and his critics, continue to pour forth...But sheer bulk does not, of course, in itself imply excellence". Ruse, The Darwin industry, *Hist. of Science*, 12 (1974), pp. 43-58. Historian of science Timothy Lenoir wrote in 1987: "The term does not refer to the sheer bulk of Darwin scholarship...More exactly, [it] is a self-styled reference coined by a select group of scholars who have in recent years concentrated their efforts on utilizing the vast resources of Darwin's unpublished notebooks and correspondence...These are the manuscript groupies". Lenoir, The Darwin industry, *Jrnl. of the Hist. of Bio.*, 20 (1987), pp. 115-30. Lenoir thus saw the term as connoting a self-styled sophisticated elite of Darwin scholars. Some scholars certainly do obsess over and insist on their 'sophistication' in this respect but the term 'Darwin industry' as it is most often used refers to either Darwin scholarship in general and especially its vastness. See also John van Wyhe, *Darwin Online* and the evolution of the Darwin industry. *Hist. of Science* 47 (4) no. 158 (2009), pp. 459-76.

Darwin, Jane Eleanor, 1823-38. 3^d child of William Brown Darwin. Died aged 15 after a protracted illness.

Darwin, Leonard, 1850 Jan. 15-1943 Mar. 26. 8th child of CD. CD called him "Bony". Nicknamed "Pouter", "Pout" or "Pouts". 12 Egerton Place, Brompton Road, London. 1862 Educated Clapham Grammar School. 1871 Royal Engineers, commissioned Dec. 1870. Became a leading member of the Photographic Society of Great Britain. 1874 & 1882 Observed transits of Venus in New Zealand. 1874 Sept. 27 arrived in New Zealand. 1878 Photographed CD at Down House. 1878 exhibited landscape photographs of Switzerland, Wales and England at the Royal Photographic Society. 1882 Jul. 11 Married 1 Elizabeth Frances Fraser s.p. 1883 Jan. 1890 Retired from army. 1890 Aug. 27 sailed to New York. Arrived in San Francisco on Sept. 23 1890. In 1891 Feb. 14, he arrived in Hong Kong. Went to see his mother in 1891 Apr. 1. 1892 Jul.-1895 Jul. MP Liberal-Unionist, for Lichfield. 1895 Stood again but not re-elected. 1900 Married 2 Charlotte Mildred Massingberd s.p. On 2^d marriage moved to Cripp's Corner, Forest Row, Sussex. 1908-11 President Royal Geographical Society. 1911-28 Chairman British Eugenics Society, succeeding Francis Galton. 1928 Honorary President British Eugenics Society. Main works: 1897 *Bimetallism*. 1926 *The need for eugenic reform*. 1929 Memories of Down House, *Nineteenth Century*, 106; pp. 108-23. Known as "Uncle Lenny". 1943 He left an estate valued at £62,924 (net personality £59,067). Biography: Margaret Keynes (niece), *Leonard Darwin, 1850-1943*. Cambridge, 1943. Photographs in DAR225 & DAR232.12.

Darwin Manuscript Catalogue, 2006-. The first and largest union catalogue of CD's handwritten manuscripts and private papers ever published, part of the core bibliographical database of *Darwin Online*. As of July 2020 it contains 76,027 records from over 25 institutions and private collections. It is intended to record all CD manuscripts in the world. The Catalogue's fully normalized records are at the level of individual items, not whole class numbers. The electronic catalogue of the Darwin Archive at CUL (which has never been made available online) forms the base of the catalogue, in the same way that Freeman 1977 forms the core of the Freeman Bibliographical Database. The Catalogue also includes the Supplement to the Darwin catalogue at the CUL. Since 2006 thousands of corrections and additions have been added to the Catalogue. It is the fastest, simplest and often the only way to locate a CD manuscript or other item in his private papers such as clippings and offprints. http://darwin-online.org.uk/MScatintro.html

Darwin, Margaret Elizabeth, 1890 Mar. 22-1974 Dec. 19. 3^d child of Sir George Howard Darwin. CD's granddaughter. 1917 Married Sir Geoffrey Keynes. 4 sons, 1 daughter. 1943 wrote biography of Leonard Darwin, privately printed.

Darwin, Marianne, 1798-1858. 1st child of Robert Waring. CD's sister. 1824 Nov. 9 Married Henry Parker, M.D. 5 children: 1 Robert, 2 Henry [II], 3 Francis, 4 Charles, 5 Mary Susan. Susan and M drank tea with the Darwins in 1839 Apr. 7. On her death the grown-up family was adopted by her sister, Susan Elizabeth, and lived at The Mount, Shrewsbury. Pattison, *The Darwins of Shrewsbury*, 2009.

Darwin, Mary Eleanor, 1842 Sept. 23-1842 Oct. 16. 3d child of CD. Born at Down House and died there. ED recorded "died" in 1842 Oct. 16. ED had moved into Down House on Sept. 14. Death notice in *The Times*, (19 Oct.): p. 7: "On the 16th inst., at Down, in Kent, the infant daughter of Charles Darwin, Esq."

Darwin, Reginald, 1818-92. Eldest son of Sir Francis Sacheverel Darwin. CD's half first cousin. Married Mary Ann Sanders, 1 son Sacheverel Charles Darwin. 1879 Lent CD documents, including a commonplace book, on Erasmus Darwin [I], which CD used for his preliminary notice in E. Krause's *Erasmus Darwin*. The commonplace book is now at Down House.

Darwin, Robert Alvey, 1826 Apr. 17-1847 Dec. 7. 3d child of William Brown Darwin. Of Elston Hall and Exeter College Oxford. Last male in senior branch of family, he left Elston Hall to his sister Charlotte Maria Cooper Darwin.

Darwin, Robert Waring [II], 1766 May 30-1848 Nov. 13. 4th child of Erasmus Darwin [I] and Mary Howard. CD's father. Named after his uncle, Robert Waring Darwin of Elston. Strictly teetotal. Known as "The father of Frankwell" by his poorer patients. Before 1785 Studied at Edinburgh before Leyden, the Netherlands. 1785 Physician, MD Leyden, Feb. 26. Lived at St. John's Hill before he built The Mount. 1788 FRS. 1796 Apr. 18 Married Susannah Wedgwood at St. Marylebone. Had 6 children: 1. Marianne, 2. Caroline Sarah, 3. Susan Elizabeth, 4. Erasmus Alvey, 5. Charles Robert, 6. Emily Catherine. 1816 Pastel chalk drawing by Ellen Sharples. Darwin Heirlooms Trust. On display at Down House. Woodall 1884, pp. 11, 14. 6' 2" tall, very corpulent, "When he last weighed himself, he was 24 stone, but afterwards increased much in weight." *Autobiography*, p. 29. "Personally of huge bulk with a very squeaky voice". F.E. Gretton *Memory's harkback through half-a-century*, 1808-58, 1889 p. 33. c.1800 Had a large practice in Shrewsbury and around, where he built The Mount (c.1800). Pattison, *The Darwins of Shrewsbury*, 2009.

Darwin, Ruth Frances (Boofy), 1883 Aug. 20-1972 Oct. 15. Second child of Sir Horace Darwin. CD's granddaughter. High Hackhurst, Abinger Hammer, Dorking. 1932 Appointed to the Brock Committee that called for forced sterilization of "mental defectives". 1938 CBE. 1948 Married W. Rees-Thomas as second wife, s.p. Her grandmother ED noted her birth in 1883 Aug. 20 "Ruth Mary born".

Darwin, Sarah Gay Forbes, 1830 Feb. 13-1889 Jun. 15. 7th child of William Brown Darwin. CD's cousin. 1848 Married Edward Noel. Had 8 children.

Darwin, Susan Elizabeth (Chucky), 1803 Aug. 3-1866 Oct. 3. 3d child of Robert Waring Darwin [II]. Unmarried. CD's sister. Lived at The Mount, Shrewsbury until her death. Henrietta Litchfield: "My father [CD] told me that anything in coat and trousers from eight years to eighty was fair game to Susan". ED1:141. c.1822 She and Jessie Wedgwood, daughter of John Wedgwood, were known as "Kitty" and "Lydia" after the Bennetts in Jane Austen's *Pride and prejudice*, because they were flirts. ED1:141. 1836 CD called her "Granny". After the death of her sister Marianne in 1858 she adopted the grown-up Parker children who lived with her. Pattison, *The Darwins of Shrewsbury*, 2009. ED recorded in 1866 Aug. 30 "Susan illness" and on Oct. 3, "Susan's death".

Darwin, William [I], died before 1542. Yeoman. Of Marton, Lincolnshire. Two sons, 1. William?, 2. John. The earliest ancestor given by Burke 1888. Tenth generation to CD in male line.

Darwin, William? [II], died before 1542. Eldest son of William [I]. 2 sons, 4 daughters. Of Marton, Lincolnshire, Yeoman. Burke 1888 is not certain of christian name. Ninth generation to CD in male line.

Darwin, William [III], -1580. Eldest son of William? [II]. Married Elizabeth ?, 3 sons. Inherited Marton from uncle John Darwin. 8th generation to CD in male line.

Darwin, William [IV], c.1573-1644. 3d son of Richard Darwin. Married as second husband Mary Healey of Cleatham, Lincolnshire. Yeoman of the Royal Armoury, Greenwich. Also held Marton. Sixth generation to CD in male line.

Darwin, William [V], 1620-75. Eldest son of William Darwin [IV]. Barrister. Recorder of Lincoln. Royalist. Erasmus became a family name through his wife. Fifth generation in male line to CD. 1653 Married Anne Earle, daughter of Erasmus Earle. 5 sons, 1 daughter.

Darwin, William [VI], 1655-82. Eldest son of William [V]. Waring became a family forename through his wife, and Elston Hall the family seat. Fourth generation in male line to CD. Portrait "at Elston shows him as a good-looking young man in a full-bottomed wig". LL1:3. 1680 Married Anne Waring, heiress of Robert Waring of Elston Hall, Newark, Nottinghamshire. 2 sons.

Darwin, William [VII], 1681-1760. Eldest son of William D [VI]. Of Cleatham and Elston Hall. 1706 Married 1 Elizabeth D (first cousin). 2 sons, 2 daughters. 1715/16 Married 2 Mary Secker. 1 son, 4 daughters. 1749 Married 3 Mary Hurst s.p.

Darwin, William Brown, 1774-1841. Son of William Alvey D [I]. Of Elston Hall. Married Elizabeth de St. Croix. 3 sons, 4 daughters. CD's first cousin once removed.

Darwin, William Erasmus, 1839 Dec. 27-1914 Sept. 8. 1st child of CD. Called "Hoddy Doddy" in infancy. Birth notice in *The Times*, (30 Dec.): 8: "On the 27th inst, at Upper Gower-street, the lady of Charles Darwin, Esq., of a son." The only one of CD's surviving sons who never grew a beard, although Leonard only did so in old age. Obituary: F. Darwin, *Christ's College Mag.*, 1914. As first-born, his activities were noted in *ED's diary*. 1840 Feb. 10, he smiled for the first time, on Sept. 29 he cut his first tooth. His weight was frequently recorded. Robert Waring Darwin to CD when William was young and supposed to be delicate, "Let him run about and get his feet wet and eat green gooseberries". B. Darwin, *Green memories*, 1928, pp. 27,

42-3. Educated at Mr Wharton's preparatory school, Rugby and Christ's College, Cambridge. 1861 Ridgemount, North Stoneham, Bassett, Southampton. 1862-1902 Partner in Grant & Maddison's Union Banking Company of Southampton, also called Southampton & Hampshire Bank. Looked after CD's financial affairs with great success. 1877 Nov. 29 Married Sarah Price Ashburner Sedgwick s.p. 1877 He is the child in CD's paper in *Mind*, 2, (F1779). 1882 Raverat, *Period piece*: "He had felt the top of his head cold at his father's funeral in Westminster Abbey and balanced his black gloves there." 1902 After death of wife, lived at 12 Egerton Street, London, next door to brother Leonard. Of the Egerton Street house "a rather tall, gaunt house, with a butler almost too perfect to live". Gwendolen Mary Darwin (later Gwen Raverat) lived with him whilst at Slade School. His detailed 1883 recollection of CD is in DAR112.B3b-B3f. A shorter one is in DAR112.A26-A27, both in *Darwin Online*. In the latter: "The point I should have liked to have made strong is the wonderful impression one got of his immense reverence for the laws of nature." and "I feel for Hen.^tta but much more for Mother & Bessy as being believers." A rare reference to unbelief. His botanical notebook with observations and notes on experiments (1862-70) is in DAR117. He made notes from botanical textbooks in a notebook in DAR234. His botanical sketchbook (1862-72) is in DAR186.43. See the appendices to latter volumes of CCD. Photographs in DAR225.

Darwin, William Robert, 1894 Aug. 22 -1970 Dec. 28 (cremated 31 Dec.). 4th child of Sir George Howard D. Married Sarah Monica Slingsby in 1894. CD's grandson. ED: 1894 Aug. 22 "Birth of G's boy". He was ill with jaundice on Sept. 4.

Darwin, William Waring, 1822-35. 2d child of William Brown Darwin. Died aged 13 after a protracted illness.

Darwinism, CD's papers on, 1871 [letter] A new view of Darwinism, *Nature*, Jul. 6, 4: pp. 180-1, refers to letter by Henry B. Howorth of same title, *ibid.*, Jun. 29, 4: pp. 161-2 (*Shorter publications*, F1754). 1872 Bree on Darwinism, *Nature*, 6: p. 279 (*Shorter publications*, F1756), relates to a review by A.R. Wallace of Bree's book, *An exposition of the fallacies in the hypothesis of Mr Darwin*, 1872.

Darwin's delay. The notion that CD held back or avoided publishing his theory of evolution from about the time of its initial inception in 1839 until its publication in 1858-9. The phrase was coined in a 1974 popular magazine article by palaeontologist S.J. Gould (*Nat. Hist.*, vol. 83. no. 10. Dec., pp. 68-70) who attributed CD's purported delay to fear of ridicule and persecution. In the ensuing 30 years the notion became one of the most widely believed aspects of CD, even though it was a new idea absent in all literature before the 1950s. Numerous reasons were then proposed for why CD supposedly held back his theory such as fear of offending ED, the church, scientific colleagues or even of providing a doctrine that would be appropriated by social radicals to challenge the status quo. Although none of these has evidence to support it, they have been accepted as common knowledge for many years. In fact, CD did not slow or withhold his theory. There is no evidence that he was afraid of the consequences to the point of altering his plans to publish, nor did he keep his belief in evolution a secret. More than 50 friends, family, neighbours and colleagues have been identified whom CD told about his belief in evolution before publication. In a 20 Jul. 1857 letter to Asa Gray, CD explicitly referred to these 20 years as spent working on many projects and that he continued to accumulate facts for a theory of evolution: "It is not a little egotistical, but I sh^d. like to tell you, (& I do not *think* I have) how I view my work. Nineteen years (!) ago it occurred to me that whilst otherwise employed on Nat. Hist, I might perhaps do good if I noted any sort of facts bearing on the question of the origin of species; & this I have since been doing." CCD6:432. In the 6th edn of *Origin* (1872), CD responded to critics who suggested that he exaggerated his originality. Before publishing *Origin* (1859), he wrote, "I formerly spoke to very many naturalists on the subject of evolution, and never once met with any sympathetic agreement." (p. 424) The word "agreement" showing that he advocated or argued for the point. See John van Wyhe, Mind the gap: Did Darwin avoid publishing his theory for many years? *Notes and Records of the Roy. Soc.* 61 (2007): 177-205 and additional evidence in John van Wyhe *Dispelling the darkness*, 2013, pp. 249-58.

Daubeny, Charles Giles Bridle, 1795-1867. Botanist. 1822 FRS. 1832 Prof. Chemistry Oxford. 1834 Prof. Botany Oxford. 1839 *Sketch of the geology of North America*. 1840 Prof. Rural Economy, attached to chair of Botany. 1860 Jun. 30 Conversazione held in his rooms after BAAS scene. 1860 D commented on *Origin* in *Report British Assoc*. CD on D's comments "very liberal & candid; but scientifically weak". CCD8:320. 1860 Remarks on final causes of sexuality of plants. 1867 *Miscellanies*.

Davidson, Thomas William St. Clair, 1817-85. Artist and palaeontologist. Specialist on brachiopods. Anti-*Origin*. 1851-86 *Monograph of British fossil Brachiopoda*. 1857 FRS. 1856-66 CD corresponded with.

Davies, Jane, (or Davis) 1807-?. Welsh cook at Down House. Census 1851. Called "Mrs" although unmarried. Known to the children as "Daydy"; she was kind to them. F. Darwin, *Springtime*, 1920, p. 55.

Davis, Charles Oliver Bond, 1816-87, Maori interpreter and writer mentioned in the newspaper article by FitzRoy and CD. (*Shorter publications*, F1640).

Davis, Richard, 1790-1863. Born in Piddletrenthide, Dorset, son of a farmer. Selected by Church Missionary Society for service in New Zealand. 1824 Arrived Bay of Islands, New Zealand. 1831 Missionary at Waimate, North Island, New Zealand. Not in orders, but ran a farm to teach the Maoris agriculture. 1835 Dec. 23-24 CD met. CD spells "Davies". *Journal* 2^d edn, p. 425; *Shorter publications*, F1640.

Davy, Dr John, 1790-1868. Army surgeon. Brother of Sir Humphry Davy. Inspector General of Army Hospitals. Friend of Sir James Mackintosh. 1834 FRS. 1855, 1856 & 1863 CD to D on salmonid eggs; two long replies printed in *Phil. Trans. of the Roy. Soc.*, 1855, and *Proc. of the Roy. Soc.*, 1856 (both transcribed in *Darwin Online*), as well as in his *Physiological researches*, pp. 251-69, 1863.

Dawes, Richard, 1793-1867. Educationalist. Older friend of CD at Cambridge. 1813-20 Trinity College, Cambridge. BA 1817. MA 1820. 1818 Mathematical tutor and bursar of Downing College, Cambridge. 1831 Spring, CD and D talked of a trip to Teneriffe with Ramsay and Kirby. 1850 Dean of Hereford. 1867 CD subscribed £2. 2s. through John Maurice Herbert for memorial to him.

Dawkins, Sir William Boyd, 1837-1929. Geologist and archaeologist. 1861-69 Member Geological Survey of Great Britain. 1867 FRS. 1869 Curator of Natural History, Manchester Museum. CD was friendly with and 1873 wrote testimonial for an application for Chair of Geology at Cambridge, which D did not get. 1874-1908 Prof. Geology, Owen's College, Manchester. 1919 Kt. See below.

Dawkins Testimonials. [1873] *Testimonials in favour of W. Boyd Dawkins...a candidate for the Woodwardian Professorship of Geology [at Cambridge]*, Cambridge, University Press printed (F1216). CD's letter p. 2. Position was formerly held by Adam Sedgwick.

Dawson, Sir John William, 1820-99. Canadian geologist. Influenced by Robert Jameson. 1855 D described *Eozoon*. Anti-*Origin*. ML1:210, 466, 468. 1855-93 Prof. Geology and Principal McGill University. 1860 D reviewed *Origin* in *Canad. Nat.* 1862 FRS. 1862 CD to Hooker, "Lyell had difficulty in preventing Dawson reviewing the Origin on hearsay, without having looked at it". CCD10:503. 1884 Kt.

Dawson, Robert, 1776-1860. Cartographer to Ordnance Survey. 1831 CD met at Llangollen when on geological tour with Sedgwick.

de Bary, Anton Heinrich, 1831-88. German botanist and physician. Prof. Botany, Strasbourg. 1879 D and CD corresponded about *Utricularia* (bladderworts, genus of carnivorous plants). D sent CD two specimens. See CCD27.

de la Beche, Sir Henry Thomas, 1796-1855. Geologist. 1819 FRS. 1835 Director Geological Survey. 1842 Kt. 1848 CD listened to D's Presidential address to Geological Society, "a very long & rather dull address". CCD4:140. 1852 Wollaston Medal, Geological Society. Only three letters to him from CD survive.

de la Rue, Warren, 1815-89. Astronomer, chemist and inventor. Son of the printer Thomas de la Rue. R invented the first envelope-making machine. 1850 FRS. 1851 Feb. CD met D at Royal Institution. 1853 CD to Henslow. R had written to CD from Brighton; the matter concerned some speculative investment. CCD5.

Decaisne, Joseph, 1807-82. French botanist. 1859 CD probably sent D copy of 1st edn of *Origin*. CCD7:536. CD cited D in *Natural selection* (F1583).

Defence of Science. 1881 Mr Darwin and the defence of science, *British Med. J.*, 2: p. 917 (*Shorter publications*, F1799). Regarding an offer to make a subscription to defend David Ferrier who had been charged with infringing the Vivisection Act by the Victoria Street Society for the Protection of Animals. Ferrier was innocent and the case was dismissed. See the *British Med. J.*, (19 Nov. 1881); *The Times* (18 Nov. 1881).

Delpino, Giacomo Giuseppe Federico, 1833-1905. Italian botanist. Prof. Botany Genoa and later at Naples. 1867- Frequent correspondent with CD. 1869 Criticised

CD's theory of pangenesis to which CD replied in *Scientific Opinion*, 2 (20 Oct.): p. 426. (*Shorter publications*, F1748b) at the request of the editor.

Denny, Henry, 1803-71. Entomologist, specialist on lice and minute beetles, was for 45 years curator of the museum of the Literary and Philosophical Society in Leeds. The BAAS in 1842 made a grant to D of fifty guineas to assist him in the study of British *Anoplura*. 1844-65 Corresponded with CD on lice. 1844 Jun. 3 CD to D about races of human lice and on a Mr Martial's observations on them. CCD3. 1871 *Descent* 1:219 mentions D's work on the lice of pigeons, fowl and dogs.

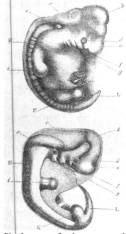

Embryos of a human and a dog from *Descent*.

Descent (book). The most widely quoted line in the book was "We thus learn that man is descended from a hairy quadruped, furnished with a tail and pointed ears, probably arboreal in its habits, and an inhabitant of the Old World." The last sentence of the work reads: "we must acknowledge, as it seems to me, that man with all his noble qualities, with sympathy for the most debased, with benevolence which extends not only to other men but to the humblest living creature, with his god-like intellect which has penetrated into the movements and constitution of the solar system— with all these exalted powers—Man still bears in his bodily frame the indelible stamp of his lowly origin". (F938). 1871 Apr. 7th thousand, with textual changes (F939), facsimile of this issue 1969 (F939). 1871 Dec. 8th thousand, with textual changes (F940). 1874 2d edn, 10th thousand (F944), in one vol. 1875 2d edn corrected, 11th thousand (F945). 1877 2d edn revised and augmented, 12th thousand (F948). First foreign editions: 1906 Bulgarian (F2504). 1930 Chinese (F1047a). 2007 Croatian (F2419). 1906 Czech (F1048). 1874-75 Danish 1870, 1871 *The descent of man, and selection in relation to sex*, 2 vols. (F936). CD's copy of vol. 1 dated 1870 was the only one known until 2000 when another copy surfaced at an auction at Sotheby's; the same set resurfacing at auction in 2001 and 2006. 1871 Feb. Normal issue of both vols., 25 errata on verso of title leaf of vol. 2, 1st issue (F937). 1871 Mar. 2d issue, text changes and no errata, works by the author on verso of title leaf of vol. 2 (F1050). 1871 Dutch (F1053), 1872 French (F1058). 1871 German (F1065), 2006 Greek (F2048). 1884 Hungarian (F1084). 1871 Italian (F1088), 1881 Japanese (F1099). 1986 Korean (F2358). 1874-76 Polish (F1101). 1910 Portuguese (F2033). 1967 Romanian (F1106). 1871 Russian (F1107), 1977 Serbo-Croat (F2421). 1950 Slovene (F1122). 1876 Spanish (F1122a). 1872 Swedish (F1136). 1968 Turkish (F1137). 1921 Yiddish (F1138). Extracts from "advance-sheets of Darwin's new work" appeared in *Appleton's Jrnl*. 5 (98) (11 Feb. 1871): 171-3. List of presentation copies is in CCD19 Appendix V, *Descent* 2d edn CCD22 Appendix IV. CD's notes are in DAR69. Materials for the 1st edn are in DAR80-85; 2d edn in DAR87-90 and DAR157. Proofs of 2d edn are in DAR213.

Descent, Theory of. 1837-8 First use of term by CD in *Notebook B*.

Devil's chaplain. Desmond and Moore, *Darwin*, 1991, claimed that CD remembered radical freethinker Richard Carlile's (1790-1843) visit to Cambridge in 1829 by this title and furthermore feared that he himself (CD) would come to be reviled by respectable society as one. In fact, Carlile was not dubbed the Devil's chaplain until two years after his Cambridge visit and the phrase had been in common use since the time of Chaucer. Hence there is no link between CD's use of the phrase in a letter to J.D. Hooker (13 Jul. [1856], CCD6:178) and Carlile or indeed of any fear of social repercussions to CD's theories. It is likely that CD thought that T.H. Huxley would make a good 'Devil's chaplain'. See van Wyhe, *Darwin in Cambridge*, 2014.

Devonport, Town and naval dockyard west of and contiguous with Plymouth, Devon. 1831 Sept. 13 CD with FitzRoy and Musters arrived after three days by packet from London. 1831 Sept. 16 CD returned to London. 1831 Oct. 30 CD back and stayed at 4 Clarence Baths until *Beagle* finally sailed Dec. 27 after two unsuccessful attempts to put to sea.

Dianths, hybrid, 1857 Hybrid Dianths, *Gardeners' Chron.*, no. 10: p. 155 (*Shorter publications*, F1693). CD recounts his experiments cross pollinating a red Carnation and a crimson Spanish Pink from this genus of 300 flowering plants. See CCD6:349-350.

Dicey, Albert Venn, 1835-1922. Barrister. 1882-1909 Vinerian Prof. English Law, Oxford. 1882 D was on "Personal Friends invited" list for CD's funeral. ED in 1868 May. 3 recorded "Mr Dicey". Other records 1870 Jul. 1, 1882 Jan. 22, 28-30. 1872 Married Elinor Bonham-Carter. Even after 1882, ED kept in contact, noting in 1888 Feb. 17 "Albert Dicey's lecture". Henrietta paid him a visit at Aberdovey in 1896 Aug. 15. DAR210.8, in *Darwin Online*, on religious part of CD's autobiography.

Dick, c.1847-50 A dog at Down House which was killed trying to jump through the flywheel of the well. F. Darwin, *Rustic sounds*, 1917, p. 12.

Dicky, 1885 A small male fox terrier of ED's widowhood, given to her by Mrs (Margaret) Vaughan Williams (née Wedgwood, Margaret Susan). Dicky, according to Henrietta was "the greatest possible pleasure to her". ED2:358.

Dieffenbach, Johann Karl Ernst, 1811-55. German physician, geologist and naturalist. Studied at university of Giessen. CD discussed evolution with before publication of *Origin*. c.1837 In London, teaching German. c.1840 First trained naturalist in New Zealand. 1844 Translated first German edn of *Journal*. (F188).

Ditchfield, Field at Down, just north of Little Pucklands.

Dickson, Augusta, 1837-? Ladies Maid at Down House in 1881 census.

Dickson, William, ?-1833 Aug. 26. 1833 Mar. D was the only Englishman at Port Louis, Falkland Islands, "now has charge of the British Flag". The British had just annexed the islands. Called "Mr. Dixon" in *Beagle diary*, pp. 138-9.

Dixie, Lady Florence Caroline, 1855-1905. British traveller and writer. Born in Scotland. 1880 wrote to CD on her observations of the tuco tuco. DAR172.

Dobbin, A pony in CD's childhood. 1838 autobiographical fragment and ML1:5.

Dobell, Horace Benge, 1828-1917. Physician and medical author. 1863 CD to D, thanking for a copy of his *On the germs and vestiges of disease*, 1861, and on regeneration. CCD9:136.

Dobrski, Konrad, 1849-1915. 1873 translated *Expression* into Polish (F1203).

Dodgson, Charles Lutwidge, 1832-98. Mathematician and, as "Lewis Carroll", author of children's books. 1851 Christ Church College, Oxford. 1852-81 Christ Church Mathematical Lectureship, Oxford. 1872 Sent a photograph of a young girl, Flora Rankin, to CD for his work on *Expression*, now in DAR53.1.C98.

Doedes, Nicolaas Dirk, 1850-1906. Studied theology but changed to history at Utrecht. Thanked CD for a photograph and sent one of himself and his friend Costerus (frontispiece to CCD21). Asked CD about his belief in God. CD replied "The safest conclusion seems to be that the whole subject is beyond the scope of man's intellect; but man can do his duty." CCD21:150. The letter was published in 1883. Photograph in DAR162.201.

Dogs. CD or his family owned the following dogs: Bob or Bobby ('Bob Darwin' in the village), Bran, Button, Butterton, Czar, Dash (CD called him Mr Dash), Dick, Dicky, Lil, Nina, Otter, Pepper, Pincher, Polly, Quiz, Sappho, Shelah, Spark, Tartar, Tony and Tyke. See also Emma Townshend, *Darwin's dogs*, 2009.

Dogs, (paper) 1882 On the modification of a race of Syrian street-dogs by means of sexual selection, by Dr [W.] Van Dyck, with a preliminary notice by Charles Darwin, *Proc. of the Zool. Soc. of London*, no. 25: pp. 367-70 (*Shorter publications*, F1803). Read Apr. 18 by the Secretary: CD died on Apr. 19. Draft notes are in DAR28.2.C1-C5.

Dohrn, Felix Anton, 1840-1909. German zoologist. 1862 Introduced to CD's work by Haeckel. 1870 Sept. 26 D visited CD at Down House, and perhaps again later. C. Gröben, Charles Darwin and Anton Dohrn, 1982, pp. 93-4, gives Dohrn's account of his visit to Down House, 1870 Feb. 26, with Ulan story (see below) in detail, spells the German "ulan" not "Uhlan", its English equivalent. 1872 Apr. 3 CD wrote to D about success of *Descent* in Germany. 1873 Founder of Zoological Station at Naples 1873, later Stazione Zoologica. 1875 CD wrote to D about Naples station and invited D and wife to visit Down House, "I have often boasted that I have had a live Uhlan in my house!" CCD23:57. 1879 CD gave D £100 for the station from his Bressa Prize money and £10 each for his sons George and Francis. 1899 Foreign Member Royal Society. Recollection of CD, in *Darwin Online* (F2090).

Don, David, 1799-1841. Scottish botanist. 1822-41 Librarian Linnean Society. 1836-41 Prof. Botany King's College, London. 1836 CD approached D about identifying *Beagle* plants.

Donders, Frans Cornelis, 1818-89. Dutch ophthalmologist and physiologist. Prof. Physiology Utrecht. 1864 *Accommodation and refraction of the eye*. 1871 D gave CD information for *Expression*. 1872 Apr. D wrote to CD to tell him of his election as foreign member of the Koninklijke Akademie van Wetenschappen. 1874 Jul. CD wrote to D, to thank him for entertaining his son George. Late 1881 At Int. Med. Congr. CD sat between D and Virchow. Brent 1981, p. 499.

d'Orbigny, Alcide Charles Victor Dessalines 1802-57. A palaeontologist sent out by the French government to South America. In the *Falkland notebook*, Chancellor & van Wyhe, *Beagle notebooks*, 2009.

Dorking, Surrey. 1876 May 6-Jun. 6 CD had family holiday there.

Double Flowers. 1843 Double flowers—their origin, *Gardeners' Chron.*, no. 36: p. 628 (*Shorter publications*, F1663). CD's first botanical publication.

Doubleday, Henry, 1808-75. Entomologist and Quaker. Correspondent with CD on insect matters. 1860 Sent CD true oxlip *Primula elatior*.

Douglas, Charles D., surveyor and pilot, long resident in Chiloe. In the *Port Desire notebook*, Chancellor & van Wyhe, *Beagle notebooks*, 2009, *Shorter publications*, F1649.

"Doveleys, The", or **Dovelies**, Nickname for sisters Frances and Emma Wedgwood in childhood and up until Frances' (Fanny's) untimely death in 1832. ED(1904)1:76. ED1 and Loy, *A Victorian life*, 2010, p. 48.

Doveton, William, 1753-1843. Magistrate and judge. Napoleon famously died after a picnic at Doveton's house in 1820. In the *Despoblado notebook*, Chancellor & van Wyhe, *Beagle notebooks*, 2009.

Dowie, Annie Chambers, daughter of Robert Chambers. Corresponded with CD in 1871, 1872 and 1875. Called on the Darwins in 1871 Jun. 20 and 1875 Sept. 11. Both times Mrs A.D. Hamilton called as well.

Down, Kent. The village was so spelt before 1842. See Down. A useful history is Howarth, [1933.] *A history of Darwin's parish: Downe, Kent*. In *Darwin Online*.

Down Bible, used to record illnesses and vaccinations of the family. 1920 Henrietta at 77 wanted to know if she ever had chicken-pox so wrote to her nephew Charles at Cambridge to fetch the record. Raverat, *Period piece*, 1952, p. 124.

Down Coal and Clothing Club. A local charity that supplied parishioners with coal and clothes in exchange for regular savings, the local gentry made "honorary" contributions. CD helped establish, run at first by J.B. Innes. CD was treasurer from 1848-69. A small notebook at Down House: The 'Down Coal Club Honorary Subscriptions 1841 to 1876 Inclusive.' (Down House MS) contains accounts in CD's hand from 1848-69. The payments by eight to ten subscribers totalled £6-£35 annually. CD regularly contributed £5.

Downey, W. & D., 1855-1940. Newcastle and London commercial photography company that printed CDVs by Wallich from 1872. See Darwin, Iconography.

Down Friendly Club. A local savings and insurance club. CD helped to found in 1850 and acted as its treasurer for 30 years. LL1:142-3. The annual general meeting was held at Down House every year, usually on Whit Monday. 1852 Mar. Rules for the Club printed at CD's expense. His accounts show: 18 Mar. 1852 Smith & Elder. Printing 250 Rules for Down Club £3.5. CD's manuscript accounts. Registered at Friendly Societies, 17 North Audley Street, W1. File number for Downe Club

F51/232, but the file passed to Public Record Office. 1877 'To members of the Down Friendly Club', a single sheet printed for CD to dissuade members from disbanding (*Shorter publications*, F1303). After the Friendly Societies Act of 1875 (38 & 39 Vict. Ch. 60), and an amending Act of 1876 (39 & 40 Vict. Ch. 22), under which the Downe Club would have been placed in Class 5 'Local Village and Country Societies', there seems to have been dissatisfaction; some members wanted to disband and share out the proceeds. The leaflet was distributed to members, in Feb. 1877, to dissuade them, successfully, from this course. ED, wrote to Francis Darwin on 3 Feb. 1879, that the band was expected that day. ED2:237.

Down House, Luxted Road, Down, Orpington, Kent. 1842 Jul. 22 CD and ED first saw. Bought from Rev. J. Drummond, Vicar of Down, for £2,020 with 18 acres of which 12 were then the paddock. ED moved in 14 Sept. CD moved in 17 Sept. Ordnance datum 565ft, the well 325ft deep, to the clay below the chalk of the North Downs. From 1844 the house and its contents were insured by Royal & Sun Alliance Insurance whose records survive. London Metropolitan Archives.

Accounts of: 1842 Jul. CD's own account of house, estate and district, written to his sister Catherine, is printed in ML1:31-36 & CCD2:323. Darwin's home. *London Daily News*, (24 Apr.), p. 3. E. von Hesse-Wartegg, 1880, Bei Charles Darwin. *Frankfurter Zeitung und Handelsblatt* (30 Jul.), transcribed in *Darwin Online* in English translation courtesy of Randal Keynes. 1882. Darwin's Heim. *Ueber Land und Meer.* no. 34: 691-2, transcribed in *Darwin Online* with an English translation. A.D. Webster, 1888. Darwin's garden. *Gardeners' Chron.* (24 Mar.): 359-60. Anon. 1888. Darwin's house. *Shields Daily Gazette* (22 Sept.) details of books, experiments and grandchildren. O.J. Vignoles, 1893. The home of a naturalist. *Good Words* 34: 95-101. [G. Newman] 1893. Darwin's house at Down. *Maidstone Jrnl. and Kentish Advertiser* (22 Jun.), p. 5 Unique details of wormstone and CD's coffin. Anon. 1909. A visit to Darwin's village: reminiscences of some of his humble friends. *Evening News* (12 Feb.): 4. D.S. Jordan, *The days of a man*, 1922, vol. 1, pp. 272-4. 1929 L. Darwin, Memories of Down House, *Nineteenth Century*, 106: pp. 118-23. At Down House, June 7th, 1929. *Cornhill Magazine*, 1929, vol. 67. 1952 G. Raverat, *Period piece*, chapter 8, from personal experience in childhood, but not in CD's lifetime as is B. Darwin, *Pack clouds away*, 1941. 1955 Keith, *Darwin revalued*, Chaps. 4 & 24. 1969 Dobson, *Historical and descriptive catalogue of the Darwin Memorial*. 1969 British Association for the Advancement of Science: Historical and descriptive catalogue of the Darwin Memorial at Down House. 1966 Dobson, *Charles Darwin and Down House*. 1981 [Titheradge,] *Charles Darwin memorial at Down House*. Atkins 1976. (The preceding all transcribed in *Darwin Online*.) 1998 Reeve, *Down House* and the English Heritage reports cited below.

Alterations to house: The house was originally bare brick but CD had it plastered and painted white. 1843 Angled bay extension to all three storeys of west front added. 1845-46 Kitchen area rebuilt and butler's pantry added, with stove, and schoolroom and two small bedrooms above. Schoolroom above butler's pantry has carved on shelf in cupboard "Darwin A 10 W. E. DARWIN 1853". 1851 Alterations in upstairs rooms and stairs. 1857 New dining-room (later the drawing room) added at north end, with two bedrooms above it, cost £500. 1859 A new drawing room added to the southwest of the house. Billiard table set up in the old dining room. 1872 Verandah added outside drawing room. 1877 New billiard room added, with a bedroom and small drawing room above and new main entrance on east side.

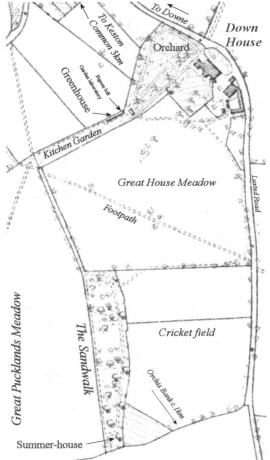

Map of Down House and grounds based on the 1869 Ordinance Survey map, also showing later features.

1881 Billiard room converted to new study. There was no interior plumbing during CD's lifetime. 1878 June, The Classed Accounts record payments for the construction of "earth closets". See Lea, *Down House...perspective reconstruction drawings*, 1998.

Outbuildings included the well house, large water tanks, a summerhouse near the sandwalk by 1846, a garden 'douche' built beside the well in 1849 "shaped something like a very diminutive church" (George Darwin) with "a tank thirteen feet deep, with a stage in the middle. [With]...a big cistern above that held six hundred and forty gallons." (John Lewis). By 1853 it was no longer in use. 1855 An elevated hexagonal wooden pigeon house built above the tool house behind the kitchen garden wall. 1856 CD: "I am building a new house for my tumblers". "Throughout 1855 and 1856, CD paid John Lewis, the Down carpenter, each half-year for work on pigeon houses (CD's Account book)". CCD6:45-6. 1858 Sept. 8 CD to Tegetmeier "I want to clear away my pigeon-houses".

CCD7:154. "the principal wooden pigeon house was erected as a sort of elevated summer house above the tool-house at the near end of the kitchen gardens...This old pigeon house became all overgrown with a splendid mass of ivy & the woodwork became very rotten...During a gale in the spring of 1882 it was bodily blown over, & it has been left lying on its side, for the sake of preserving the magnificent clump of ivy, which entirely hides the wood. (George Darwin) 1882 Apr. 22 "~~Summer house Great storm blown down~~" *ED's diary*. See *Variation* 1:180. A gardener's cottage near cow yard. There was also the Hot-house & greenhouse, stables, a small brick darkroom painted black inside (later called the laboratory) built behind the greenhouse wall in 1881, a laundry with wringing machine, servant's cottage and woodshed. At the time of CD's death the stables contained a brown mare named Lady Florence, a bay horse named Druid, a chestnut horse named Sailor, an old grey mare, an Alderney cow and four other cows, three were "in milk". Carriages: a double Brougham, a Landau, a Waggonette, a dog cart and a farm cart. A story circulated in Down in the 1930s that "the body of Darwin's carriage was acquired by a carpenter sometime during Darwin's lifetime and used as a cupboard for tools." *Aberdeen Press and Jrnl.*, (22 Jun.), p. 2. Lea, *Down House...Notes on the trellis*, 1997; Lea, *Down House, Bromley. Kent: The billiard room fireplace*, 1999; and Lea, *Down House...Notes on the original heating system installed in Darwin's greenhouse*, 1999. These important English Heritage reports are available in the Historic England website: https://research.historicengland.org.uk/.

Alterations to land: 1843 The lane in front of the house lowered and a wall built along it to block view of the house from the road. 1844 New garden wall built. There is an 11 page manuscript catalogue in CD's hand dated Jun. 1844 which lists the trees and plants on the Down House grounds. (Private Collection, California, 2020. In *Darwin Online*.) 1845 Mound under yews on west side removed; mound added at east side as wind protection. 1846 Sandwalk wood planted on land rented from John W. Lubbock (The Jan. 1846 rental agreement is in DAR210.15.) Outhouses rebuilt. 1874 sandwalk wood exchanged for a piece of pasture with John Lubbock. 1881 Strip of field beyond orchard purchased from Sydney Sales for hard tennis court, new wall built. 1959 the garden was restored by Nora Barlow and Margaret Keynes, CD's granddaughters, according to their memories of the house during the 1890s. 1960s the roadside wall to the east of the house was set back and lowered in height. 1962 the scullery was demolished. 1996 house and grounds restored more closely to condition in CD's time by moving a fence, restoring original trees and plants and removing later additions and Sandwalk paved by English Heritage. See Miele, *Darwin's garden: The estate and gardens at Down House*, 1996 and Lea, *Down House...Notes on the historic development of the greenhouse and laboratory...*, 1996.

Hot-house & greenhouse. c.1855 CD had a small greenhouse, mentioned in his Classed account book and letter to Hooker 19 Apr. [1855]. 1863 Jan.-Feb. John Horwood, gardener of neighbour G.H. Turnbull, superintended the building (by William Ledger, builder) of a hot-house in the kitchen garden. 1863 An adjoining greenhouse was built in the summer. 1864 hot-house extended a further 12.2m, doubling the size of the complex to 23.7m. This was divided into four sections, each kept at a different temperature. c.1881 a laboratory was built behind the brick wall

5.3m. 1898 the greenhouse was reduced in length and re-roofed. See CCD10-11 especially the important introduction to 'Darwin's lists of hothouse plants' Appendix VI to CCD11 and Lea, *Down House...Notes on the historic development of the greenhouse and laboratory complex*, 1996. Woodcut of the interior by Alfred Parsons, in *Century Illustrated Monthly Mag.*, Jan. 1883. (Right)

Furnishings: Today the old study and the new drawing room are furnished almost as they were when CD was alive; this includes the original study chairs, the portrait of Lyell given to CD by Lady Lyell in 1847, the photograph of Hooker given to CD by Julia Margaret Cameron, the photographer, and the print of Josiah Wedgwood [I] by Samuel Reynolds given back by Francis Darwin in 1927. The drawing room piano, bought in 1839, was bought back from the Positivist Society for £20 in 1929. Many other objects were provided by the family and are on display. For example, Henrietta donated CD's study chair and letter weighing machine. After English Heritage acquired Down House in 1996 it was further restored and reopened in 1998. See Lea, *Inventory & valuation of furniture, pictures books linen &c.* (English Heritage report).

Interior of hot-house in 1882 by Parsons.

The history of how CD's manuscripts and private papers came to be divided between Down House and CUL in the late 1940s is described in Sydney Smith, Historical preface, in *Darwin's notebooks*, 1987. *Beagle* voyage, domestic items, finances and documents on CD's health were given to Down House (a public museum since 1929) for the purpose of making a popular or public display whereas the vast bulk went to CUL for the use of scholars.

Household expenditure: See also entry under CD Finance. CD and ED kept detailed accounts from the date of their marriage. These, although preserved at Down House, have unfortunately not been published. Keith, *Darwin revalued*, 1955, pp. 221-32, and Atkins 1976, pp. 95-100, give useful extracts. In 1867 there were four dependent sons and two daughters, only the eldest son being away and employed, probably eight indoor servants and the garden staff. Atkins 1976, p. 99, gives a detailed breakdown for 1867-81:

Figures in £'s	Annual Household Expenditure 1867-1881 (Eggs and milk were produced on the estate.)														
	1867	1868	1869	1870	1871	1872	1873	1874	1875	1876	1877	1878	1879	1880	1881
Meat	250	259	275	268	272	281	290	246	221	267	330	292	251	246	223
Butter	5	10	11	9	9	5	6	8	5	9	10	9	12	12	15
Cheese	18	14	18	22	18	20	19	21	13	15	10	12	15	12	11
Candles	16	12	11	16	20	12	39	37	19	27	39	37	24	17	25
Oil	7	8	8	12	13	16									
Bacon	10	11	15	14	10	16	17	17	19	9	16	14	12	9	4
Soap	10	8	8	9	11	7	6	8	3	2	2	2	—	8	4
Grocery	53	63	76	63	67	67	77	56	57	80	70	61	91	103	79
Sugar	16	16	19	19	17	14	15	8	12						
Bread	63	22	41	47	43	59	38	50	36	42	42	33	57	51	34
Fish & game	20	30	23	23	28	36	46	31	31	36	23	31	27	27	38
Servants	71	81	57	74	93	89	100	82	86	116	68	106	108	114	117
Poultry	38	34	34	28	37	31	40	28	30	36	41	31	41	39	31
Tea, servants & best tea	27	18	26	23	14	19	11	26	11	21	13	12	17	23	21
Coffee	11	14	25	20	14	19	21	12	13	19	13	14	8	8	21
Washing	6	10	18	6	13	8	10	34	36	56	35	36	49	40	34
Dress: Emma & the girls	28	205	215	193	174	129	152	126	119	133	80	60	65	91	63
Gifts	79	102	13	87	123	105	116	110	147	120	121	126	114	170	174
Misc.	75	53	53	67	49	79	55	75	60	57	77	101	72	75	28
Dripping	3	2	2	2½	3	2½	2	2¾	3	2	4	3	—	—	—

These figures do not include those expenses which CD paid for himself, menservants wages, alcohol, snuff and later cigarettes and the clothing of the boys. Some expenses are also recorded in *ED's diary*.

Iconography, Down House: There are photographs of the interior by Leonard Darwin during CD's lifetime and others after his death as well as sketches, watercolours and an etching of CD's study shortly after his death, showing its appearance as he left it. Many more photographs were taken of the house, inside and out, after it became a museum in 1929. Not all of these are listed here.

c.1835 Pencil sketch of the front of the house by William Arthur Johnson. It shows the house with its original roof. On display at Down House.

c.1835 Pencil sketch of house, east, by William Arthur Johnson. On display at Down House.

c.1835 Watercolour of the front of the house by William Arthur Johnson, very similar to the first sketch above. On display at Down House.

1858-72 Photograph of back of house with three members of the family before construction of the verandah. Believed to be the earliest photograph of the house. DAR225.17-19.

c.1863 Photograph of family, except CD, seated at and around centre dining-room window with a dog (Bobby?) and a visitor. DAR219.12:9 and DAR225.156.

The drawing room in 1882 photographed by William England.

The new study in 1882 photographed by William England. CD's hat and cloak are draped over the chaise longue on the left. The small calendar on the wall above the mantelpiece was left on "18" [Apr] the last day it was changed by or for CD before he died. This is seen in all the posthumous images.

c.1865 Watercolour of Down House from the garden by Horace Darwin. Down House collection. Reproduced in Reeve, *Down House: The home of Charles Darwin*, 2009, p. 34.

c.1865 Photograph "Bessy taken by LD" sitting in the centre window of the dining room. DAR219.12:11.

c.1865 Photograph of "Etty" sitting in the southern dining room window with a dog, Polly, who looks at the camera. Bobby absent. Reproduced by Elliott & Fry. DAR225.73.

c.1865 Photograph of "Etty" sitting in the southern dining room window with two dogs, Polly and Bobby. Reproduced by Elliott & Fry. A copy in the Huntington Library.

c.1866 Photograph of CD astride Tommy in front of house by Leonard Darwin. DAR225.116, copy on display at Down House.

c.1873 Photograph of back (i.e. west side) of house. DAR219.12.172-3.

c.1875 Photograph of back of house with verandah, some trellis-work, climbers, and a fence across the lawn. DAR225.27.

c.1876 Photograph of old study by Leonard Darwin. DAR225.24 and Down House.

c.1877 Photograph of garden and house from south-west by Leonard Darwin. Down House.

c.1877 Photograph of the garden and house by Leonard Darwin. Down House.

c.1877-1910 Photograph, flower beds from the verandah, looking south-west. Down House.

c.1877-1910 Photograph of the rose walk, house visible through garden arch. Down House and DAR225.38-39.

1877 Photograph of rear of house from the garden. Down House.

1878 Photograph by Leonard Darwin of CD on verandah. DAR225.1 and Down House.

1880 Watercolour painting by Albert Goodwin, signed "A. Goodwin '80", back of house from south-west, in ED2:76; in colour in D. Donald & J. Munro eds., *Endless forms*, 2009, p. 28. On display at Down House, EH88202055.

1880 Photograph of new study. DAR225.2, 22-23. In Reeve, *Down House*, 2009, p. 5.

1875-82? Model of Down House made of cork by William Jackson. At Down House.

1882 Watercolour painting by Albert Goodwin, view from the north-west. On display at Down House, EH88202056. Exhibited at Christ's College in 1909 and described as "signed 'A.G. 82".

after1872 Photograph of the flower beds west of the verandah. Down House and DAR225.38-39.

c.1882 Photograph of back of house from north-west with trellises but few climbers, with William, Francis (and Amy?), and Howard(?) seated. DAR225.29a. Dated as "sometime after 1872" in CCD23:424.

c.1882 Photograph of the house from the south-west. Down House.

1882 Woodcut of rear of house, trees without leaves, a man leans against column of the verandah while a lady is seated within by J.R. Brown. *The Graphic*, (Jul.), p. 16.

1882 Woodcut of CD's new study from just inside the doorway by J.R. Brown. *The Graphic*, (Jul.), p. 16.

1882 Apr. Copperplate etching of the new study (25x36cm) by Axel H. Haig (1835-1921). Commissioned by the family a week after CD's death. Originally in a limited edn of 250 unnumbered copies signed "Axel H. Haig" in pencil in the lower right border and "Mr Darwin's Study at Down, Kent." in the lower left. Monogram "AHH 1882" in lower left of the etching. Described as not yet published in *Liverpool Mercury*, (6 Apr.) 1887. In Seward ed., *Darwin and modern science*, 1909 et al.

1882 Sketch of back of house by L. Buob. Engraved in *Ueber Land und Meer*, p. 688 & *Der Spiegel*, 1962.

1882 Two photographs of drawing room by William England (d.1896). DAR225.4 & 20. The former is annotated on verso in pencil in an elderly hand "Drawing Room at Down as it was at the time of C.D's death…" Also on display at Down House.

1882 Two photographs of the New Study by William England. One published in 1900.

1882 Photograph of dining room by William England. On display at Down House.

1883 Woodcut of new study by Alfred Parsons, *Century Illustrated Monthly Mag.*, (Jan.).

1883 Back of house from southwest, woodcut by Parsons, *ibid.*, also in LL1:320.

1883 Woodcut of front of house from field opposite east side by Parsons, *ibid.*

1884 Woodcut of the back of the house. Walford, *Greater London: a narrative of its history, its people.*

1884 Woodcut near view of part of back of house from a photograph. Woodall, Charles Darwin. A paper contributed to the Transactions of the Shropshire Archæological Society.

1886 Jun. 8 Watercolour by Julia Wedgwood of back of house showing flowerbeds, sundial and climbers. Signed "Down June 8 1886". Framed and on display at Down House.

[1886] Watercolour of CD's new study by Julia Wedgwood (13.5x9cm). Robert M. Stecher Collection, Dittrick Medical History Centre, Case Western Reserve University. Cover of this volume.

1887 Engraving of CD sitting in the new study after Parsons. *Illustrated London News*, (10. Dec.), p. 686.

1887 A fine engraving of the house from the rear with three figures standing on the lawn. *Ibid.*

1888 Woodcut of house from the back/west. *Frank Leslie's Popular Monthly*, vol. xxv.

1888 Woodcut of CD's new study. *Frank Leslie's Popular Monthly*, vol. xxv.

nd Drawing of the back of Down House from memory by George Darwin. It shows how the house looked before 1856. At Down House.

nd Photograph of the back of house by Searle Brothers, 191 Brompton Road S.W. DAR225:176 annotated on verso "[20] Photograph of Down House after 1880".

1890 "Down House—south side. View from the garden." Vignoles, *Good Words*, 1893.

1890 "Front of Darwin's house." Vignoles, *Good Words*, 1893.

1890 "Darwin's study." Unique perspective, other side of room. Vignoles, *Good Words*, 1893.

1892 "Photograph of Professor Darwin's Study, Down, Interior" copyright by Charles Essenhigh Corke, 39 London Road, Sevenoaks, Kent, on 25 Oct. 1892. National Archives, COPY 1/410/88.

1894 Photograph of house from the rear. *Bromley record*, (1 Jan.): 15.

1903 Photograph of back of the house with four children on the lawn. B. Darwin, *The world that Fred made*, 1955, facing p. 24.

1904 Colour-tinted postcard photograph of back of house and lawn. *"Downe House, Kent. Home of the Late Dr. Darwin."* by F.G. Julyan, Orpington. Registered for copyright 7 May 1904.

1909 Photograph of the front of the house. DAR233.1:7.

1909 Photographs of front and back of house. *The Sphere*, (13 Feb.), p. 148.

1909 Photograph of back of house. *Order of the proceedings at the Darwin celebrations*, 1909, plate VII.

1909 Postcard photograph of the back of Down House with Barraud 1881a in an oval frame in the upper right corner. "6747 D Charles Darwin. His house at Downe. Rotary Photo. E.C."

1912 Photograph of back of house from a distance. *Children's Mag.*, vol. 4, p. 491.

-1918 Postcard photograph of the back of house. In a white border strip: "Downe House, Kent. Home of the late Dr. Darwin."

1920 Oil on canvas, CD with pigeons behind Down House by Vatagin. SDMM. OF7667/460.

1926 Pastels on paper, CD with Ernst Haeckel by M. Ezuchevski. SDMM. OF7667/503.

1927 Postcard photograph of back of house, in the direction of the drawing room "Downe House, Downe." (in white). Handwritten message on verso reads "this is the Monkey House".

1927 Photographs of back of house, sandwalk and laboratory. *Illustrated London News*, (10 Sept.)

1927 Photograph of back of house from southwest. *The Times*. Widely syndicated.

1928 Photograph of back of house. *Sheffield Daily Telegraph*, (5 Sept.), p. 7.

1929 Photograph of back of house during opening ceremony. G.K. Chesterton, *Illustrated London News*, (15 Jun.), p. 1028. Several photographs were taken on this occasion.

1929 Jun. 7 Photograph of back of house with crowd during opening as a museum. Morris & Wilson, *Down House*, 1998, p. 3.

1929? Drawing by George Darwin of the house from the rear as it looked before 1858. In Keynes, *Annie's box*, 2001, p. 88.

1929- Postcard photograph of the front of house (from a ladder?) Daniell Bros., Lewisham.

1929 Photograph of front of house. *Staffordshire Sentinel*, (27 Sept.), p. 16.

1929- Postcard photograph through front gate (sign already in wall). BAAS.

1929- Postcard photograph of the back of house from the south-west. BAAS.

1929- Postcard photograph "The Drawing Room." (14x9cm) BAAS.

1929- Postcard photograph "The Study." (14x9cm) BAAS.

1929 Postcard photograph "THE OLD STUDY, DOWN HOUSE." With stuffed dog in basket.

1929 Postcard photograph of the front of house. "DNE-21 DOWNE, DOWN HOUSE (HOME OF DARWIN) Copyright T. Sergeant."

c.1920s Colour tinted postcard photograph of front of house from within gate. "Downe House." (in white). n.p., n.p.

1920s Postcard photograph "Downe House, Downe, Kent Front of House". A. Buchanan & Co., Surrey, No: 14727.

1930 Painting of CD at his desk in the old study by S. Uranova. Tretyakov Gallery, Moscow.

c.1930 Photograph of the front of the house after conversion to a museum. DAR233.1:6.

1934 Unique drawing of back of house with head portrait after Collier 1881. Ty-phoo Trade Cards. "Homes of Famous Men" No. 6 of 25.

c.1934 Postcard photograph of the back of "Down House, Downe." (in white script).

c.1934 Postcard photograph of the back of "Down House Downe." (in white) showing entire lawn.

1935 Oil on canvas. CD in cloak in garden behind Down House with dog by Victor Vatagin. SDMM. OF7667/110. Black and white photograph in DAR238.14:94.

1948 Oil painting of CD in the old study, by Evstafiev (40x52cm). Down House, EH88204429.

1952 Woodcut of Aunt Bessy in the drawing-room by Gwen Raverat. *Period piece*, 1952.

1952 Woodcut of playing croquet on lawn behind house by Gwen Raverat. *Period piece*, 1952.

early 20th century Watercolour of Down House when it was Downe House School by Dorothy Willis, sister of the school's founder, Olive Willis. Reeve, *Down House*, 2009, p. 49.

c.1958 "Early Spring", oil painting showing young CD on verandah and ED at her Broadwood piano by Evstafiev? Sent by the SDMM to Down House. See DAR238.14.64.

c.1958 Oil painting, "Late Autumn" by Evstafiev, showing ED at her piano playing for elderly CD. (frame size 36x47cm). Given to Down House by the SDMM. Down House EH88202062.

nd Etching of southwest front, not signed, not in CD's lifetime. Moorehead, pp. 250-1. Down House

1958 Oil on canvas of beardless CD in his old study writing with pen by Evstafiev, signed and dated, (82.5x109cm). Given to Down House by the SDMM. EH88202272.

1959 Woodcut, back of house with a female figure. Корсунская, *Великий натуралист чарлз дарвин.*

1960 Oil painting of bearded CD in his library in the old study by Evstafiev (50x70cm). Initialled and dated. Given to Down House by the SDMM. No original or copy is in the SDMM. EH88204432.

Plan of ground floor, from information supplied by Leonard Darwin. [Keith,] Down House. *[BAAS] Report of the ninety-sixth meeting…1928*, 1929.

Plan of ground floor. Keith, *Darwin revalued*, 1955, p. 46.

Plan of ground floor. Atkins 1974 & 1976, p. 22.

Plan of ground floor. [Titheradge], *Darwin memorial*, 1981.

Plans of entire house. Lea, *Down House…Notes on the inventory*, English Heritage, 1996.

Plans and drawings of entire house. Lea, *Down House…perspective reconstruction drawings…in seven phases of its historical development*, English Heritage, 1998.

Floor plans and silhouettes of house as it changed 1730s-1870s. Reeve, *Down House*, 2009.

Iconography, grounds:

1840 Tithe Map of Down, Kent. Public Record Office, Kew, JR 30/17/151.

late 1850s Photograph by William Darwin of back lawn with gardeners William Brooks and Henry Lettington with donkey-drawn mowing machine. Down House. Reproduced in several publications.

1854 Plan of grounds with sandwalk and wood in 1885. Über die Wege der Hummel-Männchen. In Krause ed. *Gesammelte kleinere Schriften von Charles Darwin*. vol. 2, pp. 84-88 and in Freeman, Darwin on the routes of male humble bees. *Bulletin of the BMNH, Historical Series*, 1968 (captions in English). Both in *Darwin Online*. The former is reproduced below.

-1872 Watercolour of sundial through the garden wall arch by Julia Wedgwood. At Down House.

1870s Photograph of sandwalk by Leonard Darwin. In Reeve, *Down House*, 2009, p. 35.

c.1880 Photograph of approach to the sandwalk from the north with Horace Darwin. Down House and DAR225.33-37. On display at Down House.

c.1880 Photograph of the north end of the sandwalk with the 'Hollow beech' with Horace Darwin. Down House and DAR225.33-37. On display at Down House.

c.1880 Photograph of the sandwalk. DAR225.33-37. On display at Down House.

c.1880 Photograph of the T-junction to sandwalk by L. Darwin. DAR225.33-37. At Down House.

c.1880 Photograph of sandwalk and Great Pucklands Meadow. DAR225.33-37 and Down House.

1882 Watercolour of the sandwalk by "Dickinson". Exhibited at Christ's College, 1909.

1883 Woodcut of garden gate "Darwin's usual walk." by Parsons, *Century Illustrated Monthly Mag.*, (Jan.).

1887 Engraving, CD "on his usual walk" after Parsons 1883. *Illustrated London News*, (10 Dec.), p. 686.

1887 Engraving of CD (drawn too small) in the hothouse after Parsons 1883. *Ibid.*

1890 "Down House—south side. View from the garden." Vignoles, *Good Words*, 1893.

1890 "Front of Darwin's house." Vignoles, *Good Words*, 1893.

Down House from the rear from an 1882 sketch by Louise Buob. *Ueber Land und Meer*, p. 688.

1890 "Summer-house at south end of the 'sand-walk'" Vignoles, *Good Words*, 1893.

1890 "The 'Sand-walk'" with Great Pucklands. Vignoles, *Good Words*, 1893.

1895/6 Ordnance Survey Map 2ᵈ edn. 1 to 2500 scale in Miele, *Darwin's Garden*, 1996.

late 19ᵗʰ-century photograph of flower garden. Morris & Wilson, *Down House*, 1998, p. 30.

early 20ᵗʰ-century watercolour of Down House when it was a school. In Reeve, *Down House*, 1998.

1909 Photograph of sandwalk. *Order of the proceedings at the Darwin celebrations*, 1909, plate VIII. Later sold as a postcard by the BAAS.

c.1910 Photograph of the beds in the rose walk with a young woman. Down House.

1920 Pastels on paper, CD in pigeon house by M. Ezuchevski. SDMM. OF7667/447.

1923 Pastels on paper, CD with orchids in his hothouse by M. Ezuchevski. SDMM. OF7667/448.

1929 Plan, from information supplied by Leonard Darwin. [Keith], Down House. *[BAAS] Report of the ninety-sixth meeting…1928*, 1929.

c.1930 Map of after conversion of Buckston Browne Farm. Down House.

1931 Leonard Darwin's sketch of the plantings at Down House made with the help of the 1868-69 Ordnance Survey. Reproduced in Miele, *Darwin's Garden*, 1996. Down House.

1868/9 Ordnance Survey Map. 1 to 2500 scale. Miele, *Darwin's Garden*, 1996.

1952 Mulberry tree and back of Down House; woodcut by Gwen Raverat. *Period piece*, 1952.

Plan. Keith, *Darwin revalued*, 1955, p. 47.

Plan. Atkins 1974 & 1976, p. 34.

Plan. [Titheradge], *Darwin memorial*, 1981, p. 27.

Plan of house, grounds and village in Chapman & Duval, *Darwin commemorative*, 1982.

Plan based on the 1868-9 Ordinance Survey map of Kent. CCD211:745, 1999.

Plan of grounds S. Morris & L. Wilson, *Charles Darwin at Down House*, 1998-2003 (title varying). Perspective drawings of grounds and greenhouse. Reeve, *Down House*, 2009.

Sketch of part of the grounds of Down House (1854). Krause, *Gesammelte kleinere Schriften*, 1886, vol. 2.

Staff of Down House (servants): The details are incomplete, especially for those of junior staff and full names and dates are often unavailable. Many have their own entry in the text where more details can be found. The following staff are known:

Butlers: Joseph Parslow, William Jackson, James Price.

Coachmen: Joseph Comfort, John Skinner, Samuel Jones.

Cooks: Margaret Evans, Jane Davies, Mrs Brummidge.

Custodians: Harold, Samuel and Sydney Robinson.

Footmen: Moffatt, George Bridgen, James Pearce, James Johnson, James Tester.

Gardeners: Comfort, Hills, Horwood, George Coote, William Brooks, Henry Lettington, Frank Skinner, Henry Wheeler/Mr Bailey and Shearman (d. c.1915), Thomas Price, William Duguid.

Governesses: Barellien, Louise Buob, Grant, Grut, Latter, Camilla Ludwig, Louisa Ludwig, Pugh, Catherine Anne Thorley, Maria Emily Thorley.

Grooms: Mark Ansell, Frederic 'Fred' Hill.

Housekeeper: Margaret Evans (also cook).

Maids: "Lucy" (CD to ED [24 Jun. 1846]). Anne, Betsy, Emily Jane, Matheson, Eddy, Emily Asborne, Elizabeth Chapman, Jane Holledge, Augusta Dickson, Elizabeth Bradford, Jane Roffey, Harriet Irvine, Hellen Parslow, Mary Wilkins, Harriet Wills.

Nursemaids/nurses: Anne Abberley, Mrs Bennett, Elisabeth Harding, Jannet Brodie, Mrs Locke, Mary Cooper, Sarah Young, Margaret Evans, Anne Parslow, Jane Asborne, Sara Griffith, Mary Anne Westwood, Harriet Wills, Pauline Badel.

Page: John Lewis.

late 1850s Photograph by William Darwin of back lawn with gardeners William Brooks and Henry Lettington with donkey-drawn lawnmower. Down House. Reproduced in Reeve, *Down House*, 2009, p. 23 and Browne, *Power of place*, facing p. 312, et al.

1878 Photograph "Down House household 1878", written on recto, including Jackson and wife, John, Fred, Price, Evans, Jane, Harriet, Mary Anne and Bernard Darwin(?) by W.J. Sell. From L.A. Nash. DAR225.191. Reproduced in Browne, *Power of place*, facing p. 312.

History: Back to 1651. Earlier given in Atkins 1976, pp. 12-17, with list of owners or tenants back to 1651. Reeve, *Down House*, 2009 has further details. 1900-06 Rented from George Darwin by a Mr Whitehead about whom nothing seems to be known except that he owned the first motor car in Down. 1907-22 Rented by Downe House School. 1924-27 Run as an unsuccessful girls school by a Miss Rain. 1927 Bought from the Darwin heirs by Sir George Buckston Browne for £4,250. 1929 After spending about £10,000 on repairs and giving £20,000 as an endowment, Buckston Browne handed it over to the BAAS in 1929. It was formally opened at a tea on Jun. 7. Admission to the museum was free. 1953 Given free to Royal College of Surgeons of England who administered it until 1996, although they attempted to transfer it to the National Trust in 1958. The Surgeons' research establishment marches with the grounds to the southwest. 1996 Acquired by English Heritage with a grant from the Wellcome Trust. Restored and reopened in 1998.

Visitors: The Darwins entertained considerably at Down House, although seldom large gatherings. Casual calling, which was customary in cities, was usually confined to near neighbours. John Lubbock, who was 8 years old when CD came to Down, was the most frequent. Visitors from London and elsewhere came for lunch, dinner, short stays, weekends or even longer. The following list omits close relatives and neighbours and it is no doubt far from complete. Nevertheless, the sheer number of visitors would seem to contradict the notion that CD was quite such a recluse as often depicted. The numbers of visitors increased in later years when the children were grown and brought their own friends. Even after CD's death, his peers continued to visit ED and the family. Those who are known to have left written recollections of CD (all in *Darwin Online*) are given in italics.

Agassiz, A. & A.R. 1869.
Airy, Hubert 1872.
Alcock, Sir Rutherford 1878.
Allen, J.S. 1846.
Allfrey, Dr Charles H. 1882.
Appleton, J.T. 1844-74.
Appleton, Thomas G. 1874.
Armstrong & family 1851-71.
Ashburner 1880.
Atkin family 1879.
Atkinson, Tindall 1880.
Aubertin, John James 1871.
Aveling, Edward B. 1881.
Badel, Pauline 1881.
Balfour, Arthur. 1873.
Balfour, Charles 1880.
Balfour, Francis M. 1870-81.
Balfour, Strutt 1871.

Barbier, Edmond 1880.
Bates, Henry Walter 1862-64.
Bell, Thomas 1844-47.
Bennett, Alfred William 1872.
Blunt, T.P. & family 1861-81.
Blyth, Edward 1868.
Bonham-Carter, E.M. 1855-77.
Bowman, William 1869-78.
Brace, Charles L. & Mrs B. 1872.
Bradshaw, Henry 1872-84.
Brand, Mr 1876.
Brinton, Dr William 1863.
Brodie, Jessie 1842-59?.
Brodie, Sir Benjamin 1872.
Brooke, Sir Victor 1872-77.
Brown, J.R. 1882.
Bryce, James 1882.
Büchner, Ludwig 1881.

Buckley, Arabella B. 1871-72.
Buob, Louise 1863-75.
Burdon-Sanderson, J.S. & Mrs G.B. 1873.
Burgess, Edward & C.L. 1877.
Busk, G. & family 1862-73.
Butler, Mary 1860.
Butler, Samuel 1860-72.
Caird, James 1878.
Cameron, Mrs Julia M. 1872.
Carlyle, Thomas 1875.
Carpenter, Miss 1861.
Carpenter, William B. 1843-61.
Caspary, Robert 1866.
Caton, Mrs 1855.
Chadwicks, Mr & Mrs 1866.
Chapman, Dr John 1865.
Clark, Dr Andrew 1873-82.

Clark, J.W. & family 1880-95.
Claus, Carl Friedrich 1871.
Clifford, William 1839.
Clive, Archer 1858.
Clough, Eves 1850.
Cobbe, Frances Power 1868.
Cockell, Edgar 1842.
Cohn, Ferdinand, Mrs P. 1876.
Collier, John 1878-81.
Conway, Moncure Daniel 1873.
Cooksons, 1873-78.
Crawley, Charles 1871-95.
Cresy, Edward, family 1842-67.
Crofton, Amy 1866-71.
Cupples, Anne Jane 1869.
D'arcy, Miss M. 1877-79.
de Candolle, A.L.P. 1880.
de Vries, Hugo 1878-79.
Dew-Smith, Mr 1880.
Dicey, Albert Venn 1882-95.
Dohrn, Anton 1870.
Donders, Frans Cornell 1869.
Dowie, Mrs 1871-75.
Downing, John 1873.
Dresser, Henry Eeles 1875.
Drummond, Miss 1879.
Duff, M.E.G. 1871-95.
Duncan, Dr Peter Martin 1868.
Durando, Gaetano 1878.
Durdík, Josef 1875.
Elliott & Fry, 1869-81.
England, William 1882.
Errera Léo Abram 1878.
Evans, Margaret 1857.
Evans, Mary A. 1873.
Falconer, H. 1844-47, 58, 64.
Farr, William 1870.
Farrer, Thomas H. 1869-78.
Ffinden, G.S. & Mrs 1871-95.
Fiske, John & Mrs 1873-80.
FitzRoy, Robert & wife 1857.
Flower, W.H. & G.R. 1877-80.
Forbes, Edward 1844-47.
Forrest, Mr 1879.
Forster, Laura Mary 1874-81.
Fox, Eliza Ann 1846.
Fox, W.D. 1843, 49, 54, 61.
Francis, Duke of Teck 1875.
Franke, Hermann, J. 1881.
Fraser, Thomas R. 1882.
Frische, Gustavus? 1875.

Fry family 1860-67.
Galton, Francis 1870.
Gaskell, Meta family 1856-83.
Gifford, G. Mrs 1845-56.
Gladstone, Miss 1880-95.
Gladstone, W.E. 1876-77.
Goodwin, Albert 1880, 1882.
Goodwin, Rev. H. 1874.
Gray, Asa & family 1868-96.
Grifith, C. 1858.
Grut, Mrs 1859.
Gulick, J.T. 1872.
Günther, A. 1870-71.
Gurney family, 1877-87.
Haeckel, Ernst 1866-76.
Hague, James Duncan 1871.
Hahn, Otto 1881.
Haig, Axel H. 1882.
Hamilton, Mrs 1875.
Hancock, Albany 1876.
Harding, Elizabeth 1841.
Harrison, Frederic 1871.
Harrison, Jane E. 1876.
Harvey, William Henry 1858.
Haweis, John O.W. 1868-69.
Haycock, John Hiram 1842.
Head, Henry A. 1872.
Hemmings, Henry 1861-63.
Henry, William 1877.
Henslow, George 1866.
Henslow, J.S. & Mrs 1857-60.
Hesse-Wartegg, Ernst von 1880.
Higginson, T.W. 1872&78.
Hoare, Arabella 1869.
Hoare, Charles 1877.
Holland, Henry 1860.
Holland, Saba? 1855.
Hooker, B.H.H. 1873, 1876.
Hooker, Joseph D. 1846-95.
Hoole, Stanley 1878-79.
Hope, Elizabeth R.C. 1881.
Horner, L. & Mrs H. 1840-55.
Horners & daughters 1859-81.
Horsman, S.J. O'Hara 1867.
Husseys & family 1840-51.
Huxley, T.H. & family 1856-96.
Hyatt, Alpheus 1874?
Innes, Brodie J. 1840-66 & family 1851-79.
Isitt, Virginia L. 1871.
James, Henry 1869.

Jeffreys, John Gwyn 1868.
Jenner, Dr William B. 1864-65.
Jones, Dr Henry B. 1865-69.
Jones, Mary 1863.
Judd, John Wesley 1880.
Kirk, Miss 1857.
Klein, Edward Emanuel 1874.
Kölliker, R. A. von 1860-64.
Kovalevsky, V.O. & Mrs K. 1867-72.
Lady L. & German Prof. 1871.
Langton, Ch. & Charlotte 1859.
Langton, Edmund 1863-65.
Lankester, R.E. 1875.
Latter, Miss 1859-61.
Laugel, Antoine Auguste 1865.
Leaf, Mr 1876.
Lewes, George Henry 1873.
Lewin, Dr. Friend 1872.
Lowe, Robert 1871.
Lowell, Mrs 1880.
Ludwig, Camilla 1862-96.
Ludwig, Louisa 1860-95.
Lushington, Vernon 1868-81.
Lyell, Ch. & family 1845-92.
Mackintosh, R. & children 1860.
Margo, Theodor nd.
Marsh, Othniel Charles 1878.
Marshall, V.A.E.G. 1880.
Marshall, W.C. family 1872-96.
Martin, Alice & Miss 1854-73.
Martin, S. 1853.
Maryanne, Miss 1876.
Maudsley, Dr Henry 1872.
McGilvray, Miss 1879.
McLennan, John F. 1878.
Meek, Mary A. 1854-64.
Meldola, Raphael 1878.
Mellersh, Arthur 1862.
Moggridge, John T. 1866-71.
Mohl, Mme. 1859-80.
Monteagle, Lrd & Ldy 1873-77.
Moore, Dr N. & Mrs 1871-96.
Morgan, Lewes Henry 1871.
Morley, John 1877.
Moseley, Henry Nottidge 1876.
Moulinié, Jean Jacques 1868.
Müller, Friedrich M. 1882.
Nash, Wallis & Mrs 1874-76.
Neale, Mrs Edward V. 1867.
Nevill, Lady Dorothy 1875.

Newton, Alfred 1870-96.
Noel, Edward 1865-1890.
Normans, H. family 1854-91.
North, Marianne 1874-81.
Norton, C. & family 1868-69.
Norton, Jane 1873.
Norton, L. Playfair 1876.
Ogle, William 1879-95.
Ouless, William W. 1874-75.
Overstone, Lord 1854.
Paget, Dr James 1869-89.
Paget, Stephen 1857-69.
Parker, Charles 1850 (5days).
Parker, Henry 1854-72.
Parslow & family 1857-96.
Parsons, Arthur G. 1882.
Parsons, Col. R.M(?) 1874.
Parsons, Laurence(?) 1877.
Pattrick, St. Reginald 1871-94.
Pearce, Mr & family 1878-96.
Perry, Matthew C.(?) 1845.
Pertz, Anna 1877.
Pertz, G.H. & family 1868-96.
Playfair, Lyon 1877.
Poulton, Sir Edward B. 1895.
Powell, Rev. H. 1869-71.
Price, John 1842.
Pryor, Marlb. Rob. 1873-88.
Pugh, Miss M.A. 1857-63.
Ramsay, Andrew C. 1848, 50.
Reade, William W. 1871-72.
Reeds & family 1858-66.
Rich, Mrs 1849.
Richmond, William Blake 1879.
Richter, Hans 1881.
Riley, Charles Valentine 1871.
Ritchie, R.T.W. family 1881-95.
Robinson, Mr & Mrs 1855.
Romanes, George J. 1874-87.

Römer, Ferdinand 1876.
Rouse, R.C.M. 1870.
Rowlands, 1851, 53, 56, 71.
Russell, A. & Lady R. 1869-78.
Rütimeyer, Ludwig Karl nd.
Sarcey, Francisque 1879.
Schomburgk, Robert H. 1848.
Scott, John 1864.
Scott, Mrs 1879.
Sedgwick, Theodora 1878.
Sellwood, Thomas 1859-67.
Semper, Karl Gottfried 1872.
Severtsov, Nikolai A. 1875.
Shaen, Margaret J. 1874-96.
Shore, Miss 1871-87.
Smith, Godwin 1868-71.
Smith, Julia 1872-75.
Sophy, Katherine E. 1865.
Sowerby, George B. 1843-61.
Spalding, Douglass A. 1874.
Spring Rice, T. & E. 1877.
Stanley, Mary C. 1871-96.
Stephen, Leslie 1873-96.
Stephens, 1860-66.
Stokes, John Lort 1864.
Strachey, Richard 1873.
Strickland, Sefton West 1863-7.
Strutt, John William 1871.
Studer, Bernhard 1847.
Sulivan, B.J. 1846, 1851, 1862.
Susan, Margaret 1865.
Sutherland, Charles L. 1872.
Swears & Wells 1881.
Swettenham, R.P.A. 1867-71.
Swinhoe, Robert 1870-71.
Sylvester, James Joseph 1875.
Symonds, Miss H. 1869.
Tait, Robert Lawson 1875-80.

Teesdale & family 1876-84.
Tegetmeier, William B. 1858-87.
Tennyson, Lord Alfred 1868.
Thackeray, A.I. 1866, 81, 82(?).
Thackeray, Harriet M. 1866-73.
Thackeray, William M. 1873.
Thiselton-Dyer family 1875-94.
Thompson & family 1860-96.
Thomson Adm. family 1867-90.
Thorley, Anne C. 1848-94.
Thorley, John 1852, 1854.
Thorley, Maria Emily 1856-93.
Timiryazev, K.A. 1877.
Tollets, 1843-90.
Trimen, Roland 1867.
Turnbull, Lizzy 1856-72.
Tylor, Alfred 1875.
Tylor, Edward Burnett 1872.
Tyndall, John 1868-78.
Virchow, Rudolf Carl 1879.
von Kölliker, R.A. 1862, 1870.
Wallace, A.R. & Mrs 1862-68.
Wallich, George Charles 1871.
Waterhouse, G.R. 1844-47.
Weir, John Jenner 1868-71, 80.
Wenger, Mlle 1856.
Wheler, Lucy 1868.
Wickham, John Clements 1862.
Williams, 1845.
Williams, Edward A. 1861.
Wills, Harriet 1881.
Wollaston, Thomas V. 1856.
Woolner, T. & W. 1866-72.
Wright, Chauncey 1872.
Yates, Edmund H. 1878.
Yonge & family 1873-86.
Young, George 1875, 1880.

Among the groups of people who visited are: members of Down Friendly Club, yearly; 1872-6 London Working Men's College (WMC), groups of 60 or 70 for the day, yearly; 1880 the Scientifics q.v.; 1880/81? J.W.C. Fegan's street boys for the day; Yorkshire Naturalists' Union (Sorby, Brook, Roebuck & 11 other officials); 1880 Lewisham and Blackheath Scientific Association, 43 visitors including J.J. Weir and Hesse-Wartegg (see DAR226.1.77); 1882 "Sunday tramps", led by Leslie Stephen.

Down or **Downe**, Village, Orpington, Kent. BR6 Post Office spelling was "Down" before 1842. Census of 1841 total population 444; of 1851 437, in 1881 555. Postal addresses, near Bromley in 1845, near Farnborough 1845-early 1855, near Bromley late 1855-late 1869, near Beckenham 1869 Sept. Present address is in the Bromley postal code. In 1855 letters arrived from Bromley at 8am and were dispatched at

1.30pm. 1786 Church: St. Mary the virgin, illustration 1786, before drastic restoration. Atkins 1976, p. 25. Inns: The George & Dragon and The Queen's Head on church side. Village hall, the one built by the Darwin's is next to the George & Dragon. Both Petley's and Trowmers are in Luxted Road. 1842 Jul. CD to his sister Catherine, "The little pot-house, where we slept is a grocers-shop & the land-lord is the carpenter—so you may guess style of village— There are butcher & baker & post-office.— A carrier goes weekly to London & calls anywhere for anything in London, & takes anything anywhere." CCD2:324. Schoolmasters: Norman, Fletcher, Skinner. Physician: Engleheart. 1933 Howarth & Howarth give a detailed description of the village and its history. 1969 Newman, in Pevsner's *Buildings of England, West Kent*, 1969, p. 251, describes the architecturally worthwhile buildings.

Down village by G.W. Smith, c.1900.   St. Mary's church, Down. c.1900 postcard.

Downe Church, St. Mary's. The building dates to the 13th century. 1786 drawing in Atkins 1976. Vicars: Drummond, Innes, Ffinden. Curates: Hoole, Horsman, Humphreys, Palin (or Ealin?), Powell, Robinson, Salin, Stephens. Churchwarden: Lovegrove. The spire had three bells. "The benefice is a perpetual curacy, in the Diocese of Canterbury; the rectorial tithes, amounting to £395 per annum, are held by Lady Carew on lease, and the vicarial tithes are cummited at £91 per annum. There is a chapel for Baptists, and a charity school for girls and infants, near the church". (*Post Office Directory of Essex, Herts, Kent…1855. Part 1.*) Churchyard has two slab tombs which are memorials to Darwins: Grave of Erasmus Alvey, also to CD and ED. Grave of Mary Eleanor and Charles Waring, but adult-sized slab, which also commemorates Henrietta, Bernard Darwin and Elinor Mary Monsell, his wife. Summary of graveyard inscriptions in *North West Kent Family Hist. Jrnl.* I, no. 1, 1978. References to the Darwins and Parslow in the minute book of the vestry meetings from 1844-82 from St. Peter and St. Paul Cudham and St. Mary Downe, https://www.pcd.org.uk/:

1844 Apr. 9: Sir J.W. Lubbock & Charles Darwin Esq. appointed Surveyors of Roads for ensuing year.
1845 Mar. 25: Sir J. Lubbock & C. Darwin Esq. appointed Surveyors of the Roads for ensuing year.
July 15: CD present at a Vestry Meeting which agreed matters concerning: expenses of the repairs for the chancel…erecting slate tablets over the communion table; the Parish contributing £10 out of £50 towards painting, whiting and other repairs.
1846 Mar. 24: CD present at a Vestry Meeting, appointing surveyors (but not CD) etc.

1848 Mar. 24: CD Present at a Vestry Meeting which agreed "that one years' rent be forgiven to the widow Osborne - on condition of her paying the remaining arrear, and in future the rent be £5 a year to be paid weekly."

Apr. 25: CD seconded a motion by Sir John Lubbock "that it is the opinion of the vestry that the rating on the small tithe be reduced £20 on the prop value, and a further reduction of £5 making the prop value £70 and the rateable value £65." carried.

1850 Mar. 28: CD present at Vestry meeting which examined and passed accounts and set a highways rate of 6*d* in the pound.

1852 Sept. 10: CD proposed a motion (passed) requesting the overseers "to ascertain from the poor law commissioners if they have the power of selling the old workhouse by private contract to be held in trust for the site of a school for poor children".

1853 May 27: CD present at Vestry meeting which consented "to the Guardian of the poor of the Bromley Union selling the premises consisting of an old house of wood & tile twenty four foot square opposite the church".

Oct. 21 Sir John Lubbock moved and CD seconded that the Guardians of the Poor of the Bromley Union convey the old Poor House and the land, and that the site be used for a school.

1855 Apr. 13: CD and four others requested to consider and report on the present state of Tromer pond and the wish of a villager to remove part of his wall abutting on the road.

1855 Apr. 24: This committee recommended fencing the pond, making a tank in which to filter the water, and repairing the road where it had been worn away by the pond, but did not agree the alteration to the wall which would have involved diverting a footpath.

1855 May 25: The vestry meeting which followed agreed the repairs to the pond, asked the committee to look at improving the supply and quality of water, but also considered that the wall could be moved, provided it was rebuilt to the same height.

1856 Mar. 25: Mr Parslow appointed one of the overseers.

1856 June 10: Because of an outbreak of smallpox, a vestry meeting requested the churchwardens and overseers "to take the most stringent steps in their power to prevent any person from Hemings house communicating with others". Also to ascertain the state of vaccination in the parish. Letter sent to Charles Heming warning that he would be punished according to law if they showed themselves outside their premises.

1856 Sept. Mrs Darwin et al subscribed £5 towards a new organ (John Lubbock £7 12*s*. 8*d*.; Lady Lubbock £5; Mrs Lubbock £2). Mr Parslow gave 5*s*.

1858 Apr./May: CD and Mr Parslow involved in a dispute between the churchwardens and Robert Ainslie about his non-payment of the church rate. CD moved that "no further legal proceedings be taken in the matter of the Church rate which Mr Ainslie refuses to pay".

1859 Mar. 25: CD present at vestry meeting.

1859 Apr. 26: Parslow present at a vestry meeting which agreed to renew churchyard gate and part of wall, and repair and whitewash church ceiling and walls. They set a rate of 9*d*. (50% increase) to do the repairs.

1860 Mar. 29: Parslow appointed an overseer for the year (and a number of years following, inc. 1871).

1871 Nov. 24: CD proposed "that it is desirable to form a School board for the parish of Downe, Kent, and that the Education Department be applied to for the necessary powers". Carried by a majority of six.

1872 Jan. 19: The previous motion was unanimously rescinded. Another motion was unanimously passed (but not apparently with CD present): "that with the view to provide the necessary funds for carrying on the Parochial Schools a voluntary rate at eight pence in the pound for the current year payable in January and June...and the following ratepayers hereby pledge themselves each to contribute their proportion of the same." (CD was not among those who pledged support.) However the Vicar proposed that certain named ratepayers be constituted a committee for the management of the School - the third name on the list after the Vicar and Sir J Lubbock was CD. The motion was carried unanimously.

1873 Jan. 21: A Vestry meeting at the Boys School Room unanimously agreed that the School Committee (including CD) be requested to retain their offices for the current year. Parslow also the meeting.
1874 Jan. 3: Vestry meeting held at the School re-appointed CD et al to the School Committee.
1875 Jan. 2: Vestry Meeting did not include CD in list of ratepayers appointed to School Management Committee.
1876 Jan. 1: Dr F. Darwin included in list of ratepayers appointed to School Management Committee.
1876 Mar. 25 Mr Joseph Parslow appointed again as one of the overseers.
1877 Jan. 6: F. Darwin again appointed to School Management Committee. The committee was given the powers to move the Infant School children to a room to be built at the National School. The Infant Schoolroom would probably be taken over by the Workmen's Club.
1878 5 Jan. Mr Joseph Parslow was added to the School Committee.
1879 4 Jan.: The present School Committee (including F. Darwin and J. Parslow) were appointed for a further year. 1880 10 Jan.: ditto.   1881 8 Jan.: ditto.   1882 5 Jan.: ditto.

Down, John Langdon Haydon, 1828-96, Physician. 1839 Nov. 19 ED recorded "Mr Haydon's". 1859 Appointed medical superintendent and resident physician at Eastwood Asylum for Idiots in Surrey. 1866 First to describe "Mongolism". Laura Lee, *The name's familiar*, 1999, p. 121. 1873 sent CD a photograph "Ear of Microcephalic" which CD noted as received in DAR87:65. CCD21:6.

Darwin's old study as a schoolroom.
Ward, *Charles Darwin*, 1927.

Downe Court. 1690 Original manor house of Down, opposite east side of Down House. 1842 Jul. CD to his sister Catherine, "There is a most beautiful old farm-house with great thatched barns and old stumps of oak trees...one field off". CCD2.

Downe House School. Always spelt with an "e". Headmistress Olive Margaret Willis was co-founder with her friend Alice Carver. Started with one girl and five mistresses, but was at once successful. 1907 Feb.-1922 Apr. 1 Occupied Down House. 1922 Moved to larger premises Hermitage Road, Cold Ash, Newbury, where it flourished. Photographs of house interior in Reeve, *Down House*, 2009.

Downes, John, 1810-90. Cambridge friend of CD. 1831 Jul. 11 CD to Henslow, "Do you by any chance recollect the name of a fly that Mr. Bird sent through Downes". CCD1:126. According to family tradition, D was offered the position of naturalist on *Beagle* before CD but declined to take Holy Orders. 1834-63 Vicar of Horton and Piddington, Northamptonshire and later Hannington, Hampshire.

Downton, Wiltshire. 1822 Jun. CD had a holiday there with his sister Caroline.

Drewe, Charlotte, ?-c. 1817. 5th child of Edward D. Unmarried. Died young.

Drewe, Edward, 1756-1810. Vicar of Broadhembury and Willand, Devon. 1793 Married Antoinette Caroline Allen. 7 children: 1 Harriet Maria, 2 Marianne, 3 Georgina, 4 Edward Simcoe, 5 Charlotte, 6 Francis, 7 Louisa.

Drewe, Edward Simcoe, 1805-77. 4th child of Edward D. c.1820 Inherited The Grange, near Honiton, Devon. 1828 Married Adèle Prévost (d. 1881). Had children.

Drewe, Francis (Frank), ?-c. 1817. 6th child of Edward D. Triplet of Charlotte and Louisa? See *Emma Darwin*, 1904, F1552.1.

Drewe, Francis Rose, 1738-1801. Half-brother of Edward Drewe. Squire of Grange, near Honiton, Devon.

Drewe, Georgina Catherine, 1791-1871. 3d child of Edward D. Mother of Lady Salisbury. 1823 Married E.H. Alderson. Darwins called on her 1840 Feb. 14.

Drewe, Harriet Maria, 1785-1857. 1st child of Edward D. 1816 Married Robert, Lord Gifford and had offspring. 1837 Was living at 1 Atholl Crescent, Edinburgh. Visited Down 1845 Sept. ? stayed till Sept. 30. *ED's diary* 1845.

Drewe, Louisa, ?-c. 1817. 7th child of Edward D.

Drewe, Marianne, 179?-1822. 2d child of Edward D. 1820 Married Rev. Algernon Langton. Died in childbed. Son Bennet L. ED1:135.

Dring, John Edward Collector of shells. 1834 Oct. Appointed acting Purser to replace Rowlett on return of *Beagle* from 2d voyage. Also acted as Clerk. In the *Sydney notebook*, Chancellor & van Wyhe, *Beagle notebooks*, 2009. Went on 3d voyage.

Dropmore, Buckinghamshire. 1847 CD visited on day trip from BAAS meeting at Oxford. 1849 May 6 CD wrote from Malvern to Henslow reminiscing "we shall not have any such charming trips as Nuneham and Dropmore." and in 1849 Oct, CD to Hooker recalled "that heavenly day at Dropmore". CCD4:36, 235.

Drosera. 1860 Irritability of Drosera, *Gardeners' Chron.*, no. 38: p. 853 (*Shorter publications*, F1813). CD posed a brief query: "In Lindley's Vegetable Kingdom (p. 433) it is stated that the leaves of Drosera lunata 'close upon flies and other insects that happen to alight upon them.' Can you refer me to any published account of the movement of the viscid hairs or leaves of this Indian Drosera?" Editor replied.

Drummond, James, 1784?-1863. Scottish botanist. Emigrated to Australia in 1829. Superintendent of the government gardens, Western Australia, until 1834. 1860 D helped CD on fertilisation of *Leschenaultia*.

Drummond, Rev. James, (1800-82). Christ Church College, Oxford, 1818; B.A. 1823; M.A. 1825. Vicar of Down before Innes. Perpetual curate of Down Feb. 6 1828-40. 1837 D purchased Down House. 1842 Sold house to CD for £2,020.

Drummond, Thomas, 1797-1840. Army officer, civil engineer and politician. Invented Drummond's light; a form of limelight for surveying purposes.

Drysdale, Lady Elizabeth, née Pew. 1787-1887. Friend of the Darwins through Moor Park Hydropathic Establishment. CD to Hooker in 1857 "The owners, Dr. Lane & wife & mother-in-law Lady Drysdale are some of the nicest people, I have ever met." 1858 Feb. 23 threw a party which the Darwins attended. *ED's diary* 1858. See Kate Summerscale, *Mrs Robinson's disgrace*, 2012.

Du Bois-Reymond, Emil Heinrich, 1818-96. German physician and electrophysiologist. 1858 Prof. Physiology Berlin. 1860 CD to Gray that D agrees with CD's views. 1876 *Darwin versus Galiani*, Berlin. 1878 D writes to CD to tell him of his election to Königlich-Preussische Akademie der Wissenschaften, Berlin, as Fellow. 1884 *Friedrich II in Englischen Urtheilen, Darwin und Kopernicus*, Leipzig.

Duberry, Amy. 1869 Sunday-school teacher at Down. CCD16:437. Listed as a 36-year-old dressmaker in 1871 census.

Dublin. 1827 CD visited on spring tour with his uncle Josiah Wedgwood.

Duck, George Francis, c.1848-1875. Carpenter, grocer, and landlord of the George & Dragon public house in Down from at least 1855. 1866 A trustee of Down Friendly Club. Stecher 1961, p. 245. 1875 May 10 CD wrote to J.B. Innes and mentioned the death of D and the need to appoint a new Trustee. CCD23:296. Parslow was one of his executors.

Duff, Mr. Of 21st Regiment. Given lift on *Beagle* to England from Tasmania by gun-room officers. *Beagle diary*, p. 410.

Duff, Sir Mountstuart Elphinstone Grant, 1829-1906. Politician. 1857-81 Liberal MP for Elgin Burghs. 1866-72 Rector University of Aberdeen; succeeded by Huxley. 1868-74 Under-Secretary of State for India. 1871 Jan. D visited Down House with Lubbock, Huxley and R. Lowe, from High Elms. 1880-81 Under-Secretary of State for the Colonies. 1881-86 Governor of Madras. 1881 FRS. 1887 GCSI. 1889-92 President Royal Geographical Society. D dined with the Darwins in 1887 Oct. 20 in company were the Sedgwicks. On Mar. 8 1880, Mr Sedley Tayor and D visited. Four more recorded visits from 1888-95. Recollections of CD in *Notes from a diary, 1873-1881*, vol. 2, pp. 283; 300, transcribed in *Darwin Online*.

Dimbola Lodge, Freshwater, Isle of Wight. A house owned by Julia Margaret Cameron. 1868 CD and Family stayed there in summer. CD was photographed.

Duncan, Andrew, 1758-1832. Prof. Materia Medica Edinburgh. 1798 FRS. 1826 CD to Caroline Darwin, "is so very learned that his wisdom has left no room for his senses. His lectures begin at eight in the morning". CCD1:25. 1847 CD to Hooker, "a whole, cold, breakfastless hour on the properties of rhubarb!" CCD4:36.

Duncan, Ethel, 1856-1927. Daughter of Andrew D. Married George J. Romanes.

Duncan, Peter Martin, 1824-91. Physician, Geologist and Invertebrate palaeontologist and writer of popular natural history. 1864-70 Secretary Geological Society. 1866-72 *A monograph of the British fossil corals*, Palæontographical Society. 1868 FRS. 1870 Prof. Geology, King's College, London. 1868 Sept. 16 visited CD. In company were Mr Flower and Mr Gwyn Jeffreys. *ED's diary* 1868. 1866? CD to D, will send coral specimens from Keeling Islands. CCD16:394. 1876-77 President Geological Society. 1876 CD to D, CD will return an overlooked coral and manuscript by William Lonsdale. CCD24:281.

Dundee and Angus, 1827 CD visited on a spring tour.

Dun horses. 1861 Dun horses. *The Field*, 17 (27 April): p. 358 (*Shorter publications*, F1960). 1861 On dun horses, and on the effect of crossing differently coloured breeds. *Ibid.*, 17 (25 May): p. 451 (*Shorter publications*, F1962). 1861 Dun horses. *Ibid.*,

17 (15 Jun.): p. 521 (*Shorter publications*, F1963). Dun horses often exhibit a stripe along the back and faint zebra striping on the upper legs.

Dunker, Wilhelm Bernhard Rudolph Hadrian, 1809-85. German palaeontologist especially of Mollusca, Conchologist. Lecturer Technical High School Cassel, later Prof. Geology Marburg. CD sent *Fossil cirripedes* to. *Lychnos*, 1948-49: pp. 206-10. 1851 D sent fossil and recent cirripedes to CD. 1854 CD sent *Living cirripedes* to D.

Duns, Rev. John, Free Church minister and dabbler in natural history. 1860 Reviewed *Origin* "its publication is a mistake", *North British Rev.* 32 (May), p. 486. Transcribed in *Darwin Online*. CD to Asa Gray "very severe review on me". CCD8:223.

du Puy, Martha (Maud) Haskins, 1861 Jul. 26-1947 Feb. 6. Daughter of Charles Meredith and Ellen Reynolds of Philadelphia. Niece of Lady Jebb (Caroline Reynolds, maternal aunt). Raverat, *Period piece*, 1952. 1884 Married George H Darwin.

Dutch. First editions in: 1891 *Journal* (F176). 1860 *Origin* (F2056.1). 1889-90 *Variation* (F910). 1871-72 *Descent* (F1053). 1873 *Expression* (F1182). 1913 Essay on instinct (F1440a). 1913 Humble bees (F1583h). 2000 *Autobiography* (F1906).

Dyck, Dr William Thomson van, 1857-1939. American physician, ornithologist and lecturer on materia medica and hygiene at the Syrian Protestant College, Beirut 1880-82. 1882 D to CD on sexual selection in Syrian street dogs. 1882 Apr. 2 CD to P.L. Sclater submitting it, with covering note, for *Proc. of the Zool. Soc. of London* 1882 Apr. 18 Read, no. 25: pp. 367-70 (*Shorter publications*, F1803); last intentional publication in CD's lifetime; he died on Apr. 19.

Dyster, Frederick Daniel, 1810-93, CD gave a copy of Maull & Polyblank 1855 to D which is now in the Old Library of Christ's College, Cambridge. Notes written on the back of the frame read: "This photograph of Darwin was presented by him to my Uncle, F.D. Dyster, of Tenby. I am informed by Francis Darwin, his son, that the photograph was probably taken in the year 1854, but he had never seen it. F. H. H. Guillemard." Below this, in a different and very faint hand is written in pencil: "NB F.D. Dyster was the microscopist after whom the genus Dysteria was named." A third pencil note in yet a different hand records: "Exhibited at the Darwin Commemoration in Christ's College – June 1909." Below the photograph was a signature of CD, which was the endorsement on verso of a cheque "to self". D was a friend of T.H. Huxley and attended ED's aunt, Jessie Sismondi, when she died on 3 Mar. 1853. No correspondence with CD known. See John van Wyhe, A Darwin manuscript at Christ's College, *Christ's College Mag.*, (2007) no. 232, pp. 66-8.

Earle, Augustus, 1793-1838. Wandering artist of some distinction; visited all five continents. Draughtsman at start of 2d voyage of *Beagle*. CD "Earle's eccentric character". FitzRoy "I engaged an artist...at £200 per year". His original *Beagle* sketches are not known to survive although other material remains. Many of the *Beagle* sketches were engraved and published in *Narrative*. Keynes, pp. 1-2, open licentiousness from CD's letters. *Narrative of a nine months' residence in New Zealand in 1827*, 1966, ed. Robert McCormick; originally published in 1832. 1832 Aug. Left *Beagle* in

Montevideo owing to continuous ill-health. His illness was rheumatism. Replaced by Conrad Martens in Dec. 1833. 1832 Watercolour caricature sketch "Quarter Deck of a Man of War on Diskivery or interesting Scenes on an Interesting Voyage" sold at auction in 2015, attributed to E. Purportedly depicts CD in high hat together with FitzRoy on board *Beagle* with nine other figures, possibly off the coast of Bahia Blanca. The officers uniforms do not match those of the Royal Navy.

Earth, Age of. 1868-77 Some of CD's correspondence on in CCD16-25. CD made and revised estimates in chapter IX of *Origin*, most famously in his discussion of the denudation of the Weald. Immense age of the Earth predates CD.

Earthworms. 1881 *The formation of vegetable mould through the action of worms, with observations on their habits,* London (F1357). CD's notes are in DAR63-65. Drafts in DAR24-25. See Keith, *Nature,* 149: p. 716, 1942. 1881-82 3d-6th thousands (F1359-1362) contain small corrections by CD. 1882 7th thousand (F1364) contains small changes by Francis Darwin. 1888 11th thousand (F1373) contains altered footnote on p. 7 by Francis Darwin. 1969 Facsimile of 2d thousand (F1400). First foreign editions: 1896 Armenian (F1402). 1954 Chinese (F1402a). 2019 Dutch (F1875). 1882 French (F1403), 1882 German (F1404), 1882 Italian (F1407), 1938 Japanese (F1407a). n.d Portuguese (F2062.18). 1882 Russian (F1408), USA (F1363).

Eastbourne, Sussex. 1853 Jul. 14 - Aug. 4 CD had family holiday there. 1860 Sept. 22 - Nov. 10 Family holiday there. *ED's diary.*

Eastbury Park, A house near Gunville, Dorset. 1800 Bought by Thomas Josiah Wedgwood. 1803 Sold to Josiah [II] Wedgwood. Until 1805 Thomas Josiah Wedgwood continued to live there with his sisters until his death. CD who was "poorly for a fortnight past" went to E from Feb. 21-23 1842.

Eaton, Bertha, 1820/21-? Sister of Dorothea Hannah E. 1848 Married Edmund Edward Allen brother of George Baugh A.

Eaton, Dorothea Hannah, c.1825-68. Sister of Bertha E. 1846 Married George Baugh Allen, brother of Edmund Edward A.

Eddowes' Newspaper, Shrewsbury. 1880 Mrs Haliburton [Sarah Owen of Woodhouse] had reminded CD of his saying as a boy that if *EN* ever alluded to him as "our deserving fellow townsman" he would be amply gratified. LL3:335. Opening sentence of a leading article in *The Times* of 1880 is given: "Of all our living men of science none have laboured longer and to more splendid purpose than Mr. Darwin."

Eddy, Miss. 1879 Kitchen maid at Down House. F. Darwin, *Story of a childhood,* 1920, pp. 12 and 27.

Edgeworth, Maria, 1767-1849. Author. Daughter of Richard Lovell Edgeworth. Friend of Erasmus Darwin [I] and Josiah Wedgwood [I]. 1840 E described the character of Erasmus D. ED2:56.

Edgeworth, Michael Pakenham, 1812-81. Son of Richard Lovell E. "A fool, Mr. Edgeworth, you know, is a man who never tried an experiment in his life", Erasmus

Darwin [I], Woodall 1884, p. 4. Half-brother of Maria E. Botanist and Indian Civil Servant. 1861 CD met at Linnean Society.

Edible fungus from Tierra del Fuego 1845 In M.J. Berkeley, On an edible fungus from Tierra del Fuego, *Trans. Lin. Soc. of London*, 19: pp. 37-43, summary in *Proc.*, 1: pp. 97-8 (*Shorter publications*, F1671). With extracts from CD's notes.

Edinburgh, Midlothian. 1838 Jun. Apart from his time at the University, CD visited on his way to Glen Roy. George Howard Darwin went in 1869 Jul. 9.

Edinburgh University. 1825 Oct.-1827 Apr. CD was at as a medical student, but did not qualify. See 1888 Feb. 16, *St. James's Gazette*, 1888 May 22 *Edinburgh Weekly Dispatch*, 1935 Charles Darwin as a student in Edinburgh, 1825-1827. *Proc. of the Roy. Soc. of Edinburgh* 55: 97-113; Shepperson, The intellectual background of Charles Darwin's student years at Edinburgh, Darwinism and the study of society, ed. M. Banton (1961), pp. 17-35; Jenkins, Henry H. Cheek and transformism: new light on Charles Darwin's Edinburgh background. *Notes and records of the Roy. Soc.* (2015) 69, 155-171. 1825 Oct. CD stayed briefly at Star Hotel, Princes Street, moving to 11 Lothian Street, lodgings run by Mrs Mackay. CD's *Edinburgh diary* is in DAR129 and his *Edinburgh notebook* is in DAR118, both transcribed by Kees Rookmaaker and published in *Darwin Online*. Other Edinburgh notes are in DAR5.

Edmonston, John. Retired servant of Dr Duncan. "A negro lived in Edinburgh, who had travelled with Waterton, and gained his livelyhood by stuffing birds...he gave me lessons for payment". *Autobiography*, p. 51. CD paid him a guinea for an hour every day during two months. Brent 1981, p. 45. Waterton, *Wanderings in South America*, pp. 153-4, 1825 identifies him as John, a slave of Charles Edmonston of Demerara (British Guyana). On coming to Scotland and being freed he took the surname of Edmonston or Edmonstone. E lived at 37 Lothian Street, CD lived at no. 11. CD mentioned him in *Descent* 1:232: "I was incessantly struck, whilst living with the Fuegians on board the 'Beagle,' with the many little traits of character, shewing how similar their minds were to ours; and so it was with a full-blooded negro with whom I happened once to be intimate." The best source is Freeman, Darwin's negro bird-stuffer. *Notes and Records Roy. Soc.*, 33: pp. 83-6, 1978.

Edmonston, Laurence, 1795-1879. Physician and naturalist. Studied at University of Edinburgh M.D. 1830. Correspondent with CD from Unst, Shetland 1856-57. Made many additions to list of British birds. Father of Thomas. CD discussed evolution with before *Origin*.

Edmonston, Thomas, 1825-46. Botanist and naturalist. Eldest son of Laurence E. Visited Galapagos Islands in H.M.S. *Herald*. Accidentally shot and killed in Ecuador. 1845 Prof. of Botany, Anderson's University (now part of University of Strathclyde).

Edward. A manservant at 12 Upper Gower Street, London. 1839 Feb. 3 "Edward is such a perfect Adonis in his best livery, that he is quite a sight". ED2:33. 1839 May E occurs in CD's accounts. 1840 E had left and Parslow had arrived.

Edward VII, 1841-1910. 1866 Apr. 27 CD presented to when Albert Edward, Prince of Wales, at Royal Society Soirée. CD said nothing because he could not hear what the Prince said, "A nice good-natured youth". 1881 Int. Congr. Med. CD sat opposite. "The Prince (of Wales) spoke only a few civil words to me". Brent 1981, p. 499. 1901-10 King of the United Kingdom of Great Britain etc.

Edwards, Mr. A resident at Down. Stecher 1961, p. 207. Not identified by CCD.

Ernest Edwards.   20 Baker St. W.
Carte de visite by Edwards, 1866.

Edwards, Ernest, 1837-1903. Photographer at 20 Baker Street, London. Photographed CD in 1865-66. 1869 E patented the heliotype process of photographic reproduction. The CDV here, from the collection of van Wyhe, is signed on the reverse by E with his address.

Edwards, Henry, 1827-91. English born American botanist, entomologist and amateur actor. 1873 Correspondent with CD. 1876 CD to E, thanking for photograph and glad E approved of Weismann's essay on dimorphism in butterflies. CCD24:66.

Edwards, William Henry, 1822-1909. American entomologist, lawyer, and businessman. 1847 *A voyage up the river Amazon* which inspired Wallace and Bates to go to the Amazon as collectors. Wallace, 1905, vol. 1, p. 264.

Edwards, Jane, maid? 1850 Nov. 7 ED paid E £7.

Edwards, Joaquin, Major-domo of copper mines at Panuncillo, Chile. For this and another anecdote see *Beagle diary*, p. 330, *Coquimbo notebook, Beagle notebooks*.

Edwards, José Maria, Anglo-Chilean, son of the owner of silver mines at Arqueros, Chile, with whom CD travelled. In *Coquimbo notebook, Beagle notebooks*, 2009.

Egan, James, fl.1850. Hungarian agriculturist of Budapest. Also known as James Egasy. 1858 CD corresponded with on colour of horses. CCD7.

Egerton Street, Westminster, London. 1882-1900 no. 12 home of Leonard Darwin. 1902-14 no. 10 or no. 14 home of William Erasmus Darwin after death of wife in 1902. Gwendolen Mary Darwin lived with him when she was a student at Slade School of fine Art.

Egerton, Sir Philip de Malpas Grey, Bart, 1806-81. Palaeontologist. 10th Bart. 1831 FRS. 1855 Oct. CD met at Shrewsbury, "He asked me why on earth I instigated you [W.D. Fox] to rob his Poultry yard". CCD5:482. E was a neighbour of Fox at the time. 1873 Wollaston Medal, Geological Society.

Ehrenberg, Christian Gottfried, 1795-1876. German Protozoologist. Prof. Zoology Berlin. 1845 Examined fine dust from *Beagle* in Atlantic for Protozoa. *Journal* 2d edn, p. 5. An account of the fine dust which often falls on vessels in the Atlantic Ocean. *Quart. Jrnl. of the Geol. Soc. of London*, 1846; 2, pp. 26-30. (*Shorter publications*,

F1672) and in four 1844-5 papers in the *Bericht über die zur Bekanntmachung geeingneten Verhandlungen der Königl. Preuss. Akademie der Wissenschaften zu Berlin*. 1837 Foreign member Royal Society. 1838 *Die Infusionstierchen*, Leipzig. 1839 Wollaston Medal Geological Society. 1854 *Mikrogeologie*, Leipzig.

Electric fish. 1881 CD to Romanes, parable about evolution of electric organs to get rid of parasites. Romanes 1896, p. 106.

Elephant. 1836 May 5 CD rode one in Mauritius from Capt. Lloyd's country house half way to Port Louis, "The circumstance which surprised me most was its quite noiseless step". *Journal* 2d edn, p. 486. The only elephant on the island.

Elephants. 1869 [Letter] Origin of species [on the reproductive potential of elephants], *Athenaeum*, no. 2174: p. 861 (*Shorter publications*, F1746). 1869 [Letter with same title], *ibid.*, no. 2177: p. 82 (*Shorter publications*, F1747). 1859 *Origin*, p. 64, estimate of population growth potential of elephants. Revised in later editions.

Elephant Tree. Large beech on sandwalk, also known as "Bismarck" and "Rhinoceros". 1969 Cut down almost dead, but main trunk preserved and a sign added.

Elevation and Subsidence in the Pacific and Indian Oceans. 1837 On certain areas of elevation and subsidence in the Pacific and Indian Oceans, as deduced from the study of coral formations. *Proc. of the Geol. Soc.*, 2: pp. 552-4 (*Shorter publications*, F1647). See also 1881 letter to Semper in F1952.

Elevation on the Coast of Chile. 1837 Observations of proofs of recent elevations on the coast of Chile... *Proc. of the Geol. Soc.* (*Shorter publications*, F1645).

Élie de Beaumont, Jean Baptiste Armand Louis Léonce, 1798-1874. French geologist. Influentially anti-*Origin*. "Damned himself to everlasting fame" by coining the term "la science moussante" for evolutionism. LL2:185. 1870 CD to Quatrefages, É calls CD's science "frothy", his own bubbles first of craters of elevation and second of direction of mountain chains according to age have "burst and vanished into thin air" everywhere but France. CCD18. 1835 Foreign Member Royal Society. 1853- Perpetual Secretary of the Académie des Sciences.

Elliot, Sir Walter, 1803-87. Indian Civil Servant and archaeologist. 1821-60 Indian Civil Service, Madras. 1855 CD met at BAAS. 1856 CD writes to E in India asking for information on variation. CCD6. 1857 E sent poultry skins from Madras. 1866 KCSI. 1873 Title of CD's 1827 contribution to Plinian Society first printed by E in *Trans. Bot. Soc. Edinburgh*, 11: pp. 1-42, 17 footnote; also in *Nature*, 9: 38. 1877 FRS.

Elliott & Fry, From 1863 commercial photographers of London. 1963 incorporated in Bassano & Vandyck Studios, now Bassano's Ltd. 1869-81 Photographed CD on five occasions. See Darwin, Iconography.

Elston, near Newark, Nottinghamshire. Elston Hall, seat of William Darwin [VI] whose wife, Anne Waring, had inherited from her mother, and seat of senior branch of Darwin family from 1680 until the middle of 20th century. Many early Darwins are buried in All Saints' churchyard. Erasmus Darwin [I] was born there.

Elwin, Whitwell, 1816-1900. Clergyman, critic and editor of *Quart. Rev.* 1849-1900 Rector of Booton, Norfolk. 1859 Read manuscript of *Origin* for John Murray. E felt that the theory would be unconvincing without the full evidence CD had accumulated, the style was not as good as *Journal* and that instead an introductory volume on pigeons with an outline of the species theory would be preferable. John Murray Archive: (https://digital.nls.uk/jma/gallery/title.cfm?id=27&mode=transcript).

Emily Jane, 1865?-79. Domestic servant at Down House.

Emma Darwin's Diary, 1824-96. 60 small pocket diaries. Diaries for 1836-38 and 1846-47 are missing. However, there are 3,200 scanned images from all 60 diaries, now fully transcribed by Christine Chua, in *Darwin Online*. The diaries contain many unique names and dates and intimate and often moving accounts of the suffering of CD and family details. (http://darwin-online.org.uk/EmmaDiaries.html).

Engleheart, Stephen Paul, 1831?-85. 1859 Fellow Royal College of Surgeons, London. 1861-70 Village surgeon/physician at Down, known to Darwin family as "Spengle". Drowned in Old Calabar, Nigeria, Africa, trying to visit a patient.

Entomological Society of London. 1833 Established. F.W. Hope and T.C. Eyton made CD founding member. CD was an Original and Life Member, not Fellow, used after 1884 when Charter granted. 1837 [Specimens of genus *Carabus*]. *Trans. of the Entomol. Soc. of London.* (F1643a) 1838 CD Council Member and Vice-President and presided at several meetings. K.G.V. Smith, *Antenna* 6: pp. 200-1, 1982. Smith says CD exhibited five species of *Carabus* from southern tip of South America, *Proc.*, II: p. xli. 1856 CD to Mrs Lyell, "You might trust Mr. Waterhouse implicitly, which I fear as rumour goes, is more than can be said for all entomologists". CCD6:32. 1867 "No body of men were at first so much opposed to my views as the members of the London Entomological Society". CCD15:271.

d'Entrecasteaux, Antoine Raymond Joseph de Bruni, 1739-93. French navigator who explored the Australian coast in 1792. *Sydney notebook, Beagle notebooks*, 2009.

Eozoon. A supposed fossil protozoan described by J.W. Dawson, *Quart. Jrnl. of the Geol. Soc.*, 21, 1865. Later shown not to be of organic but of mineral origin, but still described as a foraminiferan by Adam Sedgwick, *Student's textbook of zoology*, 1: p. 15, 1898. 1866 CD included information on E in 4th edn of *Origin*, p. 371. 1882 CD to D. Mackintosh, "As far as external form is concerned, *Eozoon* shows how difficult it is to distinguish between organised and inorganised bodies". ML2:171. See John van Wyhe, 'Almighty God! what a wonderful discovery!' *Endeavour* (2010).

Epping Field Club. Later Essex Field Club. 1880 Jan. CD to William Cole, declining joining, but sending a guinea "in aid of your preliminary expenses". 1880 Feb. CD to same, accepting Honorary Membership. *Essex Nat.*, 21: p. 14, 1927.

Erasmus Darwin. Ernst Krause's paper first appeared in German in *Kosmos*, 3, Feb. 1879. He revised the text for the English translation. This book started the one-sided row with Samuel Butler. Butler's copy with manuscript notes is in the British Library. CD discussed in *Autobiography*, pp. 134ff. Further discussion, especially of the accusations of Butler, can be found in LL3: Chapter VI and Appendix 2 by Barlow in *Autobiography*, p. 167. 1879 Krause, *Erasmus Darwin...with a preliminary notice by Charles Darwin* (F1319), CD's notice, pp. 1-127, is longer than Krause's essay on Erasmus Darwin's scientific work, even after 16% of the manuscript was cut by Henrietta. Some of the manuscript of introduction in DAR212.1-6, DAR210.11.45-47, proofs in DAR213.14. A copy with CD's annotations for revisions is in a private collection in Virginia, USA (examined by van Wyhe in 2009). CD's list of presentation copies is in CCD27 Appendix IV and a list of reviews is in appendix V. 1887 *The life of Erasmus Darwin* (F1321), sheets of the first edn with new preface and errata. Desmond King-Hele published the complete original text, before Henrietta's omissions, in an important edn: *Charles Darwin's The life of Erasmus Darwin*, CUP, 2002, (F1853). First foreign editions of CD's notice: 1880 USA (F1320). 1880 German (F1323). 1959 Russian (F1324). 1971 Facsimile (F1322).

Erichsen, **Sir John Eric**, **Bart** (1895), 1818-96. Surgeon. Prof. Surgery University College London. 1876 FRS. 1885 E was member of Vivisection Commission.

Errera, **Léo Abram**, 1858-1905. Belgian botanist. Prof. of Botany at University of Brussels. 1877 CD to and from on heterostyly especially in *Primula elatior*. CCD25. 1878 CD to and from, E visited Down House, but CD was away. CCD26. 1879 CD thanks for offprint on heterostyly with Gustave Gevaert. CCD27. 1879 E to CD sending photograph CD had asked for; E asks for one in return. CCD27.

Erratic Boulders of South America. 1841 On the distribution of erratic boulders and on the contemporaneous unstratified deposits of South America, *Proc. of the Geol. Soc.*, 3: pp. 425-30 (*Shorter publications*, F1657); *Trans. of the Geol. Soc.*, pp. 415-31, plate XL. (F1661). CD thought erratics were carried by icebergs from glaciers.

Erratic Boulders, Transportal of. 1848 On the transportal of erratic boulders from a lower to a higher level, *Quart. Jrnl. of the Geol. Soc. (Proc.)*, 4: pp. 315-23 (*Shorter publications*, F1677). They "had been transported by the ordinary currents of the sea".

Erskine, **Frances**, 1825-70. 1854 Married Sir (later Baron) Thomas Henry Farrer as 1st wife. Emma Cecilia "Ida" Farrer, their eldest daughter, married Horace Darwin.

Erskine, **William**, 1773-1852. Scottish orientalist and historian. 1809 Married Maitland Mackintosh, eldest daughter of James Mackintosh. Issue included Frances E.

Esperanto. First edition in: 2009 *Origin* (F2160).

Essay on Instinct. 1883 In G.J. Romanes, *Mental evolution in animals*, with essay on instinct by CD, pp. 355-84, index pp. 405-11 (F1434). 1975 Complete transcript of original manuscript in *Natural selection*, pp. 466-527, (F1440). First foreign editions: 1958 Chinese (F1310a). 1913 Dutch (F1440a). 1884 French (F1441), 1885 German (F1443). 1913 Hungarian (F2380). 1907 Italian (F1447). 1967 Romanian (F1448). 1894 Russian (F1449). 1983? Spanish (F2445). 1884 USA (F1435).

Essays of 1842 and 1844, see Sketches of 1842 and 1844.

Estonian. First editions in: 1949 *Journal*, (F179). 2012 *Origin* (F2514).

Etruria Hall, Staffordshire. Home of Josiah Wedgwood [I] and several of his Wedgwood relatives. 1768-71 Built by Joseph Pickford. 1769 and before Jun. 13, foundations laid before this when the section of the works for making ornamental ware was opened. Josiah Wedgwood [I] cast six black basalt vases to commemorate, later inscribed "Artes Etruriae renascuntur". 1774 Richard Wedgwood moved there, died 1780. 1795 Jan. 2. Josiah Wedgwood [II] inherited estate and works, estate then 380 acres. 1795 Spring Josiah Wedgwood [II] moved to Stoke d'Abernon, Surrey, his mother and Kitty Wedgwood remaining at the Hall. 1799 Josiah Wedgwood [II] bought Gunville, Dorset. 1804 Hall leased to Byerley. 1810 On death of Byerley, mother and Kitty lived in Hall while Parkfields was altered. 1814 Josiah Wedgwood [II] returned to Hall. 1830 Sept. Henry Allen and Jessie Wedgwood, just married, moved in. 1832 Francis Wedgwood and Frances Mosley moved in on marriage, and Jessie Wedgwood Allen and her children John and Jane moved out. 1840 Hall and most of land sold, but works failed to reach reserve. 1930-40 The old factory worked until 1930s, until a new one was opened at Barlaston, six miles away, in 1940. 1978 Hall, then an office building, remained, but nothing of works except the Round House. Later it has formed part of a hotel.

Evans, Edward, ?-1846. Robert Waring Darwin's butler at The Mount, Shrewsbury. "A faithfull friend and servant". Brent 1981, p. 18. His wife was also in Robert Waring Darwin's employ. Pattison, *The Darwins of Shrewsbury*, 2009.

Evans, Margaret (Evvy), later **Mrs Sales**, 1832-1909. Born in Shrewsbury. Cook and housekeeper for nearly forty years. Woodall 1884, p. 39. The "Mrs" was honorary, but later made a suitable marriage in the village. Listed as housekeeper in 1871 census. 1871-82 cook at Down House after being a nurse to Leonard Darwin. 1881 Wages were £36 per annum. Attended services by J.W.C. Fegan and was "brought into the light". W.Y., Fullerton, *A tribute*, 1931, p. 31. 1882 E had a ticket for Jerusalem Chamber at CD's funeral but was asked to join family in the Choir. B. Darwin, *Green memories*, 1928, p. 15. Browne, *Power of place*, 2002.

Evans, Mary Ann, 1819-80. Novelist under pseudonym "George Eliot". 1854-78 E was common-law wife of George Henry Lewes. 1873 E with Lewes visited Down House for lunch. 1874 E attended séance with CD and ED at Richard Buckley Litchfield's house. 1879 Oct. CD and ED called after Lewes's death. 1880 Married J.W. Cross, a New York Banker.

Everest, Robert, 1799-1860. Anglican priest. 1850-60 At Calcutta. 1856? CD to E on degeneration of British dogs in India in *Variation* vol. 1. Letter from CD to E in Sotheby's sale, Honeyman, part III, May 1979.

Evolution. 1830-3 Geological uses of the term, Lyell, *Principles of Geology*. 1871 First use of the word in CD's sense is in *Descent*. First use in *Origin* is in 6th edn, 1872, p. 201 twice and p. 424 three times. Evolved is the last word in all editions of *Origin*.

Ewald, Julius Wilhelm, 1811-91. German geologist and palaeontologist. Studied in Bonn and Berlin. 1878 E seconded CD's election to Berlin Academy as Corresponding Member. He was mentioned in C. Lyell's letter to CD in 1865 Jan. 16, in F.C. Donders to CD in 1872 Apr. 1 and in Hartogh to CD in [1873].

Ewart, Rev. Henry C., Anglican priest. 1882 Article by in *Sunday Mag.* on sermons preached about CD, after Westminster Abbey memorial service of 1 May. Atkins 1976, p. 50.

Ewart, James Cossar, 1851-1933. Zoologist. 1881 CD to Romanes, unable to give E a testimonial [for Edinburgh chair] because he has already given one for Edwin Ray Lankester. Thinks that E is fit for the appointment, remembers interesting interview with E on bacteria at University College London laboratory. Carroll 1976, 604. 1882-1927 Regius Prof. Zoology Edinburgh. 1893 FRS. 1899 Penycuik experiments, on telegony in horses, a theory in which CD once believed.

"Experiment Book", (or Experimental notebook), DAR157a.1-84, contains notes on experiments, such as on natural means of species dispersal, between 13 Nov. 1855 and 20 May 1868, but mostly in 1863. In *Darwin Online.*

Photograph of and by Oscar Rejlander simulating the emotion of disgust, from *Expression.*

Expression. 1872 *The expression of the emotions in man and animals.* See also *Queries about expression.* With seven heliotype plates; one of the first uses in scientific publications. Oscar Rejlander, a Swedish photographer, posed for some of the pictures, including 'surprised man'. Others taken by Duchenne de Boulogne. First issue has last signatures 2B2 2C3, only 2B1 and 2C1 signed (F1141); second issue 2B1 2C4, with 2B1, 2C1 and 2C2 signed (F1142). Second issue has misprint "htat" in th first line of page 208. Partial draft written on the backs of MS. folios of *Descent* in DAR17.1A, notes for the book are in DAR59.2. Additional photographs for *Expression*, drawings, notes and references of chapters. 2, 3, 6, 7 and introduction in DAR53, DAR189 and DAR195-6. Proofs of the book are in DAR213. First issue has plates numbered in Arabic in most copies; second issue, sometimes Arabic, sometimes Roman.

1969 Facsimile (F1175). 1890 2d edn (F1146), edited by Francis Darwin. First foreign editions: 1935 Chinese (F2254). 1964 Czech (F1181). 1873 Dutch (F1182). 2009 Finnish (F2431). 1874 French (F1184). 1872 German (F1187). 1963 Hungarian (F1199). 1878 Italian (F1200). 1921 Japanese (F1202b). 1998 Korean (F2357). 1873 Polish (F1203). 1975 Portuguese (F2038). 1967 Romanian (F1205). 1872 Russian (F1206). c.1902. Spanish (F1214). n.d Turkish (F2409). 1872 USA (F1143) List of presentation copies is in CCD20 Appendix V. See G. Radick, Darwin's puzzling expression. *Comptes Rendus Biologies* 333 (2010): 181-7.

Extinct Mammalia in the Neighbourhood of the Plata. 1837 A sketch of the deposits containing extinct Mammalia in the neighbourhood of the Plata, *Proc. of the Geol. Soc.*, 2: pp. 542-4 (*Shorter publications*, F1646). Toxodon and Macrauchenia.

Eyre, Edward John, 1815-1901. English explorer. 1862-65 (Acting) Governor of Jamaica. 1865 E put down an insurrection of local inhabitants. 1866 CD supported

John Stuart Mill's attempt to prosecute E for murder. CD subscribed to Jamaica Fund. CCD14. Although charged twice, the cases against E never proceeded.

Eyton, Thomas (Tom) Campbell, 1809-80. Ornithologist and specialist in skeletal variation. Born at Eyton Hall, inherited in 1855. CD discussed evolution with before *Origin*. Anti-*Origin*. At Cambridge with CD and shot with him on vacations. 1835 Married Elizabeth Frances Slaney. 1839 E examined birds from *Beagle* voyage for *Zoology of the Beagle*, and wrote appendix to Part III of *Birds*, pp. 147-56, not published until 1841. Corresponded with CD on skeletal variation. 1868 E sent CD his *Osteologia avium*, Wellington, 1867. 1868 CD remembered hunting and fishing with him in their youth. CCD16. An entry in Emma's diary 1834 Mar. 18 as "Eyton".

"F", after 1868 = Father, used by ED writing to her children when grown up.

Fabre, Jean-Henri Casimir, 1823-1915. French entomologist. 1880 CD to F, praising *Souvenirs entomologiques*, 1879-1907. 1880-81 CD letters to F. ML1:385-6. 1913. My relations with Darwin. *Fortnightly Rev.* 94: 661-75, in *Darwin Online*.

Fairfax, Mary, 1780-1872. Scottish science writer and polymath. Neeley, *Mary Somerville*, 2001. 1812 Married as 2^d husband to William Somerville. 1834 On the Connexion of the Physical Sciences became one of the best-selling science books of the age. 1869 *On molecular and microscopic science*. For this CD lent her woodblocks from *Orchids*. 1870 F agreed to Henry Walter Bates revising her *Physical geography*, 6th edn, but not to "infuse any Darwinism in it".

Falconer, Hugh, 1808-65. Physician and palaeontologist. 1830 Went to India as Assistant Surgeon. 1832-42 Superintendent of Botanic Garden, Saharanpur. 1845 FRS. CD discussed evolution with before *Origin*. 1847-55 Superintendent of Botanic Garden, Calcutta. Often at Down House on his return from India. 1859 Lived at Torquay for his health. 1859 CD sent 1st edn *Origin*. 1861 offered CD a live *Proteus anguinus*, a blind amphibian. 1863-64 VP, RS. 1864 F proposed CD successfully for Copley Medal. *Palæontological memoirs*, C. Murchison ed., 1868, 2 vols.

Falkland Islands, British territory in South Atlantic. 1833 Mar. 1 "The present inhabitants consist of one Englishman (Dixon) who has resided here for many years and now has charge of the British Flag, 20 spaniards and three women, two of whom are negresses". *Beagle diary*, pp. 138-9. Keynes, *Beagle record*, 1979, p. 118, writing of Port Louis. See Dickson. 1834 Mar. 10 *Beagle* at Berkeley Sound in East Falkland, Port Louis at head of Sound. CD explored and returned Mar. 19. 1834 Port Darwin, at head of Choiseul Sound, named after CD. He crossed the isthmus near to it on Mar. 17. *Zoological diary*, pp. 136-148, 203-217. *Geological diary*: DAR32.123-150, DAR33.166-222 and DAR34.65-73. P. Armstrong, *Darwin's desolate islands: A naturalist in the Falklands, 1833 and 1834*, 1992, all transcribed in *Darwin Online*.

Falkland Islands geology. 1846 On the geology of the Falkland Islands, *Quart. Jrnl. of the Geol. Soc. (Proc.)*, 2: pp. 267-79 (*Shorter publications*, F1674).

Fancourt, Mr. 1849 mentioned in *ED's diary* while the family was at Malvern Mar. 9-Jun. 29. Starting Apr. 12, F either came to them or only William to him every 2-3 days with last entry on Jun. 6. Possibly William Joseph Fancourt (c.1809-52) the curate at Malvern Priory.

Farr, William, 1807-83. Epidemiologist and medical statistician. Sent CD large volumes of printed reports on national disease statistics. Browne, *Power of place*, 2002, p. 327. 1870 CD studied F's reports and kept notes on child mortality. Keynes, *Annie's box*, 2002, p. 285. 1870 Jul. 10 Visited Down House and 1873 May 1 ED recorded "called at Mrs Hamilton & Mrs Farr".

Farrar, Frederic William, 1831-1903. Anglican priest. Rector of St. Margaret's Westminster. 1858 *Eric or little by little*. 1865 CD to F, congratulating him on *Origin of language*. 1866 FRS. CD and F exchanged several letters between 1865-71. 1882 Canon of Westminster when CD was buried there and preached the sermon. 1883 Archdeacon and Rural Dean of Westminster. CD 1883, [Extract of a letter on classical education]. In Farrar, General aims of the teacher. A lecture in Cambridge teachers' training Syndicate course. March 3, 1883, *American Jrnl. of Education* 32: 129-154 (139-40). (F2055) Recollection of CD in F2101 and *Men I have known*.

Farrer, Cecilia Frances, 1823-1910. 1882 F was on "Personal Friends invited" list for CD's funeral. 1885 Married Sir Stafford Henry Northcote, 8th Bart, 1st Earl of Iddesleigh.

Farrer, Emma Cecilia (Ida), 1854 Nov. 7-1946 Jul. 5. Only daughter of Sir Thomas H. F. 1880 Married Horace Darwin. CD liked to hear her singing Sullivan's "Will he come". LL1:124. Henrietta wrote in ED2:238, "This marriage added a great happiness to my mother's life; whether she was well or ill, she rejoiced in seeing Ida enter the room, and was always soothed and exhilarated by her presence. Ida indeed became another daughter and entered into all her joys and sorrows. "I did so enjoy our lovers, only one of them was rather sick. I had some nice sits with them. She is a sweet little wife," my mother writes Jan. 31st, 1880." ED wrote to her often.

Farrer, Mary, 1825-1905. Sister of Sir Thomas H. F. 1848 Married Hobhouse A. B (1819-1904). 1878 CD to Romanes, "Lady Hobhouse is trustworthy". CCD26.

Farrer, Sir Thomas Henry, 1st Bart (1883), 1819-99. Statistician, barrister and civil servant. Abinger Hall, Dorking, Surrey. 1873 Married 1 Frances Erskine. 3 sons, 1 daughter; Emma Cecilia ("Ida"). Married 2 Katherine Euphemia Wedgwood s.p. 1893 1st Baron. Visited Down House often. *ED's diary* 1854-95.

Fawcett, Henry, 1833-84. Political economist and statesman. Biography: Leslie Stephen, 1885. 1858 Blinded through shooting accident. 1860 F was present at Oxford BAAS meeting. 1861 F was at Manchester BAAS meeting and spoke in defence of *Origin*. 1861 F to CD, on John Stuart Mill's opinion of the logic of *Origin*. 1862 "On the method of Mr. Darwin in his treatise on the origin of species", *Report BAAS*, for 1861, p. 141. 1863-84 Prof. Political Economy Cambridge. 1865-84 MP. 1880-84 Postmaster General. 1882 FRS. 1883-84 Rector, University of Glasgow.

Fayrer, Sir Joseph, Bart, 1824-1907. Physician and toxicologist in India. 1874 F provided cobra venom for *Insectivorous plants*. CCD22. 1875 CD letters to F on cobra poison [venom] and Drosera In *Proc. of the Roy. Soc. of London* 23 (1874-5): 261-79, pp. 273-4 (F2282). 1877 FRS. 1896 1st Bart.

Fegan, James William Condell, 1852-1925. An Evangelical worker amongst poor boys in South London. In 1872 he founded Fegan's Boys Homes in Deptford. W.Y. Fullerton, *J. W. C. Fegan: A tribute, the life of Mr. Fegan*, 1931. In 1880 F brought the boys to Down for a holiday. They sang hymns for the Darwins and, much moved, CD gave each of the boys sixpence. The visit was recorded in *ED's diary*. F wrote twice to CD to ask for the use of the Reading Room which Darwin had established for the village. Transcribed in *Darwin Online*.

Fellowes, Catherine Henrietta, 1821-1900. Daughter of Isaac F, 4th Earl of Portsmouth and Lady Catherine Fortescue. 1843 Married Seymour Phillips Allen.

Felméri, Lajos, 1840-94. Hungarian writer, journalist, university Prof. 1873 thanked CD for a copy of *Expression*. F sent CD his journal on Scotland, *Úti levelek Skóciából*, 1870. 1873 Translated into Hungarian part of *Expression* (F1849). CCD21.

Fernando de Noronha or Fernando Noronha, Archipelago of oceanic islands in the Atlantic, belonging to Brazil. 1832 Feb. 20 *Beagle* anchored off and CD ashore. *Geological diary*: (2.1832) DAR32.39-40, transcribed in *Darwin Online*. Also covered in the *Cape de Verds notebook*, Chancellor & van Wyhe, *Beagle notebooks*, 2009. There is a fine sketch of it in DAR29.3.47a.

Ferrier, Sir David, 1843-1928. Scottish physician. Prof. Neuropathology King's College Hospital, London. 1863-68 Medical student, Aberdeen. Scientific assistant to Alexander Bain. 1876 FRS. 1881 F was prosecuted under Vivisection Act. CD had met at C.L. Brunton's house and offered to subscribe towards the expenses of the case. ML2:437, *British Medical Jrnl.*, 2: p. 917, 1881. (*Shorter publications*, F1799).

Fertilisation of Flowers. 1883 Hermann Müller, preface, p. vii-x, by CD 1882, Feb. 6 (F1432). Translation of *Die Befruchtung der Blumen durch Insekten*, 1873 by D'Arcy W. Thompson. 1950 Russian edition, CD's preface only (F1433). In *Darwin Online*.
Fertilisation of Orchids, see *Orchids*.
Fertilisation of Plants. 1877 *Gardeners' Chron.*, 7: p. 246 (*Shorter publications*, F1780). Reply to a note by George Henslow on p. 203-4 of same volume.
Fertilisation of winter-flowering plants. 1869 *Nature*, 1: p. 85 (*Shorter publications*, F1748a). A reply to A.W. Bennett, botanist and publisher; his letter on the fertilisation of winter-flowering plants appeared in *Nature* 1 (1869): 58. See CCD17:477.

Ffinden, Rev. George Sketchley, 1836?-1911. Anglican priest. King's College, London; Assoc. (1st class) 1859; ordained deacon 1860; ordained priest 1861. 2 Nov. 1871-1911 vicar of Down, he was generally disliked. Olive Willis described him as "that wicked man". Atkins 1976, p. 48. 1872 Jan. 6 & Sept. 16 dined at the Darwins. *ED's diary*. Between 1872-95, there were many visits either way. George Darwin in 1874 Oct. 18 writing to CD, referred to the "Ffinden affair is all moonshine". 1896 Mrs Ffinden is mentioned with nursemaid and baby in an elegant goat-carriage. *Emma Darwin*, 1904, 2:465. Interview in Anon. 1909. A visit to Darwin's village: reminiscences of some of his humble friends. *Evening News* (12 Feb.): 4: "'[CD] lived in such retirement that though I was here for eleven years before he died I hardly saw him half a dozen times, and then chiefly about affairs of business.' I gathered from Mr. Ffinden that Darwin's beliefs and theories find less than favour in his eyes. 'I confess that, perhaps, I am a bit sour over Darwin and his works. You see, I'm a Churchman first and foremost. He never came to church, and it was such a bad business for the parish, a bad example. He was, however, most amiable and benevolent and courteous, and very liberal. I remember his giving me a subscription for the church and the house—restoration or building. 'Of course,' he told me, 'I don't believe in this at all.'" Memorial in Downe church.

Fick, Heinrich, 1822-95. Swiss jurist and Prof. in Zürich. CD letter favouring competition among trades unions and the working classes, in Fick, ed. *Heinrich Fick*. 2 vols., 2:314-5 (F1943).

Field-lane refuges, 1858 [Contribution to the Field-lane refuges.] *The Times* (29 Dec.): p. 10 (*Shorter publications*, F1935). The Field Lane Refuge, at the north side of Vine Street in Clerkenwell, London, was a charitable housing and Christian mission centre for the poor and unemployed. CD had made a donation.

Fife, George, 1807-57. Physician of Newcastle-on-Tyne. Naturalist friend of CD at Edinburgh. Member of the Plinian Natural History Society. 1827 Was absent from the meeting in Mar. where CD communicated to the Society two discoveries which he had made. See Minutes of the Plinian Society recording CD's first scientific papers. Edinburgh University Library. Transcribed in *Darwin Online*.

Figueroa, Augustín, Military Administrator of the Spanish Settlement of Port Soledad, Falkland Islands, 1784-86.

Findon, Mr, Findon's son, then a schoolboy at boarding school, of Down. Atkins 1976, p. 104. ?= Ffinden.

Fine Dust which falls on Vessels in the Atlantic. 1846 An account of the fine dust which often falls on vessels in the Atlantic ocean, *Quart. Jrnl. of the Geol. Soc. (Proc.)*, 2: pp. 26-30 (*Shorter publications*, F1672). The dust was analysed for protozoan content by Ehrenberg. CD's notes are in DAR188.

Finnish. First editions in: 1913-17 *Origin* (F2023). 1924 *Journal* (F179a). 1987 *Autobiography* (F2028). 2009 *Expression* (F2431).

Fish. CD ed. 1842. *Fish Part 4 of The zoology of the voyage of HMS Beagle* by Leonard Jenyns. Edited and superintended by Charles Darwin. London: Smith Elder and Co. On CD's work on see Daniel Pauly, *Darwin's fishes*, 2004. More than 60 of CD's specimens survive at the Zoology Museum, Cambridge.

Fish, David Taylor, 1824-1901. Professional gardener and horticultural journalist. 1868 CD called F an "excellent gardener" in *Variation*. 1869 F objected to CD's views on earthworms, *Gardeners' Chron.* 17 Apr., 1869, p. 418, prompting CD's response in *Gardeners' Chron.* 15 May, 1869, p. 530 (*Shorter publications*, F1745). 1882 Apr. 29 F wrote fine obituary tribute to CD, *Garden*, transcribed in *Darwin Online*.

Allan, *Darwin and his flowers*, 1977, pp. 295-6, Britten & Boulger, *A biographical index of British and Irish botanists*, 1893, 2ᵈ edn, 1931.

Fiske, John, 1842-1901. Born as Edmund Fiske Green. American philosopher, historian and theoretical biologist. 1871 CD to F, with invitation to visit Down House when he came to England. 1874 F sent CD *Outlines of cosmic philosophy*, 2 vols. 1879 *Darwinism and other essays*, London and New York. 1883 *Excursions of an evolutionist*, Boston. 1884 *The destiny of man viewed in the light of his origin*, Boston. 1885 *The idea of God as affected by modern knowledge*. In 1880 May 21, ED recorded a visit by Mr and Mrs Fiske. F and CD would exchange many letters between 1871-80. Recollections of and letters from CD in *Darwin Online* (F2108).

Fitch, Robert, 1802-95. Geologist and pharmacist. FSA, FGS. CD described specimens F sent. CD wrote to F 1849-51. No record of F replying. CCD4 & 5.

Fitton, William Henry, 1780-1861. Irish physician and geologist. 1815 FRS. 1838 Aug. CD dined with at Athenaeum. 1839 Apr. 1 ED recorded visitors as Mr and Mrs Henslow, Dr Fitton, Mr Brown and Lyells. Darwins invited to dine at F's on Apr. 4. The F's called in 1840 Dec. 22. 1852 Wollaston Medal, Geological Society.

FitzRoy, Robert, RN, 1805-65. Naval officer, surveyor and meteorologist. Son of Lord Charles FitzRoy, 2ᵈ son of 3ᵈ Duke of Grafton, bastard descendant of Charles II. His mother was the daughter of the first Marquess of Londonderry and the half-sister of Viscount Castlereagh. F's name is variously spelt (Fitz-Roy, Fitz Roy, Fitz-Roy, Fitzroy); throughout the spelling FitzRoy is used, except where cited literally. F himself used Fitz-Roy. 1818 Entered Royal Navy College, Dartmouth, aged 12. 1828 Nov. 13-1830 Nov. F was in command of *Beagle* after suicide of Commander Pringle Stokes in Aug. 1828 until end of 1ˢᵗ voyage. Captain Philip Parker King of HMS *Adventure* had the overall command of the surveying voyage. 1828 Commander. 1831 Jun.-1836 Nov. In command of *Beagle* for whole of 2ᵈ voyage. CD's opinion of his character "Fitz-Roy's character was a very singular one, with many noble features: he was devoted to his duty, generous to a fault, bold, determined, indomitably energetic, and an ardent friend to all under his sway": "Fitz-Roy's temper was a most unfortunate one". "...whether much hot coffee had been served out this morning"—junior officers' query about F's temper. *Autobiography*, pp. 72-3. 1832 F's opinions of CD's character are given in his letters to Beaufort, 1832 Apr. 28 "Darwin is a regular trump". Aug. 15 "He has a mixture of necessary qualities which make him feel at home, and happy, and makes everyone his friend". Francis Darwin, *Nature*, 88: pp. 547-8, 1912; Barlow, *Cornhill*, 72: pp. 493-510, 1932, which also contains the best account of CD's relationship with F. 1835 Dec. Captain. 1836 1 Married Mary H. O'Brien (d. 1852), 1 son, 3 daughters. 1854 2 Married Maria Isabella Smyth, 1 daughter. 1838 Sketch by Philip Gidley King in MLNSW, ML ZC767, p. 69, c. 1835, ink and wash, reproduced in Keynes, *Beagle record*, 1979, p. 16 et al. "Dr Wallich gave me a collection of photographs which he had made and I was struck with the resemblance of one to FitzRoy; on looking at the name I found it

Ch.E. Sobieski Stuart, Count d'Albanie, illegitimate descendant of the same monarch". *Autobiography.* 1839 F edited *Narrative of the surveying voyages of...Adventure and Beagle*, and also wrote an earlier brief account of the 2^d voyage, with a little on the 1st, *Jrnl. of the Royal Geographical Soc.*, 6: pp. 311-43, 1836. 1849-50 Commanded *Arrogant*, Steam Frigate. 1857 Rear Admiral. 1863 Vice Admiral. 1843-45 Governor-General New Zealand. 1851 FRS, was proposed by CD. 1854-65 Chief Statician [Statist], Meteorological Department, Board of Trade. 1857 F visited Down House, the last time he and CD met. 1859 F wrote to CD re *Origin*, the letter does not survive. However, see CCD7:413-4. 1859 Dec. CD to Lyell, enclosing a letter printed in *The Times* signed "Senex", "It is I am sure by Fitz-Roy...It is a pity he did not add his theory of the extinction of Mastodon, etc., from the door of the Ark being made too small". "What a mixture of conceit and folly, and the greatest newspaper in the world inserts it". CCD7:413. F was indeed the author. 1859 CD sent 1st edn *Origin*. 1860 F was at Oxford meeting of BAAS to give famous paper on British storms. Strongly anti-*Origin*, he is said to have walked out of the lecture room holding a bible over his head and exclaiming "The Book! The Book!" The story comes from George Griffith and A.G. Vernon Harcourt, who were both present. Poulton, *Darwin and the Origin*, 1909, p. 66. 1863 *The weather book: A manual of practical meteorology.* 1865 Apr. 30 F committed suicide at his home at Norwood, Surrey.

Flemish. First edition in: 1958 *Origin* (F654).

Fletcher, Joseph. A schoolmaster and copyist for CD. F was sent the 1844 essay on species (DAR7) theory to make a fair copy (DAR113) for which he was paid £2. See *Foundations.* Other payments to F are mentioned in CD's Account Book at Down House. First name not previously identified.

Fletcher, Alexander Pearson, 1825-1907 General manager of the Northern Assurance Company (1865-81). F wrote Mar. 14 1874 and asked CD for a reference CD wrote in [1874 after Mar. 14] testifying to the trustworthiness of Charles Person with disclaimer. CCD22:155.

Fletcher, Harriet, 1799-1842. Of Isle of Wight. Daughter of Sir Richard F. 1834 Married William Darwin Fox.

Fletcher, Sir Richard, Bart, R.E., 1768-1813. 1796 Married Elizabeth Mudge (1739-99). Father of Harriet F. Killed at San Sebastián in Peninsular War.

Floating Ice, 1849 [Letter on Floating Ice]. In Murchison, On the distribution of the superficial detritus of the Alps, as compared with that of Northern Europe, *Proc. of the Geol. Soc. of London*, 4 (Part 1): pp. 65-9 (*Shorter publications*, F1816).

Flourens, Marie Jean Pierre, 1794-1867. French physiologist. Influential anti-*Origin*. Appointed Perpetual Secretary Academy of Sciences since 1833 at the request of the dying Cuvier. 1864 *Examen du livre de M. Darwin sur l'origine des espèces*, Paris. CD had annotated copy, CUL (see *Marginalia*, p. 234). CD to Wallace in 1864 Jun. 15 "A great gun Flourens has written a little dull book against me". CCD12:248-9.

Flower, Sir William Henry, 1831-99. Mammalogist, comparative anatomist and surgeon. 1864 CD to F, about supposed sixth toe in frogs. 1864 FRS. 1873 "On palaeontological evidence of gradual modification of animal forms", *Jrnl. of the Roy. Institution*, pp. 94-104. 1877 F to CD, he had examined a pig's foot with an extra digit sent to CD by Otto Zacharias. CCD25. 1882 F was on "Personal Friends invited" list for CD's funeral. 1884-98 Director BMNH in succession to Richard Owen. 1892 KCB. 1868 Sept. 16 visited CD. In company were Dr Duncan, Peter Martin and Mr Gwyn Jeffreys. *ED's diary* 1878 Jun. 24 and 1880 Feb. 14, F visited.

Flowers and Insects. 1877 Fritz Müller on flowers and insects, *Nature*, 17: p. 78, introducing a letter from Müller, *ibid.*, 17: pp. 78-9. Müller's letter discussed the pollination of flowers by insects and certain butterflies with scent-producing scales thought to attract females. (*Shorter publications*, F1781).

Flowers and their Unbidden Guests. 1878 Kerner [von Marilaun, Freiherr], Anton, *Flowers and their unbidden guests*, London, prefatory letter to the editor (W. Ogle) by CD, pp. v-vi (F1318); translation by W. Ogle of *Die Schützmittel der Blüthen gegen unberufene Gäste*, Innsbruck, 1876.

Flowers. 1861 Cause of variation of flowers, *Jrnl. of Horticulture*, 1, p. 211 (*Shorter publications*, F1715). 1866 "Partial change in sex in unisexual flowers", *Gardeners' Chron.*, no. 6: p. 127 (*Shorter publications*, F1735).

Flustra. CDs paper on the ova of the *Flustra* read to Plinian Society in 1827 while in Edinburgh, but anticipated by Sir John Dalyell in 1814. In Barrett 1977, F1583a.

"Flycatcher". CD's nickname used by all ranks on *Beagle*.

Flyer. A cob used for pulling the coach at Down House. c.1882 "An old white mare living in honourable retirement in the field". B. Darwin, *Green memories*, 1928, p. 13.

Foliation. 1846-56 CD's views on geological cleavage and foliation. ML2:199-210.

Forbes, David, 1828-76. Mineralogist and geologist. Geological correspondent of CD in general. CD discussed evolution with before *Origin*. Brother of Edward F. 1857-60 Visited Chile, Peru and Bolivia as metallurgist. 1858 FRS. 1860 CD to Hooker, CD praises F's work on geology of Chile. CCD8.

Forbes, Edward, 1815-54. Naturalist. Brother of David F. Often at Down House. Founder and moving spirit of the Red Lion Club, a convivial group of the BAAS. Biography: Wilson and Geikie, 1861. 1842-44 Curator museum Geological Society, London. 1843-54 Prof. Botany King's College London. 1845 FRS. 1846 published Descriptions of Secondary Fossil Shells from South America in appendix to F273. 1848 Married Emily Ashworth. 3 children; 1 Edward 1849 died at birth, 2 Edward 1850-?, 3 Jane Teare 1852-? 1849 Nov. 20 CD to Lyell, "after more doubt & misgiving, than I almost ever felt, I voted to recommend Forbes for Royal Medal, & that view was carried, Sedgwick taking the lead." CCD4:281. The Royal Medal was not awarded to F. 1853 President Geological Society. 1854 Prof. Natural History Edinburgh. 1854 CD praised his introductory lecture at Edinburgh. 1854 Died prematurely of kidney failure. 1855 CD to Hooker, "poor Forbes", "of course I shall wish to subscribe as soon as possible to any memorial". CCD5:250. CD contributed £5 towards his memorial fund. CCD23. 1856 CD to Hooker, "but I must confess (I hardly know why) I have got to mistrust poor dear Forbes". CCD5:335. 1868 CD to Hooker, "false theories, such as the Quinarian Theory & that of Polarity by poor Forbes". CCD16:645.

Forbes, James David, 1809-68. Physicist and glaciologist. CD sent specimens of volcanic rocks described in *Volcanic islands* to F. There are only three extant letters to F. from CD. CCD3. 1832 FRS. 1833-60 Prof. Natural Philosophy Edinburgh. 1843 Gold Medal Royal Society. 1859- Principal United College of St. Andrews.

Forchhammer, Johan Georg, 1794-1865. Danish geologist and chemist. 1849? CD met F at the BAAS meeting in Birmingham. Corresponded with CD 1849-50. CD called him "my very kind friend". CCD4.

Ford. 1817 CD remembered (in 1838) that at Mr Case's school, aged 8½, he walked with F "across some fields to a farmhouse on the Church Stretton road". ML1:4.

Ford, George Henry, 1809-76. South African artist. Cut most of the blocks for the second volume of *Descent*. 1837-75 Artist British Museum. 1870 CD to Albert Günther, praising quality of blocks for *Descent*.

Ford, Richard Sutton, 1785-1850? Farmer of Newstead, CD's acquaintance from Swynnerton. 1839 answered five of CD's questions about the breeding of animals.

Fordyce, John, 1879 May 7 CD to F on being a theist or agnostic. CCD27. 1883 Author of *Aspects of scepticism*, London, p. 190 (F1861) which prints the letter. See Darwin, Charles Robert, Religion for the letter.

Forel, Auguste Henri, 1848-1931. Swiss entomologist, especially of ants. 1874 CD to F, having read *Les fourmis de la Suisse*, Zurich. CCD22:480. 1921-23 *Le monde social des fourmis*. 5 vols.

Forest, The. Nickname for Woodhouse, Felton, Shropshire, home of the Owens.

Forms of Flowers. 1877 *The different forms of flowers on plants of the same species*, London (F1277). 1878 2d edn (F1279). 1884 2d edn, 3d thousand (F1281), with new preface by Francis Darwin. First foreign editions: 1996

Two forms of Cowslip
(*Primula veris*)
from *Forms of flowers*.

Chinese (F2277). 1878 French (F1296). 1877 German (F1297). 1877 Italian (F2473). 1949 Japanese (F1300). n.d. Portuguese (2062.16). 1965 Romanian (F1301). 1948 Russian (F1302). 2009 Spanish (F2077). 1877 USA (F1278). List of presentation copies is in CCD25 Appendix IV. Proofs in DAR213.

Forster, Johann Georg Adam, 1754-97, and **Forster, Johann Reinhold**, 1729-98 Father and son. Naturalists and travellers. 1772 J.R.F FRS. 1772-75 Both were naturalists on Commander James Cook's 2d voyage after the withdrawal of Joseph Banks. 1777 J.G.A. F FRS. 1857 CD's cognomen as Member of Academia Caesarea Leopoldino-Carolina Naturae Curiosorum was "Forster".

Forster, Miss Laura Mary, 1839-1924. A lifelong friend of Henrietta D. 1873 Jun. 14-17 stayed at Down. 1874 Jul. 11 came to Down. 1879 Jun.-Jul. 1. F lent her house, West Hackhurst, Abinger Hammer, Surrey, to CD for a holiday. 1881 Mar. 4-15 F stayed at Down to recuperate from an illness. 1882 Mar. 3-Apr. 4 stayed at Down. 1882 May 15, Horace and Ida stayed at West Hackhurst and again in Aug. 23 the same year. 1882 Sept. 2. 1892 Jul. F stayed at Down House. E.M. Forster (nephew) *Marianne Thornton*, 1956. 1883-5 recollections of CD in DAR112.A31-A37, DAR112.A38-A47, DAR112.A48-A49, all transcribed in *Darwin Online*.

Forster, William Edward, 1818-86. Liberal Party statesman, industrialist and philanthropist. 1861-86 Liberal MP. 1875 FRS. 1875-78 Rector University of Aberdeen in succession to Huxley. 1875 Member of Vivisection Commission. LL3:201.

Forsyth, Charles Codrington, c.1810-73. Born South Arlington, Devon. Went on 2ᵈ voyage of *Beagle*. Served in South Africa and East Indies. 1826-70 Royal Navy 1832 Apr. Joined *Beagle* as Volunteer 1ˢᵗ Class. 1834 Junior Midshipman. 1836 Oct. Midshipman on *Beagle* on return from 2ᵈ voyage. 1850 Member Royal Geographical Society. His 1833-36 log of the voyage is now in the Museo Naval de la Nación, Tigre, Buenos Aires, Argentina and is published in *Darwin Online*. http://darwin-online.org.uk/converted/pdf/1833_Forsyth_Tigre.pdf Together with an important introduction by Simon Keynes: http://darwin-online.org.uk/EditorialIntroductions/Keynes_ForsythLog.html

Fossil Mammalia. CD ed. 1844. *Fossil Mammalia Part 1 of The zoology of the voyage of HMS Beagle*, by Richard Owen. Edited and superintended by Charles Darwin. London: Smith Elder and Co. See Adrian Lister, *Darwin's fossils*, 2018.

Fossil Remains. 1838 Copy of a Memorial presented to the Chancellor of the Exchequer [Thomas Spring Rice], recommending the Purchase of Fossil Remains for the British Museum. *Parliamentary Papers, Accounts and Papers 1837-1838*, (27 Jul.): 1. Memorial signed by CD and 17 others (*Shorter publications*, F1944).

Foster, Sir Michael, 1836-1907. Physiologist. Edited Huxley's *Scientific memoirs* with E.R. Lankester. 1869-72 Fullerian Prof. of Physiology at the Royal Institution succeeding Huxley. 1869-83 Prof. Practical Physiology University College London. 1871 CD asks F for curare for experiments for *Insectivorous plants*, and inviting to Down House: F sent it. CCD19. 1872 FRS. 1872 again invited Down House. CCD20. 1875 F saw and agreed to R.B. Litchfield's draft sketch for a vivisection bill. LL3:204. ED recorded his visit on 1881 Oct. 22. 1882 F was on "Personal Friends invited" list for CD's funeral. 1883-1903 Prof. Physiology Cambridge. 1899 KCB. 1900-06 MP for University of London in succession to Sir John Lubbock.

Foundations of The Origin of species, see Sketches of 1842 and 1844.

Fowl. 1861 Influence of the form of the brain on the character of fowls. *The Field*, 17 (4 May): p. 383 (*Shorter publications*, F1961). CD asked if the bulbous skull of the Polish fowl affected their behaviour. An anonymous response, titled 'Polish fowls', appeared in the 11 May issue of *The Field*, p. 404: "I have half a dozen…and they certainly are tame or stupid. You may tread upon them—they don't seem to see well, and they seldom find the roosting-place, but crouch or perch anywhere."

Fox, Alice Augusta Laurentia Lane (later Pitt-Rivers, under "Rivers" in Burke 1888) c.1862-1947. Daughter of Augustus H. L. F. 1884 Married Sir John Lubbock.

Fox, Frances Jane, 1807-94. W.D. Fox's sister. 1852 Married Rev. J. Hughes.

Fox, Samuel, Married Anne Darwin [III]. Father of William Darwin F.

Fox Darwin, Samuel William, 1841-1918. Clergyman. Eldest son of William Darwin F. Vicar of St. Paul's, Maidstone, Kent. 1876 Married Euphemia Rebecca Bonar 1836-1917) of Edinburgh.

Fox, Victor William Darwin, 1883-1915. Son of Robert Gerard Fox and Emily Mary Fox. Grandson of Rev. William Darwin F. Army Officer, Lieutenant, 1st Battalion Irish Guards, Western Front. Killed in action in France.

Fox, Rev. William Darwin, 1805-80. Son of Samuel F and Anne Darwin [III]. CD's second cousin. At Christ's College, Cambridge, with CD and kept up correspondence. 1827 [CD] "Became acquainted with Fox and Way and so commenced Entomology". *Journal.* 1828 CD stayed at family home, Osmaston near Derby. 1834 Married 1 Harriet Fletcher. 6 children, the first stillborn. 1838-73 Vicar of Delamere, Cheshire. 1846 Married 2 Ellen Sophia Woodd. (1820-77), had 12 children. CD discussed evolution with before *Origin*. 1859 CD sent 1st edn *Origin*. 1868 CD thanks F for a return on sheep and cattle. CCD16. 1870 Nov. CD to F, will send copy of *Descent* when published. "It is very delightful to me to hear that you, my very old friend, like my other books". CCD18. First record of his visiting the Darwins was 1843 May. 27. Again on 1849 Nov. 2, 1854 May 6 and 1861 Jul. 18. *ED's diary.* See Larkum, *A natural calling*, 2009. There is a fine pencil sketch of F as a young man in the Wellcome Collection.

Franke, Hermann, 1847-1908. German music director and violinist. 1880 Married Constance Rose Wedgwood, s.p. 1881 May. 20 had lunch with the Darwins together with Hans Richter. Visited ED in 1882-85. *ED's diary.*

Frankland, Sir Edward, 1825-99. Organic chemist. Did experiments for *Insectivorous plants*. 1853 FRS. 1863- Prof. Chemistry Royal Institution, London. 1897 KCB.

Fraser, Elizabeth Frances (Bee), 1846 Dec. 15-1898 Jan. 13. Sister of Sir Thomas Fraser, a brother officer of Leonard Darwin. 1882 Jul. 11 Married Leonard Darwin, s.p. "She was elegant, fastidious, rustling in silk". B. Darwin, *Green memories*, 1928.

Fraser, Louis, 1810-66. British zoologist and collector. Curator of the Museum of the Zoological Society of London. Worked with Richard Owen. F wrote to CD in 1845 Jul. 23 informing CD about characteristics of certain species of Galapagos birds and Jul. 24? wrote to discuss the colour of *Zenaida* (doves) from Galapagos.

Fraser, Thomas Richard, 1841-1920. Brother of Elizabeth Frances. 1882 Mar. 25 visited Down with Frances (Bee) and again on Apr. 1 the same year.

Freeman Bibliographical Database. 2006- a heavily revised and greatly augmented edition of R.B. Freeman's *The works of Charles Darwin. An annotated bibliographical* *handlist*, 1977. Freeman's book is the acknowledged authority for bibliographic information about CD's publications. It records details of CD publications, in every known language, that were known to Freeman up to 1975. Building on this, the Freeman Bibliographical Database is the most comprehensive bibliography of CD's writings ever published. Freeman 1977 recorded publications up to the number F1805, in 33 languages. As of November 2020, the Database records items up to the number F2564 in 64 languages with almost a thousand more records than Freeman 1977. It also records more than 2,000 additional items not written by CD such as reviews, obituaries and the entire *Beagle* library. http://darwin-online.org.uk/Freeman_intro.html

Freke, Henry, 1818?-88. Irish. Eccentric theoretical evolutionist. 1860 CD to Henslow, "Dr Freke has sent me his paper,—which is far beyond my scope". CCD8:444. 1860 *Observations upon Mr. Darwin's recently published work*—'On the Origin of species by means of natural selection'. Dublin: printed for the author. 1861 *Origin of species by means of organic affinity*. 1861 CD to Hooker, on F's book: "read a page here & there just to see the maximum of ill-written unintelligible rubbish, which he tells the reader to observe has been arrived at by 'induction', whereas all my results are arrived at only by 'analogy'". CCD9:8-9. F claimed priority over CD, saying he had published in 1851 an account of animals and plants evolving from a single filament of organised matter. Browne, *Power of place*, 2002, p. 109.

French. First editions in: 1860 *Journal* (extracts only) (F180). 1862 *Origin* (F655). 1868 *Variation* (F912). 1870 *Orchids* (F818). 1872 *Descent* (F1058). 1874 *Expression* (F1184). 1875 *Journal* (complete) (F181). 1877 *Climbing plants* (F858). 1877 *Insectivorous plants* (F1237). 1877 *Cross and self fertilisation* (F1265). 1877 Biographical sketch of an infant (F1311). 1878 *Coral reefs* (F309). 1878 *Different forms of flowers* (F1296). 1882 *Power of movement* (F1342). 1882 *Earthworms* (F1403). 1884 Essay on instinct (F1441). 1888 *Life and letters* (F1514). 1902 *Volcanic islands* (F310).

French, Erasmus Darwin, 1822-1906. Born in New York of Anglo-Scottish origin. Served as army physician, later worked for mining prospectors in Darwin, now a ghost town in Inyo County, California, USA. Source of forenames unknown. Darwin Bench, Darwin Canyon, Darwin Creek, Darwin Falls and Darwin Falls Wilderness named after him. *Settlers of the American West*, 2015, p. 61.

Freshwater, Isle of Wight. 1865 Summer, ED wrote to her sister Elizabeth that William had taken the brothers there. The same holiday W was said to have filled "Lenny's trousers with stones" and pelting them while they were bathing. 1868 Jul. 17-Aug. 20 CD had family holiday there. They were visited by Thomas Appleton and Tennyson and called on H.W. Longfellow. Photographed by Julia Margaret Cameron there. George and Maud went in 1890, Fr(ancis) and B(essy) went in 1892.

Frog. 1879 Fritz Müller on a frog having eggs on its back—on the abortion of hairs on the legs of certain caddis-flies, etc., *Nature*, 19: pp. 462-3; introducing a letter from Müller, *ibid.*, 19: pp. 463-4, 3 woodcuts. (*Shorter publications*, F1784).

Fry, Sir Edward, 1827-1918. Judge and zoologist. 1859 Married Mariabella Hodgkin. In his 1921 memoir, among many other references to CD, "but I did not, like so many good people, feel distressed at the influence of the Darwinian theory upon my religious beliefs." p. 65. 1881 wrote to CD about worms.

Fry, Mariabella, née Hodgkin, 1833-1930. ED made a number of diary entries starting 1854 Jan. 6 "Mrs Fry's party", the Frys visited in 1855 Aug. 2 with "Reids", then in 1860 Jan. 4 the company included Mr Stephens, Mr Thompson and Reids. The Darwins called on the Frys in Mar. 19 1860, Feb. 21 1862, George and Henrietta went to their ball in 1865 Jan. 25. In 1867 Jun. 14 was an entry "Fry". Last two records were in 1872 Jan. 10.

Fuegians at Woollya from FitzRoy's *Narrative*. Drawn by FitzRoy and engraved by T. Landseer.

Fuegians. The English name for the Indian tribes of Tierra del Fuego, most of those encountered by the *Beagle* were the Yahgan. The best account of those encountered by the crew of the *Beagle* as well as the history of Fuegia Basket, Jemmy Button, Boat Memory and York Minster, the Fuegians brought to England on the first voyage, three returned on the second, is in FitzRoy's *Narrative*, vol. 2, especially pp. 1-16, 119-227. Their later history and that of Fuegians in general is in E.L. Bridges, *Uttermost part of the earth*, 1947. More up-to-date accounts are in Richard Keynes, *Fossils, finches and Fuegians*, 2002 and A. Chapman, *European encounters with the Yamana people of Cape Horn, before and after Darwin*, 2010. 1882. [Quotation from a letter on civilizing the Fuegians]. *Leisure Hour*. F2022. See South American Missionary Society. Some artefacts collected by CD and not mentioned in any of his writings, including body paint pigments, are in the British Museum.

Fuller, Harry. Captain's steward on the *Beagle*. In the *Buenos Ayres notebook*, Chancellor & van Wyhe, *Beagle notebooks*, 2009.

Fullerton, William Young, 1857-1932. Evangelist and writer. Wrote a biography of J.W.C. Fegan. See *J. W. C. Fegan, A tribute*, 1931. pp. 29-31, contains passages about Fegan visiting Down House and the activities of the day and of F writing to CD for permission to use the Reading Room at Down. Transcribed in *Darwin Online*.

Fumariaceae. 1874 Fertilisation of the Fumariaceae, *Nature*, 9: p. 460 (*Shorter publications*, F1769). Fumariaceae is a family of flowering plants.

Furneval, Elizabeth, 1837-? Staffordshire, visiting lady's maid in 1861 census for Down House.

Gadney, William, a British merchant resident in Cape Town who turned his Sea Point home into a lodging-house after a business failure. In the *Despoblado notebook*, Chancellor & van Wyhe, *Beagle notebooks*, 2009.

Gaertner (Gärtner), Carl Friedrich von, 1772-1850. German botanist and physician and pioneer in study of hybrids. 1826 Member Deutsche Akademie der Naturforscher Leopoldina 1849 *Versuche und Beobachtungen über die Bastarderzeugung im Pflanzenreich*, which CD thought highly of. Frequently referred to in *Variation*. Reprinted in A. Weinstein, "How unknown was Mendel's paper?" *Jrnl. of the Hist. of Bio.* 10: pp. 341-64 especially pp. 347-8, 1977. 1863 CD's paper "Vindication of Gärtner—effect of crossing peas", *Cottage Gardener* 29: p. 93, not in Barrett. In *Shorter publications*, pp. 327-9, F1727a; vindication is from aspersions of Donald Beaton.

Galapagos Islands, (Insulae de los Galopegos) Named after the giant tortoises living there. The importance of the fauna of these islands, especially of the mocking birds to the development of CD's early thoughts on evolution has often been stressed (indeed usually over stressed). In notes CD prepared in Nov. 1880 for Wallace's book *Island life*, CD wrote: "Galapagos. — I regret that you have not discussed plants. Perhaps I overvalue these Islds for how they did interest me & how they have influenced my life, as one main element of my attending to origin of species." NHM WP6.4.1, transcribed with introduction in *Darwin Online*. There is a large body of literature on the finches, e.g. 1940, D. Lack, *Nature*, 146, pp. 324-7; 1959, J.R. Slevin, *Occasional Papers California Acad. Sci.*, no. 25, 1959, pp. 1-150; 1963 *Occasional Papers California Acad. Sci.*, no. 44: pp. 1-154; 1967, *National Geographic Mag.*, 131: pp. 540-85. Frank J. Sulloway, 1984, *Biol. Jrnl. of the Lin. Soc.* 21: pp. 20-59; whole part is on the islands 21: pp. 1-258, and as a book, but not about CD. See Grant and Estes, *Darwin in Galapagos: footsteps to a new world*, 2009. The Charles Darwin Foundation set up in Brussels 1959, to provide science to conserve the environment and diversity of the archipelago, Julian Huxley first President. H.Q. is at first buildings put up in 1960s at Puerto Ayoro on Santa Cruz. The National Park about 8,000 sq. kilometres. Airstrip was on Baltra (South Seymour), a legacy from World War II. 1892 The whole archipelago was renamed by Ecuador, Archipélago de Colón, but the old names are still sometimes used in English writings on the islands. The equivalent names are: Abingdon = Pinta; Albemarle = Isabela; Barrington = Santa Fé; Bindloe = Marchena; Charles = Floreana, Santa Maria; Chatham = San Cristóbal; Culpepper

= Darwin; Duncan = Pinzón; Hood = Española; Indefatigable = Santa Cruz; James = Santiago, San Salvador; Jervis = Rabida; Narborough = Fernandina; South Seymour = Baltra; Tower = Genovesa; Wenman = Wolf.

CD was ashore in 1835 as follows, from a *Beagle* log: Sept. 16 arrived, landed St. Stephen's Bay, Chatham, for 1 hour. Sept. 17 Chatham, St. Stephen's Bay, ashore after dinner. Sept. 18 Chatham, long walk after dinner, top of hill. Sept. 21-22 Northeast Chatham, CD and Covington slept ashore. Sept. 23 Charles, Post Office Bay, ashore collecting. Sept. 23 Charles, Black Beach, ashore collecting. Sept. 29 Albemarle, ashore. Sept. 30 Albemarle, Tagus Cove, ashore. Oct. 1 Albemarle, Tagus Cove, ashore. Oct. 8 James, Sulivan Bay, CD, Covington, Bynoe etc. camped ashore. Oct. 17 James, Sulivan Bay, party picked up again. Oct. 20 *Beagle* sailed for Tahiti. 1835 There was a penal settlement on Charles. CD's *Zoological diary*, pp. 290-301. *Geological diary*: DAR37.716-795A, transcribed by K. Thalia Grant & Gregory B. Estes with an important introduction. CD's *Galapagos notebook* (now missing from Down House) in Chancellor & van Wyhe, *Beagle notebooks*, 2009. All in *Darwin Online*.

Galapagos Islands Finches. 1837 John Gould, *Proc. of the Zool. Soc. of London*, Pt. 5, no. 53, 1837. Members of the sub-family Geospizinae of the buntings, Emberizidae. 1837 CD, "Remarks on the habits of the genera *Geospiza, Camarhynchus, Cactornis* and *Certhidea*", *Proc. of the Zool. Soc. of London*, Pt. 5, no. 49 (*Shorter publications*, F1644). 1839 *Journal*, pp. 378-80. 1935 the term 'Darwin's finches' coined by Percy Lowe. 1946 D. Lack, *Occasional Papers California Acad. Sci.*, no. 21. 1947 D. Lack, *Darwin's finches*. The finches did not inspire CD to think of evolution nor did CD recognize that the differently shaped beaks of the finches were adapted to different diets. This was the discovery of David Lack in 1947.

Galapagos Islands Research Station. 1964 Built by Charles Darwin Foundation at Academy Bay, Indefatigable Island, dedicated 1964.

Galapagos lichen, note on. By CD in J.D. Hooker, An enumeration of the plants of the Galapagos Archipelago; with descriptions of those which are new. *Trans. of the Lin. Soc. of London* 20: 164: "[*Usnea plicata*] Hab. James Island, "hanging from the boughs of the trees in the upper damp region, where it forms a considerable proportion of the food of the large tortoise." Transcribed in *Darwin Online* (F2017).

Galician. First edition in: 2003 *Origin* (F2152).

Galton, Emma Sophia, 1811-1904. Daughter of Samuel Tertius G and Violetta G, née Darwin. Sister of Francis G. 1879 wrote to CD with information and corrections for *Erasmus Darwin* (F1319). Transcribed in *Darwin Online*.

Galton, Sir Francis, 1822 Feb. 16-1911 Jan. 17. Polymath, traveller, travel writer, eugenicist and statistician. 7th and last child of Samuel Tertius G and Frances Anne Violetta Darwin. CD's half-first cousin. G was a voluminous writer on many topics. Biography: K. Pearson, 4 vols., 1914-30; D.W. Forrest, 1974. Archive calendar: M. Merrington and J. Golden, 1976. Complete works at http://galton.org/. 1839 Late Oct. or early Nov. visited CD at Upper Gower Street when a student at King's College Hospital. 1840 Oct. Went to Trinity College, Cambridge. 1853 Married Louisa Jane Butler s.p. 1860 FRS. 1869-71 Experiments with transfusion of rabbit's blood to test CD's theory of pangenesis proved unsuccessful. 1869 *Hereditary genius.* 1873 G sent CD a questionnaire on education and background. In 1870 Jan. 8-10, ED recorded his stay. Mr. Rouse who came a day earlier on Jan. 7, left also on Jan. 10. Between 1873-81, his visits were also noted in *ED's diary.* After that G visited ED 1884-95, the last entry in 1895 May 14. 1874 *English men of science.* 1879 CD answered G's questions on the faculty of visualising for *Inquiries into human faculty,* 1883, "I am inclined to agree with Francis Galton in believing that education and environment produce only a small effect on the mind of anyone, and that most of our qualities are innate". *Autobiography,* p. 43. 1883 *Inquiries into human faculty and its development.* 1908 Autobiography. 1909 Kt. Buried at St. Michael and All Angels Churchyard, Warwickshire. Recollections of CD in DAR112.A52-A53 and 1909. *Memories of my life,* pp. 287-88, 169, both transcribed in *Darwin Online.*

Galton, **Samuel Tertius**, 1783 Mar. 23-1844 Oct. 23. Son of Samuel "John" G. Father of Francis G. 1807 Married Frances Anne Violetta Darwin (1783-1874). Taught CD how to use a vernier on a barometer at Shrewsbury.

Gardner, **Andrew**, 1790-1861. Scottish ex-soldier and ex-convict who established the Scotch Thistle Inn (1831), later Blackheath Inn, Ross, Australia. In the *Sydney notebook,* Chancellor & van Wyhe, *Beagle notebooks,* 2009.

Gardening. 1864 Ancient gardening, *Gardeners' Chron.,* no. 41: p. 965 (*Shorter publications,* F1732). "I should be very much obliged if any one who possesses a treatise on gardening or even an Almanac one or two centuries old would have the kindness to look what date is given as the proper period for sowing Scarlet Runners or dwarf French Beans." CD received several replies, see CCD12:361-2. In *Variation* 2:314, CD remarked "I have not been able, by searching old horticultural works, to answer this question satisfactorily."

Gardeners' Chronicle. One of CD's favourite magazines in which he published many articles and letters seeking information. 1863 Sept. 5, ED recorded "Gardeners Chron", when her article with CD on vermin traps was received. See F1728 and F1931. There is a bound manuscript index to CD's collection 1841-71 (in the Cory Library, Cambridge Botanic Garden), DAR222.1. It has a list of numbers of the magazine missing from the collection and a "List of the numbers of special interest to Darwin and kept by him in separate parcels". This gives the page numbers and the subject of the articles of interest to CD and which he often annotated. The index is transcribed only in *Darwin Online.*
http://darwin-online.org.uk/content/frameset?pageseq=1&itemID=CUL-DAR222.1-&viewtype=text

Garson, **John George**, (1854-1932). Assistant to William Henry Flower at Royal College of Surgeons, although never on the official staff. See correspondence between CD and Reuben Almond Blair in Carroll. 1878 Nov. 6 Flower to CD, on deformity in goose wings, gives a report on wings provided by Blair. See CCD26.

Gaskell, **Mrs Elizabeth Cleghorn**, see Stevenson.

Gaudry, Jean Albert, 1827-1908. French geologist and palaeontologist. *Calendar* & CCD lists under Albert Jean G. Published under name Albert Gaudry. 1868 CD to G, on reception of *Origin* in France and on paper in *Geological Mag.*, p. 372, 1868. 1868 G was pro-*Origin*. 1869 CD cited G's work on Greek fossil mammals that were clearly related to living mammals. 1884 Wollaston Medal, Geological Society. 1895 Foreign Member Royal Society.

Gay, Claude, 1800-73. French naturalist & traveller. Prof. of physics & chemistry, Santiago, 1828-42. In *Galapagos notebook*, Chancellor & van Wyhe, *Beagle notebooks*.

Geach, Frederick F., 1834?-90. Cornish mining engineer in Malacca until 1866, introduced (via letter) to CD by Wallace. Answered queries about expression for Malays and Chinese, see *Expression*, p. 21. CCD15 & 16.

Geddes, Patrick, 1854-1932. Scottish biologist, social scientist and town planner. Studied under T.H. Huxley. Greatly influenced by CD's evolutionary ideas. 1877-8 recollections of CD in A.J. Thompson & P. Geddes, 1931. *Life: Outlines of general Bio.*, vol. 2, pp. 1454-55, transcribed in *Darwin Online* (F2100). G recalled CD's "Panic intoxication of ecstasy" upon seeing movement under the microscope which left a "vivid and memorable lesson in biology" in him; how the "simplest spectacle of life" made CD suddenly "broke out, positively shouting for joy".

Gegenbaur, Carl (Karl), 1826-1903. 1855 Prof. Anatomy Jena, first extraordinary, in 1858 ordinary. 1858 Taught and later worked with Ernst Haeckel. 1864 An early convert to CD's views. CCD12. 1873 Prof. Anatomy Heidelberg.

Geikie, Sir Archibald, 1835-1924. Geologist. Prolific writer; biographer of Edward Forbes (with G. Wilson), Murchison and Ramsay. Brother of James G. 1865 FRS. 1871- Murchison Prof. of Geology and Mineralogy, University of Edinburgh. 1881 Murchison Medal, Geological Society. 1881-1901 Director General Geological Survey. 1891 Kt. 1907 KCB. 1908-13 President Royal Society. 1914 OM. 1924 Autobiography: *A long life's work*.

Geikie, James Murdoch, 1839-1915. Geologist. Brother of Sir Archibald G. 1862-82 Geological Survey. 1875 FRS. 1881 [1880] *Prehistoric Europe*, London, contains extracts from two letters from CD on the drift deposits near Southampton, pp. 141-2 (*Shorter publications*, F1351). 1882- Murchison Prof. Geology and Mineralogy Edinburgh in succession to his brother Archibald G.

Geological diary. CD's geological observations from the voyage of the *Beagle* and his longest voyage manuscript (DAR32-33) and geological notes (DAR34-38) are fully transcribed and edited in *Darwin Online*. It is accompanied by an important and detailed introduction by Gordon Chancellor:
http://darwin-online.org.uk/EditorialIntroductions/Chancellor_GeologicalDiary.html.
CD used these notes, together with later readers, to write his articles and books on the geology of South America between 1837-46. These earned him the reputation of one of the leading geologists in England. See J.A. Secord, The discovery of a vocation: Darwin's early geology, *Brit. Jrnl for the Hist. of Sci.*, 24, 1991, pp. 133-57.

Geological Notes on Coasts of South America. 1836 Geological notes made during a survey of the east and west coasts of South America, in the years 1832, 1833, 1834 and 1835, with an account of a transverse section of the cordilleras of the Andes between Valparaiso and Mendoza, *Proc. of the Geol. Soc.*, 2: pp. 210-12. Communicated by Prof. A. Sedgwick. (*Shorter publications*, F3 and F1642); CD's first paper under his own name alone, published without CD's knowledge.

Geological Society of London. 1836 Sept. 8 CD proposed by Sedgwick and Henslow while still on *Beagle*. Nov. 2 elected. Nov. 4 admitted. 1838 Feb. 16-1841 Feb. 19 CD was Secretary. Sir Henry Thomas de la Beche was Foreign Secretary at the time. 1859 CD awarded Wollaston Medal, which from 1846 to 1860 was made of palladium. 1859 Feb. 18 Wollaston Medal presented to Lyell for CD in CD's absence through illness. *Proc. of the Geol. Soc.* 1860, pp. xxii-iv.

Geology of the Voyage of H.M.S. Beagle. 1842, 1844, 1846 Intended as one volume in three parts, but issued as three books, *Coral reefs*, 1842, *Volcanic islands*, 1844 and *South America*, 1846 qq.v. 1851 First appearance of the three bound in one volume, a remainder from unsold sheets (F274). 1890 Ward Lock edition of the three parts printed together (F279). See Herbert, *Charles Darwin, geologist*. 2005.

Georgian. 1951 First edition in: *Journal* (F187).

Geospiza, Camarhynchus, Cactornis and Certhidea of Gould. [Remarks on the Habits of the Genera...] 1837 *Proc. of the Zool. Soc. of London*, Pt 5, (53): p. 49 (*Shorter publications*, F1644). CD's notes on habits of so-called Darwin's finches, following Gould's descriptions of CD's specimens from Galapagos. There are four other papers by Gould in part 5 on CD's South American birds, but without notes by CD.

German. CD had great difficulty in understanding the German language. See also Wien. 1880 CD to R.L. Tait, "German, which to almost all Englishmen is a great trouble and sorrow". De Beer 1968, p. 81. CD to Hooker "I have begun German". Hooker to CD "I've begun it many times". LL1:126. First editions in: 1844 *Journal* (F188). 1860 *Origin* (F672). 1862 *Orchids* (F820). 1868 *Variation* (F914). 1870 Tendency of species to form varieties (F365). 1871-72 *Descent* (F1065). 1872 *Expression* (F1187). 1876 *Coral reefs* (F311). 1876 *Climbing plants* (F860). 1876 *Insectivorous plants* (F1238). 1877 *Volcanic islands* (F312). 1877 *Cross and self fertilisation* (F1266). 1877 *Different forms of flowers* (F1297). 1877 Biographical sketch of an infant (F1312). 1878 *South America* (F313). 1880 *Erasmus Darwin* (F1323). 1881 *Power of movement* (F1343). 1882 *Earthworms* (F1404). 1885 Essay on instinct (F1443). 1887-88 LL (F1515).

Gibbs, George, 1815-73. Ethnologist of Smithsonian Institution. Geologist. Wrote on Native American linguistics. 1867 Mar. 31 G wrote to CD about *Queries about expression*, which Spencer Fullerton Baird had shown him.

Gibson, Lucie, 1864-1939. Red-haired. From Cork. 1888 Married Maj. Cecil Wedgwood. Governess to Mary Wedgwood, his half-sister. 1916 Board Director of Wedgwood after Cecil Wedgwood's death.

Gifford, Robert, Baron, 1779-1826. Judge and MP. Married Harriet Drewe, seven children. Woodchester, Stroud, Gloucestershire. 1817-24 MP for Eye. 1824 1st Baron Gifford of St. Leonard's.

Gilbert, Henry, (practiced 1850's). Dentist at Pall Mall, London, pigeon fancier. Designed chair for dental extractions. Sent CD "as a present two young Runts (a breed of pigeon) one a fine young Cock". CD to Tegetmeier 19 Jul. 1857, CCD6.

Gilbert, Sir Joseph Henry, 1817-1901. Agricultural chemist. 1843-1901 At Rothamsted Experimental Station. 1860 FRS. 1876 CD to G on soil without organic matter; CD had met G at Linnean Society. CCD24. 1893 Kt.

Gill, Mr, 1835 Apr. 5 "When at Lima I was conversing with a civil engineer, Mr Gill, about" ruins of houses in uninhabitable places. *Beagle diary*, p. 321, Keynes, *Beagle record*, 1979, p. 274. Referred to in the *Geological diary* as an architect, DAR37.698.

Gisbert, Francis J., Spanish consul in Amoy 1867-? CD cited G's observation on hedgehogs carrying strawberries to their holes on their spines. (*Shorter publications*, F1740). First name not previously identified.

Glaciers of Caernarvonshire, 1842 Notes on the effects produced by the ancient glaciers of Caernarvonshire, and on the boulders transported by floating ice, *Phil. Mag.*, 21 (Sept.): pp. 180-8 (*Shorter publications*, F1660). 1842 CD visited Caernarvonshire in May and Jun. Notes in DAR27.1.

Gladioli, 1861 Parents of some gladioli. *Jrnl. of Horticulture*, (10 Sept.): p. 453 (*Shorter publications*, F1819). A reply appeared after CD's query. See CCD9:257.

Gladstone, Helen, 1849-1925. 6th child (of eight) of William Ewart G. 1882-96 Vice-Principal Newnham College, Cambridge. 1882 G was on "Personal Friends invited" list for CD's funeral. G could be "Miss Gladstone" in *ED's diary*. She visited in 1880 Aug. 18, lunched on 1881 Oct. 22 and probably stayed a night and left at 7pm on Oct. 23.

Gladstone, William Ewart, 1809-98. Statesman. CD et al. 1870. Copy of a memorial presented to the Right Hon. the Chancellor of the Exchequer. dated May 14 1866, in Sclater, Transfer of the South Kensington Museum. *Nature.* 2 (16 Jun.): 118. [see F1766 & F869] CD et al. 1874. Memorial presented to the First Lord of the Treasury, respecting the National Herbaria. In S.C. Cavendish, Fourth report of the Royal Commission on the scientific instruction and the advancement of science. *Parliamentary Papers, Command Papers; Reports of Commissioners,* (884), vol. XXII.1, pp. 31-2, *Shorter publications*, F869. 1877 Mar. 11 called on the Darwins. ED recorded "Gladstone &c called". G visited Down House in company with Huxley, Lord Morley, and Playfair, whilst staying at High Elms, home of Lubbock. CD said how honoured he was "that such a great man should come and visit me!" Morley, *The life of William Ewart Gladstone*, 1911, new ed., vol. 2, p. 562. 1877-79 CD corresponded with, mostly on behaviour, 1866-81. 1880 G arranged Civil List pension for Wallace. 1881 FRS. 1881 Jan. G wrote to CD about Wallace pension.

Glasgow. 1827 May CD visited on a spring tour. Van Wyhe ed., 'Journal', (DAR158). 1838 Jun. CD visited at end of geological trip to Glen Roy. 1855 Sept. 12-19 CD and ED went to BAAS meeting. On Sept. 14, ED recorded "Section in morning, saw Persia & foundry", the next day they went again to "Sections", attending a soirée and an "electric light". On Sept. 16 CD to Col. Rawlinson's lecture.

Glaziou, Dr Auguste François Marie, 1828-1906. French botanist and Director of Botanic Garden, Rio de Janeiro, Brazil. 1880 G wrote to CD about graft hybrids of sugar cane. CD mistakenly referred to him as "Dr Glass". 1882 CD to Romanes, about preparing a paper by Villa Franca and G, *Proc. of the Lin. Soc. of London*, 1880-82: pp. 30-1. ML1:389. This paper, never published, was prepared by CD and G.J. Romanes. The paper can be gauged by a surviving draft in DAR207.4. 1882 An abstract of the paper was published in the *Jrnl. of Botany* vol. 20, p. 192 and elsewhere. Transcribed in *Darwin Online*, F2168. Although it was submitted under the names of Villa Franca and Dr Glass, CD's drafting of part of the paper makes it count as a CD publication. Depending on how one defines an intentional publication, this could be regarded as CD's final publication in a scientific journal.

Lithograph of Glen Roy after a drawing by Albert Way, Plate II for CD's 1839 Glen Roy paper.

Glen Roy, Lochaber, Inverness-shire. CD, drawing on his findings on the geology of parts of South America, argued that the parallel roads of Glen Roy were of marine origin. This was contested, among others by Louis Agassiz, and eventually CD had to concede that the roads were terraces formed along the shorelines of ancient ice-dammed lakes. See Martin Rudwick, Darwin and Glen Roy: a "great failure" in scientific method? *Studies in the Hist. and Phil. of Sci.* 5, 1974, pp. 97-185 which provides a description of CD's Agenda for Lochaber (from DAR50), F1909. 1838 End of Jun. CD spent "eight good days" there. Van Wyhe ed., 'Journal', (DAR158), LL1:290. CD's *Glen Roy notebook* is in DAR130 and images of it are in *Darwin Online* and it is transcribed in F1817. Further MS notes in DAR50. 1839 "Observations on the parallel roads of Glenroy, and of other parts of Lochaber in Scotland, with an attempt to prove that they are of marine origin", *Phil. Trans. of the Roy. Soc.*, 129: pp. 39-81 (*Shorter publications*, F1653). CD's only contribution to *Phil. Trans. of the Roy. Soc.* 1841-80 Full discussion and letters about. ML2:171-193. 1861 "My paper was one long gigantic blunder from beginning to end. Eheu! Eheu!". CCD9:257. 1862 "I do believe every word in my Glen Roy paper is false". CCD10:462. 1876 "This paper was a great failure...a good lesson never to trust in science to the principle of exclusion." *Autobiography*, p. 84. 1880 CD to Joseph Prestwich "I gave up the ghost with more sighs and groans than on almost any other occasion in my life". *Prestwich* 1899, p. 300. Prestwich settled the origin of the parallel roads of Glen Roy in an article in the *Phil. Trans.* in 1879. See also F2564, in *Darwin Online*.

Glenie, Rev. Samuel Owen, 1811-75. Anglican clergyman. 1834 Colonial chaplain, Colombo, Ceylon (Sri Lanka). Later also appointed chaplain at Trincomalee and at Kandy. 1868 G to CD (through George Henry Kendrick Thwaites), answering *Queries about expression*, and on weeping in elephants. *Expression*, p. 167. 1868 CD to Thwaites asking him to thank G for "excellent letter". CCD16. 1871 Retired.

Glossostigma. 1877 Fertilisation of Glossostigma. *Nature*, 17: pp. 163-4 (*Shorter publications*, F1812). CD forwarded a letter by T.F. Cheeseman from Auckland, New Zealand. *Glossostigma* is a genus of flowering plants in the lopseed family.

Glutton Club, see Gourmet Club, of which it was a nickname.

Goddard, Right Rev. Isaac, 1836-1909. Chaplain for many years to the Empress Eugenie. 1873 Priest at Chislehurst who annoyed ED by preaching about Louis Napoleon as if he were a saint. ED(1904)2:261.

Godínez y Esteban, Enrique, 1845-94. Spanish journalist and translator. 1877 made the first Spanish translation of *Origin* (F770).

Gold mines. 1849 [Remark on a South American gold mine.] In Murchison, On the distribution of gold ore over the Earth's surface... *Athenaeum. no.* 1143 (22 Sept.): p. 966 (*Shorter publications*, F1941). This was the report by the *Athenaeum's* correspondent at the 19th meeting of the BAAS held at Birmingham in 1849. A brief comment by CD on "a gold mine on the east side of the Cordillera...modern origin of the rocks was indubatible". Agua del Zorro, near Mendoza, Chile.

Gonzales, Mariano, CD's hired guide and companion on his expeditions in Chile. CD paid him £7. In three notebooks, Chancellor & van Wyhe, *Beagle notebooks*, 2009.

Goodacre, Rev. Dr Francis Burges, 1829-85. Clergyman and naturalist. 1879 G sent CD hybrids between common goose and Chinese goose which were apparently fertile. LL3:240, *Nature*, 1880, 21: p. 207. "Fertility of hybrids from the common and Chinese goose". (*Shorter publications*, F1786).

Goodwin, Albert, 1845-1932. Landscape painter influenced by Turner and the Pre-Raphaelites. Painted Down House in 1880 and 1882 "The House at Down" with CD, ED, Bernard and the terrier Polly just visible. John Ruskin took G on a tour of Europe. His oil painting "Old Walls, Winchester" (1875) hung in the drawing room at Down House during CD's lifetime and is on display there now.

Goodwin, Rev. Harvey, 1818-91. Anglican priest and mathematician. 1869-91 Bishop of Carlisle. G dined with the Darwins in 1874 Dec. 17 as recorded in *ED's diary*. 1882 May 1 G preached sermon at CD's memorial service, Westminster Abbey, in place of Archbishop of Canterbury, Archibald Campbell Tait, who withdrew at short notice. Atkins 1976, p. 49.

Goose. 1880 Fertility of hybrids from the common and Chinese goose, *Nature*, 21: p. 207 (*Shorter publications*, F1786). "The fact of these two species of geese breeding so freely together is remarkable from their distinctness…removes a difficulty in the acceptance of the descent-theory, for it shows that mutual sterility is no safe and immutable criterion of specific difference." See also Goodacre.

Gordon, Richard, 1828-95. French physician. 1877 translated *Climbing plants* (F858).

Gore, Philip Yorke, 1801-84. Chargé d'affaires in Buenos Aires, 1832-4. In the *Buenos Ayres notebook*, Chancellor & van Wyhe, *Beagle notebooks*, 2009.

Goree Roads, Eastern end of Beagle Channel, Tierra del Fuego. Mentioned as Goeree Sound in *Beagle diary*. 1833 Jan. 15-Feb. 9 and Feb. 28 *Beagle* at.

Gorringes. A former dower house on the Lubbock estate. In 1851 Mrs. Susannah Eaton recorded as living there. A Sir Hugh Lubbock and a Mrs Forrest are recorded as living there by Atkins 1976, p. 104. 1926-54 Home of Bernard Richard Meirion Darwin. See preface to Barrett et al eds., *Darwin's notebooks*, 1987.

Gosse, Philip Henry, 1810-88. Naturalist, prolific writer and Plymouth Brother (Evangelical Christian movement). CD discussed evolution with before *Origin*. Biography: Edmund Gosse (son), 1890 *Life*; 1907 *Father and son*. 1856 FRS. 1857 *Omphalos; an attempt to untie the geological knot*. Attempt to reconcile Creation with natural law, especially Lyell's *Principles of geology*. 1861 CD read some book of his, Francis Darwin suggested *Naturalist's sojourn in Jamaica*, 1851, but more likely *Letters from Alabama*, 1859. 1863 CD to G, on fertilisation of orchids, which G cultivated.

Gould, Elizabeth, née **Coxen**, 1804-41. Natural history artist. 1829 Married John Gould. 1838 All 50 plates for *Birds* are after her drawings but she is not credited.

Gould, John, 1804-81. Ornithologist. Taxidermist to Zoological Society of London. Producer of sumptuous bird books. 1827 First Curator and Preserver at the museum of the Zoological Society of London. 1837 G described CD's *Beagle* birds in *Proc. of the Zool. Soc. of London* (F1643, F1644) with notes on habits by CD and others without. There are four additional articles (1837-8) on the Galapagos birds by G without contributions by CD. Established that CD's Galapagos finches were separate species. 1838-41 *Zoology of the Beagle*, Pt. III, *Birds* (F8). 1843 FRS.

Gourmet Club. Formed by CD and friends at Cambridge University, nicknamed "Glutton Club". CD was at one time President. Members included Blane, Lovett, Cameron, Heaviside, Herbert, Lowe, Watkins and Whitley qq.v. F. Watkins recalled: the club was "so called not because its members were gluttons, but because they made a devouring raid on birds & beasts which were before unknown to human palate. Our menu was certainly a choice one but the appetite for strange flesh did not last very long & I think the Club came to an untimely end by endeavouring to eat an old brown owl which was indescribable! we tried hawk & bittern & other delicacies". DAR112.A111-A114, transcribed in *Darwin Online*. John van Wyhe, *Darwin in Cambridge*, 2014, p. 81.

Gower Street, no. 110, see Upper Gower Street no. 12.

Graham, John, 1794-1865. 1829 Examiner for Little-go at Cambridge. 1829 CD's tutor at Christ's. 1830-48 Master of Christ's College, Cambridge.

Graham, William, 1839-1911. Prof. Jurisprudence Queen's College Belfast. 1881 CD to G, on reading his *Creed of science*. LL1:315.

Grange estate. c.1830 Inherited by Edward Simcoe Drewe, near Honiton, Devon.

Grange, The, see Newnham Grange.

Grant & Maddison, Bankers, Southampton. Looked after CD's investments. 1862-1902 William Erasmus Darwin a partner. 1902 Taken over by Lloyd's.

Grant, Miss 1857 Governess at Down House for six months.

Grant, Robert Edmond, 1793-1874. Zoologist and physician. G was with CD at Edinburgh and they collected on the sea-shore together with John Coldstream in 1826/27. 1815-20 Travelled and studied in Europe. In 1815-16 G studied in Paris where he met Cuvier, Lamarck and Étienne Geoffroy St. Hilaire. 1827-74 Prof. Zoology and Comparative Anatomy University College London. 1836 FRS. 1836 G was willing to examine *Beagle* corallines. 1837-38 Fullerian Prof. Physiology Royal Institution. 1861 G dedicated his *Tabular view of the primary divisions of the animal kingdom* to CD, with a long letter about G's early views on evolution. 1861 G is mentioned in the historical sketch of 1861, but not in the USA and German versions of 1860. 1876 "He did nothing more in science, a fact which has always been inexplicable to me". *Autobiography*, p. 49. Huxley of G: "I met nobody, except Dr. Grant, of University College, who had a word to say for Evolution—and his advocacy was not calculated to advance the cause". LL2:188. 1984 Two papers stressing G's predarwinian Lamarckist views, 1984 Adrian Desmond, *Jrnl. of the Hist. of Bio.* 17: pp. 189-223, *Archives of Nat. Hist.* 11: pp. 395-413.

Grasmere, Westmorland. 1879 Aug. 14 CD visited on day trip from Coniston. Henrietta in ED2:298-99 wrote "One expedition was made to Grasmere. I shall never forget my father's enthusiastic delight, jumping up from his seat in the carriage to see better at every striking moment."

Gray, Asa, 1810-88. American botanist. Intimate friend and correspondent of CD. CD discussed evolution with before *Origin*. Biography: Jane Loring Gray (wife), 2 vols., 1894. Letters are at Gray Herbarium, Harvard. 1838 First Prof. of Botany and Zoology, University of Michigan. 1842-73 Fisher Prof. Natural History Harvard. G donated his collection of books and plant specimens to Harvard; it formed the basis of the Gray Herbarium. 1855 or before CD met at Kew. 1857 CD outlined his evolutionary views in a sketch for G. The sketch was included in the 1858 Linnean Society publication of CD and Wallace's views. CD's draft of the sketch is in DAR6.51, transcribed in *Darwin Online*. 1859 CD sent 1st edn *Origin*. G arranged publication of American edition with Appleton. 1860 "Natural selection not inconsistent with natural theology", *Atlantic Monthly*, Jul., Aug., Oct. 1861 Oct. Produced in London as a pamphlet at CD's expense. Letters on its distribution; CD presented thirty-two copies. CCD9. 1862 Hooker to CD "A. Gray knows no more of the philosophy of the 'struggle for life' than the Bishop of Oxford does". L. Huxley, *Life and letters of Hooker*, II, p. 41, 1918. The remark refers to the American civil war. 1868 Oct. 24-30 G and wife dined at Down House and stayed. Dr & Mrs Hooker were guests at the same time. In 1869 Aug. 28, ED wrote in her diary "Asa Grays came", they might have enjoyed games of croquet at Bromley Commons. 1876 *Darwiniana*, New York. 1873 Foreign Member Royal Society. 1877 CD's *Forms of flowers* is dedicated to G. 1881 Aug. 15 an entry "Grays went". Grays visited ED in 1887 Jun. 18, with Hookers and Sedgwick. In 1893 Nov. 16, Mrs Gray visited ED.

Gray, George Robert, 1808-72. Younger brother of John Edward G. Zoologist, entomologist. Assistant Natural History Department, British Museum. CD discussed evolution with before *Origin*. 1839-41 G wrote much of the text for John Gould's *Birds*, Pt. III of *Zoology of the Beagle*, when Gould was in Australia. 1845 published descriptions of CD's specimens in *Catalogue of the specimens of lizards in the collection of the British Museum*. 1865 FRS. 1869 CD declined to write testimonial for G on grounds that he did not know enough of G's work. CCD17.

Gray, Jane Lathrop, née Loring, 1821-1909. 1848 Married Asa G. 1893 edited letters of Asa G. 1868 Oct. 28 Letter on a visit to Down House in Archives of the Gray Herbarium (Box G AG-B10: 8). Transcribed by Darwin Correspondence Project: https://www.darwinproject.ac.uk/people/about-darwin/family-life/visiting-darwins#

Gray, John Edward, 1800-75. Elder brother of George Robert G. Zoologist. CD discussed evolution with before *Origin*. Biography: *Annals and Mag. of Nat. Hist.*, 15: p. 218, 1875. 1832 FRS. 1840-74 Keeper of Zoology, British Museum. 1854 CD to G regarding viewing Cirripedes at British Museum. 1856 CD to Mrs Lyell (sister-in-law to Charles Lyell through her marriage to Lyell's brother), suggesting that she offer a collection of beetles to G for the Museum.

Great Cumberland Street, London. 1830 no. 14 home of Sir James Mackintosh and his daughter, Mrs Rich (Mary Mackintosh, 1789-1876).

Great Marlborough Street, London. no. 36 CD's lodgings 1837 Mar. 13-1838 Dec. 30. From before 1837 no. 43 home of Erasmus Alvey Darwin, now no. 48, the home of Schott Music London. Mark Pallen, *The rough guide to evolution*, 2009.

Greek. First editions in: 1877 Biographical sketch of an infant (F2059). 1900 *Journal* (F206). 1915 *Origin* (F698). 2006 *Descent* (F2048). 2007 *Autobiography* (F2049).

Green, Mr, a ship owner in Valparaiso. See *Narrative* 2:559. In the *Galapagos notebook*, Chancellor & van Wyhe, *Beagle notebooks*, 2009.

Green, Rev. John Richard, 1837-83. Historian. 1860 G was present, as an undergraduate student, at BAAS Oxford meeting. He described the scene to Boyd Dawkins, then a fellow student. LL2:322. 1869- Librarian at Lambeth Palace. 1860 recollection of CD in letter to W. Boyd Dawkins, 3 Jul. in *Letters of John Richard Green*, pp. 43-45, transcribed in *Darwin Online*.

Greg, William Rathbone, 1809-81. Social essayist. 1878 CD to G, on G's son's views on and objections to CD's views on evolution. CCD26.

Gresson, Rev. John George, Of Worthing. Before 1863 Second master at St Andrew's College, Bradfield, Berkshire, "a great dandy who wore white flannel trousers, a delicately tinted shirt, a purple velvet cap with tassel and primrose gloves for football". John Blackie, "Bradfield 1850-1975", p. 37, 1976. 1863 Innes suggested G as a possible tutor to CD's sons. CCD11:694.

Gretton, Frederick Edward, 1803-90. Was at Shrewsbury School and a friend of Erasmus Alvey Darwin. Anglican priest. 1844-72 Headmaster Stamford Grammar School. 1889 Of CD: "I just remember him—a dullish apathetic lad, giving no token of his after-eminence". *Memory's harkback*, p. 33, transcribed in *Darwin Online*.

Greville, Robert Kaye, 1794-1866. Botanist, expert on cryptogams especially Scottish. Read medicine at Edinburgh but did not qualify. Philanthropist. Collected with CD on shores of Firth of Forth, including Isle of May; "He had actually to lie down on the greensward to enjoy his prolonged cachinnation" (at the cries of kittiwakes). W.F. Ainsworth, *Athenaeum*, (13 May), p. 604, 1882. 1856 MP for Edinburgh.

Grey, Sir George, 1812-98. Army officer and Australian explorer. Governor of South Australia, two terms as governor of New Zealand, and later of the Cape Colony (South Africa). 1870 Settled in New Zealand. Prime minister of New Zealand 1877-79. Correspondent of CD, mostly on geology. 1846-7 correspondence with G and Stokes in 'Stokes' charges and Darwin's letters', In Rees & Rees, 1892. *The life and times of Sir George Grey*, 2ᵈ edn, pp. 591-595 (F1835). 1902 *N.Z. Herald*, Auckland Sept. 6. William Lee and Lily Rees biography of G, 1892. 1837 Travelled to Australia in *Beagle* on 3ᵈ voyage, occupying CD's old cabin. CD discussed evolution with before *Origin*. 1855 Dec. 9 CD to G "I have during many years been collecting all the facts and reasoning which I could to the variation and origin of species" possibly CD's earliest use of this phrase, as well as to others consulted at the same time including E.L. Layard and A.R. Wallace.

Grey Powder, common name of a medicine of mercury in chalk used as a laxative etc. Used in the Darwin family up to 1896 Aug. 7 as recorded in *ED's diary*. Leonard was given in 1853 when he had vomited, CD was given when "stomach disordered". ED took when she had a headache. Horace took when he had been sick.

Grice, Jane, 1801-? Though never married was known as Mrs Grice. Housekeeper for the Darwins at The Mount. In service for more than twenty five years. Pattison, *Darwins of Shrewsbury*, 2009, p. 104. ED recorded a few recipes she shared. *Recipe book*.

Griesbach, Rev. Alexander William, 1807-62. Entomologist. *Newsletter of the Geological Curators Group* I, no. 2, pp. 49-51, 1974. 1864 Benjamin Dann Walsh to CD, G introduced W to CD at Christ's College "more than thirty years ago". CCD12:161.

Grieve, Symington, 1848-1932. Ornithologist, expert on great auk. 1882 Mar. 22 CD to G, on floating stones supporting fuci. Transcribed in *Darwin Online* (F2260).

Griffin, R. & Co., Publishers, London. 1860 CD corrected his own entry for their *Dictionary of contemporary biography* (1861). Transcribed in *Darwin Online*. See CCD8:57.

Griffith, Charles, British Consul at Buenos Aires, 1834. In the *Buenos Ayres notebook*, Chancellor & van Wyhe, *Beagle notebooks*, 2009.

Griffith, John, 1821-? In 1851 census as Erasmus Darwin's manservant. Erasmus living at Park St, sent G with message to ED that Annie Darwin "has rallied—has passed good night – danger much less imminent". Keynes, *Annie's box*, 2001, p. 209.

Griffith, Sara, 1845-? Kitchen maid at Down House in 1871 census.

Gros, near Abergele, Denbighshire. 1813 CD with family for sea bathing. 'Journal'.

Grote, George, 1794-1871. Historian, educationalist and political radical. 1832-41 MP In the 1840s CD met at Lord Stanhope's. LL1:76. 1857 FRS. 1862-71 Vice-Chancellor University of London, succeeded by John Lubbock.

Grove, The, Hartfield, Sussex. Until 1862 Home of Charles Langton.

Grove, The, Huntingdon Road, Cambridge. 1882-96 ED moved there for the winters after CD's death and the summer in 1882 Jul. 29-Aug. 3. ED2:338. 1891 census lists Hannah Bronwick, Joanna Matheson, Harriet Irvine and Elizabeth Batholomew as servants. Now part of Fitzwilliam College.

Grove, Sir William Robert, 1811-96. Physicist, barrister and judge. 1840 FRS. 1842 Developed first fuel cell, 'gas voltaic battery'. 1866 CD to Hooker, G as President of BAAS, Nottingham, "disappointed in the part about Species; it dealt in such generalities that it would apply to any view or no view in particular". CCD14:304.

Growth. 1877 Growth under difficulties, *Gardeners' Chron.*, 8: p. 805 (*Shorter publications*, F1782). CD took a branch from a plant in his greenhouse and kept it suspended in his warm and dry study where it had "sent out two fine flowering stems".

Grut, Madame, Swiss governess at Down House from 1859 Jan. 24 - Mar. 16. G was dismissed by CD. See Browne, *Power of place*, 2002 and *ED's diary*.

Guanacos. 1844 [Extracts from letters on guanacos.] In W. Walton, *The alpaca*, 20, pp. 43-44, 50-1 (*Shorter publications*, F1833). Transcribed in *Darwin Online*. "Perhaps there is no animal in the world which, in its wild state, flourishes under stations of such different, and indeed directly opposite characters, as the guanaco."

Gulick, John Thomas, 1832-1923. American missionary and naturalist from Hawaii. 1872 correspondence about extremely limited distribution of species, especially land molluscs in Hawaii. CCD20. 1872 Aug. 2 G visited CD recollection in *The American Naturalist* vol. 42, no. 493 (Jan.): 48-57, transcribed in *Darwin Online*.

Gulliver, George, 1804-82. Anatomist and physiologist. FRS. 1855 CD accepted G's offer of *The works of William Hewson*. CD sends blood of pigeons and discussed variation of blood in animals. CD mentioned G's runt in *Variation*, 1:144.

Gully, James Manby, 1808-83. Physician. In charge of cold water cure at Malvern. 1825 Medical student at Edinburgh, contemporary with CD. 1849 Mar. 9- Jun. 29 When CD first went to Malvern, G made him give up snuff. 1849-51 CD took several cures at Malvern. 1851 CD took daughter Annie to G with persistent indigestion. In 1851 Apr. 14, from letters sent by CD to ED, she recorded in her diary "Better. Dr G. said she had turned the corner". The next day, she wrote "Dr G alarmed wrote for Ch." Annie died and buried in Malvern.

Günther, Albert Karl Ludwig Gotthilf, (also Albert Charles Lewis Gotthilf Gunther), 1830-1914. German-born British zoologist and herpetologist. 1859 Staff of British Museum. 1858, 1860 and 1877 published descriptions of CD's reptile specimens. 1869 G gave CD information on sexual differences in fish. 1870 G arranged for cutting of blocks for *Descent* by Ford. 1871 Jan. 28 G at Down House. 1882 G was on "Personal Friends invited" list for CD's funeral.

Gunville House, Tarrant Gunville, Dorset. 1799 Bought by Josiah Wedgwood [II]. 1800-05 Lived there. 1802 bought Maer Hall, but continued to live at GH.

Gurney, Edmund, 1847-88. Writer on music and psychic research. 1876 CD to G on music and sounds made by insects. The Gurney's visited on 1877 Jul. 14 and 1880 Jul. 10, in company were the Dyers. 1881 G wrote on vivisection in *Fortnightly Rev.*, 30: p. 778. 1882 On same subject, *Cornhill*, 45: p. 191, referred to. 1882 G was on "Personal Friends invited" list for CD's funeral.

Haast, Sir John Francis Julius von, 1822-87. German-born New Zealand geologist. 1863 CD to H on New Zealand geology and natural history. 1866 Prof. Geology Canterbury College, New Zealand (later University of Canterbury).

Hacon, William Mackmurdo, 1821-85. CD's solicitor, although they never met. "Everything I did was right, and everything was profusely thanked for". H's feeling for CD in Francis Darwin's reminiscences. LL1:120. 1843-85 Practised. 1861 31 Fenchurch St, London (CD's Address book, Down House MS). 1870-84 His partners varied. 1879 18 Fenchurch St. Hacon & Turner, 101 Leadenhall St.

Haeckel, Ernst Heinrich Philipp August, 1834-1919. German biologist, physician and natural history artist. Second son of Karl H and Charlotte Sethe. The apostle of darwinism in Germany. H's wild, and mostly unsupported, phylogenetic speculations, combined with his popular reputation, held back experimental scientific work on evolution. His biological imagery like the 'tree of life' and embryological 'recapitulation' remain popular icons of evolutionary history. 1862 Married 1 Anna Sethe (his first cousin) d.s.p. 1862-1909 Prof. Zoology Jena. 1863 Mar. CD to Lyell, "A first-rate German naturalist (I now forget name!!)" CCD11:244. 1864 H asked CD about the origins of his theory of evolution. CD's reply was published in German in *Natürlichen Schöpfungsgeschichte* (1868). Later, the German zoologist Oskar Schmidt asked CD for an English translation of the letter and CD thus wrote a very concise account of the origins of his theory as he then recollected it: "Having reflected much on the foregoing facts, it seemed to me probable that allied species were descended from a common ancestor. But during several years I could not conceive how each form could have been modified so as to become admirably adapted to its place in nature. I began, therefore, to study domesticated animals and cultivated plants, and after a time perceived that man's power of selecting and breeding from certain individuals was the most powerful of all means in the production of new races. Having attended to the habits of animals and their relations to the surrounding conditions, I was able to realize the severe struggle for existence to which all organisms are subjected; and my geological observations had allowed me to appreciate to a certain extent the duration of past geological periods. With my mind thus prepared I fortunately happened to read Malthus's 'Essay on Population;' and the idea of natural selection through the struggle for existence at once occurred to me. Of all the subordinate points in the theory, the last which I understood was the cause of the tendency in the descendants from a common progenitor to diverge in character." Schmidt, *The doctrine of descent and Darwinism*, 1875, 2ᵈ edn, pp. 132-3 (F1916). 1866 Oct. 21 H stayed at Down House. W. Bölsche, *Ernst Haeckel*, 1909, p. 179 with recollection of CD, transcribed in *Darwin Online*. 1867 Married 2 Agnes Huschke. 1 son, 2 daughters. 1867 CD complains to Huxley of excess of neonyms in H's *Generelle Morphologie*, 1866. 1868 CD to H "your boldness sometimes makes me tremble". CCD16:850. 1869 Huxley "The Coryphaeus of the Darwinian movement in Germany". 1876 Sept. 26, 1879 Visited Down House. His recollections "I fancied a lofty world-sage of Hellenic antiquity—a Socrates or Aristotle—stood alive before me" *Nature*, 26, (1882). pp. 533-41. Main works: 1866 *Generelle Morphologie*, 2 vols. 1868 *Natürliche Schöpfungsgeschichte*. 1874 *Anthropogenie*. 1877 *Die heutige Entwickelungslehre in Verhältnisse zur Gesammtwissenschaft*. 1878-79 *Gesammelte populäre Vorträge aus dem Gebiete der Entwickelungslehre*. 1882 *Die Naturanschauung von Darwin, Goethe, and Lamarck*. 1894 *Die systematische Phylogenie*. 1904 *Kunstformen der Natur*. See di Gregorio, *From here to eternity*, 2005. 1882 recollection of CD in *The Times* (28 Sept.): 6, transcribed in *Darwin Online*.

Hague, James Duncan, 1836-1908. American geologist "resident in California". 1871 Feb. Visited Down House. Concerning *Descent*, "everybody is writing about it without being shocked". See F1810 and F1761. Recollection in 1884 *Harper's new monthly Mag.* 69, (Oct.): 759-63, transcribed in *Darwin Online*.

Hahn, Otto, 1828-1904. German lawyer and amateur palaeontologist. 1879 Sept. 1 H sent CD his article *Die Urzelle*. H claimed to have found microscopic fossils in chondritic meteorites. CD's reply is lost. *Calendar* 12211. 1880 Dec. 16 H sent CD his *Die Meteorite und ihre Organismen* (1880) and claimed to have discovered the beginning of life on earth. *Calendar* 12917. 1880 Dec. 20 CD to H: "If you succeed in convincing several judges as trustworthy as Professor Quenstedt, you will certainly have made one of the most remarkable discoveries ever recorded". 1882 Feb. 5 T.G. Bonney to CD disbelieving the story; H cannot distinguish between mineral and organic structures. *Calendar* 13663. 1889 H in a letter to George Darwin claimed to have visited CD in Jan. 1881 and CD was convinced of the organic structure of the chondrites. No proof of H's visit or CD's views is available. See John van Wyhe, 'Almighty God! what a wonderful discovery!': Did Charles Darwin really believe life came from space? *Endeavour*, 34, no. 3, (Sept. 2010): pp. 95-103.

Haig, Axel Hermann (Hägg), 1835-1921. Swedish artist and architect. 1882 engraved CD's new study a week after his death, when it had not been disturbed.

Haile, Peter, A bricklayer at Parkfields, the home of CD's aunts Sarah Elizabeth Wedgwood [I] and Catherine Wedgwood. A recollection of H was one of CD's earliest childhood memories. ML1:2.

Haliburton, Thomas Chandler, 1796-1865. Nova Scotian politician, judge and author. Married 2 Sarah Harriet Owen Williams. 1837-40 Author of *Sam Slick* (the *Clockmaker* serial). 1859-65 MP for Launceston.

Hall, Captain Basil, 1788-1844. RN Traveller, writer, naval captain, explorer and anthropologist. 1816 FRS. In the *Buenos Ayres notebook*, Chancellor & van Wyhe, *Beagle notebooks*, 2009. 1838 Athenaeum acquaintance of CD. 1840 [Letter to H on the valley of Coquimbo] F2163.

Hall, ?Jeffrey Bock, 1807-86. 1829 Cambridge friend of CD.

Halsey, Henry, Of Hanley Park, Surrey. Father of Mary H.

Halsey, Mary, Daughter of Henry H of Hanley Park, Surrey. 1848 Married Rev. Robert Wedgwood as 2ᵈ wife. 7 children.

Hamilton, Alfred Douglas, 1819-95. Barrister. Nine entries in *ED's diary* on meetings with the Hamiltons. Visits in 1869 Jan. 16 with Godfrey Lushington and Miss Sedgwick. The Hamiltons invited the Darwins on several occasions.

Hamond, Robert Nicholas, 1809-83. Mate on HMS *Druid*, loaned to *Beagle* in Nov. 1832 as mate, spent a lot of time ashore with CD. Went with CD to sacrament prior to voyage to Tierra del Fuego. 1827 Lieutenant. 1828 His elder brother Anthony married Mary Ann Musters, sister of Charles M. 1832 Jul. Joined *Beagle* to replace Musters. CD: "I have seen more of him than any other and like him accordingly". 1833 May Left *Beagle* for stammering. In the *Buenos Ayres notebook*, Chancellor & van Wyhe, *Beagle notebooks*, 2009. 1836 Married Caroline Musters, another sister of Charles M. 1882 One of CD's surviving shipmates from *Beagle*. LL1:221. 1882 recollections of CD in DAR112.A54-A55, transcribed in *Darwin Online*.

Hancock, Albany, 1806-73. Invertebrate zoologist. 1849 "On the occurrence on the British coast of a burrowing barnacle...", *Athenaeum*, no. 1143: p. 966 (*Shorter publications*, F1678), with notes by CD, read to BAAS meeting 22 Sept. 1849. 1855 CD (to Huxley) thought H a "higher class of labourer than J.O. Westwood", and suggested him for a Royal Medal of Royal Society. 1858 Royal Medal Royal Society. Visited CD 1876 May 15.

Hancock, Mary, sister of Albany H. 1883 Aug. 18 ED wrote "Mary Hancock".

Harbour, Mr, A man employed by CD to collect beetles for him around Cambridge. 1829 CD to Fox, "I have caught Mr. Harbour letting Babington have the first pick of the beettles; accordingly we have made our final adieus, my part in the affecting scene consisted in telling him he was a d------d rascal, & signifying I should kick him down the stairs if ever he appeared in my rooms again: it seemed altogether mightily to surprise the young gentleman." CCD1:81.

Harcourt, Edward William Vernon, 1825-91. Naturalist, editor, travel writer and conservative politician; MP for Oxfordshire and Henley. 1856 CD asked to borrow a copy of C.L. Brehm's work on ornithology. H duly obliged. H is referenced in *Origin*, p. 391. H was given a presentation copy of *Origin*, inscribed in a clerk's hand "E.W. Harcourt, from the Author, 1859" sold at Sotheby's in 2019 for £50,000.

Hardie, James, Physician. Founding member of the Plinian Society. Friend of CD at Edinburgh when a student, went on natural history trips together.

Harding, Elizabeth, 1826-? 1842 Nurserymaid at Down House, aged 16, from Maer, Staffordshire. Known as "Bessy". B got lost in Cudham Wood with Annie, Ernest Wedgwood and Brodie. She carried Annie for three hours while waiting to be found. The group somehow split up and Snow had returned with only Doddy (William Darwin). 1851 Mar. 30 census showed she was still with the household.

Harding, Anna, aged 14 in 1851, day school teacher.

Hardy, Francis, 1816-79. Farmer in Beesby, Lincolnshire and CD's tenant. See Worsely, *The Darwin farms*, 2017.

Hare Dene Albury, near Guildford, Surrey. The house belonged to Henry Drummond, an Irvingite. 1871 Jul. 28-Aug. 25 CD and family spent a holiday there. See Albury, near Guildford.

Harris. A gentleman farmer of Orange Court, Down. A villager, Miss Canning, recollected "My brother and I walked every day to Orange Court to fetch *our milk*. Mr. Harris was the farmer and a gentleman farmer too." Atkins 1976, p. 104.

Harris, James, A sealer of Del Carmen on Rio Negro. Acted as pilot to Wickham in *La Paz*, whilst his friend Roberts acted for Stokes in *La Liebre*. *Beagle diary*, p. 95, CCD1 and in the *Falkland notebook*, Chancellor & van Wyhe, *Beagle notebooks*, 2009.

Harris, Sir William Snow, 1791-1867. Physicist and electrical engineer. Known as "Thunder and lightning Harris". 1831 FRS. 1831 Nov. 11 CD met at Plymouth. 1831 H's type of lightning conductor was fitted to all masts of *Beagle*, long before they were adopted by the navy for all ships. 1835 Copley Medal. 1848 Kt.

Harrison, Frederic, 1831-1923. Jurist, historian and popular writer. 1871 Apr. 1 CD to H on beauty. CCD19. In 1873 Nov. 16 lunched with the Darwins. In her final diary in 1896, ED wrote "F Harrison" on the end pages.

Harrison, Jane Ellen, 1850-1928. British classicist scholar and linguist. A suffragist, also interested in CD's work, wrote essay, The influence of Darwinism on the study of religions. In Seward ed. *Darwin and modern science*, 1909. In *Darwin Online*. H visited in 1876 Sept. 22. ED wrote to her in 1877.

Harrison, Matthew James, 1846-1926. Canadian naval officer. 1874 Married Lucy Caroline Wedgwood. Had 5 children. Many visits to ED between 1883-96.

Hartfield. Village in East Sussex, on the edge of Ashdown Forest. ?1840-63 In *ED's diary* can mean Hartfield Grove, a quarter of a mile from The Ridge, home of Charles Langton and family. 1847-68 In biography usually means The Ridge, Hartfield, home of Sarah Elizabeth Wedgwood [II], built for her in 1847, left 1868. 1855 George, aged 10, was allowed to ride the 20 miles from Down alone. Atkins 1976.

Hartfield Grove, House at Hartfield, Sussex. Home of Charles Langton.

Hartley, George Justinian, 1875-1919. 1902 Married Mary Frances Wedgwood.

Hartogh Heijs van Zouteveen, Hermanus, 1841-91. Dutch naturalist and geologist. Dutch translator of *Descent, Expression, Variation* and *Journal*.

Hartung, Georg, 1822?-91. German geologist, specialist on geology of Atlantic islands. 1858 CD corresponded with, through Lyell, on Azores. CCD7.

Harvey, William Henry, 1811-66. Algologist, botanist. From 1847 CD was a friendly correspondent with. 1856- Prof. Botany Trinity College Dublin. 1858 FRS. 1860 Feb. 17 H read a "serio-comic squib" to Dublin University Zoological and Botanical Association. This was published as a pamphlet *An inquiry into the probable origin of the human animal etc.*, Dublin, 1860. CD's copy, at Cambridge, is marked "With the author's repentance, Oct. 1860". 1860 H wrote courteous but anti-*Origin* review in *Edinburgh Rev.* 1860 Aug. CD to H about Whale-bear story, "I struck it out in the 2nd Edition". CCD8:371. 1860 CD to Gray, "Even [H]...is not nearly so savage against me as...when he published his foolish pamphlet". CCD8. 1861 H wrote a review in *Dublin Hosp. Gazette*, May 15.

Hastings, Sussex. 1853 Aug. 2. CD visited for day from Eastbourne. In 1857 Apr. 9, CD went to H again, took a walk on shore and went to castle hill, went again a month later on 8 May, did some shopping and returned on May 12.

Hatherly, Baron, see Sir William Page Wood.

Hatschek, David, (also known by Hatsek, David Haek, Franz Helbing or Hans Helling), 1854-1920? Jewish-Hungarian author and editor who translated *Origin* (F2556) and *Descent* (F2557) into German.

Haughton, Rev. Samuel, 1821-97. Man of science. 1851-81 Prof. Geology Trinity College Dublin. 1858 FRS. 1859 Feb. 9 H's address to Geological Society of Dublin is one of the first comments on the CD and Wallace 1858 communications to Linnean Society: "If it means what it says it is a truism; if it means anything more, it is contrary to fact". LL2:157. 1860 CD to Gray, with footnote CD to Hooker, "A review in last Dublin Nat. Hist. Review, is the most unfair thing which has appeared,—one mass of misrepresentation." CCD8:247. "Do you know whether there are *two* Rev. Prof. Haughtons at Dublin...Can it be my dear friend?". CCD10:499.

Hawkins, Benjamin Waterhouse, 1807-94. Artist. H drew and put on stone the plates for *Fish* and *Reptiles* in *Zoology of the Beagle*. 1852-54 H made the Crystal Palace giant reptile replicas for the Dinosaur Court with advice from Richard Owen.

Hawkshaw, Sir John, 1811-91. Civil engineer. Of Hollycombe, Sussex. Father of John Clarke. 1855 FRS. 1873 Kt. 1876 Jun. CD and ED visited his home. CCD25.

Hawkshaw, John Clarke (Clarke), 1841-1921. Civil engineer. Eldest son of Sir John H. Brother of Mary H. 1865 Married Cicely Mary Wedgwood. 5 children; 1 Dorothy M 1866-1932 *ED's diary* "Cicely confined" on 31 Aug. 1866, 2 Katherine Anne 1868-1911?, 3 Oliver H 1869-1949, 4 Cicely F 1871-1925, 5 Mildred C 1875-

1927. 1870 June 18, H visited the Darwins. 1872 Jul. 12, H and Cicely dined. ED was particularly close to Mildred.

Hawkshaw, Mary Jane Jackson, 1836?-63. Daughter of Sir John H. Sister of John Clarke H. 1862 Married Godfrey Wedgwood as first wife. Died in childbed 10 days after the birth of Cecil. ED wrote in her diary 1863 Apr. 7 "Mary's death".

Hawley, Dr Richard Maddock, A "Dr Hanley" is mentioned in ML1:6 but it is a misspelling. English Lecturer in Physiology, Edinburgh and a medical author. 1807 MD Edinburgh. 1825 Oct. CD and Erasmus Alvey Darwin called on him on their arrival in Edinburgh. CCD1:18. 1827 FRCP Edinburgh.

Healey, Mary, ?-1679. Of Cleatham, Lincolnshire. Sixth generation ancestor of CD in male line. c.1600 Married William Darwin [IV] as second husband.

Health diary, (or Diary of health), 64 pages of loose foolscap sheets at Down House in which CD recorded his daily health, symptoms and water cure treatment from Jan. 1849 until 1855. His health was variously recorded some days as "poorly" or "unwell". Sometimes he was "Well <u>very</u>" with double underlining, what he called his "double dash" days. The most common symptom was an entry like this "4 or 5 slight fits of ft [flatulence]", also common were "boils" or vomiting. On 8 Jul. 1849 he recorded: "Poorly, much flat; excessive at night with slight trembling & fright". Other nights he recorded being "wakeful" or "restless". The diary is most intensively analysed and discussed in Ralph Colp Jr, *To be an invalid: the illness of Charles Darwin*, 1977 and Colp, *Darwin's illness*, 2008 (with a transcription of the diary); also cited frequently in CCD as it provides not only evidence of CD's health but sometimes his whereabouts. (Down House EH88202563) Many further records of CD's health are in *ED's diary* and the CCD. See also John Bowlby, *Charles Darwin*, 1990.

Heathcote, Maria Sophia, 1874 CD to Lyell, "I was glad to hear at Southampton from Miss Heathcote a good account of your health & strength". CCD22:434.

Heathorn, Henrietta Anne (Nettie), 1825-1914. Of Sydney. 1855 Married Thomas Henry Huxley. 1882 Was on "Personal Friends invited" list for CD's funeral. 1907 edited Huxley's *Aphorisms and reflections*. 1913 *Poems of Henrietta A. Huxley with three of Thomas Henry Huxley*.

Heaviside, Rev. James William Lucas, 1808-97. Cambridge friend of CD, member of Gourmet Club. 1833-38 Fellow of Sidney Sussex College, Cambridge. 1836 CD met in Cambridge. 1838-57 Prof. Mathematics East India College Haileybury. 1860-97 Canon of Norwich. 1882 recollections of CD in DAR112.A56-A57, transcribed in *Darwin Online*.

Hebrew. First editions in: 1930 *Journal* (F207). 1948-49 *Autobiography* (F1520). 1960 *Origin* (F700).

Heckel, Dr Édouard, 1843-1916. Physician and botanist. Translated *Cross and self fertilisation* (F1265), *Forms of Flowers* (F1296), *Power of movement* (F1342).

Hedgehogs. 1867 Hedgehogs, *Hardwicke's Science Gossip*, 3: p. 280 (*Shorter publications*, F1740). CD related a report in a letter from Robert Swinhoe of hedgehogs that purportedly carried strawberry-like fruits away stuck on their spines. See Gisbert.

Heer, Oswald, 1809-83. Swiss palaeobotanist and entomologist. 1835 Prof. Botany and Entomology Zürich. 1850 went to Madeira for his health. 1855-82 Prof. Botany Polytechnikum Zürich. 1878 Royal Medal, RS. 1878 seconded CD's election to Fellowship Königlich-Preussische Akademie der Wissenschaften, Berlin.

Hellyer, Edward H., 1811?-33. Clerk on 2ᵈ voyage of *Beagle*. 1833 Mar. 4 Drowned at Falkland Islands, collecting bird for Captain FitzRoy. *Beagle diary*, pp. 145-6.

Helmholtz, Hermann Ludwig Ferdinand von, 1821-94. German physician and physicist. 1855-58 Prof. Anatomy and Physiology Berlin. 1858- Prof. Physiology Heidelberg. 1878 H seconded CD's election to Fellowship Königlich-Preussische Akademie der Wissenschaften, Berlin.

Hemmings, Henry, 1810-? Until 1856 Manservant to Sarah Elizabeth Wedgwood [I] at Petley's, Down, until her death when he returned to Maer. Listed on 1861 census for Down House as "Emmings" a 51-year-old widowed visitor and retired servant from Maer. Visited the Darwins in 1861 Apr. 3 and 1863 Aug. 30 to Barlaston. 1872 H was alive but with a bad heart.

Henry, Isaac Anderson, 1799-1884. Lawyer and plant hybridiser, of Edinburgh. Name more commonly spelt Anderson-Henry. 1849 CD to H, on *Phlox* and *Mimulus*. CCD4. 1863 CD to H, on cross and self fertilisation and on the uselessness of the compound microscope. CCD4. 1867 CD to H to thank for offer to lend De Maillet's *Telliamed*, 1748. CCD15.

Henry Poole & Co., Tailor at No. 15 Savile Row, London. CD's tailor. Recent media reports claim that they retain CD's measurements. Writing to them in 2019, no information was forthcoming.

Henry, Samuel Pinder Tiritahi, 1800-52. Son of William H, missionary in Tahiti. 1835 Nov. 23 CD met with. *Narrative* 2:524, 546, 615. *Red Notebook*, p. 83.

Hensleigh, Elizabeth, 1738-90. Of Panteague. Origin of name Hensleigh in Wedgwood family. CD's maternal great-grandmother and ED's paternal great-grandmother. 1763 Married John Bartlett Allen as first wife.

Henslow, Anne, 1833-99. Daughter of John Stevens Henslow. Married Robert Cary Barnard in 1859. 1871 Mar. 30 H to CD, telling him of a visit to Colchester mental asylum in 1852, seeing a girl with pointed ears. CCD19. 1871 Mar. 31 CD to H, thanking her for information and praising John Stevens H. CCD19.

Henslow, Frances Harriet, 1825-74. Daughter of John Stevens Henslow. 1851 Jul. 15 Married as his first wife J.D. Hooker. 1856 CD to Hooker, on her "pedestrian feats". CCD6. 1874 Dec. 25 CD to Gray, "The death of Mrs Hooker has indeed been a terrible blow. Poor Hooker came here [Down House] directly after the funeral and bore up manfully". CCD22.

Henslow, Rev. George, 1835-1925. 3^d son of John Stevens Henslow. Botanist. Schoolmaster. Hon. Prof. to Royal Horticultural Society. VMH. Reviewed *Fertilisation*. 1865 Headmaster, Grammar School, South Crescent, Bedford Square, London. 1866 Apr. 2, ED wrote in her diary "Mr G. Henslow Mr Innes & Mr Stephens to dinner". 1873 *The theory of evolution of living things*. 1882 H was on "Personal Friends invited" list for CD's funeral. See *Shorter publications*, F1737 and F1809.

Henslow, Rev. John Stevens, 1796-1861. Clergyman, botanist and geologist. Father-in-law of Sir J.D. Hooker. 1818 FLS. 1822-27 Prof. Mineralogy Cambridge. 1823 Married Harriet Jenyns. 3 sons, 3 daughters. 1827-61 Prof. Botany Cambridge. CD, when at Cambridge, was known as "the man who walks with Henslow". 1829-31 CD attended his botany lectures. CD regularly attended his Friday evening gatherings, which continued every week in term until 1836 and were the forerunners of the Cambridge Ray Club 1837-. H became a strong personal friend of CD and looked after specimens sent back from *Beagle* voyage. 1830 Oct. 8 CD to Fox, of Mrs H, "she is a devilish odd woman, I am always frightened whenever I speak to her, yet I cannot help liking her". 1835 H edited CD's letters to him as *Letters on geology*, privately printed for members of the Cambridge Philosophical Society (*Shorter publications*, F1). 1836 CD at Sydney to H, "my master in natural history". CCD1. 1837-61 Vicar of Hitcham, Suffolk. 1839-41 ED recorded numerous dinners and visits with H, sometimes with his wife as well. CD discussed evolution with before *Origin*. 1844 Mr. Darwin's Memorandum. In H., Rust in wheat. *Gardeners' Chron.*, (28 Sept.), p. 659, (F1668a). 1854 H visited Down House when Hooker was staying for a fortnight. 1855 CD paid little girls in H's parish to collect seeds of *Lychnis* etc. 1858 H stayed Nov. 25-27. 1859 CD sent 1st edn of *Origin*. 1860 Feb. 14-16, he stayed at Down. 1860 Sat. Jun. 30 H was in the chair of Section D at BAAS Oxford scene between Wilberforce and Huxley. 1861 May 24 CD to Hooker, on H's death and the question of a biography, "The equability & perfection of Henslows whole character". CCD9:137. "His judgement was excellent and his whole mind well-balanced; but I do not suppose that anyone would say that he possessed much original genius". *Autobiography*, p. 64. 1862 Biography: 1862, Leonard Jenyns, with recollections by CD, pp. 51-5 (*Shorter publications*, F830). 1871 Mar. 31 CD to Anne Barnard (H's daughter), "To the last day of my life I shall think of your father with the deepest respect and affection, and gratitude for his invariable kindness towards me". CCD19. 1967 Barlow, *Darwin and Henslow* (F1598), transcribed in *Darwin Online*.

Herbert, John Maurice, 1808-82. County Court judge on Monmouth and Cardiff circuit. Cousin of Charles Whitley. Close friend of CD at Cambridge and member of Gourmet Club. Nicknamed "Cherbury", from Lord Herbert of Cherbury. 1828 CD collected beetles with H at Barmouth. 1839 H sent CD a silver forficula, i.e. asparagus tongs, as a wedding present. ED2:24. 1856 Jan. 2 CD to H, thanking him for a book of poetry, "I shall keep to my dying day an unfading remembrance of the many pleasant hours, (especially at Barmouth) which we have spent together". CCD6. 1867 May 7 CD invited H to Down House. CCD15. 1868 Jan. 30 H had given CD his old microscope. CCD16. 1872 Nov. 21 CD sent H 1st edn of *Expression*. CCD20. 1882 recollections of CD in DAR112.A60-A61, DAR112.A58-A59 and [at Cambridge] DAR112.B57-B76, transcribed in *Darwin Online*.

Herbert, Hon. and Rev. William, 1778-1847. Poet and expert on bulbous plants. Dean of Manchester. 1844 CD to Hooker, H in relation to heaths from Cape of Good Hope. 1845 Warden of the Collegiate Church. 1845 CD visited. 1847 Collegiate Church became a Cathedral in 1847 and H its Dean. 1847 CD visited in London and discussed hybridizing, "I...saw that he was very feeble", he died in his chair later in the same day. CD probably discussed evolution with before *Origin*. CD refers to in 1863 Vindication of Gärtner—effect of crossing peas. (F1727a)

Hermitage, house near Woking, Surrey. c.1847- Home of Henry Allen Wedgwood. Darwins stayed 1848 Apr. 7-12 and 1850 Jun. 1-5. *ED's diary*. 1853 Aug. 13-17, van Wyhe ed., 'Journal', (DAR158).

Hero. CD's name for a plant of morning glory, *Ipomoea purpurea*, of exceptional vigour. *Cross and self fertilisation*, pp. 37, 47. Allan, *The Hookers of Kew*, 1967, p. 252.

Herschel, Sir John Frederick William, Bart, 1792-1871. Astronomer and chemist. 1813 FRS. 1821 Copley Medal, Royal Society. 1831 Knight of Hanover (also referred to as Royal Guelphic Order or Hanoverian Guelphic Order). 1836 Jun. 4 CD and FitzRoy met with. Later that month CD dined with at Cape of Good Hope, at Lady Caroline Bell's house. Her comment on H "he always came into a room as if he knew that his hands were dirty, and that his wife knew that they were dirty". *Autobiography*, p. 107. CD also dined with him in London. 1838 1st Bart. 1849 H edited *Manual of scientific enquiry*, to which CD contributed the geology chapter (*Shorter publications*, F325). 1850-55 Master of the Mint. 1859 CD sent copy of 1st edn of *Origin*. 1859 10 Dec. CD to Lyell "I have heard by round about channel that Herschel says my Book 'is the law of higgledy-piggledy'.— What this exactly means I do not know, but it is evidently very contemptuous.— If true this is great blow & discouragement." CCD7:423. 1861 Jun. 5 CD to Gray, on evolution as stated in H's *Physical geography of the globe*, 1861. 1871 CD was a pallbearer at H's funeral.

Hesse-Wartegg, Ernst von, 1851-1918. Austrian-American writer and traveller. Visited CD and Down House in 1880, publishing a detailed account: Bei Charles Darwin [At Charles Darwin's]. *Frankfurter Zeitung und Handelsblatt* (30 Jul.), in *Darwin Online*, also an English translation in *Darwin Online* courtesy of Randal Keynes.

Heterogeny. 1863 [Letter] The doctrine of heterogeny and the modification of species, *Athenaeum*, no. 1852: pp. 554-5 (*Shorter publications*, F1729). "Heterogeny, as the old doctrine of spontaneous generation is now called".

Hewitt, Edward, poultry breeder of Birmingham. H is much quoted in *Variation* and *Descent*. 1868 Mar. CD to John J. Weir on sexual preferences of pheasant cocks when crossed with poultry hens. 1868 Apr. CD to the same, H says "the common hen prefers a salacious cock, but is quite indifferent to colour". CCD16.

Heywood Lodge, Heywood Lane, Tenby, South Wales. See Tenby.

Higgins, John, 1796-1872. Born in Shrewsbury. 1819 Moved to Alford, Lincolnshire, land agent to Robert Waring and Susan Darwin and CD from 1845. H expertly managed and collected rent for CD's farm in Beesby. His papers survive in Lincolnshire Archives (Higgins Deposit) and provide abundant detail. See Worsley, *The Darwin farms*, 2017.

Higginson, Col. Thomas Wentworth, 1823-1911. American Unitarian minister, abolitionist and soldier. Of Newport, Rhode Island, USA. 1872 & 1878 visited CD at Down. 1873 Feb. 27 CD to H, he had enjoyed his book *Life with a black regiment*, 1870. CD to H on this: "I always thought well of the negroes, from the little which I have seen of them; and I have been delighted to have my vague impressions confirmed, and their character and mental powers so ably discussed." CCD21. CD also had his *Atlantic essays*, 1871. Recollections of CD in 1872-8 in *Cheerful yesterdays*, 1900, pp. 283-6, transcribed in *Darwin Online*, (F2096).

High Elms. Estate of about 3,000 acres marched with Down House grounds. c.1842 Home of, and rebuilt, after burning down, by, Sir John William Lubbock, and then of his son Sir John L, Baron Avebury. Visits by the Darwin's and social occasions are recorded in *ED's diary* between 1855-95. Mansion burned to the ground in 1967. Now a public park and nature reserve.

"High Elms", Pseudonym of Edward Levett Darwin as an author.

Hildebrand, Friedrich Hermann Gustav, 1835-1915. German botanist. CD often praised H for writing German which was as clear as French. 1860-1908 Prof. Botany Freiburg. 1866 May 16 CD to H, on his papers on fertilisation of Fumariaceae and *Salvia*. CCD14. 1868 Jan. 5 CD to H, on graft hybrids. CCD16. 1881 Member Deutsche Akademie der Naturforscher Leopoldina.

Hill, The, near Abergavenny, Wales. 1830 Home of John Wedgwood IV.

Hill, Elizabeth, 1702-97. Daughter of John H. CD's great-grandmother. 1723/1724 Married Robert Darwin.

Hill, Frederic (Fred), 1861-? A groom at Down House, later on gardener. Aged 20 in 1881 census. "Fred...wore in his (tie) a metal horse-shoe which aroused unstinted admiration". Bernard Darwin, *The world that Fred made*, 1955, p. 11.

Hill, John, 1678-1717. Of Sleaford, Lincolnshire. Married Elizabeth Alvey. Father of Elizabeth H. Fourth generation ancestor of CD in male line.

Hill, Richard, 1795-1872. Born in Jamaica. Naturalist. Studied in England. H helped Philip Henry Gosse with Jamaican birds. Illustrated his own books. 1859 CD to re *Origin*. CCD7:322. CD sent 1st edn of *Origin* to, copy on market in 1981.

Hill, Major Richard Noel, (Richard Noel, Noel-Hill), 1800-61. Capt. Owen of Woodhouse's cousin. 1820s A shooting companion of CD. Took part in a joke at CD's expense. *Autobiography*, p. 54. 1848 5th Baron Berwick of Attingham.

Hindi. First edition in: 1964 *Origin* (F702).

Hindmarsh, Luke. See *Annals Nat. Hist.*, 2: pp. 274-84, 1839. 1861 May 3 CD to, about Earl of Tankerville's wild white cattle at Chillingham, Northumberland. CCD9.

"Historical sketch". Of previous studies and ideas on transmutation or evolution in *Origin*. 1860 Appeared in a shorter version, written before Feb. 20, in 1st German edn and 4th USA printing. 1861 First added to 3d English edn of *Origin*, in answer to criticisms by reviewers. The text was amended over following editions.

Hitote. Tahitian Chief. 1835 Nov. 26 CD discussed lightning conductors with H and several other Chiefs.

Hoare, Arabella, daughter of James Charles H? 1852 Feb. 3 ED recorded a visit by H, later in Aug. ED took Eliz to have tea with the Hoares. 1853 May. 30 ED wrote "went to the Hoares" and another entry was made in 1869 May 17.

Hoare, James Charles, 1781-1865. 1831 A canon residentiary of Winchester Cathedral. 1811 Married Jane Isabella Holden (d. 1865). Had 7 children. A Cambridge friend of CD. 1831 Dec. 25 CD went to church and found H was preaching there. The *Beagle* sailed two days later.

Hoare, Rev. John Newenham, 1838-1901. B.A., F.R.H.S., Vicar of Keswick. Exchanged a few letters with CD 1871, 1875. CCD19 & 23.

Hobart, Tasmania. 1836 Feb. 5-17 *Beagle* anchored in Storm Bay; CD landed. See Banks, A Darwin manuscript on Hobart Town. *Papers and Proc. of the Roy. Soc. of Tasmania* 105 (1971): 5-19, F1829 and F1821 (in *Darwin Online*). *Zoological diary*, pp. 302-4, 312-13. *Geological diary*: DAR40.97-99. All transcribed in *Darwin Online*.

Hobhouse, Arthur, 1st Baron, 1819-1904. Lawyer and judge. 1848 Married Mary Farrer.

Höchberg, Karl, 1853-85. German financier, author and socialist. 1879 CD to H, answering his queries on diet in relation to activity. In *Darwin Online*, F1984. CCD27.

Hochstetter, Christian Gottlob Ferdinand, Ritter von, 1829-84. Austrian geologist and explorer. 1860 Prof. Mineralogy and Geology, Imperial Polytechnic Institute Vienna. 1861 H wrote to Gray that Hooker informed him that evolution was making "very considerable progress" in Germany.

Hocken, Thomas Morland, 1836-1910. Ethnographer, bibliographer and book collector. Secretary of Otago Institute, New Zealand. 1862 Settled permanently in New Zealand. 1880 Institute celebrated 21st birthday of *Origin* by sending illuminated address to CD 1881, *Nature*, Feb. 24:393-4, in *Darwin Online*. 1881 Feb. 21 CD to H thanking and expressing continued interest in New Zealand. *Calendar* 13059.

Hodgson, Brian Houghton, 1800-94. Vertebrate naturalist and ethnologist of Darjeeling, India. 1862 Dec. 6 Hooker wrote to H, in succinct praise of CD. L. Huxley, *Life and letters of Sir J.D. Hooker*, 1918. CD cited H in *Variation*.

Hofmann, August Wilhelm von, 1818-92. Chemist. H helped CD with experiments for *Insectivorous plants*. 1845 Director College of Chemistry London. 1851 FRS. 1861 President Royal Chemical Society London. 1864 Prof. Chemistry Berlin. 1875 Copley Medal, Royal Society. Often cited by CD in *Descent* and *Variation*.

Holden, Rev. James Richard, 1807-76. Cambridge friend of CD. Rector of Lackford, Suffolk. Mentioned in letters to Fox 1829 and J.M. Herbert 1837 CCD1 & 2.

Holland, Edward, of Dumpleton north Gloucestershire, CD's second cousin. 1857 Oct. 4 CD to James Buckman, CD had asked "my cousin Mr. Holland of Dumpleton to make the enquiries, but as he is not on the spot, I have ventured to ask you". The enquiry was about a rare breed of pigeon. CCD6:463.

Holland, Elizabeth, 1771-1811. 1797 Married Rev. William Stevenson. Died 13 months after birth of Elizabeth Cleghorn (Mrs Gaskell). Sir Henry Holland's aunt.

Holland, Sir Henry, Bart, 1788-1873. Physician and travel writer. Physician to Queen Victoria. CD's second cousin. Son of Dr Peter H and Mary Willetts. His grandmother, Catherine E. Willett née Wedgwood, was 10th child of Thomas Wedgwood [III]; "A long and intimate friendship with whom (namely CD) I have more pleasure in recording than any family tie". 1815 FRS. Married 1 Emma Caldwell d.1830, married 2 Saba Smith d.1866. Holland, *Recollections of a past life*, 1868. Woodall 1884, p. 2. Constantly kind to the Darwin family in their illnesses. 1827 Harry Wedgwood to his mother: "Nobody shall persuade me that Dr. H. is either the most agreeable or the cleverest man in London. If he was he would not have shocked Charles Darwin by saying that a whale has cold blood". ED1:197. 1853 1st Bart. 1859 Nov. 18 CD to William Benjamin Carpenter, "I do not think (privately I say it) that the great man has knowledge enough to enter on the subject [evolution]". 1859 Oct. 25 CD to Lyell, CD hopes that H will not review *Origin* in *Quarterly Rev.* because he "is so presumptuous and knows so little". CCD7. 1859 Dec. 10 CD to Lyell, CD had "found him going an immense way with us (i.e. all Birds from one)—good". CCD7. Visits with Darwins in *ED's diary* 1824-63.

Holledge, Jane, 1827-?, Laundry maid from Kent at Down House 1851 census. 1850 Sept. 21. ED recorded "Jane Holledge came". *ED's diary*.

Holly berries. 1877 Holly berries, *Gardeners' Chron.*, 7: p. 19 (*Shorter publications*, F1774). 1877 [The scarcity of holly berries and bees], *ibid.*, 7: p. 83 (*Shorter publications*, F1775). Discussed possible reasons for the recent "scarcity of Holly-berries in different parts of the country".

Hollycombe, near Midhurst, Surrey. Home of Sir John Hawkshaw. 1876 Jun. 7-10. CD stayed there. Van Wyhe ed., 'Journal', (DAR158).

Holmgren, Alarik Frithiof, 1831-97. Physiologist, vocal opponent of vivisection. Prof. Physiology Uppsala. 1881, 1887 CD letter to H on vivisection, *The Times*, Apr. 18 (*Shorter publications*, F1352); *Nature*, Apr. 21; *British Medical Jrnl.*, 1: p. 660; *The Times* 22 Apr. (*Shorter publications*, F1793); also in a pamphlet by G. Jesse and several times in Sweden. Also in LL3:208 and Bettany 1887, pp. 160-2. (F1352-F1356).

Holwood House. The great house at Keston, 1½ miles from Down. George Bentham visited Down House from there. LL3:39. Atkins 1976, p. 103 says that the estate belonged to Earl of Derby. 1865 Home of Robert Rolfe, Baron Cranworth. There are many diary entries by ED of dinners and visits from 1854-95.

Homefield. A small house 400 yards northwest of Down House. On two acres originally part of little Pucklands field. Bought by the Darwins and in the Downe House School period a convalescent dormitory. 1930 Leased and added to by Sir Arthur Keith as his country cottage until his death in 1955.

Hooker, Harriet Anne, 1854-1945. Second child of Sir Joseph Dalton H and Frances Henslow. Married Sir William Thiselton-Dyer. Studied botany and was an accomplished illustrator, rendering c.100 for publication during the period 1878-80.

Hooker, Sir Joseph Dalton, 1817 Jun. 30-1911 Dec. 10. Second son of Sir William Jackson H. Botanist. Biography: L. Huxley, 2 vols., 1918; Turrill, 1963; Mea Allan, *The Hookers of Kew*, 1967; A. Desmond, *Sir Joseph Dalton Hooker*, 1999; J. Endersby, *Imperial nature*, 2008. H was CD's greatest personal friend and confidant, much more so than either Lyell or Huxley, and provided much plant material for CD from Kew. CD discussed evolution with before *Origin*. H preserved all CD's letters, see Janet Browne, *Jrnl. for the Soc. of the Bibliography of Nat. Hist.*, 8: pp. 351-66, 1978. Often at Down House. Records in *ED's diary* of visits spanned from 1852-96, with a final entry in 1896 Mar. 1 "wr Sir J. Hooker". 1839 MD Glasgow University. 1839 Jan. CD and H first met in company with Asa Gray at Hunterian Museum, Royal College of Surgeons. Also in Trafalgar Square in company of Robert McCormick (briefly surgeon on H.M.S. *Beagle*). 1839-43 Voyage to the Antarctic as assistant-surgeon on H.M.S. *Erebus*. 1844 Sept. CD to Lyell, "Young Hooker talks of coming [to Down House]; I wish he might meet you— he appears to me a most engaging young-man." CCD3:57. 1845 CD to Henslow, CD was disappointed that H had not got the post of Prof. of botany at Edinburgh. 1847 FRS. 1847-51 Voyage to the Himalayas and India. 1851 Jul. 15 Married 1 Frances Henslow, eldest daughter of John Stevens Henslow. 4 sons, 3 daughters. Second child Harriet Anne H. 1854 Royal Medal, Royal Society. 1858 Jul. 1 Communicated Darwin Wallace papers at meeting of Linnean Society with Charles Lyell. 1859 CD sent 1st edn of *Origin*. 1859 Nov. H accepted CD's theory in print in introductory essay to *Flora Tasmaniae*, I, Pt. 3, pp. ii-xxviii; this is vol. 3 of *Botany of H.M. discovery ships Erebus and Terror, 1839-1843*, 3 vols., 1849-60. The introductory essay was also available separately. 1865-85 Director of Royal Botanic Gardens, Kew, Surrey, in succession to his father. 1866 Aug. 27 H satirized Oxford meeting of BAAS as "the gathering of a tribe of savages who believed that the new moon was created afresh each month. The anger of the priests and medicine men at a certain heresy, according to which the new moon is but the offspring of the old one, is excellently given." LL3:48. 1869 CB. 1873-78 PRS. 1874 Frances Henslow died. 1876 Aug. Married 2 Hyacinth Symonds,

Lithograph of J.D. Hooker. Montague Chatterton.

widow of Sir William Jardine, Bart. 2 sons. 1877 KCSI. 1882 H was Pallbearer at CD's funeral. 1885 H retired to The Camp, Sunninghill, Berkshire. 1887 Copley Medal, Royal Society. 1892 Darwin Medal, Royal Society. 1897 GCSI. 1897 Victoria Medal of Honour of Royal Horticultural Society. 1897 Given gold commemorative medal by Linnean Society on H's 80th birthday. Medal was sculpted by Frank Bowcher and is also available in bronze. 1907 OM. 1908 Darwin-Wallace Medal Linnean Society. Recollections of CD in 'Unveiling the Darwin statue at the museum' *Jackson's Oxford Jrnl.* (17 Jun. 1899): 8, transcribed in *Darwin Online.* (F2169).

Hooker, Sir William Jackson, 1785 Jul. 6-1865 Aug. 12. Botanist. Father of Sir Joseph Dalton H. CD knew and met often but was not familiar with. Biography: Joseph Dalton Hooker, *Annals of Botany*, 16: pp. ix-ccxxi, 1902; Mea Allan, *The Hookers of Kew*, 1967. 1812 FRS. 1815 Married Maria Sarah Turner. 2 sons, 3 daughters. 1820-41 Regius Prof. Botany Glasgow. 1836 Kt of Hanover. 1841-65 Director Royal Botanic Gardens, Kew, Surrey.

Hoole, Rev. Arthur Stanley, 1866-1935. 1877- Curate at Downe church, presumably as a locum for G.S. Ffinden, then the vicar. Son of Stanley H.

Hoole, Stanley, 1840-1904. Underwriter for Lloyds. 1865 Married Mary Alice Swan (1846?-87), 2 daughters, 2 sons. He dined with the Darwins in 1878 Apr. 24 and 1879 Apr. 29. 1880 Aug. 23 CD to Innes "My wife has not seen poor Mrs Hoole since her return" Stecher 1961, p. 246. 1880 Nov. 29 Innes to CD "You will be glad to hear that Mrs. Hoole continues to improve..." Stecher 1961, p. 247.

Hope, Lady Elizabeth Reid Cotton, 1842-1922. Known as Lady Hope. Widow of Admiral of the Fleet Sir James Hope, writer of evangelical tracts and on temperance. "Of Northfield". H concocted the myth of CD's so-called death-bed conversion, see Atkins 1976, pp. 51-2; Moore, *The Darwin legend*, 1994. 1915 Encouraged by D.L. Moody, she told the story to one of M's schools at Northfield, Massachusetts, USA. Her story was printed in *Watchman Examiner*, Boston, on 15 Aug. 1915. Henrietta Litchfield denied the story in detail in *The Christian*, 1922 Feb. 23, "The whole story has no foundation whatever". H was not present at CD's last illness, although the possibility that H visited Down House at the end of 1881 cannot be excluded. 1922 Mr Tucker, of Salvation Army, asked H for details.

Hope, Rev. Frederick William, 1797-1862. Entomologist and print collector. Founder of Hope Chair of Zoology (Entomology) Oxford. CD gave him many insects which are now in the Hope collection, Oxford. Poulton, *Darwin and the Origin*, 1909, p. 202. 1829 Feb. H gave CD specimens of about 160 species of beetles in London. CCD1. 1829 Jun. CD visited Barmouth with H to collect beetles, but CD was ill and returned to Shrewsbury after two days. LL1:178. 1834 FRS. 1837 CD to H, about Australian insects. 1838 Aug. 9 CD to Lyell, "How much I disliked the manner [Hope] referred to his other works, as much as to say 'you must...buy everything I have written'". CCD2:96. 1838 published descriptions of CD insect specimens from the voyage.

Hope, Thomas Charles, 1766-1844. Scottish physician and chemist, discovered the element Strontium. The only teacher at Edinburgh of whose lectures CD approved. "Lectures, and these were intolerably dull, with the exception of those on chemistry by Hope". *Autobiography*, p. 47. 1799-1843 Prof. Chemistry Edinburgh. 1810 FRS. Some of CD's lecture notes "Dr Hope's Chymistry" in DAR5.A6-A11.

Hopedene. A house near Dorking, Surrey, built in 1875. Lent to Hensleigh Wedgwood. Wedgwood 1980. 1876 May 24-Jun. 10 CD stayed there. Van Wyhe ed., 'Journal', (DAR158). CD began writing his autobiography there.

Hopkins, William, 1793-1866. Mathematician and geologist. Mathematical coach at Cambridge. 1833 Introduced to geology by Sedgwick. 1837 FRS. 1851 Wollaston Medal, Geological Society. Often cited in CD's geological writings. 1860 H reviewed *Origin* in *Fraser's Mag.*, Jun., Jul., against but friendly. In *Darwin Online*.

Hordern, Ellen Frances, 1835-79. Daughter of Rev. Peter H. Memorial in Downe Churchyard gives date of birth. 1856 Married Sir John Lubbock as first wife.

Horner, Frances Joanna, 1814-94. Second child of Leonard Horner. 1844 Married Sir Charles James Fox Bunbury. 1894 Editor of memorial of her husband, London [1894], nine volumes, privately printed.

Horner, Francis [I], 1778-1817. Barrister and statesman. Elder brother of Leonard Horner. Statue by Sir Francis Chantry in Westminster Abbey.

Horner, Francis [II], 1820-24. 6th child and only son of Leonard Horner.

Horner, Joanna, 1822?-? 7th child of Leonard Horner. Unmarried. 1873 *Walks in Florence* published with Susan H.

Horner, Katherine Murray, 1817-1915. 4th child of Leonard Horner. Sister-in-law to Sir C. Lyell. 1848 Married Lt-Col. Henry Lyell, Sir Charles Lyell's younger brother. 1856 H wrote to CD about some beetles which she had. 1875 H asked CD to be a Pallbearer at Lyell's funeral. CD declined on grounds of ill-health. LL3:197. 1876 She visited Darwins as K. Lyell. 1882 H was on "Personal Friends invited" list for CD's funeral. Author of: *Life, letters and journals of Sir Charles Lyell*, 2 vols., 1881. *Memoir of Leonard Horner*, 2 vols., privately printed, 1890.

Horner, Leonard, 1785-1864. Son of John Horner. Linen draper of Edinburgh. Scottish merchant, geologist and educational reformer. Fairly frequent correspondent of CD and met when CD was in London. Member of Whig circle and friend of Erasmus D (I). CD discussed evolution with before *Origin*. Biography: Katherine Murray Lyell (daughter), 2 vols., 1890. Married Anne Susan Lloyd. 1 son, 6 daughters: 1. Mary Elizabeth, 2. Frances Joanna, 3. Susan, 4. Katherine Murray, 5. Leonora, 6. Francis, 7. Joanna. 1813 FRS. 1826 H took CD to meeting of Royal Society of Edinburgh. CCD1. 1827-31 First Warden of University College London. 1833-60 Factory Commissioner. 1845-47 and 1860-62 President Geological Society. 1846 H visited Down House with his wife. 1860 CD sent 1st edn of *Origin* to. Other visits noted by ED in her diaries from 1850-96, showed Joanna as the most frequent visitor but it is unclear if it was Joanna or Frances Joanna.

Horner, Leonora, 1818-1908. 5th child of Leonard Horner. 1839 H dined with CD and ED at Upper Gower Street. 1847 Sept. H visited Down House with the Lyells. 1854 Married Chevalier Georg H. Pertz. She probably saw CD one final time in 1882 Apr. 8, with another visit on May 20 after his death. Continued her visits up to 1896 with possibly a last visit on Jun. 6.

Horner, Mary Elizabeth, 1808-73. First child of Leonard Horner. 1832 Married Charles Lyell.

Horner, Susan, 1816-1900. 3d child of Leonard Horner. Unmarried. 1873 *Walks in Florence* published with Joanna H. Visited the Darwins with her sister Joanna on several occasions.

Hornschuch, Friedrich, 1793-1850. German botanist. CD cited his essay on the sporting of plants "Flora, 1848", in *Natural Selection*, pp. 102, 127.

Horses. The following Darwin family horses are known by name: Dandy (carriage horse, bought 1867, sold 1868), Dobbin, Flyer, Tara and Tommy. 1882: Lady Florence, Druid and Sailor.

Horsman, Samuel James O'Hara, d.1887? 1867-68 Curate of Down. H got, after a prison sentence, curacy of St. Luke's, Marylebone, London.

Horwood, John, 1823-c.1880 G.H. Turnbull's head gardener. 1862-63 H superintended building of CD's hothouse.

Hotham, Harriet, 1810-73. 1833 Married Sir John William Lubbock. Made a pencil sketch of CD in 1855.

Houseman, Emma, 1839-1929. Daughter of John H. 1871 Married Lawrence (Laurence) Wedgwood. 6 children.

Houseman, John, London bookseller. Father of Emma H.

Houseman, Laurence, So spelt in Wedgwood 1980, "Lawrence" in *Emma Darwin*.

Howard, Mary, 1740-70. Daughter of Charles H (?1706-71) of Lichfield and Penelope Foley (1708-?48). Known as "Polly". CD's grandmother. 1757 Married Erasmus Darwin [I] as 1st wife. Had 5 children. On dying was to have said "Dr Darwin; he has prolonged my days, and he has blessed them." Seward 1804, p. 14.

Howarth, Osbert John Radcliffe, 1877-1954. Geographical scholar. 1909-46 Assistant Secretary and later Secretary BAAS. 1929-54 Curator Down House. 1933 H and Eleanor Katherine H. (wife), *A history of Darwin's parish*. In *Darwin Online*.

Hubbersty, Nathan, 1803-81. 1826 CD went on walking tour in North Wales with H. 1826-28 Assistant master Shrewsbury School. 1829-51 Headmaster Wirksworth Grammar School. 1839 CD suggested to H that he should do some plant-breeding experiments. In *Notebook E* and *Notebook M*.

Hudson, William Henry, 1841-1922. Ornithologist and prolific popular writer. Born in Argentina and lived there until 1874, when he settled in England. Challenged CD's observation on Pampas woodpecker. See Pampas woodpecker.

Hughes, Charles, H helped CD and became interested in geology. 1818-19 Shrewsbury School with CD. 1832-3 Resident in Buenos Aires. 1832 Nov. 11 CD met at Buenos Aires. In the *Buenos Ayres notebook*, Chancellor & van Wyhe, *Beagle notebooks*, 2009. CCD1:277. 1833 Returned to England because of ill health.

Hughes, Frances Jane Fox, 1806?-. Sister of William Darwin Fox. 1852 Married Rev. John Hughes. 1880 CD to H, about an essay on religion and science by H which no good scientific journal would publish, "there have been too many attempts to reconcile Genesis and science". *Calendar* 12596 also 13683, Carroll 1976, 573.

Hughes, Thomas McKenny, 1832-1917. Geologist. 1862 Vice-president Geological Society. 1873-1917 Woodwardian Prof. Geology Cambridge in succession to Sedgwick. 1880 CD to H, about award to CD of a medal by Chester Natural History Society. 1880 Oct. Took tea with CD and ED in Cambridge. 1889 FRS.

Humble Bees, (bumble bees). 1841 *Gardeners' Chron.*, no. 34: p. 550 (*Shorter publications*, F1658). 1861 Is the female bombus fertilised in the air?, *Jrnl. of Horticulture*, (22 Oct.): 76 (*Shorter publications*, F1818). 1885 Ueber die Wege der Hummel-Männchen, in *Gesammelte kleinere Schriften*, (F1584). 1965 1885 paper translated as On the flight paths of male humble bees, pp. 70-3 in Freeman, *The works of Charles Darwin*, (F1580). 1968 Darwin on the routes of male humble bees, *Bulletin of the BMNH, Historical Series.*, 3:177-89. As 1965 translation but with transcript of CD's field notes added (F1581). Transcribed in *Darwin Online*. 1913 Dutch translation: F1583h.

Humboldt, Friedrich Wilhelm Heinrich Alexander, Baron von, 1769-1859. German naturalist and traveller. 1799-1804 Travelled to South and North America. His *Personal narrative* influenced CD, LL1:55. 1815 Foreign Member Royal Society. 1831 CD's copy of *Personal narrative*, 1819-29, was given him by Henslow before he sailed. It was inscribed "J. S. Henslow to his friend C. Darwin on his departure from England upon a voyage round the World." CD once met, when CD was resident in London, at Murchison's house, LL1:74. 1852 Copley Medal, Royal Society. 1874. [Letter to D.T. Gardner on Humboldt] *New York Times*. (F2283). 1881 CD to Hooker, H was "the parent of a grand progeny of scientific travellers".

Humphreys, Of 32 Sackville Street, London. c.1868 Supplied curates for Down.

Hungarian. First editions in: 1873-74 *Origin* (F703). 1884 *Descent* (F1084). 1873 *Expression* (partial). 1913 *Journal* (F208). 1913 Essay of Instinct (F2380). 1955 *Autobiography* (F1521). 1959 *Variation* (F919). 1963 *Expression* (F1199).

Hunt, Robert, 1807-87. Scientific writer. 1854 FRS. 1866 CD sent a third-person summary of his life for inclusion in Reeve & Walford eds., *Portraits of men of eminence*, 1866. (F1856) See the letter in CCD14:152. The published text is transcribed, with the photograph by Ernest Edwards (Edwards 1866b), in *Darwin Online*.

Huntsman & Sons, of Savile Row, now called Huntsman. CD's tailor. CD normally paid his bill with them each January and July. (Classed account books Down House MS). CCD25:222. It was recently reported that they maintain CD's measurements in their books. Requests for this information in 2019 were unsuccessful.

Hussey, Dr Thomas John, 1792-1866. Clergyman and astronomer & **Anna Maria H** (1805-53) mycologist, writer, and illustrator of Hayes, Kent. Her brother, George

Varenne Reed, became tutor to CD's sons. As early as 1844 Jul. 11, ED wrote in her diary of having dined at the Husseys. They were present at a dinner party in 1851 Nov. 27. The other guests included Capt. Sulivan, the Innes and the Lascelles. In 1852 Oct. 6 while at Bromley, Mrs Hussey was visited. *ED's diary.*

Hutton, Frederick Wollaston, 1836-1905. Army Officer and geologist. Curator of Canterbury Museum, Christchurch, New Zealand. 1861 Apr. 23? CD to Hooker; H reviewed *Origin* in *The Geologist*, 132. 1861 CD to H, on his review, praising it. 1866 Resigned his commission and settled in New Zealand, eventually, as geologist. 1867 Jun. 10 CD to Kingsley, "a very acute observer". CCD15. 1887 Author of *Darwinism.* 1892 FRS. 1899 Author of *Darwinism and Lamarckism, old and new.* 1905 Died on journey back to England and was buried at sea off Cape Town.

Hutton, Richard Holt, 1826-97. Unitarian clergyman, which he later abandoned. Man of letters. Anti-vivisectionist. 1861 Joint owner and editor of *The Spectator.* Opposed Huxley's agnosticism. 1875 H was a member of Vivisection Commission.

Hutton, Robert, 1784-1870. 1837-41 M.P. for Dublin. CD corresponded with. 1839 Mar. 9 ED wrote in her diary "refused Huttons President Roy. Soc." 1839 Jun. 8 CD and ED dined at the Huttons.

Huxley, Ethel Gladys, 1866-1941. Youngest child of Thomas Henry H. 1889 Married John Collier as second wife after her sister Marian's death in 1887.

Huxley, Henrietta Anne, see Heathorn.

Huxley, Leonard, 1860-1933. 4th child of Thomas Henry H. CD was his godfather. Biographer of his father and of Hooker. 1885 Married 1 Julia Frances Arnold (1862-1908). 3 sons, 1 daughter: 1. Julian Sorell, 2. Noel Trevenen, 3. Aldous, 4. Margaret Arnold. 1912 Married 2 Rosalind Bruce. 2 sons: 1. David Bruce, 2. Andrew. 1900 *Life and letters of Thomas Henry Huxley*, 2 vols. 1918 *Life and letters of Sir Joseph Dalton Hooker*, 2. vols. Recollections of CD in 1921. Home life of Charles Darwin. *R.P.A. Annual*, pp. 5-9, and 1929 *Cornhill Magazine* both in *Darwin Online.*

T.H. Huxley by Elliott & Fry. Not known as Darwin's bulldog.

Huxley, Marian (Mady), 1859-87. Artist. Studied at Slade School, London. 3d child of Thomas Henry H. 1864 Darwin's "pet" CCD12:147. 1878 H made pencil sketch of CD. 1879 Married John Collier. 1880-84 Exhibited at the Royal Academy, London.

Huxley Testimonials, [1851] *Testimonials for Thomas H. Huxley, F.R.S., candidate for the Chair of Natural History at the University of Toronto.* London, Richard Taylor printed. CD's letter at p. 4 (*Shorter publications*, F344). The Chair went to William Hincks, brother of Sir Francis Hincks, then Prime Minister of Upper Canada.

Huxley, Thomas Henry, 1825 May 4-1895 Jun. 29. 7th child of George H and Rachel Withers. Man of science and reformer of science education. Administrator, sat on numerous (Royal) commissions. CD discussed evolution with before *Origin.* Biography: Leonard Huxley (son) 1900; F. Chalmers Mitchell 1900. Mario A.

di Gregorio, *T.H. Huxley's place in natural science*, 1984. Adrian Desmond, *Huxley.* 2 vols., 1994-7. Frequent correspondent and often at Down House, but was never on such close personal terms with CD as was Hooker; see Bartholomew, M., *Annals Sci.*, 32: p. 525, 1975. 1845 MB (*Medicinae Baccalaureus*) London. 1846-50 Assistant Surgeon on HMS *Rattlesnake*, mostly in Australian waters. 1851 FRS. 1852 Royal Medal, Royal Society. 1854 Prof. Natural History School of Mines London. 1854 Apr. 23 CD to H on archetypes. 1855 Naturalist British Geological Survey. 1855-58, 1865-67 Fullerian Prof. Royal Institution. 1855 Jul. 25 Married Henrietta Anne Heathorn. 3 sons, 5 daughters: 1. Noel, 1856-60. 2. Jessie Oriana, 1858-1927. 3. Marian. 4. Leonard. 5. Rachel, 1862-1934, married 1884 Alfred Eckersley. 6. Henrietta, 1863-1940, known as "Nettie", married 1889 Harold Roller. 7. Henry, 1865-1946, married 1890 Sophia Stobart. 8. Ethel Gladys, 1866-1941, known as "Babs" and "Pabelunza", married 1889 John Collier (as deceased wife Marian's sister). 1856 May 21 CD to Hooker, about H's Royal Institution lectures "I think his tone is much too vehement". CCD6:111. 1859 CD sent 1st edn of *Origin*. 1860 Apr. H reviewed *Origin* in *The Times* and *Westminster Rev.* 1860 Sat. Jun. 30 H defended *Origin* against Bishop Samuel Wilberforce's remarks at Oxford meeting of BAAS. 1860 "Time and life: Mr. Darwin's 'Origin of species'" *Macmillan's Mag.* 1: pp. 142-8. 1863-69 Hunterian Prof. Royal College of Surgeons. 1869-70 President BAAS. 1871 Nov. 2 H to Haeckel "The dogs have been barking at his heels too much of late". *Life and Letters of Huxley*, vol. 1, p. 363. During the 20th century it became a commonplace 'fact' that H was widely known as Darwin's bull-dog because of his tenacious defence of CD and evolutionary theory. However this name is posthumous and was not used during his lifetime. See John van Wyhe, Why there was no 'Darwin's bulldog': Thomas Henry Huxley's famous nickname. *The Linnean*, vol. 35, 2019. 1873 Apr. 23 £2,100 subscribed by CD and other friends to let H have a long rest after nervous breakdown. All H's children were looked after by ED at Down House whilst he was away. ML1:72. 1875 Apr. 14 CD to Hooker. H was member of Vivisection Commission. He saw and agreed to Litchfield's draft for bill. LL3:204. 1876 Wollaston Medal, Geological Society. 1880 Apr. 9 H lectured to Royal Institution on "The coming of age of the Origin", published in *Nature*, and in *Science and Culture*. CD sorry that he could not attend. LL3:240. 1881-85 Inspector of Fisheries. 1882 CD left him £1,000 in his will. ML1:72, *Life and letters of Huxley*, vol. 1, p. 366. 1882 H was Pallbearer at CD's funeral. 1883-85 PRS. 1884-90 President Marine Biological Association. 1887 H recollected that on mastering the central idea of *Origin* in 1859-60, he felt "How extremely stupid not to have thought of that!" LL2: Chapter V, pp. 179-204. c.1887 recollections of the reception of *Origin* in DAR112.B77-B84, transcribed in *Darwin Online*. 1888 Copley Medal, Royal Society. 1890 Linnean Medal, Linnean Society. 1890 H retired to Hodeslea (a name which he invented and believed related to the origin of his surname), Stavely Road, Eastbourne, Sussex, which he designed and had built. 1891 Anthony Rich left H his house, Chappel Croft,

Heene, Worthing, Sussex, and contents. H sold house for £2,800. 1892 PC (Privy Council). 1894 Darwin Medal, Royal Society. 1908 E.R. Lankester of H "the great and beloved teacher, the unequalled orator, the brilliant essayist, the unconquerable champion and literary swordsman". *Darwin-Wallace celebrations at Lin. Soc.*, p. 29. 1909 E.B. Poulton of H: "the illustrious comparative anatomist, Huxley, Darwin's great general in the battles that had to be fought, but not a naturalist, far less a student of living nature". *Darwin and the Origin*, p. 58. Main works: 1862 *On our knowledge of the causes of the phenomena of organic nature.* 1863 *Evidence as to man's place in nature.* 1871 *Anatomy of vertebrate animals.* 1873 *Lay sermons, addresses and reviews.* 1873 *Critiques and addresses.* 1877 *Anatomy of invertebrated animals.* 1880 *The crayfish.* 1881 *Science and culture and other essays.* 1893-94 *Collected essays,* 9 vols. 1898-1903 *Scientific memoirs,* 5 vols. Many visits recorded by ED. His death was recorded by ED in 1895 Jun. 29 "Death of Mr Huxley".

Hyatt, Alpheus, 1838-1902. American zoologist and palaeontologist. H worked especially on fossil cephalopods. Pupil of Louis Agassiz and friend of Edward Drinker Cope. Proponent of Neo-Lamarckism. 1870-88 Prof. Palaeontology and Zoology Massachusetts Institute of Technology. 1870 Custodian Boston Society of Natural History. 1872 Oct. 10 CD to H about H and Cope's ideas on acceleration and retardation in evolution. CD wrote on the back of one of H's papers "I cannot avoid thinking this paper fanciful". 1877-1902 Prof. Biology Boston Uni. 1877 CD to M. Neumayr on H's views on inheritance of acquired characters. CCD25:120.

Hybrids. 1868 On the character and hybrid-like nature of the offspring from the illegitimate unions of dimorphic and trimorphic plants, [Read 20 Feb.] *Jrnl. of the Lin. Soc. of London (Bot.)*, 10: pp. 393-437 (F1742). An edited version of this paper was later published in *Forms of flowers* (1877), chapter V.

Icebergs Making Grooves. 1855 On the power of icebergs to make rectilinear uniformly-directed grooves across a submarine undulatory surface, *Phil. Mag.*, 10 (Aug.): pp. 96-8 (*Shorter publications*, F1681). Arguing against a glacier explanation.

Icelandic. First editions in: 1945 *Descent.* 2004 *Origin* (F2483).

Indonesian. First edition in: 2002 *Origin* (F2487).

Ilkley. A spa town near Otley in West Yorkshire. CD was there for the hydropathic establishment when *Origin* was published in Nov. 1859. *ED's diary* 1859 Sept. 30 "Send Sat Tue Wed Thur to Ch at Ilkley Wells House Otley York", and Oct. 17 "Came to Ilkley"; Nov. 24, "left Ilkley with Etty & Lizzy" and to Shrewsbury the next day. 1859 Oct. 2 until Dec. 7 CD to water cure there, stayed at Wells Terrace. See Mike Dixon & Gregory Radick, *Darwin in Ilkley*, 2009.

Illot, Edward. Surgeon in Bromley. 1858 June, summoned to Down House to attend to Charles Waring D who died on Jun. 28. Browne, *Power of place*, 2002, p. 34.

Impey. CD's gyp (servant) at Christ's College, Cambridge. 1858 Impey was still there when William went up to Christ's. CD to W.E. Darwin [15 Oct.] CCD7:178.

Inchkeith, Fife. Island in Firth of Forth. CD visited with Ainsworth when at Edinburgh and was benighted, took refuge in lighthouse. W.F. Ainsworth, *Athenaeum*, (13 May), p. 604, 1882.

Index Kewensis. Also known as the Kew Index of Plant-Names or *Nomenclator Botanicus Darwinianus*. Originally supervised by Hooker and carried out by B. Daydon Jackson. LL3:353, *Kew Bull.*, 29, 1896. 1882 Jan. CD sent £250 and left a letter desiring that his children should send a similar sum for four or five years. 1892-95 4 vols., with 12 subsequent supplements to 1959, and a supplement since quinquennially. List of plant genera and their species, with relevant literature. Wording of announcement in vol. 4: "The expense of preparing the work has been entirely defrayed by the members of the family of the late Charles Darwin". In *Darwin Online*.

Ingall, Margaret Rosina (Rosena), 1854-1922. Daughter of Richard Ingall of Valparaiso, Chile. 1873 Married Alfred Allen Wedgwood.

Inglis, Sir Robert Harry, Bart, 1786-1855. Politician. Inglis was legal guardian of Laura Forster's mother, Laura Thornton. 1813 FRS. 1820 2d Bart. 1829-54 MP for Oxford University. 1854 CD took breakfast with him in company. 1854 PC (Privy Council). 1839 Apr. 3 ED recorded "Sir R. Inglis party".

Inheritance, 1881 Inheritance, *Nature*, 24: p. 257 (*Shorter publications*, F1795). Discussed "The tendency in any new character or modification to reappear in the offspring at the same age at which it first appeared in the parents".

Innes, Rev. John Brodie, 1817-94. Trinity College, Oxford, 1835 BA. 1839 MA. 1842. Letters to and from CD edited by R.M. Stecher, The Darwin-Innes letters. The correspondence of an evolutionist with his vicar, 1848-84. *Annals Sci.*, 17: pp. 201-58, 1961 (F1597). They contain a lot of information about people at Down not contained in other sources. 1842 Curate of Farnborough, Kent. 1846-c.60 Perpetual curate of Down c.1860-69 Vicar of Down. 1859 CD sent 1st edn of *Origin*. CD said to Innes, "I do not attack Moses, and I think Moses can take care of himself". LL2:288. 1869 Innes retired to his ancestral home Milton Brodie, Forres, which he had inherited. In this year he also added "Brodie" to his name. Until 1871 Down was served by curates until George Skertchley Ffinden became vicar in 1871. 1871 "Brodie Innes and I have been fast friends for thirty years, and we have never thoroughly agreed on any subject but once, and then we stared hard at each other, and thought one of us must be very ill". 1882 Innes was on "Personal Friends invited" list for CD's funeral. In 1851 Jul. 15, ED wrote in her diary, "Innes dined". Later on until 1879, I and his family were often invited to dine. Last record in *ED's diary* was about her writing to Mrs Innes in 1894 Oct. 24. Recollections of CD in DAR112.B85-B92 and DAR112.A65-A66, both transcribed in *Darwin Online*.

Innes, John William Brodie, 1848-1923. Son of John Brodie Innes. Barrister, novelist and occultist. In letters between CD and his father, as a child and young man.

Insectivorous Plants. 1875 *Insectivorous plants* (F1217). 2700 copies were immediately sold. 1875 2d thousand, with 6-line errata slip (F1218). 1875 3d thousand, 6 errata corrected, but with a further 6 on slip (F1219). 1888 2ᵈ edn, revised by Francis Darwin (F1225). 1969 Facsimile 1ˢᵗ edn (F1235). First foreign editions: 1875 USA (F1220). 1876 German (F1238), Russian (F1244). 1877 French (F1237). 1878 Italian (F1242). 1965 Romanian (F1243). List of presentation copies is in CCD23 Appendix IV. CD's notes in DAR54-61, DAR190 and DAR200. Proofs are in DAR213.

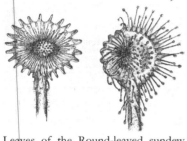

Instinct. 1873 [Letter] Inherited instinct, *Nature*, 7: p. 281, introducing a letter without title from William Huggins, *ibid.*, 7: pp. 281-2 (*Shorter publications*, F1757). 1873 "Origin of certain instincts", *Nature*, 7: pp. 417-18 (*Shorter publications*, F1760). 1873 [Introductory Letter] Instincts: Perception in Ants, *Nature*, 7: pp. 443-4 (*Shorter publications*, F1810). CD introducing a letter by James Duncan

Leaves of the Round-leaved sundew from *Insectivorous plants*.

Hague on ant behaviour. 1883 The late Mr. Darwin on instinct, *Nature*, 29: pp. 128-9 (F1804), summary, with last three paragraphs in full, of a communication by Romanes to Linnean Society of London, published in full in *Mental evolution in animals*, 1883.

Iquique, Peru. See Benchuca bug. 1835 Jul. 12-14 *Beagle* at. 1835 Jul. 13 CD landed and made short journey to saltpetre mines. *Geological diary*: DAR37.677-684. Barometer Measurements at in DAR39.158.

Ireland. 1827 May CD visited Belfast and Dublin at end of a tour in Scotland, his only visit to Ireland. Lizzy (Elizabeth Darwin, CD's daughter) went in 1866 Jun. 6 and returned a month later on Jul. 9 with Frank (Francis Darwin). Horace went and returned on Aug. 2 1872. George made two visits in Sept. 1 & Sept. 8 1882. Henrietta and R.B. Litchfield went in 1895 Oct. 5-24.

Iron, a supplement CD would take from 1841 with an entry in *ED's diary* on Sept. 2 "Ch. began Iron". Then William at under 2 years old too "began Iron". Bessy took in 1849, Leonard began in 1852, Lizzy in 1859. ED herself also took it.

Irvine, Harriet, 1859-1950? Born in Maer, Shropshire. 1876 Came as wet-nurse for Bernard Darwin at seventeen. Stayed with the family for the rest of her working life. Keynes, *Annie's box*, 2001, p. 338-39. Long description in B. Darwin, *Green memories*, 1928, pp. 15-16. 1881 census as housemaid, 1891 head housemaid. Raverat, *Period piece*, 1952, p. 151. After ED's death, stayed with Bessy Darwin until 1926.

Isitt, Virginia Lavinia, governess to the children of Emily Jesse, Tennyson's sister. ED wrote to I in reply to I's interest, via ED's niece (Snow?) to be CD's secretary. ED offered "to make the experiment for about a month". Evans, *Darwin and women*, 2017, pp. 207-08. Apparently she was not engaged.

Isle of May, Fife, Firth of Forth. CD visited with Ainsworth and Greville when at Edinburgh. William Francis Ainsworth, *Athenaeum*, (13 May), p. 604, 1882.

Isle of Wight, Hampshire. 1837 Nov. CD visited William Darwin Fox there. 1846 Sept. 12 CD and ED visited on day trip from BAAS meeting at Southampton. 1858

Jul. 17-Aug. 12 Family holiday at Sandown and Shanklin, between 21-27 Jul. 1868
Jul. 17-Aug. 20 Family holiday at Freshwater. CD photographed by J.M. Cameron.
Italian. First editions in: 1864 *Origin* (F706). 1870 *Orchids* (F2471). 1871 *Descent* (F1088). 1872 *Journal* (F211). 1874 *Coral reefs* (F2463). 1876 *Variation* (F920). 1877 *Forms of flowers* (F2473). 1878 *Climbing plants* (F863). 1878 *Expression* (F1200). 1878 *Insectivorous plants* (F1242). 1878 *Cross and self fertilisation* (F1269). 1882 *Earthworms* (F1407). 1884 *Power of movement* (F1347). 1907 Essay on instinct (F1447). 1919 *Autobiography* (F1522). 1960 On the tendency of species to form varieties (F368). 1960 *Foundations of the origin of species*. (F1562).

Jackson, Mrs. Wife of William J, she had been a nurse; "the most perfectly tidy person I ever saw, with a row of shiny black buttons down the front of her dress and an overwhelming sense of propriety". B. Darwin, *Green memories*, 1928, p. 13. ED recorded many encounters with Mrs J, starting 1882 Nov. 1- 1895 Dec. 31. A "little Jackson" was entered into her diary in 1889 Apr. 29.

Jackson, Benjamin Daydon, 1846-1927. Botanist on staff at Kew, in charge of first volume of *Index Kewensis*, 1880-1902 Secretary to Linnean Society. 1909, 1910 *Darwinia*, 1910, contains three essays published elsewhere, 1909, republished as a pamphlet with alterations; one gives a list of plants named for CD.

Jackson, Henry, 1839-1921. Classical scholar. BA, Cambridge, 1862 Fellow of Trinity, 1864 Assistant tutor, 1866 Praelector in ancient philosophy, 1875 Vicemaster, 1914. Regius Prof. of Greek, Cambridge, from 1906. CCD21. Corresponded with CD. Married, had twin sons. 1879 Dec. Francis Darwin and Bernard visited the Jacksons in Cambridge. Credited in *Expression*, 2^d edn., 1890, p. 250.

Jackson, William, a manservant at Down House. 1875 Succeeded Parslow as butler. "A little man with very red cheeks, little loose curly wisps of side whiskers; not very tidy and not at all smart, nor, I imagine, very efficient". B. Darwin, *Green memories*, 1928, p. 11. J made model of Down House in cork, once in Galton Collection at University College London, now at Down House. J also made a sentry-box in the orchard at Down House for Bernard. F. Darwin, *Story of a childhood*, 1920, p. 48. 1881 Sept. 27 served as a witness to the signing of CD's last will and testament. c.1882 Retired. 1882 J attended CD's funeral, walking in procession with Parslow behind the family mourners, but ahead of the official representatives. ED recorded some money matters in 1894 with J. J kept a scrapbook of newspaper cuttings at the time of CD's death and funeral tickets. Now in the Galton papers 6, UCL.

Jacobsen, Jens Peter, 1847-85. Considered the most important communicator of Darwinism in Denmark in the 19th century. Translation of *Origin* was based on the 5th edn (1869), 1871-72 (F643). His translation of *Descent* (F1050.1-2) was published in 13 parts from Oct. 1874 to Nov. 1875 in 1,250 copies.

Jäger [Jaeger], Gustav, 1832-1917. German zoologist and physicist of Stuttgart. [1869] Author of *Die Darwin'sche Theorie und ihre Stellung zu Moral und Religion*. 1875 Feb. 3 CD to J, thanking him for copy of his book *In Sachen Darwins ins-besondere con-*

tra Wigand, 1874. 1897 *Problems of nature*, London, translations of some of J's papers, prints two letters from CD thanking J for books sent.

James, Henry, 1843-1916. American author. At 26 visited Down House and met with the Darwins and their friends. ED recorded in 1869 Mar. 27 "Mrs C. Norton & Mr James Mrs Innes dined". The following day there was a lunch party and the Darwins found the "young Yankee" a delight with his "talkable" and straight-forwardness. J described his luncheon with CD at Down House as "beautifully be-nignant, sublimely simple". F.W. Dupee ed. 1956. *Autobiography: A small boy and others*, p. 515, transcribed in *Darwin Online*.

Jameson, Robert, 1774-1854. Mineralogist and natural historian. J edited several editions of Cuvier's *Essay on the theory of the earth*. CD studied the 5th edn of 1827 while at Edinburgh. 1804-54 Regius Prof. Natural History Edinburgh. CD found his lectures "incredibly dull". *Autobiography*. 1808 J founded Wernerian Society, Edin-burgh. 1823 J founded Plinian Society, Edinburgh. 1826 FRS. 1854 May? 29 CD to Hooker, about Forbes "I wish, however, he would not praise so much that old brown, dry stick Jameson." CCD5:105,

Jamieson, Thomas Francis, 1829-1913. Agriculturalist and geologist of Ellon, Ab-erdeen. Correspondent of CD and Lyell. 1862-77 Fordyce Lecturer in Agriculture, 1862 FGS. 1862 J was the first person to give correct solution to parallel roads of Glen Roy, *Quart. Jrnl. of the Geol. Soc.*, 19: pp. 235-59, 1863.

Jane, 1865?-79. Housemaid at Down House. Not the same person as Emily Jane. Head housemaid and leaving to get married. B. Darwin, *Green memories*, 1928, p. 15.

Jane, 1845?-61 ED made several payments to J and advanced her money.

Janet, Paul Alexandre René, 1823-99. French philosopher, writer and entomolo-gist. 1845-48 Prof. Moral Philosophy, Bourges. 1857-64 Prof. Logic Lycée Louis-le-Grand, Paris. 1863 Reviewed *Origin*. 1864-? Prof. Philosophy Sorbonne, Paris. 1866 CD to Wallace, "As for M. Janet he is a metaphyscian & such gentlemen are so acute that I think they often misunderstand common folk." CCD14:236.

Japanese. First editions in: 1896 *Origin* (F718). 1915 *Foundations* (F1563a). 1921 *Ex-pression* (1202b). 1949 *Coral reefs* (F319). 1881 *Descent* (F1099c). 1937 *Variation* (F921a) 1938 *Climbing plants* (F863a). 1938 *Earthworms* (F1407a). 1939 *Orchids* (F823a). 1949 *Forms of flowers* (F1300). (1954) *Journal* (F216). 1927 *Autobiography* (F1523c).

Jardine, Sir William, 7th **Bart**, 1800-74. Scottish cabinet naturalist, especially of birds. J's relict Hyacinth Symonds married Hooker. 1833-43 Edited *The Naturalist's Library*. 1859 CD sent 1st edn *Origin*. 1860 CD to Lyell, CD had had a letter from J who opposed CD on evolution, but his attack on CD's ornithological accuracy is worthless. CCD8. 1860 FRS. 1860 Reviewed *Origin*.

Jeanneret, Henry, 1802-86. Born in London. MD, LRCS. 1834-38 Practised dental surgery in Hobart Town, Tasmania. Treated CD in Feb. 1836. See note by CD in the *Sydney Mauritius notebook*, Chancellor & van Wyhe, *Beagle notebooks*, 2009 and J.

Hayman & J. van Wyhe, Charles Darwin and the dentists. *Jrnl. of the Hist. of Dentistry.* (2018) vol. 66, no. 1, pp. 25-35.

Jeffreys, John Gwyn, 1809-85. Lawyer. Conchologist and malacologist. 1840 FRS. 1851 J sent CD his cirripede collection. J received a presentation copy of *Origin.* 1859 J was anti-*Origin*, objecting that there were insufficient intermediate gradations amongst fossil molluscs etc. CD to J 29 Dec. 1859 CCD7:460, 1860 Jan. 4 CD to Lyell. CCD8:15. In 1868 Sept. 16, called on CD with Mr Flower & Dr Duncan.

Jenkin, Henry Charles Fleeming, 1833-85. Scottish electrician and engineer. 1865 FRS. 1866 Prof. Engineering University College London. 1867 Jun. Review of *Origin* in *North British Rev.*, vol. 46 criticizing theory on the basis that single variations could not be perpetuated; hence the term 'swamping argument'. 1867 CD to Kingsley, the review is telling and hostile, but lacking in knowledge. CCD15. 1868 Prof. Engineering Edinburgh. 1869 CD to Hooker, "Fleming Jenkins has given me much trouble, but has been of more real use to me, than any other Essay or Review." CCD7:21, 1887 Francis Darwin, "my father, as I believe, felt the review to be the most valuable ever made on his views". LL3:107.

Jenner, Sir William, Bart, 1815-98. Physician. 1837 Member Royal College of Surgeons. 1854-79 Physician at University College London. 1861 Physician to Queen Victoria. 1863 CD consulted, van Wyhe ed., 'Journal', (DAR158). 1864 FRS. 1868 1st Bart. 1877 KCB. J called in 1864 Mar. 20, Apr. 10 and May. 22. CD had one of the worst attacks of sickness that year. *ED's diary.* In 1864 Jul. 5, ED went to see J.

Jenny, c.1834-39. A female Bornean orangutan in the London Zoological Gardens, purchased in Nov. 1837. Named Lady Jane, which was shortened to Jenny. CD visited her in late March and made observations and experiments relevant to his developing theory of evolution. See CD to Susan Darwin [1 Apr. 1838] CCD:2:80. CD's notes on the visit in DAR191, transcribed in John van Wyhe & Peter C. Kjærgaard, Going the whole orang: Darwin, Wallace and the natural history of orangutans. *Studies in Hist. and Philosophy of Sci. Part C: Studies in Hist. and Philosophy of Biol. and Biomedical Sciences.* vol. 51, 2015, pp. 53-63 and in *Darwin Online.*

Jenyns, Leonard, later Blomefield, 1800-93. Anglican priest and naturalist. Vicar of Swaffham Bulbeck, Cambridgeshire. Member of many learned societies. Henslow's brother-in-law. CD discussed evolution with before *Origin.* 1826 Founding member of Zoological Society. 1831 Invited to join FitzRoy on *Beagle* but declined on grounds of health and parish duties. 1840-42 Wrote *Fish* for *Zoology of the Beagle.* His list of the specimens is in the Zoology Museum Cambridge. 1844 Founding member of Ray Society. 1859 CD sent J 1st edn of *Origin.* 1862 J wrote *Memoir of John Stevens Henslow*, with recollections by CD, pp. 51-5 (F830). 1871 J changed his surname to Blomefield on inheritance. 1876 CD in autobiography about J "At first I disliked him, from his somewhat grim and sarcastic expression...but I was completely mistaken, and found him very kind-hearted and with a good stock of humour".

And biographical note. ML1:49. 1887, 1889 *Chapters in my life*, 1889, Recollections of CD in 1882 in DAR112.A67-A68, transcribed in *Darwin Online*.

Jesse, **George Richard**, 1820-98. Civil engineer. Anti-vivisectionist. 1875 *Evidence given before the Royal Commission on Vivisection*. 1881 J had written, very politely, to CD on the subject. 1881 J's pamphlet (F1356) reprints CD's letter to Frithiof Holmgren, which had appeared in *The Times*, Apr. 18 (*Shorter publications*, F1352).

Johnson, **Charles Richardson**, 1813-82. 1832 May Joined *Beagle* for 2^d voyage. Acting mate on return of *Beagle* from 2^d voyage. 1879 Vice-Admiral. 1882 Died same week as CD.

Johnson, **Henry**, 1802?-83. Physician. Contemporary of CD at Shrewsbury school. 1826 J was at Edinburgh with CD. 1826 CD to his sister Caroline, saying that J had changed his lodgings for the third time. CCD1:25. J returned to Shrewsbury to practice. 1880 CD to J about excavations at Wroxeter and worms. De Beer 1968, p. 74.

Johnson, **James**, 1858-? Footman at Down House in 1881 census. Told Bernard about Columbus. 1882 Sept. 19, Francis wrote J "was going away to a new place" which made Bernard and Pauline cry. F. Darwin, *Story of a childhood*, 1920, pp. 44, 57.

Jones, **Henry Bence**, 1813-73. Physician and chemist. Of St. Georges Hospital. CD's physician for many years. 1846 FRS. 1865 Jul. CD began to consult. 1865 Nov. 1 saw CD. 1866 3 Jan. CD to J, he was taking muriatic acid with cayenne pepper and ginger suited for his stomach disorder which suited him "excellently". 1866 10 Feb. J. to CD, regarding his ongoing flatulence "I wish you could get a rough pony & be shaken once daily to make the chemistry go on better". CCD14. CD began riding Tommy in June 1866. 1866 Apr. 27 CD met at Royal Society soirée.

Jones, **Richard**, 1790-1855. Economist. Known as "Old Jones"; moved in scientific circles and was partial to a lot of wine, especially port; he liked to share his food and drink with young men. 1835 Chair political economy and history East India College at Haileybury as successor to Malthus. 1838 CD to Lyell, "Old Jones" was going to quarrel at the Newcastle meeting of the BAAS. CD dined with. CCD2:97.

Jones, **Samuel**, 1836-? Coachman at Down House from 1858. 25 years old in 1861.

Jones, **Thomas Rymer**, 1810-80 Dec. 10. Physician and naturalist. 1836-74 Prof. Comparative Anatomy, King's College, London. 1840-44 Fullerian Prof. Physiology, Royal Institution. 1844 FRS.

Jones, **Thomas Rupert**, 1819-1911. Geologist and palaeontologist. Authority on Foraminifera and Entomostraca. Editor of works by Mantell. 1849 Assistant secretary Geological Society London. 1854 Feb. 18 CD to Lyell, about a meeting of the Geological Society, J had told CD about Prestwich's views on red clay with flints. 1862 Prof. Geology, Royal Military College, Sandhurst. 1872 FRS. 1890 Lyell Medal, Geological Society.

Jordan, **David Starr**. 1851-1931. Regretted not calling on CD when he was in London. 1883 visited Down and heard from several persons CD's extensive and judicious charities. J was told how CD, during a drought, rode out to see who needed water. Recollection in *The days of a man*, 1922, 1:272-73 transcribed in *Darwin Online*.

Jordan, John, 1839 end of, manservant at CD's house, 12 Upper Gower Street. Mentioned in CD to ED [3 June 1844]. CCD3.

'Journal', a small 152-page notebook (DAR158.1-76) in which CD recorded, from Aug. 1838-Dec. 1881, key dates and details of his life and especially the progress of his work and other personal matters. Francis Darwin wrote "It is unfortunately written with great brevity, the history of a year being compressed into a page or less, and contains little more than the dates of the principal events of his life, together with entries as to his work, and as to the duration of his more serious illnesses." LL1:iv. It contains such important and oft quoted passages as: "In July opened first note Book on 'transmutation of Species'. — Had been greatly struck from about month of previous March on character of S. American fossils — & species on Galapagos Archipelago. — These facts origin (especially latter) of all my views." It was first published in LL but highly interleaved with other material. It appeared more fully in ML. It first appeared in its entirety in English in 1959, edited by Gavin de Beer from a copy by an unknown copyist. The original manuscript was not re-discovered until 1962. It is reproduced year by year as an appendix to each volume of the CCD. Published and edited in full by John van Wyhe in 2006 in *Darwin Online*:
http://darwin-online.org.uk/content/frameset?viewtype=side&itemID=CUL-DAR158.1-76&pageseq=3

Journal of researches, see also *Voyage of a naturalist*, *Voyage of the Beagle*. CD's first published book and probably his most widely read. "The success of this my first literary child always tickles my vanity more than that of any of my other books." *Autobiography*, p. 116. "Charm arising from the freshness of heart which is thrown over these virgin pages of a strong, intellectual man and an acute and deep observer". *Quarterly Rev.* 1845 CD sold the copyright of the 2^d edn of 1845 to John Murray for £150 and so made no profit from it or from its many subsequent printings or translations. Notes for 2^d edn which "may be useful in species theory" in DAR46.2. Selection of UK editions: 1839 As vol. 3 of R. FitzRoy, ed. *Narrative of...H.M.S. Adventure and Beagle*, sub-title *Journal and Remarks* (F10), CD's text was completed and printed in 1838. 1839 First independent issue of same text, *Journal of researches into the geology and natural history* etc. (F11). 1840 Reissue (F12). 1845 2^d edn, *Journal of researches into the natural history and geology* etc. (F13). 1860 Edn from stereos with postscript added. Final definitive edn. (F20). 1890 Edn with postscript incorporated in text. (F58). 1890 First Murray illustrated edn, with prefatory notice by John Murray (F59). 1916 English braille edn, based on 1890 (F168). First foreign editions, in whole or in part: 1844 German (F188). The 1st German is the only translation based on the 1st English. 1846 USA (F16). 1860 French (F180). 1870-71 Russian (F226). 1872 Swedish (F259), Italian (F211). 1875 German of 2^d edn (F189). 1876 Danish (F174). 1887 Polish (F223). [1891] Dutch (F176). 1900 Greek (F206). 1902 Spanish (F249). 1913 Hungarian (F208). 1930 Hebrew (F207). 1949 Armenian (F169), Estonian (F179), Serbo-Croat (F244). 1950 Slovene (F248). 1951 Georgian (F187). 1954 Japanese (F216). 1956 Czech (F171). 1958 Romanian (F225). 1963 Lithuanian (F222). 1967 Bulgarian (F170).

Judd, John Wesley, 1840-1916. Geologist. Prof. Geology Royal College of Science London. Correspondent of CD and visitor to Down House. 1877 FRS. 1891 Wollaston Medal, Geological Society. 1877 Reviewed *Geology*, *Nature*, 1 Feb., pp. 289-90.

Jukes, Joseph Beete, 1811-69. Geologist and naturalist. 1848 CD to Hooker, "the man not content with moustaches, now sports an entire beard, & I am sure thinks himself like Jupiter tonans." CCD4:140. 1850-69 Director of the Geological Survey of Ireland. 1853 FRS. 1860 J was pro-*Origin*. CCD8:112.

Jumbo. 1883 a black cat at Down House. Sat in a game of "spinning a knife" with Francis and Bernard Darwin. F. Darwin, *Story of a childhood*, 1920, p. 58-59.

Justice of the Peace. 1857 3 Jul. CD appointed magistrate for the County of Kent. Records of attendance at the bench listed in Itinerary (28 occasions). 1869 CD to T. Thompson, Secretary of the Westerham Friendly Club: "Mr W. Reeves in this place has called on me as a County Magistrate to consult me on the best means of obtaining the payment which he states is due to him from the Westerham Club." CCD17:425. Westerham is about 5 miles from Down. 1881 Jan. 28 CD to Romanes, he was, as a magistrate, giving orders daily to allow animals to cross roads, at a time of swine fever. *Calendar* 13029.

Kay, William, 1807-61. Physician of Clifton, Gloucestershire. Naturalist friend of CD at Edinburgh. 1827 Secretary and later president of Plinian Society. 1828 K and CD produced story of a 'Zoological Walk' to beach of Portobello. Manuscript in DAR5.A49-A51, transcribed in *Darwin Online*. "Together Kay and Darwin produced an account of a 'Zoological walk' to the nearby beach of Portobello—possibly completely fictitious, although it seems more likely that a real expedition served as a basis for their humour. Everything that could have gone wrong, they implied facetiously, went wrong." Janet Browne, *Voyaging*, 1995, pp. 78-80.

Kannada. First edition in: 2008 *Origin* (F2401).

Kazakh. First edition in: 1996 *Origin* (F2153).

Keen, George, 1794-1884 & **Mary Yates K**, 1802/3-72, British residents in La Virgen de los Dolores, Argentina, near Mercedes. (Spelt 'Keane' in *Journal*, p. 181) 1833 Nov. 22-26 CD visited their estancia on river Beguelo ("Berguelo") and collected a skull of "Megatherium" (actually Toxodon) from a nearby hill, Cerro Perico Flaco (CD called it Cerro del Pedro Flaco). Winslow, Mr. Lumb and Masters Megatherium: an unpublished letter by Charles Darwin from the Falklands, *Jrnl. of Historical Geography*, 1, 1975. Chancellor & van Wyhe, *Beagle notebooks*, 2009. CCD1.

Keith, Sir Arthur, 1866-1955. Surgeon, anthropologist and darwinian. K was much involved in the purchase of Down House for the BAAS and its later acquisition by the Royal College of Surgeons. K retired to Homefield, a small house on the western side of the Down House grounds. 1908-33 Conservator Hunterian Museum, Royal College of Surgeons, London. 1913 FRS. 1921 Kt. 1942 "A postscript to

Darwin's *Vegetable mould through the action of worms*", *Nature*, 149: p. 716. 1955 *Darwin revalued*, which contains a last chapter on the later history of Down House, as well as much other information which is not available elsewhere.

Kelvin, **Baron**, see Sir William Thomson.

Kemp, William. Scottish amateur geologist of Galashiels, Selkirk. "Almost a working man", "partially educated", "a most careful and ingenious observer". 1843 K sent CD seeds from a sandpit near Melrose, found under 25 feet of white sand, which germinated into a common *Rumex*, an unrecognized species of *Atriplex*, and two species of *Polygonum*. CD forwarded the seeds to Lindley. K sowed part of the seeds and sent the plants to Henslow. The case in the end not proven. CCD2, Barlow, *Darwin and Henslow*, 1967, p. 151. *Annals and Mag. of Nat. Hist.*, 13, p. 89-91, 1844 although signed by K was actually drafted by CD (*Shorter publications*, F1918). Rough draft made for K in DAR50. See CCD2 Appendix VI.

Kempson, William John (John), 1835-77. 1864 Married Louisa Frances Wedgwood and had 4 children: 1 Jessie 1867-1939; 2 Hester Louisa 1869-1930; 3 John Wedgwood 1870-1958 and 4 Lucy C 1874-1958.

Kendall, Thomas, 1778-1832. Not in holy orders but a schoolmaster. 1814 Early missionary for Church Missionary Society in New Zealand, arriving 1814. 1815 Author of the first book published in New Zealand, *The New Zealander's first book!*, Sydney printed. 1822 K was dismissed for living with a Maori girl. 1835 CD mentions K (spelling "Kendal") in "Moral state of Tahiti, New Zealand etc.", 1836, q.v. (*Shorter publications*, F1640) in company with John King, but CD did not meet.

Kennedy, Mr, Santiago, Chile. 1834 Aug. 28 CD to FitzRoy: "Corfield took me to dine with a Mr Kennedy, who talks much about the *Adventure* and *Beagle*; he says he saw you at Chiloe". Keynes, *Beagle record*, 1979, p. 235. CCD1:406.

Kennedy, Dr Benjamin Hall, 1804-89. Classical scholar. 1836-66 Headmaster of Shrewsbury School. 1867-89 Regius Prof. Greek, Cambridge. 1881 Oct. CD saw "old Dr. Kennedy of Shrewsbury" at Cambridge.

Kensington Square, London. 1883-1903 no. 31, home of Richard Buckley Litchfield and Henrietta. In 1883 Jun. 6, ED went and returned to Down with the Litchfields on the 9th. She stayed again in 1883 Oct. 3-6. 1884 Feb. 29-Mar. 4 she was visited by Lady Lyell there. There were many visits after with a last mention in 1896 Mar. 17. "The old house at Kensington Square had a very strong flavour of its own. It was a peculiar kind of earthly paradise". Raverat, *Period piece*, 1952, p. 130.

Kent, William, ?-1882. 1831 Jul. Passed as Surgeon. 1833 Dec. 7 Joined *Beagle* as Assistant Surgeon. 1836 Oct. Assistant Surgeon on return of *Beagle* from 2d voyage. 1838 Appointed Surgeon.

Keppel Island, Falklands. 1855 Mission to house Fuegians started, the building called Sulivan House after Admiral Bartholomew James Sulivan. 1860 Jemmy Button visited mission. 1898 Transferred to Tekeeneka. 1911 Old building sold.

Kerner, Anton Joseph, Ritter von Marilaun, 1831-98. Austrian botanist. 1878 CD wrote prefatory letter to translation by W. Ogle of K's book *Die Schutzmittel der Blüthen gegen unberufene Gäste*, 1876, *Flowers and their unbidden guests*, (F1318). See F2546.

Kew Gardens, see Royal Botanic Gardens, Kew.

Keynes, Sir Geoffrey Langdon, 1887 Mar. 25-1982 Jul. 5. Physician, bibliographer and bibliophile. Younger brother of John Maynard K. 1917 Married Margaret Elizabeth Darwin. 4 sons. 1955 Kt. 1981 FBA.

Keynes, John Neville, 1852-1949, economist and father of John Maynard Keynes, became a Fellow of Pembroke College, Cambridge, in 1876. 1877 diary entry on CD's honorary LLD degree in CUL Add.7831.2, transcribed in *Darwin Online*.

Keynes, Richard Darwin, 1919 Aug. 14-2010 Jun. 12. Son of Sir Geoffrey Langdon K. The first member of the generation of Darwins to carry the continuous Darwin Fellowship of Royal Society into sixth generation from Erasmus Darwin [I]. 1959 FRS. 1972- Prof. Physiology Cambridge. 1979 Editor of *The Beagle record*. Contains much unpublished material including extracts from Covington's diary, many plates mostly by Martens, list of 307 Martens watercolours. 1988 Editor of *Charles Darwin's Beagle diary* (an edited and improved version of Nora Barlow's transcription.) 2000 *Charles Darwin's zoology notes and specimen lists from H.M.S. Beagle*. 2002 *Fossils, finches and Fuegians. Charles Darwin's adventures and discoveries on the Beagle, 1832-1836*. Account of the voyage with a focus on science.

Keyserling, Alexander Friederich Michael Lebrecht Nikolaus Arthur, Graf von, 1815-91. Baltic German geologist and palaeontologist, born in Latvia, then part of Russia. CD had sent a presentation copy of *Origin*. K is referred to in Historical sketch in *Origin*. See J.A. Roger, *Isis*, 64: pp. 487-8. 1860 K wrote to CD about *Origin*, stating his belief that species change, natural selection explains well adaptation, "but thinks species change too regularly, as if by some chemical law, for natural selection to be sole cause of change". CCD8:188.

King, Sir George, 1840-1909. Physician and botanist. 1871-98 Superintendent of Royal Botanical Garden Calcutta. 1873 K sent CD *Aldrovanda* (waterwheel plant) for *Insectivorous plants*, and also helped with *Earthworms*. See CCD20-21. 1887 FRS. 1891 Linnean Medal, Linnean Society. 1898 KCIE.

King, John, 1787-1854. Missionary, but not in holy orders, a shoemaker by trade. 1810 First missionary for Church Missionary Society in New Zealand. 1835 Dec. CD met Mrs K and their son, but K was away. "Moral state of Tahiti, New Zealand etc.", p. 231. (*Shorter publications*, F1640).

King, Philip Gidley [I], 1758-1808. Captain, RN. Father of Philip Parker K, grandfather of Philip Gidley K [II] qq.v. 1800-06 3d Governor New South Wales.

King, Philip Gidley [II], 1817-1904. Son of Philip Parker K. Naval officer. Born in Parramatta, Sydney, New South Wales, Australia. Midshipman on 1st and 2d voyages of *Beagle*. CD very friendly with. 1829 pencil sketch by his father in MLNSW, ML ZC767, p. 142. Pencil sketch of FitzRoy by K in MLNSW, PXC 767/ no. 50 and Keynes, *Beagle record*, 1979, p. 16. 1832 Apr. 25 CD at Botafogo Bay to Caroline

Darwin "I believe King is coming to live here, he is the most perfect pleasant boy I ever met and is my chief companion". CCD1:226. 1836 Jan. 12 K left *Beagle* to remain with his father at Sydney. 1880- K was a member of Legislative Council of Sydney. LL1:221. 1890 K drew the diagrammatic layout of *Beagle* which first appeared in *Journal*, 1890. Section of *Beagle* by K 1890 at Hallam Murray's request, found by Geoffrey Keynes in map pocket of *Narrative*, now at MLNSW, with a letter to Capt. Fisher, reproduced in Keynes, *Beagle record*, 1979, p. 21 et al, also a drawing of quarterdeck and poop cabin at CUL, p. 39. Recollections of CD in DAR112.A74-A75 and MLNSW, FM4/6900/, both transcribed in *Darwin Online*. A copy of the latter, in another handwriting, is in DAR107.11-18.

King, **Philip Parker**, 1791-1856. Born Norfolk Island, Australia. Son of Philip Gidley K [I]. Father of Philip Gidley K [II] q.v. Naval Officer. Surveyor and geologist. Biography D.F. Branagan, 1985, *Spec. Publ. Soc. Hist. Nat. Hist.*, 3, pp. 179-93. Settled in Australia with rank of Rear Admiral. 1824 FRS. 1826-31 K commanded, as Captain, *Adventure* on 1st voyage of *Adventure* and *Beagle*. Collected plants which Robert Brown was dilatory in identifying. *Narrative* edited by FitzRoy and published as first volume of *Narrative of the surveying voyages of His Majesty's ships Adventure and Beagle*. 1836 Jan. 23 CD spent evening with K at Dunheved outside Sydney. 1836 Jan. 28 CD stayed with K 30 miles from Sydney and visited his relatives, the MacArthurs, for lunch "beautiful very large country house" which Keynes, *Beagle record* identifies as Camden Park. p. 346.

King George's Sound, Western Australia. 1836 Mar. 6-14 *Beagle* anchored there, CD landed. *Geological diary*: DAR38.858-881, transcribed in *Darwin Online*. DAR40.92

King, **William James**, **Colonel**, Of Hythe, Kent. K sent CD specimens of rock-pigeon and is acknowledged as "Colonel King" in *Variation*, 1:184.

Kingsley, **Charles**, 1819-75. Anglican clergyman. Author and naturalist. Curate and later Rector of Eversleigh, Hampshire. 1859 CD sent 1st edn of *Origin*. "That the Naturalist whom, of all naturalists living, I most wish to know & to learn from, should have sent a sciolist like me his book, encourages me at least to observe more carefully, & think more slowly." CCD7:379. 1860-69 Regius Prof. Modern History Cambridge. 1860 CD to Henslow telling him that the "celebrated author and divine" who is quoted in 2d edn of *Origin* (p. 481) was K. CCD8. 1863 *The water-babies*. 1867 CD to K about Duke of Argyll's *Reign of law* and Fleeming Jenkin's review of *Origin*. 1867 Sent K 4th edn of *Origin*. CCD15. 1873 Canon of Westminster. 1877 [CD letter on Stock Dove]. In *Charles Kingsley*. F1951.

Kinnordy House, near Kirriemuir, Forfarshire, Scotland. Home of Sir Charles Lyell's father and later his.

Kippist, **Richard**, 1812-82. Botanist. 1842-81 Librarian of Linnean Society. CD often wrote to K to borrow books from the library of the Linnean Society.

Kirby. (Possibly Henry Kirby) Cambridge friend of CD from Clare College. 1831 K was interested in going with CD to Canary Islands.

Klein, Edward Emanuel, 1844-1925. Slavonian-born British bacteriologist. MD Vienna. 1871 Moved to London. Brown Animal Sanatory Institute, University of London. Lectured St. Bartholomew's Hospital. 1874 K helped CD with *Insectivorous plants*. CCD22. 1874 Provided evidence for the Vivisection Commission showing indifference to animal suffering. This prompted CD to give evidence for the Committee on 3 Nov. CCD23; correspondence between Huxley and CD. 1875 FRS. 1889-91 Prof. Bacteriology, College of State Medicine, London.

Knight, Thomas Andrew, 1759-1838. Distinguished plant hybridizer. *A selection from the physiological horticultural papers...a sketch of his life*, London 1841. 1805 FRS. 1811-38 President London Horticultural Society. 1868 CD drew extensively on his work in *Variation*. Knight's Law, also called Knight-Darwin Law, "nature abhors perpetual self fertilisation". ML2:250. Francis D., *Annals of Botany*, 13: 1899 p. 13.

Knole Park, Sevenoaks, Kent. Seat of Baron Sackville. 1846 Sept. 22 CD, ED and Susan Darwin made day trip to. 1850 Aug. 6 ED wrote in her diary "went to Knole". 1860 Aug. 23 there was a "Review at Knole". In 1861 Jun. 14, the Darwin family went and enjoyed a hot day. Another excursion with Etty only in 1861 Oct. 3.

Koch, Eduard, 1838-97. German publisher. 1867 took over Schweizerbart'sche Verlagsbuchhandlung publishing. Published CD's works translated by J.V. Carus

Koch, Friedrich Karl Ludwig, 1799-1852. German mineralogist. CD sent him copy of *Fossil Cirripedia*. *Lychnos*, 1948-49. 1851 K sent CD fossil cirripedes.

Kölliker, Rudolph Albert von, 1817-1905. Swiss biologist. 1844 Prof. Physiology and Comparative Anatomy Zürich. 1847 Prof. Physiology, Microscopy and Comparative Anatomy Würzburg. 1860 Foreign Member Royal Society. 1860-64 At some time between 1860 and 1864 K visited Down House. 1860 CD to Huxley who had suggested K as possible translator of *Origin* into German, CD preferring Bronn. 1861 *Entwicklungsgeschichte des Menschen und der höheren Thiere*, Leipzig. 1897 Copley.

Král, Josef Jiří, 1870-1951. Born in Loužná, Czech Republic, emigrated to America. Translated *Descent* into Czech in 1906 (F1048), published in Chicago, USA.

Kollmann, Julius Constantin Ernst, 1834-1918. German anatomist, zoologist and anthropologist. 1876 K to CD on atavism, extra digits and rudimentary organs. CCD24:78, *Variation*, 1:459. 1878 Prof. Anatomy Basel, Switzerland.

Korean. First editions in: 1957 *Origin* (F732). 1965 *Autobiography* (F1525). 1986 *Descent* (F2358). 1998 *Expression* (F2357). 2006 *Journal* (F2348).

Kororareka [Russell], Town on Bay of Islands, New Zealand. 1835 "Capt. Fitz-Roy, Mr Charles Darwin and the Officers of H.M.S. 'Beagle' 15.0.0" subscription to building fund for the chapel there. Manuscript subscription list 1834-41, at Russell Centennial Museum. 1836 "Placing a church at the headquarters of iniquity, at such a notorious place as Kororadika (the older spelling), is certainly a bold trial... This little village is the very stronghold of vice". "Moral state of Tahiti" p. 231. (*Shorter publications*, F1640) 1844 Renamed "Russell". 1873 Chapel renamed Christ Church.

Koskimies, Aarno Rafael, 1882-1971. Translated *Origin* into Finnish in booklet form 1913-17, (F2023). See Engels & Glick eds., *The reception of Charles Darwin in Europe*, 2008.

Kovalevskii, Aleksandr Onufrievich, 1840-1901. Russian embryologist of Polish descent. K was the first to point out the chordate affinities shown by ascidian tadpoles. Brother of Vladimir Onufrievich K. 1885 Foreign Member Royal Society.

Kovalevskii, Vladimir Onufrievich, 1842-83. Russian palaeontologist. ("Kovalevsky" in *ED's diary* Kowalewsky). Brother of Aleksandr Onufrievich K. Married Sof'ya Vasil'yevna Krukovskaya. 1867-68 K translated *Variation* into Russian. Visited Down House 1867, 1869 stayed Sept. 30-Oct. 1. 1870. Visited CD 1871 Oct. 15 and 1872 Oct. 22. 1883 Apr. 15 Committed suicide.

Kōzu Senzaburō, first to translate CD's work into Japanese. 1881 *Descent* (F1099c).

Krause, Ernst Ludwig, 1839-1903. German botanist. Also published under the name Carus Sterne, an anagram of his name. 1879 Feb. K's biography of Erasmus Darwin [I] appeared in *Kosmos*, the number being a Gratulationsheft for CD's 70th birthday. 1879 An English translation (by W.S. Dallas), with introductory matter by CD had K's own alterations to his part (F1319). It was this edition which so offended Samuel Butler. Butler's copy with his manuscript notes is in the British Library. 1880 German translation of the 1879 English edition (F1323). 1885 *Charles Darwin und sein Verhältniss zu Deutschland*, 1885-86 *Gesammelte kleinere Schriften*, vol. I contains "Humble bees", translated from CD's manuscript by K (F1584). The sketch is reproduced on p. 233 of this volume. Portrait of K in CCD27:317.

Krohn, August David, 1803-91. Russian-born invertebrate anatomist and embryologist of Bonn. 1860 Sept. 28 CD to Lyell, K had pointed out errors in interpretation of CD's anatomy of cirripedes "with the utmost gentleness & pleasantness" in *Archiv für Naturgeschichte* 25 (pt 1), pp. 355-64. CCD8:396. CD's recanting of his views is in *Nat. Hist. Rev.*, 3: p. 115 (*Shorter publications*, F1722).

Kruell, Gustav, 1843-1907. German Artist. Emigrated to the USA in 1873. 1884 Wood engraving after Maull & Polyblank photograph of CD, the profile, for *Harper's Mag.*, Oct. LL1: frontispiece. 1887 Wood engraving of CD from Elliott & Fry photograph for LL3: frontispiece.

Krukovskaya, Sof'ya Vasil'yevna Korvin-, 1850-91. Russian mathematician. Published under the name Sophie Kowalevski. After moving to Sweden changed name to Sonya. 1868 Married Vladimir Onufrievich Kovalevskii. 1869 Visited Down House with husband. 1889 Prof. of Mathematics, Stockholm, Sweden.

Kynaston, Sir Edward, Bart, 1758-1839. Vicar of Kinnerley, Shropshire. 1822 2d Bart. 1831 Sept. 6 CD to his sister Susan, describes FitzRoy as a "dark but handsome edition of Mr Kynaston". CCD1:144.

Lacaze-Duthiers, Félix Joseph Henri de, 1821-1901. French invertebrate zoologist. 1854 Prof. Zoology Lille with the support of Milne-Edwards. 1865 Prof. Muséum nationale d'histoire naturelle, Paris. 1868 Prof. University of Paris. 1872 CD to Quatrefages, "I am gratified to hear that M. Lacaze-Duthiers will vote for me [for Académie des Sciences] for I have long honoured his name". CCD20:23. The election was for the zoology section. CD did not get in, but was elected for the botany section in 1878. 1897 Foreign Member Royal Society.

Lacy, Dyson, Australian. Of Aramao, Bacao near Rockhampton, Queensland. 1868 L answered CD's *Queries about expression* (forwarded by Edward Wilson, a neighbour of CD in Down). CCD16:162.

Lamarck, Jean-Baptiste Pierre Antoine de Monet, Chevalier de, 1744 Aug. 1-1829 Dec. 18. French naturalist and evolutionist. Widely remembered today for his theory of the inheritance of acquired characteristics, called (since 1900) Lamarckism or soft-inheritance. In fact not original or exclusive to Lamarck at all, nor was this the defining characteristic of his theory. 1801 *Système des animaux sans vertèbres*, a major work on the classification of invertebrates. 1809 L's main work, *Philosophie zoologique*, Paris. 1815-22 *Histoire naturelle des animaux sans vertèbres*, 7 vols. 1844 CD to Hooker, "Heaven forfend me from Lamarck nonsense of a 'tendency to progression' 'adaptations from the slow willing of animals' &c". CCD3:2. 1844 CD to Hooker, "Lamarck's [book] which is veritable rubbish", but praises *Animaux sans vertèbres*. LL2:29. 1845 CD to Hooker about L "in his absurd though clever work has done the subject harm, as has Mr Vestiges". CCD3:253. 1861 CD discusses L's views in "Historical sketch" in *Origin* 3ᵈ edn, "This justly celebrated naturalist". "He first did the eminent service of arousing the attention to the probability of all change...being the result of law, and not of miraculous interposition".

Lambert, Charles, 1793-1876. A Franco-British entrepreneur who made a fortune with copper and silver mining and smelting in Chile. In the *Santiago notebook*, Chancellor & van Wyhe, *Beagle notebooks*, 2009.

Lamont, Sir James, Bart, 1828-1913. Explorer, traveller, sportsman and geologist. FGS. Of Knockdow, Argyllshire. 1860 Mar. 5 CD to L in reply to his letter of 23 Feb. about *Origin*. CCD8:120. 1861 *Seasons with the sea-horses*. L sent CD a copy. CD replied about whales and bears. The book, p. 17, contains an important statement about the relationship between British red grouse and Scandinavian willow grouse, and, p. 277, quotes whale-bear story, from 1ˢᵗ edn of *Origin*, p. 184, in full, one of the few reproductions of it in CD's lifetime except in 1860 USA editions of *Origin*. CCD9:37. 1865-68 MP for Buteshire. 1910 1ˢᵗ Bart.

Lane, Edward Wickstead, 1823?-89. Physician and proprietor of Moor Park Hydropathic Establishment, near Farnham, Surrey. Later at Sudbrooke Park, Petersham, Surrey. Son-in-law of Lady Drysdale. CD liked L and visited his establishment between 1857-59. See Browne, *Power of place*, 2002, pp. 63ff. 1882 L was on "Personal Friends invited" list for CD's funeral. L gives his recollections of CD in W.B. Richardson, *Lecture on Charles Darwin* (1882), transcribed in *Darwin Online* and LL1:131 with quotation. R. Colp, Charles Darwin, Dr. Edward Lane, and the 'Singular Trial' of Robinson v. Robinson and Lane. *Jrnl. of the Hist. of medicine and allied sciences*. 1981 Apr. 36(2):205-13. Summerscale, *Mrs Robinson's disgrace*, 2012.

Lane, H.B., Australian of Belfast, Victoria, police magistrate and warden. 1868 Jun. 24 L answered CD's *Queries about expression* via Robert Brough Smyth. CCD16:675.

Lane, Mary Margaret, née Drysdale, 1823-91. Daughter of Lady Elizabeth Drysdale. 1847 Married Edward Wickstead L. CD enjoyed discussing popular fiction with her as she was an avid reader of volumes from Mudie's Lending Library. Summerscale, *Mrs Robinson's disgrace*, 2012, p. 76.

Lane, Richard James, Physician? Brother of Edward Wickstead L. 1860 L met CD at Sudbrooke Park Hydropathic Establishment, Petersham, Surrey.

Langdon, Miss, Governess to the Wedgwoods at Maer. "When I knew her in her latter days she was certainly the most unattractive old lady I ever saw, nearly stone deaf, with a harsh countenance, and a voice like a parrot's". ED2:155. 1854 L was taken in by Sarah Elizabeth Wedgwood [II] at The Ridge, Hartfield.

Langton, Algernon, 1781-? Soldier, later Anglican clergyman. Uncle of Charles L. 1820 Married Marianne Drewe. 1 son, Bennet L.

Langton, Charles, 1801-86. Anglican clergyman. Nephew of Algernon L. Had a weak chest. Lost nine siblings through consumption. Before 1831 L had been tutor to Lord Craven's children. 1832 Married 1 Charlotte Wedgwood. 1 surviving son, Edmund. 1832-41 Vicar of Onibury near Ludlow. 1839 Sept. 21 the Darwins went with L to Shrewsbury. 1841 L lost his faith and resigned living. 1841-47 L lived at Maer. 1844 Jul. 6 L went with the Darwins to Sevenoaks. They stayed for two days in 1845 May 9-10, CD was away in London. 1847-63 L lived at Hartfield Grove, Hartfield, Sussex, which he left after death of first wife. 1849 Mar. 3 L came to Petley's to meet with the Darwins who were on the way to Chevening and London. Many recorded visits from 1851-59. 1863 CD and ED stayed there. 1863 Married 2 Emily Catherine Darwin. s.p. L moved to Shrewsbury and, after death of second wife, moved into lodgings "at Mrs Tasker's".

Langton, Edmund, 1841-75. Only surviving child of Charles L and Charlotte Wedgwood. CD's second cousin. His birth brought "the most intense joy" and the Allen aunts were rapturous. Was "the delight of the Maer household." *ED's diary* in 1875 Nov. 27, "Edmund's death". 1867 Married Emily Caroline Langton Massingberd. 4 children 1. Charlotte Mildred, 2. Stephen, 3. Mary, 4. Diana.

Langton, Stephen Massingberd, 1869-1925. Son of Edmund L and Emily Caroline. 1895 Married Margaret Lushington. Was lent the Grove for their honeymoon. Judge Vernon Lushington, the bride's father was most grateful. The Langtons saw ED often. 1896 Sept. 19 ED recorded one final time, "Stephen & Margt".

Lankester, Sir Edwin Ray, 1847-1929. Zoologist and evolutionary biologist. 1870 CD's Note on the age of certain birds in L's, *On comparative longevity in man and the lower animals*, p. 58. (*Shorter publications*, F1991) 1873 Apr. 15 CD to L, about reproduction of elephants "I can clearly see that you will some day become our first star in Natural History". CCD21:174. 1874-90 Jodrell Prof. Zoology and Comparative Anatomy University College London. 1875 FRS. 1875 CD to Wallace, about L being blackballed for election to Linnean Society, "he is not my personal friend, only an acquaintance". CCD23:494. 1875 Jul. 18 visited CD, in company with Dyer. 1879 CD to L, CD is glad that L is to spend more time on original research, does "splendid work". CCD27:313. 1880 *Degeneration: a chapter on Darwinism.* 1881 CD wrote a testimonial for L's application for Edinburgh Chair. L held it briefly in plurality. *Calendar* 13446. 1891-98 Linacre Prof. Comparative Anatomy Oxford. 1898-1907 Director BMNH. 1907 KCB. 1908 Darwin-Wallace Medal Linnean Society. 1913 Copley Medal Royal Society. Recollections of CD: 'Charles Robert Darwin' in 1896 C.D. Warner ed. *Library of the world's best literature ancient and modern,* vol. 2, pp. 4835-4393, transcribed in *Darwin Online* (F2113).

Laslett, Hannah, mistress of the Girls & Infant School in Down in 1855.

Laslett, Isaac Withers, 1830-87. Builder/bricklayer and undertaker of Down. Possibly the same L that CD quoted in an 1840s MS "invariably when clay rests on solid chalk, it becomes black with ? iron ?" EH88202300, transcribed in *Darwin Online.*

Latter, Miss, 1859 Apr. – 1862 Feb. Governess at Down House. ED recorded 1859 Apr. 9 "Miss Latter came". Replaced by Camilla and Louisa Ludwig in 1862.

Latter, Robert Booth, 1804?-69. Lawyer in the firm Latter & Latter of Bromley, Kent. 1862 Dec. 29 ED wrote "Mr Latter 7 Cornwall Terrace Belmont Hill" and in 1863 Mar. 14 "[£12]". In 1881 May 28 L came to visit while the family was at Penrith region. Another visit was recorded in 1890 Jul. 5-7.

Latvian. First editions in: 1914-15 *Origin* (F736). 1935 *Autobiography* (F1526).

Laugel, Antoine-August, 1830-1914. French engineer, historian and philosopher. 1860 L gave a favourable review of *Origin* in *Revue des deux Mondes,* Apr. LL2:305.

Laurence, (also spelt **Lawrence**), **Samuel,** 1812-84. Artist. 1853 Chalk drawing of CD is at Down House. There is a study for it at Botany School, Cambridge.

Lawless, Hon. Miss Emily, 1845-1913. Irish novelist and poet. Amateur botanist, entomologist and nature writer. 1876 Aug. 9 CD to Romanes, about fertilisation of plants; she sent CD "a very good manuscript". Romanes 1896, p. 58.

Lawson, Nicholas Oliver, 1790-1851. Norwegian, born as Nicolai Olaus Lossius on Sekken Island, Moldefjord, Norway. Vice-Governor of Galapagos Islands. c.1816 L left Norway and became British Naval officer. c.1835 L commanded colony at Floriana for Ecuadorian government. CD mistook him for English and all subsequent literature has described him as an English governor. 1835 Sept. 25 Entertained CD and FitzRoy on Charles Island; "he could tell at once [from] which island any one (tortoise) was brought". *Beagle diary.* Keynes, *Beagle record,* 1979, pp. 302-3. See Grant and Estes, *Darwin in Galapagos: footsteps to a new world,* 2009.

Layard, Edgar Leopold, 1824-1900. Ornithologist. CD discussed evolution with before *Origin.* 1855 Dec. 9 CD to L asking for information, especially about birds, CCD5. 1856 Jun. 8 L provided CD with information for *Variation.* CCD6.

Leadendale. 1897 Home of Cecil Wedgwood.

Leandre, Pons i Dalmau, 1815-87. Born in Barcelona, Spain. Graduated from School of Fine Arts. Translated *Journal* from the 1875 French edition (F181) into Catalan, 1879 (F1907). 1879-81 translated *Journal* into Spanish (F2068).

Leaves. 1881 [Letter] The movement of leaves, *Nature*, 23: pp. 603-4 (*Shorter publications*, F1794). 1881 Leaves injured at night by free radiation, *Nature*, 24: p. 459 (*Shorter publications*, F1796). Referring to heat loss at night.

Lecoq, Henri, 1802-71. French botanist. 1854-58 *Études sur la géographie botanique de l'Europe*, 9 vols., Paris. LL3:301. 1861 L referred to in Historical Sketch of 3^d edn of *Origin* in support of modification of species. 1862 CD to Hooker, "Here is a good joke; I saw an extract from Lecoq. Geograph. Bot. & ordered it & hoped that it was a good sized pamphlet & my God *nine* thick volumes have arrived!" CCD9:350. 1863 CD to Bentham, L "is a believer in the change of species." CCD11:497.

Lee, Rev. Samuel, 1783-1852. Historian and orientalist. 1819-31 Prof. Arabic Cambridge. 1831-48 Regius Prof. Hebrew. 1838 CD dined with L at Trinity College.

Leeves, Mrs Elizabeth, (Spelt Reeve in LL1:378) Juliana Sabine's mother. 1849 CD sat with her in the coach to BAAS meeting at Birmingham. CCD4:271.

Leggett, William Henry, 1816-82. Botanist of New York, USA. 1864 Oct. 14 L helped CD with information on forms of flowers through Asa Gray. 1870-80 Founder and editor of *Torrey Botanical Club (Bulletin)*.

Lehr, Christian Wilhelm Jacob, 1856-1928. German-Italian Sculptor. 1883 Bust of CD, not from life, listed in LL3. Signed "C. Lehr D.J. sculpt. / Leipzig 1883". Copies at Oxford University Museum, Amsterdam, Universiteit van Amsterdam 2017 Aug. 30 and CUL.

Leidy, Joseph, 1823-91. American palaeontologist, zoologist and parasitologist. 1853- Prof. Anatomy University of Pennsylvania. Later Prof. Natural History Swarthmore College. 1861 Mar. CD to L, welcoming L's partial acceptance of CD's views on evolution, "I have never for a moment doubted, that though I cannot see my errors, that much in my Book will be proved erroneous." CCD9:39.

Leighton, Francis Knyvett, 1772-1834. Army Officer; Shropshire Militia. 1805 Married Louisa Anne Aldworth, 3 children. 1835 Apr. 23 CD at Valparaiso to Susan Darwin "I am indeed very sorry to hear of poor Col. Leighton's death. I can well believe how much he is regretted." CCD1:447.

Leighton, William Allport, 1805-89. Anglican clergyman and lichenologist. Schoolfellow of CD at Mr Case's school, Shrewsbury and fellow student at Cambridge. CCD1. 1841 *Flora of Shropshire*. Recollections of CD in 1881 (F2563) and in c.1886 DAR112.B94-B98, transcribed in *Darwin Online*.

Leith, Midlothian, the port of Edinburgh. 1838 Jun. CD went to L by boat from London on his way to Glen Roy.

Leith Hill Place, near Dorking, Surrey. 1842 Josiah Wedgwood [III] bought it, about 4,000 acres, on resigning his partnership in the family firm, home c.1847-80 Also home of Margaret Susan Wedgwood (Mrs Vaughan Williams), before 1944 and later. It was passed to Hervey Vaughan Williams, and in 1944 on his death to Ralph Vaughan Williams, who gave it to the National Trust. They leased it to Ralph Wedgwood, his cousin and close friend.

Lesquereux, Charles Léo, 1806-89. Swiss bryologist and palaeobotanist. 1847 Followed Louis Agassiz to USA and settled there. 1865 CD to Hooker, "he says that he is converted [to evolution] because my books makes the Birth of Christ—Redemption by Grace &c plain to him!!" CCD13:29.

Lessona, Michele, 1823-94. Zoologist specialized in amphibians. Translated four of CD's works into Italian. 1867 Prof. Zoology Turin. 1882 "Commemorazione di Carlo Darwin", *Atti Accad. Scienze di Torino*, 18: pp. 709-18. 1883 *Carlo Darwin*, Rome.

Lester, James, Petty Officer Cooper on 2d voyage of *Beagle*.

Letters. Correspondence to and from CD, in whole or in part, has been published in collections since 1887 with *Life and letters*, 3 vols., 1903 *More letters*, 2 vols., 1904, 1915 *Emma Darwin*, 2 vols. And these are still valuable. Freeman 1978 gave a list of other publications containing correspondence, many focusing on particular individuals or collections. But all of these have been superseded by the 1985- *The Correspondence of Charles Darwin*, 27 vols. by 2019. The definitive edition. 2006- 'Collections of Published Letters & Recollections' in *Darwin Online*. Virtually all CD letters that appeared in print during his lifetime and the years after his death. These provide evidence of which of CD's views had entered the public domain, and when.

Letters on Geology. Extracts from letters sent to Henslow by CD when on the *Beagle* voyage were read to the Cambridge Philosophical Society, Nov. 16, 1835. 1835 These extracts were printed, without CD's knowledge, for private circulation amongst members of the Society, by Cambridge University Press. The pamphlet is not dated, although the preface is dated Dec. 1, 1835 (*Shorter publications*, F1). 1960 A type facsimile, also for private distribution, was issued in 1960 (F4). 1967 The letters are printed in *Darwin and Henslow* (F5, F1598). Foreign editions: 1959 Russian (F7).

Lettington, Henry, c.1822/3-c.1910. Gardener in 1851 census. 1854-79 Gardener at Down House. L of CD "He moons about in the garden, and I have seen him standing doing nothing before a flower for ten minutes at a time. If only he had something to do I believe he would be better". Lubbock, *Darwin-Wallace celebrations of the Lin. Soc. of London*, 1908, pp. 57-8. Helped CD in his experiments on the crossing of plants. More anecdotes on L at Down House in F. Darwin, *Springtime and other essays*, 1920, pp. 56-8. 1860s Photograph of L with William Brooks by William Darwin, Down House collection. Reproduced in Reeve, *Down House*, 2009, p. 23 and Browne, *Power of place*, facing p. 312. Mrs. Amy L was draper in the village. 1882 L was on "Personal Friends invited" list for CD's funeral. 1895 Jul. Alive.

Leuckart, Karl Georg Friedrich Rudolf, 1822-98. Zoologist and parasitologist. 1850-69 Prof. Zoology Giessen. 1864 Nov. 4 CD to Falconer, L was an early convert to evolution. CCD12:396. 1869-98 Prof. Zoology Leipzig.

Lewes, East Sussex, 1853 The Darwins were there coming from Pevensey on Jul. 29. "went to Brighton & Lewes", to Hastings on Aug. 1. *ED's diary*.

Lewes, George Henry, 1817-78. Philosopher, man of letters. Many letters to and from CD. 1854-78 Common law husband of Mary Ann Evans "George Eliot". 1868 Feb. L reviewed *Variation* favourably and "gratifyingly" in the *Pall Mall Gazette*. 1873 Oct. Lunched at Down House together with George Eliot. 1874 CD and ED attended a séance at Erasmus Darwin's house in London with L and George Eliot. CD to Hooker. CCD22:25.

Lewis, John [I], 1799-1866. Carpenter of Down, born in Stone, Staffordshire. Father of John L [II]. Often worked for CD. 1849 Built hydropathic douche beside the well. *ED's diary* 1853 Oct. 7 "Lewis 1. 13. 6". 1862 With his son built hothouse.

Lewis, John [II], c.1834-1915 3ᵈ son of John L. [I]. "A short hale man with white hair and beard and a rare smile". *Zoologist*, 1909, p. 120. c.1849 Page at Down House for two years. Later village carpenter, first working with his father. 1882 Built CD's first coffin. See Charles Darwin Death and Funeral; *Maidstone Jrnl.* (22 Jun.), p. 5. Buried in a coffin he had made himself.

Lewis, Thomas, 34 in 1871 census. Labourer in Cudham, Kent, a village near Down. Secretary of the Down Friendly Society. 1891 Grocer in Down.

Lewy, Naphtali (Naphtali Hallevi), 1840-94. Rabbi and humanistic writer of Radom, Russian Poland. 1874 Pamphlet *Toledoth Adam* [The descent of man], Vienna, which is the first to introduce CD's views into rabbinical literature. 1876 L wrote to CD about *Toledoth Adam* and sent him the pamphlet. CD had the pamphlet translated through Henry Bradshaw, librarian at Cambridge University. CCD24:100. The pamphlet argued that the theory of evolution was consistent with the Old Testament account of creation. 1891 L's book *Nachlat Naphtali*, Pressburg, prints extracts from his correspondence with CD. 1894 L died at Southport, Lancashire.

Library, CD's. The library has been broken up since CD's death and is now found in multiple repositories and collections. Several important sources are available, listed below, but this is not an exhaustive list. These items, books, pamphlets, offprints and so forth total in excess of 5,800 items. A large proportion, especially annotated works, is in the 'Darwin Library-CUL' with 734 volumes. The other large collection is in the Darwin Library – Down with 779 volumes. In addition to books, CD collected offprints and pamphlets. See Pamphlet Collection. An important survey of CD's library was written by J. David Archibald:

http://darwin-online.org.uk/EditorialIntroductions/Archibald_DarwinsLibrary.html.

Anon. 1900. List of donations [books] received during the year 1899: From the executors of the late Mrs Darwin. *Cambridge University Reporter* 30 (41) (15 June), pp. 1079-1080. (Transcribed in *Darwin Online*)

Anon. 1960. *Handlist of Darwin papers at the University Library Cambridge*. Cambridge: Cambridge University Press. (Transcribed in *Darwin Online*.)

Anon. 1961. *Darwin library. List of books received in the University Library Cambridge March – May 1961*. (Manuscripts reading room, CUL and transcribed in *Darwin Online*.)

Anon. Down House Library list. (Available in manuscripts reading room, CUL).
Anon. "List of reviews of Origin of Species & of C Darwin's Books." (DAR262.8.9-18, (EH88206151-60), transcribed in *Darwin Online*).
CD & Francis Darwin. 1878-. Catalogue of Charles Robert Darwin's pamphlet collection. (DAR252.1-5).
Francis Darwin & H.W. Rutherford, Scrapbook of reviews. (DAR226.1-2) (Reproduced in *Darwin Online*, 781 images).
H.W. Rutherford, 1908. *Catalogue of the library of Charles Darwin now in the Botany School, Cambridge. Compiled by H. W. Rutherford, of the University Library; with an Introduction by Francis Darwin.* (Transcribed in *Darwin Online*)
Mario di Gregorio, 1990. *Charles Darwin's Marginalia*, with the assistance of N.W. Gill. New York: Garland Publishing.
Nick Gill, Darwin Library Unbound Material. (Available in manuscripts reading room, CUL)
P.J. Vorzimmer, 1963. A catalogue of the Darwin reprint collection at the Botany School Library, Cambridge. (Manuscripts reading room, CUL and *Darwin Online*- A1037)
T.W. Newton, H.W. Rutherford and Francis Darwin, 1875. Catalogue of the library of Charles Darwin Esq. M.A., F.R.S., &c. at Down, Kent. Bound manuscript book, with later manuscript additions. 426pp. (DAR240)

Liebre, La. Schooner. CD says 11½ tons, but FitzRoy says 9 tons "sharp built or frigate barge". Surveyed southeast coast of Argentina. 1832 Sept. 11 Hired at £140 by FitzRoy from James Harris, resident at Rio Negro, Argentina, for eight lunar months, with Schooner *La Paz*. Commanded by Stokes, who had Lieut. Bartholomew James Sulivan in *La Paz* under his command. CD travelled on her for a time and then Wickham was in charge.

Liesk, William C., Resident in Cocos Keeling Islands. 1836 Apr. 3 CD met.

Life and letters. 1887, 1888 F. Darwin, ed. *The life and letters of Charles Darwin, including an autobiographical chapter*, 3 vols.. Contains CD's "Autobiography" in 1:26-160. Three other printings in 1887 and one in 1888 have small corrections (F1452-1455, 1457). 1969 Facsimile (F1507). Foreign editions of whole work: 1887 USA (F1456). 1887-88 German (F1515). 1888 French (F1514) 1889 Norwegian (F1528). 1892 See also *Charles Darwin: his life told in an autobiographical chapter* (F1461), which is largely, but not entirely an abridged version. Transcribed in *Darwin Online*.

Lime water. 1864 In one of the worst years of CD's illness, "began lime water chalk & mag" to ease CD's suffering. *ED's diary*. W. Jenner to CD 14 Aug. 1864. The lime was not the citrus fruit but the calcium-containing mineral.

Lincecum, Gideon, 1793-1874. American physician, and naturalist. 1862 Based on L's letters, CD communicated a "Notice on the Habits of the 'Agricultural Ant' of Texas ['Stinging Ant' or 'Mound-making Ant,' *Myrmica (Atta) malefaciens*, Buckley]" to *Jrnl. of the Proc. of the Lin. Soc. of London*, (*Shorter publications*, F1938). See CCD9:39-48.

Lindley, John, 1799-1865. Botanist and prolific writer on botanical subjects. Member of Horticultural, Linnean and Geological Societies. Played a pivotal role in traditional botany in the Victorian era. 1828 FRS. 1829-60 Prof. Botany University College London. 1843 CD sent L some seeds which had been found by William Kemp under 25 feet of white sand. CCD2:355. 1853 L was in competition with CD for award of Royal Medal of Royal Society. CCD5:165. 1856 CD to Hooker, suggesting that L was worth a Copley Medal. 1857 L got a Royal Medal in 1857. CCD6:72.

Linnean Club. Dining club of Linnean Society. 1861 CD and Thomas Bell dined.

Linnean Society of London. 1854 CD Fellow. CD borrowed from the Library a great deal through its librarian Richard Kippist. 1856 CD sent £20 "with heavy groans". CCD6:171. 1858 Jul. 1 "On the tendency of species to form varieties", by CD & Wallace communicated by Lyell and Hooker. It appeared in print on Aug. 20. 1881 The Society commissioned John Collier's oil portrait of CD. It hangs in their rooms at Burlington House. 1958 Jul. 1 Organised the Darwin-Wallace Celebration.

Linum. 1863 "On the existence of two forms, and on their reciprocal sexual relation in several species of the genus *Linum*", *Jrnl. Proc. of the Lin. Soc. (Bot.)*, 7: pp. 69-83 (*Shorter publications*, F1723). 1863 French translation *Annals Sci. Natural Bot.*, 19: pp. 204-95, with CD's 1862 papers on *Primula* and *Catasetum* (F1717 & F1718). List of presentation copies in CCD11 Appendix IV.

Lion Inn, The, Wyle Cop, Shrewsbury. 1835 April, CD to his sister Susan, CD considered staying there when he got back from *Beagle* voyage, travelling by coach (Wonder), from Falmouth, to avoid waking family "in the dead of the night." In the event he reached Shrewsbury in the early morning. Now there as The Lion Hotel.

Litchfield, Henrietta Emma, see Darwin, Henrietta Emma.

Litchfield, Richard Buckley, 1832 Jan. 6-1903 Jan. 11. Scholar and philanthropist. Had 2 siblings. L worked on the legal side of the Ecclesiastical Commission. One of the founders of the Working Men's College (later Birkbeck College) London, where he became Bursar and Vice Principal. L lived at no. 4 Bryanston Square. See Gwen Raverat, *Period piece*, Ch. 7. 1871 Aug. 31 Married at Downe Church, Henrietta Emma Darwin d.s.p. After their marriage, they often visited and stayed for a few days at Down House. 1872 L and Henrietta assisted CD with *Expression*. ED2:208. 1881 ED to George Darwin "the Litches came at lunchtime". 1882 Apr. 15, they came late to Down, the night CD fainted at dinner. They left on 11 May and returned 10 days later. 1883 Moved to 31 Kensington Square. L arrived at Down on Oct. 1 1896 and was present when ED "peacefully died with no consciousness of the end". ED(1904)2:466. 1903 Biographer of Thomas Wedgwood [VI], an early experimenter in photography. Died at Cannes. Photograph in DAR225.58.

Lithuanian. First editions in: 1959 *Origin* (F738). 1959 *Autobiography* (F1527). 1963 *Journal* (F222).

Little Etruria. House on the Etruria estate, near Etruria Hall. First home of Josiah Wedgwood [II]. 1769 Josiah Wedgwood [I] and family moved in from Brick House, the Hall not being finished. 1792 Josiah Wedgwood [II] and Bessy (Sarah Elizabeth Allen) moved in on marriage.

Liverpool, Lancashire. 1818 Jul. CD visited with Erasmus Alvey Darwin. 1838 CD passed through on return from Glen Roy.

Lizard's eggs. 1855 Lizard's eggs, *Gardeners' Chron.*, no. 21: p. 360 (*Shorter publications*, F1808). CD requested from schoolboys, Sand lizard eggs in order to experiment to see if or how long they could survive contact with sea water. See CCD5:337-338.

Llangollen, Denbighshire. 1831 Aug. CD visited with Sedgwick for geology. See DAR5.B5-B14. (Images in *Darwin Online*)

Lloyd, Mary Charlotte 1819-96. Welsh landowner and artist. Studied sculpting in Rome. 1869 Jul. 24 CD to L, sending letter from Boyd Dawkins about CD's visit to Caerdeon, Barmouth. Rev W.E. Jelf, owner of house where they stayed. L neighbour to Darwins. CCD17. See P. Lucas, *Archives of Nat. Hist.*, 2007, pp. 318-45.

Lloyd, Anne Susan, 1786-1862. Daughter of Gamaliel L of Yorkshire. 1806 Married Leonard Horner.

Lloyd, Capt. John Augustus, 1800-54. British engineer and surveyor. 1830 FRS. 1831-49 Surveyor General Mauritius. 1836 May 3 L entertained CD. In the *Sydney notebook*, Chancellor & van Wyhe, *Beagle notebooks*, 2009. "So well known from his examination of the Isthmus of Panama". *Journal*, 2^d edn, p. 485. See also Elephant.

Loch Leven, Argyllshire. 1838 Jun. CD visited on way to Glen Roy.

Lock & Whitfield, 1856-94. Commercial London photography company. Photographed CD in 1878. See Darwin, Iconography.

Locke, Mrs. 1843 Servant, nursemaid for Henrietta Emma Darwin. *ED's diary*. CD observed Henrietta smiled at three weeks but Mrs L said it was two. *Notebook of observations on the Darwin children*. (DAR210.11.37) Transcribed in *Darwin Online*.

Loddiges, Conrad, Nurseryman. 1838 Sept. CD visited his garden in Hackney, saw 1,279 varieties of roses. *Darwin's notebooks*, p. 371.

London Stereoscopic & Photographic Company, 1854-1922. Company issued CDVs of CD after photographs by Ernest Edwards. They did not photograph CD themselves and did not issue actual stereoscopic photographs of CD.

London Working Men's College, founded in 1854 by Frederick Denison Maurice, later Birkbeck College. Richard Buckley Litchfield was one of the founders. 1871 WMC presented a picture by Maccallum to the Litchfields as wedding gift. 1871 Nov. 4 CD and ED attended their party. 1872 Jun. 30 WMC visited. 1873 Henrietta Litchfield: "Several times after my marriage, my father and mother invited the party to Down." WMC's arrival often brought the household a lot of joy. ED(1904)2:262. ED's last diary entry mentioning WMC was in 1876 Jul. 2.

Longfellow, Henry Wadsworth, 1807-82. American poet and educator. 1843 Married Frances Elizabeth Appleton. 1868 L called on CD at Freshwater, Isle of Wight, with brother-in-law Thomas Gold Appleton. 1868 recollection of CD in *Louis Agassiz*, 1885, vol. 2: 666, (F2088), transcribed in *Darwin Online*.

Longevity in birds. 1870 [Note on the age of certain birds.] In E.R. Lankester, *On comparative longevity in man and the lower animals*, p. 58 (*Shorter publications*, F1991). Lankester wrote: "In reply to enquiries, Mr. Charles Darwin writes that he has no information with regard to the longevity of the nearest wild representatives of our domesticated animals, nor notes as to the longevity of our quadrupeds." In a footnote L added: "Mr. Darwin very kindly furnished me with a note relative to the age of certain birds, which is quoted in the Table of Statements, which follows."

Longley, Charles Thomas, 1794-1868. The archbishop of Canterbury. 1868 Innes to CD, "I was very sorry to lose good Dr. Longley". CCD16:892.

Longueville, Cecile, 1860 May 18 Married Francis Parker, q.v.

Lonsdale, William, 1794-1871. Soldier, geologist and palaeontologist. L served at Waterloo. CD discussed evolution with before *Origin*. 1829-42 Curator and Assistant Secretary Geological Society of London. 1842 CD to Lyell, "I had long talk with Lonsdale on Friday— I have not for years seen him so cheerful, or I might say I never saw him really cheerful before. His setting to work at corals as an avowed return for the sum presented to him is a noble return & is one which will, I think, especially please you. He is evidently deeply gratified by the Present." The gift was a silver cup and subscription of £600. CCD2:336-7. 1846 Wollaston Medal, Geological Society London. Letters to in F2262.

Lothian Street, Edinburgh, no. 11. Mrs Mackay charged £1.16.0 per week for two bedrooms and a sitting room, regularly let to medical students, including Edward Forbes. 1825 Oct.-1827 Apr. CD lodged there when a medical student. 1825 Erasmus Alvey was there. John Edmonston lived at no. 37. 1888 A tablet was put up on the house commemorating CD's stay, at suggestion of Francis Darwin. The building was demolished by 1913.

Lovegrove, Charles, 1828-96. Merchant in the City of London, and churchwarden of St Mary's, Down. L corresponded with CD in 1861? and 1866? about L's "pretty pony" and a subscription to the Down Coal & Clothing Club. In 1869 Dec. 14 the Darwins called on him. From 1872-79, ED may have called on Mrs Lovegrove.

Lovén, Sven Ludvig, 1809-95. Swedish marine zoologist and malacologist. CD corresponded with L 1849-76. CCD4 & 24.

Lowe (later Sherbrooke), Henry Porter, 1810-87. Cambridge friend of CD. Brother of Viscount Robert Sherbrooke. Member of Gourmet Club.

Lowe, Rev. Richard Thomas, 1802-74. Anglican clergyman and botanist. 1832-52 Chaplain at Madeira. 1866 Hooker to CD, H had a letter from L on distribution of plants in Atlantic islands which was of interest to CD. CCD14:276.

Lowe, Robert, 1811-92. At Oxford. Statesman. Liberal MP for Kidderminster, Calne and University of London. Biography A.P. Martin, 1893: "I saw something in him (CD) which marked him out as superior to anyone I had ever met". This when they met at Barmouth. 1831 Was at Barmouth with CD, not the earlier trip. "Journal" kept by H.P. Lowe & R. Lowe during 3 months of the summer 1831. "at Barmouth. North Wales..." Nottinghamshire Record Office, NRO-DD.SK.218.1, transcribed and introduced in *Darwin Online* by Peter Lucas. 1842-50 In Australia as lawyer and politician. 1868-73 Chancellor of the Exchequer. 1871 FRS. 1871 L visited Down House from High Elms with Lubbock, Huxley and Mountstuart Elphinstone Grant Duff. 1880 Viscount Sherbrooke. Recollections of CD in *Life and letters of…Robert Lowe*, 1893, vol. 1, pp. 19-20; vol. 2, pp. 198-207, transcribed in *Darwin Online*.

Lowell, James Russell, 1819-91. American author, romantic poet and diplomat. 1857 First Editor of *Atlantic Monthly*. 1880-85 American Minister in London. 1872 Feb. 25 ED recorded in her diary "heard Lowell" when she and CD were in London. 1880 May 30, Mrs Lowell paid a visit. 1882 Pallbearer at CD's funeral.

Lubbock, Henry James, 1838-1910. Son of Sir John William L. Younger brother of Sir John L. Married Frances Mary Turton. L visited Down House with his elder brother. 1870 Jun. 29 he visited Down House. The Darwins had "Little Lubbocks" over to tea in 1871 and 1881. Mrs Francis L would also dine with ED in 1889 with a last mention in 1890. *ED's diary*.

Lubbock, Sir John, Bart, 1834 Apr. 30-1913 May 28. First child of Sir John William L. Statesman, banker and man of science. Home: High Elms near Down. L was the closest of CD's younger friends and frequent visitor to Down House from childhood. CD discussed evolution with before *Origin*. Biography: Hutchinson, *Life of Lord Avebury*, 2 vols:, 1914. 1856 Married 1 Ellen Frances Hordern. 3 sons, 3 daughters: 1. Constance Mary, 2. Gertrude, 3. Amy Harriet, 4. John Birkbeck, 5. Norman, 6. Rolfe Arthur. 1853 His first scientific paper was in *Annals and Mag. of Nat. Hist.* describing *Labidocera darwinii*, a calanid copepod, from material lent by CD. "How on earth you find time is a mystery to me." CD to L in Hutchinson, I, p. 176. Social calls between the families were recorded in *ED's diary* starting with an entry in 1848 Dec. 27 "dance at Lubbocks". There were balls, lunches and dinners. 1853, 1855 published descriptions of CD's specimens of crustaceans and copepods. 1858 FRS. 1859 CD sent 1st edn of *Origin*. 1865 4th Bart. 1865 CD to Hooker, "Many men can make fair M.P.s, & how few can work in Science like him." CCD13:210. 1870-80 MP for Maidstone. 1880-1900 MP for London University. 1882 L suggested Westminster Abbey funeral for CD and organized memorial to the Dean. Served as a Pallbearer. 1884 Married 2 Alice Augusta Laurentia Lane Fox. 3 sons, 2 daughters. 1. Ursula, 2. Irene, 3. Harold Fox Pitt, 4. Eric Fox Pitt, 5. Maurice Fox Pitt. 1900 1st Baron Avebury. Account of interactions with CD in Duff, *The lifework of Lord Avebury*, 1924, transcribed in *Darwin Online*. A microscope given to L by CD is now at Down House.

Lubbock, Sir John Birkbeck, Bart, 1858 Oct. 4-1929 Mar. 26. Eldest son of Sir John L. and Ellen Frances Hordern. 1913 5th Bart. 1913 2d Baron Avebury.

Lubbock, Sir John William, Bart, 1803 Mar. 26-1865 Jun. 21. Father of Sir John L. Banker, barrister and astronomer. Home: High Elms near Down, which he largely rebuilt after burning down. CD's neighbour, their land marching together. On friendly terms, but not close. 1829 FRS. 1833 Married Harriet Hotham. 1836-42 Vice-Chancellor London University. 1840 3d Bart.

Lubbock, Montagu, 1842-1925. Doctor of Medicine (London and Paris). Fellow of the Royal College of Physicians. John Lubbock's younger brother. 1861 L seriously injured in a carriage accident. 1862 Feb. 24 CD to Innes, "They [the John Lubbocks] gave us a good account of poor Montague". CCD10:92.

Lubbock, Ursula, 1885-1959. Second child of Sir John L and Alice Fox. 1906 Married Major Adrian Grant Duff.

Lucas, Prosper, 1808-85. French physician and writer on heredity. 1847-50 *Traité philosophique et physiologique de l'hérédité naturelle*, 2 vols. CD read in 1856 and cited extensively in *Natural selection, Origin* and especially *Variation*. R. Noguera-Solano & R. Ruiz-Gutiérrez. Darwin and inheritance: The influence of Prosper Lucas. *Journal of the History of Biology* 42, 2009, pp. 685-714.

Ludwig, Miss Camilla, 1837-1912. Sister of Karl L. 1860-63 or later Governess at Down House. Evans, *Darwin and women*, 2017, p. 252. Her arrival in 1860 Feb. 21 was recorded in *ED's diary*. From 1862-66 her comings and goings were recorded by ED. In 1867 she came with her sister Louisa. 1868 L translated for CD L. Rütimeyer, *Die Grenzen der Thierwelt. Eine Betrachtung zu Darwin's Lehre*, Basel, 1868. 1874 Married Reginald Saint Pattrick, Vicar of Sellinge, Kent. "It seems that Darwin usually sent a copy of his books to Camilla Ludwig, as if she too were a member of the family. A number of items, mostly offprints of significant reviews of Darwin's works, are listed in a Sotheby's catalogue, described as 'from the collection of C. Ludwig, Leipzig.'" Browne, *Power of place*, 2002, p. 519. 1881 Sept. 24 the sisters came with children and were joined by Mr Pattrick on 27 Sept. 1891-96 ED wrote often to her with a final letter on 1896 Sept. 25.

Ludwig, Miss Louisa, 1839-1915. Sister of Camilla L. Governess and schoolmistress. Evans, *Darwin and women*, 2017, p. 252. 1862 took over care of Horace D when Camilla Ludwig went back to Germany. 1861 Sept. 13 ED recorded "Miss Louisa Ludwig came". 1862 Aug. 1-8 Payments totalling £8. CD's Classed account book (Down House MS). She visited ED in 1890 Jul. 28 and in 1895 Aug. 17.

Ludwig, Rudolf August Birminghold Sebastian, 1812-80. German geologist and palaeontologist of Darmstadt; factory and saltworks inspector. 1877 Jul. 16 CD to L, thanking for essay dedicated to CD and referring to "*Crocodilus darwini*, Fossile Crocodiliden aus der Tertiär Formation des Mainzer Beckens", *Palaeontographica*, supplement 3, Lieferung 4 and 5. CCD25.

Luftschifferei der Spinnen, 1839 Über der Luftschifferei der Spinnen [On aeronaut spiders], *Froriep's Neue Notizen aus dem Gebiete der Natur- und Heilkunde*, 11: columns 23-24 (F1654); a translation from *Journal*, pp. 187-8.

Lumb, Edward, 1804-75. English merchant at Buenos Aires. 1833 CD stayed with. 1834 L arranged for shipment to England of a "Megatherium" skull, actually *Toxodon*, which CD had found near the estancia of Mr Keen, on river Beguelo. J.H. Winslow, *Jrnl. of Historical Geography*, 1: pp. 347-60. In the *Buenos Ayres* and *Galapagos notebooks*, Chancellor & van Wyhe, *Beagle notebooks*, 2009. [1833]. Recollection of CD in Buenos Aires in *Reminiscences of diplomatic life*, pp. 27-9, (F2097), transcribed in *Darwin Online*. See daughter Macdonell, Anne.

Lushington, Sir Godfrey, 1832-1907. 5th son of Stephen L; twin brother of Vernon L. Barrister and Civil Servant. 1865 Married Beatrice Ann Shore Smith. 1868 CD and ED gave luncheon to him and his wife in London. ED2:189. 1870 Apr. 24 called on the Darwins. 1882 L and Mrs L on "Family Friends invited" list for CD's funeral. 1892 KCB. 1899 GCMG.

Lushington, Margaret, 1867-1906. See Langton, Stephen Massingberd.

Lushington, Susan, 1870-1953. Daughter of Vernon L. Unmarried.

Lushington, Vernon, 1832-1912. 6th son of Stephen L. Twin brother of Godfrey L. 36 Kensington Square, London, and Borden, Hampshire. 1865 Married Jane Mowatt. 3 daughters. c.1869 Henrietta Emma D first met Richard Buckley Litchfield, her future husband, at the L's London house. The L's and their three daughters remained family friends. Katherine ("Kitty") married Leopold James Maxse; Margaret married Stephen Massingberd son of Edmund Langton and Susan never married. 1871 Secretary to the Admiralty. 1877-1900 County Court Judge for Surrey and Berkshire. 1871 May 21-23, Jul. 18 and 1881 Jul. L visited Down House with wife. 1882 L and Mrs L on "Family Friends invited" list for CD's funeral.

Luxan, also known as Luján de Cujo.

Lyell, Charles [I], 1767-1849. Amateur botanist and country gentleman. Of Kinnordy, Kirriemuir, Forfarshire. 1796 Married a daughter of Thomas Smith of Swaledale, Yorkshire. 10 children; 1 Charles; 2 Thomas and 3 Henry.

Lyell, Sir Charles [II], Bart, 1797 Nov. 14-1875 Feb. 22. Lawyer and Geologist. First son of Charles Lyell [I]. Family home Kinnordy, Kirriemuir, Forfarshire (now Angus); in London 16 Hart Street, Bloomsbury Square, later 53 Harley Street. CD discussed evolution with before *Origin*. Blind in old age. Biography: Mrs Katherine Murray Lyell (sister-in-law), 2 vols., 1881; Bonney 1901; F.D. Adams 1933; E. Bailey 1962; L.G. Wilson 1970; L.G. Wilson 1972. L was of independent means and worked as a geologist, the most distinguished of his age. L was a close friend and correspondent of CD, but never on the same comfortable terms as Hooker. L never stated

Sir Charles Lyell after 1862. Stereoscopic carte de visite.

unequivocally in print his views on CD's position in regard to evolution. 1820 Called to the Bar and practised until 1827. 1826 FRS. 1830-33 Prof. Geology King's College London. 1832 Married Mary Elizabeth Horner, d.s.p. 1836 Oct. 29 CD first met at L's house in London. 1839 ED to her sister Sarah Elizabeth Wedgwood [II], "Mr Lyell is enough to flatten a party, as he never speaks above his breath, so that everybody keeps lowering their tone to his". ED2:40. 1844 CD to Leonard Horner, "I always feel as if my books [the geologies] came half out of Lyell's brain".

CCD3:55. 1845 CD dedicated 2ᵈ edn of *Journal* to L. 1848 Kt. 1858 Jul. 1 Communicated Darwin-Wallace paper at Linnean Society meeting with J.D. Hooker. 1858 Copley Medal, Royal Society. 1858 CD to Hooker, sending H notes on L's excellence to help him award Copley Medal. CCD7:186. 1859 CD sent L 1ˢᵗ edn of *Origin*, copy now at Down House, presented by Sir George Buckston Browne. 1863 CD to Hooker on publication of *Antiquity of man*, "The Lyells are coming here...I dread it, but I must say how much disappointed I am that he has not spoken out on Species, still less on Man." CCD11:173. 1864 1ˢᵗ Bart of Kinnordy. 1865 Feb. CD broke 6ᵗʰ edn of *Elements of geology* into two halves in his dislike of heavy books. 1873 L's wife died. 1874 Sept. 23 CD's last letter to Lyell about Judd's views on volcanoes. CCD22:434. 1875 CD was asked by Mrs Henry Lyell to be a Pallbearer at L's funeral. CD declined on grounds of ill health. L's secretary for many years was Arabella Burton Buckley. Seven of L's scientific notebooks were edited by L.G. Wilson, 1970. His notebooks are now in the National Library of Scotland. Main works: 1830-33 *Principles of geology*, three vols. CD's copy of vol. 1 was presented to him by FitzRoy. Vols. 2 and 3 reached him in South America. Twelve editions published until 1875, the last posthumously. 1838 *Elements of geology*. Seven editions published during L's lifetime, including the Student's Lyell of 1871. 1863 *The geological evidence of the antiquity of man*. Four editions published during L's lifetime. In his will he left the die by Wyon to be cast in bronze to Geological Society and £2,000, not less than one third interest to go with gold medal annually. Lyells visits were entered in *ED's diary* from 1839 Feb. 6 onwards.

Lyell, Lt-Col. Henry, 1804-75. 3ᵈ son of Charles Lyell [I]. Indian Army Officer. 1848 Married Katherine Murray Horner. 4 children: 1 Leonard (1850), 2 Francis Horner (1852), 3 Arthur Henry (1854), 4 Rosamund Frances (1856).

Lyell, Marianne, 1801-81, One of Charles L [I]'s daughters. 1875 Feb. 23 CD to Miss Buckley mentions her. CCD23:55.

Lyell, Thomas, Naval Officer. Second son of Charles L [I].

Lynch, Richard Irwin, 1850-1924. Botanist. L was head gardener to the Earl of St. Germans. 1867-79 Foreman of the propagating department at Kew. 1878 L supplied CD with plants. CCD26. See Francis Darwin, *Jrnl. of the Lin. Soc. of London (Bot.)*, 22: p. 102. 1879-1919 Curator Botanic Garden Cambridge. 1881 Associate Linnean Society. 1901 Veitch Memorial Medal Horticultural Society, followed by the Veitch Memorial Gold Medal in 1923. 1916 Honorary M.A. Cambridge University.

Lyne, Mrs, ?-1881 Feb. 16. ED wrote to Henrietta telling her a villager at Down had died suddenly. Mrs L's family wanted Francis to go and see her, which in ED's opinion, "he wisely declined." ED(1904)2:314.

Lynn, John, 1822-1889. A policeman at Down. Later a head constable. "he made the acquintance of [CD]...The acquintance ripened into a warm friendship, and Darwin presented him with several of his works, on which Mr Lynn set great value." *Western Morning News*, (14 Jan. 1889), p. 5, transcribed in *Darwin Online*.

Lythrum salicaria. 1865 On the sexual relations of the three forms of *Lythrum salicaria*, *Jrnl. of the Lin. Soc. of London (Bot.)*, 8: pp. 169-96 (F1731). List of presentation copies is in CCD12 Appendix III. Draft of and notes in DAR27.2.

"M", "Old M", blacksmith and "a notable old drunkard in the village". Converted by J.W.C. Fegan. ED2:244. No name starting with M found in the 1881 census.

Macarthur, Hannibal Hawkins, 1788-1861. Australian colonist, politician, businessman, wool pioneer and brother-in-law of Philip Parker King. In the *Sydney notebook*, Chancellor & van Wyhe, *Beagle notebooks*, 2009.

MacArthur, James, 1798-1867. Australian politician. 4th son of John M. Brother of William. Also lived at Camden Park. 1836 Was in England at the time of CD's visit.

MacArthur, John, 1767-1834. British army officer, entrepreneur and pastoralist. Father of Sir William M and James M. A founder of New South Wales sheep farming. Encouraged wine making in Australia, importing vines from Germany. 1790 Settled in Australia. 1832 Commissioned building of Camden Park.

MacArthur, Sir William, 1800-82. 5th son of John M. Philip Parker King was a cousin by marriage. Australian sheep farmer, viticulturist and horticulturalist of Camden Park, New South Wales. Amateur botanist. Member of New South Wales Legislative Council. 1835 Finished building Camden Park. 1836 Jan. 27 CD visited Camden Park with Capt. King. 1856 Kt. 1857 CD dined with M in London. CD to Asa Gray, "a clever Australian gardener". CCD6:361.

Macaulay, Thomas Babington, 1800-59. Historian and politician. c.1842 CD met at Lord Stanhope's house in London. 1857 1st Baron. M recalled in 1856 Jul. 16 CD whom he described as "Darwin, a geologist and traveller, came to dinner." In Trevelyan, ed. 1876. *The life and letters of Lord Macaulay*, vol. 2, pp. 403-4.

Macaw Cottage, see: Upper Gower Street.

Macedonian. First edition in: 1978 *Origin* (F2158).

McCormick, Robert, 1800-90. Also spelt MacCormick, M'Cormick or Maccormick. Surgeon on 2d voyage of *Beagle*. Trained and "wished to be employed on scientific voyages." Not, as commonly believed, the official naturalist on the *Beagle*. See van Wyhe, "my appointment" 2013. E. Steele, *He is no loss: Robert McCormick and the voyage of HMS Beagle*, 2011, like all other works that mention him, perpetuates the error that M was the de facto naturalist and that CD usurped his place. FitzRoy to Beaufort of M "a sad empty headed coxcomb". Keynes, *Beagle record*, 1979, p. 77. 1832 Apr. M returned to England, ostensibly sick, but had quarrelled with FitzRoy and Wickham. J.J. Keevil, *Jrnl. Roy. Naval Med. Serv.*, 29: pp. 36-62, 1943. 1832 Apr. 25 CD to Caroline "Maccormick returning to England, being invalided, ie. being disagreeable to the Captain & Wickham.— He is no loss." CCD1:225. M turned down positions offered to him and invalided out of multiple voyages as a result of his cantankerous personality. 1839 CD met M with Hooker in Trafalgar Square. 1839-43 M was on *Erebus* and *Terror* expedition with J.D. Hooker his junior.

McDermott, Frederick, 1855-1924. Born Grove Park, Camberwell. 1876 A student of the Middle Temple. 1879 Barrister. 1880 Wrote to CD requesting a 'yes' or 'no' answer whether CD believed in the New Testament. See Darwin, Charles Robert, Religion for CD's reply. 2015 Letter sold at Bonham's for US$197,000.

M'Donnell, Robert, 1828-89. Comparative anatomist of Dublin. 1860 Nov. 24 CD to Lyell, "a first rate man". M had written to CD about the difficulties of electric organs in evolutionary theory. 1861 M's observations on homologous structures in skate and torpedo published in *Nat. Hist. Rev.*, 1:57-60, 1861.

Macdonell, Anne, née Lumb, 1850-? recalled CD's stay at Buenos Aires with her parents when the *Beagle* arrived. CD brought back a tuco tuco, left it in the lodging and departed for another excursion. The bad smell in the room during his absence was traced to the tuco tuco so her mother threw it into the fire. When her father Edward L explained the missing specimen, CD said "I will forgive Mrs. Lumb, for she is nearly as beautiful as the touca-touca." In Macdonell, *Reminiscences of diplomatic life*, 1913, pp. 27-9. Transcribed in *Darwin Online* (F2097).

Macgillivray, William, 1796-1852. Ornithologist and fine field naturalist. CD knew him in Edinburgh and later met in London. 1823 Assistant to Robert Jameson. 1830-39 Wrote large part of his friend John James Audubon's *Ornithological biographies*. 1831-41 Conservator of Museum of Royal College of Surgeons of Edinburgh. 1841-52 Prof. Natural History Marischal College, Aberdeen.

Mackay, Mrs, Landlady of 11 Lothian Street, Edinburgh, who specialized in accommodating medical students, Lothian Street being near the University. 1825 Oct.-1827 Apr. CD lodged there. Erasmus lodged there until Summer 1826.

Mackintosh, Catherine [I], see **Stuart**.

Mackintosh, Lady [Catherine II], see **Allen**.

Mackintosh, Catherine [III], 1795-? 3d child of Sir James M. and Catherine Stuart. Married 1 Sir William Saltonstall Wiseman, 7th Bart. Married 2 G.H. Turnbull.

Mackintosh, Daniel, 1815-91. Geologist. M earned his living by tuition and lecturing. Biography: *Geological Mag.*, p. 432, 1891. 1861 FGS. 1869 Author of *The scenery of England and Wales*. 1879 CD praises his work under difficulties and writes to on erratic boulders. CCD27:424, 428.

Mackintosh, Elizabeth "Bessy", 1804-23. 3d child of Sir James M and Catherine Allen.

Mackintosh, Frances Emma Elizabeth (Fanny Mack), 1800-89. 1st child of Sir James M and Catherine Allen. 1832 Married Hensleigh Wedgwood. 1851 M was a partisan of Mazzini. ED2:143. 1878 M stayed at Down House, "quite an invalid".

Mackintosh, Sir James, 1765-1832. Lawyer, philosopher and statesman. M was related to the Darwin's through second marriage and some of the children were family friends. 14 Great Cumberland Street and Ampthill Park. CD, "The best converser I ever listened to". *Autobiography*, p. 55. Biography: R. Mackintosh (son) 1836. 1789 Married 1 Catherine Stuart. 3 daughters: 1. Mary (I), 2. Maitland, 3. Catherine (III). 1798 Married 2 Catherine Allen. 2 sons, 2 daughters: 1. Frances Emma Elizabeth, 2. Robert, 3. Elizabeth "Bessy", 4. Robert James. 1803 Kt. 1813-32 MP. 1822-

24 Rector of University of Glasgow. 1827 Sept. CD visited. Van Wyhe ed., 'Journal', (DAR158). 1832 M died from a chicken bone in his throat.

Mackintosh, Mary [I], 1789-1867. First child of Sir James M and Catherine Stuart. Married Claudius James Rich s.p. 1831 On being widowed M lived with her twice widowed father. Mary M's visits were noted in *ED's diary*. 1849 CD lent her Lyell's *Principles of geology*. CCD4.

Mackintosh, Maitland, 1792-1861. Second child of Sir James M and Catherine Stuart. Married William Erskine.

Mackintosh, Robert, 1803. Second child of Sir James M and Catherine Allen. Died in infancy.

Mackintosh, Robert James, 1806-64. Colonial governor. 4th child of Sir James and Catherine Allen. Married Mary Appleton. 2 sons, 1 daughter. Emma Wedgwood wrote in her diary in 1835 Sept. 4 "R. Mack. came". 1836 M wrote biography of his father. 1846 ED called on M at a cottage near The Grange, Lord Ashburton's house, when CD went to BAAS meeting at Southampton. *ED's diary* entries showing "Macks came". His death was recorded by ED on Jun. 24 1864, "Mack's death".

Mackintosh (M'Intosh), William Carmichael, 1838-1931. Physician and invertebrate zoologist. Among many academic functions and distinctions he was Director of Gatty Marine Laboratory, St. Andrew's University. 1872-76 On *Challenger* expedition. 1877 FRS. 1881 Dec. 10 CD to Romanes on M "of whose work I have a very high opinion". CD declined to give a testimonial to M for the Edinburgh Chair of Natural History, on the grounds that he had already given one for Edwin Ray Lankester. Carroll 1976, 604. 1899 Royal Medal Royal Society. 1924 Linnean Medal.

Maclaren, Charles, 1782-1866. Scottish journalist and geologist. Founder and editor of *The Scotsman*. 1843 CD corresponded with M on M's review of *Coral reefs* in *Edinburgh New Phil. Jrnl.* CCD2. See *Shorter publications*, F1662.

Maclear, Sir Thomas, 1794-1879. 1831 FRS. 1834-70 Astronomer Royal at Cape of Good Hope. 1836 Jun. CD met at Cape of Good Hope. Keynes, *Beagle record*, 1979, p. 365. 1860 Kt.

Macleay, William Sharp, 1792-1865. Civil Servant, cabinet naturalist and entomologist. M invented the quinary system of classification. 1825-36 Havana, Cuba. 1830- British commissioner and judge. 1839 M emigrated to NSW, Australia. 1859 CD to Owen, "I have thought that perhaps my book might be a case like Macleay's Quinarian system." [i.e. laughed at]. CCD7:430.

McNab, William Ramsay, 1844-89. Scottish physician and botanist. Curator of the Royal Botanic Garden, Edinburgh. 1862 Nov. 19 CD to John Scott, "present my thanks to Mr. McNab". CCD10:538. 1863 May 23 CD to Hooker, Scott was not happy under McNab. CCD11:438. 1872-89 Prof. Botany Dublin. Scientific Superintendent Royal Botanic Gardens, Glasnevin, Dublin. Two diary entries by ED regarding M in 1848 Feb. 12-14 were crossed out.

MacNalty, Francis Charles, 1846-1914. MD. Practised at Patterdale, near Penrith, for 16 years, later at Winchester. 1870 Qualified Dublin as physician. 1880 MD. 1881 Jun. CD saw at Glenridding, diagnosed angina pectoris with signs of myocardial degeneration. Sir Arthur Salusbury MacNalty, son, *Nursing Mirror*, Dec. 4, 1964.

Madagascar Squib. Description of a spoof carnivorous plant supposed to eat humans. Crinoida Dajeeana. The man-eating tree of Madagascar. *World* (New York), 28 Apr. 1874, p. 7. 1874 Jun. 3 CD to Asa Gray, "I thought it was a false story, & did not perceive it was a hoax till I came to the woman". CCD22:275.

Madonna, The. 1868 nickname of Julia Margaret Cameron's pretty Irish maid, Mary Ryan (1848-1914), who often sat for her. CD and family met her at Freshwater, Isle of Wight. ED2:191.

Maer Hall, Maer, Staffordshire. Seven miles from Etruria and Stoke-on-Trent. Home of Josiah Wedgwood [II] (Jos). CD was a frequent visitor there in his youth especially for the shooting in partridge season, and, after his return from the *Beagle* voyage, for his courting. "The happiest of all Wedgwood houses". Description in Wedgwood 1980, p. 246. Parkfields was a cottage with about 100 acres which Josiah Wedgwood [II] added to the estate, borrowing from Robert Waring Darwin. 1802 Bought for £30,000. 1807 Moved in. 1808 May 2 ED born there. 1814 Josiah Wedgwood [II] thought of selling it because he was then having to live at Etruria. 1816 He was back at Maer. Emma Wedgwood at 16 made a diary entry in 1824 Jun. 19, "Susan & Charles came" and again on Dec. 18. Before their wedding, Charles went to MH and stayed for four days. 1839 Jan. 29 married ED at St. Peter's church from there. CD and ED returned on Apr. 26. 1846. Left, on death of his wife Bessy Allen. 1847 Sold. Watercolour by Charlotte Langton at Down House. Photographs of the house in CCD1 and Pattison, *The Darwins of Shrewsbury*, 2009, p. 81.

Magendie, François, 1783-1855. French physiologist and vivisector. 1830-55 Prof. Medicine Collège de France. 1881 Apr. 22 CD in letter to *The Times* refers to the cruelty of his experiments "some half a century ago" (*Shorter publications*, F1793).

Magistrate. See Justice of the Peace.

Maguire, Thomas Herbert, 1821-95. Irish lithographer. 1849 Portrait of CD drawn and put on stone by M, printed by M. & N. Hanhart. The only engraving of CD from life. One of the Ipswich British Association portraits.

Ma Jun-wu, 1881-1940. First to translate CD's works into Chinese. *Origin* began to receive attention only at turn of the 20th century even though Darwinian references appeared in the 1870s through the writings of Western missionaries. M incorporated non-Darwinian doctrines, particularly Lamarckian and Spencerian principles, into his translation of *Origin*. Xiaoxing Jin, Translation and transmutation: the *Origin of Species* in China. *BJHS* 52(1): 117-141, Mar. 2019. 1902 *Origin* (Historical sketch) (F1846), 1903 Chapters 3 & 4 only as separate booklets, 1904, vol. 1 (preliminaries and chaps. 1-5 only) (F634), 1918 in 3 vols. (F637), 1930 *Descent* (F1047a).

Malay. First edition in: 2017 *Origin* (F2195).

Malayalam. First editions in: 1985 *Origin* (F2149). 2009 *Autobiography* (F2486).

Malden, Bingham Sibthorpe, 1830-1906. Anglican clergyman and botanist. 1861 Jun. 15-16 CD to M on orchids and insects. CCD9. 1862 M is thanked in *Orchids*.

Maldonado, Uruguay. 1832 Jul.-Oct. CD stayed at. He used the mouth of La Plata River as base for inland expeditions. *Zoological diary*, pp. 150-179. *Geological diary*: DAR33.153-164, both transcribed in *Darwin Online*.

Maling, Thomas James, 1778-1849. Naval Officer. 1821 Married 1 Harriet Darwin d.s.p. 1825. 1828 married 2 Elizabeth Colyear. 1841 Vice Admiral.

Mallet, Robert, 1810-81, Irish geophysicist and civil engineer. Known as "father of seismology". Corresponded with CD on earthquakes. See Herbert, *Charles Darwin, geologist*, 2005, p. 214.

Maltese. First editions in: 2007 *Journal* (F2007).

Malthus, Rev. Thomas Robert, 1766-1834. Cleric and scholar, specialized in political economy. M's statements on the geometrical increase in population and its relation to the availability of resources were extremely important in CD's formulation of natural selection. It is not evidence, however, that CD took inspiration from economic theory, rather M's book was replete with examples of animal populations. 1805-34 Prof. Modern History and Political Economy, East India Company College Haileybury. 1818 FRS. 1838 Sept. 13 CD started to read *An essay on the principle of population*, 1798, in the enlarged edn of 1803. LL1:83. See Winch, Darwin fallen among political economists. *Proc. of the American Phil. Soc.* 145, 2001, pp. 415-37.

Malvern Wells, Worcestershire. A spa town where Dr James Manby Gully had his water cure establishment. 1848 Summer CD stayed for "some months". *Autobiography*, p. 117. 1849 Mar. 9-Jun. 30 CD again took water cure there staying at a house called The Lodge. 1849 Sept. 16 CD visited for day from BAAS meeting at Birmingham returned home on 20th. 1850 Jun. 11-18, see van Wyhe ed., 'Journal', (DAR158), p. 12. 1850 Jun. 11 CD went for a day before heading to London. 1851 Mar. 24 Anne Elizabeth Darwin, suffering from a fever, arrived there with CD, ED and Henrietta. Miss Thorley the governess arrived a few days later on 28th. Apr. 17 CD arrived from London. Apr. 23 Anne died. ED2:132. Apr. 24 CD returned to Down. 1863 Sept. 1-Oct. 14 CD took a house for whole family. ED2:180. The whole time he was there, he was just as ill with giddiness and sickness. After a water cure on 15th, although he was better the next day, had "bad sickness" the following day. Horace was poorly. See Janet Browne, Spas and sensibilities: Darwin at Malvern. *Medical Hist.*, Supplement no. 10, 1990, pp. 102-113.

Mammalia. CD ed. 1839. *Mammalia Part 2 of The zoology of the voyage of HMS Beagle*, by George R. Waterhouse. Edited and superintended by Charles Darwin. London: Smith Elder and Co.

Mansell, Henry Longueville, 1820-71. Anglican clergyman and metaphysician. 1855- Fellow of Magdalen College, Oxford. 1861 CD sent him Gray's *Natural selection not inconsistent with natural theology*. CCD9. 1866-68 Regius Prof. Ecclesiastical History Oxford. 1868-71 Dean of St. Paul's Cathedral, London.

Mantell, Gideon Algernon, 1790-1852. Surgeon, geologist and palaeontologist. Father of Walter Baldock. Describer of *Iguanodon* and other dinosaurs. Disliked Owen as much as CD did. 1927, biography by Spokes. 1940 *Manuscript journal* ed. E.C. Curwen. 1825 FRS 1848 Feb. 2 CD listened to on New Zealand fossil birds at Geological Society, also met at Royal Society committees. Letters with CD 1839, 43.

Mantell, Walter Baldock Durrant, 1820-95. Naturalist and politician, son of Gideon Algernon M. CD letters to M in 1854 and 1856. 1840 Went to New Zealand.

Manual of Scientific Enquiry. 1849 CD's article on Geology is Section VI in *A manual of scientific enquiry; prepared for the use of Her Majesty's Navy: and adapted for travellers in general*, edited by J.F.W. Herschel. Early copies have a serious transposition of text pp. 178-90 (*Shorter publications*, F325). Later copies are corrected (F326). CD's own copy, at Cambridge, has the correct section inserted in a pocket in back cover. CD contributed "On the use of the microscope on board ship" to Richard Owen's article on Zoology (*Shorter publications*, F1822). 1851 2^d edn (F328). 1859 3^d edn, superintended by R. Main (F329). 1871 4th edn, revised by J. Phillips (F331). 1886 5th edn edited by Sir Robert S. Ball; CD's contribution, Article X, revised by Archibald Geikie (F333). CD's article alone occurs as a pamphlet: 1849 1st edn (F327). 1859 3^d edn (F330). 1871 4th edn (F332). First foreign editions: 1860 CD's article only, Russian (F336). 1860 Whole book, Russian (F337).

Manures and Steeping Seed. 1844 Manures and steeping seed, *Gardeners' Chron.*, no. 23: p. 380. (*Shorter publications*, F1666). CD forwarded quotations from P. Vallemont, *Curiosities of nature and art in husbandry and gardening*, 1707.

Marcet, Jane, 1769-1858. Writer of popular science books. 1819 Wrote *Conversations on natural philosophy*. CD read the chapter on optics to George Darwin (8-9 years old) every day. Keynes, *Annie's box*, 2001, p. 124.

Marginalia, Charles Darwin's, Mario A. di Gregorio with the assistance of N.W. Gill, *Charles Darwin's marginalia*, vol. 1, 1990. Many of CD's books are annotated with his notes, reactions and so forth. This work not only transcribes these voluminous and often terse, scattered or obscure comments but explains his methods of note taking and working with sources such as the fact that he often returned to a work multiple times and that subsequent occasions of annotating, called layers, can often be identified and distinguished (such as through different writing medium or colour of notepaper or "slips"). The annotations are furthermore classified into numerous themes. There appear to be more than 1,400 annotated works recorded.

Marindin, Samuel, 1807-52. Captain, Life Guards. M was at Shrewsbury School and Cambridge with CD. 1821-25 At Shrewsbury School. 1829 Trinity College Cambridge BA. 1834 Married Isabella Colville of Ochiltree and Craigflower, Ayrshire. 1834 M is mentioned in letters to Whitley. CCD1:376 & 397. 1835 Rector of Penselwood, Somerset.

Markham, Thomas, Farmed twenty acres in Cudham, Kent, a village near Down. A member of the Down Friendly Society.

Marriage notes, see "This is the Question".

Marsh's wagon. Mentioned several times by CD to Henslow for shipments of specimens. Marsh and Swann's Fly Wagons travelled daily (except Sunday) between Cambridge and the Bull, Bishopsgate Street, London. CCD1, Barlow 1967, p. 123.

Marsh, Othniel Charles, 1831-99. American palaeontologist. 1866- Prof. Vertebrate Palaeontology Yale, where his uncle George Peabody endowed the Peabody Museum. 1880 Aug. 31 CD thanks M for sending *Odontornithes*, 1880. LL3:241.

Marshall, William, 1815-90. Solicitor and botanist of Ely. 1852 M wrote a pamphlet on spread of *Anacharis alsinastrum* (a supposed new British plant), London, reprinted from *Cambridge Independent Press*. CD had corresponded with. 1860 May 8 CD to Henslow, about spread of *Elodea canadensis* (an aquatic plant). CCD8, Barlow 1967, p. 203. In *Darwin Online*.

Marshall, William Cecil, 1849-1921. Architect. Cambridge friend of CD's sons. 1876 CD to M, on adding billiard room with dressing room and bedroom above at Down House. CCD24, Atkins 1976, p. 28. Many visits over the years of 1872-96 were recorded by ED. The first being in 1872 Jun. 25 and the last in 1896 Mar. 8. A presentation copy to M of *Insectivorous plants* sold at a Stockholm auction in 2015.

Marshall, Victor Alexander Ernest Garth, 1841-1928. Landowner of Monk Coniston, cousin of W.C. Marshall. 1879 Aug. 2-27 the Darwins stayed in a hotel in Coniston owned by M. CD wrote to thank him for the enjoyable stay and sent M a young oak tree. CCD27. *ED's diary*. 1880 Feb. 14 M visited the Darwins. DAR210.9.

Martens, Conrad, 1801-78. Draughtsman on 2^d voyage of *Beagle*. M replaced Augustus Earle at Montevideo. Later a distinguished landscape painter in Australia. Later assistant librarian, Legislative Council New South Wales. "A pupil of C. Fielding and excellent landscape drawer". 1833 Early Dec. At Montevideo where he met FitzRoy, who engaged him as draughtsman. 1834 Pencil sketch self portrait in MLNSW, DL Pd 279. 1834 Aug. Valparaiso, where M left *Beagle* to travel to Sydney, where he arrived in 1835. 1836 CD bought two pictures from him: Jan. 17 Ponsonby Sound, actually *Beagle* in Beagle Channel, Jan. 21 Santa Cruz river, for 3 guineas each at Sydney. CD "It is necessary also to leave our little painter, Martens, to wander about ye world." CCD1:411. Biography, Lindsay, Sydney, Revised edn, 1968. his works are listed and a surprising range of them illustrated in Keynes, *The Beagle record*, pp. 389-402. 1837 Married Jane Brackenbury Carter. 2 daughters. 1862 CD was sent a third picture depicting Brisbane River. Two of his sketchbooks are in CUL, MS Add 7983 and MS Add 7984, in the Cambridge Digital Library: https://cudl.lib.cam.ac.uk/view/MS-ADD-07984/1 and https://cudl.lib.cam.ac.uk/view/MS-ADD-07983/2. Many others in MLNSW.

Martial, Mr, Surgeon on a whaling ship. M gave CD information on races of human lice. *Descent*, 1:219, where he is not named. "Worthless and slightly educated"; CD to Henry Denny. Carroll 1976, 35, CCD3:38.

Martin, John Royle, 1871 CD to M, asking for ten shares in Artizans, Labourers, and General Dwellings Company for £100. CCD19.

Martin, Septimus, Son of the Rector of an adjoining parish to Down. 1853 Apr. M dined at Down House. M had emigrated to Melbourne before this and was visiting. M had been to Two-fold bay and told CD about it during the visit. CCD5:164.

Martineau, Harriet, 1802-76. 'Feminist', social theorist and author. CD's father Robert did not like her. ED1:276. 1831 ED met M at Hensleigh Wedgwood's in London, "She is so happy, good-humoured and conceited that she will not much mind what people say of her". ED1:257. 1841 Erasmus tried to help her when she was ill and poor. ED2:58. ED met with M on several occasions in 1833. In 1839 Mar. 16 M visited the newly wedded Darwins. Recollection of CD in *Harriet Martineau's autobiography*, 1877, vol. 1, p. 268, transcribed in *Darwin Online*.

Martineau, James, 1805-1900. Brother of Harriet M. Nonconformist minister. Unitarian pastor. 1840-85 Prof. Philosophy Manchester New College. *ED's diary* 1834 Aug. 3 "went to hear Mr Martineau". In 1854 the college moved to London. 1859-73 CD went to Little Portland Street chapel in London to hear him preach. R.V. Holt, 1938, p. 344. 1861 CD sent M Asa Gray's *Natural selection not inconsistent with natural theology*. CCD9. 1869-85 Principal Manchester College. 1885 President Manchester College.

Marx, Karl Heinrich, 1818-83. German political philosopher and communist. CD never met. See also entry for E.B. Aveling. From 1848 Living in London. 1873 CD was sent a copy of 2^d edn of *Das Kapital*, vol. 1, 1873, inscribed to CD "On the part of his sincere admirer". Now at Down House. The inscription is in the hand of Aveling; who also wanted to dedicate the English translation to CD. CD wrote to M to briefly thank him for the book, which CD never read. For many years a letter to Aveling, in which CD declined having a book by Aveling dedicated to him, was mistakenly believed to by to M. See Colp, The myth of the Darwin-Marx letter, *Hist. of Political Economy*, 14, 1982, pp, 461-82 and M.A. Fay, Did Marx offer to dedicate Capital to Darwin? *Jrnl. of the Hist. of Ideas* 39, 1978, pp. 133-46.

Massingberd, Charlotte Mildred (Mildred), 1868-1940. Granddaughter of Charles Langton. Took the name "Massingberd" as had her mother Emily Caroline M. CD's daughter-in-law. "She had a lively seriousness...she was charming to look at, with a great air of breeding and, I imagine, took more pains over her clothes than she would have confessed". B. Darwin, *Green memories*, 1928, p. 51. Many ED diary entries of her visits and often staying up to a week. M was ED's companion "or lady-in-waiting as we sometimes called it". Raverat, *Period piece*. A final diary entry in 1896 Aug. 29. 1900 Married as second wife to Leonard Darwin.

Massingberd, Emily Caroline (Lena), 1847-97. Elder daughter of Charles Langton M and niece and co-heir of Algernon Langton M. 1867 Married Edmund Langton, her second-cousin. 1875 Widowed. 1887 May 20 She assumed the name and arms of "Massingberd" by Royal Licence. Burke 1888, p. 11.

Mastodon, The position of the bones of, a 1,300 word essay written by CD in Feb. 1835 which reveals that at that time he still believed in special creation of species but was speculating about natural causes for extinctions. The purported Masto-don (actually a *Macrauchenia*) seemed to have died in a stable environment, as op-posed to a catastrophe; CD speculated that perhaps species had fixed lifetimes after which they went extinct. CD made use of the phrase "the gradual birth & death of species", as used in Lyell, *Principles of geology*, vol. 3, p. 33, thus revealing CD's early interest in questions of the disappearance and origins of species. The essay (DAR42.97-99) is transcribed with an introduction in *Darwin Online*. It is discussed in M.J.S. Hodge, Darwin and the laws of the animate part of the terrestrial system. *Studies in the Hist. of Bio.* 6, 1983, 1-106, pp. 19-20 and D. Kohn, Theories to work by. *Studies in the Hist. of Bio.* 4, 1980, pp. 67-170.

Masters, **Maxwell Tylden**, 1833-1907. Son of William M. Physician and plant tera-tologist. 1860 Apr. 7 CD to M, on papilionaceous flowers. CCD8:146. 1860 Apr. 13 CD to M, about evolution, mentioning that M had written to CD's father who was ill. CCD8:157. 1862 Feb. 26 CD to M, about M's approval of *Origin*. CCD9:95. 1865-1907 Editor of *Gardeners' Chron.* 1869 *Plant teratology.* 1870 FRS.

Masters, **William**, 1796-1874. Nurseryman in Canterbury. 1823 Founded the Can-terbury Museum, 1823-46 honorary curator. Conducted hybridisation experiments on passion flowers. Father of Maxwell Tylden M. Friendly correspondent of CD.

Matta, **Eugenio**, Spanish co-owner of the San Antonio mine. "a hospitable old Spaniard" *Beagle diary*, p. 339. In the *Coquimbo notebook* and *Copiapò notebook*, Chancel-lor & van Wyhe, *Beagle notebooks*, 2009.

Matthew, **Henry**, 1807-61. Cambridge friend of CD. He was ill and paralysed for 20 years. CD lent or gave him money. 1830 President of the Cambridge Union. 1831 wrote to CD "our friendship was the growth of a day but I trust it will bear fruit for years. Once for all I do love you and shall ever come what may to either of us." CCD1:118. 1837 After some impoverished years in London, priest in Linsdale. 1837-43 Rector of Grove. 1843-61 Rector of Eversholt, Bedfordshire.

Matthew, **Patrick**, 1790-1874. Author on political and agricultural subjects. Of Gourdiehill, Errol, Scotland. One of CD's predecessors in publishing a notion of natural selection. Biography W.J. Dempster, 1983. 1831, 1839 The main statement regarding natural selection is in an appendix to his *Naval timber and arboriculture*, Lon-don, 1831, and there are further remarks in *Emigration fields*, London, 1839. 1860 M drew attention to his priority in *Gardeners' Chron.*, Apr. 7, with an extract from *Naval timber*, and reinforced it in *Saturday Analyst and Leader*, Nov. 24. CD's reply to first paper is in *Gardeners' Chron.*, Apr. 21: pp. 362-3. (*Shorter publications*, F1705) Gracious-ly accepting that M had preceded him with that point. 1864 In his pamphlet *Schles-wig-Holstein* M put on title page "Solver of the species problem". 1865 Oct. CD to Hooker, about W.C. Wells's work read to the Royal Society in 1813, "So poor old Patrick Matthew, is not the first, & he cannot or ought not any longer put on his Title pages 'Discoverer of the principle of Natural Selection'"! CCD13:279. M sometimes inserted calling cards with this message in his books. 1912 Miss Euphemia M, daughter, visited W.T. Calman at BMNH with copies of CD-M correspondence. See Calman, *Jrnl. Bot. British foreign*, pp. 192-4, portrait of M.

Matthews, Mary Anne, 1821?-1905. Married Lawrence Ruck. Mother of Mary Elizabeth and Amy Richenda Ruck. Mother-in-law of Francis Darwin. Francis Darwin's book *The story of a childhood* contains extracts from letters addressed to M about Bernard Richard Meirion Darwin's youth. Known to Bernard as Nain, North Welsh for grandmother. Home Pantlludw, Merionethshire, Wales; picture of in B. Darwin, *Green memories*, 1928, p. 24. 1890 Was visiting ED in Cambridge "once a year". She taught ED solo whist. There are many entries on Mrs Ruck in *ED's diary*. After the death of Amy, ED wrote regularly to R and visited her often. In her later years, the correspondence increased. An 1896 Apr. 14 letter may be last.

Matthews, Richard, 1811-93. Missionary from Church Missionary Society to Fuegians. Carried there on 2d voyage of *Beagle*. A young catechist rather than a qualified missionary, also a seaman. He became an Able Seaman after the rescue from Woollya. 1834 Jan. 23 M landed at Woollya to establish a mission. 1834 Feb. 6 M was taken off again because his life was in danger. "No companion could be found in time". In the *Buenos Ayres notebook*, Chancellor & van Wyhe, *Beagle notebooks*, 2009. Finally landed at New Zealand where his brother was a missionary. *Journal*, 2d edn, p. 207. 1834 Married Johanna Sara Blomfield. 5 daughters. 1839 Farmed 3,000 acres at Te Kumi, North Island, New Zealand. 1893 Died in Auckland. 1893 Obituary *New Zealand Herald Suppl.*, Feb. 24.

Maull & Polyblank. Commercial photographers of London, 1854-65. From 1866-72: Maull, Henry & Co., from 1873-78: Maull & Co., from 1879-85 Maull & Fox. Photographed CD in 1855 and 1857 as well as ED. See Darwin, Iconography.

Mauritius, Indian Ocean. 1836 Apr. 29-May 9 *Beagle* at Port Louis. CD made several inland trips including one to Captain J.A. Lloyd's house on May 4. CD returned part of the way on an elephant, the only one on the island. 1874 [Memorial to Gordon, Governor of Mauritius, requesting the protection of the Giant Tortoise on Aldabra]. *Shorter publications*, F2006. *Geological diary*: DAR38.882-901, in *Darwin Online*. *Sydney Mauritius notebook*, Chancellor & van Wyhe, *Beagle notebooks*, 2009.

Maw, George, 1832-1912. Manufacturer, geologist, botanist and antiquarian. Of Benthall Hall, where he had a well-known garden. M provided *Drosophyllum* for *Insectivorous plants*. 1861 Jul. M reviewed *Origin* in *Zoologist*. CD to Lyell, "evidently a thoughtful man". CCD9:12. 1871 Travelled to Morocco and Tunisia with Hooker and John Ball. 1886 *A monograph on the genus Crocus*.

May, Jonathan, 1800-? Petty Officer. Carpenter on 2d voyage of *Beagle*. Boat builder, built several and maintained all *Beagle*'s boats. In the *Buenos Ayres notebook*, Chancellor & van Wyhe, *Beagle notebooks*, 2009.

Mazhar, Ismail, 1891-1962. Born in Egypt into a wealthy family. Prolific writer, editor and translator. 1918 the first to translate CD's work into Arabic, *Origin* (F1928a), first five chapters only, adding four more in 1928 (F1928b) with the complete translation in 1964 (F1928c). See Ziadat, *Western science in the Arab world: The impact of Darwinism 1860-1930*, 1986 & Muzaffar Iqbal, *Science and Islam*, 2007.

Mayzel, Waclaw, 1847-1916. Born in Kunów, Poland. 1873 translated *Origin* into Polish (F739).

Medicago lupulina. 1865-6 Note on *Medicago lupulina*. In G. Henslow, Note on the structure of *Medicago sativa... Jrnl. of the Lin. Soc. of London (Bot.)* 9:328. (*Shorter publications*, F1809). CD's experiments suggested that the flowers of Black medick or Hop clover were adapted to ensure pollination by insects rather than self-fertilisation.

Meehan, Thomas, 1826-1901. American botanist; born in England. 1848 Settled in Germantown, Philadelphia, USA, and worked for Bartram's Garden. Started his own nursery. Reviewed *Origin* and *Fertilisation*. 1874 Oct. 9 CD to M, about colours of dioecious flowers. CCD22. A copy of an Elliott & Fry 1874b.3. CDV signed on the verso "Mr Meehan from Ch Darwin" sold at Christie's in 1998 for £322.

Meek, Mary Ann, 1817-75. School teacher in Birkenhead. A payment of £3 was recorded in 1854 Nov. 16 in CD's account book. In 1858 ED wrote in her diary "Miss Meek 11 Oxton Rd Birkenhead" and in 1864 Jul. 11, "Hen. came from London with Miss M." and the next day took her back.

Meldola, Raphael, 1849-1915. Chemist and entomologist. Educated in chemistry at the Royal College of Chemistry, London. 1873 Aug. 13 CD to M, about saltations. CCD21. 1882 M translated August Weismann, *Studien zur Descendenz-Theorie*, Leipzig, 1875-76, as *Studies in the theory of descent*, London, with prefatory note by CD, pp. v-vi (F1414). 1885 Prof. Chemistry, Finsbury Technical College, London. 1886 FRS. 1912-15 Prof. Organic Chemistry University of London. Recollections of CD in Unveiling the Darwin statue at the museum. *Jackson's Oxford Jrnl.* (17 Jun. 1899): 8, transcribed in *Darwin Online* (F2169).

Mellersh, Arthur, 1812-94. Volunteer 1st class on 1st voyage of *Beagle*. 1832 Apr. Midshipman/Mate's warrant on 2^d voyage of *Beagle*. 1862 Oct. 21 visited CD with Wickham & Sulivan. CD was unwell that day. 1878 Vice-Admiral. 1882 recollections of CD in DAR112.A83, transcribed in *Darwin Online*. 1884 Retired as Admiral.

Melastomaceae. This group of flowering plants has, in some species, two forms of stigmata. 1862-81 CD worked on them, but never published results. ML2:292-302 summarizes his work & quotes correspondence on subject. Notes in DAR205.8.

Memorials. CD signed numerous memorials and testimonials in support of individuals applying for a position, the National Sunday School League (1860), educational reform, against war with Russia and in Afghanistan (1878), the purchase of fossils by the BM (1838), copyright for British authors in the USA (1872), the protection of the giant tortoise on Aldabra (1874), for the right of black South Africans with property to vote (1877), on the persecution of Jews in Russia (1882) and other matters he found important. See F324a, F345b, F869, F1702, F1766, F1831, F1910, F1926, F1937, F1939, F1942, F1944, F1945, F1954, F1954a, F1957, F2003, F2006,

F2030, F2080, F2159, F2166, F2280, F2288, F2449, F2527, F2529, F2534, F2547 and F2561 (National Sunday League). *Shorter publications* and *Darwin Online*.

Mendel, Johann Gregor, 1822-84. Augustinian monk at Brno, Moravia. CD never heard of M and, although his now famous 1865 paper on inheritance in peas, "Versuche über Pflanzenhybriden", *Verh. Naturforsch. Verein Brünn*, 4, was available at the Royal Society and at the Entomological Society, it was ignored until 1900 and CD did not have a copy. M thought hybridization a superior theory to natural selection.

Mendoza, Argentina. 1835 Mar. 27 CD visited from Valparaiso, crossing the Andes by the southern, Portillo, pass. 1835 Mar. 29 Returned by northern Uspallata or Aconcagua, pass, crossing the Incas' bridge on Apr. 4. *Geological diary*: DAR36.462-501, transcribed in *Darwin Online*.

Mental Evolution in Animals. 1883 George John Romanes, *Mental evolution in animals*, London, contains posthumous Essay on instinct by CD, q.v. (F1434).

Mephistopheles, Henrietta Darwin's cat. Also known as "Phisty". ED2:202.

Meteorites. CD 1881. [Letter] In G.W. Rachel, Mr. Darwin on Dr. Hahn's Discovery of Fossil Organisms in Meteorites. *Science* 2 (61) (27 Aug.): 410. Rachel claimed, incorrectly, that CD had been convinced that extraterrestrial fossils had been found in meteorites. See John van Wyhe, 'Almighty God! what a wonderful discovery!' *Endeavour* 34, no. 3, (2010): 95-103.

Meteors. "There have been two bright meteors passing from East to West." Jan. 21 1832 *Beagle diary*, p. 27. ED recorded in her 1866 Nov. 13 diary entry "meteors" which was the Leonid meteor showers; she recorded meteors again in 1867 Nov. 14.

Meteyard, Miss Elizabeth, 1816-79. Daughter of a surgeon to Shropshire Militia. Spent her early years in Shrewsbury. Woodall 1884, p. 1. Biographer of the Wedgwoods. 1871 *A group of Englishmen (1795-1815) being records of the younger Wedgwoods and their friends*, London, is an important sourcebook, including information about CD's mother and of Darwins and Allens.

Miall, Louis Compton, 1842-1921. Zoologist. 1876-1907 Lecturer Yorkshire College of Science, Leeds, later Prof. Zoology there. 1881 FLS. 1883 *The life and works of Charles Darwin; a lecture delivered to the Leeds Philosophical and Literary Society on February 6th, 1883*, Leeds, the first biography after the obituaries. 1892 FRS.

Microscopes. CD was given (by John Maurice Herbert) a small microscope when a student at Christ's College. The microscope now in the old study at Down House is a portable in original case, by Cary, London, used during the *Beagle* voyage. 1863 May 2 CD to Isaac Anderson Henry, "I have, as yet, found no exception to the rule, that when a man has told me he works with the compound alone his work is valueless". "Experience, however, has fully convinced me that the use of the Compounds without the simple microscope is absolutely injurious to progress of Natural History (excepting, of course, with *Infusoria*)". CCD11:373. For his barnacle research, CD purchased an 1847 achromatic compound microscope, made by James Smith, London, for £36. Now in the Whipple Museum, Cambridge with a heavily annotated instruction manual and notes on the preparation of specimens and making measurements under the microscope. (Wh.3788) CD used an old silk handkerchief as a dust cover (one at WSM). See B. Jardine, Between the *Beagle* and the barnacle: Darwin's microscopy, 1837-1854. *Studies in Hist. and Philosophy of Sci.* 40 (2009): 382-395.

Middleton, John, Royal Marine on 2^d voyage of the *Beagle*. See *Narrative* 2:516 and FitzRoy & Darwin 1836, *Shorter publications*, F1640.

Midhurst, Sussex. 1876 Jun. 7-9 CD visited Sir John Hawkshaw there.

Mill, John Stuart, 1806-73. Philosopher. 1823-58 In service of East India Company until dissolution. 1861 Henry Fawcett to CD, "he considers that your reasoning throughout is in the most exact accordance with the strict principles of Logic. He also says, the Method of investigation you have followed is the only one proper to such a subject." CCD9:203. 1865-68 Whig MP for Westminster.

Miller, Hugh, 1802-56. Geologist, stonemason, author and evangelical Christian. CD never knew this remarkable man, but he borrowed Lady Lyell's copy of M's *Footsteps of the Creator*, 1849, and then bought one himself.

Miller, William Hallowes, 1801-80. Mineralogist and crystallographer. 1832-70 Prof. Mineralogy Cambridge after Whewell. 1836 M helped CD with examination of rock specimens from *Beagle*. 1838 FRS. 1859 M and CD corresponded on structure of cells of honeybee comb. CCD7. 1843 M mentioned in Extracts...on the analogy of the structure of some volcanic rocks with that of glaciers, F1670.

Milman, Henry Hart, 1791-1868. Anglican clergyman and author. Dean of St. Paul's Cathedral, London. 1846 CD met Sydney Smith at M's house in London. CCD3:367. *ED's diary* 1839 Jan. 10 has a list of ladies and Mrs Milman was one of them. In 1864 Jul. 15 "Milmans & Lyells" were recorded in *ED's diary*.

Milne, David, (Milne-Home, Milne-Hume), 1805-90. Geologist and mineralogist. 1847 M was against CD's interpretation of Glen Roy. Frequent correspondent. *Trans. Roy. Soc. Edinburgh*, 16: p. 395, 1849. 1852 Name changed upon his wife inheriting the 'Home' estates in Berwickshire.

Milne-Edwards, Henry (Henri) Milne, 1800-85. Zoologist. Born in Bruges, the Netherlands, when the Netherlands was part of France; since 1830 Bruges is in Belgium, then newly founded. Of British father and French mother. Frequent correspondent with CD. 1841- Prof. Entomology Paris. 1848 Foreign Member Royal Society. CD was inspired by M's concept of the "physiological division of labour". 1854 CD sent *Living Cirripedia* to. 1856 Copley Medal Royal Society. 1859 CD sent 1st edn *Origin*. 1862 Prof. Zoology, Paris.

Milner, Sir William Mordaunt Edward, 5th **Bart**, 1820-67. Of Nunappleton, Tadcaster, Yorkshire. 1859 CD to Hooker on M having found nuts in young petrels' crops at St. Kilda. CCD7:252.

Milnes, Richard Monckton, 1809-85. Politician, poet and patron of literature. In youth M was a member of Cambridge Apostles Debating Society. Known by Sydney

Smith as "the cool of the evening". ED2:114, 121. late 1830s CD met at Lord Stanhope's house. 1837-63 MP for Pontefract. 1863 1st Baron Houghton.

"Minerva". Nickname for Athenaeum Club, London, from bust on top of façade. 1838 CD to Lyell "I did not even taste Minerva's small beer today". CCD2:98.

Miranda, Commandante, was a subordinate of General Rosas. In the *Falkland notebook*, Chancellor & van Wyhe, *Beagle notebooks*, 2009.

Missionaries. 1836 Jun. 28 CD and FitzRoy, "On the whole...we are very much satisfied that they thoroughly deserve the warmest support, not only of individuals, but of the British Government". *South African Christian Recorder*, 2: p. 238 (*Shorter publications*, F1640). See also Moral state of Tahiti. See Richard Matthews. This was CD's first intentional publication, although it consisted of quotations from his *Beagle diary* together with a longer letter by FitzRoy. CD later followed the activities of missionaries in Tierra del Fuego. In 'Philosophical Tracts' CUL.

Mitten, Annie, 1846 Feb. 12-1914 Dec. 10. Eldest daughter of botanist and an authority on mosses. William M and Anne M. 1866 Married Alfred Russel Wallace. 2 sons, 1 daughter. In 1868 visited the Darwins with Wallace. *ED's diary*.

Mivart, St. George Jackson, 1827-1900. Barrister and biologist. Lecturer in Biology, St. Mary's Roman Catholic College, Kensington. Roman Catholic anti-Darwinian who was initially an adherent of the theory of natural selection. Biography: J. Gruber, 1960. 1862 FLS; later Secretary and President. 1867 FZS. 1869 FRS. 1871 *The genesis of species*. 1871 CD to Wallace, "but he was stimulated by theological fervour". CCD19:60. 1871 CD to Wallace, "I conclude with sorrow that though he means to be honourable, he is so bigoted that he cannot act fairly". CCD19:505. 1900 Excommunicated. M's other evolutionary works: 1873 *Man and apes, an exposition of structural resemblances bearing upon questions of affinity and origin*. 1876 *Contemporary evolution; an essay on some recent social changes*. 1882 *Nature and thought*.

Moffatt, or **Moffat**, 1858-78 Liveried and later senior footman at Down House.

Moggridge, John Traherne, 1842-74. Naturalist. M was tubercular and lived in Mentone, South of France. M sent orchis *Neotina intacta* to CD. Allan, *Darwin and his flowers*, 1977. 1865 Oct. 13 CD to M, about fertilisation of bee orchis. 1866 Jun. 23-5 M visited CD at Down with Joseph Dalton and Frances Hooker. 1871 Jun. CD to M, about habits of ants and about orchids. CCD19. 1871 *Contributions of the flora of Mentone*. 1872 Oct. 9 CD to M, about trap-door spiders. CCD20. 1874 a writer in *Nature* described M as "one of our most promising young naturalists". *Nature*, 11: p. 114. 1873 Author of *Harvesting ants and trap-door spiders*, 1873[-74], which rediscovered the habits of *Atta*, described in *Proverbs*, vi. 6.

Mohl, Madame, unclear if she was related to Hugo von Mohl. 1859 Jul. 29, ED recorded "Mme Mohl came". 1880 Aug. 23 Came with Mme von Schmidt for lunch.

Mohl, Hugo von, 1805-72. German biologist. Prof. of physiology and botany. Foreign member, RS, 1868. D. Oliver and J.D. Hooker discussed M's work with CD.

Mongolian. First foreign edition: 1980 *Origin* (F2156).

Mojsisovics von Mojsvár, Johann August Georg Edmund, 1839-1907. Austro-Hungarian geologist and palaeontologist. Vice-Director Imperial Geological Institute, Vienna. Sent CD his *Die Dolomit-Riffe von Südtirol und Venetien*, 1879, and the accompanying *Geologische Übersichtskarte*, 1878. CD replied praising it. CCD26.

Monk, Thomas James, 1830-99. Merchant and brewer. Lived at 111 High Street, Lewes, Sussex. Kept an aviary. 1873 J.J. Weir informed CD that two species of the genus *Motacilla* [Wagtails] produced a fertile hybrid. CCD21:500 & CCD25:18.

Monkeys. 1876 Sexual selection in relation to monkeys, *Nature*, 15: pp. 18-19 (*Shorter publications*, F1773). CD suggested that sexual dimorphism in monkeys was the result of sexual selection. Reprinted in *Descent*, 12th thousand, 1877 and onwards.

Monro, Alexander, 1773-1859. Scottish anatomist. Prof Anatomy Edinburgh, in succession to his father and grandfather. M is said to have lectured from his grandfather's notes. 1826 Jan. 6 CD to sister Caroline Sarah, "I dislike him & his Lectures so much that I cannot speak with decency about them." CCD1:25. "Made his lectures on human anatomy as dull as he was himself, and the subject disgusted me". *Autobiography*, p. 47. CD's lecture notes are in DAR5.A12 and DAR5.A13-A23, transcribed in *Darwin Online*.

Monsell, Elinor Mary, 1878-1954. Irish engraver and portrait painter. 1906 Married Bernard Richard Meirion Darwin. Memorial in Downe churchyard.

Montevideo, Uruguay. 1832-33. 1832 Jul. 26 CD took several inland trips from here and from Buenos Aires when *Beagle* was based on La Plata river, until 1833 Dec. 6 when *Beagle* left for Patagonia. *Zoological diary*, pp. 65-68, 107-110. *Geological diary*: (11.1833) DAR32.83-84 and DAR34.3-6, both transcribed in *Darwin Online*.

Montgomery, Wales. 1822 Jul. CD visited for holiday with sister Susan Elizabeth.

Moor Park, near Farnham, Surrey. A water cure establishment, run by E.W. Lane, which CD visited from 1857-59. According to his recollection (F2024), Alexander Bain recommended that CD visit there. 17th century was home of Sir William Temple and Esther Johnson, Jonathan Swift's "Stella". ED wrote in her diary in 1857 Apr. 22, "Ch went to Moor Park." Many visits were made between 1857-59 with a last entry in 1859 Jul. 26 when CD returned. 1859 "Dr. Lane's delightful hydropathic establishment". *Autobiography*, p. 122.

Moore, Aubrey Lackington, 1848-90. Anglican clergyman. Argued that Darwinism was not in conflict with Christianity. "The clergyman who more than any other man was responsible for breaking down the antagonism towards evolution then widely felt in the English Church". E.B. Poulton, *Darwin and the Origin*, 1909, p. 11. 1876-81 Rector of Frenchay, Bristol. 1881- Fellow of Keble College Oxford.

Moore, David, née Muir, 1807-79. Botanist. Changed surname on moving to Ireland. M provided *Drosophyllum* for *Insectivorous plants*, and information on *Pinguicula*. 1838-79 Director of Glasnevin Botanical Gardens, Dublin.

Moore, Sir Norman, Bart, 1847-1922. Physician and antiquary. Of St. Bartholomew's Hospital. M attended CD in his last illness. Atkins says that CD had no confidence in him. Atkins 1976, p. 38. "He [CD] once remarked to Dr. Norman Moore that one of the things that made him wish to live a few thousand years, was his desire to see the extinction of the Bee-orchis". LL3:276. ED started recording Dr M's visits from 1874 Mar. 14. M and Dr Allfrey were in attendance in 1882 Apr. 12. 1882 M was on "Personal Friends invited" list for CD's funeral.

Moral state of Tahiti, New Zealand &c. 1836 Jun. 28 A letter, containing remarks on the moral state of Tahiti, New Zealand &c., *South African Christian Recorder*, 2: pp. 221-38, by FitzRoy and CD (*Shorter publications*, F1640). CD's contributions are suffixed "D". CD's first intentional publication.

More, Alexander Goodman, 1830-95. Botanist. Lived in Isle of Wight. 1860 helped CD with orchid work. CCD8. 1867-80 Assistant Natural History Museum, Royal Dublin Society. 1881 Curator of Botany.

More letters. 1903 Francis Darwin and A.C. Seward, eds., *More letters of Charles Darwin. A record of his work in a series of hitherto unpublished letters*, 2 vols. (F1548). 1972 Facsimile (F1550). Foreign editions: 1903 USA (F1549). 1959 Russian, autobiographical fragment and account of Down House (F1551). 2019 Autobiographical fragment translated into Portuguese by Pedro Navarro. (F2230) In *Darwin Online*.

Morley, John, Viscount, 1838-1923. Statesman and man of letters. 1871 Reviewed *Descent* in *Pall Mall Gazette*. 1871 Mar. 24 CD wrote to anonymous reviewer through editor. CCD19. 1877 Visited Down House with Gladstone, Huxley and Playfair, whilst staying at High Elms. Browne, *Power of place*, 2002, p. 440. 1902 OM.

Morlot, Charles Adolf von, 1820-67. Swiss geologist and archaeologist. 1844 CD advised him on how to rewrite a paper on glaciers. CCD3.

Morgan, Lewis Henry, 1818-81. American anthropologist. Corresponded with CD and paid a visit to Down House in 1871 recounted in *Extracts of Lewis Henry Morgan's European travel journal*, 1937, pp. 338-9, transcribed in *Darwin Online*.

Moresby, Sir Fairfax, 1786-1877. Naval Officer. 1845 Apr. 6 "Captain Moresby informs me that in the Chagos archipelago in the same ocean, the natives, by a horrible process, take the shell from a living turtle. *Journal*, p. 459. 1865 GCB.

Morrey, Martha, 1861 census as 40-year-old retired servant visiting Down House.

Morrey, Mrs. Sister of Martha. Cook to Sarah Elizabeth Wedgwood [I] at Petley's, Down until 1856 when Sarah E Wedgwood died. Keynes, *Annie's box*, 2001.

Morris, George, clerk to 'Mr Bacon'. 1881 Sept. 27 served as a witness to the signing of CD's last will and testament, transcribed in *Darwin Online*.

Morse, Edward Sylvester, 1838-1925. American zoologist and Japanophile. 1873 Sept. 16 CD to M on supposed relation of brachiopods to annelids. *Proc. of Boston Nat. Hist. Soc.*, 15; *Proc. of American Soc. Advancement Sci.*, 19: p. 272, 1870; *Annals and Mag. of Nat. Hist.*, 6: p. 267, 1870. 1877 Apr. 21 CD to M, on his Presidential Address to American Association for the Advancement of Science, on the advance of evolutionary work in USA, published in *Proc. of American Assoc. Advancement Sci.*, 25, 1876. 1880-1914 Director Peabody Museum Salem.

Moseley, Henry Nottidge, 1844-91. Zoologist. 1872-76 Naturalist on *Challenger*. 1877 FRS. 1879 Feb. 4 CD to M, about M's book *Notes of a naturalist on the "Challenger"*, London, 1879, which is dedicated to CD. LL3:237. 1880 FLS. 1881- Linacre Prof. Human and Comparative Anatomy Oxford in succession to Rolleston.

Mosley, Frances (Fanny Frank), 1807-74. Daughter of John Peploe M. 1832 Married Francis Wedgwood. "Blonde and beautiful and frivolous". Wedgwood 1980, p. 217. Died when nightdress caught fire in an hotel in Guernsey. 7 children.

Mosley, Rev. John Peploe, 1766-1833. Rector of Rolleston, Staffordshire. 1790 Married Paget, Sarah Maria (1770-?). 6 children. Father of Frances.

Mosquitoes. 1881 Sept. 15 Mr. Darwin on mosquitoes. *The Times,* p. 10, in response to a letter from Astley Paston Price (1826-86), consulting chemist in London. (*Shorter publications*, F1948).

Motley, John Lathrop, 1814-77. American historian and diplomat. 1840's CD met M at Lord Stanhope's house. 1856 Author of *The rise of the Dutch Republic*, 3 vols. 1860-67 Author of *History of the United Netherlands*, 4 vols.

Mould. 1838 "On the formation of mould", *Proc. of Geol. Soc.*, 2: pp. 574-6 (*Shorter publications*, F1648). 1840 "On the formation of mould", *Trans. of the Geol. Soc.*, 5: pp. 505-9 (*Shorter publications*, F1655). 1844 "On the origin of mould", *Gardeners' Chron.*, no. 14: p. 218 (*Shorter publications*, F1665). 1869 "The formation of mould by worms", *Gardeners' Chron.*, no. 20: p. 530 (*Shorter publications*, F1745). 1881 see *Earthworms*.

Mount, The, Parish of St. Chad, Shrewsbury, Shropshire. Home of Dr Robert Waring Darwin [II]. 1797 Built by him c.1800. late Georgian, red brick, 5 bays and 2½ storeys, quite plain, deep Tuscan porch; lower wings of different length and height, that on the left of four bays, one-storeyed with windows in blank arches. Pevsner, *Buildings of England, Shropshire*, 1958, p. 289. 1809 CD was born there. Until 1866 After CD's father's death, Susan Elizabeth D lived there until her death 1866. 1869

The Mount, Shrewsbury, c.1860.

CD visited, then owned by Spencer Phillips. LL1:11. 2004 The Mount is the District Valuer and Valuation Office in Shrewsbury. Plan of the grounds in Pattison, *The Darwins of Shrewsbury*, 2009. A forgotten photograph from 1903 is in *The Sphere* (18 Jul.), p. 57.

Movement in plants. See also "Movements of Leaves" and "Movements of Plants". 1880 *The power of movement in plants*, London, two-line errata slip p. x, assisted by Francis Darwin (F1325). Advertised as "The circumnutation of plants." Drafts in DAR19-23, proofs in DAR213. CD notes on in DAR209. 1880 2d thousand, errata corrected (F1326). 1882 3d thousand, preface slightly altered (F1328). 1966 Facsimile of 1st edn (F1339). 1969 Facsimile of 2d thousand (F1340). First foreign editions: 1881 German (F1343). USA (F1327). 1882 French (F1342). Russian (F1349). 1884 Italian (F1347). 1970 Romanian (F1348).

Movements of Leaves. 1881 The movements of leaves, *Nature*, 23: pp. 603-4, observations on a manuscript letter from Fritz Müller (*Shorter publications*, F1794).

Movements of Plants. 1881 Movements of plants, *Nature*, 23: p. 409, observations on a manuscript letter from Fritz Müller (*Shorter publications*, F1791).

Mowatt, Jane, 1836-84. Daughter of Francis Mowatt. Born in Sydney, Australia. 1865 Married Vernon Lushington.

Moxon, Walter, 1836-86. Physician of Guy's Hospital. 1882 Apr. 19 M was sent for to Down House, but CD was dying when he arrived. M was on "Personal Friends invited" list for CD's funeral. Browne, *Power of place*, 2002, p. 495.

Mudie's lending library, A lending and subscription library founded by Charles Edward Mudie (1818-90). The library bought 500 copies of the first edition of *Origin* in Nov. 1859. ED got second-hand books from M. Her diary has entries for "Mudie", "Mudie box" 1852, 54, 91.

Müller, Sir Ferdinand Jacob Heinrich von, 1825-96. Botanist of German origin. 1847 Moved to Australia. 1852-96 Government Botanist of Victoria, Australia. 1857-73 Director Royal Botanic Gardens Melbourne, Australia. 1861 FRS. 1861 M answered CD's *Queries about expression*.

Müller, Friedrich Max, 1823-1900. Better known as Max M. German philologist and orientalist living in England. Curator of Bodley's Library. Friendly correspondent with CD. 1850-68 Deputy Taylorian Prof. modern European languages Oxford. 1868- Corpus Christi Prof. Comparative Philology Oxford. 1870 onwards M criticised CD's theories of the origin of language. 1882 Visited Down House.

Müller, Fritz, see Johann Friederich Theodor M.

Müller, Heinrich Ludwig Hermann, 1829-83. German botanist. Younger brother of Fritz M. 1855- Science teacher at Lippstadt, Germany. 1872 CD sent M manuscript of "On the flight paths of male humble bees", which was translated by Ernst Krause as "Über die Wege der Hummel-Männchen", *Gesammelte kleinere Schriften von Charles Darwin*, 1885-86 (F1602). 1872 *Anwendung der Darwinischen Lehre auf Bienen*, Berlin. 1873 Author of *Die Befruchtung der Blumen*, Leipzig. 1873 CD to M, saying that he is reading the German edition slowly. CCD21. 1883 *Die Befruchtung der Blumen* translated by James d'Arcy Wentworth Thompson, *The fertilisation of flowers*, 1883, with prefatory note by CD (*Shorter publications*, F1432). 1950 *The fertilisation of flowers*, foreign edition: Russian, CD's preface only (F1433). *Darwin Online*.

Müller on Kitchen Middens, 1876 Fritz Müller on Brazil kitchen middens, habits of ants, etc. *Nature*, 13: pp. 304-5 (*Shorter publications*, F1811). The letter was forwarded by CD but none of his words were quoted in the article.

Müller, Johann Friederich Theodor, 1821-97. Elder brother of (Heinrich Ludwig) Hermann M. Known as and writing as "Fritz". German schoolmaster in Brazil and naturalist. CD and M never met, but "of all his unseen friends Fritz Müller was the one for whom he had the strongest regard". "Uninterrupted friendship and scientific comradeship". "He had for Müller a stronger personal regard than that which bound him to his other unseen friends". Francis Darwin, *Annals of Botany*, 13: p. xiii, 1899. CD to Hermann M, "One of the most able naturalists living". Photograph ML2:344. Biography ML1:382. Married and had 7 daughters and 1 son, who died young. One daughter, Rosa, observed circumnutation in *Linum usitatissimum*. 1852 Emigrated to Brazil. Teacher of mathematics at Gymnasium, Blumenau, Santa Catarina. Lived in Itajaí, Desterro (later Florianópolis) on the island Santa Catarina, and Itajaí again. Dismissed because he refused to live in Rio de Janeiro. 1864, 1869 M was author of *Für Darwin*, translated by W.S. Dallas, at CD's expense on commission, 1869, *Facts and arguments for Darwin*. It contains one of the earliest statements of the recapitulation theory and Haeckel took the theory from here without acknowledgement. It also contains a joke classification of the Crustacea. 1865-81 Many letters, to and from M, first 1865 Aug. 10, last 1881 Dec. 19. 1874-81 CD wrote introductory notes to six short papers by M in *Nature*. 1880 M was nearly drowned in a flood of the Hajahy river. CD to Hermann M, offering financial help to replace books etc. (£100), but not needed. See *Shorter publications*, F1811, F1781, F1784.

Mumford, John, 1822-82. 1851?- Master of the free school for boys in Down. One of three known copyists for CD along with Mr Fletcher and Ebenezer Norman.

Muñiz, Francisco Javier, 1795-1871. Of Lujan, Argentina. Physician, politician and palaeontologist. 1845 M had discussed Niata cattle, the pug-faced breed, with CD. *Journal*, p. 145. 1845 CD to Owen on bones of *Machairodus* sp. which M offered for sale and which British Museum bought. CCD3. 1845 M described it as *Muñi-Felis bonarensis* in *Le Gaceta Mercantil*, Oct. 9. Letter to M in *Nature*, 14 Jun. 1917, F2481.

Murchison, Sir Roderick Impey, Bart, 1792-1871. Geologist and geographer. CD knew M fairly well during London period, calling him Don Roderick. "He was very far from possessing a philosophical mind". "The degree to which he valued rank was ludicrous". *Autobiography*, p. 102. Biography: A. Geikie 1875. 1826 FRS. 1839 *The Silurian system*. Passages refer to CD collecting shell fragments from drift at Little Madeley, Staffordshire, and near Shrewsbury, between the town and village of Moele-Brace, pp. 352, 533. 1846 Kt. 1849 Copley Medal Royal Society. 1855 Director General British Geological Survey, Director Royal School of Mines and Museum Practical Geology London. 1858 Jun. 19 CD to M, about British Museum enquiry. CCD7. 1863 KCB. 1866 1st Bart. CD wrote 1849. [Remark on a South American gold mine]. In M, On the distribution of gold ore over the Earth's surface... *Athenaeum*, no. 1143 (22 Sept.): 966. (F1941).

"Murder, like confessing". 1844 Jan. 11 CD to J.D. Hooker "I am almost convinced (quite contrary to opinion I started with) that species are not (it is like confessing a murder) immutable." CCD3:2. Since the 1950s this passage, in print since LL in 1887, has come to be interpreted to mean that CD was terrified to reveal that he believed in evolution. Following this interpretation, this quotation has become immensely popular in recent years, perhaps one of the most famous attributed to CD. In fact the language is typical for CD's humour and no one, including all those who knew CD, interpreted this sentence as an expression of fear, until the 1950s. When work on his books felt overwhelming CD would write "the descent half kills me" or "I am ready to commit suicide". CD to Hooker 10 Feb. [1875] CCD23:62. When Hooker was preparing to travel overseas, CD wrote to him "I will have you tried by a court martial of Botanists & have you shot." 17 Mar. [1869] CCD17:135. CD once playfully remarked to Wallace "may all your theories succeed, except that on oceanic islands, on which subject I will do battle to the death". 22 Dec. 1857 CCD6:514. Even the word murder was typical CD hyperbole: "if [the plant] dies, I shall feel like a murderer." to Hooker 24 Nov. 1873 CCD21:520. "You ought to have seen your mother she looked as if she had committed a murder & told a fib about Sara going back to America with the most innocent face." To Henrietta 4 Oct. [1877] CCD25:400. An 1871 letter to O. Salvin shows the same sort of humour and vastly stronger language than the famous "like confessing a murder": "When I saw your bundle of observations, I felt as if I had committed theft, arson or murder." CCD19. See John van Wyhe, *Dispelling the darkness*, 2013, p. 254-6.

Murray, **Andrew**, 1812-78. Advocate and naturalist. 1857 FRSE. 1860 M was anti-*Origin*, paper in *Proc. of Roy. Soc. Edinburgh*, 4: pp. 274-91. 1860 Jan. 10 CD to Lyell, "the entomologist and dabbler in Botany". CCD8. 1861 FLS. 1867 Mar. 17 CD to Hooker, CD had bought a second-hand copy of M's *The geographical distribution of mammals*, 1866. 1867 "It is clear to me that the man cannot reason", "He seems to me conceited". CCD15. 1876 Jun. 5 CD to Wallace, "utter want of all scientific judgement". CCD24:185. 1877 CD to Thiselton-Dyer, "what astonishing nonsense Mr. Andrew Murray has been writing in Gardener's Chronicle about leaves & carbonic acid." CCD25:297.

Murray, **John [II]**, 1808-92. Publisher of 50 Albemarle Street, London. CD's main publisher. 1845 M bought copyright of the 2^d edn of *Journal* from Colburn, for inclusion in his Home and Colonial Library, for £150. 1859 CD and M were on personal terms from the first publication of *Origin*, 1859. 1877 Published *Scepticism in geology and the reasons for it* as Verifier a pseudonym of M. Verifier casts doubts on the principle of "causes now in action" as adopted by Lyell and CD et al. 1882 M was on "Personal Friends invited" list for CD's funeral. M published 1st and subsequent editions of ten of CD's books, as well as: 1875 2^d edn of *Climbing plants* (F836). 1879 *Erasmus Darwin* (F1319). 1869 F. Müller, *Facts and arguments for Darwin*. 1887 *Life and letters* (F1452). 1903 *More letters*, 2 vols. (F1548). 1915 *Emma Darwin* (F1553). Recollections of CD in "Darwin and his publisher John Murray." *Science progress in the twentieth century* 3 (1909): 537-542 and *John Murray III*, 1919, p. 18, both in *Darwin Online*.

Murray, **Sir John [III]**, 1841-1914. Scottish marine biologist and oceanographer. 1868-72 Studied geology at Edinburgh University under Alexander Geikie. 1872-76 Chief Naturalist on *Challenger*. 1881 CD to Alexander Agassiz, on M's firm views on origin of coral reefs, in which CD was right and M wrong. LL3:183, ML2:197. 1882-96 Editor of *Challenger expedition reports*.

Musters, Charles, 1817-32. 4th son of John M of Coldwick Hall, Nottinghamshire, and Mary Anne Chaworth, Lord Byron's Mary. Volunteer 1st Class Royal Navy, on 2d voyage of *Beagle*. 1831 Sept. 11-14 M sailed with CD and FitzRoy from London to Plymouth. Died of malarial? fever at Rio de Janeiro. Buried at Bahía.

Musters, Chaworth George, 1841-79. Royal Navy, known as the "King of Patagonia". 1871 Wrote *At home with the Patagonians, a year's wanderings on untrodden ground from the Straits of Magellan to the Rio Negro.* CD in 1870 in a letter to Sulivan said he had not heard of M's journey but looked forward to see some account of it in the *Jrnl. of Geogr. Soc.* In Nov. 1870 M's paper on his travels in Patagonia was read at the *Roy. Geogr. Soc.* 1872 Oct. ED in her letter to Aunt Fanny Allen said CD was intensely interested in the book; having been there and knowing the country. CD said the natives there are very kind to wife and children. ED2:260.

Nägeli, Carl Wilhelm von, 1817-91. Swiss botanist. 1839-40 Studied botany at Geneva under Alphonse de Candolle. 1865 *Entstehung und Begriff der Naturhistorischen Art*, which was given as a lecture, Mar. 28, to Königlich-Bayerische Akademie der Wissenschaften, Munich. 1866 Jun. 12 CD to N, praising *Entstehung*. "many of your criticisms are the best which I have met with". CCD14:203.

Nancy, CD's nurse at Shrewsbury. CD sent greetings to her in *Beagle* letters via his sisters. LL1:254. CCD1:143, 354, 382, 420, 467.

Narrative of the surveying voyages of His Majesty's Ships Adventure and Beagle. 1836 An earlier narrative by FitzRoy is in *Jrnl. Roy. Geogr. Soc.*, 6. 1839 3 vols. and appendix to vol. 2, edited by Robert FitzRoy and published by Henry Colburn, London. Vol. 1 is narrative of 1st voyage, 1826-30, under Captain Philip Parker King; vol. 2 (and appendix) is narrative of 2d voyage, 1831-35, under FitzRoy, with tables of data in the appendix; vol. 3, entitled *Journal and remarks*, is by CD and is 1st issue of *Journal*, which was issued separately in the same year (F10). The whole work has never been translated. 1972 Facsimile (F166).

Nash, Louisa A'hmuty, née Desborough. 1838-1922. Wife of Wallis N. c.1875 N drew fine head and shoulders of CD in brushed india ink. Still in the Nash family. Ran a temperance group 'Band of Hope.' 1874 Aug. 31 ED called at High Elms and Mrs N. ED last wrote to her in 1894 Jul. 19. 1910 N gave reminiscences of CD in *Overland Monthly* (San Francisco), (Oct.), pp. 404-8, transcribed in *Darwin Online*.

Nash, Wallis, 1837-1926. Lawyer, builder and writer. One of the founding fathers of Oregon State University. Married Louisa A'hmuty in 1871 as 2d wife. One of his five sons named L. Darwin Nash. See K.G.V. Smith & R.E. Dimick, *Jrnl. for the Soc. of the Bibliography of Natural Hist.*, pp. 78-82, 1976. 1873-79 N took George Wood's house, The Rookery, at Down and became friendly with the Darwins. 1879 Emigrated to USA upon death of four of his children caused by scarlet fever. In 1873 Dec. 8, *ED's diary* "Went to Nashes". Had dinner at Down House 1874 Dec. 19.

1885 Board of Regents Oregon Agricultural College, forerunner of Oregon State University. Recollections of CD in *A lawyer's life on two continents*, 1919, pp. 130-8, transcribed in *Darwin Online*.

Natural History Collections. 1858 *Public natural history collections. Copy of a memorial addressed to the Right Honourable the Chancellor of the Exchequer*, signed by CD and eight others (F371). Reprinted in *Gardeners' Chron.*, no. 48: p. 861 (F372, *Shorter publications*, F1702). See Memorials.

Natural History Review. 1860 Quarterly, founded in 1854 by Edward Perceval Wright; it was no success. In 1860 it was taken over by Huxley and others. First number appeared in 1861. By 1862 taken over by Hooker. The last number appeared in 1865. CD published one article in *Nat. Hist. Rev.*: "On the so-called 'auditory-sac' of cirripedes", vol. 3, pp. 115-16 (*Shorter publications*, F1722). 1860 CD to Huxley, warning him not to waste his energies editing a review, but to get on with original work; a warning which Huxley did not heed. CCD8.

Natural selection, (book). When Wallace's 'Ternate essay' reached CD on 18 Jun. 1858, he was well over half finished with his "big book" on the species theory having composed a quarter of a million words across ten and a half chapters. After the Linnean Society reading, CD was urged to publish an overview of his theory for an eager readership rather than make them wait for a longer period before he completed the "big book". He spent thirteen months condensing his longer work into the single-volume *Origin*. Even after the appearance of *Origin* he still intended to complete the larger work. The first two chapters were expanded to become *Variation*. The remainder was never completed as he wrote about other topics in *Orchids, Descent, Expression* and so forth. The remaining eight and half chapters were carefully transcribed, edited and published in 1975 by R.C. Stauffer as *Charles Darwin's natural selection; being the second part of his big species book written from 1856 to 1858*. Cambridge: Cambridge University Press. It is transcribed in *Darwin Online*.

Natural Selection, (phrase) originally used in CD's MS notes and was the intended title of the "big book". 1858 Phrase first published in the CD portion of the Darwin-Wallace article. 1859 Used in title of *Origin*. Ch. 3 "I have called this principle, by which each slight variation, if useful, is preserved, by the term Natural Selection". 1860 "Natural selection", *Gardeners' Chron.*, no. 16: pp. 362-3 (*Shorter publications*, F1705). In this paper CD recognizes Patrick Matthew's claim to priority in the idea, but not the expression. 1860 CD to Lyell, "I doubt whether I use the term Natural Selection more as a Person, than writers use Attraction of Gravity as governing the movements of Planets &c but I suppose I could have avoided the ambiguity". CCD8. 1873 "Natural selection", *Spectator*, 46: p. 76 (*Shorter publications*, F1758). 1880 "Sir Wyville Thomson and natural selection", *Nature*, 23: p. 32 (*Shorter publications*, F1789). 1975 R.C. Stauffer ed. *Charles Darwin's Natural Selection*. The MS is in DAR8-15.1, notes for chapter on 'Variation under Nature' DAR45, notes for chapter on

'Struggle for Existence' DAR46.1, Notes for 'Laws of Variation' in DAR47, Scraps & notes for "Transitions of Organs", i.e. chapter 8 of *Natural selection* in DAR48.

Natural Selection, speed of, "Major Leonard Darwin in conversation with E.B. Ford noted that his father once said that by choosing the right material it might be possible to detect evolutionary changes in natural populations within a period of fifty years." T.H. Hamilton, The Darwin-Wallace concept of evolution by natural selection. In: *Process and pattern in evolution*, 1967, p. 4, transcribed in *Darwin Online*.

Naudin, Charles Victor, 1815-99. French botanist. 1846 Prof. Zoology, Chaptal College, Paris, France. 1854 Muséum d'Histoire Naturelle Paris. 1861 N is referred to in Historical sketch in *Origin*. CD to Lyell that he is unable to follow his arguments in *Rév. Horticole*, 1852. 1859 CD to Armand de Quatrefages that he does not consider N to anticipate CD. CCD7:456. 1861 CD to Gray, N writes to say that he is going to publish on peloric flowers in *Pelargonium*. CCD9. 1862 *Mémoir sur les hybrides du règne végétal* advances his theory of transmutation by hybridism. 1864 CD to N, about N's work on Cucurbitaceae (cucurbits and the gourd family). CCD12. 1868 *Variation* refers to N's work. 1880 CD to Romanes, "Naudin, who is often quoted, I have much less confidence in", about plant hybrids. Romanes, *Life and letters of George John Romanes*, 1896, p. 102, in *Darwin Online*. 1863 letter to in F2291.

Neale, Edward Vansittart, 1810-92. Barrister, co-operative reformer. Paper in *Proc. of Zoological Soc. London*, for 1861: pp. 1-11. 1861 CD to Hooker, "a Mr. Neale has read a paper before the Zoological Society on 'Typical Selection'; what it means I know not". CCD9:9. ED recorded Mrs Neale visit in 1867 Feb. 12.

Nectar-secreting organs of plants. 1855 Nectar-secreting organs of plants, *Gardeners' Chron.*, no. 29: p. 487 (*Shorter publications*, F1684). On this letter and CD's notes and observations on the habits of bees see CCD5:383.

Nelson, Richard John, 1803-77. Geologist. Major-General Royal Engineers. 1854 CD to Owen, CD had corresponded with on coral formations in Bermuda. CCD5.

Netherlands, The. 1877 [Letter of thanks to A.A. van Bemmelen] in P. Harting, "Testimonial to Mr Darwin. Evolution in the Netherlands", *Nature*, 15: pp. 410-12 (*Shorter publications*, F1776, CCD25). CD received an album of 217 portrait photographs for his 68th birthday. 1874 Jun. 18-Jul. 11, George Darwin went to Holland.

Netley Abbey, Hampshire. 1846 Sept. 14 CD visited on a day trip from BAAS meeting at Southampton. Summer of 1865, William Darwin took the boys to Netley and Winchester and then to Isle of Wight. ED2:207.

Neumayr, Melchior, 1845-90. German Palaeontologist. An enthusiastic Darwinian. 1873 Prof. Palaeontology, Vienna, a position created for him. 1877 Mar. 9 CD to N, on inheritance of acquired characters and on his work with Carl Maria Paul, "Die Congerien- und Paludinenschichten Slavoniens und deren Faunen; ein Beitrag zur Descendenz-Theorie", *Abhandl. K.-K. Geol. Reichs-Anstalt*, 7, p. 3, 1875. 1878 Jun. 27 CD to Judd, praising N's work and with brief biography. CCD26.

Nevill, Reginald Henry, 1807-78. Of Dangstein, Rogate, Hampshire. Justice of the Peace for Sussex. 1848 Married Lady Dorothy Frances Walpole.

New Forest, Hampshire. 1848 Jul. CD and family visited on return from holiday at Swanage. See CCD4.

Newington, Samuel, 1815-83. Physician and botanist, of Hawkhurst, Sussex. N was joint proprietor of Ticehurst Private Asylum for Insane and Nervous Patients. Ticehurst House Hospital was founded by his grandfather, also named Samuel Newington (1739-1811). Corresponded with CD in 1875. CCD23.

Newnham College, Cambridge, Ellen Wordsworth Crofts, wife of Francis Darwin, was a lecturer on History there.

Newnham Courtney, Oxfordshire. 1847 Jun. CD visited on day trip from BAAS meeting at Oxford.

Newnham Grange, Cambridge. Home of Sir George Howard Darwin and his wife Maud du Puy. He bought it in Mar. 1885. The house had no name. The district from the Silver Street Bridge to Barton Road was also known as Newnham, so George Darwin named the house Newnham Grange. Raverat, *Period piece*, 1952, p. 32. All his five children born there. Now part of Darwin College. 2003 a blue plaque was unveiled to commemorate Gwen Raverat, George's eldest daughter.

Newport, George, 1803-54. Surgeon and insect anatomist. Francis Darwin writes of CD watching this brilliant anatomist dissect a humble bee "getting out the nervous system with a few cuts of a fine pair of scissors". LL1:110. 1843-44 President Entomological Society. 1846 FRS.

Newton, Alfred, 1829-1907. Ornithologist. Biography: A.F.R. Wollaston, *Life of Alfred Newton*, 1921. 1858 the retrospective account by N that he was converted by the Darwin-Wallace paper is contradicted by contemporary letters. 1860 H.B. Tristram to N, "The infallibility of the God Darwin and his prophet Huxley". 1865 Oct. 29 CD to N declining to write a testimonial for the Cambridge Chair on the grounds that N knew only about birds. CCD13. 1866-1907 First Prof. Zoology and Comparative Anatomy Cambridge. 1870 FRS. 1870 Feb. 9 CD to N; N, Swinhoe, Hooker and Günther spent Sunday 23 Jan. at Down House. ED recorded in 1870 Jan. 22 "Frank Mr Newton Gunther Swinhoe". 1870 May 23 CD visited N at Cambridge Museum. 1881 CD and ED took tea with N at Cambridge. 1882 N was on "Personal Friends invited" list for CD's funeral. The next record in *ED's diary* was in 1883 Dec. 1 where N, Huxleys, M. Forsters and A. Sedgwick dined. From 1884-96, he would have further dinners with ED with the last visit on 1896 May 25.

New York Entomological Club. 1881 [Letter from CD to Club] in H. Edwards, Obituary. *Papilio*. 2 (5) (May): 81. Transcribed in *Darwin Online*. (F2009).

New Zealand. 1835 Dec. 21-30 *Beagle* at Bay of Islands, North Island. CD landed and was entertained especially by missionaries. 1835 Dec. 30 *Beagle* left for Sydney, "I believe we were all glad to leave New Zealand. It is not a pleasant place". *Journal*, 2d edn, p. 430. 1836 FitzRoy, "An Englishman may now walk alone...where, ten years ago, such an attempt would have been a rash braving of the club and the oven". *Jrnl. Roy. Geogr. Soc.*, 6. 1836 "A letter, containing remarks on the moral state of Tahiti, New Zealand &c", *S. Afr. Christian Recorder, (Shorter publications*, F1640), by FitzRoy and CD. *Geological diary*: DAR37.802-811, transcribed in *Darwin Online*. 1843-45 FitzRoy was Governor-General. 1874 Sept. 27 Leonard Darwin arrived as photographer on an expedition to observe the transit of Venus. See Tee, Darwin's contacts with New Zealand, *N.Z. Genetical Soc. Newsletter*, no. 4, 1978, pp. 45-52.

Nichols, 1764-? Retired postman at Down, aged 87 in 1851. Atkins 1976, p. 103.

Nicols, Robert Arthur, 1840-91. Writer and traveller. Corresponded with CD 1871-80. A 20 Mar. 1873 letter from CD discussed the purported ability of cats to return home from long distances by smell, which N doubted. Nicols, *Natural history sketches*, 1885, p. 52 (F2281). See CCD21:132 and Perception below.

"Nigger". ED's nickname for CD. Keith, *Darwin revalued*, p. 275. Henrietta in ED2:96 explains; "'My dearest N,' means 'My dearest Nigger.' He called himself her 'nigger' meaning her slave, and the expression 'You nigger,' as a term of endearment, is familiar to our ears from her lips."

Nihill, Mr, One of the dinner guests at Clives along with the Darwins, Mr Russell and Professor Henslow. *ED's diary* 1839 Mar. 2.

Nilsson, Sven, 1787-1883. Swedish naturalist and anthropologist. 1832-56 Prof. Natural History Lund University. 1868 N provided CD with information about growth of reindeer antlers. *Descent*, 1:288. S. Lindroth, *Lychnos*, 1948, pp. 144-58.

Nitre, as a medication it was sometimes used as an antispasmodic. ED recorded in her diary on 11 Mar. 1864 "began nitre".

Nitroglycerin. CD took as medication. Often prescribed for angina. When cough and cold began in 1864 Mar. 11, he took and left off a week later. The next day he was well enough to go to the hot house but got sick again on 20th and Dr Jenner came to see him. *ED's diary*.

Nixon, Mr. 1834 Sept. 13 CD stayed four days at Yaquil near Rancagua at a gold mine owned by N, an American. *Beagle diary*, pp. 296-8; Keynes, *Beagle record*, 1979, p. 236; Chancellor & van Wyhe, *Beagle notebooks*, 2009.

Noel, Edward, 1825-99. Magistrate and Deputy Lieutenant of Derbyshire. 1849 Married Sarah Gay Forbes Darwin, of Elston Hall. Had 8 children. Visited CD in 1865 Sept. 24, 1871 Jun. 16 and 1890 Dec. 22. *ED's diary*.

Nomenclature of Zoology. CD one of 8 people appointed to BAAS committee on Feb. 11 1842; 5 more names were added on Jun. 29th. H.E. Strickland was appointed to act as reporter. This is the first appearance of the Strickland, or Stricklandian, Code or Rules. The whole Committee, except William Ogilby, signed the Report on 27 Jun. and the whole document is dated 29 Jun. *Report of a Committee appointed "to consider the rules by which the nomenclature of Zoology may be established on a uniform and permanent basis."* F1661a & F1661b. In *Darwin Online*.

Norgate, Francis, 1844?-1919. Naturalist, ornithologist. Of Sparham, Norfolk. His diary, containing field notes, is kept at the Norfolk Record Office. 1881 Mar. 8 N to CD, about dispersal of fresh-water bivalve molluscs by water beetles, *Nature*, 25: pp. 529-30, 1882. (*Shorter publications*, F1802.)

Norman, Ebenezer, 1835/6-1923. 1854- Schoolmaster at Down and from 1856 and many years thereafter copyist for CD. 1856 Aug. 17 First payment for copying in CD's Account book (Down House MS). Many thereafter. CCD6:444. 1857 CD to Hooker, "I am employing a laboriously careful Schoolmaster". CCD6:443. 1858 CD to Hooker, "I can get the Down schoolmaster to do it [i.e. transcribe] on my return". CCD7:130. 1871 Banker's clerk in Deptford.

Norman, George Warde, 1793-1882. Writer on finance. Resident at The Rookery, Bromley Common near Down. Biography by D.P. O'Brien & J. Creedy, *Darwin's clever neighbour*, 2010. 1821-72 Director at Bank of England. From 1850's Dec. 30 entry in *ED's diary* "Normans & Lubbocks came to tea", also dances, lunches and dinners until 1895 Jul. 10. His children were also in frequent contact with the Darwins. 1860 CD to Hooker, "My clever neighbour, Mr. Norman, says the article [*Edinburgh Rev.* on *Origin*] is so badly written, with no definite object, that no one will read it". CCD8:162. 1874 CD on increase of numbers of starlings "Mr. Norman a well-known man in Kent". CD to Alfred Newton 12 Mar. [1874], CCD22. 1876 CD to N, thanking for condolences on death of Amy Richenda Darwin. CCD24. 1881 Romanes to his sister, recounts an episode about CD and N's liking for snuff. Romanes 1896, p. 129. 1882 N was on "Personal Friends invited" list for CD's funeral.

North, Marianne, 1830-90. Naturalist, botanical artist and traveller. 1874 May 5 and 1881 Jul. 16. visited CD. Both times the Lushingtons were there too. *ED's diary*. 1880 On suggestion of CD travelled to Australia and New Zealand. 1882 N was on "Personal Friends invited" list for CD's funeral. [1880-82] recollections of CD in Symonds ed. *Recollections of a happy life being the autobiography of Marianne North*, 1894, vol. 2, pp. 87, 214-6 (F2002), transcribed in *Darwin Online*.

Norton, Andrews, 1786-1853. Influential American Unitarian theologian. Father of Charles Eliot N. 1813-21 Harvard College Librarian. 1819-30 Dexter Prof. Sacred Literature Harvard. 1848 CD and ED read *The evidences of the genuineness of the gospels* by N. See Randal Keynes, *Annie's box*, 2002.

Norton, Charles Eliot, 1827-1908. Son of Andrews N. Married Susan Ridley Sedgwick. 1868 Summer, N spent four months staying at Keston Rectory near Down. 1875-98 Prof. History of Art Harvard. 1876 CD to Gray, two detachments of Nortons had visited Down House, "I then verified a grand generalisation, which I once propounded to you, that all persons from the U States are perfectly charming". CCD24. Recollections of CD (1868) in *Letters of John Ruskin to Charles Eliot Norton*, 1905, vol. 1, pp. 194-5 and (1873) in *Letters of Charles Eliot Norton*, 1913, vol. 2: 476-7, transcribed in *Darwin Online*.

Norwegian. First editions in 1889: *Life and letters* (F1528); *Origin* (F2164).

Notebook of observations on the Darwin children. 1839-56, DAR210.11.37. Includes measurements, habits and charming anecdotes of the children. Some of CD's observations in this notebook were later used in *Expression* (1872) and his article: A biographical sketch of an infant (F1305, 1877). He began recording observations of his first-born child, William Erasmus, from his birth on 27 Dec. 1839 until Sept. 1844. ED added entries from Jan. 1852 recording amusing statements by the children and their activities. CD continued adding observations from 20 May 1854 until Jul. 1856. The latter entries record the children's logic and evidence of self-perception. Many of the most amusing anecdotes about the Darwin children come from this notebook. On William "When little under five weeks old, smiled, but certainly not from pleasure, but merely a chance movement of muscles, without a corresponding sensation.". On Anne, "at 2 months & four days had a very broad sweet smile & a little noise of pleasure very like a laugh." On Henrietta, "just a year old has only 3 teeth & cannot walk alone though she is very near it. She does not understand very much of what is said to her though she looks very wise." By ED, George to Frank "Are you glad there is such a boy as me in the world — F. Yes I am tremendously glad. I shall never do any thing without you." On Lizzy "4 1/2 years old. She has always had the oddest pronunciation possible." On Leonard, "Lenny 2 years old speaks perfectly correct except that he lisps. He says Thiff for fish." On Horace, "3 yrs old walking with Bessy in London saw a very little boy smoking a cigar "Look at that fellow how well he does it." The notebook was exhibited at Christ's College in 1909 and first published in CCD4, Appendix III along with an important introduction. It is also transcribed in *Darwin Online*.

Notes/note taking see: Abstracts, *Marginalia*, Notebooks on geology, transmutation and metaphysical enquiries and Portfolios.

Notebooks on geology, transmutation and metaphysical enquiries. These were published first in the 1960s by Gavin de Beer and team. However, the many pages CD had excised from the notebooks in later years after he changed from a notebook to a portfolio or file system for organizing his notes were unknown and not located at that time. After their identification within the Darwin Archive and a great deal of additional research, the definitive edition appeared in 1987 transcribed and edited by Paul H. Barrett, Peter J. Gautrey, Sandra Herbert, David Kohn and Sydney Smith. That volume contains eleven notebooks and four related manuscripts covering 1836-44. These are the *Red Notebook* [1836-37] the geological *Notebook A* [1837-39] and *Glen Roy Notebook* [1838] the *Edinburgh Notebook* [1837-39] and the best-known, those on the transmutation of species: *Notebook B* [1837-1838], *Notebook C* [1838], *Notebook D* [1838], *Notebook E* [1838-39], the so-called *Torn Apart Notebook* [1839-41]. Other notes included are known as 'Summer 1842', Zoology Notes and Questions & Experiments [1839-44]. Those categorized by the editors as mostly about metaphysical enquiries are *Notebook M* [1838], *Notebook N* [1838-1839], Old & Useless Notes [1838-1840], and the Abstract of Macculloch [1838]. The seven notebooks labelled with letters are of most general interest as they reveal the gradual and complex development of CD's evolutionary theorizing and research and other broad research directions that CD's lifelong work would develop. Some were exhibited at Christ's College in 1909 and at several subsequent exhibitions.

Nott, Henry, 1774-1844, of the London Missionary Society, arrived in Tahiti in 1797. CD met in 1836. Mentioned in *Shorter publications*, F1640.

Oakley, village in Aylesbury district, Buckinghamshire. 1868 Apr. 16 ED wrote "Called at Oakley & Holwood". 1870 Nov. 7, "Girls to Oakley".

Oakley, Mr, "A joiner with red hair". The agent of Sir Woodbine Parish. 1833 CD met at Montevideo. O provided at least one fossil bone. *Buenos Aires and the provinces of Rio de la Plata*, pp. 175-7, 1839. In the *Buenos Ayres notebook*, Chancellor & van Wyhe, *Beagle notebooks*, 2009.

Obituaries of CD. after CD's death on 19 Apr. 1882 an unusually large number of obituaries appeared around the world and in many languages. No complete list has ever been attempted. By far the largest collection is in the Darwin Archive CUL (DAR133, DAR138, DAR140.4-5, DAR215, DAR216). *Darwin Online* includes all of these and also many transcribed and scanned obituaries from elsewhere (http://darwin-online.org.uk/recollections.html).

O'Connell, Daniel, 1775-1847. Irish statesman. In the *Sydney notebook*, Chancellor & van Wyhe, *Beagle notebooks*, 2009.

Ogilby, William, 1808-73. Irish barrister who studied Stonesfield slate. 1839-46 Honorary Secretary Zoological Society London. Often cited in *Darwin's notebooks*.

Ogle, William, 1827-1912. Physician and naturalist. Superintendent of Statistics to the Registrar General. 1868 Mar. 6 O to CD on Hippocrates's views on pangenesis. [1874-78?] CD to O, CD had called on him in London, invites him to lunch. 1878 O translated A. Kerner *Flowers and their unbidden guests*. It contains a prefatory letter by CD (*Shorter publications*, F1318). 1879-95 O visited Down House. *Nineteenth Century*, 106, 1929, pp. 118-23. 1879 Nov. 22 ED recorded in her diary "Ogles came". 1881 Aug. 15 they left. 1882 Feb. 22 O sent CD his translation of Aristotle on the parts of animals. 1882 O was on "Personal Friends invited" list for CD's funeral. 1883 Aug. 18-20 they stayed with ED. 1884 onwards O was addressed as Dr Ogles. They visited ED every two years or so, with the last record in 1895 Jul. 13-15.

Oldfield, Henry Ambrose, 1822-71. Surgeon and Painter. 1845 MD St. Andrews. 1846 Assistant surgeon Bengal Medical Service. 1850-63 Doctor British Residency Kathmandu. 1856 May 10 CD to O on breeds of dogs. CCD6:107. 1880 Author of posthumously published *Sketches of Nepal*.

Oldham, Rev. Joseph, 1821?-96. 1845 St. John's College, Cambridge (B.A.). 1848-51 Curate of Down, Kent, who lived with his wife and mother. 1851 census.

Oliphant, Margaret, 1828-97. Scottish novelist born in Wallyford near Edinburgh. CD enjoyed her stories and purportedly remarked "people do not quite appreciate Mrs Oliphant yet." *Shields Daily Gazette* (16 Jan. 1883) F1886. CD cited her in *Expression*. See his 'Books Read' DAR128. ED made two records of her in 1860 and 1887.

Oliver, Daniel, 1830-1916. Botanist at Kew. O provided material for CD's botanical work and was a long-standing and important correspondent. 1860 Dec. 17 CD to Hooker, "Remember me kindly to Oliver. He must be astonished at not getting a string of questions". CCD8:532. 1860-90 Librarian Herbarium Kew. 1861-88 Prof. Botany University College London. 1861 Jan. 15 CD to Hooker, "How capitally Oliver has done the résumé of botanical books. Good heavens how he must have read"; in reaction to the Bibliography of botanical literature published in the *Nat. Hist. Rev.*, vol. I, pp. 85-120. CCD9:8. 1862 Sept. 27 CD to Hooker, "the all-knowing Oliver". CCD10:486. 1862 CD to Hooker, "Oliver the omniscient". CCD10:246. 1863 FRS.

Omori shell mounds. Omori is in Japan, near Tokyo. 1880 [CD letter] The Omori shell mounds, *Nature*, 21: p. 561, introducing a letter from Edward S. Morse, *ibid.*, pp. 561-2 (*Shorter publications*, F1788).

Orange Court, Down, a house owned by gentleman farmer Mr Harris. Atkins 1976, p. 104.

Orchard, The. Built by Horace Darwin on spare land close to ED at The Grove. Mason, *Cambridge minds*, 1994, p. 124. From 1884-96 ED wrote of many happy times there. 1962 given by Darwin family to the now Murray Edwards College.

Orchids, *Angraecum sesquipedale*, "Comet" or "Star of Bethlehem", is the orchid for which CD predicted an insect with a 30cm proboscis. *Xanthopan morgani praedicta* is the Madagascan race of an African sphingid which pollinates the white flowers at night. It was discovered in 1904 and its vernacular name is Morgan's sphinx moth. *Angraecum sesquipedale* is now frequently referred to as Darwin's orchid. 1860 CD to Lyell, "I showed the case [of Orchids] to Elizabeth Wedgwood, and her remark was "now you have upset your own book, for you won't persuade me that this could be effected by nat. selection." CCD8:263. CD to Gray, "It really seems to me incredibly monstrous to look at an orchid as created as we now see it. Every part reveals modification on modification". CCD9:302.

Orchids, Fertilisation of. (book) 1861 Sept. 24 CD to John Murray, "I think this little volume will do good to the '*Origin*' as it will show that I have worked hard at details". CCD9:279. 1862 *On the various contrivances by which British and foreign orchids are fertilised by insects, and on the good effects of intercrossing*, London (F800). 1862 Asa Gray suggested that if it had appeared before *Origin*, the author would have been canonized rather than anathematized by the natural theologians. A reviewer in *Literary Churchman*, Oct. found only one fault, that Mr Darwin's expression of admiration at the contrivances of orchids is too indirect a way of saying "O! Lord, how manifold are thy works". Review by Duke of Argyll, *Edinburgh Review*, Oct., is in much the same vein. 1877 2^d edn, *The various contrivances by which orchids are fertilised by insects* (F801). First foreign editions: 1965 Chinese (817c). 1862 German (F820). 1870 French (F818) (see papers below 1869). 1877 USA (F802). 1883 Italian (F823). 1939 Japanese (F823a). 1900 Russian (F825). 1964 Romanian (F824). List of presentation copies of *Orchids* is in CCD10 Appendix IV, *Orchids* 2ᵈ edn is in CCD25 Appendix IV. Materials for the 2ᵈ edn are in DAR70 and DAR196. Corrected proofs of the 2ᵈ edn are in DAR140.6.

Orchids, Fertilisation of, (papers) See also *Catasetum* and *Cypripedium*. 1860 Fertilisation of British orchids by insect agency" *Gardeners' Chron.*, no. 23: p. 528 (*Shorter publications*, F1706). 1860 Fertilisation of British orchids by insect agency, *Ent. Wkly Intelligencer*, 8: pp. 93-4, 102-3 (*Shorter publications*, F1707). 1861 Fertilisation of British orchids by insect agency, *Gardeners' Chron.*, no. 6: p. 122 (*Shorter publications*, F1710). 1861 Fertilisation of orchids, *Gardeners' Chron.*, no. 37: p. 831 (*Shorter publications*, F1712). 1869 Notes on the fertilisation of orchids, *Annals and Mag. of Nat. Hist.*, 4: pp. 141-59 (*Shorter publications*, F1748). 1870, 1877 The last two inserted in the French translation, 1870, and occur in the 2ᵈ English edn, 1877 (F801).

Dissected Early-purple orchid from *Orchids*.

Orchis Bank, Downe Bank, now a nature reserve. CD called it orchis bank because many wild orchids grew there. Francis Darwin recalled: "Another favourite place was 'Orchis Bank,' above the quiet Cudham valley, where fly- and musk-orchis grew among the junipers, and Cephalanthera and Neottia under the beech boughs". LL1:116. Henrietta recalled "Just on the other side of the narrow steep little lane leading to the village of Cudham, perched high above the valley, was 'Orchis bank,' where bee, fly, musk, and butterfly orchises grew. This was a grassy terrace under one of the shaws of old beeches, and with a quiet view across the valley, the shingled spire of Cudham church shewing above its old yews." ED2:376. Many believe, presumably because of the shared use of the word 'bank' that it inspired the famous conclusion of *Origin*: "It is interesting to contemplate an entangled bank..." See Randal Keynes, *Annie's box*, 2001 and Kathryn Tabb, Darwin at Orchis Bank: Selection after the *Origin*. *Studies in Hist. and Philosophy of Sci. Part C: Studies in Hist. and Philosophy of Biol. and Biomedical Sciences.* Feb. 2016, pp. 11-20.s

Origin, (book). The text of each of the first six editions is much altered. The changes are given in detail in the variorum edition, 1959, listed below. A full variorum by Barbara Bordalejo is in *Darwin Online*. The whole of LL2 is devoted to the preparation, publishing and reception of *Origin*. Janet Browne, *Darwin's Origin of species. A biography*, 2006, provides information on the genesis and impact of *Origin*. See Shillingsburg, The first five English editions of Charles Darwin's On the Origin of Species. *Variant*, 2006. A. Ellegård, *Gothenburg Studies in English*, 8: pp. 1-394, Ellegård 1958, 1990, covers reviews in popular journals in detail. The appendices of the CCD often give lists of reviews. Hundreds of reviews, in multiple languages, are reproduced in *Darwin Online*.

1859 Nov. 24 *Origin*, London, John Murray, 1,250 copies (F373). This is the first of CD's books for which details of author's presentation copies are available. CD's list is in DAR201.18 and reveals that ninety copies of the first edition were sent out. See CCD7 Appendix VIII and CCD8 Appendix III. No copy inscribed by CD himself is known. The whereabouts of the following copies are known: Agassiz (Harvard); Buist (sold in auction 2019), Butler (St. John's College, Cambridge); Robert Caspary (sold at Bonham's in 2019); Dana (Yale); Harcourt (Sotheby's 2014); Horner (BMNH); Herschel (Texas); Innes (University of London); Jenyns (in the family); Lyell (Down House); Owen (Shrewsbury School); Prestwich (University Library, Cambridge); Sedgwick (Trinity College, Cambridge); Tegetmeier (BMNH); Wallace (Keynes Room, CUL), Linnean Society of London, Royal Society. Another presentation copy is in a private collection in Virginia, USA. This copy belonged to Francis Darwin. As he was 11 years old when the *Origin* was published it seems unlikely that he received his copy at the time. The book bears a handwritten '92' on its title page which corresponds to the number given in the 'Catalogue of the Library of Charles Darwin' in DAR240, which demonstrates that this copy originally belonged to CD.

ON

THE ORIGIN OF SPECIES

BY MEANS OF NATURAL SELECTION,

OR THE

PRESERVATION OF FAVOURED RACES IN THE STRUGGLE FOR LIFE.

By CHARLES DARWIN, M.A.,
FELLOW OF THE ROYAL, GEOLOGICAL, LINNÆAN, ETC., SOCIETIES;
AUTHOR OF 'JOURNAL OF RESEARCHES DURING H. M. S. BEAGLE'S VOYAGE ROUND THE WORLD.'

LONDON:
JOHN MURRAY, ALBEMARLE STREET.
1859.

The right of Translation is reserved.

Title page and spine of the first edition of *Origin*.

Copies were sent to the following, but their present whereabouts are unknown: Bronn, Bunbury, de Candolle, Cresy, Erasmus, Falconer, FitzRoy, Fox, Asa Gray, Henslow, Hooker, Huxley, Jardine, Kingsley, Lubbock, Milne-Edwards, John Phillips, Pictet de la Rive, Ramsay, Herbert Spencer and Caroline Susan Wedgwood.

Further presentation copies may have been sent to: Nils Johan Andersson, Joachim Barrande, William Benjamin Carpenter, Joseph Decaisne, Thomas Campbell Eyton, Richard Hill, Henry Holland, Alexandr Andreevich Keyserling, Armand de Quatrefages de Bréau, Karl Theodor von Siebold, Hewett Cottrell Watson, Edward A. Williams and Hensleigh Wedgwood. Galton's copy, at University College London, is said to be an author's presentation copy, but is not inscribed. CD's own copy is in CUL. The list of presentation copies of *Origin* is in CCD7 Appendix VIII, CCD8 Appendix III, CCD9 Appendix VII (3^d edn), CCD14 Appendix IV, and CCD20 Appendix IV (6th edn). CD's notes for revisions and translations are in DAR69. Proofs for the 6th edn 33^d thousand are in DAR213. A list of 1859-60 reviews is in CCD8 Appendix VII.

The print run was 1,250 without covers; CD had twelve free copies, five were for copyright and forty-nine were sent out for review. If CD bought another thirty-five for presentation, then the number available for purchase was 1,156, or 1,144 if the uncertain presentation copies are included. 1859 CD's 'Journal' (DAR158) entry reads Nov. 24 "all copies ie 1250 sold first day." The story that the book sold out on publication day may stem from a letter from CD to Huxley on Nov. 24, "I have heard from Murray today that he sold whole edition of my book the first day and he wants another instantly". CCD:7:393. These statements have often been construed as meaning that all copies were bought by the public on the first day. What these mean is that the booksellers took up the whole printing available to them as soon as it was offered by John Murray.

1859 [2^d edn, 1st issue], 1st thousand may be present on title page, three quotations on p. [ii], (F375). 1860 Jan. 5 5th thousand [2^d edn, 2^d issue] (F376), 3000 copies. 1861 Apr. 3^d edn, 7th thousand, with historical sketch added (F381), 2000 copies. 1862 CD to John Scott, "The majority of criticisms on the 'Origin' are in my opinion not worth the paper they are printed on." CCD10:583. 1866 4th edn, 8th thousand (F385), 1500 copies. 1868 CD to W.D. Fox, "I must prepare a new edn of that everlasting *Origin*. I am sick and tired of correcting". CCD16. 1869 5th edn, 10th thousand (F387), 2000 copies. 1872 6th edn, 11th thousand (F391), title changes to *Origin of species etc.*, 3000 copies. 1876 6th edn (with additions and corrections), 18th thousand (F401). The final definitive text as CD left it. All subsequent issues printed by Murray are 6th edn. The book came out of copyright in Dec. 1901. 1880 Apr. 9 Huxley address to Royal Institution 'On the coming of age of The origin of species', *Nature*, 22: pp. 1-4; *Pop. Science Monthly*, 17: pp. 337-44. 1934 English Braille edn (F629). 1959 Variorum edn, Philadelphia, edited by Morse Peckham (F588). 1964 1st edn facsimile (F602). 1969 1st edn facsimile (F614), etc.

First foreign editions: 1918 Arabic (F1928a). 1936 Armenian (F630). 1994 Basque (F2147). 1946 Bulgarian (F632). 1903 Chinese (part) (F634). 1904 Chinese (F1848) 2000 Croatian (F2150). 1914 Czech (F641). 1872 Danish (F643). 1860 Dutch (F2056.1). 2009 Esperanto (F2160). 1913-17 Finnish (F2023). 1958 Flemish (F654); first Belgian edn; the language is Dutch. 1862 French (F655). 2003 Galician (F2152). 1860 German (F672). 1915 Greek (F698). 1960 Hebrew (F700). 1964 Hindi (F702). 1873 Hungarian (F703). 1864 Italian (F706). 1896 Japanese (F718). 2008 Kannada (F2401). 1996 Kazakh (F2153). 1957 Korean (F732). 1914-15 Latvian (F736). 1959 Lithuanian (F738). 1978 Macedonian (F2158). 2017 Malay (F2195). 1985 Malayalam (F2149). 1980 Mongolian (F2156). 1889-90 Norwegian (F2164). 19xx Persian (F2522). 2004 Punjabi (F2407). 1873 Polish (F739). 1910-12 Portuguese (F1104). 1950 Romanian (F746). 1864 Russian (F748). 1952 Sanskrit (F2490). 1878 Serbian (F766). 2006 Slovak (F2154). 1951 Slovene (F768). [1877] Spanish (F770). 1869 Swedish (F793). 2000 Tibetan, central (F2148). 1970 Turkish (F796). 1936 Ukrainian (F797). 1962 Vietnamese (F2063). 1860 USA (F377).

Origin of Species, (papers) 1863 [Letter] Origin of species, *Athenaeum*, no. 1854, p. 617 (*Shorter publications*, F1730). 1869 [Letter] Origin of species [on the reproductive potential of elephants], *Athenaeum*, no. 2174: p. 861 (*Shorter publications*, F1746). 1869 Same title, *ibid.*, no. 2177: p. 82 (*Shorter publications*, F1747).

Ornithological Notes, see Darwin, Charles Robert, Manuscripts.

Orton, James, 1830-77. American naturalist, specialized on South America and the Amazon basin. 1869 Prof. Natural History Vassar College, New York. 1870 Dedicated his *The Andes and the Amazon*, to CD "by permission".

Osborn, Henry Fairfield, 1857-1935. American palaeontologist who studied under T.H. Huxley 1879-80. A student's reminiscences of Huxley. *Biological lectures delivered at the Marine Biological Laboratory*, 1896, vol. 4, p. 32. O is the inadvertent source of the modern myth that Huxley was widely known as "Darwin's bulldog" during his lifetime. O recalled that Huxley once called himself this. See Darwin's bull-dog. Recollections of CD in O's *Impressions of great naturalists*, 1928. In *Darwin Online*.

Osborne, Christopher, ?-1860 (CD and Henrietta spelled Osborn). A resident and bricklayer at Down. 1860 CD to Innes about death of O. CCD8:540. 1885 ED helped Mrs O when she was stone deaf and being looked after by another cottager, Alice Carter, who was partially blind. In 1894 Apr. 9 ED recorded she gave Mrs O £6. ED in a letter to Henrietta in 1885 Sept. wrote "there was an unmistakeable look of pleasure at the sight of me." ED also would see that a grate be put in for Mrs O. The last entry mentioning Mrs O was 1894 Apr. 13.

Osmaston Hall, near Derby. Home of Samuel Fox and his son William Darwin Fox. 1828 Sept. CD visited. CCD1.

Otter hounds, 1866. Feet of otter hounds. *Land and Water* (6 Oct.): 244. In *Variation* 1:39-40, CD referred to a response to this query from Charles Otley Groom-Napier, *Land and Water* (13 Oct. 1866), p. 270. Regarding incipient webbing on toes.

Otway, Rear Admiral Sir Robert Waller, 1770-1846, Commander-in-chief of the South American Station. In *St. Fe notebook*, Chancellor & van Wyhe, *Beagle notebooks*.

Ouless, Walter William, 1848-1933. Painter. Visited CD in 1874 Dec. and in 1875 Jan. 23, Mar. 11-13. Painted CD in oils, the 2d portrait in oils. Original in family, now at Darwin College, Cambridge, copy (1883) in the Hall at Christ's College, Cambridge. Also painted ED and Francis in Feb. 1875. ED to Leonard who was in New Zealand "He cannot nearly fill up his time with F. [Father], so it was a convenient time for me to sit. Both portraits are unutterable as yet; but he puts in the youth and beauty at the very last." ED2:272-3. 1877 Associate RA. 1881 RA.

Overton-on-Dee, Wrexham, Wales. Home of the Parker family. 1838 Jul. CD visited for a night on return from Glen Roy; van Wyhe ed., 'Journal', (DAR158).

Owen, **Arthur Mostyn**, 1814-96. Second son of William Mostyn O. [I]. 1832-48 Indian Civil Service. 1876 High Sherriff of Shropshire.

Owen, **Charles Mostyn**, 1817-94. 4th son of William Mostyn O. [I]. Trinity College, Oxford. Army Officer. Chief Constable, Oxfordshire. 1845-47 Kaffir War.

Owen, **Frances Myddelton Mostyn**, 1807-87. Painter. Second daughter of William Mostyn O [I], sister of Sarah Harriet Mostyn O. Nicknamed "Housemaid" to CD's "Postillion". CCD1. 1830 Belle of the ball at Woodhouse. Keith, *Darwin revalued*, p. 6. 1832 CD calls her "poor dear Fanny". CCD1:220. 1832 Married Robert Myddelton Biddulph. 3 sons, 3 daughters: eldest child Frances. 1837 Jan. 14 wrote and thanked CD for sending her flowers. CCD1.

Owen, **Henry Mostyn**, 1820-43. Youngest son of William Mostyn O. [I]. Army Officer. 1834 Became a dandified young man. Brent, p. 186. 1843 Died in India.

Owen, **Sir Richard**, 1804-92. Palaeontologist and zoologist. The most distinguished vertebrate zoologist and palaeontologist of Victorian England, but, according to many, a most deceitful and odious man. "The English Cuvier." Prolific writer on anatomical and palaeontographical subjects. *Nautilus*. 1834 FRS. 1835 Married Caroline Amelia, daughter of William Clift, Conservator of the Hunterian collection at the Royal College of Surgeons. 1836-56 Conservator and Hunterian Prof. Royal College. CD discussed evolution with before *Origin*. Biographies: Rev. R. Owen (grandson), 2 vols., 1894. N.A. Rupke, *Richard Owen. Victorian naturalist*, 1994 (revised edn 2009). 1832 *Memoir on the pearly nautilus*. 1836 Oct. 29 CD and O first met at Lyell's house in London. 1838-40 *Zoology of the Beagle*, Part I, *Fossil Mammalia*, 4 numbers 1838-40, by Owen (F8 and F9 in part). 1858-62 Fullerian Prof. of Physiology. c.1840 CD [Note on a *Toxodon* skull] In Owen ed. *The life of Richard Owen*, 1894, F2032. Until 1859 CD was on friendly terms with O until the publication of the *Origin*; after that, O was probably the only man that CD hated, if he could hate. 1859 Nov. 11 CD to O and O's reply on sending a presentation copy of *Origin*, both in friendly manner. 1859 Dec. 10 CD to Lyell: "repeat nothing. Under garb of great civility, he was inclined to be most bitter and sneering against me". "He was quite savage and crimson at me". "A degree of arrogance I never saw approached". "He is the most astounding creature I ever encountered". CCD7. 1859 Dec. 13 CD to O, before O had shown his hand in public, "I should be a

Richard Owen from Reeve & Walford, *Portraits of men of eminence*, 1866.

dolt not to value your scientific opinion very highly". 1860 Apr. O reviewed *Origin*, anonymously, in *Edinburgh Rev.*, pp. 487-532. 1860 Apr. 10 CD to Lyell, "It is painful to be hated in the intense degree with which Owen hates me". CCD8:154. 1860 May CD to Henslow, "Owen is indeed very spiteful". "The Londoners says he is mad with envy because my book has been talked about: what a strange man to be envious of a naturalist like myself, immeasurably his inferior!" CCD8:195. 1860 Jun. CD to Asa Gray, "No one fact tells so strongly against Owen...that he has never reared one pupil or follower". CCD8:247. 1863 The editors of ML discuss CD's relationship with O and instance his conduct in relation of Falconer's fossil elephants. ML1:226. 1863 CD to Hooker, "There is an Italian edition of the *Origin* preparing...Owen will not be right in telling Longmans that Book w^d be utterly forgotten in ten years—Hurrah!" CCD11:688. 1863 Feb. 4 CD to Lyell, "He ought to be ostracised by every naturalist in England". CCD11:114. 1866-68 *Anatomy of vertebrates*, 3 vols. 1867 CD said to Trimen, about O's review in *Edinburgh Rev.* "The internal evidence made me almost sure that only Owen could have written it: but when I taxed him with the authorship and he absolutely denied it—then I was quite certain". Trimen told this story to Poulton. *Quarterly Rev.*, 1909, pp. 4-6. 1868 Dec. 5 CD to Hooker, "Owen pitches into me & Lyell in grand style in last chapt of vol. 3 of Anat. of Vertebrates.—— He is a cool hand— he puts words from me in inverted commas & alters them." CCD16:907. 1878 Clarke Medal Royal Society. 1881 First Director of BMNH. 1884 KCB. 1887 "Mrs Carlyle said that Owen's sweetness reminded her of sugar of lead". Huxley to Tyndall, L. Huxley's, *Life and letters of Thomas Henry Huxley*, 1900, 2:167, ML1:309. 1887 When LL was published in 1887 O was alive and very little was printed on the matter. ML, 1903, contains a lot, and more recent publications have added to it. 1892 Dec. 28 Huxley to Flower, "Gladstone, Samuel [Wilberforce] of Oxford, and Owen belong to a very curious type of humanity, with many excellent and even great qualities and one fatal defect—utter untrustworthiness". *Life and letters of Thomas Henry Huxley*, 1900, 2:341.

Owen, Sarah Harriet Mostyn, 1804-86. Eldest daughter of William Mostyn O [I]. Sister of Frances Mostyn O. A romantic interest of CD's before *Beagle* voyage. 1831 Pinned a lock of her hair to a letter to CD before *Beagle* voyage. CCD1 1831 Married 1 Edward Hosier Williams. 1833 CD to Emily Catherine Darwin at Maldonado "one of the kindest (letters) I ever received. I was very sorry to hear...that she has lost so much of the Owen constitution: I am very sure that with it none of the Owen goodness has gone". CCD1:311. 1856 Married 2 Thomas Chandler Haliburton. 1872 Nov. 1 CD to O, "for old times sake", sending photograph and copy of *Expression*. 1880 Nov. 22 CD to O, "My dear Sarah, see how audaciously I begin". "I have always loved and shall ever love this name". O had reminded him of his old ambition about *Eddowes' Newspaper*. They had met at Erasmus Darwin's house in London. 1882 O was on "Personal Friends invited" list for CD's funeral.

Owen, William Mostyn [I], c.1770-1849. Squire of Woodhouse, Rednal, Shropshire, 13 miles northwest of Shrewsbury. Had been in Royal Dragoons. Married Harriet Elizabeth Gordon-Cumming. 10 children: 1 William [II], 2 Arthur, 3 Francis, 4 Charles, 5 Henry, 6 Sarah Harriet, 7 Frances, 8 Caroline, 9 Emma, 10 Sobieski. CD used to shoot on his estate in the 1820s.

Owen, William Mostyn [II], 1806-68. Eldest son of William Mostyn O. [I]. Unmarried. Major, Royal Dragoons. Of Woodhouse, Shropshire. 1820s Then Captain, shooting companion of CD who records in *Autobiography* how O helped to play a trick on him, preventing CD from knowing how many birds he had shot. 1865 and again in 1881 CD's accounts show interest on a mortgage to Major O (?the same man). Atkins 1976, p. 96.

Oxalis bowei. 1866 *Oxalis bowei, Gardeners' Chron.*, no. 32: p. 756 (*Shorter publications*, F1736). CD enquired if there were different forms of flowers in *Oxalis bowei* or Bowie's wood-sorrel. Notes on *Oxalis* are in DAR109.

Oxenden, George Chichester, 1797-1875. Poet and orchid enthusiast. Provided CD with specimens, "I am indebted to Mr. Oxenden for a few spikes of this rare species". *Orchids,* pp. 31, 43, 61, 72 and 154. Also mentioned in Notes on the fertilization of orchids, 1869, F1748. Given a copy of *Journal* (F20), which was inscribed by CD (sold at Bonhams in 2015 for £27,926). O received a presentation copy of *Orchids*. CCD20:153.

Oxford. 1847 Jun. 22-30 CD visited for BAAS meeting. 1860 meeting of the BAAS where Huxley and Wilberforce had their famous exchange. CD was not present.

Oxford University. 1870 CD declined Hon. D.C.L. on grounds of not being able to attend the ceremony due to ill-health. *Oxford University Gazette*, Jun. 17. [Letter of CD]. *The Times.* (*Shorter publications*, F1940) It was offered at the instigation of the Marquis of Salisbury on his installation as Chancellor. His list was opposed by Hebdomadal Council. 1899 On erection of statue of CD by Hope-Pinker at University Museum: 1899. [Recollections of CD by Hooker, Meldola & Tylor.] Unveiling the Darwin statue at the museum. *Jackson's Oxford Jrnl.* (17 Jun.), p. 8. 1909 Feb. 12 The University celebrated the centenary of CD's birth. William, George, Francis and Leonard Darwin were present. Main speeches were by George and Francis Darwin and Poulton. Poulton, *Darwin and the Origin*, 1909, pp. 78-83, 1909.

Packard, Alpheus Spring, 1839-1905. American entomologist, palaeontologist and founder of *American Naturalist*, 1867. 1880 [CD on purportedly carnivorous bees]. In Packard, Moths entrapped by an Asclepiad plant…killed by honey bees. *American Naturalist* 14 (1) (Jan.), pp. 48-51. (*Shorter publications*, F1953) 1881 Jul. 8 CD to Gray, saying that he had invited P to Down House, but he may not have got letter.

Paget, Sir George Edward, 1809-92. Physician. Had dinner with CD in Cambridge after CD's return from *Beagle* voyage. Recollected a story CD told about vultures' sense of smell. 1882 Recollections of Darwin, transcribed in *Darwin Online*.

Paget, Sir James, Bart, 1814-99. Surgeon and pathologist. St. Bartholomew's Hospital. 1851 FRS. 1857 Feb. 13, CD accompanied ED to see P for a busted lip. 1869 Apr. 21, P came to see CD after CD's fall from Tommy. 1871 CD to W. Turner, "he is so charming a man", and notes that he had been seriously ill of a post-mortem infection. 1871 1st Bart. 1872 P gave CD information for *Expression*. 1875 CD to Hooker; P probably agreed to Litchfield's draft sketch for a vivisection bill. 1875 CD thanked P for sending his *Clinical lectures and essays*. CCD23. 1880 CD to Hooker, on P's work on growth in plants and on galls. ML2:425. 1881 CD met P at breakfast party for International Medical Congress in London. 1882 P was on "Personal Friends invited" list for CD's funeral. 1882 Nov. 29, ED saw P again. 1883-95 Vice-Chancellor University of London. 1882 recollections of CD in DAR112.A86-A93, transcribed in *Darwin Online*.

Paget, Stephen, 1855-1926. Surgeon and author. 4th son of Sir James P. Surgeon Middlesex Hospital. 1882 P was on "Personal Friends invited" list for CD's funeral.

Paihia, Bay of Islands, New Zealand. 1835 CD spent Christmas Day there at house of William Colenso.

Paley, William, 1743-1805. Theologian. DD. 1763 Senior Wrangler, Cambridge. 1782 Archdeacon of Carlisle. 1785 *The principles of moral and political philosophy*. 1802 Author of *Natural theology*, which is largely a crib from John Ray's *Wisdom of God*, 1691. ED aged 16 in 1824 read P's sermons and *Moral philosophy*. According to recent tradition CD and P had the same set of rooms at Christ's College, Cambridge. However, there is no evidence for this. See van Wyhe, *Darwin in Cambridge*, 2014. CD: "The logic of this book [*Evidences of Christianity*] and as I may add of his *Natural Theology* gave me as much delight as did Euclid". "I did not at this time trouble myself about Paley's premises". *Autobiography*, p. 59. Former was not a required reading.

Palin, Vernon, (or Ealin?) 1868 Acting Curate at Down. CCD16:591.

Pampas woodpecker, *Colaptes campestris* (Campo flicker). 1870 Notes on the habits of the pampas woodpecker (*Colaptes campestris*), *Proc. of Zool. Soc. London*, no. 47: pp. 705-6 (*Shorter publications*, F1750). CD replied to the sceptical criticisms of ornithologist and popular writer William Henry Hudson (1841-1922) *Proc. of Zool. Soc. London* near Buenos Aires that CD's observations that the pampas woodpecker did not nest in trees but instead mud banks. The last sentence in CD's paper reads "I should be loath to think that there are many naturalists who, without any evidence, would accuse a fellow-worker of telling a deliberate falsehood to prove his theory". CD had noted in his *Banda Oriental notebook*, p. 6 on 15 Nov. 1833 "woodpecker nest in hole"; see also *Zoology notes*, p. 154. CD referred to ground nesting woodpeckers in *Origin* pp. 184, 186, 204 and 471. "on the plains of La Plata, where not a tree grows, there is a woodpecker, which in every essential part of its organisation, even in its colouring, in the harsh tone of its voice, and undulatory flight, told me plainly of its close blood-relationship to our common species; yet it is a wood-pecker which never climbs a tree!" In the 6th edn of *Origin* (1872): "in certain large districts it does not climb trees". The bird nests both in mud banks and holes in trees.

Pampean formation. 1863 On the thickness of the Pampean formation near Buenos Ayres, *Quart. Jrnl. of the Geol. Soc. (Proc.)*, 19: pp. 68-71, 2 figs. (*Shorter publications*, F1724). Also in *Geological Observations* 2d ed., pp. 363-7. CD was sent "some excellent sections of, and specimens from, two artesian wells lately made at Buenos Ayres".

Pamphlet Collection (CD's). From about 1878 CD began to arrange the articles, papers, and reprints he received into a numbered collection. He maintained this until his death, when it was taken over by Francis Darwin who continued the collection, adding new items, the catalogue numbers assigned running consecutively from those of CD. There is a catalogue of the pamphlet collection prepared by CD and Francis Darwin from 1878 (see ED to Henrietta, [Jun. 1878] DAR219.9.175) in DAR252.1-5. The catalogue is in two sections, a list of the quarto pamphlets (c.340 items) and one for the general collection (c.1550 items). Both sections are alphabetically arranged with the entries pasted on scraps of paper in a loose-leaf folder. These were transcribed and edited by Peter J. Vorzimmer in 1963 in a privately printed typescript as the "Darwin reprint collection." (Transcribed in *Darwin Online*) Vorzimmer's own copy (now in a private collection) is reproduced in *Darwin Online* and includes many corrections and additions by Vorzimmer and his wife. In the CCD this reprint collection is referred to as the "Darwin Pamphlet Collection-CUL." Nick Gill has prepared a new catalogue of the collection.

Pander, Heinz Christian (Christian Heinrich), 1794-1865. Russian, German-Baltic. Embryologist and palaeontologist. 1861 CD attributed P's ideas to Josef W. E. d'Alton in a footnote to the historical sketch in 3d edn of *Origin*, p. xviii.

Pangenesis. CD's theory of heredity, first published in 1868 as 'Provisional Hypothesis on Pangenesis' in *Variation*, Chapter XVII, vol. II. MS notes in DAR51. See also R.C. Olby, Charles Darwin's manuscript of Pangenesis, 1963, (F1578, in *Darwin Online*). 1860 Jul. Elizabeth Barrett Browning's poem first published, "What was he doing the great god Pan, Down in the reeds by the river". 1867 CD to Gray, sending clean sheets of *Variation*, "What I call Pangenesis will be called a mad dream...I think it contains a great truth". CCD15. 1868 The term was coined by CD and first appears in print in *Variation*. He thought that the idea was new although it was not. 1868 CD to Hooker, "You will think me very self-sufficient, when I declare that I feel sure if Pangenesis is now stillborn it will, thank God, at some future time reappear, begotten by some other father, and christened by some other name". CCD16:163. 1868 CD to Wallace, "It is a relief to have some feasible explanation of the various facts, which can be given up as soon as any better hypothesis is found". "I had given up the great god Pan as a stillborn deity". CCD16:195. 1868 CD to

Lyell, "An untried hypothesis is always dangerous ground". CCD16:280. 1871 "Pangenesis", *Nature*, 3: pp. 502-3, a letter criticising a paper by Francis Galton, *Proc. of Roy. Soc.*, 19: pp. 393-410 (*Shorter publications*, F1751) who had concluded pangenesis could not work based on transfusing the blood of rabbits of different colours. 1880 CD to Paget, "To anyone believing in my pangenesis (if such a man exists)". ML2:427. First foreign editions: 1869 Italian (F2071). 1898 Russian (F2511).

Papilionaceous Flowers. 1858 On the agency of bees in the fertilisation of papilionaceous flowers, and on the crossing of kidney beans, *Gardeners' Chron.*, 13 Nov. 1858, no. 46: pp. 828-9 (*Shorter publications*, F1701). (Reprinted in *Annals and Mag. of Nat. Hist.*, 2: pp. 459-65). Papilionaceous flowers (Latin *papilion*=butterfly) often have a butterfly-like flower. CD observed that their fertilisation seemed to depend on visits by bees. A reply appeared: J.B.W. 1858. Accidental fertilisation of papilionaceous plants. *Gardeners' Chron.* (20 Nov.), p. 845.

Parfitt, Edward, 1820-93. Entomologist and botanist. 1860 CD to Stainton, mentions P as a correspondent about orchids. CCD8:250. Cited in *Orchids*, p. 36.

Paris. 1827 Spring CD visited with his uncle Josiah Wedgwood [II], his only visit to continental Europe. 1879 Oct. 29 ED wrote "Sara & Wm from Paris".

Parish, Sir Woodbine, 1796-1882 Diplomat, traveller and geologist. 1824 FRS. 1825-32 Consul General Buenos Aires. 1832 Sent back to London an almost complete skeleton of *M. americanum*. Lister, *Darwin's fossils*, 2018, p. 42. CD knew him in London. *Red Notebook*, p. 106. 1837 Kt Commander of the Guelphic Order of Hanover. CD letter to P (1839?) in *A life of Sir Woodbine Parish*, 1910, p. 412 (F2294).

Park Street, London. 1845-52 no. 7 home of Erasmus Alvey Darwin.

Parker, Down village cobbler. "A short, sallow man with the bristling chin." Provided leather for making slings. F. Darwin, *Springtime*, 1920, p. 60.

Parker, Mr. 1837 P forwarded to CD a chart of Diego Garcia, Indian Ocean, which related to *Coral reefs*, see 3d edn, 1889, pp. 90-5. Barlow 1967, p. 130. CCD2:21.

Parker, The Misses, Two illegitimate daughters of Erasmus Darwin [I], Susanna and Mary Jr., by Mary Parker, a governess. CD's aunts. 1790s Erasmus Darwin set up a school for them at Ashbourne, Derbyshire, in the 1790s. His *A plan for the conduct of female education in boarding schools*, 1797, relates.

Parker, Rev. Charles, 1831-1905. 4th child of Henry Parker [I] and Marianne Darwin. Unmarried. CD's nephew. 1850 Apr. 1, ED wrote in her diary "Charles Parker came". He stayed six days. Wrote to CD in 1864 about needing funds to buy land to build a school house. 1864 Jun. 8 ED wrote him. 1884 Living in Shrewsbury.

Parker, Francis, 1829-71. 3d child of Henry Parker [I] and Marianne Darwin. CD's nephew. 1860 Married Cecile Longueville. 3 sons.

Parker, Henry [I], 1788-1856. Physician and surgeon. CD's brother-in-law. Overton-on-Dee, Flint. 1824 Married Marianne Darwin. 5 children: 1 Robert, 2 Henry [II], 3 Francis, 4 Charles, 5 Mary Susan.

Parker, Henry [II], 1827-92. 2d child of Henry Parker [I] and Marianne Darwin. CD's nephew. Classical Fellow of Oriel College, Oxford. 1862 Reviewed *Orchids* in *Saturday Rev.* 1862 Dec. 29 Visited Down House. Unmarried.

Parker, Mary Susan, 1836-93. 5th child of Henry Parker [I] and Marianne Darwin. CD's niece. 1866 Married Edward Mostyn Owen of Woodhouse. 5 children.

Parker, Marianne, see Darwin, Marianne.

Parker, Robert, 1825-? 1st child of Henry Parker [I] and Marianne Darwin. CD's nephew. Story about his idleness. *Autobiography*, p. 33.

Parkfields, Staffordshire. 1803 A cottage adjoining the Maer Hall estate which Josiah Wedgwood [II] bought for his mother and two sisters. Home of Mrs Josiah Wedgwood [I] until her death in 1815. 1823 Home of Sarah Elizabeth Wedgwood [I] and her sister Catherine. When the latter died in 1823, Sarah Elizabeth Wedgwood [I] went to Camphill. 1847 Sold with the Maer Hall estate.

Parle, North Wales. 1826 Oct. 30 CD visited on riding tour with his sister Caroline.

Parr, Thomas, "An old miserly squire" of Lythwood Hall, 2 miles from Shrewsbury. ED2:54.

Parry. A leading merchant at Montevideo. 1833 *Beagle diary*, pp. 82-3, 119. FitzRoy to CD. P's wife dies. Son, Robert, sent to England to school, daughters sent to Buenos Aires. Keynes, *Beagle record*, 1979, p. 72.

Parslow, Anne, Under nurse at Down House in 1861 census.

Parslow, Arthur, 1846?-95. Son of Joseph and Eliza P. Carpenter and served as CD's butler. Married Mary Anne Westwood, also employed in the Darwin household. She was Bernard Darwin's wet nurse. ED in 1895 Jan. 19 wrote to Joseph P about Arthur. On Feb. 4 1895, ED wrote "wrote to Parslow on Arthur's death."

Parslow, Mrs Eliza, née Richards, 1812-81. 1845 Married to Joseph P. Mrs P was ED's personal maid before marriage. Later she ran a dressmaking school. ED gave E pounds of tea in 1840. *ED's diary*.

Parslow, Hellen, 1852-? Servant at Down House in 1871 census.

Parslow, Jane, 1865?-79. Nursery maid at Down House. Joseph Parslow's daughter.

Parslow, Joseph, 1812-98. CD's butler/manservant. Hooker described him as "an integral part of the family, and felt to be such by all visitors to the house". LL1:318. Known by the family as "the venerable P" after "the aged Parslow" in Dickens' *Great expectations* (1860-1). ED once told Henrietta they were not fortunate in their men-servants till Parslow came. c.1840 Manservant at 12 Upper Gower Street. 1841-42 Wages £25 per annum all round. 1857 he received £44 per annum. By 1871 P was living out at Home Cottage, Back Lane, Down. Until 1875 Butler at Down House. 1881 Wages £60. ED recorded payments to P in 1867 May 22 £5, 1882 Nov. 5 £10. 1845 Married Eliza Richards. Two sons, Arthur and Ernest (died in 1856 of smallpox) and a daughter Jane. 1881 Arthur "married comfortably". Stecher 1961, p. 251. 1882 P was at CD's funeral, walking in procession with William Jackson, behind the family mourners, then seated in Jerusalem Chamber. After CD's death P had a pension of £50 per annum and the rent of his house. CD's memorandum on the pension is in DAR210.10.26, transcribed in *Darwin Online*. 1885 June. 9 P went to unveiling of CD statue at BMNH at South Kensington. In 1890 Sept. 16, ED wrote "Little Parslows", presumably grandchildren visited. 1891 little Ps came again. They played games and had tea and ED wrote P annuities. Interview with P in D.S. Jordan, *The days of a man*, 1922, vol. 1, pp. 272-4, transcribed in *Darwin Online*. Gravestone Downe churchyard: "In memory of Joseph Parslow, who died 4th. October 1898 aged 86 years. The faithful servant and friend of Charles Darwin of Down House, in whose household he lived for upwards of 36 years. Also of Eliza his wife, who died 12th Jul. 1881, aged 69 years."

Parson, Arthur George, 1845?-1907. Of Haslemere, Surrey. Solicitor. 1880 Married Mabel Frances Wedgwood s.p. 1885-96 visited ED. Last record in 1896 May 11.

Partington, Eliza, 1806-?. Landlady at Montreal House in Great Malvern. See Malvern Wells. 1849 Annie Darwin wrote to her mother to say they were "lodging with Miss Partington". 1851 As landlady, she had to report Annie's death for the official record. In the 1861 census she was 55 years old. 1863 Sept. CD and ED with help from P and W.D. Fox's information, located Annie's headstone in the Priory churchyard. CCD11; Randal Keynes, *Annie's box*, 2002.

Pasteur, Louis, 1822-95. French chemist and bacteriologist. 1863 May 23 CD to Bentham, "I was struck with infinite admiration at his work". CCD11:433. 1869 Foreign Member Royal Society.

Patagonian Fossils. 1853 [CD on Patagonian fossil beds.] In Richard Owen, Description of some species of the extinct genus *Nesodon...Phil. Trans. of the Roy. Soc.*, vol. 143: p. 309 (*Shorter publications*, F1820). In his introductory comments, Owen referred to CD as "my friend".

Patten, John Wilson, Baron Winmarleigh, 1802-92. Politician. 1830-31 and 1832-74 Conservative MP. 1866 PC (Privy Council). 1874 1st Baron. 1875 P was member of Vivisection Commission to which CD gave evidence. LL3:201.

Patterson, Robert, 1802-72. Naturalist and author of zoological works in Belfast. 1847-60 letters to P in corresponding vols. of CCD. 1859 FRS.

Pattrick, Francis, 1837-96. Classical scholar. Identification uncertain. 1876-96 Magdalene College Cambridge, President. 1882 P was on "Personal Friends invited" list for CD's funeral.

Pattrick, Saint Reginald 1834-97. Clergyman. BA, Oxford. 1874-97 Vicar of Sellindge, Kent. In 1871 Jul. 18, ED wrote "Mr Pattrick." Records of his visits began again from 1881-94 with a letter written to him in 1895 Jun. 23.

Pawson, Iris Veronica, 1887-1982. Daughter of Albert Henry P. 1906 Married Ralph Lewis Wedgwood. 3 children, 1 son died in infancy.

Payne, George, 1841?-1924. Sir Thomas Farrer's gardener at Abinger Hall from April 1870 until at least 1914, trained at Kew. P helped CD on *Mimosa*. 1873 CD to Farrer, "As he is so acute a man, I should very much like to hear his opinion" on water damage to leaves. CCD21:321.

Paz, La. Schooner hired by FitzRoy during *Beagle* voyage to accelerate survey. Fitz-Roy says 15 tons burthen "ugly and ill built craft", "soaked with rancid seal oil". 1832 Sept. 11 Schooner hired from James Harris, resident at Rio Negro, Argentina, for eight lunar months by FitzRoy with schooner *La Liebre*. Commanded by Lieut. Bartholomew James Sulivan under Lieut. John Clements Wickham. Surveyed southeast coast of Argentina.

Peacock, George, 1791-1858. Anglican clergyman, mathematician and astronomer. P wrote to Henslow about post of naturalist on *Beagle*. 1818 FRS. 1837-58 Lowndean Prof. Astronomy Cambridge. 1837 P was present at interview of CD by Thomas Spring Rice [I] about £1,000 grant for publishing *Zoology of the Beagle*. CCD1-2, Barlow 1967, p. 134. 1839-58 Dean of Ely.

Pearce, Elizabeth, Of Maer, Staffordshire. 1871 servant to Erasmus Alvey Darwin.

Pearce, James, 1774?-1852. Surgeon. Father of Joseph Chaning P (1811-47). P had a large collection of fossils which he bequeathed to his family in 1847. CD wrote to P for access to the collection. CCD24:47.

Pearce, James, 1833-? Footman from Yorkshire at Down House in 1851 census. 1882 P was on "Personal Friends invited" list for CD's funeral.

Pearson, Charles, 1845?-1914. Schoolteacher, innkeeper, and organist. Schoolmaster at Down National School from 1867-c.1881. 1875 P was elected a trustee of Down Friendly Club. CCD23:185. 1881 Shopkeeper, wine and spirit merchant, and proprietor of the George & Dragon, Down.

Peas. 1862 Peas, *Gardeners' Chron.*, no. 45: p. 1052 (*Shorter publications*, F1719). "Will any one learned in Peas have the kindness to tell me whether Knight's Tall Blue and White Marrows were raised by Knight [1799] himself?" CD discussed Knight's pea varieties in *Variation* 1: 326, 329-30, vol. 2: 129. See CCD10:510-11. Many notes on P throughout the Darwin Archive. DAR46.2, 60, 69, 77, 78, 205, 206 and 209.

Penally, near Tenby, South Wales. 1846 Home of CD's aunts Frances Allen and her sister Madame [Jessie] Simonde de Sismondi.

Pengelly, William, 1812-94. Geologist. Explorer of Devon caves. 1861 Jul. CD met at Torquay. 1863 FRS. 1873 corresponded with CD.

Penguin Ducks. *Jrnl. of Horticulture*, 30 Dec. 1862, 3:797 (*Shorter publications*, F1825). CD enquired if Penguin ducks (a breed of domestic duck with a striking upright posture) could run faster than other breeds. Discussed in *Variation* 1:281-6.

Pennethorne, Deane Parker, 1835-1917. Barrister of Lincoln's Inn and School Inspector. 1868 May 22 CD to P, acknowledging letter on descent of man. CCD16.

Pepper. A dog belonging to George Darwin which bit gardeners. P was taken over by William but bit gardener again; then to Sir Leslie Stephen in London, where it bit children; finally to Archbishop A.C. Tait at Addington Palace, Surrey. Atkins 1976, p. 80. 1890 ED to Henrietta "I am vexed about Pepper". ED2(1904):402.

Pepsine (pepsin), a medication that was meant to help with digestion. CD began taking in 1862 Feb. 7 when he had "hardly an hour comf in day". He took again on

Feb. 22 and left off it Mar. 4 only to pick up again on 7th. ED recorded in her diary 1863 Dec. 30 "left off pepsine". 1864 Jan. and Apr. it had to be taken again. See CD to W.W. Baxter 18 Mar. 1882.

Perception. 1873 [Letter] Perception in the lower animals, *Nature*, 7: p. 360 (*Shorter publications*, F1759), supporting a letter from Wallace, *ibid.*, 7: p. 303, *Zoologist*, 8: pp. 3488-9. CD cited examples, some from his own experience, of animals (including his horse Tommy) that seemed to orient their movements by smell.

Period piece. 1952 *Period piece: a Cambridge childhood*, London, by Gwen Raverat (née Darwin). The most important source of information on CD's children in their adult day-to-day lives, and on ED in old age, written as through the eyes of Gwen as a child. A most interesting and amusing book. Extracts in *Darwin Online*.

Pernambuco, Brazil. 1836 Aug. 12-17 *Beagle* at. CD visited old city of Olinda and studied the sandstone bar off the harbour; now called Recife. *Beagle diary*, pp. 434 ff. 1841 "On a remarkable bar of sandstone off Pernambuco, on the coast of Brazil", *Philosophical Mag.*, 19: pp. 257-60. (*Shorter publications*, F1659).

Perristone, or **Perrystone**, near Ross, Herefordshire. Home of William Clifford, family friend of Wedgwoods. 1824-48 Several family letters are addressed from there. 1849 Mar. 29 ED wrote "I went to Perristone". Henrietta wrote in ED2:124-125, "her visit must have been a great pleasure as well as change in her home-keeping life. My father's health now quite unfitted him for visiting or for her to leave him with any comfort, so that practically she never left home."

Persian. First edition in: 1944 *Origin* (F2517).

Pertz, Miss Anna Julia, 1855-1928. First daughter of Georg Heinrich P. 1877 P drew a leaf of *Trifolium resupinatum* (Persian clover) reproduced in CD letter to Thiselton-Dyer in ML2:412, not the same as the drawing in CCD25:344. The last record of her visit to ED was in 1895 May 22 where she came with her sister Mme Pertz (below).

Pertz, Dorothea Frances Matilda (Dora), 1859-1939. Second daughter of Georg Heinrich P. British botanist, co-authored several papers with Francis Darwin. Paid several visits to elderly ED with a last call in 1896 Jun. 22.

Pertz, Georg Heinrich, 1795-1876. 1842 Royal Librarian Berlin. 1854 Married Le-onora Horner. 2 daughters, Anna and Dorothea. Records of his visits either alone or with his family to ED began in 1885. *ED's diary*.

Peters, Wilhelm Carl Hartwig, 1815-83. German zoologist, palaeontologist and traveller. 1878 P seconded CD's election as Corresponding Member of Königlich-Preussische Akademie Berlin.

Peterson, John, 1787-?. Quartermaster on 2d voyage of the *Beagle*. 35 years at sea. Shetlander. In *Buenos Ayres notebook*, Chancellor & van Wyhe, *Beagle notebooks*, 2009.

Petleys, House at Luxted Road, Down, north of Down House. The Petley family came to Down in the 13th century. 1847-56 Home of Sarah Elizabeth Wedgwood [I] until her death. Leased from Charles Lovegrove. CD wrote to his sons William and George after the funeral of Sarah Wedgwood (Nov. 12 1856) that he took a walk down to Petleys with the Wedgwoods. ED in 1861 Oct. 17 walked with Henrietta "round Green Hill & Petleys".

Philippi, **Rudolph Amandus**, 1808-1904. German, later Chilean zoologist and palaeontologist. Prof. Natural History Technical High School Kassel. 1851 P sent fossil cirripedes to CD. CD sent P *Fossil cirripedes*. *Lychno*s, 1948-49: pp. 206-10. CCD5.

Phillips, **George Lort**, 1811-66. Of Laurenny Park. 1840 Married Isabella Georgina Allen. 1861-66 MP for Pembrokeshire.

Phillips, **John**, 1800-74. Geologist. 1834 FRS. 1844 Prof. Geology Dublin University. 1845 Wollaston Medal Geological Society. 1854-70 Keeper of Ashmolean Museum Oxford. 1856 Reader Geology Oxford University. ?1856 CD to P on foliation and offers copies of three vols. of geology of *Beagle*. CCD6. 1859 P to CD, to acknowledge award of Wollaston Medal of Geological Society. 1859-60 President Geological Society London. 1859 CD sent 1st edn of *Origin*. Sollas, *The age of the earth*, pp. 251-3, 1905, J.M. Edmonds, *Proc. of Ashmolean Nat. Hist. Soc.*, for 1948-50, pp. 25-9, 1951. 1860 P gave Rede lectures at Cambridge, anti-*Origin*, but very fair. *Life on earth*, Cambridge 1860, contains substance of Rede lectures, CD wrote to Hooker that they were "unreadably dull". 1869 P sent CD his *Vesuvius*, Oxford 1869. CCD17. 1870 CD to Herschel, recommending that P be asked to revise 4th edn of *Manual of scientific enquiry*, 1871, which he did. CCD18.

Phillips, **Mary**, 1822-69. 1840 Married as first wife to Darwin Galton.

Phillips, **William Walker & Mrs Phillips**, 1860 Residents at Down Hall farm. 1868 "Old Phillips" would not sell land to Innes to build a vicarage on. ? A farmer, John Phillips of Orange Court, perhaps the son, would not either. Orange Court seems to have been owned by a Mr Harris. CCD7:28, 893. P had been a subscriber to the National School in Down and to the Down Friendly Club. CD discussed information P provided on the stripes of asses in *Variation* 1:62-4, but P is not named.

Philoperisteron Club, 1855 A pigeon-fancy club of which CD was a member and which met at the Freemasons' Tavern in London. CCD5. See also Columbarian Society. Jim Secord, Nature's fancy: Charles Darwin and the breeding of pigeons. *Isis*, vol. 72, no. 2 (Jun., 1981), pp. 162-186.

"Philos". CD's youthful nickname for his brother Erasmus. Also CD's nickname on board the *Beagle* by officers and men, short for philosopher.

Philosophical Club of Royal Society. A dining club of forty-seven members. It met on Thursdays at 6pm and chair quitted at 8.15pm for members to attend meetings of the Society. 1847 Founded. 1854 CD elected. 1855 Jan. 25 ED recorded "Phil club." in her diary. 1855 Dec. 20 CD attended. 1864 CD resigned.

Phosphorus, a medicinal supplement that was given on two occasions, taken with iron when both Horace and CD were ill. It is unclear if it was given to both or to only one of them on the first occasion but on the second it was CD. *ED's diary*.

Physiognomy. 1874 In *Notes and Queries on Anthropology, for the use of travelers and residents in uncivilized lands. (Drawn up by a Committee appointed by the British Association for the Advancement of Science.)*, pp. 12-13, no. IX. (*Shorter publications*, F1832). The table of contents, under "Part I.—Constitution of man" reads: "IX. Physiognomy. By C. Darwin, Esq., F.R.S.—Questions as to the expression of the countenance, natural gestures, blushing &c." Earlier versions of these queries about expression were printed from 1867. See Freeman and Gautrey who examined the five then known in 1972 and 1975. F1832 was then unknown and it has additions not present in earlier versions. See F1739 and F874. Answers to the pre-1872 queries were used in the writing of *Expression* where a list of 16 queries is given on pp. 15-16. See CCD15.

Physiological Society. 1876 Founded, partly as a result of the anti-vivisection movement. At the beginning "dinners with discussion" constituted the activities of the Society. 1876 Jun. 1 CD elected the first, and at that time the only, Honorary Member. William Sharpey was elected the second Honorary Member shortly after. From 1880 Scientific Meetings were held.

Piano. 1839 ED was given a piano from Broadwoods by her father shortly after her marriage. ED wrote in her diary in 1839 Jan. 31, "went to Broadwoods for P.F" (piano forte). It had belonged to the Rev. Thomas Stevens, who had married Caroline Tollet. 1858 22 Feb. A new piano selected. 31 Mar. CD recorded payment to Wornum & Sons, 16 Store Street, London for £75 5s. Account book (Down House MS) 1873 Henrietta recalled ED said how sad she felt to see Elizabeth Wedgwood doing nothing due to her failing sight. "To fill up some of her weary useless time my mother gave her our old Broadwood grand-piano and helped her to learn by heart simple airs to play without seeing the notes." ED2:262-263. Thus the piano used to test the hearing of worms (*Earthworms*, pp. 26-7) was the later instrument. 1929 The Broadwood was bought for Down House, for £20, from the Positivist Society. Atkins 1976, p. 116. It remains on display at Down House.

Pictet de la Rive, François Jules, 1809-72. Swiss zoologist and palaeontologist. 1835-59 Prof. Zoology and Comparative Anatomy Geneva in succession to Alphonse P. de Candolle. 1859 CD sent 1st edn *Origin*. 1860 P was courteously anti-*Origin*, review in *Arch. Sci. Bibliothèque Universelle*, Mar: 233-55. In *Darwin Online*.

Pigeons. The races of domestic pigeon, *Columba livia*, are extensively discussed in *Variation* and CD kept all the breeds he could acquire as well as getting material from other breeders. Much information on pigeons (and poultry) was provided by W.B. Tegetmeier. CD was a member of the pigeon fancying clubs the Columbarian and Philoperisteron Societies qq.v. In 1855 CD had a pigeon house built, it was a hexagonal structure, with a high pitched roof, approximately 4.8x3 and 2.7 meters high. 1855 Nov. 8 CD to Hooker, "I love them to that extent that I cannot bear to kill & skeletonise them." CCD5:497. 1859 Dec. 13 CD to Huxley, offering drawings of pigeons from his portfolio. J.A. Secord, Nature's fancy: Charles Darwin and the breeding of pigeons, *Isis*, 72 (1981), pp. 163-86.

Pigott, Katherine Gwendoline, c.1868-1958. Eldest daughter of Rev. Edmond Vincent P, Vicar of Trentham. 1902 Sept. 11 Married Francis Hamilton Wedgwood.

Pimienta, Sebastián, ["Sebastian Pimiento" by CD], estate owner in Maldonado district. See *Beagle diary*, pp. 158-9 and *Falkland notebook*, Chancellor & van Wyhe, *Beagle notebooks*, 2009. 2008 Stone monument erected Pan de Azúcar, Uruguay.

Pinguicula. 1874 [Irritability of Pinguicula], *Gardeners' Chron.*, 2: p. 15 (*Shorter publications*, F1767). Summary of a report by CD read to Royal Horticultural Society on 1 Jul. 1874 the leaves of the common butterwort "possess a power of digesting animal matter similar to that shown by the Sundews". See *Insectivorous plants*, chapter 16.

Pinker, Henry Richard Hope-, 1849-1927. Sculptor. c.1860 Statue in stone of CD in University Museum Oxford is by P, model for it at Down House.

Pistyll Rhaeadr, Denbigh, Wales. Waterfall. 1820 Jul. CD and Erasmus went on a riding tour from there. "Pistol Rheyadur" in van Wyhe ed., 'Journal', (DAR158).

Plas Edwards, near Towyn, Merioneth, on the coast of Wales. 1819 Jul. CD went on holiday for three weeks. See *Autobiography*, p. 45 and Smith, *Darwin's insects*, p. 5.

Plant notes, from the voyage of the *Beagle*. These were prepared for Henslow with the intention that he would identify the dried specimens, mostly vascular plants, ferns and flowering plants as well as algae and fungi. The surviving copy is transcribed and edited in Duncan M. Porter, Darwin's notes on *Beagle* plants. *Bulletin of the BMNH, Historical Series*, 14, no. 2 (1987), pp. 145-233. (Transcribed in *Darwin Online*) About 2,700 of CD's plant specimens are preserved at the Cambridge University Herbarium.

Playfair, Sir Lyon, Baron, 1818-98. Chemist, administrator and politician. 1843 Prof. Chemistry Royal Manchester Institution. 1845 Chemist to Geological Survey and Prof. School of Mines London. 1848 FRS. 1868-92 MP. 1876 P visited Down House whilst staying at High Elms in company of Huxley, Morley and Gladstone. 1883 KCB. 1892 1st Baron. 1895 GCB.

Plinian Society of Edinburgh, officially the Plinian Natural History Society. 1823-c.1848 A student society founded by Robert Jameson. 1826 Robert Edmund Grant Secretary, but gave up the post to join the Council of the Wernerian Natural History Society. 1826 Nov. CD elected. Proposed by John Coldstream. He attended eighteen out of a possible nineteen meetings up until 1827 Apr. 3. 1827 Mar. 27 CD made a communication to it, not "at beginning of the year 1826" as stated in *Autobiography*, p. 39. Title was: 1 "That the ova of *Flustra* possess organs of locomotion." 2 "That the small black globular body hitherto mistaken for the young of *Fucus loreus* is in reality the ovum of *Pontobdella muricata*." CD was wrong in both these assertions; the "ova of *Flustra*" were pilidium larvae, and the "ovum of *Pontobdella*" was an egg case full of eggs. CD was disappointed that Grant had presented the results to the Wernerian Society on 24 Mar. without acknowledging CD. The Minutes of the Plinian Society recording CD's first scientific papers are in Edinburgh University Library and transcribed in *Darwin Online*. Barrett, *Collected papers*, 1977, vol. 2, p. 285

gives a full transcript of CD's original notes, now at CUL (transcribed in *Darwin Online*). 1870 [Note on CD's papers 27 Mar. 1827] In W. Elliot, Opening address by the President. *Trans. of the Botanical Soc. of Edinburgh* 11: 1-42, (*Shorter publications*, F1749). See 1888 *Edinburgh weekly Dispatch*, May 22; J.H. Ashworth, Charles Darwin as a student in Edinburgh, 1825-27. *Proc. of Roy. Soc. Edinburgh*, 55 (1935): pp. 97-113 [In *Darwin Online*]; P.H. Charles Darwin and Dr. Grant, *Lychnos*, 1948-1949: pp. 159-167. Jenkins, Henry H. Cheek and transformism: new light on Charles Darwin's Edinburgh background. *Notes and records of the Roy. Soc.* (2015) 69, pp. 155-171.

Plumptre, Charles John, 1818-87. Barrister and lecturer on elocution at University of Oxford. CD 1881. [Letter on the expression of the eye]. In Plumptre, *King's College lectures on elocution...*pp. 290-1. (*Shorter publications*, F1994).

Podophyllin, an extract or resin from the root of *Podophyllum peltatum* or May-apple. CD was prescribed it to take when he was ill in the stomach. He was given "pod." every 2-7 days from Mar. 24 1864 until Jul. 11. None was administered in Jun, 1864. ED took in 1864 Sept. 27. Two more intakes were recorded in 1867 Feb. 7 and 8 but it is unclear who took it. *ED's diary*. See E.A. Darwin to ED 30 [Mar. 1864?] CCD12.

Polish. First editions in: 1873 *Origin* (F739). 1873 *Expression* (F1203). 1874 *Descent* (F1101). 1887 *Journal* (F223). 1888-89 *Variation* (F922). 1891 *Autobiography* (F1529). 1959 *Cross and self fertilisation* (F1270).

Pollen. 1861 Effects of different kinds of pollen, *Jrnl. of Horticulture*, (9 Jul.): pp. 280-1 (F1823). 1863 Influence of pollen on the appearance of seed, *Jrnl. of Horticulture*, (27 Jan.): p. 70 (*Shorter publications*, F1828).

Pollock, Sir Frederick, Bart, 1845-1937. Jurist. 1882 Jan. 8 P came to Down House on a "Sunday tramp". ED(1904)2:329. 1902 3ᵈ Bart. FBA.

Pollock, George Frederick, 1821-1915. Master of the King's Bench and King's Remembrancer. 1858 Read manuscript of *Origin* for John Murray and advised printing of 1,000 copies, not 500 as Murray had suggested. E.S.P. Haynes, *Cornhill Mag.* 41 Aug. 1916, p. 233. L. Huxley, *Charles Darwin*, 1921; p. 57. P said *Origin* would be widely discussed in the sense that it would be a financial success. Thought it "brilliantly written" and that CD surmounted the "formidable obstacles which he was honest enough to put in his own path." Guido Jozef Braem, *Darwin*, 2012.

Polly. Henrietta's white rough-haired female fox terrier. 1870 ED to Henrietta, description of behaviour after her litter of puppies had been removed and illustration of Haeckelian joke on phylogeny. ED2:198. Drawn from a photograph by Rejlander in *Expression*, p. 43. Caricature sketch as "Ur-hund" by Huxley in ED2:[199]. 1871 Attached herself to CD when Henrietta married. 1882 Apr. 20 P was put down at ED's request because of "bodily pain" shortly after CD's death and buried under the Kentish beauty apple tree in the orchard. 1927 A stuffed replica was placed in the restored old study by Buckston Browne, curled up in her basket. Visible in some photographs. It soon became moth eaten and was thrown out. Atkins 1976, p. 115.

Pomare IV, 1813-77. Queen of Tahiti. "Pomare" was a lineal name, real name Aimata Pōmare IV Vahine-o-Punuatera'itua, "Aimata" meaning eye-eater. She signed herself "Pomare Vahine", "vahine" meaning "woman". 1835 Nov. 25 P was entertained on board *Beagle*. "A large awkward woman without any beauty, grace or dignity". *Journal*, 2ᵈ edn, p. 416. (*Shorter publications*, F1640).

Pontobdella muricata, Marine leech. 1827 R.E. Grant, *Edinburgh Jrnl. Sci.* 7, pp. 160-2, acknowledged "my zealous young friend Mr Charles Darwin of Shrewsbury". First appearance of CD's name in print. Preceded, however, by Sir John Dalyell.

Poole, Dorset. 1847 Jul. CD visited on way home from family holiday at Swanage.

Poole, Skeffington, 1803-76. Lieutenant-colonel in the first regiment light cavalry of the Bombay Military Establishment. 1850 Retired. CD corresponded with P about horses. P's information is cited in *Origin*, 1859, p. 163.

Port, Georgina Mary Ann, 1771-1850. Married Benjamin Waddington. Mother of Frances, Baroness von Bunsen. Grand-niece of Mrs Mary Delany. Mme D'Arbley described her as "the beautiful Miss Port". P was a friend of the Allens, especially of Lancelot Baugh A. ED1:48. 1817 Mrs Josiah Wedgwood to her sister Emma Allen "the inconceivable Mrs Waddington". ED1:110.

Port Darwin, East Falkland Island. Named after CD. 1834 Mar. 17 CD crossed the isthmus near it.

Port Desire (now Deseado, Puerto, Patagonia, Argentina). 1833 Dec. 23 *Beagle* at, when it was a deserted Spanish settlement. *Zoological diary*, pp. 186-88. *Geological diary*: DAR33.229-244 and DAR34.29-35a, both transcribed in *Darwin Online*.

Port Famine, Patagonia. (Puerto del Hambre) On Magellan Straits, south of Punta Arenas. 1834 Feb. 2-10, Jun. 1-8 *Beagle* there. *Zoological diary*, pp. 232-3. *Geological diary*: DAR34.125-128 and DAR34.153-156, both transcribed in *Darwin Online*.

Port Jackson, New South Wales, Australia. 1836 Jan. 12 *Beagle* arrived and anchored in Sydney Cove.

Port Louis, Berkeley Sound, East Falkland Island. 1833 Mar. 1-Apr. 6 *Beagle* at or near. 1834 Mar. 10-Apr. 7 *Beagle* at or near.

Port Louis, Mauritius. 1836 Apr. 29-May 9 *Beagle* at. CD made several excursions. See the *Sydney Mauritius notebook*, Chancellor & van Wyhe, *Beagle notebooks*, 2009.

Porter, George Richardson, 1790-1852. Statistician. 1834- Secretary to the Board of Trade. 1849 CD went to BAAS meeting at Birmingham with P. CCD4:270. Member of council of the BAAS and vice-president of section F (statistics).

Portfolios. After his use of notebooks to record information and ideas, CD changed to a topical portfolio system to organize notes according to subject. Most of the portfolios still extant were relevant to his theory of evolution research programme and each was numbered. This allowed him to write just a number on diverse sheets of paper and these could always be filed back where they belonged. In Dec. 1856 CD excised many pages from his notebooks and filed them by subject in portfolios. For example, his portfolio 3 was for the subject of variation and varieties.

Others included: 10 Rudimentary & abortive organs; 11 Divergence; 12 Embryology; 16 Hybridism; 18 Migration; 19 Island endemism: animals; 20 Island endemism: plants; 21 Palaeontology: extinction; 23 Instinct: change in habits. For later projects he employed portfolios that were organized according to the chapter of the book he was preparing, for example there was a portfolio 5 for *Origin* chap. 5 laws of variation. See the table of excised pages in *Darwin's notebooks*, 1987, pp. 643ff.

Porto Praya, St Jago (now Praia, Santiago), Cape Verde Islands. 1832 Jan. 16-Feb. 8 *Beagle* at and CD landed. 1836 Aug. 31-Sept. 5 *Beagle* visited again. Sketch of the cliffs and mountains from the sea in DAR29.3.45-46a.

Portobello, Edinburgh. 1826 CD wrote an unpublished paper 'Zoological walk to Portobello', perhaps intended for Plinian Society. DAR5.A49-A51. Transcribed in *Darwin Online*, 2009. Colp, *N.Y. State Jrnl. Med.*, (21 Dec. 1979), p. 36.

Portsmouth, Hampshire. 1846 Sept. 12 CD visited on way to Isle of Wight on day trip from BAAS meeting at Southampton. 1858 Jul. 16 CD stopped at on way to family holiday in Isle of Wight.

Portuguese. First editions in: 1900? *Journal* (F2520). 1904 'Bar of sandstone off Pernambuco' (F268). 1910 *Descent* (F1104). [?192-] *Origin* (F743). 1975 *Expression* (F2038). 2004. *Autobiography* (F2039). nd *Variation* (F2062.11), *Insectivorous plants* (F2062.14), *Climbing Plants* (F2062.09), *Different forms of flowers* (F2062.16), *Power & Movements* (F2062.17), *Earthworms* (F2062.18). See A.L. Pereira, The reception of Darwin in Portugal (1865-1914). *Revista Portuguesa de Filosofia* (2010): 643-60, 2019 Autobiographical fragment (F2230).

Pouchet, Félix Archimède, 1800-72. French naturalist and biologist. 1868 CD quotes P in translation "variation under domestication throws no light on the natural modification of species". *Variation* 1:2. Review of *Variation* in *Athenaeum*, Feb. 15.

Poulton, Sir Edward Bagnall, 1856-1943. Entomologist and evolutionary biologist. Specialist on mimicry in butterflies and author of many papers on evolution. Lifelong advocate of natural selection. 1889 FRS. 1893-1933 Hope Prof. Zoology (Entomology) Oxford. 1908 *Essays on evolution*, Oxford. 1909 *Charles Darwin and the Origin of species*. 1914 Darwin Medal Royal Society. 1922 Linnean Medal Linnean Society. 1935 Kt.

Powell, Rev. Baden, 1796-1860. Mathematician. Three times married and had 14 children. Father of Lord Baden Powell, Chief Scout. 1824 FRS. Correspondent of CD. 1827-60 Savilian Prof. Geometry Oxford. 1855 *Essays on the spirit of inductive philosophy*, is referred to in Historical sketch in 3d edn of *Origin*, 1861, p. xviii. 1860 P was one of seven contributors to pro-natural selection *Essays and reviews*.

Powell, Rev. Henry, 1841-1901. 1869-71 P was Vicar at Down from 12 Apr. 1869-71. CCD17:330. 2d child from Rev. Baden Powell's 2d marriage (Charlotte Pope d. 1844). In 1869 Nov. 20 dined with Darwins. Called again in 1871 Jun. 16.

Power, Mrs, a resident of Port Louis, Mauritius. Possibly the wife of Colonel James Power, Royal Artillery. She is mentioned in the *Red notebook*, p. 17: "Mrs Power at

Port Louis talked of the extraordinary freshness of the streams of Lava in Ascencion known to be inactive 300 years?" and the *Sydney notebook*, Chancellor & van Wyhe, *Beagle notebooks*, 2009. Also a reference to a Dr Power of Mauritius in *Descent* 1:335.

Prehistoric Europe. 1881 James Geikie, *Prehistoric Europe, a geological sketch*, London, extracts from two letters from CD, pp. 141-2 (*Shorter publications*, F1351). Published late in 1880, although dated 1881.

Prestwich, Sir Joseph, 1812-96. Geologist and wine merchant. 1853 FRS. 1859 CD sent 1st edn of *Origin*. 1859 CD to Lyell, "I wish there was any chance of Prestwich being shaken; but I fear he is too much of a catastrophist". CCD7. 1865 Royal Medal Royal Society. 1870-72 President Geological Society. 1879 Settled Glen Roy question.

Prévost, Jane Susan Adèle, 1803?-81. 1828 Married Edward Simcoe Drewe.

Preyer, Thierry William, 1841-97. English physiologist and child psychologist. 1857 Moved to Heidelberg to study chemistry. 1862 P wrote dissertation on great auk, *Plautus impennis* (now *Pinguinus impennis*), along darwinian lines, almost the earliest piece of special work based on *Origin*. 1868 Mar. 31 CD to P, that he is glad to hear that P is pro-*Origin*. CCD16:349. 1869-88 Prof. Physiology Jena. 1879 Feb. P compiled a list of darwinian papers in Gratulationsheft number of *Kosmos* for CD's 70th birthday. 1891 CD letters to P in: Briefe von Darwin. Mit Erinnerungen und Erlaeuterungen. *Deutsche Rundschau* 17, no. 9 (Jun.): 356-390. (F6).

Price, James, c.1882- Butler at Down House. 1885 Oct. 17, ED recorded in her diary "Pay Price full wages". 1891 ED "Parslow wants me to raise Price's wages again". Atkins 1976, p. 74.

Price, John, 1803-87. Welsh scholar, botanist. 1818-22 At Shrewsbury School with Erasmus and CD. 1826 BA Cambridge. CD called him "Old Price". 1826-27 Assistant Master at Shrewsbury; private tutor at Chester. Brent 1981, p. 28. 1874 CD to P thanking for sending *Utricularia* from Cheshire for *Insectivorous plants*. CCD22. Recollections of CD in DAR112.B101-B117 and *Llandudno*, 1875, in *Darwin Online*.

Price, Mrs Mary, neighbour at Down Lodge. 1842 Sept. 16 ED having just moved into Down House called on her. Keynes, *Annie's box*, 2001, p. 79.

Price, Mrs Sara, Dr Robert Waring Darwin's housekeeper. Brent 1981, p. 18.

Price, Thomas, Gardener at Down House. Known as "The Dormouse". Said to be a deserter from the army. Drank too much beer. Unmarried. Died young. Francis Darwin, *Springtime*, 1920, p. 57; B. Darwin, *Green memories*, 1928, p. 14.

Prichard, James Cowles, 1786-1848. Physician and ethnologist. Physician to Bristol Infirmary. 1813, 1826 Some hesitant ideas about evolution in *Researches into the physical history of mankind*, 2d edn. 1827 FRS. 1844 CD to Hooker. CCD3:79. 1897, 1908 Poulton, *Science Progress*, 1, Apr. 1897, and *Essays on evolution*, 1908, pp. 173-92, stresses importance of 2d edn.

Primula. 1862 "On the two forms, or dimorphic condition, in the species of *Primula*, and on their relations", *Jrnl. Proc. of Lin. Soc. of London (Bot.)*, 6: pp. 77-96 (*Shorter publications*, F1717). French translation of this paper with CD's

"On *Catasetum*" and "On *Linum*" *Annals Sci. Natural Bot.*, 19: pp. 204-95 (F1723). 1868 "On the specific differences between *Primula veris*, Brit. Fl. (var. *officinalis* of Linn.), *P. vulgaris* Brit. Fl. (var. *acaulis*, Linn.) and *P. elatior* Jacq.; and on the hybrid nature of the common oxslip. With supplementary remarks on naturally produced hybrids in the genus *Verbascum*", *Jrnl. of the Lin. Soc. of London (Bot.)*, 10: pp. 437-54 (F1744). 1874 "Flowers of the primrose destroyed by birds", *Nature*, 9: p. 482, 10: pp. 24-5 (*Shorter publications*, F1770, F1771). Material on illegitimate progeny of dimorphic and trimorphic plants, including notes, tables, memoranda and press-cuttings, mainly on *Primulae*, are in DAR108-111.

Pringsheim, Nathanael, 1823-94. German botanist. 1878 P seconded CD's election to Königlich-Preussische Akademie as Corresponding Member. CCD26.

Pritchard, Rev. Charles, 1808-93. Astronomer and educationalist. 1834-62 Founder and Headmaster of Clapham Grammar School. 1840 FRS. All CD's [surviving] sons, except William, went to Clapham Grammar School, but only George and Francis were taught by P. 1866 letter from CD in *Charles Pritchard...memoirs of his life.* (F2289). 1870 Savilian Prof. Astronomy Oxford University.

Pritchard, George, 1796-1883. Missionary at Tahiti. 1835 Nov. CD met and attended his church on Nov. 22. 1837-44 British Consul in Tahiti, and advisor to **Queen Pomare IV**, 1844-56 British consul in Samoa. (*Shorter publications*, F1640).

Proctor, George, ?-1858. Cambridge friend of CD. "His uncle, Robert Proctor, had lived and travelled in Peru and Chile in 1823-4." CCD1:394. 1831 Christ's College BA. 1834 Jul. 20 CD to Emily Catherine Darwin regarding P. Keynes, *Beagle record*, 1979, p. 218. 1846-58 Vicar of Stroud, Gloucestershire.

Prothero, Sir George Walter, 1848-1922. Historian and writer. Cambridge friend of CD's sons. 1876-94 Fellow of King's College Cambridge. 1882 P was on "Personal Friends invited" list for CD's funeral. 1894-99 Prof. Modern History Edinburgh University. 1903 P drew Francis Darwin's attention to Baden Powell's article in favour of natural selection in *Essays and reviews*, 1861, pp. 138-9; quotation from it. ML1:174. 1920 KBE. FBA. From 1884-93 the Protheros came often to teas and dinners with the Darwins.

Pryor, Marlborough Robert, 1848-1920. Fellow of Trinity College Cambridge. Man of business. Cambridge friend of CD's sons. Received presentation copy of *Origin* 6th edn.? 1871 CD to Francis Darwin giving views on Mivart, refers to P. CCD19:31-32. 1882 P was on "Personal Friends invited" list for CD's funeral. 1873 Mar. 1 Pryor called on the Darwins. He visited again on May 10. 1884 Jan. 24 P came with Messrs S. H. Vines and W. Ward. In Jun. P went with Francis Darwin to Germany. His next visit was in 1888 Apr. 29. 1895 ED wrote twice to P, in second letter Mr Pryor became Prof. Pryor.

Public natural history collections. 1858 *Public natural history collections. Copy of a four page memorial addressed to the Right Honourable the Chancellor of the Exchequer* [Benjamin Disraeli]; signed by CD and eight others, dated 18 Nov. 1858. (*Shorter publications*,

F371, F1702). "on the subject of the proposed move of the British Museum's natural history collections to a new site". See Memorials.

Pucklands, Great and Little. Two fields to west of Down House, 19½ acres together. Great Pucklands was known as "Stoney field" by the Darwins. ED seemed to enjoy going out to "Stony field", going there 1884-94. 1931 Bought by Sir George Buckston Browne, Royal College of Surgeons research station built on Little Pucklands. Buckston Browne gave £100,000, of which £83,000 was invested after purchase and building.

Pugh, Mary Ann, ?-1895. 1857 Jan. 8 - 1859 Jan. 27 Governess at Down House. ED recorded her first visit in 1857 Jan. 8. She replaced Catherine Thorley. On Apr. 1, P & Lizzy went to Mrs Hawkshaw. Dec. 12, the family took her to Crystal Palace. 1859 Jan. 27 ED wrote "Miss Pugh went to Brighton". Three days prior, ED went to London and engaged Mrs Grut. P visited in 1859 Mar, Jul. and Dec. 1860 Dec. 10 P came and stayed a month. 1861-63 made several visits staying up to a week sometimes. 1862 P was replacement governess for Elizabeth Darwin [VI] for one month. P later went mad and was in an asylum, paid for by Sir John Hawkshaw whose children she had taught. CD paid £30 a year for her to have a holiday. "apparently suffered from melancholia" CCD7:263. 1866 ED visited P. ED2:185. In 1895 Jul. 20 ED recorded in her diary "Miss Pugh's death". Evans, *Darwin and women*, 2017.

Pulleine, Robert, 1806-68. Cambridge friend of CD. 1845-68 Rector of Kirkby-Wiske, Wensleydale, where William Darwin Fox visited him.

Pumilio argyrolepis. 1861 Notes on the achenia of *Pumilio argyrolepis*, *Gardeners' Chron.*, no. 1: pp. 4-5 (*Shorter publications*, F1709). The achene "a small dry one-seeded fruit which does not open to liberate the seed." (OED) of the Australian *Pumilio argyrolepis* (an orchid) were observed to affix themselves to a surface if damp and then become oriented upright. CD called this "a pretty adaptation to ensure the attachment of the seed to the first damp spot on which it may be blown".

Punjabi. First edition in: 2004 *Origin* (F2407).

Punta Alta, Argentina, about 15 miles southeast of Bahia Blanca. CD found very large fossils there. See *Beagle diary*, pp. 176ff., *Journal* and *Geological diary* DAR32.61-72. There is a hand-drawn and water-coloured stratigraphical section map in DAR44.22 based on one in the *Rio notebook*, which has further details of his discoveries here. Chancellor & van Wyhe, *Beagle notebooks*, 2009.

Pyt House, Wiltshire. More commonly spelt Pithouse. 1866 Temporary home of Charles Langton. 1866 Oct. 1, ED went with Lizzy, ED(1904)2:212. 1866 Francis and George went first on Dec. 24 with the rest following two days later. Last record in *ED's diary* was of girls returning in 1868 May. 9 after a six day stay.

Quatrefages de Bréau, Jean Louis Armand de, 1810-92 Jan. 12. French naturalist and anthropologist. CD discussed evolution with before *Origin*. 1850 Prof. Natural History Lycée Napoléon. 1859 CD sent 1st edn of *Origin*. 1862 Jul. 11 CD to Q, about French translation of *Origin*. CCD10:313. See 1863 CD letter to Q in F1837.

1868 CD to Stainton, CD plans to write to Q about silk moths. CCD16:226. 1869 Q to CD, opposes CD on evolution, but hopes that their differences of opinion will never alter their good relationship. CCD17. 1870 *Charles Darwin et ses précurseurs Français: étude sur la transformisme*, Paris. 1879 Foreign Member Royal Society.

Queen Anne Street, Cavendish Square, London. 1852- no. 57, later no. 6, home of CD's brother Erasmus. 1859 no. 14. Miss G. Tollet there in Apr. no. 31 Hensleigh Wedgwood's house. In 1874 Jan. 10 the family visited.

Quekett, John Thomas, 1815-61. Histologist. Co-founded Royal Microscopical Society. FRS. 1848 CD consulted his expertise. At his passing, CD donated five guineas to his widow. CCD4 and 9.

Queries about expression, distributed by CD, probably originally in manuscript to people in contact with other races, to discover what expressions were used in different circumstances. See also *Bulletin of the BMNH, Historical Series*, 4 (1972), pp. 205-19: *Jrnl. for the Soc. of the Bibliography of Natural Hist.*, 7 (1975), pp. 259-63. They were printed as follows: 1867 [No copy known], title probably *Queries about expression for anthropological enquiry*, Cambridge or Boston, Mass., printed for Asa Gray before Mar. 26, fifty copies (F871); this was the first edn anywhere, see no. 4. 1867 "Signs of emotion among the Chinese", *Notes and Records for China and Japan*, 1: p. 105, Aug. 31, anonymous, submitted by Robert Swinhoe from manuscript received from CD (F872). 1867 *Queries about expression*, single sheet, [?London], printed for CD late in the year (F873). No. 3 is printed in all editions of *Expression*, 1873-, in which the answers are analysed. 1868 (actually 1872) "Queries about expression for anthropological enquiry", *Report Smithson. Instn*, for 1867; [p. 324], text perhaps printed from a copy of 1 (*Shorter publications*, F874). First edn in 1868 German (F876a).

Queries respecting the human race. 1841 Queries respecting the human race, to be addressed to travellers and others. Drawn up by a Committee of the BAAS, appointed in 1839. *Report of the British Association for the Advancement of Science, at the Glasgow meeting, August 1840*. 10: 447-458 (F1975). The committee consisted of CD, James Cowles Prichard, Thomas Hodgkin, James Yates, John Edward Gray, Richard Taylor, Nicholas Wiseman, and William Yarrell. See Browne, *Voyaging*, 1995, p. 421. Offprints were distributed and a revised version also appeared in the report of the Association in 1842 (F1976).

Query to Army Surgeons. 1862 CD circulated a questionnaire to army surgeons about health of troops in the tropics. No copy known (F799), but text is printed in *Descent*, 1:244-245.

Questions about the Breeding of Animals. [1839] 8 pp., [London], probably late Apr., certainly before May 5 (*Shorter publications*, F262). A printed questionnaire distributed by CD. First reply on 6 May 1839. See CCD2. 1968 Facsimile of the above, wrongly dated 1840 (F263). With introduction by Sir Gavin de Beer. See also *Jrnl. for the Soc. of the Bibliography of Natural Hist.*, pp. 5220-5, 1969.

Questions for Mr. Wynne. An early set of questions in manuscript about the breeding of animals. Transcribed by Barrett in Gruber, *Darwin on man*, pp. 423-5, 1974 (F1582) and in *Darwin Online*. Mr Wynne was Rice Wynne. See also "Mr Wynne Doubts about Irish Horses hereditarily jumping" (DAR205.11.37).

Quiz. A deerhound belonging to John William Brodie Innes (Johny), son of John Brodie Innes, then 15. 1862 Jan. Taken over by Down House. CCD9:372. 1862 May shot because it "bit a child & two days after a beggar woman". CCD10:177.

Rade, Emil, 1832-1931. German civil servant. Member of Münster Zoological Society. 1877 Feb. 12 R sent CD an album of 165 portraits of German and Austrian men of science for his 68th birthday. R originated the idea. The album is finely bound and title page states "Dem Reformator der Naturgeschichte, Charles Darwin." and decorated by Arthur Fitger who also contributed a dedicatory poem. Now at Down House (EH88202652). 1877 Feb. 16 CD thanks R and writes to Haeckel on the subject. See CCD25 Appendix V.

Radovanović, Milan M. 1849-78. Serbian doctor and writer. 1878 Translated *Origin* into Serbian (F766). CCD26.

Rae, John, 1813-93. Surgeon, fur trader, explorer and author. 1880 FRS. CD used R's information on seed distribution in Stauffer, ed. *Natural selection*, 1975, p. 562.

Rain, yellow. 1863 [Letter on yellow rain], *Gardeners' Chron.*, no. 28: p. 675 (*Shorter publications*, F1727). CD recounted that ED gathered flowers outside after a light shower and found "the drops of water appeared yellowish". CD examined the droplets under the microscope and found they contained "numerous brown spherical bodies". CD sent a sample. The editor, M.J. Berkeley, suggested that the "larger bodies" were "pollen grains of some Thistle or Centaurea".

Rájon, Paul Adolphe, 1843-88. French painter and print-maker. 1875 R engraved on copper the Ouless portrait of CD.

Raleigh, Sir Walter Alexander, 1861-1922. Scholar, poet and author. 1881 Oct. CD and ED took tea with R in Cambridge. 1900-04 Regius Prof. English language and literature. 1904- Merton Prof. English literature Oxford. 1911 Kt.

Ralfs, John, 1807-90. Surgeon and botanist. R lived at Penzance, Cornwall. 1874 R sent CD *Pinguicula* and observations for *Insectivorous plants* (described pp. 390-1).

Ramirez. CD gave him £12 for horses(?) in the *Santiago notebook*, Chancellor & van Wyhe, *Beagle notebooks*, 2009.

Ramsay, Sir Andrew Crombie, 1814-91. Geologist. Biography: A. Geikie, *Memoir* 1895. 1846 CD to Lyell, R was in favour of sudden elevations. CD scoffs. 1848 Prof. Geology University College London. 1848 R visited Down House for weekend. ED in 1848 Feb. 12 recorded "Lyells Owen Forbes & Ramsay came" but the line was crossed out. The Feb. 14 record of "Messrs Owen Forbes & Ramsay went" was also crossed out. CCD4. R's own diary recorded the visit on Feb. 13. He found CD an enviable man, their home pleasant and ED a nice wife. 1859 CD sent 1st edn of *Origin*. 1859 CD to Lyell, "I infer from a letter from Huxley that Ramsay is a convert". CCD7:407. 1862 FRS, R et al published descriptions of CD's rock specimens. 1872- Director General Geological Survey in succession to Murchison. 1871 Wollaston Medal. 1880 Royal Medal Royal Society. 1881 Kt. Recollections of CD in Geikie, *Memoir of Sir Andrew Crombie Ramsay*, 1895, pp. 123, 130, 276-77, transcribed in *Darwin Online*.

Ramsay, Marmaduke, 1799-1831. 5th son of Sir Alexander R, brother of Sir Andrew Crombie R. Cambridge friend of CD and Henslow. R intended to go on a projected trip to Canary Islands with CD when he died. 1818-31 Fellow and tutor Jesus College, Cambridge.

Ramsgate, Kent. 1850 ED wrote in her diary Oct. 3 "children & Miss T. to Ramsgate." 1850 Oct. 18 CD visited for the day from Hartfield, Sussex; ED recorded "came to Ramsgate". On the 21st Annie was feverish. The next day she wrote that the others went home but she and Annie stayed until the 24th.

Ransome, George, 1811-67. Agricultural instrument maker of Ipswich. 1849 or 1850 R commissioned set of 60 Ipswich Museum portraits for BAAS meeting there in 1851. CD to R, happy to promote R's project and subscribes £1 towards portrait of "the Bishop". There are two bishops in the set, both of Norwich, Edward Stanley, died Sept. 1849, and Samuel Hinds, appointed Oct. 1849. CCD4. 1850 R gave CD a set which includes CD by T.H. Maguire.

"Ras", Family nickname for Erasmus Alvey Darwin; also for Erasmus Darwin [III].

Rats, see **Casks**.

Rattan, Volney, 1840-1915. Californian botanist and schoolmaster. Published on botanical subjects. 1880 Letters with CD on germination of *Echinocystis*. See 1881, *Movement in plants*, p. 82.

Raverat, Gwendolen Mary, (Gwen Raverat) see Darwin, Gwendolen Mary.

Raverat, Jacques Pierre Paul, 1885-1925. French painter. 1911 Married Gwendolen Mary Darwin. 2 daughters; Elisabeth and Sophie Jane.

Raverat, Sophie Jane, 1919-2011. Daughter of Gwendolen and Jacques R. 1940 R married entomologist Mark Gillachrist Marlborough Pryor, who died in 1970. 1973 Married Henry Charles Horton Gurney.

Ray Society. 1844 Instituted, for the publication of biological monographs, in honour of John Ray. 1851, 1854 (1855) CD's *Living Cirripedia*, two vols. (F339). Since 1964 several facsimile reprints have appeared. Only in Russia parts have appeared in translation in 1936. A planned Portuguese translation was announced in 2009. 1856 CD to Hooker, "I profited so enormously by its publishing my Cirripedia, that I cannot quite agree with you on confining it to translations". CCD6:171.

Reade, Thomas Mellard, 1832-1909. Civil engineer and geologist. 1881 R corresponded with CD about the success of *Earthworms*. LL3:217. 1881 CD to Hooker, about R's views on permanence of continents. LL3:247.

Reade, William Winwood, 1838-75. Historian, traveller and controversialist. 1868 CD sent *Queries about expression* to. CCD16. 1868-69 R gave CD information on Africa for *Expression*. 1872 *The martyrdom of man*.

Reading notebooks. CD's lists of books he planned to read or had read. 'Books to be read' & 'Books Read' (1838-51) (DAR119), 'Books Read' & 'Books to be Read' (1852-60) (DAR128) and 'Books [read]' (1838-58) (DAR120). The latter notebook is the most useful. Transcribed with important introductory matter in CCD4: Appendix IV. All are also transcribed in *Darwin Online*.

Reale Accademia delle Scienze di Torino, Scientific academy originally founded in Turin, Italy, in 1757. See Bressa Prize.

Reclus, Jacques Élisée, 1830-1905. French geographer, writer and anarchist. 1871 Taken prisoner during the Siege of Paris and sentenced to transportation. CD was amongst the 100 or so signatories to a memorial asking for a reprieve. The sentence was commuted in Jan. 1872 to perpetual banishment from France. Remitted 1879.

Recollections. In addition to publications and contemporary manuscripts about or mentioning CD, there is a considerable body of later recollections. Although these need to be treated with more caution than contemporary documents, recollections by his family and others who knew or met him often provide information available nowhere else. Many were solicited by Francis Darwin in preparation for *Life and letters* (1887). Several important recollections and fragments were published there. Francis Darwin's own 'Reminiscences of my father's everyday life' was incorporated there. The preliminary draft is in DAR140.3 and transcribed and edited in *Darwin Online* with an introduction by Robert Brown:
http://darwin-online.org.uk/EditorialIntroductions/Brown_IntroFrancisRecollections.html

The majority of the recollections in the Darwin Archive are found in DAR262 with further recollections in DAR107.1-4, DAR107.42-47, DAR112.A10-A12, DAR210.8.36-41, DAR210.9 and DAR251.1106-7. More than 150 published and manuscript recollections of CD are transcribed in *Darwin Online*, http://darwin-online.org.uk/recollections.html, including those by: Abbot, F.E., Allen, G., Allingham, W., Aveling, E.B., Bailey, Bain, A., Balfour, A.J., Bentham, G., Blomefield, L., Bowen, E., Brace, C.L., Bryce, J., Büchner, L., Bunbury, C.J.F., Buob, L., Butler, S., Butler, T., Calman W.J., Cameron, J.H.L., Campbell, G., Duke of Argyll., Candolle, A. de., Carlyle, T., Clark, J.W., Claus, C., Clemens, S., Cobbe, F.P., Cohn, F., Colenso, W., Collier, J., Conway, M.D., Cox C.F., Cradock, E.H., Darwin, Bernard, Darwin, Emma., Darwin, F., Darwin, G.H., Darwin, L., Darwin, W.E., Derby, Ld. [Stanley, E.H.], de Vries, H., Dicey, A.V., Dohrn, A., Duff, G.M.E., Duff, U.G., Durdík, J., Fabre, H., Farrar, F.W., Fegan, J.W.C., Fiske, J., Forbes, E., Forster, L.M., Fraser, G.D., Galton, F., Geddes, P., Geldart, E.M., Gray, A., Gray, J., Green, J.R., Gretton, F.E., Griggs, E.L., Gulick, A., Gulick, J.T., Haeckel, E., Hague, J.D., Hamilton, T.H., Hamond, R.N., Harting, P., Heaviside, J.W.L., Herbert, J.M., Hesse-Wartegg, E. von., Higginson, T.W., Holder, C.F., Hooker, J.D., Horner, L., Huxley, T.H., Innes, J.B., James, H., Jordan, D.J., Judd, J.W., Keith, A., Keynes, N., King, P.G., Krause, E., Lane, E., Lankester, E.R., Leighton, W.A., Lettington, H., Lewins, R., Lewis, J., Litchfield, H.E., Longfellow, H.W., Lowe, R., Lumb, E., Macauley, T.B., Macdonell, A., Mantell, G., Martineau, H., Maxwell, G.S., Mellersh, A., Minching, W., More, A.G., Morgan, L.H., Morley, J., Murray, J., Nash, L.A., Nash, W., Nevill, D., North, M., Norton, C.E., Osborn, H.F., Paget, G.E., Preyer, W.T., Price, J., Pym, H.N., Ramsay, A.C., Raverat, Gwen., Rees, W.L., Richter, H., Riley, C.V., Ritchie, H., Rodwell, J.M., Romanes, G.J., Romilly, Bury, J.P.T., Ruskin, J., Russell, A., Skinner, A.J., Smalley, G.W., Stephen, L., Stokes, J.L, Sulivan, B.J., Sully, J., Tegetmeier, W.B., Tennyson, H., Thackeray, J.C., Timiriazev, K., Tyndall, J., Usborne,

A.B., Vaughan, E.T., Vignoles, O.J., Wallace, A.R., Watkins, F., Webster, A.D., Wedgwood, C.S., Wedgwood, K.G.S., Wheeler, H., Whitley, C.T., Wolf, J., Wright, C., Yates, E.H., and Youmans, E.L.

Red notebook, a notebook used by CD during and after the voyage of the *Beagle*. It is particularly noteworthy because post-voyage notes contain some of his first speculations on evolution. Published by Sandra Herbert under this title in 1980 (F1583e and in *Darwin Online*), in *Darwin's notebooks*. At Down House EH88202322.

Reed, Rev. George Varenne, 1816-86. Anglican clergyman. R was tutor to George (1856), Francis (1858), Leonard (1859) and Horace (1864) before they went to Clapham Grammar School. 1854-86 Rector of Hayes, Kent. 1859 R gave CD a cutting of "carrion-smelling Arum". J.R. Moore, *Notes and Records Roy. Soc.*, 32: pp. 51-70, 1977. 1882 R was on "Personal Friends invited" list for CD's funeral. CD once wrote to R complimenting him for his "kind labours" which kept George top of his class. In 1864 Apr. 19, ED wrote "Ho to Mr Reed". CD's account book showed R was paid £12 12s for tutoring Horace for three months. 1872 Sept. 16 the Reeds and Ffindens dined. Last record in *ED's diary* 1878 Oct. 28 Mrs Reed was called on.

Reeks, Trenham, 1823-79. 1851 Curator of the Museum of Practical Geology including the Geological Survey Library. Had great knowledge of pottery. 1845 exchanged several letters with CD. CCD3. 1858 CD sent some 30 slabs of Wedgwood slate to be appraised. CD thanked him for his assistance with the Barberini Vase which was sold for £75. CCD7:22, CCD7:45.

Reeve Brothers. London printers. 1846 CD gave instructions on the printing of coloured plate for *South America*. Made a payment of £10. 2s.

Reeve, Lovell Augustus. Conchologist. CD letter to L supporting his election to the Royal Society. In J.C. Melvill, Lovell Reeve: a brief sketch of his life and career, with a fragment of an autobiography, excerpts from his diary (1849), and correspondence. *Jrnl. of Conchology* 9 (Jul.): 344-357, p. 352. (F1997).

Reeves, William, 1836?-1915. Blacksmith and farrier of Down, Reeves, grandfather, father and, in 1951, son. ?Successors to "Old M" q.v. 1885 Dec. 30, ED noted in her diary "Mr Reeve 92 West Road".

Referee reports, CD's 12 referee reports for the Royal Society of London and the Geological Society of London are transcribed and published in *Darwin Online*.

Regent Street, London. 1833 no. 24 home of Erasmus Alvey Darwin.

Reinwald, Charles Ferdinand, 1812-91. 1873- Publishers of Paris as C. Reinwald et Cie; published 1st French editions of eleven of CD's books, as well as editions of *Origin* from 1873 onwards and *Life and letters* in 1888.

Rejlander, Oscar Gustave, 1813-75. Swedish-born professional photographer of Victoria Street, London. 1871-2 photographed CD. 1872 Some of the photographs (heliogravures) in *Expression* are by R. Several of these depict R himself. Darwin family members were also photographed by R, e.g. William and R.B. Litchfield.

Religious views. 1871 A letter from Mr. Darwin, *The Index, a weekly paper devoted to free religion* (Ohio), 2: p. 404 (*Shorter publications*, F1753). Letter is addressed to F.E.

Abbott. CD agrees to allow his previous letter on the contents of *The Index*, to be published as "I agree to almost every word". See Darwin, Charles Robert, Religion.

Rendel, Emily Catherine, 1840-1921. Daughter of James Meadows R, FRS. 1866 Married Clement Francis Wedgwood. 6 children. 1883 Nov. 16-17 R and Clement dined and stayed. They visited again in 1888 Feb. 18-21. *ED's diary.*

Rengger, Johann Rudolph, 1795-1832. Swiss naturalist and physician. CD cited R in *Journal*, *Darwin's notebooks* and *Natural selection*. CD cited R's observations on monkeys in *Descent*, vol. 1, pp. 11, 40-57 and 110-118. R's *Naturgeschichte der Saugethiere von Paraguay*, 1830 was one of the books CD had in the *Beagle* library. CD's copy replete with marginalia is in *Darwin Online*.

Renous, Juan, German collector of insects etc. 1834 Sept. 17 CD met at Yaquil, near Nancagua, house of Mr Nixon, an American who owned a gold mine there. *Beagle diary*, pp. 245-8. Keynes, *Beagle record*, 1979, p. 236. In the *Santiago notebook*, Chancellor & van Wyhe, *Beagle notebooks*, 2009. CD called him Don Pedro R.

Reptiles. CD ed. 1843. *Reptiles Part 5 of The zoology of the voyage of HMS Beagle*. by Thomas Bell. Edited and superintended by Charles Darwin. London: Smith Elder and Co. CD's List of reptiles and amphibians from the *Beagle*: "Reptiles in spirits of wine" is in the NHM and is transcribed in *Darwin Online*.

Bibron's Tree Iguana, (now *Liolaemus bibronii*) collected by CD at Port Desire, Patagonia.
Plate 3 from *Reptiles*.

Reviews of CD's works. Darwin family lists include: "List Reviews of Origin of Sp & of C. Darwins Books" EH88206151-60 but kept in CUL at DAR262.8.9-18 with c.350 references. It is transcribed in *Darwin Online*. There is a scrapbook of reviews: DAR226 (2 vols.). Reproduced in *Darwin Online*, 781 images for about 335 items. DAR226.1 is labelled on the spine "Reviews of C. Darwin's works" and lists reviews of *Origin*, *Orchids* and others. DAR226.2 is labelled "Reviews. Descent. Expression. Insect. Pl. Eras. D." There are lists of reviews in appendices to the CCD: CCD8 VII. Reviews of *Origin*, 1859-60; CCD10 VII. Reviews of *Orchids*; CCD16 VI. Reviews of *Variation*; CCD19 V Reviews of *Descent*; CCD23 V Reviews of *Insectivorous* *plants* (see DAR139.18); CCD24 IV Reviews of *Cross and self fertilisation*; CCD26 IV. Reviews of *Forms of flowers*; CCD27 Reviews of *Erasmus Darwin*. By far the largest collection of reviews (over 900) is in *Darwin Online*: http://darwin-online. org.uk/reviews.html and in Darwin's private papers in CUL. The *Darwin Online* project has so far recorded over 1,000 reviews. A. Ellegård, *Darwin and the general reader*, 1958, surveys reviews in the press, in relation to popular rather than scientific opinion, in great detail with full references. There have been several important studies since.

Reynolds, Caroline Lane, 1840-1930. American. Aunt of Maud du Puy. Married Sir Richard Claverhouse Jebb in 1874 as second husband. 1952 Aunt Cara in Rave-rat, *Period piece*. 1884 Apr. 12 Mrs Jebb went to Abinger. Maud, George, Ras, Ruth and the Litches also were present. Up until 1896 Apr. 6 they frequently called and dined with the Darwins. *ED's diary*.

Reynolds, Susanna, Of Ketley Bank, Arleston. CD liked her very much and thought her very pretty. CD was thirteen. CCD13:344.

Rhea. The correct name for *Rhea darwinii* is *Rhea pennata*, with synonym *Pterocnemia pennata*. Often called Darwin's rhea. 1837 [Notes on *Rhea americana* and *Rhea darwinii*], *Proc. of Zool. Soc. London*, Part V, no. 51: pp. 35-6, follows John Gould's original description of *R. darwini* (*Shorter publications*, F1643). 1841 Described by John Gould in Part XV of *Zoology of the Beagle*, and depicted on Plate 47 as drawn by Elizabeth Gould. CD's specimen was famously eaten during the voyage.

Rhodes, Francis, later **Darwin**, 1825-1920. 1849 Married Charlotte Maria Cooper Darwin. 1850 R inherited Elston Hall under will of his brother-in-law, Robert Alvey Darwin, and changed his name. 1882 R was present at CD's funeral as head of the senior branch of Darwin family.

Rice, Thomas Spring [I], 1790-1866. Statesman. 1820-39 MP. 1835-39 Chancellor of the Exchequer. 1837 Aug. R authorized £1,000 grant for publishing scientific results of *Beagle* voyage. 1839 1st Baron Monteagle of Brandon. 1841 FRS. CD et al. 1838. Copy of a Memorial presented to the Chancellor of the Exchequer, recommending the Purchase of fossil remains for the British Museum. *Parliamentary Papers, Accounts and Papers 1837-1838*, (27 Jul.): 1. (*Shorter publications*, F1944).

Rice, Thomas Spring [II], 1849-1926. Grandson of Thomas Spring [I] R. Cambridge friend of CD's sons. Irish resident landlord, of Foynes, Co. Limerick. 1866 2d Baron Monteagle of Brandon. Recorded in *ED's diary* as visiting the Darwins 1875 May 22, 1877 May 26 and 1879 Apr. 26. All the entries appear as "Monteagles". 1882 R was on "Family Friends invited" list for CD's funeral.

Rich, Anthony, 1804-91. Chapel Croft, Heene, Worthing, Sussex. Honorary Fellow of Caius College, Cambridge. 1878 R made a will leaving nearly all his property to CD, on death of himself, then 74, and his sister; at that time it included some property in Cornhill, London, with income above £1,000. As CD predeceased R, CD's children wrote to R in May 1882 expressing how grateful they were for his generous intention of leaving his property to CD. They also said R was free to alter his will. In the return letter R wrote "nothing could induce me to alter it in that respect. It is a source of pleasure and pride to me to think that it could have been in my power to do anything which would give him ever so small an amount of gratification, and I am equally pleased to think that, when my course is also run, property which belonged to me will descend to the worthy children of so noble a man". ED2:259. 1879 May 6 and 1881 Sept. 8 CD visited R at Worthing. 1882 CD to R about success of *Earthworms*. LL3:217. 1882 After CD's death R left his estate to the Darwin children, except house and contents which went to Huxley who immediately sold it. Net value of estate £15,083. The gift was in recognition of CD "to whose transcendent genius and subtle investigation, extending over a long period of years, the discovery and practical proofs of the law of evolution is due". ML2:445-448.

Rich, Claudius James, 1787-1821. Orientalist, East India Company lived in Baghdad. 1807 Married Mary Mackintosh [I] s.p. CCD2:534. Died at Shiraz of cholera.

Richardson, John, 1787-1865. Arctic explorer and naturalist. CD, as a member of the Council, nominated R for a Royal Medal and wrote the announcement of the award: *Proc. of the Roy. Soc.*, 8: pp. 257-8 (*Shorter publications*, F1936). See CCD6:86.

Richardson, Sir Benjamin Ward, 1828-96. Physician. Writer on medical history. 1867 FRS. 1876 CD to Romanes, R's letter to *Nature* is capital. "Experimentation on animals for the advancement of practical medicine", *Nature*, 14: pp. 148-52. R's letter was forwarded by CD to *The Times* on 23 Jun. 1876 but not published. See *Calendar* 10546. 1893 Kt.

Richmond, George, 1809-96. Artist. RA. Especially portrait painter in water colour. In his youth he was a member of The Ancients, followers of William Blake, a close friend of R. 1839 Mar. Watercolour by R, unsigned, of CD, painted in London, note on back says 1840 Mar. Pencil sketch for this found in Botany School, Cambridge, 1929, now on display in CUL. 1839 Watercolour of ED. ED2:31, 33.

Richmond, Sir William Blake, 1842-1921. Son of George R. Artist. 1879 Jun. CD sat for him in LL.D. robes, exhibited RA 1881. £400 subscribed by members of Cambridge Philosophical Society. Now hangs in the Department of Zoology, Cambridge. A copy in the family. R recollected, CD "told me that he was confident that his heavy brow, his frontal sinus, had become developed during his lifetime." *Manchester Evening News*, (5 Sept. 1915). 1881 CD and ED went to see it in the Society's Library. ED: "the red picture, and I thought it quite horrid, so fierce and so dirty". ED2:248. Francis Darwin "according to my own view, neither the attitude nor the expression are characteristic of my father". LL3:222. 1895 RA. 1897 KCB.

Richter, Hans, 1843-1916. Austrian-Hungarian conductor and pianist. Admirer of Wagner and later Sir Edward Elgar. 1876 Conducted the first complete performance of Wagners's *Ring des Nibelungen* in Bayreuth. CD wrote to Henrietta in May 1880 calling R "a most jolly, happy-looking man, who made some gracious joke in German to Rose." ED2:304-05 1881 May 20 R and Frankes lunched at Down House.

In notes written after CD's death, ED recalled CD much enjoyed R's playing. LL3:223. R recollected his visit in *Neues Wiener Tagblatt*, republished in O. Zacharias, *Charles R. Darwin*, Berlin, 1882. (F2094), transcribed in *Darwin Online*.

Ridge, The, House at Hartfield, near Tunbridge Wells, Sussex, on border of Ashdown Forest. Quarter of a mile from Hartfield Grove, home of Charles Langton. 1849-68 Home of Sarah Elizabeth Wedgwood [II].

Ridgemount, House at Bassett, North Stoneham, Southampton, Hampshire. 1862-92 Home of William Erasmus Darwin.

Ridley, Henry Nicholas, 1855-1956. Botanist and geologist. Discovered a method to tap rubber without damaging the trees seriously. Visited A.R. Wallace in later years. 1878 Nov. 28 CD to R, about Dr E.B. Pusey, an important letter, "Dr Pusey's attack will be as powerless to retard by a day a belief in evolution as were the virulent attacks made by divines fifty years ago against Geology, & the still older ones of the Catholic church against Galileo, for the public is wise enough always to follow scientific men when they agree on any subject; & now there is a most complete unanimity amongst Biologists about Evolution, tho' there is still considerable difference as to the means, such as how far natural selection has acted & how far external conditions, or whether there exists some mysterious innate tendency to perfectibility." CCD27:458, 1888-1911 First Scientific Director Singapore Botanical Gardens.

Riley, Charles Valentine, 1843-95. English born American entomologist. 1868 State Entomologist to Missouri. 1871 CD to R, "our Parliament would think any man mad who should propose to appoint a State Entomologist". 1875 CD to Weismann, R supports John Jenner Weir's views on caterpillars. 1878-94 Entomologist to US Department of Agriculture. 1882 recollection and letters of CD in *Darwin's work in entomology* (F2104), transcribed in *Darwin Online*.

Ring, Abraham, & Mrs Charlotte Ring, Villagers at Down. A. was a gardener. 1862 R's wife ill. J.B. Innes to CD 16 Dec. [1862]. CCD10:606-7.

Rio de Janeiro, Brazil. 1832 Apr. 4 *Beagle* arrived at. 1832 Apr. 8-23 CD travelled inland. 1832 Jul. 5 *Beagle* left. *Zoological diary*, pp. 36-64. *Geological diary*: (4-6.1832), DAR32.51-60. *Rio notebook*, Chancellor & van Wyhe, *Beagle notebooks*, 2009. All transcribed in *Darwin Online*.

Rio Negro [I], Patagonia, Argentina. 1833 Aug. 3 *Beagle* at. CD travelled from there overland, about 850km, to Buenos Aires, arriving Sept. 20. *Geological diary*: DAR34.17-24, transcribed in *Darwin Online*.

Rio Negro [II], Entre Rios, Uruguay. 1833 Nov. 22-26 CD stayed with Mr Keen at his estancia on Rio Beguelo, a tributary, and collected fossils nearby.

Ritchie, Sir Richmond Thackeray Willoughby, 1854-1912. Born in India and worked as civil servant in the India Office. Married Anne Isabella Thackeray. 1881 Feb. 27 "R. Ritchies to lunch". The R's stayed 1882 Jan. 21-23. Visited ED in 1888 and stayed 1895 May 4-5. 1882 R was on "Personal Friends invited" list for CD's funeral. 1907 KCB. Recollections of the 1882 visit in *Letters of Anne Thackeray Ritchie*, 1924, pp. 183-4, transcribed in *Darwin Online* (F2175).

Rivers, Thomas, 1797-1877. Nurseryman, of Sawbridgeworth, Hertfordshire. R is repeatedly referred to in *Variation*. 1866 CD to R, on bud variation. 1866 CD to R, on plant variation in general. 1874 CD to Alfred Newton, R had reported great increase in number of birds in his garden.

Rivière, Briton, 1840-1920. British painter of French descent. RA. 1870 CD sent copy of *Journal* to R. 1872 CD to R about two drawings of dogs to show expressions. R made two drawings for *Expression*, depicted on pages 52 and 53.

Robarts, Curtis & Co., Bankers. 15 Lombard Street, London. CD drew bills on his father's account via R during *Beagle* voyage. 1834 Nov. "I hope not to draw another bill for 6 months". CCD1:419.

Roberts. Sealer. Pilot to John Lort Stokes in *La Liebre*. Friend of James Harris. Lived at Del Carmen on Rio Negro. A very large man who was used to trim the boat; "he did harm one day by going up to look out, and breaking the mast". Fitz-Roy, *Narrative* 2:120-2; CCD1:336.

Robertson, George Croom, 1842-92. 1866- Prof. Philosophy of Mind and Logic University College London. 1877 Apr. 27. CD sent R his manuscript Biographical sketch of an infant, as editor of *Mind*, with explanatory letter. LL3:234 (*Shorter publications*, F1305, F1779). 1882 Jan. 20 CD to Romanes, indicating that R was involved in helping Charles G. B. Allen in his financial difficulties. Carroll 1976, 612.

Robinson, James, 1813-62. Dentist at Gower St, London. Surgeon-dentist to Prince Albert. First president of the College of Dentists. CD and ED as well as the children were patients, as was Hooker. In a letter to Hooker, CD spoke of 'Robin-sophobia'. CCD7:254. Pioneer in the use of ether as an anaesthetic. 1858 Feb. 16 Henrietta had her tooth out. *ED's diary*.

Robinson, Rev. John Warburton, 1837?-? Clergyman. 1859 Trinity College, Dublin, (B.A.). See CCD17, correspondence with J.B. Innes. 1868, 30 Aug.-4 Feb. 1869 Curate of Down, unsatisfactory and walking at night with village girls, among whom was Esther West. 1868 Sept. Esther's mother had forbidden him to visit the cottage.

Robinson, Rev. Thomas Romney, 1792-1882. Also referred to as John Thomas Romney R. Astronomer. Director of Armagh observatory. 1846 CD met R at BAAS meeting. 1849 CD met R at BAAS meeting, where R was President. 1856 FRS.

Rock seen on an iceberg, 1839 Note on a rock seen on an iceberg in 61° South latitude, *Jrnl. Geogr. Soc.*, 9: pp. 528-9, one woodcut. (*Shorter publications*, F1652).

Rodwell, John Medows, 1808-1900. Orientalist. Translator of the Ku'ran (Koran) in 1861. R was nephew of William Kirby, entomologist. Cambridge contemporary of CD. Recollections of CD in Cambridge are in DAR112.A94-A95 and DAR112.B118-B121, transcribed in *Darwin Online*. 1843 Rector of St. Ethelburga's, Bishopsgate, London. 1860 R to CD, about *Origin*. CD said to R when students at Cambridge "It strikes me that all our knowledge about the structure of our earth is very much like what an old hen would know about a hundred acre field, in a corner of which she is scratching". LL2:348 (Taken from the first recollection cited above).

Roffey, Jane, 1837-? Kitchen maid at Down House in 1861 census. 1859 Sept. 7, ED recorded "Jane Roffey came £9". 1875 Jun. 12 ED wrote "Jane went home".

Rogers, Henry Darwin, 1808-66. Born in USA. His middle name was given in honour of Erasmus Darwin. Geologist. 1857 Regius Prof. Geology Glasgow. 1858 FRS. 1860 CD to Lyell, "He goes very far with us". 1861 Like CD in 1838 R published on the parallel roads of Lochaber (Glen Roy) attributing them incorrectly to the grooving influence of floods. CCD9:438.

Rolfe, Robert Monsey, 1790-1868. Lawyer and statesman. 1832-40 MP. 1850 1st Baron Cranworth. 1852-58 Lord Chancellor. 1865-66 Lord Chancellor. 1865 R lived at Holwood, near Down.

Rolle, Friedrich, 1827-87. German palaeontologist and dealer in fossils. Early follower of CD. 1862-68 Corresponded with CD. 1862, 1863, *Darwin's Lehre von der Entstehung der Arten*, p. iv thanks CD for "briefliche Ausdruck".

Rolleston, George, 1829-81. Comparative anatomist, physician and zoologist. Friend and protégé of Huxley. 1860-81 First Linacre Prof. Anatomy and Physiology Oxford. 1860 R was present at the Oxford meeting of the BAAS where Huxley and Wilberforce clashed. 1861 CD had heard R speak at Linnean Society. 1862 FRS. 1871 CD to George Busk, R had pointed out error about supracondyloid foramen in *Descent*. CCD19. 1875 R to CD, on primitive man. CD 1881. Rolleston memorial. *The Times* (5 Aug.): 9.

Romanes, Charlotte Elizabeth, 1853-1911. Younger sister of George John Romanes. 1880-81 R kept a Tufted capuchin on behalf of her brother George John R. 1878 Aug. 20 CD to G.J. Romanes had suggested keeping a monkey. *The life and letters of George John Romanes*. 1896, p. 74-76. Romanes, *Animal intelligence*, pp. 483-4.

Romanes, George John, 1848-94. Canadian-born British biologist and zoologist who also wrote on psychology and physiology. R worked at University College London and at Oxford. R coined the term Neo-Darwinism. Biography: Ethel Romanes (wife) 1896. Joel S. Schwarz, *Darwin's disciple; George John Romanes*, 2010. R was the most important of CD's younger biological friends, frequent correspondent and more than once at Down House. Francis Darwin records a conversation with R telling of a discussion with CD about recognition of natural beauty and its relation to natural selection. LL3:54. 1874 Dec. 7 CD would like to meet R and asks to lunch. CCD22. 1874 CD first met R in London. Romanes 1896, p. 14. 1874 CD to R, "How glad I am that you are so young". 1874 CD introduces R to Hooker. CCD22. 1874 R to CD, on disuse of organs. 1877 CD to R, pleased to propose R for Royal Society. CCD25. 1877 CD to R, astonished that R has not been elected. CCD25. 1878 CD to R, "Frank says you ought to keep an idiot, a deaf mute, a monkey, and a baby in your house", i.e. to study psychology. 1879 FRS. 1879 Married Ethel Duncan. 5 sons, 1 daughter. 1880 Dec. 17 "I have now got a monkey. Sclater let me choose one from the Zoo". 1881 Apr. CD to R, about letter from Frances Cobbe on vivisection in *The Times*. LL3:206. See also CD's letter on vivisection to Frithiof Holmgren published in *The Times* of Apr. 18, reproduced in LL3:207-8. 1882 R was on "Personal Friends invited" list for CD's funeral. 1882 *Animal intelligence*, (F1416), contains many extracts from CD's notes. 1883 *Mental evolution in animals*, London (F1434), contains CD's essay on instinct, pp. 355-84. 1888-91 Fullerian Prof. Physiology Royal Institution. 1890 Moved to Oxford, partly for health reasons. 1892 Founded the Romanes Lectures, held at Oxford. 1892-97 *Darwin and after Darwin*, 3 vols. 1893 *An examination of Weismannism*. 1875-81 correspondence and recollection of CD in *Life and letters of George John Romanes*, 1896. (F2111).

Romanian. First editions in: 1950 *Origin* (F746). 1958 *Journal* (F225). 1962 *Autobiography* (F1532). 1963 *Variation* (F924). 1964 *Orchids* (F824). 1964 *Cross and self fertilisation* (F1271). 1965 *Insectivorous plants*. (F1243). 1965 *Different forms of flowers* (F1301). 1967 *Descent* (F1106). 1967 *Expression* (F1205). 1970 *Climbing plants* (F864). 1970 *Movement in plants* (F1348).

Römer, Carl Ferdinand von, 1818-91. German geologist and palaeontologist. CD sent R *Fossil cirripedes*. *Lychnos*, 1948-49: pp. 206-10. 1851 R sent fossil cirripedes to CD. 1855-91 Prof. Mineralogy and Geology Breslau.

Romilly, Caroline Jane, 1781-1831. 1807 Married Lancelot Baugh Allen, 1st wife.

Roots. 1882 The action of carbonate of ammonia on the roots of certain plants, *Jrnl. of the Lin. Soc. of London (Bot.)*, 19: pp. 239-61 (*Shorter publications*, F1800). Drafts in Francis Darwin's hand (corrected by CD) in DAR28.1-2. See CD to ? 22 Feb. 1882 on instructions to engraver for illustrations. *Calendar* 13698.

Rorison, Gilbert, 1821-69. Episcopalian clergyman of Peterhead, Aberdeenshire. 1861 Anonymous author of *The three barriers: notes on Mr. Darwin's 'Origin of species'*, Aberdeen, preface signed G.R., anti-evolution. In *Darwin Online*. The barriers are the breast, the backbone and the brain. 1861 CD to Huxley, "In a little Book, just published, called the Three Barriers (a theological hash of old abuse of me) Owen gives to the Author a new resume of his Brain doctrine". CCD9:325

Rosas, Juan Manuel José Domingo Ortiz de, 1793-1877. Politician, army officer, gaucho, cattle rancher and Dictator of Argentina. 1829-52 Ruled the Argentine Confederation. 1833 R helped CD with horses and safe conducts on inland journeys. *Journal*. 1834 CD to E. Lumb "The Caesar-like Rosas". In the *Falkland notebook*, Chancellor & van Wyhe, *Beagle notebooks*, 2009. 1852 R was overthrown and retired to Swaythling, Hampshire. 1862? CD met at Southampton. Portrait at Down House.

Rose, Sibyl Renée, 1896-? 1917 Married Charles John Wharton Darwin.

Rothrock, Joseph Trimble, 1839-1922. American botanist. Studied under Asa Gray at Harvard. R answered CD's queries for *Expression* on American Indians. 1862 Aug. 22 CD to Gray, refers to R's work on *Houstonia*. CCD10. Reviewed *Expression*.

Rouse, Henry William, c1798-1870. Appointed vice consul at Concepción in 1823 and consul in 1827. Was present when the earthquake hit in 1835. Described to CD his escape from his home. See *Journal*, pp. 304-6. R visited Down House in 1870 Jan. 6-10.

Roux, Wilhelm, 1850-1924. German zoologist and embryologist. Studied under Ernst Haeckel at Jena. 1881 R sent CD a copy of his *Der Kampf der Theile im Organismus*, 1881. CD to Romanes, thought the book important, especially on the struggle of cell against cell within the body. LL3:244.

Rowland, David & Mrs R. Solicitor of Rowland and Hacon (William Mackmurdo Hacon). Friends from as early as 1833, many records in *ED's diary* when the Rowlands either dined at Maer, Down or were called on. Mrs R was one of the ladies whom ED recorded in her diary in Jan. 1840. All perhaps invited to see the three-week old baby, William.

Rowlett, George, 1796-1834. On 1st voyage in *Adventure*. Purser on 2d voyage of *Beagle*. He was, in his late 30s, the oldest officer aboard. 1834 Jun. R died aged 38 at sea. In the *Rio notebook*, Chancellor & van Wyhe, *Beagle notebooks*, 2009.

Royal Botanic Gardens, Kew, Surrey. First Director Sir William Jackson Hooker appointed in 1841; 2d Sir Joseph Dalton Hooker, his son, appointed in 1865. CD visited and received much plant material from, for his botanical work. 1874 Memorial presented to the First Lord of the Treasury [Gladstone], respecting the National Herbaria. In Fourth report of the Royal Commission on the scientific instruction and the advancement of science. *Parliamentary Papers*, paper number (884), vol. XXII.1, pp. 31-2 (*Shorter publications*, F1954). Signed by CD and 53 others.

Royal College of Physicians, England. 1879 Awarded CD the Baly Medal (medal established by F.D. Dyster).

Royal Medical Society of Edinburgh. 1826-27 CD Member whilst a medical student. 1861 CD Honorary Member.

Royal Society, See also **Philosophical Club**, **X Club**. Details of Darwin related Fellows in Freeman, *Darwin pedigrees*, 1984, p. 66. 1839 Jan. 24 CD was elected Fellow. 1849-50, 1855-56 CD Member of Council. 1853 CD Royal Medal. 1864 CD Copley Medal. 1877 CD to Romanes, who had failed to be elected in that year. Hooker (then President) had implied that age and position in scientific society weighed heavily, as did having been proposed many times. Youth is a disqualification. CCD25. 1882, 1903 There were three living Darwin Fellows briefly in the spring of 1882 and again 1903-12. 1890 Darwin Medal instituted, with residual funds from Darwin Memorial appeal. The effigy of CD is reduced from medallion by Allan Wyon. Until 1962 the Darwin family is the only one in the history of the Society to have had a continuous succession from father to son of Fellows, with no year without at least one fellow, from Erasmus Darwin [I], elected 1761, to Sir Charles Galton Darwin, died 1962 (*Darwin pedigrees*, p. 66). 1959 The succession continued, through the female line, to Richard Darwin Keynes, elected 1959 (d. 2010).

Royer, Mademoiselle Clémence-August Audouard, 1830-1902. French self-taught scholar who lectured and wrote on economy, philosophy and science. 1862 R translated *Origin* into French, adding a long preface and her own footnotes. 1862 CD to Gray, R "must be one of the cleverest & oddest women in Europe". CCD10:241. 1862 CD to Quatrefages, "I wish the Translator had known more natural history". CCD10:314. 1867 CD to Lyell, "Nevertheless with all its bad judgement and taste it shows, I think, that the woman is uncommonly clever". CCD15:355.

Royle, John Forbes, 1798-1858. Surgeon and naturalist. Originally in Medical Service in India. Secretary of Geological Society before CD. 1836-56 Prof. Materia Medica and Therapeutics King's College London. 1837 FRS. 1847 Sept. 1 CD to R, thanking for a book, perhaps *Illustrations of the botany...of the Himalayan mountains*, [1833-]1839[-1840]; "Long may our rule flourish in India." CCD4:63.

Ruck, Amy Richenda, 1850 Feb. 9-1876 Sept. 11. Daughter of Lawrence R. CD's daughter-in-law. Portrait in B. Darwin, *Green memories*, 1928, p. 14 and Evans ed., *Darwin and women*, 2017, p. 39. 1874 Jul. 23 Married Sir Francis Darwin as first wife. 1876 Sept. 11 A died in childbed. CD wrote "Poor Amy died; a most dreadful blow to us all", van Wyhe ed., 'Journal', (DAR158). R was staying at Holy Trinity, Corris, Gwynedd, Wales. 1876 Sept. 15 CD to G.W. Norman, "she was sweet and gentle". Francis Darwin had gone to North Wales for the funeral. CCD24:275.

Ruck, Lawrence, 1819-96. Of Pantlludw, near Machynlleth, Wales. Father of Mary Elizabeth and Amy Richenda. Magdalen College, Oxford. Married Mary Anne Matthews. The 3 sons who were at Clapham Grammar School with CD's sons were: 1 Col. Arthur Ashley (1847-1939), father of Amy Roberta; 2 Sir Richard Matthews (1851-1935); 3 Laurence Ethel, (1854-88), died young.

Ruck, Mary Elizabeth, 1842-1920. Daughter of Lawrence and Mary Anne R and sister of Amy Richenda. 1864 1 Married Robert Travers Atkin. 3 sons; 1 James Richard (Dick), 2 Walter S.G.D. (Dor), 3 Robert L, cousins of Bernard Darwin. 1868 spent time with ED. ED2:220. 1879 Jun 17-Jul 21, 3 sons stayed at Down. 1880 stayed also Jun. 19-Jul. 2. See Geoffrey Lewis, *Lord Atkin*, 1999. CCD27.

Rucker, Sigismund, 1809-75. Orchid grower of West Hill, Wandsworth, Surrey. R lent CD *Mormodes ignea*, "goblin orchid". Allan, *Darwin and his flowers*, 1977, p. 205.

Rudd, Sophia Helena, 1857-99. 1883 Married Rowland H Wedgwood as 1st wife.

Rüdinger, Nicolaus, 1832-96. German anatomist. 1876 CD to Lawson Tait, R had written to CD about regeneration of digits.

Rugby, Warwickshire. 1839 Jan. 29 CD and ED took a train to London from R after their wedding at Maer as far as train went. 1845 Sept. 15 slept at Rugby and the next day to Shrewsbury. 1852 Mar. 24 CD and ED brought along Henrietta and George, visited William Erasmus at Rugby School. Edward Meyrick Goulburn (1818-97) was headmaster when William Erasmus Darwin was there. 1855 Sept. 25 CD and ED stopped there on return from BAAS meeting at Glasgow.

Ruskin, John, 1819-1900. Art critic, draughtsman and social reformer. In a 22 Apr. 1837 letter to his father: "While we were sitting over our wine after dinner, in came Dr.[C.G.B.] Daubeny, one of the most celebrated geologists of the day—a curious little animal, looking through its spectacles with an air very *distinguée*—and Mr. Darwin, whom I had heard read a paper at the Geological Society. He and I got together, and talked all the evening." Collingwood, *The life of John Ruskin*, 1902, p. 61. 1879 Aug. CD met again in the Lake District. 1879 Aug. 12 lunched at his home, Brantwood, Coniston, but could not understand the Turners in R's bedroom. CD considered R's mind clouded. ED2:238. 1879 CD to Romanes, "We saw Ruskin several times, and he was uncommonly pleasant". CCD27. R also saw Dr Woodhouse, the Darwin's family dentist.

Russell, John Edward Arthur, 1825-92. Liberal Party politician. Member of Athenaeum. 1869 Aug. 29 called on the Darwins. 1878 Aug. 9 Lord and Lady A Russell called. 1869 Sept. 9 R wrote to Kate R, "We saw a great deal of the great Darwin at Holwood—he lives at Down not far off—a charming man, but with wretched health. Laura painted his picture—He is the greatest name in English science at this moment—the most universally known abroad." Recollection in: B. Russell & P. Russell eds., 1937. *The Amberly papers*, vol. 2, p. 450. Transcribed in *Darwin Online*.

Russell, Lord John, 1792-1878. 1848 CD signed a memorial to on reform at Oxford and Cambridge Universities, one of 224 signatories. F2080. See CCD18:348-9.

Russell, Laura, 1816-85. Daughter of Jules, vicomte de Peyronnet. 1865 married Arthur Russell MP. 1869 Painted portrait of CD in oils. The portrait is in private hands. Exhibited at Cambridge in 2009. Reproduced as frontispiece CCD17. See Randal Keynes, Portrait of Charles Darwin by Laura Russell. CCD17: Appendix V.

Russian. First editions in: 1846 *Coral reefs* (summary only, F320). 1860 Manual of scientific enquiry (CD's only, F336). 1864 *Origin* (F748). 1867-68 *Variation* (F925). Publication of the first parts preceded the English first edn. 1865 *Journal* (2383.1-2). 1871 *Descent* (F1107). 1872 *Expression* (F1206). 1876 *Insectivorous plants* (F1244). 1877 "Biographical sketch of an infant" (F1314). 1882 *Earthworms* (F1408). 1882 *Power of movement* (F1349). 1894 Essay on instinct (F1449). 1896 *Autobiography* (F1533). 1898 *Pangenesis* (F2511). 1900 *Orchids* (F825). 1900 *Climbing plants* (F865). 1932 *Foundations of the origin of species* (F1564). 1936 'Bar of sandstone off Pernambuco' (F270). 1936 *Coral reefs* (complete, F321). 1936 *Living Cirripedia* (F341). 1936 *Volcanic islands and South America* (F323). 1938 *Cross and self fertilisation* (F1272). 1935-59 *The collected works*, edited by S.L. Sobol' is by far the most comprehensive in a foreign language. 1939 "On the tendency of species to form varieties" (F370). 1948 *Different forms of flowers* (F1302). 1959 *Letters on geology* (F7). 1959 Memoir of Professor Henslow (CD's recollections only, F832). 1959 *Erasmus Darwin* (CD's notice only, F1324). 1971 On routes of the male humble bees (for children) (F1591).

Ruthin, Denbigh, Wales. 1831 Aug. CD visited with Sedgwick on geology trip. Van Wyhe ed., 'Journal', (DAR158), Sandra Herbert, *Charles Darwin, geologist*, 2005, p. 45.

Rütimeyer, Karl Ludwig, 1825-95. Swiss zoologist and palaeontologist. 1855-94 Prof. Comparative Anatomy Basel. 1867 CD to Lyell, R had sent him his pamphlet *Über die Herkunft unserer Thierwelt*, 1867, but CD had not read it. CCD15. 1868 R author of *Die Grenzen der Thierwelt: eine Betrachtung zu Darwin's Lehre, Zwei in Basel gehaltene Vorträge*. See DAR85.A84. CD had this pamphlet translated by Camilla Ludwig.

Ryan, Mary, 1848-1914. Julia Margaret Cameron's pretty Irish maid who often sat for her, known as "The Madonna". Married John Henry Stedman Cotton, who had fallen in love with Cameron's portrait of R. Cotton was knighted in 1902, making R Lady Cotton. 1868 Aug. 1 at Freshwater, CD and family met R. ED2:191.

Ryle, Jane Harriet, (Harriot), 1794-1866. 1815 Married Sir Francis Sacheverel Darwin. They had ten children.

Sabine, Sir Edward, 1788-1883. Anglo-Irish astronomer and physicist. Educated at the Royal Military Academy. Attained the rank of General Royal Artillery in 1870, saw little action, but went on several expeditions as naturalist. S was anti-Darwinian. 1818 FRS. 1849 CD to Hooker, CD had travelled with S to BAAS meeting at Birmingham. 1861-71 PRS. 1864 CD to Falconer, S had asked him to attend Royal Society to receive Copley Medal; CD did not attend. 1864 S's Presidential address to Royal Society about CD's Copley Medal, "Speaking generally and collectively, we have expressly omitted it [*Origin*] from the grounds of our award": a remark which caused much offence. 1866 Apr. 28 CD attended S's evening reception. 1869 KCB.

Sagitta. 1844 Observations on the structure and propagation of the genus *Sagitta*, *Annals and Mag. of Nat. Hist.*, 13: pp. 1-6 (*Shorter publications*, F1664) French, 1844 *Annals Sci. Natural Zool.*, 1: pp. 360-5. (F1664a). *Sagitta* are a genus of carnivorous marine arrow worms (Chaetognaths). CD's notes in DAR27.1.

St. Andrews, Fife, Scotland. 1827 CD visited on spring tour. 'Journal' (DAR158)

St. Chad's Church, Shrewsbury, Parish Church where CD was baptized and went as a child after the death of his mother. Built 1790-92.

St. Croix, Elizabeth de, 1790-1868. 1817 Married William Brown Darwin.

St. Helena, Atlantic Ocean. 1836 Jul. 7-14 *Beagle* at. CD stayed ashore four days "within a stone's throw of Napoleon's tomb". *Geological diary*: DAR38.920-935. *Despoblado notebook*, Chancellor & van Wyhe, *Beagle notebooks*, 2009. List of St Helena Coleoptera, Homoptera, Lepidoptera and Orthoptera in DAR29.3.34. See CD's notes on the "St Helena model" (9.1838) in DAR44.30[.1]. See Chancellor, Charles Darwin's St. Helena model notebook. *Bulletin BMNH Historical Series* (F1839). All of these transcribed in *Darwin Online*.

St Jago, (Santiago), Cape Verde Islands, Atlantic Ocean. 1832 Jan. 17-Feb. 8 *Beagle* at Porto Praya and CD landed. 1836 Aug. 31-Sept. 5 *Beagle* again at and CD landed.

St. Paul's Rocks, Atlantic Ocean. Uninhabited islets 590 miles from the coast of Brazil. With St. Peter form the St. Paul and St. Peter Archipelago, belonging to Brazil. 1832 Feb. 16-17 *Beagle* at and CD landed. *Geological diary* DAR32.37-38 & DAR34.196, transcribed in *Darwin Online*.

Sales, Sydney, c.1844-. 1881 Sept. 20 ED wrote to Henrietta "Frank and Leo went to Sydney Sales on Sunday about buying a bit of his field. He does not seem willing to sell, which will be a disappointment to me more than to them I think". ED2:318.

Sales, William, 1808-80. Grocer and victualler, proprietor of The Queen's Head and landowner at Down, west and north of Down House. A £50 cheque to S from CD, Sotheby's, 1979, Jun. 18, lot 467, Union Bank of London. 1843 CD bought an acre and a bit from him adjoining the Down House grounds. CD called it Sales field. 1845 CD wrote his sister Susan about "making a new walk in the kitchen-garden; and removing the mound under the yews, on which the evergreens we found did badly". 1861 Nov. 12 ED recorded "from Sales 2/6". 1872 "Mr. Sales would be sure to build some more ugly houses on it if he got the land".

Saliferous Deposits. 1838 Origin of saliferous deposits: salt lakes of Patagonia and La Plata, *Jrnl. Geol. Soc.*, 2: pp. 127-8 (F1651), pre-publication extract from *Geological observations on South America*, pp. 73-5. 1846 Origin of saliferous deposits. Salt lakes of Patagonia and La Plata, *Quart. Jrnl. of the Geol. Soc. (Proc.)*, 2: pp. 127-8 (F1673).

Salt, 1847 Salt, *Gardeners' Chron.*, no. 10 (*Shorter publications*, F1676). CD's reply to W.B.N. 1847. Native Patagonian salt. *Gardeners' Chron.* no. 8 (20 Feb.), p. 117.

Salt on Carbonate of Lime, 1844 What is the action of common salt on carbonate of lime?, *Gardeners' Chron.*, no. 37: pp. 628-9 (*Shorter publications*, F1668). The only response was by T.P.; Manures and drainage. *Gardeners' Chron.* no. 40 (5 Oct.): 675.

Salt-Water and Seeds, see Seeds, vitality of.

Salter, John William, 1820-69. Naturalist and palaeontologist. Apprenticed to James De Carle Sowerby in early career. 1846-63 Palaeontologist to Geological Survey, until 1854 under Edward Forbes. 1860 S showed CD a series of brachiopods from different geological epochs arranged in an evolutionary sequence at Museum of Practical Geology, Jermyn Street, London. CCD8 & 9.

Salter, Thomas Bell, 1814-58. Physician and botanist of Ryde, Isle of Wight. Nephew of Prof. Thomas Bell. 1855 S was sending seeds to CD for hybrid studies. Barlow 1967, p. 175, as Dr I. Bell Salter. CCD5:357; 360.

Sandown, Isle of Wight. 1858 17 Jul.-Aug. CD and family visited. CD began writing 'abstract' there which became *Origin*.

Sandwalk or **sand walk**. Path running around a small wood on the southeast corner of the grounds of Down House where CD took his daily walk. It took this name from a sandpit at the south end used for dressing the path. Also known as CD's "thinking path", however a source for CD calling it this has not been found. He did call it the "path". 1846 CD rented the 1½ acres from J.W. Lubbock for £1 12s per annum. See CD to J.W. Lubbock [16 Jan. 1846], CCD3:276. CD planted it with hazel, alder, lime, hornbeam and birch, with a dogwood privet and fence along the exposed side. 1855 His observations on plants growing along the sandwalk led to 'Vitality of Seeds' (*Shorter publications*, F1686). There was eventually a summer-house at far southeast end. ED2:76. An outing to the sandwalk by the convalescing Henrietta was noted in *ED's diary* in 1861. CD's number of circuits when he was ill was also noted by ED, with "4 times Sandwalk" meaning CD was feeling better. 1874 CD purchased it at the high price of £300. After his death, May 7 ED went in bath chair. The next day she walked there with Horace. ED may have taken her last walk in 1896 Jul. 19. CD took his last walk there on 7 Mar. 1882 when he experienced a seizure and made it back to the house with difficulty. LL3:357. Gwen Raverat's recollection of it is very detailed, Raverat, *Period piece*, 1952, p. 156. See Atkins 1974, pp. 25-6. 1996 paved by English Heritage to support larger numbers of visitors.

Sandys, **Sir John Edwin**, 1844-1922. Classical scholar. 1876-1919 Public Orator Cambridge. 1877 Nov. 17 S gave oration on CD's Honorary LL.D. "Tu vero, qui leges naturae tam docte illustraveris, legum Doctor nobis esto". LL3:222. 1911 Kt.

Sandstone, 1846 [Note on Sandstone and query on coral reefs.] In J.L. Stokes, *Discoveries in Australia*, 1:108, 331 (*Shorter publications*, F1915).

Sanskrit. First edition in: 1952 *Origin* (F2490).

 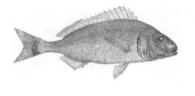

Darwin's fox [now *Lycalopex fulvipes*], found by CD on Chiloe. Plate 6 of *Mammalia*. "I was able, by quietly walking up behind, to knock him on the head with my geological hammer. This fox, more curious or more scientific, but less wise, than the generality of his brethren, is now mounted in the museum of the Zoological Society." *Journal*, p. 341.

Pristipoma cantharinum [now *Orthopristis cantharinus*] collected by CD in the Galapagos. Plate X from *Fish*.

Santa Fé, "St Fe" (de la vera Cruz), Argentina. 1833 Oct. 2-15 CD at. See *Beagle diary*, pp. 172, 194 and *St. Fe notebook*, Chancellor & van Wyhe, *Beagle notebooks*, 2009.

Santiago, Chile. 1835 Mar. 13 CD visited on his way from Valparaiso to cross the cordilleras to Mendoza. 1835 Apr. 10 CD returned through. See *Beagle diary*.

Santa Cruz, Rio, FitzRoy led a party from the *Beagle* including CD up this river in a vain attempt to discover its source in April 1834. During this expedition CD became convinced of the continental scale of the geological uplift he had discovered. See the *Beagle diary*, pp. 232ff, *Zoological diary*, pp. 230-1, *Geological diary*: DAR34.122-124. All transcribed in *Darwin Online*.

Saporta, Louis Charles Joseph Gaston, Marquis de, 1823-95. French palaeobotanist. S was a fairly frequent correspondent on botanical matters. 1863 CD to Lyell, S was pro-*Origin*. 1868 CD to S, about the growth of belief in evolution in France. 1878 CD to S, about his election to Académie des Sciences.

Sara. 1881 Nurse to Bernard Richard Meirion Darwin. ED2:246-47.

Sarcey, Francisque, 1827-99. French dramatic critic and journalist. Also known as "l'oncle" (the uncle). 1880 Lunched at Down House in summer with Edmond Barbier. S. 1879. Lettres de Londres: X. *Le XIX Siècle* (19 Jun.). Translated in F2001.

Savile, Bourchier Wrey, 1817-88. Church of England clergyman and theological writer. 1882 Extract from a letter by CD on the origin of mammals, *Shorter publications*, F2031, see *Calendar* 13366.

Scarlet fever, records of family infection: CD 1818 Sept., ED 1862 Aug. 12, William 1855 May 18, Annie 1849 Nov. 12, Henrietta, Bessy, 1849 Nov. 19, Charles Waring 1858 Jun., Leonard 1862 Jun. 12. *ED's diary.* CD's 'Journal', DAR158.

Schaaffhausen, Hermann Joseph, 1816-93. German anthropologist and paleoanthropologist, Prof. Anatomy University Bonn. 1853 Über Beständigkeit und Umwandlung der Arten, *Verhandl. Naturhist. Vereins, Bonn,* an evolutionary forerunner. 1858 Published on the anatomy of the Neanderthal discovered the previous year by Johann Carl Fuhlrott. 1860 S sent a copy of his 1853 article to Lyell.

Scherzer, Carl Heinrich, Ritter von, 1821-1903. Austrian ethnologist and printer. 1861-62 S edited *Reise der Österreichischen Fregatte Novara um die Erde.* 1868 CD to S, addressing him as "Ministerial Rath", thanking for translating *Queries about expression,* "and inserting". S inserted the *Queries about expression* in a pamphlet containing instructions for participants in expeditions to East Asia and South America. (F876a) CCD16. 1879 CD to S, "What a foolish idea seems to prevail in Germany on the connection between Socialism and Evolution through Natural Selection". LL3:237.

Schlagintweit-Sakünlünski, Hermann Rudolph Alfred von, 1826-82. German botanist, geologist and traveller. 1857 CD to Hooker, "I believe he is returned to England, and he has poultry skins from W. Elliot of Madras". CD spells "Schlagenheit". CCD6. 1857 Paper with his brother Robert von S-S., *Report British Association Advancement Science,* 27: pp. 106-8, 1857. CCD6; *Natural selection,* p. 438. 1860 CD to Lyell, "Do not trust Sclangenweit", about yak-cattle crosses. CCD8:403.

Schleiden, Matthias Jakob, 1804-81. German jurist, botanist and, with Theodor Schwann, founder of the cell theory. 1850 Prof. Botany Jena. 1863 Prof. Plant Chemistry Dorpat. 1864 CD to Benjamin Dann Walsh, S was, with other Germans, coming round to belief in natural selection.

Schmerling, Philippe-Charles, 1790-1836. Dutch/Belgian geologist and palaeontologist. 1829 Discovered first Neanderthal fossils. 1863 CD to Hooker, concerning antiquity of man, "Falconer...does not do justice to...Schmerling". CCD11:31.

Schomburgk, Sir Robert Hermann, 1804-65. German born traveller, naturalist and diplomat. Travelled widely in South America, the West and East Indies. 1837 Discovered the *Victoria regia* (Queen Victoria's water lily) in British Guyana, described and named in honour of Queen Victoria in 1837. 1844 Kt. 1848 S visited Down House for weekend. CCD4, letter to George Robert Waterhouse. 1859 FRS. 1862 S told CD about the three forms of *Catasetum tridentatum* which had been described as belonging to three different genera. *Jrnl. Lin. Soc. (Bot.)*, 6: pp. 151-7. (F1718) 1864 Retired form public service and returned to Germany.

Schultze, Fritz Karl August Julius, 1846-1908. German philosopher. Prof. Technische Hochschüle Dresden. 1875 CD to S, thanking him for copy of *Kant und Darwin*, Jena 1875. CCD23.

Schweizerbart'sche Verlagsbuchhandlung, Publisher of Stuttgart, Germany. 1860-82 S published first German editions of eleven of CD's books. Also 2^d edn of *Journal*, 1875, and *Life and letters*, 1887-88, and many other editions of CD's works including the 1876-77 and 1899 *Gesammelte Werke* (Collected Works).

Science Defence Association. 1881 CD to T. Lauder Brunton. The Science Defence Association was formed as a result of prosecution of David Ferrier. CD's name was put forward as President, but he declined. ML2:439.

Scientifics, Perhaps a Working Mens' Club. Atkins 1976, p. 85. 1880 Fifty members visited Down House and were entertained with claret-cup, wine and biscuits. Francis Darwin talked to them, but CD did not appear.

Sclater, Philip Lutley, 1829-1913. Lawyer, ornithologist and animal geographer. 1859 Founder and editor of the journal *The Ibis*. 1860-1902 Secretary to Zoological Society of London. 1860 CD to S, thanking for list and notes on Galapagos Islands birds. CCD8. 1861 FRS.

Scoresby, Rev. William, 1789-1857. Anglican clergyman, whaler and arctic naturalist. 1824 FRS. 1839 DD. 1855 CD corresponded with S on seed transport.

Scotland. 1825 Oct. 22-1827 Apr. c.24 CD at Edinburgh University. Apart from his time at university CD made two tours in Scotland: 1827 Apr.-May On leaving University, Dundee, St. Andrews, Stirling, Glasgow and from there to Belfast. 1838 Jun. By boat from London to Leith, Edinburgh. Loch Leven, Glen Roy, Glasgow on a geological trip. 1855 CD and ED went to BAAS meeting in Glasgow.

Scott, John, 1838-80. Botanist and gardener. Correspondent with CD from 1862 on botanical matters. "The only naturalist who can be described as a pupil of Darwin's" [sic]. Poulton, *Darwin and the Origin*, 1909, p. 53. 1859-64 On staff at Royal Botanic Garden Edinburgh. He left Edinburgh "at what...he considered discouragement and slight". 1862 CD to Hooker, "I have been corresponding largely with him: he is no common man". CCD10:598. 1863 CD to S, "I cannot help doubting...whether you fully comprehend what is meant by Natural Selection". CCD11:213. S's name appears on the presentation list for 'Two forms...*Linum*' (F1723). 1864 CD to Hooker, "I have paid the poor fellow passage out to India". CCD12:319. CD had thought of employing him at Down House, and suggested that Hooker take him on at Kew. S left for Calcutta at the end of the year without having secured a post. 1864 S visited Down. 1865- Curator Botanic Garden Calcutta. 1867 Brief biography by Sir George King, "shy and modest almost to being apologetic", "almost morbidly modest". ML1:217. 1867 CD to Hooker, he had had a nice letter from S on acclimatization. 1871 S offers to repay his fare. CD replies strongly that he is "a rich man" had given it as a present, not as a loan. 1873 FLS. S helped with *Expression*, p. 21, "The habit of accurate observation, gained by his botanical studies, has been brought to bear on our present subject". 1877 Became an expert on opium husbandry, *Manual of opium husbandry*, Calcutta. See CCD11-24.

Seale, Robert F., geologist, who had a museum on St. Helena. 1834 *Geognosy of the island of St Helena* read by CD. In the *Sydney notebook*, Chancellor & van Wyhe, *Beagle notebooks*, 2009. Gordon Chancellor, Charles Darwin's St. Helena model notebook. *Bulletin BMNH Historical Series* (F1839). MS notes in DAR44.30.

Sea-Water and Seeds. CD carried out many experiments to prove that many plant seeds remain viable after prolonged submersion in salt water in order to explain the occurrence of related species in different islands and continents. See also Seeds. 1855 Does sea-water kill seeds?, *Gardeners' Chron.*, no. 15: p. 242 (*Shorter publications*, F1682). 1855 Does sea-water kill seeds?, *ibid.*, no. 21: pp. 356-7 (*Shorter publications*, F1683). 1855 Effect of salt-water on the germination of seeds, *ibid.*, no. 47: p. 773 (*Shorter publications*, F1687). 1855 Effect of salt-water on the germination of seeds, *ibid.*, no. 48: p. 789 (*Shorter publications*, F1688). 1857 On the action of sea-water on the germination of seeds, *Jrnl. Proc. of Lin. Soc. of London (Bot.)*, 1: pp. 130-40 (*Shorter publications*, F1694). Notes in DAR27.1.

Sedgwick, Rev. Adam, 1785-1873. Geologist. Defined the Devonian and Cambrian periods. Biography: Clark and Hughes, 1890. 1818-73 Fellow of Trinity College and Woodwardian Prof. Geology Cambridge. 1821 FRS. 1831 Aug. CD made geological tour in North Wales with S. CD's notes of the tour are in DAR5. Walking tour in North Wales. In *Life and letters of the Reverend Adam Sedgwick*, vol. 1, pp. 379-81 (*Darwin Online*). 1833 Wollaston Medal Royal Society. 1859 CD sent S 1st edn of *Origin*. 1859 S to CD, "I have read your book with more pain than pleasure. Parts of it I admired greatly; parts I laughed at till my sides were almost sore; other parts I read with absolute sorrow; because I think them utterly false & grievously mischievous—You have *deserted*—after a start in that tram-road of all solid physical truth—the the true method of induction—& started up a machinery as wild I think as Bishop Wilkin's locomotive that was to sail with us to the Moon." CCD7:396. 1859. 1860 S spoke to Cambridge Philosophical Society, reported in *Cambridge Chron.*, May 19: "Darwin's theory may help to simplify our classifications...but he has not undermined any grand truth in the constancy of natural laws, and the continuity of true species". 1863 Copley Medal Royal Society. 1870 CD called on S at Cambridge in May. Letter from S, "I was overflowing with joy when I saw you". CCD18.

Sedgwick, Sarah Price Ashburner, 1839-1902. Of Cambridge, Massachusetts, USA. Sister of Theodora S. CD's daughter-in-law. Friend of Chauncey Wright. LL3:165. Letters of Chauncey Wright, pp. 246-8. "She was the kindest of the kind but a little formidable...Sedgwicks, Eliots and Nortons are not to be lightly encountered". B. Darwin, *Green memories*, 1928, p. 42. 1877 Nov. 29 Married William Erasmus Darwin. ED wrote often to S with the last letter perhaps 1896 Sept. 19.

Sedgwick, Theodora Price Ashburner, 1851-1916. Sister of Sarah Price Ashburner S. 1878 S visited Down House and Bassett (home of William Erasmus Darwin). 1884 S visited The Grove, Cambridge.

Seeds. CD's collections of seeds were deposited in the Botany School, Cambridge, now the Herbarium of the Department of Plant Science, University of Cambridge.

Seeds. 1855 Vitality of seeds, *Gardeners' Chron.*, no. 46: p. 758 (*Shorter publications*, F1686). 1855 Effect of salt-water on the germination of seeds, *ibid.*, no. 47: p,773 (*Shorter publications*, F1687). 1855 Effect of salt-water on the germination of seeds, *ibid.*, no. 48: p. 789 (*Shorter publications*, F1688). 1855 Longevity of seeds, *ibid.*, no. 52: p. 854 (*Shorter publications*, F1689). 1857 On the action of sea-water on the germination of seeds, *Jrnl. Proc. of Lin. Soc. of London (Bot.)*, 1: pp. 130-40 (F1694). 1857 Productiveness of foreign seed, *Gardeners' Chron.*, no. 46: p. 779 (F1698).

Seedling Fruit Trees. 1855 Seedling fruit trees, *Gardeners' Chron.*, no. 52: p. 854 (*Shorter publications*, F1690). Refers to French botanist C.T.A. Jordan (1814-1897).

Self-Fertilisation. See also *Cross and self fertilisation*, 1876. 1865 Self-fertilisation, *Hardwicke's Sci. Gossip*, 1: p. 114. Only a quotation from *Orchids*, p. 359.

Selling, Alfred Mauritz, 1847-1932, Swedish doctor, medical writer. 1871 translated *Origin* into Swedish (F793). See Eve-Marie Engels and Thomas F. Glick eds., *The reception of Charles Darwin in Europe*, 2008.

Semper, Carl (or Karl) Gottfried, 1832-93. German zoologist and ethnologist. Prof. Zoology Würzburg. 1876 *Der Haeckelismus in der Zoologie*, Hamburg. 1878 CD to S, on speciation in relation to isolation, "I shd. think nearly perfect separation would greatly aid in their "specification,"—to coin a new word." CCD26:454. 1878 CD to S, on variation; S was strongly in favour of direct action of environment. 1879 CD to S on Coral formations in Pellew Islands. CCD27. 1878 CD gave S a typewriter. 1879 S sent CD proof sheets of *Die natürlichen Existenzbedingungen der Thiere*, 1880. 1881 Translated to English. 1881 CD to S, on variation, "the even still kinder manner in which you disagree with me". ML1:391.

Serbian. First editions in: 1878 *Origin* (F766). 1937 *Autobiography* (F1542).

Serbo-Croat. First editions in: 1949 *Journal* (F244). 1977 *Descent* (F2421).

Settegast, Hermann Gustav, 1819-1908. German agronomist. Boschetti, Federico, *Darwin-Settegast (trasformisti), Linneo-Sanson (non Trasformisti), e le Leggi dell' Ereditarieta*, Turin, 1890. 1870 S sent CD a copy of his book *Die Thierzucht*, Breslau 1868.

Sevenoaks, Kent. 1844 Jul. 6-17 the family went with the Langtons and Sarah Elizabeth W. 1862 Jul. 19 ED wrote "children went to seven oaks". CD and ED stayed home as CD was ill. Aug. 1 ED went there. 1877 Oct. 5-26 ED persuaded CD to leave home for a few weeks. She sent her 3 sons in "different directions to look for a house". Horace found an "uncommonly pretty" one with Knole Park close at hand.

Seward, Anna, 1747-1809. Poet and author. Known as the Swan of Lichfield. Generally considered that S wanted to marry Erasmus Darwin after death of his first wife and was chagrined when he married Elizabeth Chandos Pole. 1754-1809 Lived at The Swan, Lichfield. 1804 S was author of *Memoirs of the life of Dr. Darwin*. 1810 Sir Walter Scott edited her *Poetical works*. 1879 CD to Reginald Darwin, his cousin, he had written his introduction to Krause's biography of Erasmus Darwin [I] "to contradict flatly some calumnies by Miss Seward". CCD27:136.

Sex Ratios. 1868 [Inquiry about sex ratios in domestic animals], *Gardeners' Chron.*, no. 7, p. 160 (*Shorter publications*, F1743). CD asked for information on "proportional number of males and females born to our various domestic animals".

Sexual selection. Evelleen Richards suggested that CD's earliest datable use of the phrase was in 1856 reading notes. Richards, *Darwin and the making of sexual selection*, 2017. 1858 phrase first published in Darwin-Wallace paper. 1871 "Sexual selection" forms Part II of *Descent*, chapters VIII-XXI, more than half the book. 1876 Sexual selection in relation to monkeys, *Nature*, 15: pp. 18-19 (*Shorter publications*, F1773), reprinted in *Descent*, 12th thousand, 1877, and later editions. 1880 The sexual colours of certain butterflies, *ibid.*, pp. 21, 237 (*Shorter publications*, F1787) 1882 "On the modification of a race of Syrian street-dogs by means of sexual selection", by Dr. [W.] Van Dyck, with a preliminary notice by Charles Darwin, *Proc. of Zool. Soc. London*, no. 25: pp. 367-70 (*Shorter publications*, F1803). One of CD's last publications. It was read by Sclater, secretary of the Zool. Soc. on 18 April; CD died the next day. R.A. Richards, Sexual selection. In *Cambridge encyclopedia of Darwin*, 2013.

Seymour (St. Maur), Edward Adolphus, Duke of Somerset, 1775-1855. Landowner and amateur mathematician. S was influential in obtaining money from the Exchequer for publication of scientific results of *Beagle* voyage. 1797 11th Duke. 1797 FRS. 1834-37 President of the Linnean Society. 1837 KG.

Seymour, Gertrude, 1784-1825. 1812 Married John Hensleigh Allen [I].

Shaen, Margaret Josephine, 1852-?. Daughter of William S. Family friend. Photographed ED, aged 88, in drawing room at Down House. ED2, list of illustrations.

Shaen, William, 1821-87. Solicitor and educationalist. Unitarian. 1851 Sept. 2 Married Emily Winkworth. Two sons, John and Godfrey, one daughter Margaret. 1882 S was on "Family Friends invited" list for CD's funeral.

Shanklin, Isle of Wight. 1858 Jul. 21-Aug. 19 CD and family stayed in the village after the death of Charles Waring to avoid infection with scarlet fever. CD's sister Marianne Parker died three days before their trip. CD started writing *Origin* there.

Sharpe, Daniel, 1806-56. Geologist. S was in Portuguese mercantile business. 1846-51 CD to S, on cleavage and foliation. 1849 CD to Lyell, CD had been discussing mica schist with S. 1850 FRS. 1856 described CD's specimens of Palaeozoic Mollusca *Trans. of the Geol. Soc. of London* (2) 7:206-215. In *Darwin Online*.

Sharples, Ellen née Wallace, 1769-1849. English painter specialized in pastel and watercolour portraits. Mother of Rolinda S. 1787 Married to James S (1751/2-1811). 1816 Pastel of CD with sister Emily Catherine is always said to be by "Sharples".

Shaw, Henry, 1812-87. Notable taxidermist of Shrewsbury. 1829 CD to William Darwin Fox. CD shot a Dun diver and had it stuffed by Shaw. CCD1.

Shaw, James, 1826-96. Scottish writer and schoolmaster in Tynron, Dumfriesshire. A supporter who corresponded with CD during the 1860s. The editors of the CCD singled out a letter from S as "probably the most enthusiastic encomium of the year [1865]". CCD13:xxi. 20 Nov. S to CD: "Newton's ocean of wonder on whose shore he gathered shells seems no longer a dark unnavigable sea crested with howling breakers. Here and there through its awful bosom the strong vision with which you have supplied us enables us to see at least that there are other islands whose conditions are explicable by our own." CCD13:312. CD to S "I quite agree with what you say on the beauty of birds, and the same view may be extended to butterflies and some other beings. I think I can show that the beauty of flowers and of many kinds of fruit is solely to attract, in the former case, insects for their intercrossing, and in the latter case, to attract birds for the dissemination of the seed." CCD13:316.

Shaw, Joseph, 1786-1859. 1803 Admitted as a Sizar at Christ's College, Cambridge. 1807 B.A. 1810 M.A. 1807-49 Fellow of Christ's. Tutor until 1829. Liked hunting and the Newmarket races. 1827 Oct. 15 CD "Admissus est pensionarius minor sub Magistro Shaw". See van Wyhe, *Darwin in Cambridge*, 2014.

Shaw-Lefevre, John George, 1797-1879. British barrister, Whig politician and civil servant. A founder of the Athenaeum Club. FRS 1820. Among founders of UCL. *ED's diary* for 1871 Jun. 18 shows he paid a visit.

Sheep. 1880 Black sheep, *Nature*, 23: p. 193 (*Shorter publications*, F1790), extract from a letter by John Sanderson about selective value of black sheep in Australian flocks.

Shell Rain. 1855 Shell rain in the Isle of Wight, *Gardeners' Chron.*, no. 44: pp. 726-7 (*Shorter publications*, F1685). CD asked for particulars of shells falling during rain.

Shells. 1878 Transplantation of shells, *Nature*, 18: pp. 120-1, introducing a letter from Arthur H. Gray, *ibid.*, p. 120 (*Shorter publications*, F1783). 1880 The Omori shell mounds, *Nature*, 21: p. 561, introducing one from Edward S. Morse, *ibid.*, p. 561 (*Shorter publications*, F1788) 1882 On the dispersal of freshwater bivalves, *Nature*, 25: pp. 529-30 (*Shorter publications*, F1802).

Shrewsbury School, 1798-1836 Samuel Butler [I] was Headmaster. 1818 Summer term-1825 Jun. 17 CD there. He boarded even though the school was hardly more than a mile from The Mount, his home. The old school is now the borough library. 1882 The new school was first occupied. The school owns Richard Owen's copy of 1st edn of *Origin*, a copy of *Coral Reefs* 2d edn given by CD to Leonard Darwin inscribed "Leonard Darwin From his Father." with a note by Leonard: "The above being in my father's handwriting, I am glad to present this volume to the library of his school on the occasion of the opening of the new laboratories Leonard Darwin. Aug. 1938" and CD's annotated 1822 school atlas Dr Butler's *Ancient geography*. See Pattison, *The Darwins of Shrewsbury*, 2009.

Shuttleworth, Sir James Phillips Kay-, Bart, 1804-77. Physician, politician and educationalist. CD knew him (then James Phillips Kay) at Royal Medical Society Edinburgh. 1849 1st Bart.

Simcox, S is mentioned as someone CD rode with at Cambridge. CCD1. S is not in Venn, 1922-54. *Alumni Cantabrigienses* 1752-1900. According to CCD1 "Possibly a familiar name for George Simpson (see letters to William Darwin Fox, [18 May 1829], 'Simpson', [3 Jan. 1830], 'old Simpcox')".

Simon, Sir John, 1816-1904. Surgeon and pathologist. Medical Officer to Privy Council. 1845 FRS. 1875 CD to Hooker; S saw and agreed to Litchfield's draft sketch for a vivisection bill. CCD23. 1879-80 PRS. 1881 CD to Romanes praising his address on vivisection to International Medical Congress. LL3:210.

Simon's Bay, Cape of Good Hope, South Africa. 1836 May 31 *Beagle* anchored at.

Simonde de Sismondi, Jean Charles Léonard, 1773-1842. Swiss historian. Home at Chêne, near Vevey, Canton of Geneva, Switzerland. 1819 Married Jessie Allen d.s.p. 1840 Jun. 2-18 S and Mrs S stayed at 12 Upper Gower Street, CD and ED away most of the time. They stayed again Aug. 22-28. 1842 Sept., Jessie S said S wished ED to have his Miltons. ED2:54.

Simonde de Sismondi, Jessie, see Allen, Jessie.

Simpson, George, 1809?-88. M.A. Curate of Luddenham, Kent. 1831 wrote to congratulate CD on his *"very very good degree"*. CCD1:113.

Skertchly, Sydney Barber Josiah, 1850-1926. Naturalist, geologist and traveller. Assistant curator Geological Society London. 1878 S sent CD copy of his *Geology of the Fenland* (1877). CCD25. 1878 CD sent S copy of *Origin* "with my autograph". CCD25. 1891 Settled in Brisbane, Australia.

Sketches of 1842 and 1844. CD's earliest essay drafts of his evolutionary views. Neither was published in his lifetime, except a fragment from the 1844 essay published in the 1858 Darwin-Wallace paper. 1887 Sketch of 1842 was not known to Francis Darwin when he edited *Life and letters*. 1896 It was found in a staircase cupboard after ED's death. 1909 Both were published in *Foundations* edited by Francis Darwin. A recent myth has arisen that CD hid one or both of the essays under the stairs at Down House. This is incorrect, they were part of his working library before the publication of *Origin* and would only have been stored away when no longer needed after *Origin* or after his death. The museum at Down House even has a parcel in the cupboard under the stairs with a note written on it: "Only to be opened in the event of my death." a reference to the extremely common misreading of the memo to ED in which CD asked her to publish the 1844 essay in case he died unexpectedly, not that he wished it only published after his death. The MS for the sketches are in DAR6-7, the fair copy of the 1844 essay is in DAR113.

1. 1909 Francis Darwin, ed. *The foundations of The origin of species, a sketch written in 1842*, Cambridge University Press, not published, issued to delegates to the anniversary celebrations at Cambridge (F1555). Facsimile 1969 (F1559). In *Darwin Online.*

2. 1909 Francis Darwin, ed. *The foundations of The origin of species, two essays written in 1842 and 1844*, pp. 1-53, the sketch of 1842, from same setting of type as no. 1 (F1556). Transcribed in *Darwin Online.* 1958 G.R. de Beer, ed. *Evolution by natural selection*, contains both drafts (F1557), issued for the XVth International Congress of

Zoology and for the Linnean Society (Darwin Centenary). 1971 Facsimile (F1560). First foreign editions: 1911 German (F1561). 1925 French (F1560a). 1915 Japanese (F1563a). 1932 Russian (F1564). 1960 Italian (F1562).

Skimp. Family nickname of Horace Darwin at the age of ten. ED in a letter to Henrietta 1879 Jan. "I am sorry to say that it is a severe case of 'capers'...however Skimp [Horace] shall order the smoking-room one at Maple's." ED(1904)2:294-5

Skinner, John, 1841-1903. Coachman at Down House, late 1880s. 1867 Married Helen Nightingale. His son Frank worked in the gardens. Came to Cambridge with ED for the winters; "soothing and tranquil rather than exciting company, as tranquil as the horses he drove", "John's kingdom was the harness room and his carriage whips were magic wands" B. Darwin, *Green memories*, 1928, p. 13. 1884 Aug. at Down, Bernard Darwin had a long game of cricket with Frank S. and Arthur.

Skinner, Albert J., son of John S. Letter of reminiscences about CD and Down House In L.F. Abbot, *Twelve great modernists*, 1927, pp. 247-9, in *Darwin Online*.

Slaney, Elizabeth Frances, 1813-70. Eldest daughter of Robert Aglionby S. 1835 Married Thomas Campbell Eyton.

Slaney, Robert Aglionby, 1791-1862. Barrister and politician. Father of Elizabeth Frances S. 1826-62 MP for Shrewsbury four times. 1854 High Sheriff of Shropshire.

Slavery. CD, like his grandfather Erasmus Darwin [I] and all educated Whigs, was against slavery, CD especially so from his experience of it in South America. 1791 Josiah Wedgwood [I]'s cameo of a kneeling slave in chains, with inscription "Am I not a man and a brother" is illustrated in Erasmus Darwin [I], *The botanic garden*, Pt. 1, facing p. 87, with note "a Slave in chains, of which he distributed many hundreds, to excite the humane to attend to and to assist in the abolition of the detestable traffic in human creatures". 1826 Anti-slavery agitation by Josiah Wedgwood [II] and his family detailed. ED1:181. 1833 CD to his sister Caroline, "What a proud thing for England if she is the first European nation which utterly abolishes it". CCD1:312. 1833 CD to Herbert, "Hurrah for the honest Whigs! I trust they will soon attack that monstrous stain on our boasted liberty, Colonial Slavery". CCD1:320. 1833 CD at Rio de Janeiro, "On such fazêndas as these, I have no doubt the slaves pass happy and contented lives". *Journal* 2ᵈ edn, p. 24. "This man had been trained to a degradation lower than the slavery of the most helpless animal". *ibid.* "I thank God, I shall never again visit a slave country", followed by two pages of description of its horrors. *ibid.* p. 499. 1845 CD to Lyell, "this odious deadly subject". CCD3:242. 1861 Many of CD's letters to Asa Gray refer to slavery in relation to the American civil war, e.g. 1861 "If abolition does follow with your victory, the whole world will look brighter in my eyes & in many eyes." CCD9:266. Anti-slavery feelings played no part in CD's evolutionary theorizing or motives for theorizing.

Sliding board, a long slide of polished wood with sides and a small flange at one end by which it can be hooked to any step of the staircase. Built by John Lewis for CD's children to play on rainy days. Bernard Darwin at five, conquered the fear of sliding down six steps standing up. F. Darwin, *Story of a childhood*, 1920, p. 42. Keynes, *Annie's box*, 2001, pp. 127-28, Browne, *Power of place*, 2002, p. 492. Photograph in Reeve, *Down House*, 2009, p. 8. Now at Down House.

Slingsby, Sarah Monica, 1900-87. 1926 Married William Robert Darwin.

"Slip-slop, Little Miss", Nickname of Emma Wedgwood (ED) in childhood.

Slovak. First editions in: 1959 *Autobiography* (F2426). 2006 *Origin* (F2154).

Slovene. First editions in: 1950 *Journal* (F248). 1950 *Descent* (F1122). 1951 *Origin* (F768). 1959 *Autobiography* (F1543).

Smith, Albert George Dew-, née Dew, 1848-1903. Physiologist of Trinity College Cambridge. Skilled photographer and collector of jewels. 1895 Married Alice Lloyd s.p. Friend of Horace Darwin and Director of Cambridge Scientific Instrument Company, a company S founded with Horace Darwin. 1873 CD to S, about physiology of *Dionaea*. CD had given all his best specimens to J.S.B. Sanderson. CCD21. 1882 S was on "Personal Friends invited" list for CD's funeral.

Smith, Sir Andrew, 1797-1872. Physician, naturalist and explorer. Army surgeon and zoologist. 1831 CD to Henslow, asking for an introduction to. 1836 CD met at Cape of Good Hope. CCD1. S provided CD with rock samples he had collected in 1835; these samples are now at the Sedgwick Museum in Cambridge. In the *Despoblado notebook*, Chancellor & van Wyhe, *Beagle notebooks*, 2009. 1837 CD met in London and "took some long geological rambles". *Beagle diary*, p. 409. 1838-50 *Illustration of the Zoology of South Africa*, 5 vols. 1849 CD to Strickland, about use of author's names in nomenclature which others, including S in conversation, were against. 1853-58 Director General Army Medical Services. 1857 FRS. 1858 KCB.

Smith, Beatrice Ann Shore, 1835-1914. 1865 Married Godfrey Lushington.

Smith, Edgar Albert, 1847-1916. Zoologist. Son of Frederick Smith. 1867-1913 Assistant Keeper Zoological Department BMNH. 1869 CD thanks S for proofs of excellent woodcuts for *Descent*. CCD17.

Smith, Catherine Amelia, née Pybus, 1800 Married Rev. Sydney Smith. Called in *ED's diary* Old Mrs Smith. Fanny Allen described her as affectionate but very unwell and infirm.

Smith, Edmund, 1804-64. Physician and surgeon. 1858-64 Proprietor of Ilkley Wells Hydropathic Establishment. 1859 CD felt that "he constantly gives me impression, as if he cared very much for the Fee & very little for the patient". CCD7.

Smith, Elder & Co., Publishers of London. Founded by George Smith and Alexander Elder. 1838-43 S published *Zoology of the Beagle*. 1842-44-46 published the three parts of CD's *Geology of the voyage of the Beagle*, and later editions until 1891.

Smith, Frederick, 1805-79. Hymenopterist at British Museum. Specialist on ants. Friendly correspondent of CD. 1849 Zoology Department British Museum. 1872 S gave CD information on copulation of bumble-bees. *Bulletin BMNH. Historical Series*, 3: p. 179, 1969. 1875 Assistant Keeper Zoology Department British Museum.

Smith, Gilbert Nicholas, 1796-1877. Vicar of Gumfreston near Tenby. 1840 CD questions S on remains found in caves on Caldy and requested bird's beaks. CCD2.

Smith, Goldwin, 1823-1910. Historian and journalist. 1858-66 Regius Prof. Modern History Oxford. 1861 Jan. ED recorded "Goldwin Smith lectures on the study of History". 1868 CD to Hooker; S had lunched at Down House with the Nortons.

Smith, James, 1782-1867. Geologist, merchant, man of letters and sailor. Of Jordanhill, Glasgow. 1830 FRS. 1848 CD to Lyell says S had a poor opinion of Chambers's *Ancient sea margins*, 1848. CCD4:151-2.

Smith, John, 1798-1888. Botanist. Specialist on ferns. 1822 Started at Kew as stove boy but was soon promoted to take responsibility of tropical glass houses. 1841- First curator Royal Botanic Gardens, Kew. 1873 CD to Hooker, to ask S about watering plants during sunshine.

Smith, John, a farmer at Down Court with Josiah S. 1854 Dec. 31 ED recorded payments to S, 1861 Nov. and 1868 Sept. 2 CD to J.B Innes informing him of S's delicate health. CCD16. 1869 Dec. 14 ED recorded "called on... & Smith D. Court".

Smith, Julia, aunt of Elinor Mary Bonham-Carter. Lived near Down. 1872 Apr. 19 visited Down House with Elinor. Again in 1875 Apr. 13.

Smith, Saba, Lady Holland, 1802-66. Daughter of Sydney [I] S. Wrote memoir of her father in 1855. 1834 Married as 2ᵈ wife Henry Holland.

Smith, Rev. Sydney [I], 1771-1845. Writer, Anglican clergyman and wit. Canon of St. Paul's Cathedral, London. Member of Holland House set and friend of Wedgwoods, Allens and E.A. Darwin. Knew the Wedgwoods well and had visited Maer. Many references to in *Emma Darwin*. CD "I once met Sydney Smith at Dean Milman's house". *Autobiography*, p. 110.

Smyth, Robert Brough, 1830-89. Australian geologist and mining engineer. Correspondent of Adam Sedgwick. 1852 Emigrated to Australia. 1867 S answered CD's *Queries about expression*, p. 20.

Smyth, Capt. William Henry, 1788-1865. RN, Vice-President of the Royal Society, Late of the Hon. East India Company's Bengal Civil Service. CD met once and mentioned S in *Coral Reefs*, p. 158.

Snelgar, Margaretta, 1818-1919. c.1840 Married John Hensleigh Allen [II].

"Snow", Nickname for Frances Julia Wedgwood.

Snow, Mrs, Listed 1882 amongst "Personal Friends invited" to CD's funeral.

Snow, George, Coal dealer and carrier from London to Down, from Nag's Head public house, Borough of Westminster, at least from 1849-55.

Snow, George, 1811-85. Road surveyor. Lived in Down for 30 years. 1863 A framed letter in saloon bar of George & Dragon signed by CD and others recommending him for post of District Surveyor, 1863. Buried Downe churchyard.

Snowdon, Mountain, North Wales. 1826 Jan. CD climbed on walking tour.

"Soapy Sam". Nickname of Rev. Samuel Wilberforce.

Somerville, Mary, see Fairfax, Mary.

Somerville, William, 1771-1860. Physician and army surgeon. 1812 Married as second husband Mary Fairfax. 1817 FRS.

Soper, William, c.1831-. A constable at Down between 1858 and the mid-1860s? Listed as "police" in 1861 census. Randal Keynes suggested (*Annie's box*, 2001, p. 312) that S may have been the constable interviewed by G. Foote, *Darwin on God*, 1889, p. 20: "I was informed by the late head constable of Devonport, who was himself an open Atheist, that he had once been on duty for a considerable time at Down. He had often seen Darwin escort his family to church, and enjoyed many a conversation with the great man, who used to enjoy a walk through the country lanes while the devotions were in progress."

Sorby, Dr Henry Clifton, 1826-1908. Geologist and microscopist. 1857 FRS. 1874 Royal Medal Royal Society. 1880 S presented address to CD from Yorkshire Naturalists' Union. LL3:227. (See *Shorter publications*, F1969).

Sorrell, Thomas, 1797-? Was on all three *Beagle* voyages. 1832 Jul. Acting Boatswain on 2ᵈ voyage of the *Beagle*.

South America, Part 3 of Geology of the Voyage of the Beagle. 1846 *Geological observations on South America. Being the third part of the geology of the voyage of the Beagle, under the command of Capt. FitzRoy, R.N., during the years 1832-1836,* London (F273). Notes for this are in DAR42. The book contains descriptions of tertiary fossil shells by G.B. Sowerby, and descriptions of secondary fossil shells by Edward Forbes. The scarcest of the three geology volumes. 1851 Combined edn with the two other parts from unsold sheets (F274). 1876 2ᵈ edn, combined with *Volcanic islands* (F276). 1891 3ᵈ edn (F282), only a reprint of 2ᵈ edn. 1972 Facsimile from second USA edn of 1896 (F307). First foreign editions: 1877 German (F312). 1891 USA (F283). 1906 Spanish (F324). 1936 Russian (F323). See Herbert, *Charles Darwin, geologist.* 2005. List of presentation copies of *Geological observations* 2ᵈ edn is in CCD24 Appendix III.

South American Missionary Society. The mission stems from the station on Keppel Island, West Falkland Islands. LL3:127. See also Thomas Bridges. 1885 Apr. 24 Admiral Sulivan, to *Daily News*, CD subscribed to their orphanage at the Mission Station, Tierra del Fuego, 1867-81 and saw the *Missionary Jrnl.* for 1867, although he had at first regarded the task as hopeless. 1882. [Quotation from a letter on civilizing the Fuegians]. *Leisure Hour.* F2022.

Southampton, Hampshire, 1846 Sept. 9-16 CD and ED attended BAAS meeting at. The whole family would go to S starting 1861. 1868-80 CD and ED visited their son William at Bassett, outside Southampton, in most years. Rough notes on Southampton gravel, 1874, in DAR52. See "drift deposits near Southampton". F1351.

Southey, Robert, 1774-1843. 1813-43 Poet Laureate. 1839 CD met S with Thomas Butler on stage coach from Birmingham to Shrewsbury, after BAAS meeting. Jones, *Life of Samuel Butler,* vol. 1, p. 13 and Silver 1962. *Family letters of Samuel Butler,* p. 209.

Sowerby, George Brettingham [I], 1788-1854. Second son of James S. Biological artist, author and conchologist. 1836 CD to Henslow, "Also about fossil shells. Is

Sowerby a good man? I understand his assistance can be purchased". CCD1:513. 1839 *A conchological manual.* 1844 S wrote appendix on fossil shells for CD's *Volcanic islands.* 1846 S wrote appendix on Tertiary Fossil Shells for CD's *South America.*

Sowerby, George Brettingham [II], 1812-84. Son of George Brettingham S [I]. Naturalist, illustrator and conchologist. Published on molluscs. 1861 Oct. 5-6 S was at Down House drawing orchids for *Orchids,* 1862 (F800).

Sowerby, James, 1757-1822. Naturalist and illustrator. First of the S dynasty. Father of George Brettingham [I] S and James de Carle S. Perfected naturalistic depictions of natural history specimens, tradition carried forward by his sons and grandson. Provided many illustrations for *Botanical Mag.* 1790-1814 *English botany,* 36 vols. 1812-29 *Mineral conchology of Great Britain.* Finished by G. Brettingham [I] S.

Sowerby, James de Carle, 1787-1871. First son of James S, elder brother of George Brettingham [I] S. CD discussed fossil molluscs of Falkland Islands with S. *Journal,* p. 253. 1839 Co-founded Royal Botanical Society and Gardens. 1851-54 S drew illustrations for all CD's work on cirripedes (F339, F342).

Spanish. First editions in: ?1877 *Origin* (F770), with two unique CD letters. 1879 *Journal* (F2068). 1876 *Descent* (F1122a). 1900 *Insectivorous Plants* (F2444). ?1902 *Expression* (F1214). 1902 *Autobiography* (F1544). 1906 *South America* (F324). ?1983 Essay on Instinct (F2445). 2006 *Coral reefs* (F2074). 2007 *Orchids* (F2050). 2009 *Different forms of flowers* (F2077). 2009 *Climbing plants* (F2079). 2011 *Earthworms* (F2135).

Spencer, Herbert, 1820-1903. First child of William George S and Harriet Homes, eight other children all died in infancy. Philosopher, biologist, anthropologist and writer. Unmarried. Biography: Duncan, 1908; Medawar, 1964, *Encounter,* 21: pp. 35-43. 1859 CD sent 1st edn *Origin.* 1860 CD to Lyell, CD had read S's essay on population, *Westminster Rev.,* 57: pp. 468-501, "such dreadful hypothetical rubbish on the nature of reproduction". CCD8. 1865 CD to Lyell upon reading *The principles of biology,* "somehow I never feel any wiser after reading him, but often feel mistified [sic.]". CCD13. 1866 CD to Hooker, "If he had trained himself to observe more, even if at the expence, by the law of balancement, of some loss of thinking power, he wd. have been a wonderful man." CCD14:418. 1870 CD to Lankester, "I suspect that hereafter he will be looked at as by far the greatest living philosopher in England". CCD18. 1873 CD to S, on receiving S's *The study of sociology,* 1873, "Those were splendid hits about the P. of Wales & Gladstone. I never before read a good defence of Toryism." CCD21:475. 1874 CD to Romanes, "I have so poor a metaphysical head that Mr. Spencer's terms of equilibration &c always bother me and make everything less clear". CCD22:392. 1874 CD to Fiske, "with the exception of special points I did not even understand H. Spencer's general doctrine; for his style is too hard for me". CCD22:560. 1882 S was on "Personal Friends invited" list for CD's funeral. Works: 1862 *First principles.* 1864-67 *The principles of biology.* 1862-93 The whole body of his work in *The synthetic philosophy,* 9 vols.. Autobiography: 1889 Privately printed. 1904 Published.

Spencer, John Poyntz, 1835-1910. Whig statesman. 5th Earl Spencer of Althorp. Also known as Viscount Althorp and the Red Earl. 1882 S, as Lord President of the Council, represented the Queen in Council at CD's funeral.

"Spengle", Darwin family's name for Dr S.P. Engleheart

Spearman, Sir Alexander Young, 1793-1874. 1840 Baronet, Assistant secretary to the Treasury. 1824-37 CD corresponded with S through Smith, Elder & Co. CCD1.

Spiders, collecting mentioned in *Beagle notebooks*, *Beagle diary* and *Zoological diary*. 1839 Über der Luftschifferei der Spinnen, *Neue Notizen aus dem Gebiete der Natur- und Heilkunde*, 11: cols 505-509 (F1654), translated from pp. 187-8 of *Journal*, 1839. 1873 Aeronaut spiders, *Gardeners' Chron.* (F1765) only a quote from *ibid.*

Spottiswoode, William, 1825-83. Physicist, mathematician and President of the Royal Society 1878-83. 1853 FRS. Died of typhoid fever and was buried in Westminster Abbey. 1861 Nov. 27 Married Eliza Arbuthnot. Two sons: William H (1864-1916) and Cyril A (1867-1915). 1880 Jun. 26 dined at High Elms with the Darwins and the Hookers. 1882 S was, as PRS, a Pallbearer at CD's funeral. 1882-83 Chairman, Darwin Memorial Fund.

Sprengel, Christian Konrad (or Conrad), 1750-1816. German botanist, theologist and teacher. Rector of Spandau, but dismissed for neglecting his duties. See J.C. Willis, *Natural science*, 2, 1893. 1793 *Das entdeckte Geheimniss der Natur im Bau und in der Befruchtung der Blumen*, Berlin. Neglected at the time but seen by CD as most important. In retrospect foundation of pollination ecology. 1841 CD read on Robert Brown's recommendation; "full of truth" although "with some little nonsense"; "It may be doubted whether Robert Brown ever planted a more fruitful seed than in putting such a book into such hands". LL3:258. 1873 CD to Hermann Müller, "it is a great satisfaction to me to believe that I have aided in making his excellent book more generally known". CCD21:204. "Wonderful book". *Autobiography*, p. 127.

Spring Gardens, London. no. 17. 1831 Sept. CD lodged there whilst preparing for *Beagle* voyage. Wrote letters to Susan, Henslow, FitzRoy and Whitley, while there.

Squirrels. The story of young red squirrels mistaking CD for a tree on one of his rounds of the sandwalk is in Francis Darwin's reminiscences, "their mother barked at them in agony from a tree". LL1:115. An illustration, entirely imaginary, is in Holder 1892 (see Darwin Iconography). ED in a letter to Henrietta in 1870 Mar. 19 wrote "Polly has had a great deal to suffer in her mind from the squirrels, and sits trembling in the window watching them by the hour going backwards and forwards from the walnut to the beds where they hide their treasures." ED2:198.

Stack, James West, 1835-1919. Missionary in New Zealand. Published on Maori subjects. 1867 S answered *Queries about expression* on Maoris, receiving the sheets from Haast. *Expression*, p. 20. 1873 Feb. S received inscribed copy of the book.

Stafford, Staffs. 1869 Jun. 30 CD stopped at on way back from Barmouth holiday.

Stainton, Henry Tibbats, 1822-92. Entomologist, especially of the microlepidoptera. 1855-81 CD to and from S, a series of letters on entomological subjects. CCD. FRS.

Stanhope, Philip Henry [I] Mahon, 4ᵗʰ Earl, 1781-1855. Eccentric. Chiefly known for his involvement with the psychotic youth Kaspar Hauser. 1806-16 MP. 1807 FRS. 1849 "Long ago I dined occasionally with the old Earl...He said one day to me 'Why don't you give up your fiddle-faddle of geology and zoology, and turn to the occult sciences'". *Autobiography*, p. 112.

Stanhope, Philip Henry [II] Mahon, 5ᵗʰ Earl, 1805-75. Historian and politician. Known by courtesy title of Viscount Mahon. 1830-52 MP. c.1842 CD dined with S in London and met Macaulay. 1849 Mar. 5 CD dined with S at his seat, Chevening, Kent. 1856 CD to Hooker; CD and Lyell dined with S in London.

Stanley, Edward Henry, 15ᵗʰ Earl of Derby, 1826-93. Statesman. Of Knowsley Hall. 1848-69 MP. 1859 FRS. 1882 S was a Pallbearer at CD's funeral. Recollection of CD in *The Times*, 1 Oct., 1883. Transcribed in *Darwin Online*.

Stanley, Edward Smith, 13ᵗʰ Earl of Derby, 1775-1851. Politician, landowner and naturalist. S kept a large private menagerie at Knowsley Hall. 1796-1832 MP. 1828-33 President of Linnean Society. 1837 CD to Henslow; S supported CD's application for a Treasury grant for publishing zoological results of *Beagle* voyage.

Star Hotel, Princes Street, Edinburgh. 1825 Oct. CD and Erasmus stayed there briefly before moving into lodgings in Lothian Street. CCD1:19.

Steadman, Mr, a bookseller in Buenos Aires. In the *Buenos Ayres notebook*, Chancellor & van Wyhe, *Beagle notebooks*, 2009. See CCD1:319 note 2.

Stebbing, George James Jr, 1803-60. Instrument maker and librarian on *Beagle*. Private assistant to FitzRoy. S was a supernumerary, at FitzRoy's expense, on 2ᵈ voyage of *Beagle*. S was responsible for the maintenance of the 22 chronometers. Worked later at Meteorological Office as optician.

Stebbing, Rev. Thomas Roscoe Rede, 1835-1926. Anglican clergyman and naturalist. A distinguished marine naturalist, specialist on amphipod crustaceans. Schoolmaster and master at Wellington College. 1859 Ordained by Samuel Wilberforce. 1869 Feb. 1 S lectured on Darwinism to Torquay Natural History Society. 1869 Mar. 3 CD thanked S, "a clergyman in delivering such an address does...much more good by his power to shake ignorant prejudices". CCD17:108. 1870 S lectured to same society on Darwinism and the Noachian flood. 1871 CD to S, thanking him for a copy of *Essays on Darwinism*. CCD19. 1881 CD to S, thanking him for a letter on Samuel Butler affair, *Nature*, 23: p. 336. Carroll 1976, 583. 1896 FRS.

Stedman, Rev. Thomas, 1745-1825. Vicar of St. Chad, Shrewsbury. 1809 Nov. 15 Baptised CD. The Parish Register of Christenings and Burials gives the following entry for 17 Nov. 1809: "Darwin Chasᵖ. Robᵗ. Son of Dr. Robᵗ. & Mʳˢ. Susannah his wife/born Febʳ. 12ᵗʰ".

Steenstrup, Johannes Japetus Smith (Japetus), 1813-97. Danish zoologist and biologist. CD wrote many letters to S from 1849-81. 1875? CD thanked him for a photograph of S. S provided CD with fossil and living cirripede specimens. CCD4 & 23. CD returned 77 specimens to S with a list of their names. The list was found in the Natural History Museum of Denmark, allowing the specimens to be identified. CD's list is reproduced and transcribed in *Darwin Online*.

Stephen, Sir Leslie, 1832-1904. Biographer, author and critic. Editor of DNB. Founder of Sunday Walking Club, nicknamed the "Sunday Tramps". Father of Virginia Woolf and Vanessa Bell. 1880 S was amongst the friends who advised CD to ignore Samuel Butler's attack on him. 1878 Nov. 25 ED recorded "Leslie Stephen to lunch". 1881 May 14-16 S stayed. 1882 Jan. 8 S came to Down House on a Sunday tramp. CD enjoyed their visit heartily. 1882 S was on "Personal Friends invited" list for CD's funeral. 1902 KCB. Recollections of CD and ED in *Darwin Online*.

Stephens, Catherine, Countess of Essex, 1794-1882. Vocalist and actress. 1825 CD to Robert Waring Darwin; CD heard her in Edinburgh. CD on S "She is very popular here, being encored to such a degree that she can hardly get on with the play" CCD1:19. 1835 Retired. 1838 Married 5th Earl of Essex, who died in 1839.

Stephens, James Francis, 1792-1852. Entomologist. [1827]-46 Author of *Illustrations of British entomology*, [1827-]1828-35[-45], supplement 1846, 12 vols., Baldwin and Cradock, London, which contains a number of beetle records, and one of a moth, bearing CD's name, mostly from Cambridge and North Wales. The first appearance of CD's name in print. 1829 Feb. 23 CD to Fox; CD took tea with S, "he appears to be a very good-humoured pleasant little man". CCD1:76. 1832 S sued James Rennie for infringement of copyright; his legal costs of £400 were raised by friends. Barlow 1967, p. 79. 1839 *Manual of British beetles*. 1880 CD to Sarah Haliburton (née Owen), "I remember the pride which I felt when I saw in a book about beetles the impressive words 'Captured by C. Darwin'". LL3:335. In F1968, *Shorter publications*.

Stephens, Rev. Thomas Sellwood, 1825-1904. Worcester College, Oxford; B.A. 1847; M.A. 1850, Clergyman. 1862 Innes to CD mentions him in relation to Tegetmeier's design for beehives. CCD10:90. 1859-67 Curate of Down. 1865 S children beating their parents at billiards. ED2:182.

Stevens, Thomas, 1809-88. Warden and founder of Bradfield College, Berkshire. 1839 Married Caroline Tollet. 1839 Josiah Wedgwood [II] gave a Broadwood piano to ED, as a wedding present, which had been S's property. 1839 Jan. 28, a day before her wedding, Emma Wedgwood recorded in her diary "Tollet Mr Stevens Darwin came". On 31st, the newlyweds "went to Broadwoods for P.F". 1839 Feb. 4 Mrs Josiah Wedgwood to ED, "Mr. Stevens is now below strumming upon our old affair". ED1:30. 1842 Lord of the Manor of Bradfield upon the death of his father. 1850 Founded Bradfield College. 1863 Innes to CD mentions. CCD11:694.

Stevenson, Elizabeth Cleghorn (Mrs Gaskell), 1810-65. Novelist. CD's second cousin. Daughter of Rev. William S and Elizabeth Holland (d. 1811). 1832 Married William Gaskell, a Unitarian minister. 1857 Wrote biography of Charlotte Brontë.

Stevenson, Rev. William, 1772-1829. Unitarian minister and writer. 1797 Married Elizabeth Holland. Daughter Elizabeth Cleghorn (Mrs Gaskell).

Stewart, Lady Caroline, ?-1865. 2d daughter of 1st Marquis of Londonderry. Fitz-Roy's aunt. 1801 Married Col. Thomas Wood. 2 sons; Alexander C and Thomas W.

Stiles, John, American botanist residing in Valparaiso. "the sea 70 years ago reached foot of Dr Stiles house" *Santiago notebook*, Chancellor & van Wyhe, *Beagle notebooks*.

Stirling, Scotland. 1827 CD visited on a spring tour.

Stoke d'Abernon, Elmbridge, Surrey. 1795-1800 Home of Josiah Wedgwood [II].

Stokes, Francis Griffin, 1853-1949. Historian, bibliographer and Shakespeare scholar of Windsor. 1878 CD to S, on intonations of young children. CCD26:293.

Stokes, Sir George Gabriel, 1st Baronet, 1819-1903. Mathematician and physicist. Corresponded with CD. 1849-1903 Lucasian Prof. of mathematics, Cambridge. 1851 FRS. 1854-85 Secretary of the Royal Society of London. 1869 President of the BAAS. 1885-90 PRS. 1887-91 Conservative MP for Cambridge University. 1889 Kt.

Stokes, John Lort, 1812-85. Naval Officer. S served on all three voyages of *Beagle*. Was mate and Assistant Surveyor on 2d voyage. After 18 years, a record, on *Beagle*, surveyed New Zealand and the English Channel. Nearly became Hydrographer. 1838 CD saw in London. 1841 Commanded *Beagle* at end of 3d voyage when Wickham was invalided. During the voyage he was speared by aborigines. 1846 Author of *Discoveries in Australia*. 1863 Rear Admiral. 1871 Vice-admiral. 1877 Admiral. 1882 Apr. 27 S letter in *The Times*, printed immediately after report on CD's funeral, about CD's seasickness. CD would say "Old fellow I must take the horizontal for it". "It was distressing to witness this early sacrifice of Mr. Darwin's health, who ever afterwards seriously felt the ill-effects of the *Beagle*'s voyage". 1883 Apr. 25 S letter in *The Times* on CD "marvelous persevering endurance in the cause of science of that great naturalist, my old and lost friend, Mr. Charles Darwin". LL1:224, transcribed in *Darwin Online*. 1864 Jul. 14. ED recorded "Admiral Stokes called". 1888 Feb. 13 Mrs & Miss Stokes came to see ED. A few more visits either way until 1895. 1882 recollections of CD in DAR112.A97-A98, recalled CD's favourite expressions were "by the Lord Harry" and "beyond belief". Transcribed in *Darwin Online*.

Stonehenge, Wiltshire. 1877 Jun. CD visited from Southampton. CD and his sons carried out a number of observations of the soil on the chalk, many of which, such as those at Stonehenge and other "Druidical stones" on Salisbury Plain, are described in detail in *Earthworms*. (F1357)

Stoney field, Great Pucklands, a field west of Down House, so-called by the Darwins because of the large number of surface flints, due to recent ploughing. See Pucklands. ED sometimes spelt as "Stony". She loved going there to sit.

Stowe, Darwin, fl. 1638. Named after his great-grandfather Henry Darwin. 1667 Married Ann Brown of Gainsborough, Lincolnshire.

"Strawberries". 1862 "Cross-breeds of strawberries", *Jrnl. of Horticulture*, 3: p. 672 (*Shorter publications*, F1720). CD asked about successfully crossing different varieties. 1862 Jul. 25 CD who had been very unwell managed some strawberries and cream. 1863 Jul. also not doing very well had strawberries. *ED's diary*.

Streets, Mr, 1876 May. 29 an entry in *ED's diary* "Mr Streets". They were in Holmbury and went to Abinger on Jun. 1.

Strickland, Hugh Edwin, 1811-53. Geologist and naturalist. CD discussed evolution with before *Origin*. 1842 Author of *Strickland code of zoological nomenclature*. 1844 One of the founders of Ray Society. 1848 *The Dodo and its kindred*, with Alexander Gordon Melville. 1849 CD to S, on difficulties in nomenclature in relation to his barnacle work. 1850 Deputy reader geology, Oxford. 1852 FRS. 1853 S was killed by a train while examining geological strata. 1858 Letter from CD on zoological nomenclature in *Memoirs of Hugh Edwin Strickland, Shorter publications*, F1983.

Strickland, Sefton West, 1839-1910. Barrister. Cambridge friend of William Erasmus Darwin. S was often at Down House with the first recorded visit in 1863 Jul. 22-25 and again in Aug. 1867 paid three visits. 1882 S was on "Personal Friends invited" list for CD's funeral.

Stringer, Mary, c.1695-1766. Daughter of Rev. Samuel Stringer. c1711 Married Thomas Wedgwood [III]. CD's maternal great-grandmother.

Stringer, Samuel, Unitarian minister of Newcastle under Lyme. Father of Mary S.

Strong, Edward, 1809-? Printer and stationer on High Street, Bromley, Kent. 1863 Printed the joint appeal of CD and ED on "Vermin and traps" (F1931 and F1728).

"Struggle for existence", 1859 "We will now discuss in a little more detail the struggle for existence". *Origin*, Chapter 3. The phrase was used by Malthus in relation to competition between different tribes. An expanded sense of the phrase as applied to all species was used by Lyell in *Principles of geology*.

"Struggle for Life". 1855 Phrase first used by CD in print in 1855 "Does sea-water kill seeds?" *Gardeners' Chron.*, no. 21 (26 May), p. 357. (*Shorter publications*, F1683).

Strutt, Joseph, 1765-1844. Businessman in textile and philanthropist. Emma Wedgwood on coming to Derby in 1834 Feb. 14, drank tea with them the next day. On the 17th attended a dinner party. CD and ED called on them in 1839 Mar. 18 and were invited to dinner on Apr. 13.

Strutt, John William, 3d Baron Rayleigh, 1842-1919. Physicist. Cambridge friend of CD's sons. 1871 May 20, visited the Darwins. 1873 3d Baron. FRS. 1879-84 Cavendish Prof. of Experimental Physics, Cambridge. 1882 Royal Medal Royal Society. 1882 S was on "Personal Friends invited" list for CD's funeral. 1887-1905 Prof. Natural Philosophy Cambridge. 1902 OM. 1908-19 Chancellor University of Cambridge. 1904 Nobel Prize Physics. Married Evelyn Balfour. 1905-8 PRS.

Strzelecki, Sir Paul (Paweł) Edmund de, 1797-1873. Polish explorer and geologist. Titled Count from his Polish ancestry. 1839-43 Travelled extensively in Australia. 1845 British subject. 1853 FRS. 1856 CD to Hooker, S was on election committee of Athenaeum, and CD proposed to speak to him about election of Huxley. CCD6:103 (misspelt "Strezlecki"). 1869 CMG.

Stuart, Catherine, ?-1797. 1789 Married as first wife to Sir James Mackintosh.

Stuart, Charles Edward Sobieski, Count d'Albanie, 1799-1880. With his brother, John Sobieski S, falsely claimed descent from Stuart kings and hence from Charles II. Actual name Charles Manning Allen. CD in *Autobiography*, pp. 71-82, "Dr Wallich gave me a collection of photographs he had made, and I was struck with the resemblance of one to FitzRoy; on looking at the name I found it Ch.E. Sobieski Stuart, Count d'Albanie, illegitimate descendant of the same monarch [Charles II]".

Studer, Bernhard, 1794-1887. Swiss geologist, Prof. at the University of Berne 1834-73. 1879 Wollaston Medal Geological Society. 1847 CD was much interested in his work and thanked him for offer of his pamphlet. S invited to Down. CCD4.

Stutchbury, Samuel, 1798-1859. Naturalist and geologist. 1846 Provided CD with fossil specimens for *Cirripedia*. CCD3. 1820 Assistant curator Hunterian Museum, Royal College of Surgeons. 1831-50 Curator museum Bristol Philosophical Institution. 1851-55 Geological surveyor in Australia.

Sudbrooke Park, Petersham, Surrey. Water cure establishment run by Dr Richard J. Lane with Edward W. Lane as physician. ?Moved from Moor Park. Clubhouse of Richmond Golf Club, since 1891. 1860 Jun. 28 CD went but suffered a fever fit the next day. CCD8. 1862 Feb. 24 Henrietta went. *ED's diary*.

Suess, Eduard, 1831-1914. Austrian palaeontologist and geologist. Postulated former existence of Gondwanaland. 1856 Prof. Palaeontology Vienna University. 1861 Prof. Geology Vienna University. 1871 CD to S, on his election as Foreign Corresponding Member of Kaiserliche Akademie der Wissenschaften, Vienna. CCD19.

Sulivan, Sir Bartholomew James, 1810-90. Naval Officer. Joined FitzRoy on 1st voyage of *Beagle*. 2d Lieutenant on 2d voyage of *Beagle*. Chief Naval Officer at Board of Trade. Biography 1896: H.N. Sulivan (son) *Life and letters of S*. S made enquiries for CD on feral cattle and horses. *Journal* 2d edn, Chap. ix. 1840 S called on CD. J.A. Sulivan, great grandson, *Mariner's Mirror*, 1979, 65: p. 76. Lois Darling, *Mariner's Mirror*, 1978, 64: p. 325. 1849 S ranched and traded in Falkland Islands. 1850 S was visited in Falkland Islands by Huxley. 1855 CB. 1863 Rear Admiral. 1865 Upon suicide of FitzRoy S convinced British government to provide his widow and daughters with £3,000. CD contributed £100. 1867 S persuaded CD to subscribe to South American Missionary Society's orphanage in Tierra del Fuego. 1869 KCB. 1870 Vice Admiral. 1877 Admiral. 1851 Nov. 27 ED wrote in her diary "Cap Sulivan came dinner party." His next visit was in 1862 Oct. 21, Capt. Wickham and Arthur Mellersh were also present. 1885 Jun. 9 S was present at unveiling of statue of CD in BMNH. Recollections of CD in H.H. Sulivan, *Life and letters of the late Admiral Sir*

Bartholomew James Sulivan (1896), pp. 40-6, 381-2, DAR107.42-7 and DAR112.A99-A108, transcribed in *Darwin Online*.

Sumner, John Bird, 1780-1862. Theologian and writer. CD read S's *Evidence of Christianity* in 1827 before deciding to go up to Cambridge to gain a BA with a view to becoming a clergyman. Reading notes survive in DAR91.114-118, transcribed in *Darwin Online*. 1816 *Treatise on the records of creation*. 1821 *The evidence of Christianity derived from its nature and reception*. Howard Gruber, *Darwin on man*, pp. 124-8. 1828-48 Bishop of Chester. 1848 FRS. 1848- Archbishop of Canterbury.

Sunday Lecture Society. 1875 Aug. 6 Founded. CD founder member and one of a long list of Vice Presidents which included Huxley, Erasmus Alvey Darwin and Herbert Spencer. CD donated £2.2.0 but did not subscribe. Object "to obtain the opening of museums, art galleries, aquariums and gardens on Sundays". *Sunday Rev.*, 1876, Oct. 1, p. 68. It offered lectures at St. George's Hall, Langham Place, London to working-class audiences in direct competition to Sunday sermons. Lectures cost £1 annual subscription, 1*d* or 6*d* at the door or 1*s* for reserved seats.

"Sunday Tramps". An intellectual walking club, technically called the Sunday Walking Club, organized by Leslie Stephen. 1882 Jan. 8 They ate at Down House, and perhaps on other occasions. 1882 May 2, ED on her birthday reflected on some precious memories. One such was the Sunday Tramp's visit during which "C. was delightful to them and enjoyed their visit heartily".

Surman, F.W. 1881 Butler to Erasmus A. Darwin for 20 years. Left a legacy by Erasmus, which was doubled by CD. After Erasmus' death, 1881 Aug. 26, CD wrote to Albert Günther about some post for S at BMNH, unsuccessfully. *Calendar* 13569; Carroll 1976, 607, 608. 1882 S applied to CD for reference. *Calendar* 13656. CD obliged.

Surtees, Rev. Matthew, 1755-1827. Rector of North Cerney, Wiltshire. "The family greatly disliked Mr. Surtees, and he appears to have been jealous, ill-tempered, and tyrannical". ED1:4. 1799 Married Harriet Allen d.s.p. 1816 Mrs Josiah Wedgwood [II] to her sister Frances Allen, "the most incomparably disagreeable man I ever saw". ED1:86. 1824 Mrs Josiah Wedgwood [II] to her sister Madame Sismondi, "Harriet is positively very much attached to him incredible as it may seem...he is a dying man". ED1:158. 1834 ED recorded two visits by Mrs Surtees.

"Survival of the Fittest". Expression coined by Herbert Spencer in his *Principles of biology*, 1864. CD was persuaded by Wallace to use the phrase in *Origin*. 1869 "The expression often used by Mr. Herbert Spencer of the survival of the fittest is more accurate and is sometimes equally convenient". *Origin*, 5th edn, Chapter 3, pp. 91-2. It is now widely considered to be a misleading shorthand for CD's theory.

Sutcliffe, Thomas, 1790?-1849. Adventurer in South America. 1822-38 Held various military and administrative positions in Chile. 1834 Governor of Juan Fernandez. 1834 Aug. 28 CD to FitzRoy "I have met a strange genius a Major Sutcliffe" who had sent a book of old voyages of the Straits of Magellan to Mr Caldcleugh for FitzRoy. Keynes, *Beagle record*, 1979, p. 235-236. 1838 Returned to England.

Sutterton Fen, a 49 acre farm in Lincolnshire that CD inherited from his father. John Abbott held the tenancy for 36 years. After 1876 at a rent of £62.10*s.* per annum. Inherited by ED who left it to Leonard Darwin. See Worsley, *The Darwin farms*, 2017.

Sutton, Seth, 1828-1902. Keeper at zoological gardens. 1867 Corresponded with CD about expression in apes. 1870 Jan. 5 CD to Abraham Dee Bartlett, S was a keeper at the Zoological Society of London's Gardens, Regent's Park, who made many observations on monkeys for *Expression*. CCD18.

Swainson, William, 1789-1855. Cabinet naturalist, ornithologist and traveller. S was a proponent of the quinary system of William Sharp Macleay in classification. 1815 FLS. 1820 FRS. 1841 Emigrated to New Zealand. 1844 CD to Hooker, "I feel a laudable doubt & disinclination to believe any statement of Swainson's". CCD3:24.

Swale, William, 1816-75. Gardener from Norfolk. 1857 Went to Christchurch, New Zealand, and became prosperous nurseryman. 1858 CD to S on introduced plants. To CD with four honeybees stuck to letter. CD sent it to *Annals and Mag. of Nat. Hist.* (3ᵈ series, 2 (1858), pp. 459-65) and *Gardeners' Chron.* extracts on pp. 828-9 after CD's paper "On the agency of bees" (*Shorter publications*, F1701).

Swanage, Dorset. 1847 Jul. CD had family holiday there.

Swedish. First editions in: 1870 *Origin* (F2429). 1872 *Journal* (F259). 1872 *Descent* (F1136). 1959 *Autobiography* (F1546).

Swettenham, Richard Paul Agar, 1845-99. A contemporary of George's at Trinity College Cambridge. School inspector, studied Mathematics. 1864 George Howard D mentioned him in letters to CD. CCD15. Many visits between 1867-73. *ED's diary.* ED to Aunt Fanny Allen in May 1867 that the family has a nice time in Cambridge where they had "a most elegant breakfast at Frank's, and ditto lunch at Swetten-ham's—claret-cup and all sorts of elegancies from the Trinity kitchen." ED2:214.

Swift, Rev. Benjamin, 1819-82. 1862 Married Georgina Elizabeth Darwin. Father of Francis Darwin S. 1857-74 Vicar of Birkdale, Lancashire.

Swift, Francis Darwin, 1864-? Elder son of Benjamin S and Georgina Elizabeth Darwin. CD's half-cousin. Matriculated at Queen's College, Oxford on 1883 BA in 1887. c.1920 S compiled and had printed *Some collateral ancestors of Erasmus Darwin*; this takes the ancestry back to Isaac II Angelus (1156-1204), Eastern Emperor 1185-1204, in skeleton form. The male Darwin line only goes back to 1644.

Swinhoe, Robert, 1836-77. Ornithologist and consular official in China. 1866 CD to Sclater, S had written to CD about common domestic duck of China. CCD14. 1867 CD sent S *Queries about expression*, which S had printed in *Notes and Queries for China and Japan*, 1: p. 105. F1739. 1871 Jan. 28 S visited Down House. 1876 FRS.

Sydney, Australia. 1836 Jan. 12 *Beagle* arrived at Port Jackson and anchored in Sydney Cove. CD made short expedition to Bathurst. 1836 Jan. 30 left for Tasmania. *Beagle diary*, pp. 395ff; Geological diary: DAR38.812-836. Both in *Darwin Online*.

Sykes, William Henry, 1790-1872. Indian army officer and naturalist. 1834 FRS. 1849 Sept. 11 CD travelled with S to BAAS meeting. 1857-72 MP. 1859 CD to S, recommending Edward Blyth for position as naturalist on China expedition. CCD7.

Symonds, Hyacinth, 1842-1921. Daughter of William Samuel S. 1871 Married 1 Sir William Jardine, Bart. 1876 Married 2 Sir Joseph Dalton Hooker.

Symonds, Mary Anne Theresa, 1783-1850. Writer and landowner. Breeder of and expert on silkworms. Referred to in *Variation*. Daughter of Capt. Thomas S. 1802 Married Capt. John Whitby. 1846 CD met S at meeting of BAAS, Southampton. 1847 and 1848 CD to S. CCD4. 1848 *A manual for rearing silkworms.* Colp, *Bulletin N.Y. Acad. Med.* 1972, 48 (6), pp. 870-6. 1869 Jan. 29 Miss S and Lyells called.

Symonds, Sir William, 1782-1856. Naval Officer. Brother of Mary Anne Theresa S. 1832-47 Surveyor of the Navy. 1835 FRS. 1837 Kt. 1848 CB. 1848 Jul. CD went in S's yacht from Swanage to Poole. 'Journal', (DAR158). 1854 Rear-Admiral. 1839 Jan. 11 Lady Symonds was in the list of names Emma Wedgwood wrote in her diary. 1839 Feb. 24 the Darwins paid a call, Apr. 19 they were invited to visit. Apr. 25 they dined at Sir S. 1840 Jan. 2 Lady S visited, perhaps to see the new baby.

Symonds, Rev. William Samuel, 1818-87. Anglican clergyman and geologist. Father of Hyacinth S. Rector of Pendock, Worcestershire. Passed the later years of his life at Sunningdale, the house of Hooker, his son in law. 1860 CD to Lyell, refers to letter from S on imperfections of geological record.

Tahiti, formerly known as Otaheite. Society Islands archipelago, British Colony. French since 1880. 1835 Nov. 15/16 *Beagle* arrived, having crossed the dateline as it was then and lost a day. Anchored in Matuvai Bay. Missionaries were most hospitable. 1835 Nov. 18 CD landed at the capital city Papeete and had a short inland expedition, returned Nov. 20. 1835 Nov. 25 Queen Pomare IV entertained on board. 1835 Nov. 26 *Beagle* sailed for New Zealand. Copy of FitzRoy's official account made for Rev. G. Pritchard in MLNSW, MLDOC 1499. 1836 "A letter containing remarks on the moral state of Tahiti, New Zealand &c.", *S. Afr. Christian Recorder,* 2: pp. 221-38 (*Shorter publications,* F1640). CD's first publication, except for beetle records in J.F. Stephens. *Geological diary:* DAR37.798-801, transcribed in *Darwin Online.*

Tait, Robert Lawson, 1845-99. Surgeon and gynaecologist. There are eight letters from CD to T at Shrewsbury School. De Beer 1968, pp. 79-82. 1871-93 Surgeon and gynaecologist at Hospital for diseases of women, Birmingham. 1875 CD to T, about use of tails for sensory purposes by mice. CCD23:224. 1875 Apr. 18 CD to Romanes; T stayed at Down House. CCD23. CD 1875. [Letter on animal tails.] in Tait, The uses of tails in animals. *Hardwicke's Sci. Gossip* 11, no. 126 (1 Jun.): 126-127, p. 127, *Shorter publications,* F2126. 1875 T reviewed *Insectivorous plants* in Spectator. 1876 T reviewed 2d edn of *Variation* in *Spectator,* Mar. 4. CCD24:82. 1880 Jul. 19 CD sent T £25 "for your scientific fund in Birmingham". De Beer 1968, p. 82. 1881 T to CD, T had spoken strongly in favour of *Origin* in his physiology lectures at Midland Institute, and inviting CD to visit Birmingham. 1875 Apr. 17 Huxleys, Romanes and T were at Down House. 1880 Mar. 4 while CD was in London, T visited.

Tait, William Chester, 1844-1928. Merchant, landowner and botanist resident at Oporto, Portugal. He provided CD with specimens of a primitive and rare insectivorous plant *Drosophyllum* for *Insectivorous plants*, after Hooker had been unable to get them. 1869 CD wrote to thank T for *Drosophyllum*. CCD17.

Talandier, Pierre Théodore Alfred, 1822-90. French jurist. Prof. French Royal Military Academy Sandhurst. Active in politics. 1851-70 Republican exile in England. 1860 CD to Quatrefages, T wanted to translate *Origin* into French. CCD8.

Talbot, Charles, 1801-76, commanded HMS *Warspite*, 1830-42, rescued the Brazilian Imperial Family from an insurrection on 6 Apr. 1831. In the *Rio notebook*, Chancellor & van Wyhe, *Beagle notebooks*, 2009.

Talbot, Emily. 1881. [Letter to T on the mental and bodily development of infants]. *Social science.-Infant education. The Jrnl. of Speculative Philosophy* 15, no. 2 (April): 206-7. (*Shorter publications*, F1995). Transcribed in *Darwin Online*.

Tanks and Hose. 1853 *Gardeners' Chron.*, 19: p. 302 (*Shorter publications*, F1807). Query about connecting large water tanks in the garden with a hose to fill them.

Tara. A cob at Down House, used for the coach. "Was only seen to be moving by reference to the hedges." B. Darwin, *Green memories*, 1928, p. 13. The coachman pronounced him "tearer". T died when ED was 87.

Tasker, Harriet Mrs, Let lodgings in Down. 1871 was suspected by the housemaid Jane of turning out Cinder the kitten at night. Consequently losing the kitten for two days. 1873 Charles Langton took lodgings at Mrs T. ED(1904)2:248-49. 1890 Nov. 8 called on ED. 1894 Jul. 13 wrote to T.

Tasker, Tom, 1872 worked at Down House. Helped Elizabeth with gardening but she found it quite tiresome so mostly did things herself. ED(1904)2:260-61.

Tasmania. 1836 Feb. 5-17 *Beagle* at Storm Bay. CD made short inland journeys. *Beagle diary*, pp. 408ff.

Tasmanians, the aboriginal peoples of Tasmania. The 'Darwin's bodysnatchers' myth that CD wrote asking for the skulls of full-blooded Tasmanians in the 1870s is exposed and refuted in van Wyhe, Darwin's body-snatchers? *Endeavour* (Dec.), 2016.

Taylor, Elizabeth, 1744-85. Daughter of John T. 1765 Married Thomas Wedgwood [V]. Had 8 children.

Taylor, Jane, 1783-1824, Novelist and poet. With her sister Ann wrote poems on goodness and grace. ED read to her children. R. Keynes, *Annie's box*, 2001, p. 124.

Taylor, John, Master potter of Hill Top works, Burslem. Father of Elizabeth T.

Tearle, William, Of St. Neot's, Cambridgeshire. 1880 CD in reply to T, cannot help with his religious doubts. *Calendar* 12578, 12579, Carroll 1976, 572.

Teasel. 1877 Note to Mr Francis Darwin's paper, *Quart. Jrnl. of Microscopical Sci.*, 17: p. 272 (*Shorter publications*, F1777). F. Darwin, On the protrusion of protoplasmic filaments from the glandular hairs on the leaves of the common teasel (*Dipsacus* *sylvestris*), *ibid.*, 17: pp. 245-77. 1877 [Letter] The contractile filaments of the teasel, *Nature*, 16: p. 339 (*Shorter publications*, F1778). The wild teasel.

Teesdale, John Marmaduke, 1844-1928. Solicitor. Assistant solicitor to the Treasury. 1875 ED to Innes; T took Down Hall. CCD23:536. 1875 "In order to shew how severe our weather has been I may mention that Mr Teesdale went to Orpington for 10 days on a sledge". CD to Innes CCD23:536. 1876 Jan. 15 CD & ED called at T. Jan. 29, Ts came for lunch. Mar. 5 T and one of his sons had lunch with CD at Down House. Aug. 2, Mrs T called. Sept. 2, Ts came for lawn tennis. 1877 Aug. 8 Miss T came for lunch. 1880 Romanes to CD mentions him twice in relation to death of Sarah Elizabeth Wedgwood. Romanes 1896, pp. 99, 100. Mrs T called in Nov. 26. 1881 Jul. 23 Romanes with T at Down Hall and called on CD, the last time they met. 1881 Nov. 5 a call was made to T. Miss T called in 1883 Jun. 14 and the last record in *ED's diary* in 1884 Jul. 25 Ts came to lunch. 1882 T was on "Personal Friends invited" list for CD's funeral.

Tegetmeier, William Bernhard, 1816-1912. Ornithologist and poultry fancier. T helped CD extensively with *Variation*, especially on poultry. Helped CD with pigeon fancying and introduced him to Columbarian Society. Over the years T provided CD with a vast store of information on many subjects. T is mentioned in *Origin* and *Descent*. T also read and corrected some chapters of *Variation* before publication. Biography: E.W. Richardson, son-in-law, 1916, especially Chapter X. Corresponded with CD from 1855-81. 1855 CD visited T at Wood Green, Middlesex. 1859 T received presentation copy of 1st edn of *Origin*; now at BMNH. 1867 *The poultry book*. 1887 T called on ED at Down House. Recollections of CD in Richardson, *A veteran naturalist*, 1916, pp. 100-2, 111-12. Photograph in *ibid*. Transcribed in *Darwin Online*. Another in DAR193.22. 1906 135 letters from CD to T sold at Sotheby's for £128. Presentation copies of CD books were also sold.

Tenby, South Wales. South Cliff House. Home of four Allen sisters: Harriet Surtees after death of husband 1827. Jessie Sismondi after death of husband 1842. Emma and Frances Allen after death of John Hensleigh Allen 1843. 1856 Jul. 29, ED recorded "Willy came home from Tenby". 1859 Feb. 9 Elizabeth Darwin travelled there upon hearing of the death of Sarah Bayly, Baugh Allen's second wife.

Tendency of Species to Form Varieties. The first appearance in print of the theory of evolution by means of natural selection. The details of the preparation and publication of this paper are given in LL1:115-138 and many other publications. It has been reprinted many times. 1858 "On the tendency of species to form varieties, and on the perpetuation of varieties and species by natural means of selection", by Charles Darwin...and Alfred Wallace, *Jrnl. Proc. of Lin. Soc. of London*, 3, no. 9: pp. 1-62 (*Shorter publications*, F350). Communicated by Lyell and Hooker, and read 1 Jul. in the absence of CD and Wallace. Reprinted (without letter by Hooker and Lyell) in *Zoologist*, 16 (F349) and CD's contribution alone in [Extract from an unpublished Work on Species]. *Gardeners' Chron.* (2 Oct.): 735 (F352a). First foreign editions:

1870 German (F365). 1939 Russian (F370). 1960 Italian (F368). 1913 Dutch (Letter to Asa Gray only) (F364c). 1983 Chinese (F1844). 2008 Spanish (F2018).

Tenerife/Teneriffe, Canary Islands. See also Canary Islands. 1832 Jan. 7 *Beagle* anchored there, but CD could not land because of quarantine regulations.

Tennyson, **Alfred Lord**, **1st Baron**, 1809-1892. Poet. 1849 In memoriam *A.H.H.* contains the idea of a struggle for existence, inspired by *Vestiges*. 1850 Poet Laureate. 1865 FRS. 1868 Jul. 19 Summer T called on CD several times at Freshwater, Isle of Wight. T's wife, Emily Sellwood, recorded this brief exchange in her diary (1897): "Aug. 17th. Farringford. Mr. Darwin called, and seemed to be very kindly, unworldly, and agreeable. A. said to him, 'Your theory of Evolution does not make against Christianity': and Darwin answered, 'No, certainly not.'" (F2171) According to Henrietta, T "did not greatly charm or interest either my father or mother". ED2:190. 1884 1st Baron. c.1885 T to Dr Grove, "I don't want you to go away with a wrong impression. The fact is that long before Darwin's work appeared these ideas were known and talked about". Poulton, *Darwin and the Origin*, 1909, p. 9. *ED's diary* 1892 Oct. 12: "Lord Tennyson's funeral."

Termites. 1874 Recent researches on termites and honey bees, *Nature*, 9: pp. 308-9 (*Shorter publications*, F1768), introducing a letter from Fritz Müller.

Terrestrial Planariae. 1844(-45) Brief descriptions of several terrestrial planariae and of some remarkable marine species, with an account of their habits, 14: pp. 241-51 (*Shorter publications*, F1669), flatworm specimens from the *Beagle* voyage.

Thackeray, **Anne (Annie) Isabella**, 1837-1919. Novelist. Eldest daughter of William Makepeace T. 1877 Married Richmond Ritchie, her first cousin. 1866-82 Visited Down House, "a most amusing and pleasant person". (letter to Anthony Rich) 1882 Was on "Personal Friends invited" list for CD's funeral. [Recollection of an 1882 visit to CD] *Letters of Anne Thackeray Ritchie*. F2175, transcribed in *Darwin Online*. Brodie, the Darwin's Scottish nurse was once with the Thackerays.

Theory of Descent, Studies in the, 1875-76 August Weismann, *Studien zur Descendenz-Theorie*. The original does not contain CD's notice. 1882 August Weismann, *Studies in the theory of descent*, 2 vols., London; prefatory notice by CD, p. v-vi (*Shorter publications*, F1414); translated from German by Raphael Meldola, with notes and additions by the author. 1939 Foreign edn, of CD's notice only: Russian (F1415).

Thiel, **Dr Hugo**, 1839-1918. German agronomist and civil servant. Prof. agricultural sciences Darmstadt and Berlin. 1869 CD to T, thanking for pamphlet *Über einige Formen der Landwirtschaftlichen Genossenschaften*. T was at Agricultural Station, Poppelsdorf. "You apply to moral and social questions analogous views to those which I have used in regard to modification of species". CCD17:101.

Thierry, **Charles Philip Hippolytus**, **Baron de**, 1793-1864. Son of French refugee. Traveller and colonist. Phantast. 1845 Settled in New Zealand after travelling widely in the South Seas. 1845 CD to Henslow, prematurely, on T's death, "King of

Nukahiva and Sovereign Chief of New Zealand. I wonder what has become of his wretched wife". CCD3:228, Barlow 1967, pp. 154-5.

"This is the Question". 1838 CD's notes on whether or not to marry. (DAR210.8.2, transcribed in *Darwin Online*). Text printed in *Autobiography*, pp. 231-4 and in many other publications. Sydney Smith [II] suggested that they may have been scribbled down in ED's presence, whilst flirting; if so, before Nov. 12, when they became engaged. Another manuscript is headed "It being proved necessary to marry When? Soon or Late" with advice from the "Governor" [his father] that it be sooner rather than later. DAR210.8.2, transcribed in *Darwin Online*.

Thiselton-Dyer, Sir William Turner, 1843-1928. Botanist. Married Harriet Anne Hooker. 1876 D helped CD with experiments for *Insectivorous plants*. 1877 Reviewed *Fertilisation*. 1879 D helped CD with botanical material from Kew, *e.g.* 1879 CD to D, on a species of *Oxalis*. 1880 FRS. 1882 D was on "Personal Friends invited" list for CD's funeral. Visited CD with Ray Lankester in 1875 Jul. 18. Came to Down with Balfour in 1878 Jan. 26-28, came with Crawleys Jan. 18 1879. One more visit before CD's death was made in Jan. 7-9. Many more visits until 1894 Apr. 27. 1885-1905 Director of Royal Botanical Garden, Kew, in succession to Hooker.

Thom, John Pringle, 1840?- Tutor. 1854 was at Moor Park. CD knew him. T was caught in a scandal in which Edward Lane too was involved. T decided to migrate to Australia. CD contributed £20 towards his travel costs. 1863 T left for Queensland. Summerscale, *Mrs Robinson's disgrace*, 2012, p. 212.

Thompson, Miss, from 1868 Apr. 17, Miss T was first mentioned in *ED's diary* then 1890 Aug. 21. She came yearly until 1896 Sept. 3. Letter to Henrietta in 1892 Sept. 8, they "talked of such old times." ED2:422.

Thompson, Mr, 1880 A resident at Down, "affected by the creeping palsy". ED to Innes, Stecher 1961, p. 248.

Thompson, William, c.1823-1903. Nurseryman of Ipswich. Provided CD with plants and seed. 1855 T issued first catalogue. 1881 CD to Thiselton-Dyer, on plants with different-coloured anthers. CD had written to T for seed of *Clarkia elegans*. 1896 Victorian Medal of Honour Royal Horticultural Society. 1885 ED made a shopping list of flowers among other things "Annual from Thompson".

Thompson, Sir Harry Stephen Meysey-, Bart, 1809-74. Brother of Thomas Charles T. Agriculturalist. 1834 CD to Whitley; "The two Thompsons of Trin". Cambridge friends of CD. CCD1:397. 1859-65 MP for Whitby. 1874 1st Bart. 1860 Jan. 4 was guest along with the Frys, Mr Stephens and Reids. 1861 Apr. 1, came again with Mr Stephens for dinner. 1867 Feb. 22, CD and ED called on Reids and Thompsons. 1873 Dec. 1 they called on Thompson and Reeds.

Thompson, D'Arcy Wentworth, 1860-1948. Biologist, mathematician and classical scholar. 1883 T translated Hermann Müller, *Die Befruchtung der Blumen durch Insekten*, Leipzig 1873, as *The fertilisation of flowers*, London, with preface by CD p. vii-x (*Shorter publications*, F1432); see LL3:275. 1884-1917 Prof. Biology, later Natural History

University College Dundee. 1916 FRS. 1917 *On growth and form*. 1917-48 Chair Natural History St. Andrews. 1937 Kt. 1946 Darwin Medal, Royal Society. 1950 Foreign translation, CD's preface only: Russian (F1433).

Thompson, Thomas Charles, 1811-85. Brother of Harry Stephen Meysey Thompson. 1834 "The two Thompsons of Trin", Cambridge friends of CD. CCD1:396-7. 1848-85 Rector of Ripley, Surrey.

Thompson, William, 1805-52. Irish naturalist. Son of linen merchant of Belfast. 1849 CD to Hugh Strickland, T "who is fierce for the law of priority". CCD4:206. 1849-51, 1856 *The natural history of Ireland*, 4 vols. 1851 CD in introduction to *Living Cirripedia*, "The distinguished Natural Historian of Ireland".

Thomson, Allen, 1809-1884. 1830 Fellow of Royal College of Surgeons. 1839-41 Prof. of anatomy at Aberdeen. A 1st edn *Origin* was sold in 2019 inscribed to "Dr Allen Thomson with John Murray's kind regards Nov. 10".

Thomson, Sir Charles Wyville, 1830-82. Biologist. Held several chairs in Ireland. 1869 FRS. 1870 Prof. Natural History Edinburgh after various other academic functions since 1850. 1872-76 Director of scientific staff on the *Challenger* expedition and edited results. 1876 Changed name from Wyville Thomas Charles Thomson to Charles Wyville Thomson. 1876 Kt. CD et al. 1877. [Memorial] Zoology of the 'Challenger' Expedition. *Nature* 16, 14 Jun., 118. 1880 Wrote anti-evolution introduction to the *Challenger* results: "The character of the abyssal fauna refuses to give the least support to the theory which refers the evolution of species to extreme variation guided only by Natural Selection". ML1:388. 1880 Letter by CD, Sir Wyville Thomson and natural selection, *Nature*, 23: p. 32 (*Shorter publications*, F1789), in which CD severely castigates T, "standard of criticism not uncommonly reached by theologians and metaphysicians", see also ML1:388. CD omitted, on advice from Huxley, "for, as Prof. Sedgwick remarked many years ago, in reference to the poor old Dean of York, who was never weary of enveighing against geologists, a man who talks about what he does not in the least understand is invulnerable".

Thomson, Thomas, 1817-78. Physician and botanist in India. Superintendent of the botanic garden of Calcutta. 1855 FRS. 1855 With Hooker wrote first volume of *Flora Indica*, no further volumes were published. 1858 CD to Lyell, received a letter from T about "what heat our temperate plants can endure". CCD7:83, spelt "Thompson". The letter has not been found. 1860 T to CD, anti-*Origin*, but kindly. 1860 May 13 CD to Hooker; "He is evidently a strong opposer to us". CCD8:207. 1863 CD to Hooker, about T's views on inheritance of acquired characters; CD wrote on "foreign paper" for forwarding to T in Calcutta. CCD11:35.

Thomson, Sir William, Baron Kelvin, 1824-1907. Physicist. T was amongst the most distinguished astronomical physicists of his day, but quite wrong about the age of the earth. Established correct value of absolute zero temperature. 1846-99 Prof. Natural Philosophy Glasgow. 1851 FRS. 1856 Royal Medal Royal Society. 1866 Kt. 1869 CD to Hooker, "I feel a conviction that the world will be proved rather older than Thomson makes it". CCD17:339. 1883 Copley Medal Royal Society. 1890-95 PRS. 1892 1st Baron Kelvin of Largs. First scientist to be elevated to the House of Lords. 1902 OM. 1904-07 Chancellor University Glasgow. 1880 May, T and Mrs paid a visit to Down where the "two men had a good deal of talk, and parted with mutual respect and admiration." Thompson, *Life of William Thomson, Baron Kelvin of Largs*, 2011, p. 758. Corresponded mostly with George Darwin.

Thorley, Anne Catherine Miss, 1829?-1911. From Tarporley, Cheshire. 1848-56 Governess at Down House. Came as 19-year-old when ED was pregnant with Francis Darwin. 1849 Nov. 2 took the children and went to London. T was paid £50 per annum until 1 Jan. 1857. 1851 Mar. 28 went to London then on to Malvern. T was present at Malvern when Anne died. The Darwins, just a day after Anne's death, had in mind to send T home in order for her to "recover her cheerfulness". 1851 Apr. 26, CD wrote the most complimentary letter to Mrs Thorley, full of praise and appreciation of her daughter's admirable conduct throughout their ordeal. 1852 Aug. 12 came with her brother John T. 1854 Aug. 12 Mrs T visited. 1855 T helped CD with studies of wild plants. She left 1856 Mar. 20 and returned in July. She was replaced by Miss Pugh. 1882 T was on "Personal Friends invited" list for CD's funeral. ED and T became lifelong friends. In her old age, ED wrote often to the sisters. She wrote her last letter to T in 1896 Jun.

Thorley, Maria Emily, 1832?-1917. Listed as Governess at Down House in 1856. 1861 Census. Sister of Catherine T. Came in as interim governess 1856 Oct. 23 after Catherine T left in July. Left in 1857 Jan. 6. After that she and her sister would come and go even when Miss Pugh became the governess. *ED's diary*. Went to Ilkley with the Darwins in Oct. 1859. Evans ed., *Darwin and women*, 2017, p. 199.

Thorley, Sarah, a sister? of Catherine and Emily. ?1848 Anne Darwin wrote to T. Keynes, *Annie's box*, 2001, p. 234. ED recorded S coming to stay 1854 Aug. 12-18.

Thwaites, George Henry Kendrick, 1812-82. Botanist, entomologist and microscopist. A frequent correspondent, especially on dimorphic flowers. CD discussed evolution with before *Origin*. See also *Bulletin of the BMNH, Historical Series*, 4, 1972. 1847 CD met at BAAS meeting in Oxford. CCD4. 1849- Director of Botanic Garden Peradeniya, Ceylon. 1860 T was originally anti-*Origin*, but was coming round. CCD8. 1865 FRS. 1868 Jan. 31 and Oct. 26 CD sent T printed *Queries about expression* (F876). CCD16. T provided information about elephants for *Expression*. CCD16.

Tibetan. First edition in: 2000 *Origin* (F2148).

Tierra del Fuego, Argentina/Chile. Group of islands at southern tip of South America. See also Boat Memory, York Minster, Jemmy Button, Fuegia Basket, Richard Matthews, Thomas Bridges. 1834 Feb. 12-Mar. 12, Jun. 9-12 *Beagle* surveyed there, CD several times ashore. *Zoological diary*, pp. 123-135, 192-202. *Geological diary*: DAR32.85-122 and DAR34.157-191ff, both transcribed in *Darwin Online*.

Times, Mail coach from London to Cambridge. 1829 CD to William Darwin Fox, CD had travelled by. CCD1.

Times, The, newspaper. 1859 CD to Lyell, "the greatest newspaper in the world". CCD7:413. 1859 CD to Huxley, "For how could you influence Jupiter Olympus & make him give 31/2 columns to pure science? The old Fogies will think the world will come to an end." CCD7:458. 1863 CD to Gray, "The Times is getting more detestable,—but that is too weak a word,—than ever [on slavery]. My good wife wishes to give it up; but I tell her that is a pitch of heroism, to which only a woman is equal to.. To give up the 'Bloody Old Times' as Cobbett used to call it, would be to give up meat drink & air." CCD11:167. 1873 Jul. The Darwins began to take T again instead of the *Daily News*. ED to William Erasmus Darwin DAR219.1.87.

Timiryazev, Kliment Arkadeevich (Timiriazev), 1843-1920. Russian botanist and plant physiologist. Wrote several books on Darwinism. 1877 T visited Down House and had a two-hour talk with CD (F2093). 1878 Prof. Botany Moscow. 1878 CD to Dyer, suggesting that Dyer should get in touch with T about equipping physiological laboratories "who seemed so good a fellow". 1909 Honorary doctorate Cambridge University. 1911 Foreign Member Royal Society. 1920 U. Darvina v Daune, *Nauka demokratiia*, p. 105. T translated *Origin* into Armenian (F631, 1963). Wrote introduction to Bulgarian *Origin* (F632, 1946) Lithuanian *Origin* (F738, 1959) and Ukrainian *Origin* (F798, 1949). In Russian, translated *Origin* (F752, 1896), *Autobiography* (F1534, 1896), On the tendencies to form varieties (F370, 1939). Edited *Insectivorous plants* (F1245, 1900), *Journal* (F233, 1908). Recollection of CD K.A. Timiriazev: A visit to Darwin. *Archipelago* 9 (2006): 47-58. In *Darwin Online*.

Tineina. 1860 Do the Tineina or other small moths suck flowers, and if so what flowers?, *Entomologist's Weekly Intelligencer*, 8: p. 103 (*Shorter publications*, F1708). The editor replied that they do. See CCD8:261-2.

Tissot, James Jacques Joseph, 1836-1902. French painter and illustrator. In French better known as Jacques-Joseph T. Usually signs his caricatures "Coïdé". 1871 Settled in London. Returned to Paris in 1882. 1871 Sept. 30 Drew the Darwin caricature "Natural selection" for *Vanity Fair*. The caricature is not signed and usually, but incorrectly ascribed to Carlo Pellegrini ("Ape"). T had drawn more caricatures for *Vanity Fair* before establishing himself in London.

Tollet, Caroline Octavia, 1816?-40. Daughter of George T. 1839 Married Rev. Thomas Stevens.

Tollet, Charles, 1796-1870. Son of George T. T changed his name to Wicksted on inheriting Shakenhurst, Worcestershire, in 1814. Sporting and family friend of CD.

Tollet, Ellen Harriet, 1812-90. Daughter of George T. She was a life-long friend of ED. Emma Wedgwood in 1833 recorded "Ellen & Georgina came". In 1839 Mar. 14, T dined with the newly-weds. All through her life, she would visit and stay with the Darwin family. 1890 "this death cuts off my last link with past life". ED2:287. In her diary, ED wrote "Ellen T. death".

Tollet, George, née Embury, 1767-1855. Agricultural reformer of Betley Hall, Staffordshire. Close friend of Josiah Wedgwood [II]. T assumed the surname under a settlement, and inherited Betley Hall in 1796 from his cousin Charles T who had died that year. T's wife, Frances Mary Jolliffe, was a very strict Calvinist. Six or seven daughters, one son. The children were personal friends of ED and CD from childhood. 1816 John Wedgwood lost his fortune in a crisis at Davison's Bank, of which he was a partner. T let him have a house on his estate at a low rate "for the pleasure of their society". ED1:102. 1839 May 10 T answered Questions about the breeding of animals. *Jrnl for the Soc. of the Bibliography of Nat. Hist.*, 5: pp. 220-5, 1969, CCD2.

Tollet, Georgina, 1808-72. Daughter of George T. Records of her friendship with Emma Wedgwood go as far back as 1813. 1826 Lost an arm through an abscess. 1859 CD asked John Murray to send manuscript of chapters 1-3 of *Origin* for her to read. She finally read whole manuscript. T was then of 14 Queen Anne Street, London, close to home of Erasmus. "The lady, being an excellent judge of style, is going to look out for errors for me". CCD7:278. "one lady who has read all my M.S. has found only 2 or 3 obscure sentences." CCD7:296.

Tollet, Marianne, c.1798-1841. Third daughter of George T. 1829 Married William Clive. 1839 Mar. 2 CD and ED dined at the Clives with Mr Nihill, Mr Russell and Mr Henslow. 1839 Mar. 19 ED called on T. Lost her baby girl in 1839 Mar. 25. ED wrote "Marianne's baby died", two days after T took to bed.

Tomes, Sir Charles Sissmore, 1846-1928. FRCS, FRS. Wrote the *Manual of dental anatomy, human and comparative* (1876). Corresponded with CD on matters relating to inherited dental abnormalities and the enamel organ. CCD22:82.

Tommy. A quiet cob (a small and stout breed of horse) which CD rode for his health on Bence Jones's Feb. 1866 advice. ED recorded on 4 Jun. 1866 "Ch. began riding". 1868 CD took T to Isle of Wight by train. Mentioned in 1868 Perception in the lower animals, (*Shorter publications*, F1759). 1869 Apr. 8 or 9 T stumbled and rolled on CD on Keston Common, bruising him badly. However, it is incorrect, as Desmond and Moore wrote "his riding days came to an end". *Darwin*, 1991, p. 568. 1869 Sept. 26 CD had been riding T when Anton Dohrn visited Down House. ED2:195. CD was still riding Tommy, against the better judgement of his family, in 1870. See CCD18. CD was photographed astride Tommy in front of Down House.

Tony. A male dog owned by Sarah Elizabeth Wedgwood [II]. 1880 Nov. When Sarah Elizabeth Wedgwood [II] died, T was taken over by CD. ED2.

Torbitt, James, c.1822-95. Wine merchant, grocer and agriculturalist of Belfast. 1876-1882 Exchanged more than 100 letters with CD, mostly regarding a project to produce potatoes resistant to blight by selection. 1876-78 CD, with Farrer and Caird, subscribed to keep his work on potato blight going. CD to Farrer on the matter. 1878 CD to T, pessimistic on same subject. See F1978, F1979 and F2532. Notes and letters concerning T's experiments on potato disease, 1876- in DAR52.

Torquay, Devon. 1861 Jul. 1-Aug. 26 CD had family holiday at. CD's brother also joined the family but left on Aug. 22. CD made observations on flight paths of male humble bees there. CD wrote letters to many people from there. CCD9. CD began writing *Orchids* while there. See van Wyhe ed., 'Journal', (DAR158).

Town, Daniel, 1830-?. Carpenter in Down. 1875 Made repairs to Infant school. CD payment of £1 for estimate not covered by the school committee. CCD23.

Travers, William Thomas Locke, 1819-1903. New Zealander, born in Ireland and educated in France. Lawyer, magistrate, politician, botanist and explorer. T to CD on natural hybridization in plants. S.H. Jenkinson, *N.Z. Centennial Surveys*, 1940, xii, p. 121. 1849 Went to Nelson, New Zealand. Later moved to Christchurch. 1853-81 Intermittently MP for several terms.

Treat, Mrs Mary Lua Adelia, 1830-1923. American botanist, entomologist and popular science writer of New Jersey. 1871-76 Corresponded with CD on botanical subjects. 1873 Provided information on *Dionaea* for *Insectivorous plants*. *American Naturalist*, (Dec.), 1873, p. 715. CCD21. Portrait in Evans, *Darwin and women*, 2017, p. 80.

Trimen, Roland, 1840-1916. Entomologist. Civil servant in South Africa since 1860. Poulton, *Darwin and the Origin*, 1909, pp. 213-46. 1859 T's reminiscences of CD, "I...saw Darwin in the Bird Galleries...A clerical friend with me, also a naturalist...echoed White's warning by indicating Darwin as 'the most dangerous man in England'". 1863 "On the fertilisation of *Disa grandiflora*", *Jrnl. Proc. of Lin. Soc. of London (Bot.)*, 7: p. 144, written by CD from T's notes. CCD11. 1863-71 letters from CD to T, on orchids and on evolutionary problems in the *Lepidoptera*. 1867 Dec. 28-29 T stayed at Down House. 1868 Mar. T lunched with CD at 4 Chester Terrace, London, home of Sarah Elizabeth Wedgwood [II]. 1872-93 Curator South African Museum Cape Town. 1883 FRS. 1910 Darwin Medal, Royal Society.

Tristram, Henry Baker, 1822-1906. Anglican clergyman, traveller and ornithologist. 1859 T accepted a limited role for natural selection for the formation of varieties based on the 1858 Darwin-Wallace papers, but an opponent of the theory once clearly explained in *Origin*. R. England, Natural Selection Before the Origin', *Jrnl. of the Hist. of Biol.*, 1997. 1860 T to A. Newton, "The infallibility of the God Darwin and prophet Huxley". Wollaston, *Life of Alfred Newton*, 1921, p. 122. 1868 FRS.

"Trotty", or **"Trotty Veck"**, Childhood nickname of Henrietta Darwin.

Trowmers, House at Luxted Road, Down, north of Down House, earlier known as Trowmer or Tromer Lodge, and later as Tower House, named after original family who owned it. 1868-80 Taken by Sarah Elizabeth Wedgwood [II] and where she died in 1880. Howarth, *A history of Darwin's parish: Downe, Kent*, 1933, p. 15.

Truelove, Edward, 1809-99. Publisher, bookseller and socialist. 1870 Published Robert Dale Owen's *Moral physiology*, New Edn, on Contraception; originally published in the USA in 1830 or 1831. 1878 T was convicted in High Court for publishing *Moral physiology*. 1878 CD to a son of T, unable to sign a memorial against the conviction because he had not heard of T before the trial. CCD26.

"Truttle's" (Truettel's & Würtz and Richter), A London bookseller, 30 Soho Square. 1833 Henslow to CD, "I will ask your brother to enquire at Truttle's for Cuvier, *Anatomie des mollusques*, Paris, 1817. CCD1:293, Barlow 1967, p. 67.

Tuckwell, William, 1829-1919. Anglican clergyman and schoolmaster. Advocate of teaching science at schools. 1860 T was present at Oxford BAAS meeting and wrote it up in *Reminiscences of Oxford*, p. 50, 1900.

Turkish. First editions in: 1968 *Descent* (F1137). 1970 *Origin* (F796). 2000 *Autobiography* (F2537). nd *Expression* (F2409).

Turnbull, George Henry, 1819-80. Building contractor who resided at the Rookery, Down. Became a subscriber to the Down Coal and Clothing Club in 1854-55 until 1872. His gardener, John Horwood, oversaw the construction of CD's hothouse in 1863. Three letters from CD to T survive.

Turnbull, Lizzy, b.1846? ED called on Mrs Turnbull in 1855 Feb. 22. T visited Down House in 1856 Oct. 18. Calls to each other at least until 1872. *ED's diary.*

Turner, Dawson, 1775-1858. Banker, naturalist and botanist. Father of Maria Sarah T, grandfather of Joseph Dalton Hooker. 1802 FRS.

Turner, Edward Francis, 1850-1933. T was for many years solicitor to the Darwin family. See Turner, *The duties of solicitor to client*, 1893. See W. M. Hacon.

Turner, James Farley, ?-1841. 1826 Christ's College, Cambridge there a friend of CD: "I remember one of my sporting friends, Turner, who saw me at work on my beetles, saying that I should some day be a Fellow of the Royal Society, and the notion seemed to me preposterous." *Autobiography*. 1834-41 Vicar of Kidderminster.

Turner, Maria Sarah, 1797-1872. Eldest daughter of Dawson Turner. Mother of J.D. Hooker. 1815 Married Sir William Jackson Hooker. 2 sons, 3 daughters.

Turner, Sir William, 1832-1916. Physician. CD met at Royal Society. CD sent T 4th edn of *Origin*. 1866 T supplied information for *Descent*. 1867-1903 Prof. Anatomy University of Edinburgh. 1871 T to CD, pointed out CD's confusion of intercondyloid foramen in the humerus with the supracondyloid foramen, in *Descent*, 1:28. CCD19. 1877 FRS. 1901 KCB.

Twain, Mark [Samuel Clemens], 1835-1910. American author. "Professor Norton…said: 'Mr. Clemens…Mr. Darwin took me up to his bedroom and pointed out certain things…he said: 'The chambermaid is permitted to do what she pleases in this room, but she must never touch those plants and never touch those books on that table by that candle. With those books I read myself to sleep every night.' Those were your own books." Howell ed. 1907. *Mark Twain speeches*, pp. 33-35, (F2102). In 1876 the Norton's visited the Darwins. See Norton, Charles Eliot.

Tweedie, John, 1775-1862. Scottish gardener and plant collector at Buenos Aires. Owned a garden at Retiro, apparently visited by CD. Chancellor & van Wyhe, *Beagle notebooks*, 2009. See Ollerton, Chancellor and van Wyhe, John Tweedie and Charles Darwin in Buenos Aires. *Notes and records of the Roy. Soc.* (2012) 66(2): 115-124.

Tyke. 1881 A male family dog at Down House.

Tyler, Anne Maria, 1789-1855. Daughter of Sir George T of Cottrels, Glamorganshire. 1836 Married Thomas (Tom) Josiah Wedgwood.

Tyler, Helen Mary, 1866 Married John Darwin Wedgwood.

Tylor, Sir Edward Burnett, 1832-1917. Anthropologist. 1871 FRS. 1871 CD to T, on receiving a copy of T's book *Primitive culture and anthropology*. 1872 Jun. 6 visited CD, *ED's diary*. 1883 Keeper University Museum Oxford. 1884-95 Reader in Anthropology Oxford. 1896 First Prof. Anthropology Oxford. 1912 Kt. Recollections of CD in Unveiling the Darwin statue at the museum. *Jackson's Oxford Jrnl.* (17 Jun. 1899): 8, transcribed in *Darwin Online*, (F2169).

Tyndall, John, 1820-93. Physicist. A distinguished scientific populariser. Member of the X Club. Born in Ireland, moved to England in 1842. 1852 FRS. 1853-87 Prof. Physics Royal Institution. 1864 CD to Hooker, "I am sorry to hear that Tyndall has grown dogmatic. H. Wedgwood was saying the other day that T.'s writings & speaking gave him the idea of intense conceit; I hope it is not so, for he is a grand man of science." CCD12:391. 1867-87 Superintendent of Royal Institution. 1868 Oct. 24-31 T stayed the night at Down House with Gray and the Hookers. T returned the following year on Apr. 25. Hooker was present. 1874 Lyell to CD, congratulating him on T's Presidential Address to BAAS at Belfast, "you and your theory of evolution may be fairly said to have had an ovation". CCD22:431. The Address with additions published London, 1874 and in *Darwin Online*. 1875 Oct. 16 visited with the Huxleys and stayed the night. 1878 Dec. 7 Huxleys and Ts visited again. 1882 T was on "Personal Friends invited" list for CD's funeral. [1875] Recollections of CD in T, 1898, *New fragments*, p. 388. Transcribed in *Darwin Online*.

Ukrainian. First editions in: 1936 *Origin* (F797). 1936 *Autobiography* (F2423).

Ullswater. 1881 May 27-Jul. 4 CD with ED, William, Henrietta and Bernard Darwin spent holiday at Glenrhydding (Glenridding) House, near Patterdale, Cumberland, on shores of Ullswater; CD's last holiday. H.P. Moon, *Archives of Nat. Hist.* 10, 1982, pp. 509-14. 1881 May 27 ED wrote "came on to Glenred". Went to visit Mrs Ruck who joined in the holiday. July 4 went to Penrith on the way home.

Union Bank of London. 1873 CD banked with. 1979, Sotheby's, Jun. 18, a £50 cheque to Sydney Sales. 2006 An 1872 cheque from this bank was discovered inside a photograph frame at Christ's College, Cambridge. See John van Wyhe, A Darwin manuscript at Christ's College, *Christ's College Mag.* (2007) no. 232, pp. 66-8.

Unione Tipografico-Editrice Torinese. Publisher of Turin. 1871-83 Published 12 first Italian editions of CD's works. 1875 Also 2d Italian edn of *Origin* (F707). ?1870s *Forms of flowers* (F2472). 1871 *Origin* (F1088). ?1872 *Journal* (F211). 1874 *Coral reefs* (F2463). 1876 *Variation* (F920). 1878 *Expression* (F1200). 1878 *Cross and self-fertilization* (F1269). 1878 *Movements and habits* (F863). 1878 *Insectivorous plants* (F1242). 1882 *Descent* (F1088a). 1882 *Earthworms* (F1407). 1883 *Orchids* (F823).

Unitarian Chapel, Shrewsbury. Founded 1662. Rev. George Case was pastor 1797-1831. As their Wedgwood mother was a liberal Unitarian, CD and siblings were taken to church there, until her death in 1817. Now displays a sign "Charles Darwin worshipped here when young." Pattison, *The Darwins of Shrewsbury*, 2009.

United States of America. CD never visited. First editions published in: 1846 *Journal* (F16). 1860 *Origin* (F377). 1868 *Variation* (F879). 1871 *Descent* (F941). 1872 *Expression* (F1143). 1875 *Insectivorous plants* (F1220). 1876 *Climbing plants* (F838). 1877 *Orchids* (F802). 1877 *Cross and self fertilisation* (F1250). 1877 *Different forms of flowers* (F1278). 1880 *Erasmus Darwin* (F1320). 1881 *Movement in plants* (F1327). 1882 *Earthworms* (F1363). 1883 *Animal intelligence* (F1419). 1884 Essay on instinct (F1435). 1887 *Life and letters* (F1456). 1889 *Coral reefs* (F278). 1891 *Volcanic islands and South America* (F283). 1903 *More letters* (F1549). 1908 *Autobiography* (F1478). 1956 Biographical sketch of an infant (F1306). 1975 *Zoology of the Beagle*, Part V, *Reptiles* only (F9a). Most American first editions were published by Appleton. Notable exceptions: *Journal* (Harper & Brothers) and *Variation* (Orange Judd & Co.).

Upper Gower Street, London. No. 12, later 110 Gower Street, first home of CD and ED on marriage in 1839. CD called it "Macaw Cottage" from the gaudy colours of the walls and furniture. ED2:19. It was rented furnished, with a long thin garden backing on to Gower Mews North, later Malet Place. Staff: Gardener, Williams; Menservants, Edward, Jordan, Parslow. 1838 Dec. 31 CD moved in. 1839 Jan. 29 ED moved in. William and Anne were born there. 1842 Sept. 14 ED left for Down House. 1842 Sept. 16 CD left. For many years the house was part of Messrs Shoolbred's warehouse system. 1904 Dec. 13 A London County Council blue plaque was put up. 1929 Woodcut of the house in *Lancashire Evening Post*, (30 Sept.): 4. 1940 Sept. It was bombed and not repaired. A photograph of the bombed house is in Barlow, *Charles Darwin and the voyage of the Beagle*, 1945. 1961- Site now part of University College London, which bore the London County Council blue plaque to "Charles Darwin Naturalist", which was originally on the house. The present plaque with different wording was put up 1961 by London County Council.

Upper House, Barlaston, Staffordshire. 1845 Built as home of Francis Wedgwood, grandson of Josiah Wedgwood [I].

Usborne, Alexander Burns (Jimmy), 1808-1885 Master's Assistant on 2d voyage of *Beagle*. Mount Usborne on West Falklands was named after him. Went on 3d voyage. Surveyed New Zealand for FitzRoy. FitzRoy, *Jrnl. Roy. Geogr. Soc.*, 6: pp. 311-43, 1836. 1835 Took command of small schooner *Constitution* and surveyed the whole coast of Peru, after *Beagle* had left for Galapagos Islands. 1836 Oct. Returned to England via Cape Horn. 1836 before Nov. the schooner was then sold. 1840 Called on CD in London. 1867 Captain. 1887 Was alive LL1:221.

Valdivia, Chile. 1835 Feb. 8-22 *Beagle* at. 1835 Feb. 20 Earthquake. CD was on board and FitzRoy in the town. *Geological diary*: DAR35.343-353, in *Darwin Online*. *Beagle diary*, pp. 286ff. CD there spells "Baldivia".

Valparaiso, Chile. 1834 Jul. 23 *Beagle* arrived. 1834 Aug. 14-Sept. 27 CD stayed ashore and made expedition inland. CD then ill until end of Oct., when *Beagle* returned and set out for Chiloe. 1835 Mar. 11 *Beagle* at again. *Zoological diary*, pp. 238-45. *Geological diary*: DAR35.218-231, DAR35.371-376 and DAR36.419-461. *Beagle diary*, pp. 249ff. All transcribed in *Darwin Online*.

"Van John". University slang for *vingt-et-un*, "twenty-one", or blackjack, a card game. 1829 CD to William Darwin Fox, from Cambridge, "A little of Gibbon's History in the morning, and a good deal of *Van John* in the evening". CCD1:79. 1880 CD to John Maurice Herbert mentions *V*, again in italics. De Beer 1968, p. 73.

The English pouter, from *Variation*.

Variation. 1867-69 Title in Russian [*On the origin of species. Section I. The variation of animals and plants under domestication. The domestication of animals and the cultivation of plants*], St. Petersburg, translated from English corrected proofs by V.O. Kovalevskii; issued in 7 parts, of which parts 1-4 appeared in 1867, preceding the 1st English edn. (F925, F926).

1868 *The variation of animals and plants under domestication*, 2 vols., (F877), 1st issue Jan., 4 lines errata in vol. 1, 7 in vol. 2. 2d issue with corrections (F878), 1 line erratum in vol. 1. 1875 2d edn. (F880). 1969 Facsimile of 2d issue (F908). CD's longest book. First foreign editions: 1957 Chinese (F909a.1). 1880 Dutch (F2372). 1868 French (F912), German (F914.1), USA (F879). 1876 Italian (F920).1888-89 Polish (F922). 1959 Hungarian (F919). 1937 Japanese (F921a). nd Portuguese (2062.11). 1963 Romanian (F924). 1867-68 Russian (F925). List of presentation copies is in CCD16 Appendix IV, *Variation* 2d edn is in CCD24 Appendix III.

Variations. 1862 Variations effected by cultivation, *Jrnl. of Horticulture and Cottage Gardener*, 3: p. 696 (*Shorter publications*, F1721), CD sent queries for his research for *Variation*. 1873 G.H. Darwin, Variations of organs, *Nature*, 8: p. 505 (*Shorter publications*, F1763), gives his father's views.

Variegated Leaves. 1844 Variegated leaves, *Gardeners' Chron.*, no. 37: p. 621 (*Shorter publications*, F1667). "Mr. Groom has stated in last Number that the leaves of some of his Pelargoniums have become regularly edged with white in consequence of his having watered the plants with sulphate of ammonia".

Vaux, William Sandys Wright, 1818-85. Antiquary and ancient geographer. 1841-Department of Antiquities, British Museum. 1855-77 President Royal Numismatic Society. 1856 CD to Henry Ambrose Oldfield. CD had consulted V about *Variation* but V referred him to Oldfield. CCD6:107. 1861-70 Keeper Department Coins and Medals, British Museum. 1868 FRS.

Vaynor, Wales. 1826 Oct. 30 CD visited on riding tour with Caroline. 'Journal'.

Vegetable Mould and Earthworms see *Earthworms* and Wormstone.

Vegetarianism. CD to Karl Höchberg (1853-85), 1880. Darwin's reply to a vegetarian. *Herald of Health and Jrnl. of Physical Culture* n.s. 31: 180. (*Shorter publications*, F1984). CD had seen vegetarian miners in Chile working extremely hard, but the gauchos of Argentina ate only meat and were also extremely active.

Veitch, James, 1792-1863. Nurseryman, with his son James Jr., of Royal Exotic Nursery, King's Road, Chelsea, London. The firm was started by John V sometime before 1808 on the Killerton estate, which was extended several times. 1861 V supplied orchids for CD's work, "Mr James Veitch has been *most* generous". CCD9:321.

Veitch, James Jr., 1815-69. Nurseryman, with his father James V, of Royal Exotic Nursery, King's Road, Chelsea, London. J.H. Veitch, *Hortus Veitchii*, 1906.

Venables-Vernon-Harcourt, Colonel Francis 1801-80. CD called "Colonel Vernon", met at Buenos Aires. 1832 Nov. 11 CD to Caroline "a brother-in-law of Miss Gooch" who CD had met at Montevideo who was doing a tour, on to Lima then overland to Mexico. CCD1:277. "great Traveller". *Buenos Ayres notebook*, Chancellor & van Wyhe, *Beagle notebooks*, 2009. See *Beagle diary*, pp. 115-6.

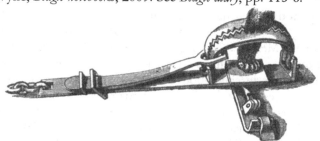

Steel vermin trap from "An appeal" (1863) by CD and ED.

"Vermin and Traps". 1863 Vermin and traps, *Gardeners' Chron.*, no. 35: pp. 821-2 (*Shorter publications*, F1728). 1863 An appeal [against steel vermin traps]. By ED and CD. The appeal on the subject in *Gardeners' Chron.* was sent to the *Bromley Record* by ED and privately printed as a pamphlet (F1931). See also CCD11, Appendix IX, pp. 776-81 and *Darwin Online*, which reproduce the appeal. Henrietta recalled how hard ED worked in 1863 "to have some humane trap substituted for the cruel steel trap in common use in game-preserving." A prize was offered if the invention was both good and humane. According to Henrietta, "No trap was invented which was portable, cheap, and effective." ED2:196. See Atkins 1976, pp. 78-84. See especially the introduction to Appendix IX CCD11.

Verrel, James, 1814-86. Newsagent in Bromley High Street. CD made monthly payments for stationery, stamps and newspapers. Browne, *Power of place*, 2002, p. 12.

Vestiges. 1844 *Vestiges of the natural history of creation*. An anonymous popular work on a progressive form of evolution, by Robert Chambers. V did not, as commonly believed, discourage CD from publishing his own theory. 1845 CD to Hooker, calls author "Mr Vestiges", although he identified the author correctly. CCD3:261. CD letter on in F2263. CD's notes on reading V in DAR205.1.37 and DAR205.9.215, transcribed in *Darwin Online*. On V see Secord, *Victorian sensation*, 2000.

Vieweg, Friedrich, und Sohn, Publisher of Brunswick, Germany. 1844 Published *Journal* (F188), the 1st translation or printing abroad of any of CD's books, and the only translation of the 1st edn of that work.

Villa Franca, Ignacio Francisco Silveira da Motta, Baron de, 1815-85. A Brazilian politician and farmer. 1881 CD to Romanes, V "wrote to me from Brazil about two years ago" on sugar-cane varieties. ML1:390. 1882 CD to Romanes, R would prepare paper on sugar-cane hybrids for the press, see "Villa Franca and Dr. Glass", *Proc. of Lin. Soc. of London*, pp. 80-1. ML1:389 (misdated 1881); *Calendar*. The paper was read at the Linnean Society meeting of May 4th, 1882. See F2168 *Darwin Online*.

Vincas. 1861 Fertilisation of Vincas, *Gardeners' Chron.*, (15 Jun.): p. 552 (*Shorter publications*, F1836). 1861 Vincas, *ibid.*, no. 37 (14 Sept.): pp. 831-2 (*Shorter publications*, F1716). 1869 [Extract of a letter on fertilisation of *Vinca* by insects] In Bennett, Fertilisation of winter-flowering plants. *Nature*. 1 (11 Nov.): 58. (*Shorter publications*, F1971). *Vinca* is a genus of flowering plants which includes Periwinkles.

Vinchuca or **winchuka**. American Spanish, *Quechua wihchuykuk*. CD spelled it "benchuca". A bloodsucking insect, *Triatoma infestans*. CD recorded in his *Beagle diary*, p. 315, in 1835 near Mendoza "At night I experienced an attack, & it deserves no less a name, of the Benchuca, the great black bug of the Pampas. It is most disgusting to feel soft wingless insects, about an inch long, cawling [sic] over ones body; before sucking they are quite thin, but afterwards round & bloated with blood, & in this state they are easily squashed." This entry was based on the initial notes in his *St. Fe notebook*, p. 152a, Chancellor & van Wyhe, *Beagle notebooks*, 2009.

Vines, Sydney Howard, 1849-1934. Botanist. Reader in Botany Cambridge. 1881 Oct. 20 CD and ED took tea with in Cambridge. 1881 Nov. CD to V on plant chemistry. LL3:346-347. 1883 May 12 called on ED. 1884 Jan. 24 ED had Messrs Vines, Ward & Pryor in her diary. 1885 Apr. 23 Vines dined. 1885 FRS. 1888-1919 Sherardian Chair in Botany Oxford University. 1900-04 President Linnean Society.

Virchow, Rudolf Ludwig Carl, 1821-1902. Pathologist, anthropologist and politician. Advocate of social reform. 1849 Chair Pathological Anatomy University Würzburg. 1856- Prof. Pathological Anatomy Charité Berlin. Director Pathological Institute for 20 years. 1877 V gave an address at Munich connecting evolution with socialism, published as *Die Freiheit der Wissenschaft im modernen Staat*, Berlin. 1878 Translated into English. Haeckel replied to it. 1878 V seconded CD's election to

Königlich-Preussische Akademie der Wissenschaften, Berlin. 1881 CD sat between V and Donders at the Int. Congr. Med. 1884 Foreign Member Royal Society.

Vivisection. CD's part in the agitation and Commission on this subject are considered in LL3:199-210 and ML2:435-441. See also Sir David Ferrier. 1875 CD warned Romanes not to discuss experiments on animals in front of Darwin ladies, since it would horrify them. CCD23. 1881 CD letter to Frithiof Holmgren, *The Times*, Apr. 18, *British Medical Jrnl.*, 1: p. 660, *Nature*, 23: p. 583 (F1792). Also in anti-Vivisection pamphlet by George Jesse (F1356), also in LL3:205-206, and in Sweden. 1881 "Mr Darwin on vivisection", *The Times*, Apr. 22, also in LL3:207-208 (F1793). 1881 CD to Romanes, about *The Times* letter, "I thought it fair to bear my share of the abuse poured in so atrocious a manner on all physiologists". Romanes 1896, p. 116. See also F1352, F1354, F2146 (*Shorter publications*). 1875 Apr. 9-12, staying at Bryanston, CD feeling rather well was "hard at work about vivisection". Nov. 1 travelled to London on account of the subject of vivisection. 1892 Oct. 26 in a letter to Henrietta, ED asked "Have you read any of the vivisection? The incredible carelessness of the accusations calls for any amount of reprobation." See F1356 for pamphlet of letters. CD's notes on vivisection are in DAR139.17.

Vivisection Commission. Members: Viscount Cardwell as Chairman, Baron Winmarleigh, W.E. Forster, Sir J.B. Karslake, T.H. Huxley, Prof. J.E. Erichsen, R.H. Hutton, with Nathaniel Baker, Secretary. 1875 Nov. 8 CD gave evidence before it in London. Viscount Cardwell, the Chairman, came to the door to receive him. 1876 *Report of the Royal Commission on the practice of subjecting live animals to experiments for scientific purposes*, London, HMSO Command 1397; CD's evidence p. 234, paragraphs 4662-4672 (*Shorter publications*, F1275). 1876 *Digest of evidence* etc., Command 1397, I, CD's evidence p. 34 (F1276). See F2146, in *Darwin Online*.

Volcanic islands. 1844 *Geological observations on the volcanic islands visited during the voyage of H.M.S. Beagle, together with some brief notices of the geology of Australia and the Cape of Good Hope. Being the second part of the geology of the voyage of the Beagle, under the command of Capt. Fitz-Roy, R.N. during the years 1832 to 1836*, London (F272), appendices by George Brettingham Sowerby [I] and William Lonsdale. Originally advertised in 1838 by Smith Elder & Co. as "Geological Observations on Volcanic Islands and Coral Formations". 1851 Combined edn of the three parts from unsold sheets, with new preliminaries (F274). 1876 2ᵈ edn, combined with Part 3 *South America* (F276). 1891 3ᵈ edn, combined with Part 3 as 2ᵈ edn (F282). 1972 Facsimile of an 1896 issue (F307). First foreign editions: 1877 German (F312). 1891 USA (F283). 1902 French (F310). 1936 Russian (F323).

Volcanic Phaenomena and Mountain Chains. 1838 On the connexion of certain volcanic phaenomena, and on the formation of mountain-chains and volcanos, as the effects of continental elevations, *Proc. of Geol. Soc.*, 2: pp. 654-60 (*Shorter publications*, F1649). 1840 On the connexion of certain volcanic phenomena in South America; and on the formation of mountain chains and volcanos, as the effect of the same powers by which continents are elevated, *Trans. of the Geol. Soc.*, 5: pp. 601-31 (*Shorter publications*, F1656). One of CD's most important geological papers in which he argued for the progressive long-term changes to the geology of South America due to incremental, non-catastrophic causes.

Volcanic Rocks and Glaciers. 1845 Extracts from letters…on the analogy of the structure of some volcanic rocks with that of glaciers, *Proc. of Roy. Soc. Edinburgh*, 2: pp. 17-18 (*Shorter publications*, F1670); letters from CD to Edward Forbes. CCD5:44.

Voyage of the Beagle, see *Journal of researches*. This title was first used on the title page in the Harmsworth Library edn, 1905 (F106), in *Darwin Online*.

Vries, Hugo de, 1848-1935. Dutch botanist and one of the first geneticists. Recollection of and CD letters to de Vries in W. Peter van der Pas, 1970, *Janus* 57: 173-213 (F2106, in *Darwin Online*) "the topics about which de Vries and Darwin were corresponding were not at all topics on evolution or heredity".

Waddington, Frances, Baroness von Bunsen, 1791-1876. Daughter of Georgina Mary Ann Port and Benjamin Waddington. Welsh painter and author. 1817 Married Christian Charles Josias baron von Bunsen. ED2:438.

Wagner, Johann Andreas, 1797-1861. German zoologist and geologist. Prof. Zoology Munich and Curator Zoologische Staatssammlung. 1861 Author of Zur Feststellung der Artbegriffes, *München Sitzungb.*, p. 301. 1863 CD to Falconer, "Poor old Wagner always attacked me in a proper spirit and sent me two or three little brochures, and I thanked him cordially." CCD11:12.

Wagner, Moritz Friedrich, 1813-87. German traveller and naturalist. Brother of Rudolph W. With his migration theory pioneered the concept of geographical speciation in evolution. 1836-39 Explored Algiers. 1852-55 With Carl Scherzer travelled through North and South America and the Caribbean. 1868 CD to Weismann on W's views about evolution in his *Die Darwin'sche Theorie und das Migrationsgesetz*, English translation, 1873, in *Darwin Online*. 1872 CD to Weismann refers to W's views. 1876 CD to W, about his evolutionary essay in *Das Ausland*, May 31, 1875.

Wagner, Rudolph, 1805-64. German anatomist and physiologist. Brother of Moritz Friedrich W. 1832-40 Prof. Zoology and Comparative Anatomy Erlangen. 1840-64 Prof. Zoology Göttingen. 1860 Abstract of Louis Agassiz, *Essay on classification*, 1857, "Louis Agassiz's Principien der Classification…mit Rücksicht auf Darwins Ansichten", *Göttingischen Gelehrten*, 1860. 1860 CD to Huxley, W had sent CD a copy of his "Abstract". 1862 letter on accepting evolution. Naturforschung und Theologie. *Jahrbücher für deutsche Theologie* 7: 167 (F2170).

Waite, George Derby, 1804-80. Surgeon-dentist; extracted CD's molars under chloroform. Published influential dental texts, including *The surgeon-dentist's anatomical and physiological manual* (1826). President of College of Dentists. See Hayman & van Wyhe, Charles Darwin and the dentists. *Jrnl. of the Hist. of Dentistry*. (2018) vol. 66.

Wales. CD visited Wales on ten occasions. 1813 Summer Family holiday at Gros, Abergele. 1819 Jul. Family holiday, Plas Edwards, Towyn. 1820 Jul. Riding tour with

Erasmus, Pistyll Rhaeadr. 1822 Jul. Holiday with Susan Elizabeth Darwin, Montgomery. 1828 Summer Reading party under George Ash Butterton, Barmouth. 1829 Jun. Beetle collecting with Frederick William Hope, Barmouth. 1830 Aug. Beetle collecting North Wales. 1831 Aug. Geology trip with Adam Sedgwick, Llangollen, Ruthin, Conway, Bangor, Capel Curig, then Barmouth alone. 1842 Jun?-Jun. 29 For geology, Capel Curig, Bangor, Caernarvon. Members of the family go over the years with Bessy most frequently. Henrietta 1865, "Boys" 1866, George and Henrietta 1868 etc. 1869 Jun. 12, returning home 30 Jul. Family holiday, Caerdeon, Barmouth. Van Wyhe ed., 'Journal', (DAR158). *ED's diary*.

Walker, Francis, 1809-74. Entomologist and traveller. Assistant at British Museum. 1838-43 W described CD's chalcid material from *Beagle* in eight articles and vol. 2 of *Monographia Chalciditum*, 2 vols., 1851-56 *Insecta Britannica. Diptera*, 3 vols., Reeve and Benham, London. In *Darwin Online*.

Walker, James, 1784-1856. Former army officer and owner of Wallerawang, Australia. In the *Sydney notebook*, Chancellor & van Wyhe, *Beagle notebooks*, 2009.

Wallace, Alexander, 1829-99. Physician and lepidopterist of Colchester. W is often referred to in *Descent* as an expert on silk moths. 1868 CD to John Jenner Weir, giving W's views on sexual selection in *Bombyx mori*. 1868 CD to Henry Tibbats Stainton, giving W's views on sex ratio in *Bombyx cynthia*.

Wallace, Alfred Russel, 1823 Jan. 8-1913 Nov. 7. 8th child (of nine) of English parents Thomas Vere W and Mary Ann Greenell. Born in Llanbadoc, a Welsh village near Usk, Monmouthshire (then part of England). W was not ethnically Welsh and never regarded himself as such. Traveller and naturalist. W's first employment was, by his eldest brother William, as a land surveyor starting in 1837 as apprentice. CD discussed evolution with before *Origin*. Although best known for his independent discovery of principle of evolution by natural selection, W made important contributions to science after the joint publication with CD in 1858. Most notable are his contributions to the study of the geographical distribution of species and biogeography. He was not, as commonly believed, the father of biogeography, a discipline that had already existed for decades. Hooker called W "Darwin's true knight" in 1868. Although CD and W were always on friendly terms and W visited Down House, there was never the intimacy that there was with Hooker, Falconer or Huxley; nor did they fully understand each other's scientific views. Biographies: J. Marchant, ed. 1916, *Alfred Russel Wallace, letters and reminiscences*, 2 vols.; 1972, H. Lewis McKinney, *Wallace and natural selection*, 1983; Peter Raby, *Alfred Russel Wallace. A life*, 2001; Ross Slotten, *The heretic in Darwin's court: the life of Alfred Russel Wallace*, 2004. John van Wyhe, *Dispelling the Darkness: Voyage in the Malay Archipelago and the discovery of evolution by Wallace and Darwin*, 2013. Wallace's complete publications are available freely online in John van Wyhe ed., 2012-. *Wallace Online* (http://wallace-online.org/).

The most important periods of his life were, as a collector of natural history specimens: 1848-52 In the Amazons, initially with Henry Walter Bates. W and Bates did not go to the Amazon, as popular legend has it, to solve "the problem of the origin of species", a phrase published by Bates in 1861 which was a modification of an original letter from W in 1847 which survives and makes no mention of a voyage or a problem. See John van Wyhe, A delicate adjustment: Wallace and Bates on the Amazon and "the problem of the origin of species". *Jrnl. of the Hist. of Bio.* (2014). 1852 Personal collection of specimens lost due to ship catching fire on way home. 1854-62 In Southeast Asia, then sometimes known as the Malay Archipelago. 1855 "On the law which has regulated the introduction of new species", *Annals and Mag. of Nat. Hist.*, 16: pp. 184-96. 1857 CD to W, "You say that you have been somewhat surprised at no notice having been taken of your [1855] paper in the Annals: I cannot say that I am; for so very few naturalists care for anything beyond the mere description of species." van Wyhe & Rookmaaker, *Wallace letters*, 2013, p. 139. Contrary to popular belief, this 1855 paper made no mention of evolution. 1858 Jun. 18 Friday. CD received letter from W, written at Ternate, Mollucas, enclosing his paper "On the tendency of varieties to depart indefinitely from the original type" in manuscript. CD wrote to Lyell the same day "Your words have come true with a vengeance that I shd be forestalled". CCD7:107. The date of sending and receipt are confirmed in van Wyhe & Rookmaaker, A new theory to explain the receipt of Wallace's Ternate essay by Darwin in 1858. *Biol. Jrnl. of the Lin. Soc.* 105 (1): 249-52, 2012. 1858 Jul. 1 Tuesday. Hooker and Lyell communicated CD and W's papers to Linnean Society, "On the tendency of species to form varieties and species by natural means of selection", *Jrnl. Proc. of Lin. Soc. of London, Zool.*, 3: pp. 45-62 (*Shorter publications*, F346). 1858 Jul. 25 CD sent "Some half dozen copies" of the offprint to W and "I have many other copies at your disposal" *Wallace letters*, 2013, p. 198. The whole episode is considered in detail in LL2:115-140 and van Wyhe, *Dispelling the darkness*, 2013. 1860 Dec. 24 W to Bates about *Origin*, "I do honestly believe that with however much patience I had worked up & experimented on the subject, I could never have *approached* the completeness of his book,—its vast accumulation of evidence, its overwhelming argument, & its admirable tone & spirit." *Wallace letters*, 2013, p. 239, CCD8:221. 1862 Returned home. 1866 Married Annie Mitten. 2 sons, 1 daughter: 1. Herbert Spencer W, 1867-74; 2. Violet Isabel W, 1869-1945 unmarried; 3. William Gore W, 1871-1951. 1868 Royal Medal Royal Society. 1868 CD to W, "I fear we shall never quite understand each other". CCD16:703. 1871 "I then applied to Mr Wallace, who has an innate genius for solving difficulties". *Descent*, 1:416. 1879 Dec. 17 CD and Hooker first raised the matter of W's income. CCD27.

On his return to England, his only income until 1881 was from sale of specimens, marking exams, editorial work for Lyell, investments and from authorship. 1881 Jan. 7 Granted a civil list pension of £200 per annum as successful outcome of lobbying by CD and friends. A draft of a sketch of W's career and other notes relating to the matter are in DAR91. See Colp, "I will gladly do my best": How Charles

Darwin obtained a civil list pension for Alfred Russel Wallace', *Isis* vol. 83, no. 1, Mar. 1992, pp. 3-26. 1882 W was Pallbearer at CD's funeral.

1889 in his *Darwinism*, pp. 46, 69, 79-89, W published extracts from unpublished notes by CD on variation in nature from *Natural selection* chapter IV, F2105. 1889 DCL Oxford. 1890 Darwin Medal, Royal Society (first recipient). 1892 Linnean Society Gold Medal. 1892-93 Offered FRS on proposal by Hooker and Thiselton-Dyer but delayed acceptance too long. Offered again in 1893 and accepted. 1908 Copley Medal Royal Society. 1908 Linnean Society Darwin-Wallace Medal (first recipient). 1908 OM. Up to 1913 W moved house often and had four houses built to his own design, the last Old Orchard, Broadstone, Wimborne, Dorset, where he died. The house was destroyed in 1964. A road through modern development called Wallace Court, now the name of a block of flats. 1913 Buried Broadstone. 1915 Nov. 1 Wall medallion to in Westminster Abbey, next to that of CD's. Main works: 1853 *Palm trees of the Amazon* [250 copies]. 1853 *Travels on the Amazon and Rio Negro*. 1869 *Malay archipelago*, 2 vols. 1870 *Contributions to the theory of natural selection*. 1876 *Geographical distribution of animals*, 2 vols. 1882 *Island life*. 1889 *Darwinism*. 1905 Autobiography: *My life*, 2 vols. Recollections of CD in The debt of science to Darwin. *Century Mag.* 25, 3 (Jan. 1883): 420-2 and The dawn of a great discovery "My relations with Darwin in reference to the theory of natural selection". *Black and White* 25: (17 Jan. 1903): 78. Since the 1960s W has been increasingly portrayed as "forgotten", overly disadvantaged and cheated or wronged by CD, his friends and by history in general. At the same time his accomplishments have come to be ever more exaggerated. He is now commonly described as "the greatest field biologist of the 19th century", the "father of biogeography" and one of or the most famous scientist in the world at the time of his death. None of these are correct. The latter error stems solely from hyperbolic language in interview reports by journalists and the obligatory panegyric of obituaries. No historians of science who specialise on the period agree with this assessment (originated by non-historians). Indeed, many scientists were *also* called the most famous scientist or greatest living naturalist etc. in the same years, e.g. Huxley (1884, 1893, 1895), Lord Kelvin (1887, 1896, 1897, 1900, 1905, 1922), Rudolf Virchow (1889, 1902), Richard Owen (1892), Tyndall (1893), William Crookes (1894), Louis Pasteur (1895, 1903, 1912), Louis Agassiz (1900), Haeckel (1901, 1909), Luther Burbank (1905), Nicola Tesla (1906), Marcellin Berthelot (1907), Élie Metchnikoff (1909), Oliver Lodge (1911) and Madamme Curie (1912). Unlike many other men of science at the time, W was not frequently caricatured in the press as he was insufficiently recognizable to the public. Yet scores of other men of science were. Most of them are today forgotten, unlike W.

Wallich, George Charles, 1815-99. Physician, marine biologist and professional photographer of Danish descent, born in India. Army surgeon and botanist in India. Superintendant Botanical Gardens Calcutta for several years. DSB, 1859-60 Naturalist on H.M.S. *Bulldog*. W was proposed by Huxley and Murchison. 1860 W sent CD a copy of his pamphlet *Notes on the presence of animal life at vast depths in the sea*. CD thanks W for. CCD8. 1861 CD met W at Linnean Society. CCD9. 1871 W provided CD with the photograph of the little girl with the hat in *Expression*, see DAR53.1.45. 1871 W visited CD at Down House. 1871 W made three CDV photographs of CD. 1882 Mar. 28 CD to W, on deepwater organisms and asking for a copy of his lecture on Protista. De Beer 1959, p. 59. 1898 Linnean Medal, Linnean Society.

Walpole, Lady Dorothy Frances (Fanny) Nevill(e), 1826-1913. Elder daughter of Horatio W, 3ᵈ Earl of Orford. 1875 Oct. W called on CD at Down House but he was ill. CD called on W several times in London. Biography: Ralph Nevill (son), *Life and letters of Lady Dorothy Nevill*, London, 1919, pp. 56-8, has reminiscences and one letter. CD signed W's birthday book, which was illustrated by Kate Greenaway. Told CD about her Siamese cat which was the colour of an otter and perhaps the first in England. Biography: Guy Nevill, *Exotic groves: portrait of Lady Dorothy Nevill*, 1984. 1847 Married Reginald Henry Nevill. 1851-78 W was an enthusiastic gardener at Dangstein, Rogate, Hampshire. 1861 CD to Hooker; W helped CD with *Orchids*, "responded in wonderfully kind manner & has sent a lot of treasures." CCD9:344. 1874 CD to W, thanking her for providing plants for *Insectivorous plants*, especially *Utricularia montana*, which lives in moss on trees, unlike the usual species which are aquatic. CCD22:465. 1875 Oct. 5-7, went with CD to Cat Show at Crystal palace as patrons. 1910 *Autobiography*, edited by Ralph Nevill: *Under five reigns*, pp. 106-12 (F2109 and F2284), has recollections of and five letters from CD, both transcribed in *Darwin Online*. Photograph in Evans ed., *Darwin and women*, 2017, p. 68.

Walpole, Colonel John, 1787-1859. Soldier and diplomat. Son of Horatio Walpole, 2ᵈ Earl of Orford. 1822-31 MP. 1833-41 Consul General at Santiago, Chile. 1834 Aug. 28 CD to FitzRoy mentions calling on "but he was in bed—or said so". Keynes, *Beagle record*, 1979, p. 235. 1841-49 Chargé d'Affaires, Chile.

Walsh, Benjamin Dann, 1808-69. Brother of John Henry W. Entomologist. C.V. Riley described W as "one of the ablest and most thorough entomologists of our time". ML1:248. 1838 Fellow of Trinity College, Cambridge. 1838 W emigrated to USA. 1864 W to CD, reintroducing himself; they had met in CD's rooms at Cambridge. W comments on *Origin*, "The first perusal staggered me, the second convinced me, and the oftener I read it the more convinced I am of the general soundness of your theory". 1867-69 W was State Entomologist of Illinois. 1868 CD to W, on 13- and 17-year cycles in cicadas. 1868 W to CD, he could not answer CD's *Queries about expression*. CCD16:314. 1869 Killed in a railway accident.

Walton Hall, near Pontefract, Yorkshire. Home of Charles Waterton 1845 Sept. CD visited and stayed the night. CCD3:58-60.

Walton, William, 1784-1857. Writer on Spain and Portugal. 1844 wrote *The Alpaca*, which included CD's comments and observations on guanacos. Walton, *The Alpaca*, 1844, pp. 20, 43-5, 50-1. (*Shorter publications*, F1833). Transcribed in *Darwin Online*.

Ward, **Nathaniel Bagshaw**, 1791-1868. Physician and botanist who invented Wardian cases, glass cases for transporting plants. 1861 CD sent W Gray's pamphlet *Natural selection not inconsistent with natural theology*. CCD9.

Wareham, Dorset. 1847 Jul. CD and family visited on way to holiday at Swanage.

Waring, **Anne**, 1662-1722. Daughter of Robert W. CD's great-great-grandmother. W inherited manor and hall of Elston, near Newark, Staffordshire, from George Lassels or Lascelles, her mother's second husband. See also Brass Close. 1680 Married William Darwin [VI, d. 1682].

Waring, **Robert**, ?-1663. Father of Anne W and origin of the forename Waring in the Darwin family. CD's great-great-great-grandfather.

Warrington, **George**, 1840-74. Author on religious topics. 1867 CD to W praising his paper on Darwinism in the *Transactions of the Victoria Institute*. CCD15:388-9. 1867 CD to Wallace, "Mr Warington has lately read an excellent & spirited abstract of the 'Origin' before the Victoria Inst. & as this is a most orthodox body he has gained the name of the Devil's Advocate. The discussion which followed during 3 consecutive meetings is very rich from the nonsense talked." CCD15:395

Warter, **John Wood**, 1806-78. Church of England clergyman and antiquary. Shrewsbury school friend of CD. 1824 Believed CD inclined to idleness, wrote to warn him against it. CCD1:10.

Waterhouse, **George Robert**, 1810-88. Mammalogist and entomologist. Keeper of Mineralogy and Geology at BMNH. A friend of CD and often at Down House. CD discussed evolution with before *Origin*. 1833 Founder of Entomological Society of London with F.W. Hope. 1837-45 W wrote 17 articles describing mammals and insects collected by CD during *Beagle* voyage. 1838-39 W wrote *Zoology*, Part II, *Mammalia*. 1843 CD to Lyell, "if Waterhouse is hired he will enjoy his seven shillings a day from the British Museum, as much as most men would ten times the sum!" CCD2:388-389. 1837 [An Australian insect] in F2015. 1840 [Chilean beetles] in F2010. 1841 [Notes on South American insects] in F2016. 1841 [Note on a ground-beetle found off the Straits of Magellan] in F2012. 1842 [Notes on South American beetles] in F2013. 1843 CD to W, "I believe...that if every organism which ever had lived or does live were collected together...a perfect series would be presented, linking all...into one great quite indivisible group". CCD2:377-379. 1844 Aug. 14 ED recorded in her diary "Mr Waterhouse came". 1847 CD reviewed W's *A natural history of the Mammalia* vol. 1, *Marsupialia*, 1846, in *Annals and Mag. of Nat. Hist.*, 19: pp. 53-56, unsigned, (*Shorter publications*, F1675).

Waterton, **Charles**, 1782-1865. Naturalist, traveller and eccentric. Walton Hall, near Pontefract, Yorkshire. 1804-c. 1824 W travelled in British Guiana and Brazil. 1825 Author of *Wanderings in South America*. ?1826 CD met W in Edinburgh with John Edmonston. 1828 W married Anne Edmondstone, granddaughter of an Arawak Indian. 1845 CD visited W at Walton Hall. 1845 CD to Lyell, "He is an amusing strange fellow; at our early dinner, our party consisted of two Catholic priests & two Mulattresses!" [W's sisters-in-law]. CCD3:258.

Watkins, Frederick, 1808-88. Anglican clergyman. Cambridge friend of CD, member of the Gourmet Club. 1860 CD to W on evolution, "I think the arguments are valid, showing that all animals have descended from four or five primordial forms; and that analogy and weak reasons go to show that all have descended from some single prototype". CCD8:308. 1874-88 Archdeacon of York. 1887 W gives memories of CD collecting beetles and talking of the beauty of the Brazilian forests. LL1:168-169, original letter in DAR112.A111-A114, transcribed in *Darwin Online*.

Watson, Hewett Cottrell, 1804-81. Botanist and phrenologist, specialist in distribution of British plants. 1836 Published the pamphlet: *An examination of Mr. Scott's attack upon Mr. Combe's 'Constitution of man.'* which focused on W's belief in a version of evolution. W "suggested testing the question of speciation by breeding a few organisms most unlike the main stock amongst themselves in isolation for many generations." John van Wyhe, *Phrenology and the origins of Victorian scientific naturalism*, 2004, p. 147. W sent a copy of the pamphlet to CD. CD discussed evolution with before *Origin*. 1847-59 *Cybele Britannica*, 4 vols. 1857 CD to Hooker, W had marked up a Flora for CD to show which he considered to be good species. 1859 CD sent 1st edn of *Origin*. W accepted evolution by natural selection. 1861 CD to Hooker, W accuses CD of egotism, "in first 4 paragraph of the Introduction, the words 'I', 'me', 'my', occur 43 times!" CCD9:70.

Watson, William, 1882 17 Apr. one of last letters CD wrote. A letter from Mr. Charles Darwin. *The Academy* 21 (527) (10 Jun.): 417. (F1949).

Way, Albert, 1805-74. Antiquary. Cambridge friend of CD. They collected beetles together. W drew two ink sketches of CD riding beetles (DAR204.29) and a mock coat of arms for CD. Drew the illustration of Glen Roy for CD. "Became acquainted with Fox and Way and so commenced Entomology". Van Wyhe ed., 'Journal', (DAR158). 1839 Fellow Society of Antiquaries London; director from 1842 to 1846. 1843-65 W edited *Promptorium parvulorum*. 1845 Principal founder of the Royal Archaeological Institute. 1860 Apr. CD to W about antiquarian information on breeds of horses, "Eheu, Eheu, the old Crux Major days are long past" [*Panagaeus cruxmajor*, a beetle collected by CD at Cambridge]. CCD8.

Weale, James Philip Mansel, 1838-1911. English naturalist and farmer in South Africa c.1860-78. Interested in the pollination of flowers, particularly orchids. 1867 CD to Lyell, W had sent seeds from locust dung from Natal. 1868 CD to Hooker, the grasses from the seeds had flowered. See *Origin*, 5th edn, p. 439, 1869. 1869 CD's article "Notes on the fertilization of orchids", *Annals and Mag. of Nat. Hist.*, vol. 4, p. 157, mentions W's publication on the fertilization of *Bonatea darwinii* by a skipper-butterfly (*Shorter publications*, F1748).

Webb, Mr. A carrier of parcels for the Ipswich Museum. 1855 CD to Henslow, CD was sending cirripedes to care of W. CCD5:283.

Webster, Mary Louisa Bell, 1854-1953. Botanist. Known as "Mary Ernest"; "an orphaned young woman ...who was estranged from her adopted parents", "stupid, dull and small-minded". Wedgwood 1980, pp. 327 and 333. 1 Married a Webster 1887 Married 2 Ernest Hensleigh Wedgwood, 1 son, Allen.

Wedgwood, Aaron [I], 1666-1743. Father of Richard W [I].

Wedgwood, Aaron [II], 1722-67. 7th child of Thomas [I] and Mary Stringer. Fat and stupid, known as "The Alderman" from his pomposity.

Wedgwood, Abner [I], 1699-1766. Younger brother of Thomas [I] W. Potter. CD's great-great-uncle.

Wedgwood, Abner [II], 1727-? 3d child of Aaron [I] and Hannah Malkin. d.s.p.

Wedgwood, Abner [III], 1779-1835. Son of "Useful" Thomas [V] and Elizabeth Taylor. Developed transferware etc. 1810 Married Amelia Hill. 3 children: 1 Elizabeth, 2 Abner [IV], 3 Josiah.

Wedgwood, Alfred Allen (Tim), 1842-1892. 5th child of Hensleigh and Fanny Mackintosh. 1866 Gave up being a Midshipman. 1873 Married Margaret Rosina Ingall. 3 children: 1 Bertram Hensleigh, 2 James "Jem" Ingall, 3 Margaret Olive 1884 Living at Horsley; "no sign of either ability or ambition". 1887 Separated.

Wedgwood, Allen [II], 1893-1915. Only child of Ernest Hensleigh and Mary Louisa Bell Webster's 2d son (from her 2d marriage). Killed at Suvla Bay, Gallipoli. His mother donated the Wedgwood Herbarium to Marlboro college in his memory.

Wedgwood, Amy, 1835-1910. 2d child of Francis and Frances Mosley. ED in a letter to Fanny H 1836 Dec. 17, "We had a very nice visit from Godfrey. It was pleasant to see how fond he is of his little maid, he always saved some dessert or asked for some for her." She was 17 months old. ED1:387. Godfrey was her elder brother. "Tiresome, selfish, narrow-minded spinster". Wedgwood, 1980, p. 3.

Wedgwood, Ann, 1712-82. Daughter of Thomas [I]. 1748 Married Philip Clark, 1 daughter, Mary.

Wedgwood, Anne Jane, 1841-78. 4th child of Henry Allen and Jessie. 1870 Married Ralph Edward Carr. 2 sons.

Wedgwood Arms, Wedgwood 1980 state that the original arms must have been gules four mullets argent. They illustrate the arms of Wedgwood of Heracles, Staffordshire: Gules, four mullets and a canton (plain) argent; a crest coronet and a lion passant, armed and langued gules, argent; motto Obstanta discendo (I split asunder obstacles) of 16-18th centuries. Most of the potter Ws of 18-19th centuries were not armigerous. Debrett gives arms for recent Barony and Baronetcy, but four mullets in cross; for the Barony "on either side a lion rampant queue fourchee argent supporting a staff raguly gales"; the Baronetcy has no supporters; in both cases the crest is not armed and langued gules.

Wedgwood, Arthur, 1843-1900. 5th child of Henry Allen and Jessie. Secretary to Charity Organization Society. 1843 Mar. 2 ED wrote "Arthur born". Unmarried.

Wedgwood, Arthur Felix (Felix), 1877-1917. 6th child of Clement Francis and Emily C. Rendel. Civil engineer, mountaineer and writer. Capt. North Staffordshire

Regiment. 1883 Nov. 13 ED wrote "Clem. Emily & Felix". 1885 visited ED with his parents. 1895 Oct. 15 visited Down House. 1910 *The Shadow of a Titan*, a novel. 1910 Married Katherine Longstaff. 3 children: 1 Frances Katherine (1912-2004), 2 Felicity Emily (1913-2003), 3 Cecil Felix Nivelle (1916-96). 1917 Killed at Bucquoy.

Wedgwood, Bertram Hensleigh (Berry), 1876-1951. 1st child of Alfred Allen and Rosina Ingall. Shipbroker of Liverpool. Served as Officer in Boer War. Wedgwood 1980, p. 274. 1905 Married 1 Winifred Eyre Heriz-Smith. 3 children: 1 Margaret Eyre Hope (1906-?), 2 Hensleigh Cecil (1908-91), 3 Geoffrey Vidal Allen (1911-86). 1921 Divorced. 1922 Married 2 Andrée Marie Perrier 1899-?

Wedgwood, Camilla Hildegarde, 1901-55. Anthropologist. 5th child of Josiah Clement [IV] and Ethel K. Bowen.

Wedgwood, Caroline Elizabeth, 1836-1916. 2d child of Henry Allen and Jessie. Nicknamed 'Carry'. ED1 and 2. *ED's diary*. Not mentioned in Wedgwood, 1980; but in pedigree in *Emma Darwin*, although not in Index. Unmarried.

Wedgwood, Catherine E., 1726-1804. 10th child of Thomas [III] and Mary Stringer. 1754 Married Rev. William Willett or Willets (1697-1779). Grandmother of Sir Henry Holland and Elizabeth Cleghorne Stevenson (Mrs Gaskell).

Wedgwood, Catherine Louisa Jane, 1799-1825. 4th child of John Wedgwood [IV] and Jane L. Allen. Unmarried.

Wedgwood, Major Cecil, 1863-1916. Only child of Godfrey and Mary Jane Jackson Hawkshaw. Major in North Staffordshire Militia in South African war. "Looked like a viking". Wedgwood, 1980, p. 128. Mother died at his birth. 1879 Joined pottery. 1884-1916 Partner Josiah Wedgwood and Sons Ltd. 1888 Married Lucie Gibson daughter of William E. Gibson of Cork. 2 daughters: 1 Phoebe Sylvia 2. Doris Audrey. 1891 Moved to Leadendale. 1902 DSO. Fought in the Boer War. ?1905 Chairman. 1910-11 First Mayor of Stoke-on-Trent. 1916 Killed at La Boiselle.

Wedgwood, Charles, 1800-c.1823. 5th child of John [IV] and Jane L. Allen. In East India Co. service. "An undisciplined and adventurous young man had died of fever in India". Wedgwood, 1980, p. 225. Unmarried

Wedgwood, Charlotte (Lotty), 1797-1862. 4th child of Josiah [II] and Sarah E. Allen. 1824 "Her beautiful fair hair reached to her knees". ED1:155. 1832 Feb. Married Charles Langton as first wife.

Wedgwood, Cicely Mary, 1837-1917. 3d child of Francis and Frances Mosley. *Emma Darwin* reads "Cicely". Wedgwood, 1980 reads "Cecily". Visited Down regularly from 1850. 1865 Married John Clarke Hawkshaw. 1866 Aug. 31 ED wrote "Cicely confined", 1st child Dorothy M. 1867 W was in Cambridge when the Darwins were and they had tea. Remained close to ED with letters up to 1895.

Wedgwood, Dame Cicely Veronica, 1910-97. Historian. 3d child of Sir Ralph Lewis and Iris V. Pawson. 1956 CBE. 1968 DBE. 1969 OM.

Wedgwood, Clement Francis, 1840-89. 4th child of Francis and Frances Mosley. Potter. 1866 Married Emily Catherine Rendel. 6 children: 1 Francis Hamilton (1867-

1930), 2 Clement Henry (1870-71), 3 Josiah Clement (1872-1943), 4 Ralph Lewis (1874-1956), 5 Cicely Frances (1876-1904), 6 Arthur Felix (1877-1917). Visited the Darwins often. 1850 Jun. 22 a list of children's names recorded in *ED's diary* "Willy Erny Effie Ronny Amy Cicely & Clem came". 1883 Nov. 13-17 came with Felix. 1888 Feb. 18-21 paid last visit. 1889 Died of cancer.

Wedgwood, Clement Henry, 1870-71. 2d child of Clement Francis and Emily C. Rendel. Died in infancy.

Wedgwood, Constance Rose, 1846-1903. 6th child of Francis and Frances Mosley. 1880 Married J.H. Franke. 1881 May 20 came for lunch, stayed Aug. 28-Sept. 2. Frankes and Arthur Felix Wedgwood called on ED: 1885 Feb. 7 & Sept. 5.

Wedgwood, Doris Audrey, 1894-1969. 2d child of Cecil and Lucie Gibson. 1928 Married Geoffrey T.R. Makeig-Jones. Secretary to the pottery in WWI.

Wedgwood, Edith Louisa, 1854-1935. Daughter of Rev. Robert and Mary Halsey. 1877 Married Clement Francis Romilly Allen.

Wedgwood, Elizabeth, 1702-74. Daughter of Thomas [I] and Margaret Shaw. Married Samuel Astbury.

Wedgwood, Elizabeth Julia, 1907-93. 6th child of Josiah Clement W and Ethel Kate Bowen.

Richmond 1840.   Maull & Polyblank c.1855.   Barraud 1881.   *Illus. London News* 1896.

Wedgwood, Emma [I], 1808 May 2-1896 Oct. 2. 9th and last child of Josiah Wedgwood [II], named after her aunt Emma Allen. CD's first cousin and wife. Biography: 1904 Litchfield, privately printed (F1552); very similar, published edn, 1915 (F1553). 1952 Gwen Raverat (granddaughter), *Period piece*, Chapter 8. 2001 Healey, *Emma Darwin*. 2010 Loy & Loy, *Emma Darwin*. Her diaries (1824-96) are all available and transcribed by Christine Chua and Kees Rookmaaker (1839) in *Darwin Online*. A recipe book which W started in 1839, labelled "Mrs Charles Darwin's Recipe Book Down" transcribed by Christine Chua is also in *Darwin Online*. Some entries in CD's *Notebook of observations on the Darwin children* are by ED. Nicknames, "The Dovelies" with Frances Wedgwood [II] in childhood, "Little Miss Slip-Slop" in childhood, "Titty" and "Em" by CD in early years of marriage, "Mammy" later. "A beautiful needlewoman, a good archer, and she rode, danced and skated". "She played delightfully on the piano". "She had lessons from Moscheles and a few from Chopin".

Autobiography, pp. 96-8, ED1:62. She read French, German and Italian. "Her brown hair kept its warm tint almost to the end of her life with hardly a grey hair in it." "In 1824 she could sit on her hair". ED1:155. CD's opinion of ED is omitted from the autobiography in LL1:69, which was published whilst she was alive.

"You all well know your mother, and what a good mother she has ever been to all of you. She has been my greatest blessing, and I can declare that in my whole life I have never heard her utter one word which I would rather have been unsaid. She has never failed in the kindest sympathy towards me, and has borne with the utmost patience my frequent complaints from ill-health and discomfort. I do not believe she has ever missed an opportunity of doing a kind action to anyone near her. I marvel at my good fortune that she, so infinitely my superior in every single moral quality, consented to be my wife. She has been my wise adviser and cheerful comforter throughout life, which without her would have been during a very long period a miserable one from ill-health. She has earned the love and admiration of every soul near her". ML1:30, *Autobiography*, pp. 96-8.

On her religious views, "In our childhood and youth she was not only sincerely religious—this she always was in the true sense of the word—but definite in her beliefs. She went regularly to Church and took the Sacrament. She read the Bible with us and taught us a simple Unitarian Creed, though we were baptised and confirmed in the Church of England". ED2:173, *Autobiography*, p. 238. ED's religious views are stated in two letters to CD. 1. ?1839, soon after marriage. CD appended a note "When I am dead, know that many times, I have kissed & cryed over this C.D." ED2:173 omitting note, CCD2:172, *Autobiography*, p. 237. 2. 1861 Jun., CD appends a note "God bless you". ED2:175, *Autobiography*, p. 238. 1818-37 Before marriage, ED travelled on the continent with her family: 1818 Apr. visited Paris; 1824-25 Paris, Geneva, Florence, Sorrento, Rome, Milan; 1826 Geneva, to visit her aunt Jessie and husband Simonde de Sismondi; 1827 Cologne; 1838 Paris. She also made a number of visits in British Isles, sometimes to relatives; 1823 Scarborough; 1828 Clifton; 1837 Edinburgh. 1822-23 ED was at school at Greville House, Paddington Green, London, with Frances [II] Wedgwood. 1824 Sept. 17 Confirmed at Maer church although brought up Unitarian. A reading book, containing four stories she wrote (The plumb pie; The little foal; The snowy night and Market) and had printed for use in the Maer school, is preserved in DAR219.11.1, bequest of Sir Geoffrey Keynes 1981, the only known copy. (An 8-page 1985 facsimile is *The pound of sugar*. DAR185.101. Transcribed in *Darwin Online*.) 1836 Oct. "We are getting impatient for Charles's arrival" [on return of *Beagle*]. ED1:272. 1836 Nov. "We enjoyed Charles's visit uncommonly". ED1:273. 1838 Nov. 11 CD proposed, at Maer, and was accepted. 1839 Married Jan. 29, by Rev. John Allen Wedgwood, CD, at St. Peter's Church, Maer. After marriage ED devoted her life to CD and to bringing up the children. 1843 Hensleigh being rather ill, ED took charge of his children. 1876 when Amy Ruck Darwin died in childbed, ED "took up the old nursery cares as if she was still a young woman" with Bernard, her first grandchild. In 1879 she donated funds for a prize for kindness to animals through Julia Goddard. After CD's death ED spent the summers at Down House and the winters at The Grove, Huntingdon Road, Cambridge. ED knitted and did Peggywork, another form of knitting involving long strips of knitting in thick wool, by pulling the wool over the pegs of a wooden frame. Raverat, *Period piece*, 1952, p. 148. 1896 ED died and is buried in Down churchyard. Her personal estate was valued at £35,077 2*s* 6*d*. She bequeathed £100 each to Catherine Thorley and Emily Thorley, £1000 each to her daughters, £550 to Mrs Pattrick (Camilla Ludwig) and numerous other gifts. 1933 recollection of ED by Bernard Darwin in Jekyll, *Children and gardens*. 2^d edn, pp. xi-xxiii & Darwin, *Pack clouds away*, 1941, all transcribed in *Darwin Online*.

Iconography of ED:

1840 Watercolour by George Richmond (20.3x25.4cm). Darwin Heirlooms Trust. On display at Down House, EH88202067.

c.1855 Half-length three-quarter left profile photograph with Leonard Darwin by Maull & Polyblank. ED2(1904). The CCD and others have dated this to 1853 or c.1853 and attributed it to Maull & Fox but Maull & Polyblank were established in 1854. As CD was photographed by them in 1855, perhaps this photograph was done on the same occasion. His payment to them was far more than required for his own photograph and for joining the 'club'. See CD Iconography above. A copy in CUL is annotated on verso "45" and in modern ink "?Leonard / Emma Darwin" and bears a label from the successor firm Maull & Fox, which existed from 1879-85. This is presumably the source of the ED2(1904) dating of the photograph to ED aged 45. DAR257.17.

1858 Jun. Photograph by William Darwin(?) ED holding the deceased baby Charles Waring Darwin. Randal Keynes. Reproduced in Browne, *Power of place*, 2002.

-1859 CDV studio photograph, full figure, seated, three-quarter left profile by Bassano. DAR225.79. Bassano had a studio at the address on this CDV until 1859.

1861 CDV three-quarter left profile by Elliott & Fry. DAR225.78. In her diary, ED noted on 19 Aug. 1861 "went to photogr" with no indication of which one.

c.1863 Albumen photograph of family, except CD, seated at and around drawing room (later dining room) window of Down House. ED is seated in window, three-quarter length left profile, reading a book. DAR219.12:9 & DAR225.156.

c.1871 ED is just visible seated in the drawing room at the extreme left of the CDV of Henrietta with Polly. Huntington Library, California. Cropped out in other variants.

1873-80 CDV three-quarter right profile of head and shoulders only by Barraud & Jerrard. Their partnership lasted 1873-80. DAR225.77.

1875 Oil painting by W.W. Ouless. Not found and presumably lost or never completed.

1880 Watercolour of Down House from the south-west by Albert Goodwin, includes CD and ED seated on the verandah. Down House EH88202055.

1881 Three-quarter right profile carbon print cabinet card, seated with hands clasped, by Barraud. Printed in ML1: facing p. 30 and ED2(1904): facing p. 320, and dated on p. [x] and in ML1:xv to 1881. A copy in CUL has an annotation on verso "Done by mistake in the carbon printing". DAR257.18. Three variants seen, one, a carbon print, cropped to chest and in an oval. A framed copy is above the mantelpiece in CD's old study at Down House.

1881 Full face cabinet card, seated with hands clasped, by Barraud. Same sitting as above. DAR257.19.

1881 Painting by John Collier reported to be in the Grosvenor Gallery in 1882. *Heywood Advertiser*, (5 May). Not traced.

1887 Oil on canvas by Charles Fairfax Murray, signed "C. F. M." (90.2x68.6cm). Darwin Heirlooms Trust, on loan to Darwin College, Cambridge.

1887 Pastel on linen lined brown paper by Charles Fairfax Murray. Darwin Heirlooms Trust, on display at Down House.

1895 Full-length, seated, left profile photograph in sitting room of Down House with her 'peggy,' by Miss M.J. Shaen, at Down House. Exhibited at Cambridge in 1909.

1896 Woodcut after Shaen 1895. *Illustrated London News*, (10 Oct.), p. 458.

1896 Woodcut, a very poor imitation of the above. *The Bryan Daily Eagle*, [Texas] (17 Dec.).

1948 Pastels on paper. Evening at Maer by Evstafiev. CD with ED and Robert Waring Darwin. SDMM. NVF1265/1085. Copy sent by the SDMM to Down House in 1959.

c.1958 Oil painting "Early Spring", by Evstafiev, showing young CD on verandah and ED at her Broadwood piano. Sent by the SDMM to Down House. Photograph of it sent by the SDMM to Sir Charles Darwin in 1963 now in DAR238.14.64.

c.1958 Oil painting, "Late Autumn" by Evstafiev, showing ED at her piano playing for elderly CD (frame size 36x47cm). Sent to Down House by the SDMM. Down House EH88202062.

1959 Plaster bust after Richmond 1840 by Victor Evstafiev. SDMM OF7667/1782. 1960 copy made and sent to Down House, EH88204645. A copy may be in the NHM.

1961 Sketch of ED playing piano by M.TS. Rabinovich. Korsunskaya, *Tales of Charles Darwin*.

1961 CD consoles ED after death Charles Waring. *Ibid.*

1967 Great men of science. *Star-Phoenix* (Canada), (Aug.): CD in top hat showing ED the back of Down House; Bearded CD sits while ED plays the piano.

Printed works: c.1825 ED wrote a reading book for her Sunday School class at Maer; the class was taught by the family and held in the laundry; "these she had printed in large type; the book contained four little stories, one about a 'plumb pie' [sic]. We, her own children were taught to read out of this little book, and were fond of these stories". ED1:142. DAR185.101.

1863 Together with CD, wrote and privately printed at Bromley "An appeal [against steel vermin traps]". Their only joint publication or writing. (F1931) Also printed as "Vermin and traps" in *Gardeners' Chron.*, no. 35 (29 Aug.): 821-2, by CD (*Shorter publications*, F1728) and published anonymously in the *Bromley Record* by ED.

1868 Letter on benefit of spinal ice-bag. Sea-sickness and how to prevent it...p. 101.

1877 Letter to editor of *Spectator* (6 Jan., p. 15) on vivisection and animal cruelty.

1887 ED wrote a four-line preface to the 1st edn in book form of Henry Allen Wedgwood's *The bird talisman*, which she had printed for the benefit of her grandchildren: "The following Fairy Tale was written and illustrated by the late Henry Allen Wedgwood for his children. It was so much liked by them that it is now reprinted by his sister for a second generation. / E. DARWIN. / CAMBRIDGE, / Oct. 1887." Reproduced in *Darwin Online*.

ED almost certainly inserted advisements and announcements in newspapers.

Wedgwood, Ernest Hensleigh (Ernie/Erny), 1837-98. 3d child of Hensleigh and Fanny Mackintosh. ED wrote her sister Eliz in 1839 "He really is the most beautiful child I ever saw, now when his eyes and cheeks are bright, and looks something like Puck. He is not a bit shy." 1863 Took a minor post in Colonial Office. 1866 Lost this post but got another. 1887 Married Mary L W Bell. 1 son: Allen.

Wedgwood, Frances [II] "Fanny", 1806-32. 8th child of Josiah [II] and Sarah Elizabeth Allen. Unmarried. "Freckled plain-faced faithful Fanny". With ED, known as "The dovelies", also as "Mrs Pedigree" from her passion for making lists, which ED kept amongst her treasures until her death. Went with Emma to Mrs Mayer's finishing school at Paddington. 1832 Died suddenly, perhaps from cholera.

Wedgwood, Frances Julia, 1833-1913. Novelist and biographer. 1st child of Hensleigh and Fanny Mackintosh. Unmarried. Known as "Snow" because she was born in a snowstorm or because it was snowing. Brent 1981, p. 176 says "Snow" was short for "Snowdrop". Her most important works were *Framleigh Hall*, 1858, a novel, *John Wesley*, 1870, *The moral ideal*, 1888, also *An old debt*, 1856, a novel under pseudonym of "Florence Dawson". CD discussed evolution with before *Origin*. 1861 The boundaries of science, a dialogue, *Macmillan's Mag.*, 1861 Jul. CD's comments on, "I could not clearly follow you in some parts, which is in main part due to my not being at all accustomed to metaphysical trains of thought". CCD9:201. 1867 "I do find myself so wicked for finding Snow such a dreadful bore...begging to discuss fate and free will...so tactless a woman I never came near and gets worse". E.M. Forster, *Marianne Thornton*, 1956, p. 223.

Wedgwood, Francis, 1800-88. Of Barlaston. 6th child of Josiah [II] and Elizabeth Allen. CD's first cousin and brother-in-law. 1827 Partner; 1844-75 Senior Partner in Josiah Wedgwood and Sons Ltd. 1832 Married Frances Mosley. 7 children: 1 Godfrey, 2 Amy, 3 Cicely Mary, 4 Clement Francis, 5 Laurence, 6 Constance Rose, 7 Mabel Frances. 1878 Jun. CD and ED visited. 1879 W visited Down House and again in 1885. 1884 W visited ED at The Grove, Cambridge.

Wedgwood, Francis Charles Bowen, 2d Baron, 1898-1959. 3d child of Josiah Clement [IV] and Ethel K. Bowen. Artist. 1920 Married Edith May Telfer. 1 son, Hugh, 3d Baron Wedgwood (1921-70).

Wedgwood, Francis Hamilton (Frank/Franky), 1867-1930. 1st child of Clement Francis and Emily C. Rendel. Justice of the Peace. Deputy Lieutenant. Major in North Staffordshire Regiment. Served in South African War and WWI. 1888 Apprenticed to the Wedgwood firm. 1893 Partner. 1916 Chairman, Managing Director Wedgwood. 1923 Director London, Midland and Scottish Railway. 1902 Married Katherine Gwendoline Pigott. 3 children: 1 Frances Dorothea Joy (1903-96), 2 Cecily Stella (1905-95), 3 Clement Tom (1907-60). Died suddenly of throat infection.

Wedgwood, Geoffrey (Geoff), 1879-97. 5th child of Laurence and Emma Houseman W.

Wedgwood, Gilbert [I], 1588-1678, Potter of Dale Hall, Burslem. This, the first Burslem Wedgwood. 3d son of Richard and Margaret Boulton. Married Margaret Burslem. 8 children: 1 Joseph, d.s.p., 2 Burslem, 3 Thomas who became the ancestor of Josiah, 4 William, s.p., 5 Moses, s.p., 6 Aaron, 7 Mary, 8 Sarah.

Wedgwood, Gilbert Henry [II], 1876-1963. 3ᵈ child of Laurence and Emma Houseman. Lieutenant-Colonel Yorkshire and Lancashire Regiment. Fought in Boer War and WWI. 1918 DSO. 1920 Married Dorothy Salmond.

Wedgwood, Gloria, 1909-74. 7th child of Josiah Clement [IV] and Ethel K. Bowen. Married 1 Heinz P. M. Married 2 Paul August Oppenheim. 1 son, Paul Felix Wedgwood (1938-2004).

Wedgwood, Godfrey, 1833-1905. 1st child of Francis and Frances Mosley. Justice of the Peace. 1859 Partner; 1875-91 Senior partner Josiah Wedgwood and Sons Ltd. Lived in Idlerocks, Staffordshire. 1834 Mar. visited Maer and left on 27. 1862 Married 1 Mary Jane Jackson Hawkshaw. 1 son, Cecil (1863-1916). 1876 Married 2 Hope Elizabeth Wedgwood. 1 daughter, Mary Euphrasia (1880-1952). 1898 Right leg amputated. Many visits over the years. ED last wrote to them in 1896 Jan. 3.

Wedgwood, Helen Bowen, 1895-1981. 1st child of Josiah Clement [IV] and Ethel K. Bowen. 1920 Married Michael S. Pease (1890-1966). 1 son, 1 daughter.

Wedgwood, Henry Allen (Hal/Harry), 1799-1885. 1837 Seabridge, Staffordshire. c.1847 The Hermitage, Surrey. 5th child of Josiah [II] and Sarah Elizabeth Allen. CD's first cousin and brother-in-law. Barrister. Author of *The bird talisman: An eastern tale*. 1821 Married 1 Caroline White. 1830 Married 2 his double first cousin Jessie Wedgwood. 6 children: 1 Louisa Frances, 2 Caroline Elizabeth, 3 John Darwin, 4 Anne Jane, 5 Arthur, 6 Rowland Henry. 1841 Jul. 24, ED wrote "went to the Hermitage". Visits by H & co. were recorded by ED as "Hermitage came".

Wedgwood, Hensleigh (Hen), 1803-91. 7th child of Josiah [II] and Sarah Wedgwood. Barrister, etymologist and philologist. In CD's London years he saw much of W. CD discussed evolution with before *Origin*. 1829-30 Finch Fellow of Christ's College. 1832-37 Police Magistrate. 1832 Married Frances E. "Fanny" Mackintosh. 6 children: 1 Frances Julia, 2 James Mackintosh, 3 Ernest Hensleigh, 4 Katherine Euphemia, 5 Alfred Allen, 6 Hope Elizabeth. 1839 Feb. 5 Dined at Upper Gower with Erasmus. H told of his getting registrarship of Hackney Cabs. 1859-65 Wrote *A dictionary of English etymology*, 3 vols. Came regularly with the family. His last visit may have been in 1889 Jul. 26. 1891 Jun. 1 ED wrote "Hensleighs death"

Wedgwood, Hope Elizabeth, 1844-1935. 6th child of Hensleigh and Fanny Mackintosh. 1876 Married Godfrey Wedgwood as 2ᵈ wife. 1879-96 kept in close contact with visits and correspondence with ED.

Wedgwood, James Ingall (Jem), 1883-1951. 2ᵈ child of Alfred Allen and Rosina Ingall. Brought up by Snow Wedgwood after parents separated. Bishop of the Old Catholic Church. Theosophist. Unmarried.

Wedgwood, James Mackintosh (Bro/Mack), 1834-1874. 2ᵈ child of Hensleigh Wedgwood and Fanny Mackintosh. Unmarried.

Wedgwood, Jessie, 1804-72. Of Seabridge. 6th child of John [IV] and Louisa Jane Allen. CD's first cousin. 1830 Married Henry Allen Wedgwood. 3 sons, 3 daughters.

Wedgwood, John, ?-1589. Of Mole in Biddulph. Elder son of Richard of Harracles and Jane Shirrot (Sherratt). 1563 High collector of subsidy. Married Anne Bowyer (d. 1582). 8 children.

Wedgwood, John [I] (Long John), 1705-79. 5th child of Aaron [I] and Mary Hollins. Potter of the Big House, Burslem. Married Mary Alsop. 6 children.

Wedgwood, John [II], 1721-67. 4th child of Thomas [III] and Mary Stringer. 1760 Represented Josiah [I] London at the sign of the Artichoke, Cateaton Street. Courtier and arranged sale of Queen's ware to Queen Charlotte. Unmarried. Drowned in the Thames.

Wedgwood, John [III], 1732-74. Son of Richard [I] and Susan Irlam. Ran the cheese factory. Brother of Sarah Wedgwood. d.s.p.

Wedgwood, John [IV], 1766-1844. Horticulturalist. 3d child of Josiah [I] and Sarah Wedgwood. CD's uncle. Partner in the Wedgwood firm. 1790-93 and 1800-12. 1829 Lived in The Hill, Abergavenny. 1794 Married Louisa Jane Allen. 7 children: 1 Sarah Elizabeth [III], 2 John Allen, 3 Thomas Josiah, 4 Catherine Louisa Jane, 5 Charles, 6 Jessie, 7 Robert.

Wedgwood, John [V] (Jack), 1877-1954. 4th child of Laurence and Emma Houseman. 1902 Married Violet Douglas. 2 children 1 Godfrey Josiah, 2 Eileen.

Wedgwood, Rev. John Allen, 1796-1882. 2d child of John [IV] and Jane L. Allen. Boarded at Westminster School. "So withdrawn that his parents were concerned over his mental stability". Wedgwood, 1980, p. 166. Consumptive in youth and later an invalid. Rector of Maer. 1821 Deacon Church of England, ordained as priest in 1822. 1825 Vicar of St. Peter's Church, Maer Hall. 1832 Mar. 22 Married Charles Langton to Charlotte Wedgwood. 1839 Jan. 29 Married CD to Emma Wedgwood at St. Peter's. Unmarried.

Wedgwood, John Darwin, 1840-70. 3d child of Henry Allen and Jessie. Army officer. 1866 Married Helen Mary Tyler. 2 children, died in infancy. 1868 Jun. 8 "John D. W." visit in *ED's diary* also "John W's death". Drowned in boating accident.

Wedgwood, Sir John Hamilton, 2d Baronet, 1907-89. Politician and industrialist. 1st child of Ralph Lewis and Emily C. Rendel.

Wedgwood, Joseph, 1757-1817. Son of Aaron [III] and Mary Stringer. Potter. Married Mary Clark. 1768 Leased Churchyard Works from Josiah Wedgwood [I].

Wedgwood, Josiah [I], 1730-95. 11th child of Thomas [III] and Mary Stringer. Born at Churchyard House, Burslem. 1770 lived in Etruria Hall, Staffordshire. Founder of the Josiah Wedgwood and Sons Ltd. 1783 FRS. CD's maternal grandfather. Close friend of Erasmus Darwin [I]. "Patient, steadfast, humble, simple, unconscious of half of his own greatness". Woodall 1884, p. 7. Meteyard, *The life of Josiah Wedgwood*, 1865. 1764 Married Sarah Wedgwood, a 3d cousin, common ancestor being Gilbert [I], his great-great-grandfather. 8 children: 1 Susannah (CD's mother), 2 Richard [III], 3 John, 4 Josiah [II], 5 Thomas [VI], 6 Catherine, 7 Sarah

Elizabeth. 8 Mary Anne. Oil painting attributed to Joshua Reynolds, Darwin Heirlooms Trust, on display at Down House

Wedgwood, Josiah [II] (Jos), 1769-1843. 4th child of Josiah [I] and Sarah Wedgwood. 1795-1841 Senior partner of Josiah Wedgwood and Sons Ltd. 1795 Stoke d'Abernon, Surrey, on his father's death. 1799 Bought Gunville, House, Tarrant Gunville, Dorset. 1800 Moved to Little Etruria. 1802 Bought Maer Hall, Staffordshire. 1803 elected Sheriff of Dorset. 1807- Lived permanently at Maer Hall. 1832-35 MP. Sydney Smith of W, "an excellent man—it is a pity he hates his friends". ED1:7. CD's uncle and father-in-law. CD was on close terms with and it was he who persuaded CD's father to let him go on the *Beagle* voyage. CD "I used to apply to him...the well known ode of Horace, now forgotten by me, in which the words 'nec vultus tyranni' etc come in". *Autobiography*, p. 56. [Justum et tenacem propositi virum/Non civium ardor prava jubentium/ Non vultus instantis tyranni/Mente quatit solida. The just man and firm of purpose not the heat of fellow citizens clamouring for what is wrong, nor presence of threatening tyrant can shake his rocklike soul. *Odes* III, iii, 1]. 1792 Married Sarah Elizabeth Allen. 9 children: 1 Sarah Elizabeth (1793-1880), 2 Josiah [III] (1795-1880), 3 Mary Anne (1796-98), 4 Charlotte (1797-1862), 5 Henry Allen (1799-1885) 6 Francis, "Frank" (1800-88), 7 Hensleigh (1803-91) 8 Frances (1806-32) 9 Emma (1808-96). ED wrote in her diary in 1843 Jun. 17, "my father worse", Jun. 22 "my father better" and finally on Jul. 12 "Died".

Wedgwood, Josiah [III] (Joe), 1795-1880. 2^d child of Josiah [II] and Sarah Elizabeth Allen. Doubly CD's brother-in-law. Potter. 1837 Married Caroline Sarah Darwin. 4 daughters: 1 Sophia Marianne, 2 Katherine Elizabeth Sophia, 3 Margaret Susan, 4 Lucy Caroline. On marriage first lived at Clayton near Etruria. 1841-44 Senior partner Josiah Wedgwood and Sons Ltd. 1841 Moved to Leith Hill Place, Surrey, 4000 acres, and sold his pottery interest to his brother Francis. 1880 CD to Hensleigh Wedgwood on his death, "there never existed a man with a sweeter disposition". *Calendar*, Carroll 1976, 573. 1880 Mar. 11 ED wrote "Death of Jos".

Wedgwood, Josiah [V], 1899-1968. 4th child of Josiah Clement [IV] and Ethel K. Bowen. 1967 Managing Director of Josiah Wedgwood and Sons Ltd. 1919 Married Dorothy Winser. 3 children: 1 John, 2 Josiah [VI] "Ralph", 3 Jennifer.

Wedgwood, Josiah Clement [IV], 1st Baron Wedgwood, 1872-1943. Of Barlaston. 3^d child of Clement and Emily C. Rendel. Director of Josiah Wedgwood and Sons Ltd. 1906-42 MP for Newcastle under Lyme. MP for 36 years, first Liberal, second Labour, then Independent. 1908 *History of the Wedgwood family*. 1915 DSO. 1924 Served in WWI attaining rank of Colonel PC. 1942 1st Baron Wedgwood of Barlaston. Took up seat in House of Lords. 1894 Married 1 Ethel Kate Bowen. 7 children: 1 Helen Bowen, 2 Rosamund, 3 Francis Charles Bowen, 4 Josiah [IV], 5 Camilla Hildegard, 6. Elizabeth Julia, 7. Gloria. 1913 Separated; 1919 Divorced. 1919 Married 2 Florence Ethel Willett, s.p.

Wedgwood, Julia, see **Frances Julia Wedgwood**.

Wedgwood, Katherine Euphemia (Effie), 1839-1931. 4th child of Hensleigh and Fanny Mackintosh. 1870 Spring stayed at Down House. 1873 Married Sir Thomas Henry Farrer as 2[d] wife. In 1840 May 16, she was weighed by ED "15 lb 11" (William Darwin "14 lb 15"). ED recorded in 1848 Nov. 9 "Effie's birthday". There are 34 records of "Effie" in *ED's diary* from 1840-82. She continued to visit after 1882 and the last entry on her was in 1895 May 12. 1925 Lived with her sister Hope Elizabeth at Idlerocks.

Wedgwood, Kennard Laurence, 1873-1950. 1st child of Laurence and Emma Houseman. Potter. Chairman Wedgwood. Established Wedgwood's American headquarters. Served in South African war. 1908 Married Kathleen Wright. 1 daughter Esmé Alice.

Wedgwood, Laurence, 1844-1913. 5th child of Francis and Frances Mosley. Potter. 1871 Married Emma E. Houseman. 6 children: 1 Kennard Laurence (1873-1950), 2 Mary Frances "Molly" (1874-1969), 3 Gilbert Henry (1876-1963), 4 Clement John (1877-1954), 5 Geoffrey Walter (1879-97), 6 Unnamed daughter died at birth (1882).

Wedgwood, Louisa Frances, 1834-1903. 1st child of Henry Allen and Jessie. 1864 Married William John Kempson. 2 sons, 2 daughters. Emma Wedgwood (later Darwin) recorded her birth. Kempsons visited in 1866 Nov. 29, 1868 Oct. 13-17, 1878 Jul. 3, 1882 Sept. 9-15. She also appeared as Louisa K on visits to Down in 1869 Aug. 30-Sept. 4 and Mar. 71. (There are 31 other entries in *ED's diary* until 1882 bearing just Louisa).

Wedgwood, Lucy Caroline, 1846-1919. 4th child of Josiah [III]. 1874 Married Capt. Matthew James Harrison RN

Wedgwood, Mabel Frances, 1852-1930. 7th child of Francis and Frances Mosley. 1880 Married Arthur George Parson (d. 1907). d.s.p.

Wedgwood, Margaret Olive Rosalind (Olive), 1892-1960. 3[d] child, Alfred Allen and Rosina Ingall. 1920 Married Verus Calvin Montgomery.

Wedgwood, Margaret Susan, 1843-1937. 3[d] child of Josiah [III] and Caroline Sarah Darwin. 1868 Married the Rev Arthur Charles Vaughan Williams. Son: Ralph Vaughan Williams. 1885 W gave ED her dog Dicky

Wedgwood, Mary Anne [I], 1778-86. 8th child of Josiah [I] and Sarah Wedgwood. Mentally retarded. Died in childhood. The pedigree in *Emma Darwin* gives the 7th and last child Sarah Elizabeth (1778-1856). Wedgwood 1980 gives Sarah Elizabeth as 1776-1856, but does not give Mary Anne in pedigree at all, only in index and text.

Wedgwood, Mary Anne [II], 1796-98. 3[d] child of Josiah [II] and Sarah Elizabeth Allen. In *Emma Darwin*, but Wedgwood, 1980 do not mention her.

Wedgwood, Mary Euphrazia, 1880-1952. Only child of Godfrey W and Hope Elizabeth W, stepsister of Cecil W. CD's first cousin twice removed. Wedgwood 1980 spell "Euphrasia" with "s", *Emma Darwin* with "z". Frances Julia Wedgwood made a scrapbook of family papers for her. Wedgwood 1980, p. 355. 1937 Married William Mosley, a cousin, after her mother's death.

Wedgwood, Mary Frances (Molly), 1874-1969. 2^d child of Laurence and Emma Houseman. 1902 Married Ernald George Justinian Hartley (1875-1919).

Wedgwood, Phoebe Sylvia, 1893-1972. Elder daughter of Cecil W. Unmarried.

Wedgwood, Sir Ralph Lewis, 1st Bart, 1874-1956. 4th child of Clement Francis and Emily C. Rendel. Trinity College, Cambridge. Established Leith Hill musical festival. 1916 Brigadier General. 1924 Kt. 1942 1st Bart of Etruria. 1944 Chairman of wartime Railway Executive. 1944 Rented Leith Hill Place from National Trust after Ralph Vaughan Williams had given it to them. Frequent visitor with brother Felix to George Darwin at Cambridge. Raverat, *Period piece*, p. 233. Railway administrator. 1906 Married Iris V. Pawson (1887-1982). 3 children: 1 John Hamilton (1907-89), 2 Ralph Pawson, died in infancy (1909), 3 Cicely Veronica (1910-97).

Wedgwood, Ralph Pawson, 1909 died in infancy. Second child of Ralph Lewis Wedgwood and Iris Veronica Pawson.

Wedgwood, Richard, c.1545-1626. Of Mowle in Biddulph. Married Margaret Boulton of Biddulph in 1567. 3 sons; 1 Richard, 2 Randle, 3 Gilbert.

Wedgwood, Richard [I], 1700-82. 1st child of Aaron [I] and Mary Hollins. Married Susan Irlam. Daughter Sarah (married Josiah [I]), 1 son John, d.s.p.

Wedgwood, Richard [II], 1725-78. 6th child of Thomas [III] and Mary Stringer. Started as a potter, but joined army, took to drink and was lost to the family. Wedgwood 1980, p. 71.

Wedgwood, Richard [III], 1767-82. 3^d child of Josiah [I]. Sarah Wedgwood. CD's uncle. Died in childhood.

Wedgwood, Rev. Robert, 1806-81. 7th child of John [IV] and Jane L. Allen. Priest at Tenby. Rector of Dumbleton. 1833 Married 1 Frances Crewe s.p. 1848 Married 2 Mary Halsey. 1 son, 6 daughters.

Wedgwood, Rosamund, 1896-1960. 2^d child of Josiah Clement [IV] and Ethel K. Bowen. 1920 Married Janos Békassy (1894-?). Children unknown.

Wedgwood, Rowland Henry (Harry), 1847-1921. 6th child of Henry Allen and Jessie. A Roman Catholic. 1883 Married 1 Sophia Helen Rudd. 1907 Married 2 Agnes Harley. Children unknown.

Wedgwood, Sarah, 1734-1815. Daughter of Richard [I] and Susan Irlam. CD's maternal grandmother. ED's paternal grandmother. The only grandparent alive in their lifetimes. 1764 Married Josiah Wedgwood [I] (a cousin). 1803 Moved to Parkfields on Maer estate. Died at Maer. 8 children.

Wedgwood, Sarah Elizabeth [I], 1776-1856. 7th child of Josiah Wedgwood [I] and Sarah Wedgwood. Of Parkfields. Unmarried. CD's aunt. Popular with CD's children and at Down House almost every day. "Tall and stately, most spartan in habits, fastidious, upright and solemn"; "kept several pairs of gloves beside her so as not to soil her hands", black cotton for shaking hands with children, lighter colours for cleaner occupations such as reading books. Wedgwood, 1980, p. 245. 1823 On death of her sister Catherine, W moved from Parkfields to Camphill on Maer Heath.

1827 Moved to Camphill which took three years to build for her. c.1829 Wrote a biography of Thomas Josiah Wedgwood, printed for the family. 1847 W moved to Petley's, Down. Petley's was leased from Sir John Lubbock. 1856 Nov. 6 Died at Down House. ED2:161-2 describes her funeral in CD's letter to his sons William and George. Present were Elizabeth, Jos, Harry, Frank, Hensleigh and all Allens. Her will was read immediately after the service presided by Mr Innes. She had, in the letter CD wrote "desired her funeral to be as quiet as possible and that no tablet should be erected to her." ED's absence was probably due to being heavily pregnant with her last child, Charles Waring, who would be born a month later.

Wedgwood, Sarah Elizabeth [II] (Elizabeth or Bessy), 1793-1880. 1st child of Josiah [II] and Sarah E. Allen. Unmarried. ED's sister. CD's first cousin. "I think none of us felt quite at ease with our aunt". ED2:105. c.1818-39 Ran Sunday school in laundry room at Maer Hall. Built a school on Caldy Island, near Tenby, built the Ridge at Hartfield. 1860 CD to Lyell, "I showed the case [of orchids] to Elizabeth Wedgwood, and her remark was 'Now you have upset your own book, for you won't persuade me that this could be effected by Natural Selection'". "The last twelve years of her life, happy with her garden, her little dog Tony, her devoted servants". ED2:106. 1880 CD to Romanes, "As good and generous a woman as ever walked this earth". Romanes 1896, p. 101, CCD28.

Wedgwood, Sarah Elizabeth [III] (Sally/Eliza), 1795-1857. 1st child of John [IV] and Jane L. Allen. Constant companion of her parents. Deeply religious. 1840 Lived with sister Jessie, husband Henry Allen and four children at Seabridge. 1843 moved to Tenby, Robert's home, with their father who died there. Visited Darwins regularly from 1840. ED recorded 1857, Sept. 11 "Eliza's death". Unmarried.

Wedgwood, "Snow", see **Frances Julia Wedgwood**.

Wedgwood, Sophia Marianne (Mary Ann), 1838-39. 1st child of Josiah [III].

Wedgwood, Susannah (Sukey), 1765-1817. Born at The Brick House, Burslem. 1st child of Josiah [I] and Sarah Wedgwood. CD's mother, ED's aunt. "She seems never to have been very strong". Meteyard, *The life of Josiah Wedgwood*, 1865, p. 357. 1796 Married Robert Waring Darwin. 6 children. 1807 W to her brother Josiah [II], "Everyone seems young but me". 1817 CD about his mother, "My mother died in July 1817, when I was a little over eight years old, and it is odd that I can hardly remember anything about her except her deathbed, her black velvet gown, and her curiously constructed work-table". *Autobiography*, p. 22. W is buried St. Chad, Shrewsbury, Shropshire, in chancel. Inscribed "Susan" on her husband's tombstone. A portrait miniature of her by Peter Paillou (the younger) from 1793 was the property of Milo Keynes, now in the Fitzwilliam Museum, Cambridge. A b/w photograph of the portrait is in CUL (DAR232.22) with an inscription on the back of the frame "Susannah wife of Robert Waring Darwin of Shrewsbury after a miniature in possession of W.E. Darwin. 1908 G.H.D." Framed by G.E. Hardwick, Cambridge.

Wedgwood, Thomas [I], 1615?-78. Potter of Churchyard House, Burslem. 4th child of Gilbert of Mole and Margaret Burslem. 1653 Married Margaret Shaw. 16 children.

Wedgwood, Thomas [II], 1660-1716. 4th child of Thomas [I] and Margaret Shaw. Potter of Churchyard House, Burslem. 1684 Married Mary Leigh (d. 1718) of Burslem. 9 children. 1 Thomas [III].

Wedgwood, Thomas [III], 1687-1739. 1st child of Thomas [II] and Mary Leigh. CD's great-grandfather. Amongst the best of Wedgwood potters of Churchyard House, Burslem. c.1711 Married Mary Stringer. 13 children. Died insolvent.

Wedgwood, Thomas, 1703-86. 3d child of Aaron [I] and Mary Hollins. Married Martha?. 2 children: 1 Mary (1715-?), 2 Aaron (1717-?).

Wedgwood, Thomas [IV], 1716-72. 4th child of Thomas [III] and Mary Stringer. Potter but not a good one of Churchyard House, Burslem. Josiah [I] was apprenticed to him for five years. 1742 Married 1 Isabel Beech, 3 children: 1 Sarah, 2 Mary, 3 Thomas. Married 2 Jane Richards, 3 children: 1 Jane, 2 John, 3 William. Died of dropsy and in debt.

Wedgwood, Thomas [V], 1734-88. 4th child Aaron [II] and Hannah Malkin. Cousin of Josiah Wedgwood [I]. Known as "Useful Thomas" because he made useful Queen's ware. Josiah Wedgwood [I] took him into partnership for this purpose. d.s.p. Died from falling into canal.

Wedgwood, Thomas [VI], 1771-1805. 5th child of Josiah [I] and Mary Stringer. Unmarried. Biography: R.B. Litchfield, 1903. Known as the first photographer. "To England belongs the honour of first producing a photograph". Wedgwood, 1980. "An account of the method of copying paintings upon glass and of making profiles by the agency of light upon nitrate of silver...", *Jrnl. of the Royal Institution*, Jun., 1807.

Wedgwood, Thomas Josiah (Tom), 1797-1860. 3d child of John [IV] and Jane L. Allen. Colonel in Scots Fusiliers. Aged 17, fought as an Ensign at Waterloo. ED1:68. 1836 Married Anne Maria Tyler, daughter of Admiral Sir Charles Tyler. d.s.p. Visited the Darwins in 1856, stayed 4 days from 1-4 Aug.

Weir, Harrison William, 1824-1906. Book illustrator. Known as "the father of the cat fancy". Breeder of poultry and pigeons. Brother of J. Jenner W. Member of Philoperisteron Club as CD. 1852 CD sent him *Journal*. 1867 Illustrated Tegetmeier's *The poultry book*. 1871 Organized the first cat show in England at the Crystal Palace. ED recorded 1871 Jul. 13, "Girls to Cat show, Crystal". CD to W (not in CCD): "Your mother ought indeed to feel proud that she had two sons such true naturalists as you and your brother." *Kent & Sussex Courier*, (25 Apr. 1884), p. 7.

Weir, John Jenner, 1822-94. Naturalist, entomologist, ornithologist and accountant. FLS, FZS and active as Fellow in the Entomological Society as treasurer and twice as vice-chairman. Controller-General H.M. Customs. 1868 Sept. 12 Sat. W stayed at Down House, with the Wallaces and Blyth. Bates was hoped for but probably was not present. Hookers came for Sunday lunch; "A very good man". 1868 CD to W,

"I read over your last ten (!) letters this morning, and made an index of their contents for easy reference; and what a mine of wealth you have bestowed on me". The letters are on selection especially in caterpillars. 1875 CD to Weismann, on W's work on selection in caterpillars.

Weismann, Friedrich Leopold August (August), 1834-1914. Physician, entomologist and student of inheritance. Known for his germ plasm theory. 1868 *Über die Berechtigung der Darwin'schen Theorie*. Leipzig. 1872 CD to W, W was having trouble with his eyes, "eyesight is somewhat better". 1872 CD to W, having read *Über den Einfluss der Isolierung auf die Artbildung*, Leipzig 1872. 1873-1912 First Prof. Zoology Freiburg and Director Zoological Institute. 1875 CD to W on selection. 1879 W sent CD his work on *Daphnia*, CD thanks for and refers to Raphael Meldola's slow progress of translation of *Studien*. ?1881 CD to W, praising *Studien*, "excited my interest and admiration in the highest degree". LL3:231. 1882 *Studien zur Descendenz-Theorie*, Leipzig, 1875-76; translated by Raphael Meldola as *Studies in the theory of descent*, London 1882, with prefatory notice by CD, p. v-vi (F1414). 1892 *Das Keimplasma: eine Theorie der Vererbung*. 1902 *Vorträge über Deszendenztheorie*, 2 vols.

Wells. 1852 Bucket ropes for wells, *Gardeners' Chron.*, no. 2: p. 22 (*Shorter publications*, F1680). CD asked if wire would be suitable for his 325 foot deep well. A cautionary reply by a C.L.C appeared in the next issue of *Gardeners' Chron.*, no. 3, 17 Jan. 1852, p. 38. 1857 The subject of deep wells, *Gardeners' Chron.*, no. 30: p. 518 (*Shorter publications*, F1696). Could a lighter material replace his 12-gallon oak bucket? The Down House well was located outside the kitchen. It has since been capped and sealed off.

Wells, Leonard Henry (Luke), ?-1903. Artist. CD commissioned W through Tegetmeier to draw six breeds of domestic pigeons, the English pouter, English carrier, English barb, English fantail, short-faced English tumbler and African owl as well as the heads of the Polish, Spanish and Hamburgh fowl for *Variation*. CCD13.

Wells, William Charles, 1757-1817. Born Charleston, South Carolina, USA, of Scottish parents. Physician and man of science. Sent to school in Dumfries and studied at Edinburgh University. See K.W. Wells, Charles Wells and the races of man, *Isis*, 1973, 64: pp. 215-25. 1785 Settled in London. 1793 FRS. 1813, 1818 Author of *Two essays*, 1818, a posthumous work which contains reprints of his two previously published and fundamental papers "On dew" and "On binocular vision", with an appendix about a black and white woman, Harriet Trets, which contains the rudiments of the idea of natural selection. There is an excellent summary by Thomas Thomson, *Annals Philosophy*, 1: p. 383, May 1813, of the paper as read to the RS, Apr. 1 and 8, 1813. The matter is referred to in "Historical sketch" in 3d edn *Origin*, 1861. 1865 CD to Hooker, "a Yankee has called my attention to a paper attached to Dr Well's famous Essay on Dew, which was read in 1813 to RS but not printed, in which he applies most distinctly the principle of N. Selection to the races of man.— So poor old Patrick Matthew, is not the first". CCD13:279.

Welsh, Jane (Jenny) Baillie, 1801-66. 1826 Married Thomas Carlyle. CD met the Carlyles on several occasions in London. 1838 CD to ED, "I cannot think that Jenny is either quite natural or lady-like". CCD2:128.

West Hackhurst, House at Abinger, Surrey. 1879 Jun. CD and ED were lent the house from Saturday to Tuesday. Horace and Ida went in 1882 May 15. 1882 Aug. 23 Richard and Henrietta Litchfield went. 1889 Bessy came home from there.

West, Esther, Mrs Frederick Allen's maid at Down. 1868 Was seen talking to Rev. John Robinson even though W's mother had written to R forbidding it. CD discussed with Innes R's conduct and was vexed by the scandal. CCD16.

West, Lady Mary Catherine Sackville-, 1824-1900. Second daughter of George John Sackville-West, 5th Earl de la Warr. Holwood House, near Down. 1847 Married 1 James Brownlow William Gascoyne-Cecil, 2d Marquess of Salisbury. 1870 Married 2 Edward Henry Stanley, 15th Earl of Derby. 1882 Jul. W called on ED at Down House from London and straight back again. ED2:260.

Westcroft. A house in Kent which CD considered buying before he saw Down House. "I am going to Westcroft (the name of one place) on Friday with a valuer & then mean to make an offer". CCD2:305.

Westwood, John Obadiah, 1805-93. Solicitor and entomologist. FLS. 1852-53 President Entomological Society. 1855 CD proposed W successfully for Royal Medal of Royal Society. De Beer 1959, p. 65. 1860 W's anti-evolutionary views discussed. W "proposed to the last University Commission the permanent endowment of a lecturer to combat the errors of Darwinism". Poulton, *Fifty years of Darwinism*, 1909, p. 15. 1861-91 1st Hope Prof. Zoology (Entomology) Oxford. Named after F.W. Hope, who donated his entomological collection to Oxford University.

Westwood, Mary Anne/Ann (Nanna), 1854-1942. 1876 Nurse to Bernard Darwin at Down House. Daughter of Edward Westwood, cheesemonger and later cab proprietor, and Jane Westwood, of Finsbury, Middlesex. ED wrote in 1876 Nov. 27 "Maryanne came". 1881 Left to get married to Arthur Parslow. ED2:282-3, 317.

Whale-Bear Story, 1859, 1860, 1861. Occurs in its full form at p. 184 of 1st edn of *Origin*, first four USA printings 1860, and in J. Lamont, *Seasons with the sea-horses*, 1861. "...I can see no difficulty in a race of bears being rendered, by natural selection, more and more aquatic in their structure and habits, with larger and larger mouths, till a creature was produced as monstrous as a whale". Referring to S. Hearne, *A journey from Prince of Wales Fort in Hudson's Bay, to the Northern Ocean*, 1795. 1859 Owen asked CD for the reference of the story. Owen attacked the story in his anonymous review of *Origin* in the *Edinburgh Rev.* published the following year. 1860 2d edn of *Origin* reads "...swimming for hours with widely open mouth, thus catching, almost like a whale, insects in the water." The rest is omitted. 1860 CD to William Henry Harvey, "As it offended persons, I struck it out in the second edition; but I still maintain that there is no special difficulty in a bear's mouth being enlarged to any degree useful to its changing habits". 1863 The full version is reprinted in *The*

Press, Canterbury, New Zealand, in a letter from T.W. Leys, Bishop of Wellington, in controversy with Samuel Butler [II]. 1881 CD to R.G. Whiteman, "This sentence was omitted in the subsequent editions, owing to the advice of Prof. Owen, as it was liable to be misinterpreted; but I have always regretted that I followed this advice, for I still think the view quite reasonable". ML1:393.

Whatford, William Starr, 1795-1887. Dentist to the Brighton and Hove Dispensary. Source of the anecdote about a dog visiting a dentist, 1871, CCD19:648. CD did not use the anecdote in *Expression*.

Wharton, Henry James, 1798-1859. Headmaster of William Erasmus Darwin's preparatory school. 1852 Feb. 24 CD wrote to William telling him to write and tell W about his examination. CD advised him to start with "My dear Sir, and to end by saying "I thank you and Mrs Wharton for all the kindness you have always done me. Believe me, Yours truly obliged." ED2:145. CCD5:81.

Wharton, Mary Dorothea, 1870-1947. Only child of Rt Hon. J. Lloyd Wharton. 1894 Married Colonel Charles Waring Darwin of Elston.

Wetherell, Nathaniel, Thomas, 1800-75. Surgeon and geologist. CD took W's *Loricula pulchella* (a fossil barnacle) specimen to be drawn for *Fossil Lepadidæ* (1851, fig. Tab V) by James de Carle Sowerby. CCD4:348.

Wheeler, Henry. b.1861 Assistant to the head gardener at Down House near the end of CD's life. 1927 Unique recollections of CD and Down House, grounds, greenhouse and darkroom and photograph of W in *Sunday Post* (4 Sept.), p. 3. In *Darwin Online*. In 1881 census a Henry Wheeler was listed as a lodger and male bakers assistant in Westerham.

Whewell, William, 1794-1866. Astronomer and philosopher. Polymath. Coined the words "scientist" and "physicist". 1820 FRS. 1832-36 Prof. Mineralogy Cambridge. 1837 *History of the inductive sciences*, 3 vols. 1838-55 Prof. Moral Theology and Casuistical Divinity Cambridge. 1841-66 Master of Trinity College, Cambridge. 1860 W to CD, "I cannot, yet at least, become a convert. But there is so much of thought and of fact in what you have written that it is not to be contradicted without careful selection of the ground and manner of the dissent". W refused, for some years, to allow a copy of the *Origin* into the Library of Trinity College. LL2:261.

Whitby, Capt. John, RN, ?-1806. Of Milford, Hampshire. 1802 Married Mary Anne Theresa Symonds.

White, Adam, 1817-78. Writer on natural history topics. 1835-63 Assistant in the Zoology Department of British Museum. 1841 described spiders collected by CD in Descriptions of new or little known Arachnida. *Annals and Mag. of Nat. Hist.* 7 (Jul.): 474, 476, (F2011). 1846 FLS. 1854 W applied for Chair of Natural History Edinburgh with printed testimonials, one by CD, but withdrew them on hearing that Edward Forbes had applied. 1859 Roland Trimen's reminiscences to Poulton, "I was at work in the next compartment to that in which Adam White sat, and heard someone come in and a cheery mellow voice say 'Good-morning Mr. White;—I am afraid you won't speak to me any more'...Ah, Sir! if ye had only stopped with the *Voyage of the Beagle!*". Poulton, *Darwin and the Origin*, 1909, p. 214. 1863 W retired from BM with mental illness. 1864 W reprinted testimonials, including CD's, with additions, to obtain paid lecturing in Edinburgh, his native town to which he had retired. 1877 CD to Günther, "that poor mad creature". CCD25:107.

White, Nicholas, 1806-? Second "Master" on 2[d] voyage of *Beagle*.

Whitehead, J., 1900-06 The first tenant of Down House after ED's death, leasing it. W owned the first motor car in Down. "Shadowy figure". Atkins 1976, p. 102.

Whiteman, R.G., 1881 CD to W, explaining why he omitted the whale-bear story from 2[d] and subsequent editions of *Origin*. ML1:392-393.

Whitley, Rev. Charles Thomas, 1808-95. Cousin of John Maurice Herbert. Hon. Canon of Durham. Intimate friend of CD at Cambridge and had been at Shrewsbury School. Member of Gourmet Club. 1833-55 Reader in Natural Philosophy and Mathematics Durham. 1838 W invited CD to Durham. 1849 Honorary canon Durham. 1854-95 Vicar of Bedlington, Northumberland. 1864-72 President College of Medicine and Surgery Newcastle-upon-Tyne. 1894 "The greatest naturalist of our time was no doubt my friend Charles Darwin. (Hear, hear)". Recollection in *The Morpeth Herald*, transcribed in *Darwin Online*.

Wilbury House, Wiltshire. 1865 A house taken by Charles Langton.

Wickham, John Clements, 1798-1864. Naval Officer. W was on all three voyages of *Beagle*. 1st Lieutenant on 2[d] voyage. Captain commanding on 3[d] voyage until invalided. 1832 CD in letter to his father: "Wickham is a glorious fine fellow". CD got on better with W than with any other officer. CCD1:206. 1834 "Although Wickham always was growling at my bringing more dirt on board than any ten men, he is a great loss to me in the *Beagle*. He is by far the most conversible being on board". CCD1:393. 1841 Resigned command *Beagle*. 1843-53 New South Wales Police Magistrate, Australia. 1853-59 W was first Government Resident at Moreton Bay (now Brisbane), Queensland, Australia. 1859 Retired to France. Biography: Barrie Jamieson, *John Clements Wickham: Charles Darwin's glorious fellow*, 2019. In *Darwin Online*.

Wien = Vienna, Austria. 1856 There is a tradition that CD once asked Hooker where "this place Wien is, where they publish so many books". It is substantiated by CD to Hooker in 1856, "to write to 'Wien' (that unknown place)". CCD6:170.

Wiesner, Julius von, 1838-1916. Austrian botanist. 1873-1909 Prof. Botany Vienna. 1881 CD to W, about movement in plants and thanking him for sending *Das Bewegungsvermögen der Pflanzen*, Vienna, 1881. LL3:336. See letters to in F2525.

Wilberforce, Rev. Samuel, 1805-73. 3[d] son of William Wilberforce, campaigner against the slave trade. Known as "Soapy Sam" from his habit of "washing his hands", whilst preaching or talking. Nickname derives from a comment by Benjamin Disraeli who described W's manner as "saponaceous". 1860 W of *Origin*, "the most unphilosophical work he ever read". CCD8:81; another version "the most illogical book ever written". Lyell, *Life*, ii, p. 358. 1845 Dean of Westminster. 1845-69

Bishop of Oxford. 1845 FRS. 1860 Sat. Jun. 30 W spoke anti-*Origin* at BAAS meeting Oxford, answered by Huxley with a remark about descent from a monkey on his grandmother or grandfather's side. 1860 Jul. W reviewed *Origin* in *Quarterly Rev.*, primed by Owen. 1860 Jul. 20 CD to Huxley, "I would give five shillings to know what tremendous blunder the Bishop made; for I see that a page has been cancelled and a new page gummed in" [pp. 251-2]. CCD8:294. 1860 CD to Innes, "Did you see the *Quarterly Rev.*, the B. of Oxford made really splendid fun of me and my grandfather". CCD8:540 1869-73 Bishop of Winchester. 1874 *Essays contributed to the Quarterly Rev.*, 2 vols., London, review of *Origin*, vol. 1, pp. 52-103.

Wilhelm, Crown Prince of Germany, 1859-1941. 1881 late Int. Med. Congr.; CD was presented to: "he looks a very nice and sensible and fine man". Brent 1981, p. 499. 1888 Became Wilhelm II, the last Emperor of Germany and King of Prussia.

Wilkes, Lieut. Charles, 1798-1877. USA Naval Officer and explorer. 1836 W was in London fitting out US Exploring Expedition to the Southern Ocean of 1838-42. CD arranged to call on W, but meeting is unconfirmed. CCD1:517.

Wilkins, Mary, Kitchen maid at Down House in 1881.

Wilkinson, Rev. Henry Marlow, 1827-1908. Clergyman living in Milford. 1874 W examined *Utricularia* for CD for *Insectivorous plants*.

Williams, 1839 A gardener, employed by CD at 12 Upper Gower Street, mentioned in CD's personal manuscript accounts.

Williams & Norgate, 1817-91. Publishers and book importers in London and Edinburgh, specialised in foreign and scientific literature. CD ordered books through them 1847-80. 1865 *Climbing Plants* sold also by them.

Williams, E. Charles, An American spiritualist medium favoured by Hensleigh Wedgwood. 1877 CD to Romanes, "a very clever rogue". CCD25. 1878 CD to Romanes, about W's exposure in *Spiritualist Newspaper*, 13, Sept. 2. CCD26. 1874 Jan. 16 at Erasmus Alvey Darwin's home, a séance of 20 persons was arranged with W. CD did join in albeit briefly and found it was worth the exertion if only to be convinced it was all an imposture. CD wrote to Hooker "The Lord have mercy on us all, if we have to believe in such rubbish". CCD22:26.

Williams, Rev. Arthur Charles Vaughan, 1834-75. Vicar of Down Ampney. Son of Sir Edward Vaughan W. Father of Ralph Vaughan W. 1868 Married Margaret Susan Wedgwood. 1875 Feb. 9 ED wrote in her diary "Arthur Wms death".

Williams, Edward Augustus, ?-1875. 1835 Sept. 3 Physician in Bromley. *London Gazette*, Part 2, p. 1701. In 1845 Aug. 12 W was sent for. He saw George in 1854, Elizabeth in 1857, Leonard in 1858, Henrietta in 1860. *ED's diary*; CCD8:436. He had a 1st edn *Origin* signed "Edwd A Williams 1859". Now in a private collection.

Williams, Edward Hosier, ?-1844. Solicitor of London. Of Eaton Mascott, near Shrewsbury. 1831 1st husband of Sarah Harriet Mostyn Owen. 1833 CD to Emily Catherine Darwin at Maldonado "one of the kindest (letters) I ever received. I was very sorry to hear...that she has lost so much of the Owen constitution: I am very sure that with it none of the Owen goodness has gone". CCD1:311.

Williams, Henry, 1792-1867. Missionary in New Zealand. Leader of the Church Missionary Society in Aotearoa. Formerly a Naval Officer who served in the Napoleonic Wars. 1823 Arrived at Waimate, Bay of Islands, North Island, New Zealand. 1835 Dec. 23-24 CD stayed at his house, "He is considered the leading person among the missionary body". (*Shorter publications*, F1640), *Journal* 2ᵈ edn, pp. 426-8.

Williams, Ralph Vaughan, 1872-1958. Composer. Son of Reverend Arthur Charles Vaughan W. and Margaret Susan Wedgwood. When young asked his mother about CD's *Origin*, she replied, "The Bible says that God made the world in six days. Great Uncle Charles thinks it took longer: but we need not worry about it, for it is equally wonderful either way". *RVW: A biography*, 1964, p. 13. 1944 Gave Leith Hill Place to National Trust.

Willis, William, "My hairdresser (Willis) says..." Near Great Marlborough Street. Comments on growth of hair and breeding of small dogs. *Notebook C*, p. 232 and *Notebook D*, p. 24. *Darwin's notebooks*, 1987 and in *Darwin Online*. First name not previously identified in the Darwin literature.

Willis, Rev. Robert, 1800-75. Mechanical engineer and historian. 1830 FRS. 1837-75 Jacksonian Prof. Natural Experimental Philosophy Cambridge. 1853-68 Lecturer Applied Mechanics, Government School of Mines. Member of the Council of the Royal Society of London. 1864 Nov. 3 W was one of twenty present at the Athenaeum where votes were cast for the Copley medal, the RS highest award. It was awarded to CD on 30 Nov. 1864. Burkhardt, Darwin and the Copley Medal. *Proc. of the American v Soc.*, vol. 145, no. 4, 2001, pp. 510-18.

Willott, Rev. John, 1813?-46. St John's College, Cambridge, 1831 BA, 1835 MA. Perpetual curate of Down, 13 Jan. 1841-Mar. 1846.

Wills, Harriet, 1857-? Visiting ladies maid at Down House in 1881 census.

Wills, William, Petty Officer, Armourer on 2ᵈ voyage of *Beagle*, on *Adventure* on 1ˢᵗ voyage. 1832 W is mentioned in the *Beagle diary* as Armourer.

Wilmot, Rev. Darwin, 1845-1935. Son of Emma Elizabeth Darwin, 2ᵈ daughter of Sir Francis Sacheverel Darwin, CD's half-second cousin. Headmaster of Macclesfield Grammar School. 1930 W had Erasmus Darwin [I]'s commonplace book which he lent to Hesketh Pearson for *Doctor Darwin*, 1930. At Down House.

Wilmot, Sacheverel Darwin, 1885-1918. 2ᵈ son and 4ᵗʰ child of Rev. Darwin W. Captain, Royal Garrison Artillery. Buried in New Delhi, India.

Wilson, Rev. Charles, 1770-1857. 1835 Nov. 15 CD met at Waugh Town, Matavai, Tahiti, "the missionary of the district". *Journal* 2d edn, p. 403.

Wilson, Belford Hinton, 1804-58. 1832-7 British Consul in Lima. 1837-41 Chargé d'Affaires. 1842-52 Consul General in Venezuela. In the *Galapagos notebook*, Chancellor & van Wyhe, *Beagle notebooks*, 2009.

Wilson, Captain James, 1760-1814. 1797 A missionary in Tahiti for more than 30 years, except for a short period when the missionaries had to flee to New South

Wales, Australia. W arrived on London Missionary Society's ship *Duff* in 1797. See FitzRoy & Darwin 1836; *Shorter publications*, F1640, In the *Galapagos notebook*, Chancellor & van Wyhe, *Beagle notebooks*, 2009. (*Shorter publications*, F1640.)

Wilson, Alexander Stephen, 1827-93. Agricultural botanist of Edinburgh. 1878-80 letters from CD to W, on races of Russian wheat.

Wilson, Colonel Sir Belford Hinton, 1804-58. 1832-41 Consul General Lima. 1835 Aug. 3 "Mr Wilson, most exceedingly obliging: having been Aide de Camp to Bolivar he has travelled much of South America". *Beagle diary*, p. 331. 1842-52 Consul General in Venezuela.

Wilson, Edmund Beecher, 1856-1939. American embryologist and cytologist. 1881 CD to W, thanking him for information on *Scyllaea*, a nudibranch mollusc found on *Sargassum* which it closely mimics (F2021). W to Poulton, "His extraordinary kindness and friendliness towards an obscure youngster who had of course absolutely no claim on his time or attention". Poulton, *Darwin and the Origin*, 1909, pp. 106-8. 1894-97 Prof. Invertebrate zoology Columbia. 1897-1928 Prof. Zoology Columbia. 1921 Foreign Member Royal Society.

Wilson, Edward, 1813-78. Anglo-Australian politician, journalist and philanthropist. 1842 Went to Australia. 1864 Settled in Hayes Place, near Bromley, Kent. 1871 Jan. 28 ED recorded "Mr Wilson dined". 1873 "Owing to the great kindness and powerful influence of Mr Wilson...I have received from Australia no less than thirteen sets of answers to my queries" i.e. *Queries about expression*, 1867. These included one from Dyson Lacy in Queensland, who was a relative of W. *Expression*, p. 20.

Winchester, Hampshire. 1846 Sept. 13 CD and ED visited Winchester and St. Cross on day trip from BAAS meeting at Southampton. Summer 1865 writing from Down to Henrietta, ED said William had taken his brothers to Netley and Winchester "on their own hook", then to the Isle of Wight where they "wandered all about". In 1875 Sept. 3, ED wrote in her diary "party went to Winchester."

Winkler, Tiberius Cornelius, 1822-97. Born in Leeuwarden, the Netherlands. Anatomist, zoologist and natural historian. 1860 translated the first edn of *Origin* into Dutch (F2056.1-2). 1889-90 translated *Descent* (F1056).

Winkworth, Catherine, 1827-78. Deeply involved in promoting women's education. Sister of Emily W. Shaen. Among the audience of Symonds lecture. ED read *Letters and Memorials of Catherine Winkworth*, edited by their niece Margaret Shaen.

Winter-flowering plants, 1869 Fertilisation of winter-flowering plants, *Nature*, 1: p. 85 (*Shorter publications*, F1748a). See *Cross and self fertilisation*, pp. 10-11.

Wiseman, Captain Sir William Saltonstall, 7th Baronet, 1784-1845. First husband of Catherine [III] Mackintosh.

Wolf, Josef, 1820-99. German-born painter and illustrator of animals. W had a studio, together with Johann Baptist Zwecker, at 59 Berners Street, London. CD contacted W to draw animals for *Expression*. W also drew the orangutans in Wallace's

Malay archipelago (1869). Letter to W and recollections of CD in A.H. Palmer, *The life of Joseph Wolf*, 1895, pp. 192-8 (F2087), transcribed in *Darwin Online*.

Wollaston, Thomas Vernon, 1822-78. Naturalist. W wintered in Madeira and other Atlantic islands, due to ill-health, and was a specialist in their invertebrate fauna, especially beetles. CD discussed evolution with before *Origin*. 1847 FLS. 1855 CD to Hooker, "Wollaston's 'Insecta Maderensia': it is an *admirable* work". CCD5:279. W visited and stayed 4 days at Down House from Apr. 25-28 1856. CD also had the Huxleys, J. Lubbock and Hooker those same days. 1857 published descriptions of some beetles collected by CD during voyage of the *Beagle*. 1860 W wrote hostile review of *Origin* in *Annals and Mag. of Nat. Hist.*, 5: pp. 132-43. 1868 CD to Stainton, "I have been sincerely grieved to hear about poor Wollaston's affairs, in which, I am told, you have taken so kind an interest". CCD16:135.

Wonder. From 1825, a coach that ran daily from Shrewsbury to London covering the 154 miles in 13 ½ hours. 1835 CD to Susan Darwin mentions it. See CCD1.

Wood, Alexander Charles, 1810-? Son of Thomas W and Lady Caroline Stewart. Nephew of FitzRoy. British colonialist. Matriculated Trinity College, Cambridge, 1831. A colonial land and emigration commissioner. Justice of the Peace for Middlesex 1831 Jan. Went up to Trinity.

Wood, Searles Valentine, 1798-1880. Palaeontologist and banker. 1848-56 *A Monograph on the Crag Mollusca*. 1860 W was pro-*Origin*. LL2:293. 1860 Wollaston Medal Geological Society. 1861-71 *A Monograph on the Eocene Bivalves of England*.

Wood, Col. Thomas, 1777-1860. Of Oxford. Politician. 1801 Married Lady Caroline Stewart. 1806-47 MP for Breckonshire. 1809-10 High sheriff of Breckonshire.

Wood, Thomas W., 1839-1910. Zoological illustrator. Introduced to CD by Wallace. W had prepared illustrations for Wallace's *Malay archipelago*. 1870 W drew figures 9, 10 and 14, of cats and a snarling dog for *Expression* and 2ᵈ edn of *Descent*. 1870 CD to Abraham Dee Bartlett, CD knew W personally in London and asks Bartlett to give him facilities at Zoological Gardens.

Wood, Sir William Page, 1st Baron Hatherley, 1801-81. Barrister and statesman. 1824-79 Fellow of Trinity College Cambridge. 1831 CD to Susan Darwin; "Captain Fitzroy (probably owing to Wood's letter) seems determined to make me [as] comfortable as he possibly can". CCD1:141. 1831 CD to Susan Darwin, "Wood (as might be expected from a Londonderry) solemnly warned Fitz-Roy that I was a whig". CCD1:146. 1831 CD to Henslow, "If you see Mr Wood remember me very kindly to him". CCD1:503. 1833 Henslow to CD, "Wood and I had intended writing by the Decr packet". CCD1:292. 1851 Kt. 1868 1st Baron Hatherley. 1868-77 Lord Justice of Appeal. 1868-72 Lord Chancellor.

Woodd, Ellen Sophia, 1820-87. 1846 Married as second wife to Rev. William Darwin Fox. Buried in Christ Churchyard, Isle of Wight.

Woodhouse, Shropshire. Home of William Mostyn Owen and his children, "a beautiful mansion of white freestone". Known as The Forest, in Rednal, 13 miles northwest of Shrewsbury on the Holyhead Road. CD was often there for shooting and social occasions, both before *Beagle* voyage and on his return. Photograph of the house in CCD1.

Woodhouse, Alfred James, 1824-1906. Dentist at Hanover Square, London and member of the Odontological Society of Great Britain. Darwin family dentist. CD wrote to him (1867?) asking about inheritance of peculiarities of deciduous ('milk') teeth. CCD24:475-6. There are 14 records of different members of CD's family going to see W between 1862-70. *ED's diary*. 1865 Aug. 1, ED to Elizabeth Wedgwood, "I seized the opportunity of [CD's] being so well to rush up to Mr Woodhouse, and he saw me at once and did not keep me waiting." ED(1904)2:207.

Woodward, Samuel Pickworth, 1821-65. Geologist and malacologist. 1845- Prof. Geology and Natural History Royal Agricultural College Cirencester. 1848-65 Assistant Department of Geology and Mineralogy British Museum. 1851, 1853 and 1856 *Manual of the Mollusca*. CD discussed evolution with before *Origin*. 1856 Jun. CD to W, had read his *Manual of the Mollusca* with "much solid instruction and interest". CD hoped to see him in London in about a fortnight. CCD6. 1856 Jun. CD to Lyell and to Hooker, on W's views on extended continents. CCD6. 1856 Jul. CD to W, on species. CCD6. 1860 CD to W, on volcanoes. CCD8. CD mistakenly thought W the anonymous author of the first review of *Origin* in the *Athenaeum* (no. 1673 (19 Nov.): 659-660 actually by J.R. Leifchild). "For years [CD] carried a grudge against the anonymous reviewer, whom he wrongly identified as [W] on the basis of something that Hooker let slip. For years, too, the unfortunate [W] did not understand why Darwin had suddenly turned so frosty." Browne, *Power of place*, 2002, p. 87.

Woodyeare, John Fountain, née Fountain-John, 1809-80. Cambridge friend of CD. CD to Fox 1829 Apr. 1 wrote that he, Jeffry Hall, James Turner and W "rode like incarnate devils" there to witness "a terrible fire…a most awful sight". CCD1:81. 1851-80 Domestic Chaplain to Dowager Countess of Cavan.

Woollya. Settlement at Tierra del Fuego, now known as Wulaia. 1834 Jan. 23 Richard Matthews, missionary, landed there from *Beagle*. 1834 Feb. 6 Matthews taken off again. See Fuegians for an illustration.

Woolner, Thomas, 1825-92. Sculptor. Founder-member of the Pre-Raphaelite Brotherhood. W called in 1866 Mar. 4, 1868 Nov. 19. Mrs Woolner (Alice Gertrude Waugh) on Nov. 28 in 1868. 1868 upon the suggestions of Erasmus Alvey Darwin, CD sat to W for bust which was finished in 1869, now in Botany School Cambridge. Francis Darwin; "It has a certain air of pomposity, which seems to me foreign to my father's expression". LL3:105-106. 1869 The Wedgwood jasperware medallion of CD is by W. 1871 CD to W, "One reviewer ('Nature') says that they ought to be called, as I suggested in joke, *Angulus Woolnerianus*". CCD19:271. *Nature*, Apr. has "Angulus Woolnerii". W had discovered this small cartilaginous lobe in the human pinna, which is more usually called "Darwin's peak". It is referred to in *Descent*, 1:22, with woodcut. 1877 May W visited Down House. 1872 May 19, ED wrote in her diary "Woolners & Mr Butler." CD wrote to Henrietta May 13 1872 that the Ws would be coming to see him on the 19th. He also would be asking Samuel Butler. CCD20. 1882 W was on "Personal Friends invited" list for CD's funeral.

Worms. 1838 On the formation of mould. *Proc. of the Geol. Soc. of London* 2: pp. 574-6 (*Shorter publications*, F1648 and F1655). 1869 The formation of mould by worms, *Gardeners' Chron.*, no. 20: p. 530 (*Shorter publications*, F1745). 1880 CD to H. Johnson, "My heart and soul care for worms and nothing else in the world just at present". nd Letter to Peter Price, *Western Mail* (6 Apr. 1883): Issue 4337, p. 4 (F2173). De Beer 1968, p. 74. "Darwin had none but kindly feelings for worms". L. Stephen, *Biography of Swift*, see also *Earthworms*.

Wormstone, (or worm stone). The modern name for a flat circular stone 460mm in diameter, c.57mm thick, c.23kg, with a hole in the centre (a mill stone?) through which a two-pronged copper rod was driven 2.6m into the ground, placed just over the fence separating Great House Meadow and Great Pucklands Meadow, under a large Spanish chestnut tree at Down House. The rods would remain stationary and the stone would slowly sink due to the actions of earthworms. CD only mentions "the large flat stone" in *Earthworms*, pp. 119-20. The stone and apparatus were prepared by Horace Darwin in 1877. It was originally called a "wormograph" (Francis Darwin to CD [28 Oct. 1877?], CCD25:443. Details of the stone and the measurements made with it are given in Horace Darwin, On the small vertical movements of a stone laid on the surface of the ground. *Proc. Roy. Soc.*, vol. 68, 1901, along with a woodcut of the vertical micrometer that sat on the stone. The stone now in the gardens of Down House is a replica by Horace's Cambridge Scientific Instrument Company in 1929, when Down House opened as a public museum. Atkins 1976, p. 118. The micrometer that sat atop the stone is no longer in place but can be seen in a photograph in Huxley & Kettlewell, *Charles Darwin*, 1956, p. 122, S. Morris & L. Wilson, *Charles Darwin at Down House*, 2003; on display at Down House.

Wray, Leonard Hume, 1816-1901. Sugar planter, fruit grower and horticultural writer. CD corresponded with on trimorphic flowers of strawberries, *Different forms of flowers*, 1877, p. 293. (F1277).

Wright, Chauncey, 1830-75. American philosopher and mathematician. Computor in American National Almanac Office, Cambridge, Massachusetts, USA. 1870- CD corresponded with on phyllotaxy after he had read W's papers in *Astronomical J.*, no. 99, 1856 and *Math. Monthly*, 1859. 1871 W reviewed Mivart's *The genesis of species* in *N. American Rev.*, Jul. 1871 Oct. 23 CD arranged to have it published as a pamphlet, with additions, *Darwinism: being an examination of Mr St. George Mivart's 'Genesis of species'*. 1871 Sept. CD to Hooker, describes W's review as "a very clever, but ill-written review". 1872 W wrote in *N. American Rev.* in reply to an article by Mivart in *ibid.*, Apr. 1872 Sept. W stayed at Down House. ED made a diary record of W on 1872 Sept. 4. W to Sarah Sedgwick (William's wife), "I was never so worked up in my life, and did not sleep many hours under the hospitable roof". LL3:165, complete letter in *Letters of Chauncey Wright*, 1878, pp. 246-8. Transcribed in *Darwin Online*. 1877 W's essays were published by Charles Eliot Norton. (F1917).

Wrigley, Alfred, 1818-98. Leonard and Horace's tutor. CCD15:363. 1861-82 Headmaster of Clapham Grammar School after C. Pritchard. 1868 Jan. ED in a letter to Henrietta said W had on hearing of George D's success in winning second place for the Smith's prize, made an announcement to the boys at school, "was going to cry, and gave a half-holiday and sent them all to the Crystal Palace." ED2:217.

Wychfield, 80 Huntingdon Road, Cambridge. 1883-1904 House of Sir Francis Darwin. Sold after death of 2d wife Ellen Crofts (d.1903). A stone inscription inside the house reads "Francis Darwin built this house for Francis & Frances Cornford 1910." Now called Conduit Head.

Wyman, Jeffries, 1814-74. American palaeontologist and anatomist. Friend of Asa Gray. 1841-42 Studied in Europe; also with Richard Owen. 1847-74 Hervey Prof. Anatomy Harvard. 1860 CD to Lyell, W had written to CD about brains of rodents. 1861-66 Corresponded with CD about *Origin*. 1866 Curator Peabody Museum of Archaeology and Ethnology, Yale University.

Wynne, Rice, 1777-1846. (not "Wynn") Apothecary and surgeon. Friend of CD's father. 1822 Mayor of Shrewsbury. Bred horses and Malay fowl. Before 1839 CD addressed some questions on animal breeding to W. A rough copy in CD's hand was transcribed by Barrett in Gruber, *Darwin on man*, 1974, pp. 423-6 (F265). See *Darwin's notebooks*, 1987, and *Darwin Online*.

X Club, 1864-1911. Founded 1864 during a dinner at Brown's and St George's Hotel (now Brown's Hotel, Albemarle St.) and met regularly until 1892, then sporadically until 1911 on Hooker's death. A small scientific dining club in London. Initiated by Huxley. First meeting was on 3 Nov. 1864. Members were Busk, Hooker, Spencer, Edward Frankland, Huxley, Spottiswoode, Thomas Archer Hirst and Lubbock. All except Spencer were FRS. They dined before Royal Society meetings, discussing its affairs. CD was not a member and may never have dined with them, but he was on intimate terms with several. The last surviving member was Lubbock, who died in 1913. 1909 *Leicester Daily Post*, (15 Feb.) reported 240 meetings and the club dissolved in 1893, also claiming that CD was an occasional guest. See J.V. Jensen, The X Club: fraternity of Victorian scientists, *British Jrnl. for the Hist. of Science*, 5, 1970, pp. 63-72 and Ruth Barton, *The X club*, 2018.

Yahgan. Indian nomadic tribe of eastern Tierra del Fuego, to which the four Indians taken to England by FitzRoy on 1st voyage of *Beagle* belonged. Full name Yahgashagalumoala ("the people from the mountain channel"), shortened by Thomas Bridges. Also known as Yamana. Called by CD and his contemporaries 'Fuegians'. See Keynes, *Fossils, finches and Fuegians*, 2002.

Yardley, Rev. John, 1805-88. Educated at Shrewsbury School 1814-24. 1836-80 Vicar of St. Chad, Shrewsbury; speaks of CD at Shrewsbury School as "cheerful, good-tempered and communicative". Woodall 1884, p. 16. Buried in old St. Chad's.

Yarrell, William, 1784-1856. London stationer and naturalist. 1825 FLS. 1831 CD to Susan Darwin, Y had helped with buying equipment for *Beagle* voyage. "But one friend is quite invaluable...he goes to the shops with me and bullies about prices". CCD1:147. 1836 *History of British fishes*, 1843 *History of British birds*. CD discussed evolution with before *Origin*. Tegetmeier claimed that Y introduced him to CD.

Yavuz, Erkoçak, First to translate CD's work into Turkish, from a German translation. 1968 *Descent* (F1137).

Yates, Edmund Hodgson, 1831-94. Journalist. Visited CD at Down House in 1878 and published an account in: Mr. Darwin at Down. *Celebrities at home*, 1878, pp. 223-30, transcribed in *Darwin Online*. (F1996).

Yiddish. First edition in: 1921 *Descent* (F1138).

Yonge, John Arthur, 1838-1922. 1873 Aug. 12 ED recorded Mrs Y called. Two days later the Yonges dined. 1876 Sept. 17, ED wrote to Mrs Yonge. Sept. 19 Mrs Y called. 1879 May 17 was another record of the "Yonges called" when they were in Basset. William Darwin wrote to his mother in May 1879 that he went to call on the Yonges. 1880 May 29 they visited. 1886 Aug. 14, they dined.

Yonge, Mary Charlotte, 1823-1901. Novelist with over a hundred books written, her admirers included W.E. Gladstone, Charles Kingsley and A. Tennyson. 1887 Dec. ED wrote to Henrietta that she read Miss Yonge's books. ED2:379.

York Minster, Fuegian man, taken to England by FitzRoy on 1st voyage of *Beagle*. Returned on 2d voyage. Named after a peak on Waterman Island near Cape Horn Island. Name in Alikhoolip language is El'leparu. 1830 Y was aged about 26. Before 1872 he was killed in a quarrel. Two engraved portraits in *Narrative* 2, facing p. 324.

York Minster, Tierra del Fuego. Southernmost peak of Waterman Island. Named by Cook who described it as "a wild-looking rock".

York Place, no. 27, off Baker Street, London. 1855 Jan. 18 - Feb. 15 CD rented this house for a family holiday. See van Wyhe ed., 'Journal', (DAR158).

Youmans, Edward Livingston, 1821-87. American science writer and publisher. Account of 1871 Jul. 12 meeting with CD in London at Erasmus Darwin's home. Also present were ED, "Miss D", likely Henrietta and "Master D", likely William. Fiske, *Edward Livingston Youmans*, 1894, p. 276, transcribed in *Darwin Online*. (F2091).

Young, George, 1819-1907. 1856-74 MP. 1869-74 Lord Advocate of Scotland. 1874-1905 Judge of the Court of Session, with title "Lord Young". CCD23:180. 1875 Y lunched at Down House. 1888 and 1889, ED recorded "Mr Young"

Young, Sarah, 1836-? Nurserymaid from Kent at Down House in 1851 census.

Zacharias, Emil Otto, 1846-1916. German freshwater biologist of Geestemünde. Journalist. Corresponded with Haeckel. 1877 CD to Z, had sent him a pig's foot with an extra digit, which William Henry Flower examined. CCD25. 1877 CD to Z, on the development of his belief in evolution, "When I was on board the Beagle, I believed in the permanence of Species, but, as far as I can remember, vague doubts occasionally flitted across my mind." CCD25:105. 1876 *Zur Entwicklungstheorie*, Jena.

1882 *Charles R. Darwin und die Kulturhistorische Bedeutung seiner Theorie vom Ursprung der Arten*, Berlin. 1892 Founded Biological Station for Limnology at Lake Großer Plöner, inspired by Anton Dohrn's Zoological Station in Naples. 1882 [Obituary of CD] *Vom Fels zum Meer* 2: 348-53. Transcribed in *Darwin Online*.

Zoological diary or **Zoology notes**. From the voyage of the *Beagle* (DAR30-31). The largest (c.200,000 words and 368 pages) and single most important of CD's records of his animal collections, experiments and observations during the voyage. It also includes a great deal of entries on plants, fungi and lichens. The entries were made chronologically according to the collection locality so that it is straightforward to consult alongside writings on other topics from the same time and place, such as the *Beagle diary* and *Geological diary*. Transcribed and edited by Richard Keynes as *Charles Darwin's zoology notes & specimen lists from H.M.S. Beagle* (2000, F1840) in *Darwin Online*. It also includes CD's lists of "Specimens in Spirits of Wine" and "Specimens not in Spirits" and the species and much else have been identified by Keynes. CD used these notes to inform specialists about his observations of the various animals and to write the introductions and various notes on the geology, geography and habits of the animals in *Zoology of the Beagle*.

Zoological Society of London, Regent's Park. 1826 Founded. 1831 CD Corresponding Member. 1839 Fellow. 1882 Apr. 3 CD to W. Van Dyck, "the Zoological Society which is much addicted to mere systematic work". LL3:253.

Zoology of the Beagle. With support from the Duke of Somerset, President of the Linnean Society, the Earl of Derby and William Whewell, the Lord Commissioners of the Treasury granted £1,000 towards the cost of publication of the *Zoology of the Beagle*. A botanical counterpart of the *Zoology* was contemplated but never undertaken. 1838-43 *The zoology of the voyage of H.M.S. Beagle, under the command of Captain Fitzroy, during the years 1832 to 1836, edited and with notes by Charles Darwin*; 19 numbers making up 5 parts. Published by Smith Elder (F8 and F9). 1838-40 Part I, *Fossil Mammalia*, 4 numbers, by Richard Owen. 1838-39 Part II, *Mammalia*, 4 numbers, by George Robert Waterhouse. 1838-41 Part III, *Birds*, 5 numbers, by John Gould [and George R. Gray and Thomas Campbell Eyton]. 1840-42 Part IV, *Fish*, 4 numbers, by Leonard Jenyns. 1842-43 Part V, *Reptiles* [and Amphibia], 2 numbers, by Thomas Bell. 1975 Facsimile Part V only (F9a). 1979 Facsimile of whole (F9b). 1994 Facsimile after Darwin's copy, C.I.L. Ltd., Peterborough, UK (F1914).

"Zoophilus" see Blyth, Edward.

BIBLIOGRAPHY

Almost all of the sources cited in *Companion* are given only in the text as they could not possibly be repeated here for lack of space. Only sources frequently cited in author date format are given below.

Atkins, Hedley. 1976. *Down: the home of the Darwins; the story of a house and the people who lived there*. London: Royal College of Surgeons [Phillimore]. Revised impression. First 1974.

Autobiography. Nora Barlow ed. 1958. *The autobiography of Charles Darwin 1809-1882, with the original omissions restored: edited with appendix and notes by his grand-daughter Nora Barlow*, London: Collins. [In *Darwin Online*]

Beagle diary. Richard Darwin Keynes ed. 1988. *Charles Darwin's Beagle diary*. Cambridge: Cambridge University Press. [In *Darwin Online*]

Brent, Peter. 1981. *Charles Darwin. A man of enlarged curiosity*. London: William Heinemann.

Browne, Janet. 1995. *Charles Darwin, volume 1: Voyaging*. London: Jonathan Cape.

———. 2002. *Charles Darwin, volume 2: The power of place*. London: Jonathan Cape.

Burke, H.F. 1888. *Pedigree of the family of Darwin*. [In *Darwin Online*]

Calendar. A calendar of the correspondence of Charles Darwin, 1821-1882. American Council of Learned Societies, 1985. 2d edn, with Supplement, Cambridge University Press, 1996. Superseded by the *Darwin Correspondence Project Online Database*. http://www.darwinproject.ac.uk/.

Carroll, P. Thomas. 1976. *An annotated calendar of the letters of Charles Darwin in the Library of the American Philosophical Society*, Wilmington: Scholarly Resources Inc. (Numbers given refer to the numbers of the letters and not to pages.) [In *Darwin Online*]

CCD: F. Burkhardt, Sydney Smith et al eds. 1985-. *The Correspondence of Charles Darwin*. Cambridge University Press.

Climbing plants. C.R. Darwin. 1875. On the movements and habits of climbing plants, *Jrnl. Linnean Society*, 9: pp. 1-118, London: Longman, Green, Longman, Roberts & Green; and Williams and Norgate, 1865. 2d edn, as a book with same title, London: John Murray. [In *Darwin Online*]

Cross and self fertilisation. C.R. Darwin. 1876. *The effects of cross and self fertilisation in the vegetable kingdom*. [In *Darwin Online*]

Darling, L. 1984. HMS *Beagle*, 1820-1870: voyages summarized, research and reconstruction. *Sea History*, Spring, no. 31, pp. 27-38.

Darwin, Francis. 1917. *Rustic sounds and other studies in literature and natural history*. London: John Murray. [In *Darwin Online*]

———. 1919. *Springtime and other Essays*. John Murray. [In *Darwin Online*]

Darwin, Leonard. 1929. Memories of Down House. 1929, *The Nineteenth Century*, 106. [In *Darwin Online*]

Darwin Online. John van Wyhe ed. 2002-. *The Complete Work of Charles Darwin Online* http://darwin-online.org.uk.

Darwin's notebooks. P.H. Barrett; Gautrey, P.J.; Herbert, S.; Kohn, D. and Smith, S. eds. 1987. *Charles Darwin's notebooks, 1836-1844: geology, transmutation of species, metaphysical enquiries*. London: British Museum.

de Beer, G.R. ed. 1960. Further unpublished letters of Charles Darwin, *Annals of Science*, 14: pp. 83-115, (for 1958). [In *Darwin Online*]
——. ed. 1959. Some unpublished letters of Charles Darwin, *Notes and Records Royal Society*, 14: pp. 12-66. [In *Darwin Online*]
——. ed. 1968. The Darwin letters at Shrewsbury School, *Notes and Records Royal Society*, 23: pp. 68-85. [In *Darwin Online*]
Descent: C.R. Darwin. 1871. *The descent of man, and selection in relation to sex*. London: John Murray. [In *Darwin Online*]
Earthworms: C.R. Darwin. 1881. *The formation of vegetable mould, through the action of worms, with observations on their habits*. London: John Murray. [In *Darwin Online*]
ED: H.E. Litchfield ed. *Emma Darwin, wife of Charles Darwin: a century of family letters*, 2 vols. (1904 privately printed edn. 1915 published edn.)
Emma Darwin's diary: Christine Chua and John van Wyhe eds., 2019. *Emma Darwin's diary*. http://darwin-online.org.uk/EmmaDiaries.html
Expression: C.R. Darwin. 1872. *The expression of the emotions in man and animals*. [In *Darwin Online*]
Geological diary: Kees Rookmaaker, John van Wyhe & Gordon Chancellor eds., *Charles Darwin's geological diary from the voyage of the Beagle*. http://darwin-online.org.uk/EditorialIntroductions/Chancellor_GeologicalDiary.html
Freeman, R.B. 1968. Charles Darwin on the routes of male humble bees, *Bulletin of the BMNH, Historical Series.*, 3: pp. 177-89. [In *Darwin Online*]
——. 1977. *The works of Charles Darwin. An annotated bibliographical handlist*. 2ᵈ edn., Revised and enlarged. [In *Darwin Online*]
——. 1978. *Charles Darwin: A companion*. Folkestone: Dawson. [In *Darwin Online*]
——. 1984. *Darwin pedigrees*. London: the author. [In *Darwin Online*]
Freeman, R.B. and P.J. Gautrey. 1969. Charles Darwin's Questions about the breeding of animals, with a note on Queries about expression, *Journal for the Society of the Bibliography of Natural History*, 5: pp. 220-5.
—— and ——. 1972. Charles Darwin's Queries about expression, *Bulletin of the BMNH, Historical Series.*, 4: pp. 205-19. [In *Darwin Online*]
—— and ——. 1975. Charles Darwin's Queries about expression. *Journal for the Society of the Bibliography of Natural History*, 7: pp. 259-63.
Insectivorous plants: C.R. Darwin. 1875. *Insectivorous plants*. London: John Murray. [In *Darwin Online*]
Journal 2ᵈ edn.: C.R. Darwin. 1845. *Journal of researches into the natural history and geology of the various countries visited by H.M.S. Beagle*, etc., 2ᵈ edn., London: John Murray. [In *Darwin Online*]
Journal: C.R. Darwin. 1839. *Journal of researches into the geology and natural history of the various countries visited by H.M.S. Beagle*, etc., 2ᵈ edn, London: Henry Colburn. [In *Darwin Online*]
Keynes, Richard Darwin. 1979. *The Beagle record. Selections from the original pictorial records and written accounts of the voyage of H.M.S. Beagle*. Cambridge University Press.
——. ed. 2000. *Charles Darwin's zoology notes & specimen lists from H.M.S. Beagle*. Cambridge: University Press. [In *Darwin Online*]
——. 2002. *Fossils, finches and Fuegians. Charles Darwin's adventures and discoveries on the Beagle, 1832-1836*.

LL: Francis Darwin, ed. 1887. *The life and letters of Charles Darwin, including an autobiographical chapter*, 3 vols., London: John Murray. [In *Darwin Online*]

ML: Francis Darwin and A. C. Seward, editors, More letters of Charles Darwin: a record of his work in a series of hitherto unpublished letters, 2 vols., London: John Murray, 1903. [In *Darwin Online*]

Narrative: Robert FitzRoy, ed. 1839. *Narrative of the surveying voyages of His Majesty's Ships Adventure and Beagle*, etc., 3 vols. and appendix vol. to vol. III, London: Henry Colburn. vol. III is Charles Darwin, *Journal and remarks*, the first printing of *Journal of researches*, 1839. [In *Darwin Online*]

Natural selection: Robert C., Stauffer ed. 1975. *Charles Darwin's Natural selection: being the second part of his big species book written from 1856 to 1858*, Cambridge: University Press. [In *Darwin Online*]

ODNB: A. Desmond, J. Browne & J. Moore. 2004. Darwin, Charles Robert (1809-1882). *Oxford Dictionary of National Biography*. Oxford: Oxford University Press.

Orchids: C.R. Darwin. 1862. *On the various contrivances by which British and foreign orchids are fertilised by insects, and on the good effects of intercrossing*. [In *Darwin Online*]

Origin: C.R. Darwin. 1859. *On the origin of species by means of natural selection, or the preservation of favoured races in the struggle for life*, London: John Murray. [In *Darwin Online*]

Power of movement: C.R. Darwin. *The power of movement in plants*, London: John Murray, 1880. [In *Darwin Online*]

Raverat, Gwen. 1952. *Period piece: a Cambridge childhood*. London: Faber & Faber.

Shorter publications: John van Wyhe ed. 2009. *Charles Darwin's shorter publications 1829-1883*. Cambridge University Press.

Stecher, Robert M. ed. 1961. The Darwin-Innes letters. The correspondence of an evolutionist with his vicar, 1848-84. *Annals Science*, 17: pp. 201-58, (F1597). [In *Darwin Online*]

Wedgwood, Barbara and Hensleigh Wedgwood. 1980. *The Wedgwood circle, 1730-1897: four generations of a family and their friends*. London: Studio Vista.

Woodall, Edward. 1884. *Charles Darwin. A paper contributed to the Transaction of the Shropshire Archaeological Society and Natural History Society*, vol. VIII, part 1, pp. 1-64. [In *Darwin Online*]

Wyhe, John van. ed. 2006. Darwin's 'Journal' (1809-81). http://darwin-online.org.uk/content/frameset?viewtype=side&itemID=CUL-DAR158.1-76&pageseq=1

——. 2013. "my appointment received the sanction of the Admiralty": Why Charles Darwin really was the naturalist on HMS Beagle'. *Studies in History and Philosophy of Biological and Biomedical Sciences* vol. 44, issue 3, Sept., pp. 316-26. [In *Darwin Online*]

——. 2014. *Charles Darwin in Cambridge: The most joyful years*. World Scientific Publishing.

Wallace letters: John van Wyhe and Kees Rookmaaker. 2013. *Alfred Russel Wallace: Letters from the Malay Archipelago*. Foreword by Sir David Attenborough. Oxford University Press.

Zoology of the Beagle: C.R. Darwin ed. 1838-43. *The zoology of the voyage of H.M.S. Beagle*. 5 vols. London: Smith Elder and Co. [In *Darwin Online*]

Zoology notes: Richard Darwin Keynes ed. 2000. *Charles Darwin's zoology notes & specimen lists from H.M.S. Beagle*. Cambridge: University Press. [In *Darwin Online*]

Printed in the United States
by Baker & Taylor Publisher Services